Application of Optimization in Production, Logistics, Inventory, Supply Chain Management and Block Chain

Application of Optimization in Production, Logistics, Inventory, Supply Chain Management and Block Chain

Special Issue Editors

Biswajit Sarkar
Mitali Sarkar

MDPI • Basel • Beijing • Wuhan • Barcelona • Belgrade • Manchester • Tokyo • Cluj • Tianjin

Special Issue Editors

Biswajit Sarkar
Yonsei University
South Korea

Mitali Sarkar
Seoul National University
South Korea

Editorial Office
MDPI
St. Alban-Anlage 66
4052 Basel, Switzerland

This is a reprint of articles from the Special Issue published online in the open access journal *Mathematics* (ISSN 2227-7390) (available at: https://www.mdpi.com/journal/mathematics/special_issues/Application_Optimization_Production_Logistics_Inventory_SCM_Block_Chain).

For citation purposes, cite each article independently as indicated on the article page online and as indicated below:

LastName, A.A.; LastName, B.B.; LastName, C.C. Article Title. *Journal Name* **Year**, *Article Number*, Page Range.

ISBN 978-3-03928-522-8 (Pbk)
ISBN 978-3-03928-523-5 (PDF)

© 2020 by the authors. Articles in this book are Open Access and distributed under the Creative Commons Attribution (CC BY) license, which allows users to download, copy and build upon published articles, as long as the author and publisher are properly credited, which ensures maximum dissemination and a wider impact of our publications.

The book as a whole is distributed by MDPI under the terms and conditions of the Creative Commons license CC BY-NC-ND.

Contents

About the Special Issue Editors

Biswajit Sarkar (Prof.) is currently an Associate Professor in the Department of Industrial Engineering, Yonsei University, South Korea. He has completed his B.S. and M.S. in Applied Mathematics in 2002 and 2004, respectively, from Jadavpur University, India. He has received his Master of Philosophy in the application of Boolean Polynomials from Annamalai University, India, in 2008, Ph.D. from the Jadavpur University, India, in 2010 in Operations Research, and post-doctorate from Pusan National University, South Korea (2012–2013). He has provided teaching and conducted research at various universities including Hanyang University, South Korea (2014–2019); Vidyasagar University, India (2010–2014); and Darjeeling Government College, India (2009–2010). Under his supervision, 15 students have been awarded their Ph.D. and three students were awarded their M.S. Since 2010, he has published 166 journal articles in reputed journals in Applied Mathematics and Industrial Engineering, and has published one book. He is an Editorial Board member for some reputed international journals of Applied Mathematics and Industrial Engineering. He is the Topic Editor of the SCIE indexed journal *Energies*, and is the Section Editor of the Scopus-indexed journal *Inventions*. He has served as the Guest Editor of three Special Issues of two SCIE indexed journals *Mathematics* and *Energies*. He is a member of several learned societies. In 2014, his paper was selected as the best research paper in an international conference in South Korea. He has presented several research papers in international conferences as an Invited Speaker and has chaired several sessions at several international conferences. He received a bronze medal for his capstone achievement from Hanyang University in 2016. He was the recipient of Bharat Vikash award as a young scientist from India in 2016. He received an International award from the Korean Institute of Industrial Engineers in 2017 at KAIST, Daejon, South Korea. He was the recipient of the Hanyang University Academic Award as one of the most productive researchers in 2017 and 2018, consecutively.

Mitali Sarkar (Dr.) is currently a Post-Doctoral Fellow in the Department of Industrial Engineering, Seoul National University, South Korea. She has completed her post-doctorate from Yonsei University, South Korea (2017–2020) and Hanyang University, South Korea (2017–2018). She has completed her Ph.D. in 2017 at Hanyang University, South Korea. She has received her B.S. in Mathematics in 2002 and M.S. in 2006 in Applied Mathematics from Jadavpur University, India. She has completed her Bachelors' in Education in 2009 from the University of Calcutta, India. She has dedicated her teaching abilities in Jaynagar Institution for Girls', India in 2003–2014. She was hired as an Adjunct Professor of Hanyang University, South Korea (2018–2020). Since 2013, she has published 23 research papers in various reputed SCI, SCIE, and SSCI journals and 7 conference papers. She worked as an Assistant Guest Editor of a SCIE indexed journal and has acted as a reviewer of different SCI and SCIE journals.

Preface to "Application of Optimization in Production, Logistics, Inventory, Supply Chain Management and Block Chain"

Rapid evolution in the technology sector is one of the massive revolutions of the 21st century. The chronological improvement in technology is accelerating the fourth industrial revolution. The use of advanced technologies within the manufacturing and production sector has revolutionized human–machine interactions with the development of artificial intelligence and the Internet of Things. The supply chain management and logistics sectors have developed using technology that helps to increase the speed and security of the entire. Smart factories, smart production, and smart homes have changed the definition of how we live along with business strategy. Autonomation is the main feature of this revolution, which supports automated machinery systems for humans. An optimized system in inventory control and supply chains must incorporate technological improvement. The fourth industrial revolution is characterized by big data analysis, cyber-physical systems, machine-to-machine interaction, smart production, the circular economy. Along with this revolution, industry is moving toward the eco-friendly development of technology and products. To improve the environmental conditions for good health from all aspects, green development is necessary.

The readers can learn about the revolution and development of the technology in the inventory, manufacturing, production, supply chain management, logistics, and optimization sectors. All articles provide new ideas and development strategies for the improvement of economy, society, and the environment. The aim is to introduce the new research trends in the above-mentioned streams. The articles are divided into three categories:

- Inventory, manufacturing, and production;
- Supply chain management and logistics; and
- Optimization, numerical methods, and simulation

Thanks to all the contributors of the Special Issue "Application of Optimization in Production, Logistics, Inventory, Supply Chain Management, and Block Chain" of the SCIE-indexed journal *Mathematics*. All the ideas and results of the research articles enrich the literature on Operations Research (Applied Mathematics) and Industrial Engineering.

Biswajit Sarkar, Mitali Sarkar
Special Issue Editors

Article

Imperfect Multi-Stage Lean Manufacturing System with Rework under Fuzzy Demand

Muhammad Tayyab [1], Biswajit Sarkar [1,*] and Bernardo Nugroho Yahya [2]

[1] Department of Industrial & Management Engineering, Hanyang University, Ansan Gyeonggi-do 155 88, Korea; mtayyabntu@yahoo.com

[2] Industrial and Management Engineering Department, Hankuk University of Foreign Studies, 81, Oedae-ro, Mohyeon-myeon, Cheoin-gu, Yongin-si, Gyeonggi-do 17035, Korea; bernardo@hufs.ac.kr

* Correspondence: bsbiswajitsarkar@gmail.com; Tel.: +82-107-498-1981

Received: 19 November 2018; Accepted: 19 December 2018; Published: 24 December 2018

Abstract: Market conditions fluctuate abruptly in today's competitive environment and leads to imprecise demand information. In particular, market demand data for freshly launched products is highly uncertain. Further, most of the products are generally manufactured through complex multi-stage production systems that may produce defective items once they enter the out-of-control state. Production management of a multi-stage production system in these circumstances requires robust production model to reduce system costs. In this context, this paper introduces an imperfect multi-stage production model with the consideration of defective proportion in the production process and uncertain product demand. Fuzzy theory is applied to handle the uncertainty in demand information and the center of gravity approach is utilized to defuzzify the objective function. This defuzzified cost objective is solved through the analytical optimization technique and closed form solution of optimal lot size and minimum cost function are obtained. Model analysis verifies that it has successfully achieved global optimal results. Numerical experiment comprising of three examples is conducted and optimal results are analyzed through sensitivity analysis. Results demonstrate that larger lot sizes are profitable as the system moves towards a higher number of stages. Sensitivity analysis indicates that the processing cost is the most influencing factor on the system cost function.

Keywords: Lean manufacturing; multi-stage production system; fuzzy demand; imperfect products; reworking

1. Introduction

Lean manufacturing concepts are becoming rapidly popular in global production industries subjected to growing market competitions and increased manufacturing costs. Initiated by Toyota's engineer Taiichi Ohno, credited with developing the principles of lean production after World War II; eliminating unwanted activities, empowering workforce, cut down inventories, and improving productivity are the focused themes of lean philosophy. Unlike Henry Ford, who maintained resources in expectation of what might be required for future manufacturing, the management team at Toyota built partnership with suppliers and consequently became made-to-order. Thus, they can easily handle quick changes, and opposite to what their competitors could, they became able to respond quicker to market demands.

Highly inflexible and having high volume/low variety products of automatic machinery are the constraints in the textile industries. Due to this nature of textile industry, implementation of lean manufacturing is a challenge in this industry. Therefore, value stream mapping, kaizen, 5S, kanban, poka-yoke, and visual controls are practiced in combination with each other to improve the textile manufacturing process [1]. Most of the textile manufacturing processes include more than one production stages, and among all other concerns, optimum work in process inventory levels and scrap control are the premier ones. Thus, managers are interested in working out ways to minimize total system costs through optimal lot sizing in the production system. This builds importance of economic production quantity model in this sector to escape high inventory levels and cut down production costs. Figure 1 shows a general multi-stage production flow of an apparel manufacturing industry.

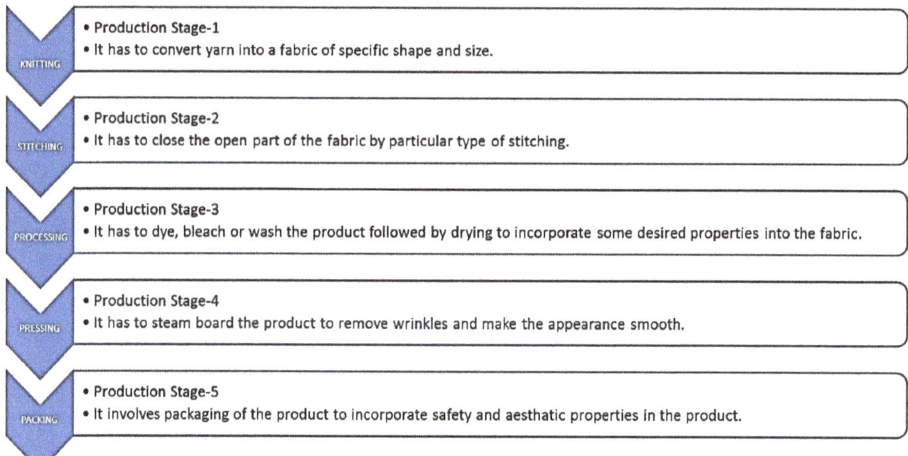

Figure 1. General production flow of an apparel manufacturing industry.

Since Harris [2] and Taft [3] provided Economic Order Quantity (EOQ) and Economic Production Quantity (EPQ) formulas, respectively, economic lot size models are broadly developed by the researchers. Due to their simplicity, these inventory models are used in many applied situations. In the inventory management research, Lee [4], Gupta and Chakraborty [5], and Tayi and Ballou [6] studied reprocessing of imperfect items, in which shortages due to reworking were not taken into account by the last two. Lee et al. [7] and Glock and Jaber [8] provided production models for imperfect production systems without considering rework opportunity. C.S Lin [9] studied the production process for its out-of-control state along with raw material resource constraints. Salameh and Jaber [10] provided an inventory model for imperfect quality items where these items are withdrawn from stock resulting in lower holding cost per unit time. Their model was extended by Khan et al. [11], and Jaber et al. [12] considers strategies for handling these imperfect quality items. Ouyang et al. [13] and Cárdenas-Barrón [14] formulated inventory models in which backordering is allowed and Eroglu and Ozdemir [15] considered shortages in their studies. Jaber and Guiffrida [16] studied Wright's learning curve along with preventive maintenance, and Daya [17] likewise extended the inventory model with planned maintenance schedules.

Most of the researchers developed inventory models for single-stage imperfect production systems, where numerous of them focused on analysis of the defective rates. Reworking in single-stage inventory models with constant defective rates has been established by Jamal et al. [18], Biswas and Sarker [19], Chung [20], and Taleizadeh et al. [21], in which the latter two included backordering in their models. Cárdenas-Barrón [22] extended the inventory model provided by Jamal et al. [18] with the addition of the algebraic solution approach. Ma et al. [23] analyzed imperfect production system without providing reworking alternatives in their model, and Barzoki et al. [24] presented discounted sales of non-reworkable items. Other researchers modified inventory model with random-defective rate in single-stage manufacturing system, where the reworking option was assumed by Chiu et al. [25], Chiu et al. [26], Noorollahi et al. [27], Haider et al. [28], and Chiu et al. [29] using the renewal reward theorem. Taleizadeh et al. [30] allowed a backordering alternative along with the rework prospect in their inventory model, whereas Chiu et al. [31] and Taleizadeh et al. [32] developed inventory models, ignoring the reworking facility. Random defect-rate contemplation was further established by Sana [33], Sarkar and Moon [34], Yoo et al. [35], Chakraborty and Giri [36], Wee et al. [37], Chiu et al. [38], and Sarkar et al. [39] using a number of other shop-floor conditions.

In real situations, determining the precise value of product demand is very difficult. To deal with such conditions, managers need to gather demand data from the experts. If, according to the experts, demand about some randomly chosen quantity is fixed; their opinion is imprecise and then the expected annual demand is vaguely expressed. This explanation of demand is through a fuzzy number. As defined by Dang and Hong [40], fuzziness is some parametric property of the demand captured when its sharp boundaries cannot be determined by the decision makers. Zadeh [41] introduced fuzzy terminologies in 1978. After him, several researches including Petrovic and Sweeney [42], Roy and Maiti [43], Yao and Su [44], Chang and Ouyang [45], Mahata and Goswami [46], and De and Goswami [47] formulated inventory models considering parameters like demand, total average cost, storage space, and holding cost as fuzzy numbers. Numerous other researches solved inventory models considering fuzzy demand. Chang and Yeh [48], Sadeghu and Niaki [49], Lin et al. [50], Cosgun et al. [51], and Sadeghi et al. [52] treated demand as a trapezoidal fuzzy number in their inventory models. Some other researchers including Rong and Maiti [53], Zhang et al. [54], Moghaddam [55], and Huang et al. [56] considered demand as a Triangular Fuzzy Number (TFN). In addition, Taleizadeh et al. [57] presented a multi-constraint EOQ model with incremental discounts and uncertain item cost under fuzzy environment, and Jana et al. [58] considered storage space and available budget as fuzzy variables. Das et al. [59] assumed credit period in a supply chain as a fuzzy number, and Kumar and Goswami [60] developed a production inventory model under fuzzy shifting time to out-of-control state and fuzzy defective proportion.

In fact, in real situations, production processes are imperfect, thus producing defective items at a certain rate. That is why many researchers and practitioners have formulated inventory models expressing different real world circumstances [21–24]. In addition, parametric information of the real world production systems is not always known with precision [48]. This imprecise data can be random or uncertain in nature. Probability theory can handle the situations with random information, whereas the computational methods provided by the fuzzy theory are required to tackle the uncertain conditions. Several approaches including the signed distance method, min-max approximation, weighted average method, the center of gravity approach (centroid method), and center of the largest area are available in the uncertainty control literature to model the uncertain environment. This research work has used the center of gravity approach to grasp the uncertainty in demand information because of its simplicity and ease of use.

Vast research has been done on single-stage production system, in which machine breakdowns and defective proportion are considered as constant, random, and occasionally fuzzy in nature [22,34,60]. Regarding multistage production systems, only a few economic batch quantity models are available with limited real world production scenarios to provide succor in decision making for the managers dealing with multi-stage production systems [61,62]. This paper attempts to add in the literature by formulating an economic lot size model for multi-stage production facility with setup time requirement, uncertain annual demand rate, and imperfection constraints. The aim of this research work is to initiate a multi-stage production model by considering a real world shop-floor situation of defective production and uncertain market conditions regarding demand information.

Literature analysis indicates that no such model exists in the current research stream of the inventory and production management. This fact provides motivation for the development of multi-stage production model with the aforementioned attributes. Thus, the novelty of this study is the introduction of imperfect production proportion at each stage of the multi-stage production system and imprecise product demand. The objective of the research is to minimize the total cost of the imperfect multi-stage production system by determining optimal lot size under uncertain environments. Fuzzy theory is applied to grasp the uncertainty in demand information and analytical optimization technique is applied to obtain global optimum results of the model. The model is analyzed through sensitivity analysis and significant conclusions are obtained that verifies the substantial contribution of the proposed model in the inventory research literature.

Framework of the paper is as follows: The mathematical model is developed in Section 2. Section 3 provides experimental study and optimal results. Section 4 presents discussion and analysis of the proposed production model. Finally, in Section 5, concluding remarks and future research directions of this paper are discussed.

2. Materials and Methods

This research work studies an n-stage imperfect production system where defective items are produced along with the perfect quality items. These defective items are reworked at an additional processing cost, and product demand is fulfilled at the nth production stage. Figures 2 and 3 show inventory behavior of production stage-k and production stage-n, respectively. The objective of the model is to minimize the total system cost by determining optimal lot size.

2.1. Assumptions

Following assumptions are considered in the development of multi-stage production model.

1. A single type of item is produced in an n-stage production system.
2. Production runs are fairly consistent, i.e., system is assumed to be a lean manufacturing system where inventory behavior follows a similar trend in repetitive production cycles.
3. Uncertain product demand is measured as a Triangular Fuzzy Number (TFN).
4. Inline inspections provides effective results, determines the defective items immediately, and helps the managers to make immediate decisions over it. This model assumes that defective items are detected during the hundred percent inline inspection.
5. In textile industries, defective items are generally reworked. This model assumes reworking of defective items at each production stage. The reworking process is considered as perfect and no item is scrapped.
6. Defective rate is constant at each production stage, but it may vary from stage to stage.
7. Setup time of each production stage is ρ percent of the total time of that production stage.
8. Transportation time and cost among the production stages is assumed to be negligible.

2.2. Model Formulation

Keeping in view the above-mentioned assumptions, a mathematical model is formulated.

Figure 2 shows the inventory behavior of kth production stage. Maximum inventory level (I_{3k}) of the kth production stage ($k = 1, 2, 3, \ldots, n-1$) is computed as $I_{3k} = I_{1k} + I_{2k}$, where I_{1k} is the inventory level after production phase and I_{2k} is the inventory level after reworking phase.

Production time of stage-k is $T_{1k} = \frac{Q}{P_k}$, and its reworking time is $T_{2k} = \frac{Q\alpha_k}{P_k}$. It is obvious from triangle *abc* in Figure 2, that $\tan \omega_1 = \frac{I_{1k}}{T_{1k}}$, which gives $I_{1k} = Q(1 - \alpha_k)$. Similarly, from triangle *bef*, one can observe that $\tan \omega_2 = \frac{I_{2k}}{T_{2k}}$, which provides $I_{2k} = Q\alpha_k$. From above the maximum inventory level of kth production stage is obtained as $I_{3k} = Q$.

Total inventory of kth production stage is the sum of the triangular areas *abc*, *bcd*, and *bdf* (Figure 2). One can simply compute these areas and the total inventory of kth production stage is now formulated as:

$$I_k = \frac{Q^2}{2P_k}\left(1 + \alpha_k - \alpha_k{}^2\right) \tag{1}$$

Total production time of stage-k can be obtained by the sum $T_k = T_{sk} + T_{1k} + T_{2k}$, where the setup time of stage-k, $T_{sk} = \rho(T_{1k} + T_{2k})$. Total time of stage-$k$ (T_k) is obtained as:

$$T_k = \frac{Q}{P_k}(1+\rho)(1+\alpha_k) \tag{2}$$

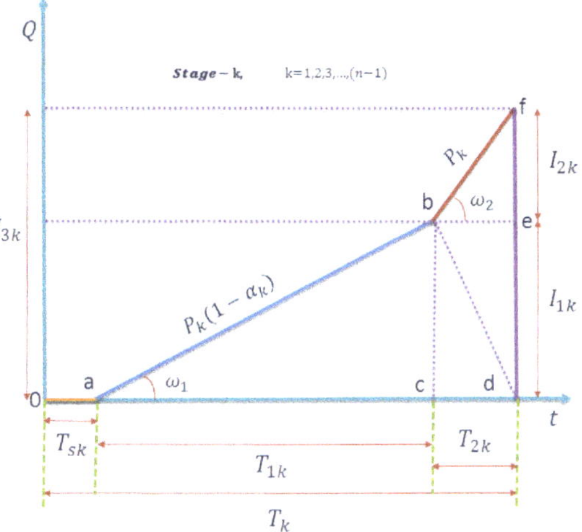

Figure 2. Inventory diagram of kth production stage.

Applying similar method for production stage-n (Figure 3), one can simply obtain production time $T_{1n} = \frac{Q}{P_n}$, reworking time $T_{2n} = \frac{Q\alpha_n}{P_n}$, pure consumption time $T_D = \frac{Q}{D}\left(1 - \frac{D}{P_n}(1+\alpha_n)\right)$, inventory level after production phase $I_{1n} = Q\left(1 - \alpha_n - \frac{D}{P_n}\right)$, and inventory level after reworking phase as $I_{2n} = Q\alpha_n\left(1 - \frac{D}{P_n}\right)$ for production stage-n. Finally, maximum inventory level of stage-n is computed as $I_{\max(n)} = Q\left(1 - \frac{D}{P_n}(1+\alpha_n)\right)$.

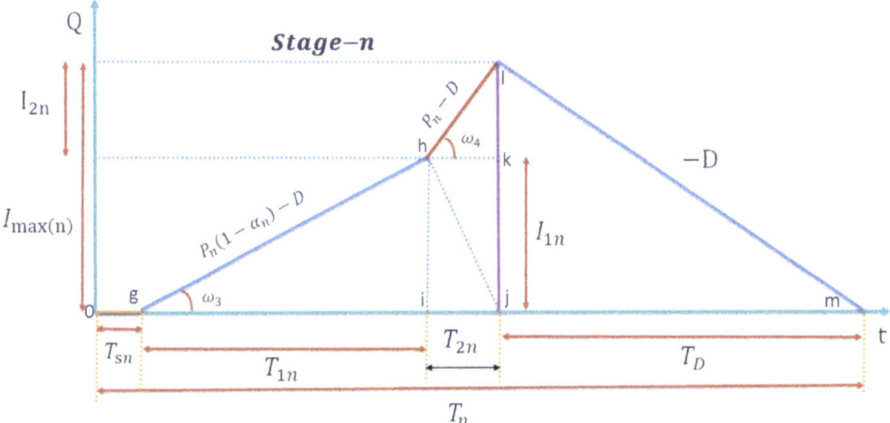

Figure 3. Inventory diagram of nth production stage.

Proceeding towards determination of total inventory level of finished items (I), we have

$$I = \frac{Q^2}{2D}\left(1 - \frac{D}{P_n}(1+\alpha_n+\alpha_n^2)\right) \tag{3}$$

Total time of production stage-n is obtained as

$$T_n = (1+\rho)\frac{Q}{D} \tag{4}$$

Total time of the complete production system can be determined by the sum $T = \sum_{k=1}^{n-1} T_k + T_n$, and is given by

$$T = \frac{Q(1+\rho)}{D}\left(1 + D\sum_{k=1}^{n-1}\left(\frac{1+\alpha_k}{P_k}\right)\right) \tag{5}$$

Average inventory level of the finished items $(\bar{I})$ in the system is calculated as $\bar{I} = \frac{I}{T}$, which provides

$$\bar{I} = \frac{Q\left(P_n - D\left(1 + \alpha_n + \alpha_n{}^2\right)\right)}{2P_n(1+\rho)\left(1 + D\sum_{k=1}^{n-1}\left(\frac{1+\alpha_k}{P_k}\right)\right)} \tag{6}$$

Total cost of the system by considering setup cost, order processing cost, inspection cost, and inventory holding cost is formulated as follows:

Setup cost (K_S) ($K/lot/stage)

Setup cost is a major part of the total cost in a production system. It is required to be incurred prior to the production of every lot. For an n-stage production system, the total setup cost is

$$K_s = \sum_{i=1}^{n} K_i \tag{7}$$

Order processing cost (K_P) ($C/item/stage)

Order processing cost and reworking cost are considered as the same in this model. The total order processing cost is the sum of processing cost of good items as well as the reworking cost of defective items:

$$K_P = Q\sum_{i=1}^{n} C_i(1 + \alpha_i) \tag{8}$$

Inspection cost (K_I) ($J/item/stage)

Inspection cost is incurred on both good quality items as well as on reworked items and is assumed to be the same. Total inspection cost:

$$K_I = Q\sum_{i=1}^{n} J_i(1 + \alpha_i) \tag{9}$$

Inventory holding cost (K_C) ($H/item/year)

Under linear assumptions, total inventory holding cost is proportional to the holding cost of average inventory of finished items in the cycle as

$$K_C = H\bar{I} = H\frac{Q\left(P_n - D\left(1 + \alpha_n + \alpha_n{}^2\right)\right)}{2P_n(1+\rho)\left(1 + D\sum_{k=1}^{n-1}\left(\frac{1+\alpha_k}{P_k}\right)\right)} \tag{10}$$

Total system cost function by considering aforementioned cost components is given by

$$TC(Q) = \frac{\sum_{i=1}^{n} K_i + Q\sum_{i=1}^{n} C_i(1 + \alpha_i) + Q\sum_{i=1}^{n} J_i(1 + \alpha_i)}{\frac{Q(1+\rho)}{D}\left(1 + D\sum_{k=1}^{n-1}\left(\frac{1+\alpha_k}{P_k}\right)\right)} + H\frac{Q\left(P_n - D\left(1 + \alpha_n + \alpha_n{}^2\right)\right)}{2P_n(1+\rho)\left(1 + D\sum_{k=1}^{n-1}\left(\frac{1+\alpha_k}{P_k}\right)\right)} \tag{11}$$

and after some simplifications

$$TC(Q) = \frac{\left(Q^2\left(HP_n - DH(1 + \alpha_n + \alpha_n{}^2)\right) + 2DP_n\sum\limits_{i=1}^{n} K_i + 2DQP_n\left(\sum\limits_{i=1}^{n} C_i(1 + \alpha_i) + \sum\limits_{i=1}^{n} J_i(1 + \alpha_i)\right)\right)}{2QP_n(1 + \rho)\left(1 + D\sum\limits_{k=1}^{n-1}\left(\frac{1 + \alpha_k}{P_k}\right)\right)} \tag{12}$$

Cárdenas-Barrón [14] viewed annual demand as a constant parameter. However, in several real world scenarios, some uncertainties may alter annual demand slightly. Thus, annual demand may be considered as a fuzzy number. To replace the annual demand D by a fuzzy number $\widetilde{D}$, consider the fuzzy number $\widetilde{D}$ as a TFN $\widetilde{D} = (D - \Delta_1, D, D + \Delta_2)$, where $0 < \Delta_1 < D$ and $0 < \Delta_2$ as shown in Figure 4. The Fuzzy Membership Function (FMF) of $\widetilde{D}$ is formulated as

$$\mu_{\widetilde{D}}(x) = \begin{cases} \frac{x - D + \Delta_1}{\Delta_1} & if \quad D - \Delta_1 \leq x \leq D \\ \frac{D + \Delta_2 - x}{\Delta_2} & if \quad D \leq x \leq D + \Delta_2 \\ \quad 0 & otherwise. \end{cases}$$

The centroid of $\mu_{\widetilde{D}}(x)$ is given by $D^* = D + \frac{1}{3}(\Delta_2 - \Delta_1)$. This result is considered to obtain the annual demand in fuzzy sense. For any $T > 0$, let $TC(Q)\,(x) = y(>0)$. Using the extension principle of Kaufmann and Gupta [63] and Zimmermann [64], the membership function of the fuzzy cost $TC(Q)(\widetilde{D})$ is formulated as

$$\mu_{TC(Q)\widetilde{D}}(y) = \begin{cases} \sup_{x \in TC(Q)^{-1}(y)} \mu_{(\widetilde{D})}(x) & if \quad TC(Q)^{-1}(y) \neq \varnothing \\ 0 & if \quad TC(Q)^{-1}(y) = \varnothing \end{cases}.$$

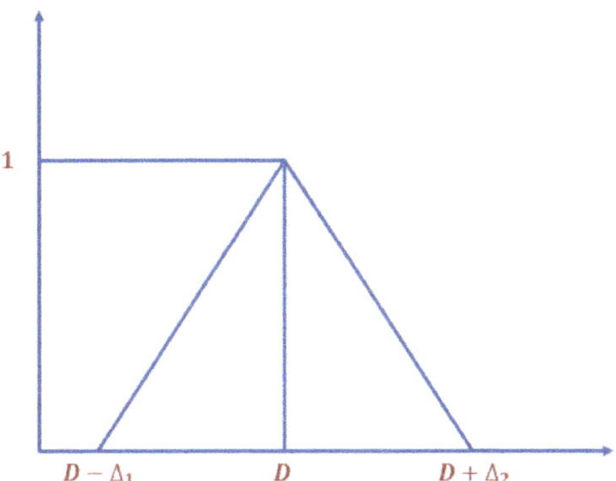

Figure 4. Triangular fuzzy number $\widetilde{D}$.

From $TC(Q)\,(x) = y$ and the system cost function, we obtain

$$y = \frac{\left(Q^2\left(HP_n - DH(1 + \alpha_n + \alpha_n{}^2)\right) + 2DP_n\sum\limits_{i=1}^{n} K_i + 2DQP_n\left(\sum\limits_{i=1}^{n} C_i(1 + \alpha_i) + \sum\limits_{i=1}^{n} J_i(1 + \alpha_i)\right)\right)}{2QP_n(1 + \rho)\left(1 + D\sum\limits_{k=1}^{n-1}\left(\frac{1 + \alpha_k}{P_k}\right)\right)} \tag{13}$$

For simplification purpose, let us assume
$B = HP_n,$
$F = H\left(1 + \alpha_n + \alpha_n{}^2\right),$
$L = 2P_n\sum\limits_{i=1}^{n} K_i,$

$$G = 2P_n \left(\sum_{i=1}^{n} C_i(1+\alpha_i) + \sum_{i=1}^{n} J_i(1+\alpha_i) \right),$$
$$M = 2P_n(1+\rho),$$

and

$$R = 2P_n(1+\rho) \sum_{k=1}^{n-1} \left(\frac{1+\alpha_k}{P_k} \right).$$

This provides

$$y = \frac{BQ^2 + D(L + Q(G - FQ))}{Q(M + DR)} \tag{14}$$

and

$$x = \frac{Q(BQ - My)}{FQ^2 + QRy - L - GQ} \tag{15}$$

From the standard triangular FMF and values of x and y, the FMF of $TC(Q)(\tilde{D})$ is given by

$$\mu_{TC(Q)\tilde{D}}(y) = \begin{cases} \dfrac{Q(BQ-My)}{(FQ^2+QRy-L-GQ)\Delta_1} - \dfrac{D-\Delta_1}{\Delta_1} & if \ \ y_1 \leq y \leq y_2 \\[4mm] \dfrac{D+\Delta_2}{\Delta_2} - \dfrac{Q(BQ-My)}{(FQ^2+QRy-L-GQ)\Delta_2} & if \ \ y_2 \leq y \leq y_3 \end{cases}.$$

Figure 5 represents the FMF of $TC(Q)(\tilde{D})$.

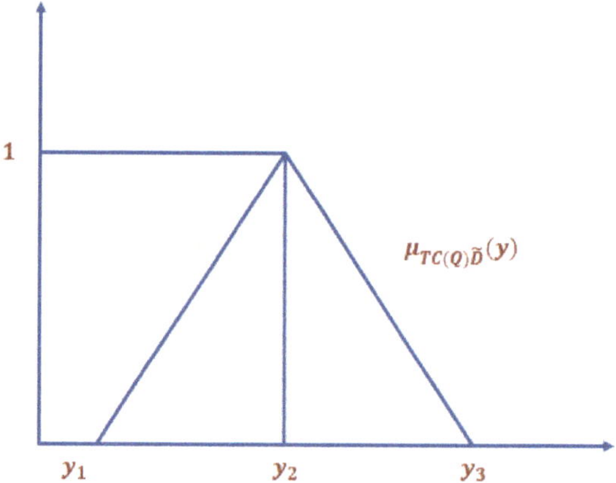

Figure 5. Triangular fuzzy number $TC(Q)(\tilde{D})$.

Now centroid of $\mu_{TC(Q)\tilde{D}}(y)$ is formulated as

$$TC(Q) = \frac{\int_{-\infty}^{\infty} \left[y\mu_{TC(Q)\tilde{D}}(y) \right] dy}{\int_{-\infty}^{\infty} \left[\mu_{TC(Q)\tilde{D}}(y) \right] dy} \tag{16}$$

After defuzzification, crisp total cost function becomes

$$TC(Q) = \frac{\left(\frac{BQ^2+D(L+Q(G-FQ))}{M+DR} + \frac{BQ^2+(L+Q(G-FQ))(D-\Delta_1)}{M+R(D-\Delta_1)} + \frac{BQ^2+(L+Q(G-FQ))(D+\Delta_2)}{M+R(D+\Delta_2)} \right)}{3Q} \tag{17}$$

As lot size Q is a decision variable, we differentiate the cost function (17) with respect to Q. Thus, from the necessary conditions, we have

$$\frac{d(TC)}{dQ} = \frac{\left(\Delta_1\left((M+DR)\left(L(M+3DR)+Q^2\left(\begin{matrix}FM-2BR\\+3DFR\end{matrix}\right)\right)+R\left(L(2M+3DR)+Q^2\left(\begin{matrix}2FM-BR\\+3DFR\end{matrix}\right)\right)\Delta_2\right)-(M+DR)\left(\begin{matrix}3(-BQ^2+D(L+FQ^2))(M+DR)\\+\left(\begin{matrix}L(M+3DR)\\+Q^2(FM-2BR+3DFR)\end{matrix}\right)\Delta_2\end{matrix}\right)\right)}{3Q^2(M+DR)(M+DR-R\Delta_1)(M+DR+R\Delta_2)} = 0 \quad (18)$$

After some simplification, Equation (18) can be written in standard quadratic form as

$$\frac{d(TC)}{dQ} = \frac{\left(\begin{matrix}L(-(M+dR)(3d(M+dR)+(M+3dR)\Delta_2)+\Delta_1((M+dR)(M+3dR)+R(2M+3dR)\Delta_2))\\+Q^2(\Delta_1((M+dR)(FM-2BR+3dFR)+R(2FM-BR+3dFR)\Delta_2)+(M+dR)(3(B-dF)(M+dR)\\+(2BR-F(M+3dR))\Delta_2))\end{matrix}\right)}{3Q^2(M+dR)(M+dR-R\Delta_1)(M+dR+R\Delta_2)} = 0 \quad (19)$$

Quadratic equation (19) provide two roots, one is negative and the other is positive. It is a well-known fact that quadratic equation has only one positive root. Thus, the two solutions for the optimal lot size are obtained as below

$$Q^* = \pm\sqrt{\left(\frac{L((M+DR)(3D(M+DR)+(M+3DR)\Delta_2)-\Delta_1((M+DR)(M+3DR)+R\Delta_2(2M+3DR)))}{\left(\Delta_1\left(\begin{matrix}(M+DR)(FM-2BR+3DFR)+\\R\Delta_2(2FM-BR+3DFR)\end{matrix}\right)+(M+DR)\left(\begin{matrix}3(B-DF)(M+DR)\\+(2BR-F(M+3DR))\Delta_2\end{matrix}\right)\right)}\right)}$$

Negative root seems quite impossible candidate solution for optimal lot size problem, and is thus discarded. The positive root is taken as the optimal lot size Q^*,

$$Q^* = \sqrt{\left(\frac{L((M+DR)(3D(M+DR)+(M+3DR)\Delta_2)-\Delta_1((M+DR)(M+3DR)+R\Delta_2(2M+3DR)))}{\left(\Delta_1\left(\begin{matrix}(M+DR)(FM-2BR+3DFR)+\\R\Delta_2(2FM-BR+3DFR)\end{matrix}\right)+(M+DR)\left(\begin{matrix}3(B-DF)(M+DR)\\+(2BR-F(M+3DR))\Delta_2\end{matrix}\right)\right)}\right)} \quad (20)$$

To satisfy sufficient conditions,

$$\frac{d^2(TC)}{(dQ)^2} = \frac{2L\left(3+M\left(-\frac{1}{M+DR}-\frac{1}{M+DR-R\Delta_1}-\frac{1}{M+DR+R\Delta_2}\right)\right)}{3Q^3R} > 0 \quad (21)$$

As the solution to the above expression is positive, the sufficient condition is also satisfied. Thus the system cost attains minimum value in the interval $(0, T)$ for Q^*.

3. Numerical Experiment and Results

This section analyzes the impact of various numerical values of parameters on the optimal cost of the system by taking appropriate units. Three numerical experiments are carried out for a five-stage production system ($n = 5$).

3.1. Example 1

Parameters $K_i = \$100/\text{stage}$, $D = 50{,}000$ items/year, $P_n = 200{,}000$ items/year, $H = \$5/\text{item}/\text{year}$, and $\rho = 2.00\%$ per stage are taken from Wee et al. [37] and Biswas and Sarker [19]. Parameters $\alpha_i = 1.00\%$ for each stage, $C_i = \$3/\text{item}$ for each stage, $J_i = \$0.02/\text{item}$ for each stage, $P_{n-1} = 210{,}000$ items/year, $P_{n-2} = 220{,}500$ items/year, $P_{n-3} = 231{,}525$ items/year, $P_{n-4} = 243{,}102$ items/year, and $\Delta_1 = 8000$, and $\Delta_2 = 12{,}000$ are considered randomly.

3.2. Example 2

Numerical values of the parameters in this example are taken from Tayyab and Sarkar [61] as K_i = \$400/stage, D = 15,000 items/year, P_n = 70,000 items/year, H = \$4/item/year, ρ = 2.00% per stage, C_i = \$35/item for each stage, J_i = \$1/item for each stage, P_{n-1} = 73,500 items/year, P_{n-2} = 77,175 items/year, P_{n-3} = 81,033 items/year, P_{n-4} = 85,085 items/year, and α_i = 1.00% for each production stage is the same as in Example 1. Parameters Δ_1 = 3000 and Δ_2 = 4000 are considered randomly.

3.3. Example 3

Numerical values of the parameters in this example are taken from Tayyab and Sarkar [61] as K_i = \$200/stage, D = 12,000 items/year, P_n = 60,000 items/year, H = \$20/item/year, ρ = 2.00% per stage, C_i = \$100/item for each stage, J_i = \$0.5/item for each stage, P_{n-1} = 63,000 items/year, P_{n-2} = 66,150 items/year, P_{n-3} = 69,457 items/year, P_{n-4} = 72,930 items/year, and α_i = 1.00% for each production stage is same as in Example 1 and 2. Parameters Δ_1 = 2000 and Δ_2 = 3000 are considered randomly.

Optimal results of the above numerical examples are provided in Table 1.

Table 1. Optimal results of the numerical experiments.

Production Stages	Example 1		Example 2		Example 3	
	Q* (items)	TC* ($)	Q* (items)	TC* ($)	Q* (items)	TC* ($)
Single-stage	1664.93	159,552	1984.44	552,648	557.94	1,236,016
Two-stage	2346.56	252,080	2795.5	905,477	786.79	2,051,893
Three-stage	2867.7	316,070	3414.82	1,160,906	961.72	2,652,656
Four-stage	3306.35	364,808	3935.67	1,360,773	1108.92	3,127,405
Five-stage	3692.53	404,441	4393.98	1,525,950	1238.47	3,522,058

4. Discussion

Decision makers of the multi-stage production systems can get the advantage of the proposed production model to reduce system cost in the presence of defective production and imprecise information. The proposed production model has provided global optimal results for lot size decision to achieve the minimum cost of the multi-stage production system. The detailed procedure of fuzzy theory application provided for uncertainty control can enable the managers and decision-makers to conveniently address the uncertain conditions of respective shop-floors. Each numerical example is solved for five-stage production systems and results are summarized in Table 1. Figures 6–8 illustrate the convexity of minimum cost objective in each case, which verifies the robustness of the model. Optimal results indicate that the larger lot sizes provide more benefit to the production systems comprising of higher number of stages. For instance, in Example 3, optimal lot size for single-stage production system is 557.94 items and 1238.47 items for five-stage production system. Consistency in the obtained results assures that this research work has achieved the desired goal of analyzing product imperfection and imprecise information in multi-stage production modeling.

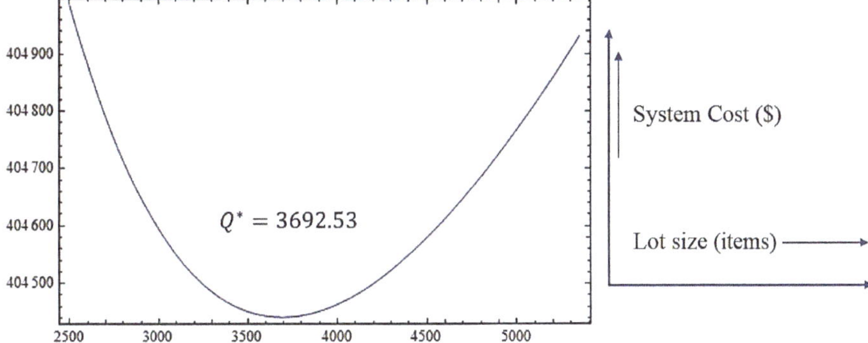

Figure 6. Objective function convexity diagram for Example 1 (Five-stage production system).

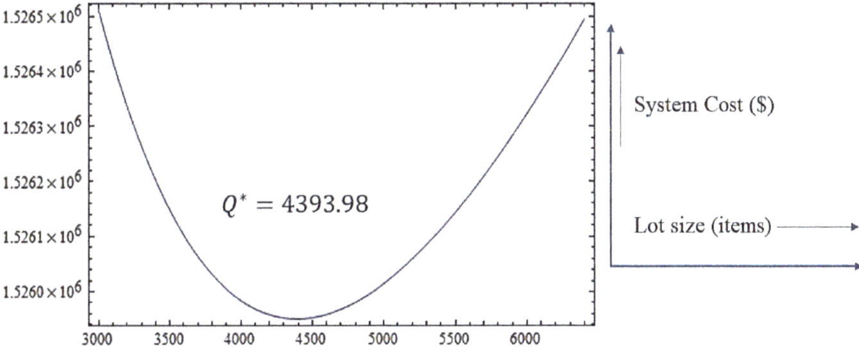

Figure 7. Objective function convexity diagram for Example 2 (Five-stage production system).

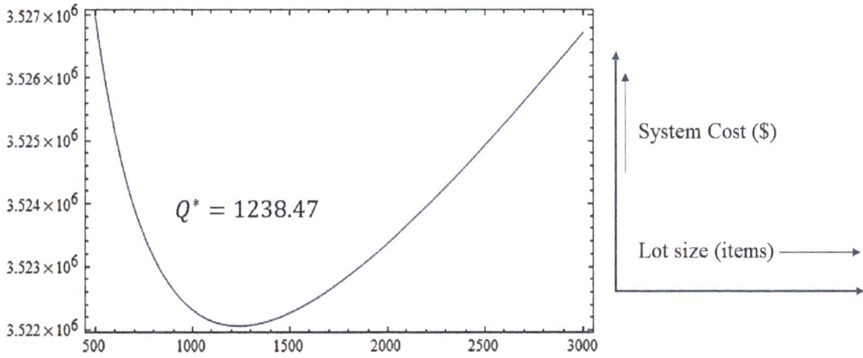

Figure 8. Objective function convexity diagram for Example 3 (Five-stage production system).

Optimal results are further evaluated for the possible variations in key parameters of the given examples. The effect of key parameters on the optimal cost function for up-to five-stage production system is summarized in Tables 2–4 for Example 1 to Example 3, respectively. The values of the parameters are varied from −50% to +50%.

1. It is evident from the change in optimal cost function that the order processing cost (C_i) is the most sensitive factor than other parameters. One can observe that with each percentage increase in C_i there exists almost similar amount of percentage increase in the system cost. For instance, regarding four-stage production system in Example 2, system cost increases by 24.17% by the

    increment of 25% in order of processing cost, and 48.34% increase in system cost is observed by a 50% rise in the order of processing cost.

2. Next to the effect of order processing cost is the impact of production rate (P_i) on the system cost. It can be noted that the effect of production rate on the production system with higher number of stages is more than the system with a lower number of production stages. This illustrates the necessity of wisely adjusting production rate at each production stage to keep the system cost at its minimum value.

3. An interesting observation from the sensitivity analysis of all the numerical examples reveal that the impact of setup cost (K_i) and the inventory holding cost (H) is same on the system cost for all the production systems with any number of stages. Thus, the decision makers have to make intelligent decision in putting efforts to reduce one of these costs first. Here arises the importance of lean philosophy, as Single Minute Exchange of Dyes (SMED) is the best possible solution for setup cost reduction. Therefore, managers should make efforts to apply SMED for setup cost reduction first, as it is much more beneficial than cutting down stock management equipment and inventory holding staff.

4. Inspection cost (J_i) and defective proportion (α_i) bear trivial impact on the system cost that gradually varies as the system moves towards a higher number of stages.

    The sensitivity analysis provides important inferences for the managers to develop optimal policies that can achieve the best outcomes at their shop floors.

Table 2. Sensitivity analysis of key parameters for Example 1.

Parameters	Change (%)	Percentage Change in $TC(Q)$*				
		Single-Stage Process	Two-Stage Process	Three-Stage Process	Four-Stage Process	Five-Stage Process
P_i	−50	−0.73	−17.09	−24.92	−29.53	−32.57
	−25	−0.23	−6.41	−9.94	−12.24	−13.85
	+25	+0.13	+4.27	+7.09	+9.13	+10.67
	+50	+0.21	+7.33	+12.40	+16.20	+19.15
K_i	−50	−1.11	−0.80	−0.65	−0.57	−0.51
	−25	−0.51	−0.36	−0.30	−0.26	−0.23
	+25	+0.45	+0.32	+0.26	+0.23	+0.21
	+50	+0.85	+0.61	+0.50	+0.44	+0.39
C_i	−50	−47.79	−48.32	−48.56	−48.70	−48.80
	−25	−23.89	−24.16	−24.28	−24.35	−24.40
	+25	+23.89	+24.16	+24.28	+24.35	+24.40
	+50	+47.79	+48.32	+48.56	+48.70	+48.80
J_i	−50	−0.32	−0.32	−0.32	−0.32	−0.33
	−25	−0.16	−0.16	−0.16	−0.16	−0.16
	+25	+0.16	+0.16	+0.16	+0.16	+0.16
	+50	+0.32	+0.32	+0.32	+0.32	+0.33
H	−50	−1.11	−0.80	−0.65	−0.57	−0.51
	−25	−0.51	−0.36	−0.30	−0.26	−0.23
	+25	+0.45	+0.32	+0.26	+0.23	+0.21
	+50	+0.85	+0.61	+0.50	+0.44	+0.39
α_i	−50	−0.47	−0.38	−0.32	−0.28	−0.25
	−25	−0.24	−0.19	−0.16	−0.14	−0.12
	+25	+0.24	+0.19	+0.16	+0.14	+0.12
	+50	+0.47	+0.38	+0.32	+0.28	+0.25

Table 3. Sensitivity analysis of key parameters for Example 2.

Parameters	Change (%)	Percentage Change in $TC(Q)^*$				
		Single-Stage Process	Two-Stage Process	Three-Stage Process	Four-Stage Process	Five-Stage Process
P_i	-50	-0.17	-15.12	-22.75	-27.42	-30.57
	-25	-0.05	-5.61	-8.95	-11.20	-12.81
	$+25$	$+0.03$	$+3.70$	$+6.28$	$+8.20$	$+9.69$
	$+50$	$+0.05$	$+6.33$	$+10.94$	$+14.48$	$+17.28$
K_i	-50	-0.32	-0.23	-0.19	-0.16	-0.15
	-25	-0.15	-0.10	-0.09	-0.07	-0.07
	$+25$	$+0.13$	$+0.09$	$+0.08$	$+0.07$	$+0.06$
	$+50$	$+0.25$	$+0.18$	$+0.14$	$+0.12$	$+0.11$
C_i	-50	-48.08	-48.23	-48.30	-48.34	-48.37
	-25	-24.04	-24.12	-24.15	-24.17	-24.18
	$+25$	$+24.04$	$+24.12$	$+24.15$	$+24.17$	$+24.18$
	$+50$	$+48.08$	$+48.23$	$+48.30$	$+48.34$	$+48.37$
J_i	-50	-1.37	-1.38	-1.38	-1.38	-1.38
	-25	-0.69	-0.69	-0.69	-0.69	-0.69
	$+25$	$+0.69$	$+0.69$	$+0.69$	$+0.69$	$+0.69$
	$+50$	$+1.37$	$+1.38$	$+1.38$	$+1.38$	$+1.38$
H	-50	-0.32	-0.23	-0.19	-0.16	-0.15
	-25	-0.15	-0.10	-0.09	-0.07	-0.07
	$+25$	$+0.13$	$+0.09$	$+0.08$	$+0.07$	$+0.06$
	$+50$	$+0.25$	$+0.18$	$+0.14$	$+0.12$	$+0.11$
α_i	-50	-0.49	-0.40	-0.35	-0.31	-0.27
	-25	-0.24	-0.20	-0.17	-0.15	-0.14
	$+25$	$+0.24$	$+0.20$	$+0.17$	$+0.15$	$+0.14$
	$+50$	$+0.49$	$+0.40$	$+0.35$	$+0.30$	$+0.27$

Table 4. Sensitivity analysis of key parameters for Example 3.

Parameters	Change (%)	Percentage Change in $TC(Q)^*$				
		Single-Stage Process	Two-Stage Process	Three-Stage Process	Four-Stage Process	Five-Stage Process
P_i	-50	-0.10	-14.38	-21.92	-26.61	-29.80
	-25	-0.03	-5.31	-8.57	-10.79	-12.41
	$+25$	$+0.02$	$+3.48$	$+5.97$	$+7.84$	$+9.31$
	$+50$	$+0.03$	$+5.95$	$+10.37$	$+13.80$	$+16.55$
K_i	-50	-0.21	-0.15	-0.12	-0.10	-0.09
	-25	-0.09	-0.07	-0.05	-0.05	-0.04
	$+25$	$+0.08$	$+0.06$	$+0.05$	$+0.04$	$+0.04$
	$+50$	$+0.16$	$+0.11$	$+0.09$	$+0.08$	$+0.07$
C_i	-50	-49.40	-49.50	-49.55	-49.58	-49.59
	-25	-24.70	-24.75	-24.77	-24.79	-24.80
	$+25$	$+24.70$	$+24.75$	$+24.77$	$+24.79$	$+24.80$
	$+50$	$+49.40$	$+49.50$	$+49.55$	$+49.58$	$+49.59$
J_i	-50	-0.25	-0.25	-0.25	-0.25	-0.25
	-25	-0.12	-0.12	-0.12	-0.12	-0.12
	$+25$	$+0.12$	$+0.12$	$+0.12$	$+0.12$	$+0.12$
	$+50$	$+0.25$	$+0.25$	$+0.25$	$+0.25$	$+0.25$
H	-50	-0.21	-0.15	-0.12	-0.10	-0.09
	25	-0.09	-0.07	-0.05	-0.05	-0.04
	$+25$	$+0.08$	$+0.06$	$+0.05$	$+0.04$	$+0.04$
	$+50$	$+0.16$	$+0.11$	$+0.09$	$+0.08$	$+0.07$

Table 4. *Cont.*

Parameters	Change (%)	Percentage Change in $TC(Q)$*				
		Single-Stage Process	Two-Stage Process	Three-Stage Process	Four-Stage Process	Five-Stage Process
α_i	−50	−0.49	−0.41	−0.35	−0.31	−0.28
	−25	−0.25	−0.20	−0.18	−0.16	−0.14
	+25	+0.25	+0.20	+0.18	+0.16	+0.14
	+50	+0.49	+0.41	+0.35	+0.31	+0.28

5. Conclusions

Agreeing to lean philosophy, high inventory levels are considered as one of the major non-value added aspect of the production system. For true implementation of lean culture, inventory as a form of non-value added activity should be reduced to its minimum. Therefore, there must be a defined optimal lot size in accordance with the input requirements of each production stage at minimum system cost. Cárdenas-Barrón [14] provided interpretation of considering defective rate in his inventory model to achieve optimal cost at the single-stage production facility. This inference is carried forward in this paper for its implication on a multi-stage production system.

This study is made-up to support production and planning managers, especially in the textile production sector where defective production and uncertain product demand are faced. It introduces an imperfect multi-stage production model with the consideration of defective proportion in the production process and uncertain product demand. Thus, the novelty of this study is the introduction of imperfect production proportion at each stage of the multi-stage production system and imprecise product demand. Fuzzy theory is applied to handle the uncertainty in demand information and center of gravity approach is utilized to defuzzify the objective function. This defuzzified cost objective is solved through the analytical optimization technique and closed form solution of the optimal lot size is obtained.

Numerical experiment comprising of three examples is conducted and optimal results are analyzed through sensitivity analysis. Optimal results indicate that as the number of stages increase, lots of larger sizes are cost effective. Sensitivity analysis indicates that the order processing cost bears the highest effect on the system cost and its effect increases slightly as the system moves towards higher number of stages. Creditably, this study can work as an elementary support in implementing lean culture in the production system. Analysis of the model illustrates the major advantage of this study by highlighting the importance of lean activities to reduce the system cost, as the results suggest application of SMED for setup cost reduction in the multi-stage production system prior to making efforts for inventory holding cost reduction.

Despite various benefits, this study has some limitations. It has considered constant defective proportion in the production system, whereas defective rate becomes imprecise in long run production processes. Secondly, the proposed model has incorporated a perfect reworking opportunity that is very rare in the real life production systems, as defective products are also produced during reworking. The analysis of the results state that the effect of order processing cost on the total system cost is highest among all other system parameters. However, this study has not provided order processing cost reduction policy to minimize system cost for future runs. Further, the model can be extended with several ideas including consideration of shortages backordered [39,65,66], variable production rate [67–69], trade credit policies [70,71], allocation problem [72,73], and different rework options [19]. The immediate possible extension to this paper can be the incorporation of uncertain defective rate [61,62].

Author Contributions: Conceptualization, M.T. and B.S.; Methodology, M.T.; Software, M.T.; Validation, B.S.; Formal analysis, M.T. and B.S.; Investigation, M.T.; Resources, M.T. B.S. and B.N.Y.; Data curation, M.T. and B.S.; Writing—original draft preparation, M.T.; Writing—review and editing, B.S. and B.N.Y.; Visualization, M.T. B.S. and B.N.Y.; Supervision, B.S.

Funding: This research received no external funding.

Conflicts of Interest: The authors declare no conflict of interest.

Nomenclature

Indices

k	production stages ($k = 1, 2, 3 \ldots (n-1)$)
i	production stages ($i = 1, 2, 3 \ldots n$)

Decision variable

Q	lot size (items)

Parameters

n	number of production stages
P_k	production rate for k^{th} production stage (items per unit time)
P_n	production rate for n^{th} production stage (items per unit time, where $P_k > P_n > D$)
$\widetilde{D}$	demand rate (fuzzy, and fulfilled at n^{th} production stage)
α_k	defective proportion at k^{th} production stage
α_n	defective proportion at n^{th} production stage
T_k	cycle time of kth production stage (years)
T_n	cycle time of nth production stage (years)
T	time between production runs (total cycle time of n-stages, years)
K_i	setup cost of ith production stage (\$/setup)
H	inventory holding cost of finished items (\$/item/year)
C_i	order processing cost at i^{th} production stage (\$/item/stage)
J_i	inspection cost at i^{th} production stage (\$/item/stage)
ρ	percentage setup time per stage
$\bar{I}$	average inventory of finished items in the system
$TC(Q)$	total cost of the production system (\$ per unit time)

Abbreviations

Below enlisted abbreviations are used for the multi-stage production model development in this research work.

EOQ	Economic order quantity
EPQ	Economic production quantity
TFN	Triangular fuzzy number
FMF	Fuzzy membership function
SMED	Single minute exchange of dyes

References

1. Hodge, G.L.; Goforth Ross, K.; Joines, J.A.; Thoney, K. Adapting lean manufacturing principles to the textile industry. *Prod. Plan. Control* **2011**, *22*, 237–247. [CrossRef]
2. Harris, F.W. How many parts to make at once. *Fact. Mag. Manag.* **1913**, *10*, 135–136. [CrossRef]
3. Taft, E.W. The most economical production lot. *Iron Age* **1918**, *101*, 1410–1412.
4. Lee, H.L. Lot sizing to reduce capacity utilization in a production process with defective items, process corrections, and rework. *Manag. Sci.* **1992**, *38*, 1314–1328. [CrossRef]
5. Gupta, T.; Chakraborty, S. Looping in a multistage production system. *Int. J. Prod. Res.* **1984**, *22*, 299–311. [CrossRef]
6. Tayi, G.K.; Ballou, D.P. An integrated production-inventory model with reprocessing and inspection. *Int. J. Prod. Res.* **1988**, *26*, 1299–1315. [CrossRef]
7. Lee, H.H.; Chandra, M.J.; Deleveaux, V.J. Optimal batch size and investment in multistage production systems with scrap. *Prod. Plan. Control* **1997**, *8*, 586–596. [CrossRef]
8. Glock, C.H.; Jaber, M.Y. Learning effects and the phenomenon of moving bottlenecks in a two-stage production system. *Appl. Math. Model.* **2013**, *37*, 8617–8628. [CrossRef]
9. Lin, C.S. Integrated production-inventory models with imperfect production processes and a limited capacity for raw materials. *Math. Comput. Model.* **1999**, *29*, 81–89. [CrossRef]

10. Salameh, M.K.; Jaber, M.Y. Economic production quantity model for items with imperfect quality. *Int. J. Prod. Econ.* **2000**, *64*, 59–64. [CrossRef]

11. Khan, M.; Jaber, M.Y.; Guiffrida, A.L.; Zolfaghari, S. A review of the extensions of a modified EOQ model for imperfect quality items. *Int. J. Prod. Econ.* **2011**, *132*, 1–2. [CrossRef]

12. Jaber, M.Y.; Zanoni, S.; Zavanella, L.E. Economic order quantity models for imperfect items with buy and repair options. *Int. J. Prod. Econ.* **2014**, *155*, 126–131. [CrossRef]

13. Ouyang, L.Y.; Chen, C.K.; Chang, H.C. Quality improvement, setup cost and lead-time reductions in lot size reorder point models with an imperfect production process. *Comput. Oper. Res.* **2002**, *29*, 1701–1717. [CrossRef]

14. Cárdenas-Barrón, L.E. Economic production quantity with rework process at a single-stage manufacturing system with planned backorders. *Comput. Ind. Eng.* **2009**, *57*, 1105–1113. [CrossRef]

15. Eroglu, A.; Ozdemir, G. An economic order quantity model with defective items and shortages. *Int. J. Prod. Econ.* **2007**, *106*, 544–549. [CrossRef]

16. Jaber, M.Y.; Guiffrida, A.L. Learning curves for imperfect production processes with reworks and process restoration interruptions. *Eur. J. Oper. Res.* **2008**, *189*, 93–104. [CrossRef]

17. Ben-Daya, M. The economic production lot-sizing problem with imperfect production processes and imperfect maintenance. *Int. J. Prod. Econ.* **2002**, *76*, 257–264. [CrossRef]

18. Jamal, A.M.; Sarker, B.R.; Mondal, S. Optimal manufacturing batch size with rework process at a single-stage production system. *Comput. Ind. Eng.* **2004**, *47*, 77–89. [CrossRef]

19. Biswas, P.; Sarker, B.R. Optimal batch quantity models for a lean production system with in-cycle rework and scrap. *Int. J. Prod. Res.* **2008**, *46*, 6585–6610. [CrossRef]

20. Chung, K.J. The economic production quantity with rework process in supply chain management. *Comput. Math. Appl.* **2011**, *62*, 2547–2550. [CrossRef]

21. Taleizadeh, A.A.; Jalali-Naini, S.G.; Wee, H.M.; Kuo, T.C. An imperfect multi-product production system with rework. *Sci. Iran* **2013**, *20*, 811–823.

22. Cárdenas-Barrón, L.E. Optimal manufacturing batch size with rework in a single-stage production system—A simple derivation. *Comput. Ind. Eng.* **2008**, *55*, 758–765. [CrossRef]

23. Ma, W.N.; Gong, D.C.; Lin, G.C. An optimal common production cycle time for imperfect production processes with scrap. *Math. Comput. Model.* **2010**, *52*, 724–737. [CrossRef]

24. Barzoki, M.R.; Jahanbazi, M.; Bijari, M. Effects of imperfect products on lot sizing with work in process inventory. *Appl. Math. Comput.* **2011**, *217*, 8328–8336. [CrossRef]

25. Chiu, S.W.; Tseng, C.T.; Wu, M.F.; Sung, P.C. Multi-item EPQ model with scrap, rework and multi-delivery using common cycle policy. *J. Appl. Res. Technol.* **2014**, *12*, 615–622. [CrossRef]

26. Chiu, S.W.; Chiu, Y.S.; Yang, J.C. Combining an alternative multi-delivery policy into economic production lot size problem with partial rework. *Expert Syst. Appl.* **2012**, *39*, 2578–2583. [CrossRef]

27. Noorollahi, E.; Karimi-Nasab, M.; Aryanezhad, M.B. An economic production quantity model with random yield subject to process compressibility. *Math. Comput. Model.* **2012**, *56*, 80–96. [CrossRef]

28. Moussawi-Haidar, L.; Salameh, M.; Nasr, W. Production lot sizing with quality screening and rework. *Appl. Math. Model.* **2016**, *40*, 3242–3256. [CrossRef]

29. Chiu, S.W.; Gong, D.C.; Wee, H.M. Effects of random defective rate and imperfect rework process on economic production quantity model. *Jpn. J. Ind. Appl. Math.* **2004**, *21*, 375. [CrossRef]

30. Taleizadeh, A.A.; Cárdenas-Barrón, L.E.; Mohammadi, B. A deterministic multi product single machine EPQ model with backordering, scraped products, rework and interruption in manufacturing process. *Int. J. Prod. Econ.* **2014**, *150*, 9–27. [CrossRef]

31. Chiu, S.W.; Ting, C.K.; Chiu, Y.S. Optimal production lot sizing with rework, scrap rate, and service level constraint. *Math. Comput. Model.* **2007**, *46*, 535–549. [CrossRef]

32. Taleizadeh, A.A.; Wee, H.M.; Jalali-Naini, S.G. Economic production quantity model with repair failure and limited capacity. *Appl. Math. Model.* **2013**, *37*, 2765–2774. [CrossRef]

33. Sana, S.S. A production–inventory model in an imperfect production process. *Eur. J. Oper. Res.* **2010**, *200*, 451–464. [CrossRef]

34. Sarkar, B.; Moon, I. An EPQ model with inflation in an imperfect production system. *Appl. Math. Comput.* **2011**, *217*, 6159–6167. [CrossRef]

35. Yoo, S.H.; Kim, D.; Park, M.S. Lot sizing and quality investment with quality cost analyses for imperfect production and inspection processes with commercial return. *Int. J. Prod. Econ.* **2012**, *140*, 922–933. [CrossRef]
36. Chakraborty, T.; Giri, B.C. Joint determination of optimal safety stocks and production policy for an imperfect production system. *Appl. Math. Model.* **2012**, *36*, 712–722. [CrossRef]
37. Wee, H.M.; Wang, W.T.; Yang, P.C. A production quantity model for imperfect quality items with shortage and screening constraint. *Int. J. Prod. Res.* **2013**, *51*, 1869–1884. [CrossRef]
38. Chiu, S.W.; Chou, C.L.; Wu, W.K. Optimizing replenishment policy in an EPQ-based inventory model with nonconforming items and breakdown. *Econ. Model.* **2013**, *35*, 330–337. [CrossRef]
39. Sarkar, B.; Cárdenas-Barrón, L.E.; Sarkar, M.; Singgih, M.L. An economic production quantity model with random defective rate, rework process and backorders for a single stage production system. *J. Manuf. Syst.* **2014**, *33*, 423–435. [CrossRef]
40. Dang, J.F.; Hong, I.H. The Cournot game under a fuzzy decision environment. *Comput. Math. Appl.* **2010**, *59*, 3099–3109. [CrossRef]
41. Zadeh, L.A. Fuzzy sets as a basis for a theory of possibility. *Fuzzy Set Syst.* **1999**, *100*, 9–34. [CrossRef]
42. Petrovic, D.; Sweeney, E. Fuzzy knowledge-based approach to treating uncertainty in inventory control. *Comput. Integr. Manuf. Syst.* **1994**, *7*, 147–152. [CrossRef]
43. Roy, T.K.; Maiti, M. Multi-objective inventory models of deteriorating items with some constraints in a fuzzy environment. *Comput. Oper. Res.* **1998**, *25*, 1085–1095. [CrossRef]
44. Yao, J.S.; Chang, S.C.; Su, J.S. Fuzzy inventory without backorder for fuzzy order quantity and fuzzy total demand quantity. *Comput. Oper. Res.* **2000**, *27*, 935–962. [CrossRef]
45. Chang, H.C.; Yao, J.S.; Ouyang, L.Y. Fuzzy mixture inventory model involving fuzzy random variable lead time demand and fuzzy total demand. *Eur. J. Oper. Res.* **2006**, *169*, 65–80. [CrossRef]
46. Mahata, G.C.; Goswami, A. Fuzzy inventory models for items with imperfect quality and shortage backordering under crisp and fuzzy decision variables. *Comput. Ind. Eng.* **2013**, *64*, 190–199. [CrossRef]
47. De, S.K.; Goswami, A.; Sana, S.S. An interpolating by pass to Pareto optimality in intuitionistic fuzzy technique for a EOQ model with time sensitive backlogging. *Appl. Math. Comput.* **2014**, *230*, 664–674. [CrossRef]
48. Chang, S.Y.; Yeh, T.Y. A two-echelon supply chain of a returnable product with fuzzy demand. *Appl. Math. Model.* **2013**, *37*, 4305–4315. [CrossRef]
49. Sadeghi, J.; Niaki, S.T. Two parameter tuned multi-objective evolutionary algorithms for a bi-objective vendor managed inventory model with trapezoidal fuzzy demand. *Appl. Soft Comput.* **2015**, *30*, 567–576. [CrossRef]
50. Lin, J.; Hung, K.C.; Julian, P. Technical note on inventory model with trapezoidal type demand. *Appl. Math. Model.* **2014**, *38*, 4941–4948. [CrossRef]
51. Coşgun, Ö.; Ekinci, Y.; Yanık, S. Fuzzy rule-based demand forecasting for dynamic pricing of a time company. *Knowl. Based Syst.* **2014**, *70*, 88–96. [CrossRef]
52. Sadeghi, J.; Mousavi, S.M.; Niaki, S.T. Optimizing an inventory model with fuzzy demand, backordering, and discount using a hybrid imperialist competitive algorithm. *Appl. Math. Model.* **2016**, *40*, 7318–7335. [CrossRef]
53. Rong, M.; Maiti, M. On an EOQ model with service level constraint under fuzzy-stochastic demand and variable lead-time. *Appl. Math. Model.* **2015**, *39*, 5230–5240. [CrossRef]
54. Zhang, B.; Lu, S.; Zhang, D.; Wen, K. Supply chain coordination based on a buyback contract under fuzzy random variable demand. *Fuzzy Set Syst.* **2014**, *255*, 1–6. [CrossRef]
55. Moghaddam, K.S. Fuzzy multi-objective model for supplier selection and order allocation in reverse logistics systems under supply and demand uncertainty. *Expert Syst. Appl.* **2015**, *42*, 6237–6254. [CrossRef]
56. Huang, M.; Song, M.; Leon, V.J.; Wang, X. Dentralized capacity allocation of a single-facility with fuzzy demand. *J. Manuf. Syst.* **2014**, *33*, 7–15. [CrossRef]
57. Taleizadeh, A.A.; Niaki, S.T.; Nikousokhan, R. Constraint multiproduct joint-replenishment inventory control problem using uncertain programming. *Appl. Soft Comput.* **2011**, *11*, 5143–5154. [CrossRef]
58. Jana, D.K.; Das, B.; Maiti, M. Multi-item partial backlogging inventory models over random planninghorizon in random fuzzy environment. *Appl. Soft Comput.* **2014**, *21*, 12–27. [CrossRef]
59. Das, B.C.; Das, B.; Mondal, S.K. An integrated production inventory model under interactive fuzzy credit period for deteriorating item with several kets. *Appl. Soft Comput.* **2015**, *28*, 453–465. [CrossRef]

60. Ku, R.S.; Goswami, A. A fuzzy random EPQ model for imperfect quality items with possibility and necessity constraints. *Appl. Soft Comput.* **2015**, *34*, 838–850.
61. Tayyab, M.; Sarkar, B. Optimal batch quantity in a cleaner multi-stage lean production system with random defective rate. *J. Clean. Prod.* **2016**, *139*, 922–934. [CrossRef]
62. Kim, M.S.; Sarkar, B. Multi-stage cleaner production process with quality improvement and lead time dependent ordering cost. *J. Clean. Prod.* **2017**, *144*, 572–590. [CrossRef]
63. Kaufman, A.; Gupta, M.M. *Introduction to Fuzzy Arithmetic*; Van Nostrand Reinhold Company: New York, NY, USA, 1991.
64. Zimmermann, H.J. Fuzzy control. In *Fuzzy Set Theory—And Its Applications*; Springer: Dordrecht, The Netherlands, 1996; pp. 203–240.
65. Sarkar, B.; Moon, I. Improved quality, setup cost reduction, and variable backorder costs in an imperfect production process. *Int. J. Prod. Econ.* **2014**, *155*, 204–213. [CrossRef]
66. Sarkar, B. A production-inventory model with probabilistic deterioration in two-echelon supply chain management. *Appl. Math. Model.* **2013**, *37*, 3138–3151. [CrossRef]
67. Sarkar, B.; Sana, S.S.; Chaudhuri, K. Optimal reliability, production lotsize and safety stock: An economic manufacturing quantity model. *Int. J. Manag. Sci. Eng. Manag.* **2010**, *5*, 192–202. [CrossRef]
68. Moon, I.; Shin, E.; Sarkar, B. Min–max distribution free continuous-review model with a service level constraint and variable lead time. *Appl. Math. Comput.* **2014**, *229*, 310–315. [CrossRef]
69. Shin, D.; Guchhait, R.; Sarkar, B.; Mittal, M. Controllable lead time, service level constraint, and transportation discounts in a continuous review inventory model. *Rairo-Oper. Res.* **2016**, *50*, 921–934. [CrossRef]
70. Sarkar, B.; Ahmed, W.; Kim, N. Joint effects of variable carbon emission cost and multi-delay-in-payments under single-setup-multiple-delivery policy in a global sustainable supply chain. *J. Clean. Prod.* **2018**, *185*, 421–445. [CrossRef]
71. Sarkar, B.; Sana, S.S.; Chaudhuri, K. An inventory model with finite replenishment rate, trade credit policy and price-discount offer. *J. Ind. Eng.* **2013**, 1–18. [CrossRef]
72. Ahmed, W.; Sarkar, B. Impact of carbon emissions in a sustainable supply chain management for a second generation biofuel. *J. Clean. Prod.* **2018**, *186*, 807–820. [CrossRef]
73. Cárdenas-barrón, L.E.; Sarkar, B.; Treviño-garza, G. Easy and improved algorithms to joint determination of the replenishment lot size and number of shipments for an EPQ model with rework. *Math. Comput. Appl.* **2013**, *18*, 132–138. [CrossRef]

© 2018 by the authors. Licensee MDPI, Basel, Switzerland. This article is an open access article distributed under the terms and conditions of the Creative Commons Attribution (CC BY) license (http://creativecommons.org/licenses/by/4.0/).

 *mathematics*

Article

Sustainable Lot Size in a Multistage Lean-Green Manufacturing Process under Uncertainty

Muhammad Tayyab [1], Biswajit Sarkar [1,*] and Misbah Ullah [2]

[1] Department of Industrial & Management Engineering, Hanyang University,
    Ansan Gyeonggi-do 155 88, Korea; mtayyabntu@yahoo.com
[2] Department of Industrial Engineering, University of Engineering & Technology, Peshawar 25120, Pakistan;
    misbah@uetpeshawar.edu.pk
*   Correspondence: bsbiswajitsarkar@gmail.com; Tel.: +82-107-498-1981

Received: 10 November 2018; Accepted: 21 December 2018; Published: 25 December 2018

Abstract: Optimal lot sizing is the primary tool applied by lean practitioners to reduce inconsistency in the manufacturing system to cut down inventories, which are often considered as a type of waste in the lean culture. Managers attempt to consider environmental impacts of the manufacturing system and find ways to reduce these effects while making efforts to achieve environmental protection. From a sustainability standpoint, carbon emissions are the major source of environmental contamination and degradation. In this context, this research provides an economic production quantity model with uncertain demand and process information in a multistage manufacturing process. This imperfect manufacturing process produces defective products at an uncertain rate, and is reworked to convert them into perfect quality products and reduce wastages. To control this uncertainty in the manufacturing process, the decomposition principle and the signed distance method of fuzzy theory are applied. The manufacturing process is analyzed with regard to environmental concerns, and a sustainable lot size is obtained through an interactive Weighted Fuzzy Goal Programming (WFGP) approach for the simultaneous achievement of economic and environmental sustainability. An experimental study is performed to verify the practical implication of the model, and results are evaluated through a sensitivity analysis. Important managerial insights and graphical illustrations are provided to elaborate the model.

Keywords: lean manufacturing; sustainability; process imperfection; uncertain information; decomposition principle; triangular fuzzy number

1. Introduction

When analyzing any industry on a global scale, presently, it is identified that the rapidly growing competitions in the global markets and frequently increasing raw material costs have led the manufacturing industries to adopt lean manufacturing policies. Taiichi Ohno, an engineer in Toyota recognized as a master of lean philosophy, provided the basis of various lean manufacturing techniques. His focused policies involved labor empowerment, inventories reduction, productivity improvement, and waste minimization. In this way, Toyota managed to accommodate frequent changes in market demands by applying a made-to-order strategy. Currently, almost all types of manufacturing industries have adopted lean culture in an appropriate proportion to minimize resource utilization, improve process reliability, enhance employee skills, and minimize system costs to achieve the ultimate goal of remaining noticeable in the sturdy market competition.

Productivity improvement is among the top application of lean culture due to its tangible and operative benefits to a manufacturing business. In order to focus on productivity improvement in a manufacturing industry, four strategic pillars of lean are simultaneously applied, namely; Poka-yoke, 5S, visual controls, and Kanban [1]. Poka-yoke is a fool-proof system, which prevents the occurrence of a mistake or defect [2]. 5S is a five-phase program of each phase starting with an alphabet "S"; (a) sort, (b) set in order, (c) shine, (d) standardize, and (e) sustain. 5S ensures workplace safety, waste minimization, and improved plant efficiency [3]. Posters, diagrammatic presentations, pictures, color codes, and symbols are effective visual tools for productivity improvement of the manufacturing facility [4]. Kanban is a production control tool for optimum utilization of the workforce capacity in order to achieve just-in-time (JIT) production [5].

With the formation of lean culture, a cleaner production approach is mandatory to reduce the environmental impacts of the manufacturing industries, which are combinedly termed as lean-green strategies. This task can be efficiently performed by applying waste control policies. According to a project carried out by Fresner [6], a cleaner production approach has the potential to minimize 0.5%–1.5% of the system cost by eradicating nonvalue added activities. Similarly, Ozturk et al. [7] analyzed implementation of 22 different environmental protection working techniques and achieved a significant reduction in resource utilization, while achieving comparable plant efficiency and product quality. They identified that the contribution of motivated decision-makers and leaders is mandatory to achieve desired results. Otherwise, lean-green strategies provide little to no benefit to the manufacturing process.

In order to attain economic sustainability, managers attempt to decide the optimal lot size for the manufacturing process. Since the first development of Economic Production Quantity (EPQ) model by Taft [8], a broad number of researchers have studied and extended the model to various real-life production scenarios. The elementary shortcoming of the basic EPQ model is the non-consideration of process imperfections. None of the real-world manufacturing processes are perfect in nature. Hence, they produce defective products due to their "out-of-control" states. Therefore, many researchers have extended the model by considering defective proportions in the manufacturing systems. Some researchers have considered a constant defective rate, whereas various others have studied a random defective rate in their production models. In addition, to convert defective products into perfect quality products by incurring additional reworking cost, several researchers have devised rework opportunity.

Sarkar et al. [9] determined the optimal reliability of a manufacturing process with random imperfection using control theory. Chiu et al. [10] developed a production model with the aim of reducing suppliers' carrying cost by considering random imperfections in the production process. Sarkar et al. [11] provided a production model with a random defective rate and provided optimal strategies for setup cost reduction and process improvement of the system. Tayyab and Sarkar [12] considered random defective proportion following beta distribution function in a manufacturing process and obtained the optimal lot size through the analytical optimization technique. Kim and Sarkar [13] further considered random imperfections in their production model and provided optimal investment policies for process improvement in a manufacturing process. They found that the spread of randomness in defective proportion data has a direct effect on the system cost.

In most of the scenarios, determining precise distribution function of the product demand and the random defective rate is not possible. In these situations, managers need to apply fuzzy theory, which can grasp the uncertainty involved in demand and the defective proportion information of the system. Zadeh introduced fuzzy theory in 1978 to handle the uncertain conditions, after which various researchers utilized the fuzzy approach to solve production models with imprecise parametric values. Chang [14] considered product demand and defective proportion as a fuzzy number in a single-stage manufacturing process and found that the uncertainty in defective proportion can be better dealt with using fuzzy theory on a cost of additional expenditure in the process. Priyan and Manivannan [15] considered defective proportion as a fuzzy number, along with inspection errors, in a manufacturing process and provided optimal delivery policies for the vendor-buyer integrated system in a supply chain.

Along with economic policies of the manufacturing processes, various other researchers have considered environmental policies to reduce detrimental impacts of the manufacturing systems on the outer environment. For this, carbon emission reduction is measured as an environment-friendly policy. Efforts are made through lot size adjustment, process improvement, better recourse utilization, and renewable energy consumption to reduce the carbon emissions of the system. Zeballos et al. [16], Sarkar et al. [17], and Xu et al. [18] considered the effects of carbon emissions in their production models and provided optimal production policies for the development of a sustainable manufacturing process. Our research work also considers carbon emission as an indicator of environmental influence of the manufacturing process analyzed in this study. Variable carbon emissions are considered during each operational activity of the manufacturing process, and the optimal lot size of the system is determined, which reduces the adverse environmental effects of the process.

The current focus of researchers is toward combinedly targeting operational and environmental scopes of the production environment through the integration of lean manufacturing and green manufacturing strategies [19]. Adopting lean practices provides green benefits, and further green policies often pave paths for lean benefits [20]. Diaz-Elsayed et al. [21] found that the collective implementation of lean and green practices has significantly reduced the production cost up to 10.80% in an automotive manufacturing firm, which is evidently a noticeable realization of implementing lean-green policies together. Recently, Thanki and Thakkar [22] developed a value-value load diagram (VVLD) tool to access the operational (lean) as well as environmental (green) performance of the organizations in combination with each other. This approach evaluates the lean-green performance of the firm established on various factors including resource utilization, value addition, and various others.

The above considerations promote the simultaneous implementation of lean and green policies to improve economic, as well as environmental, sustainability of the manufacturing industries. In view of the extensive association among lean and green practices and their combined benefits, this research work studied the simultaneous implementation of lean-green policy. Economic sustainability is aimed toward the implementation of the lean manufacturing strategy and economic lot size is obtained to attain the minimum cost of the system. Then, environmental sustainability is focused on the implementation of the green policy, considering emissions at each activity of the complete production system and the optimal lot size for minimal environmental impacts is achieved. A sustainable lot size is then obtained through the combined implementation of lean and green policies to improve eco-environmental performance of the multistage production system.

Most of the researchers have provided production models for a single-stage process, whereas almost all the products are manufactured through multistage manufacturing systems. Few researchers have studied multistage production processes. Among these researchers, Jaber and Khan [23] analyzed effects of learning and forgetting in multistage production process. Tayyab and Sarkar [12] developed a multistage lean manufacturing model by considering random imperfections at each stage of the system to obtain the optimal lot size through the analytical optimization technique. They found that the lean culture implementation has great potential to improve the economic sustainability of the multistage production process. Recently, Kim and Sarkar [13] provided optimum investment policies for setup cost reduction at all the process stages in an imperfect multistage production process. Their results showed that the reduction in imperfect production stabilizes the manufacturing process. Figure 1 shows a general flow of a multistage manufacturing process.

Literature review indicates that there exists a significant research gap in the field of the multistage manufacturing process, specifically under uncertain process conditions. Therefore, this research work extends the study of Tayyab and Sarkar [12] by taking initiative to consider highly uncertain product demand and defective proportion at all stages of the multistage manufacturing process. A Triangular Fuzzy Number (TFN) represents the product demand and the defective proportion at each manufacturing stage. The decomposition principle and signed distance method [14] of the fuzzy theory are applied to handle this uncertainty and de-fuzz the fuzzy objective function of the developed model. As this is a multi-objective multistage manufacturing model with conflicting objectives of cost-minimization and carbon emissions minimization, both the analytical optimization technique and the metaheuristic approach are applied to solve the model. The aspiration level of each objective function is determined by the individual solution of each objective function, then the Weighted Fuzzy Goal Programing (WFGP) model is developed for linearization of the multi-objective model, which is solved through the metaheuristic approach.

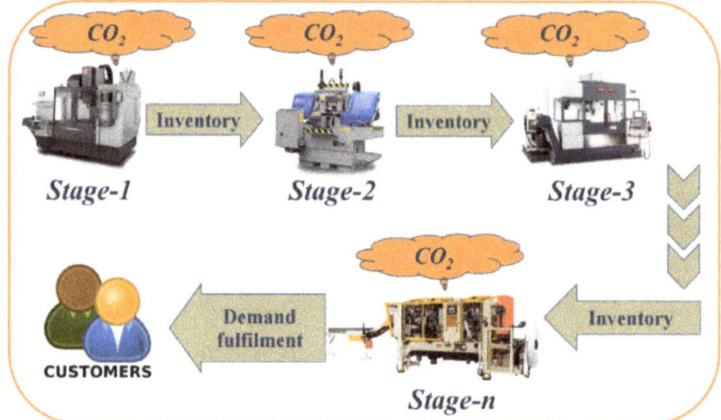

Figure 1. General flow of a multistage manufacturing process.

Structure of the paper is as follows: Section 2 provides problem definition, assumptions, and mathematical model of this research; Section 3 provides solution methodology of the model. Numerical experiment, results analysis and important managerial insights are provided in Section 4. Finally, Section 5 presents concluding remarks and future research opportunities related to the proposed model.

2. Model Formulation

This section provides problem definition, notation, assumptions, and the mathematical model of the proposed imperfect multistage lean manufacturing process under an uncertain environment.

2.1. Problem Definition

An imperfect multistage manufacturing process is studied in this paper. A single type of item is produced through an n-stage manufacturing process [12]. Due to out-of-control process conditions, imperfect quality items are produced, along with the perfect quality items at a highly uncertain rate [14]. A rework opportunity is allowed to convert these imperfect items to perfect quality items at additional reworking cost. To handle the uncertain information regarding product demand and the imperfect production proportion at each manufacturing stage, product demand and the imperfect proportion of the production are taken as TFN. The decomposition principle and signed distance method of fuzzy theory are applied to grasp the situation [14]. Customer demand is met at the final production stage [12]. Each stage emits a certain quantity of carbon emissions to the outer environment during the manufacturing process. The manufacturing system makes an effort to implement the lean philosophy of waste minimization and environmental protection by adopting the lean-green approach.

Two objectives of the model are to minimize the system cost and carbon emissions of the system. A multi-objective model for the simultaneous minimization of these two conflicting objectives (economic and environmental) is developed. To convert the multi-objective model into a single-objective model, WFGP is applied with a single objective of improving satisfaction levels of each objective function. The model is solved by an interior point metaheuristic approach and results are analyzed to illustrate the robustness and practical applicability of the proposed model.

2.2. Assumptions

For model formulation, the below assumptions are considered:

1. The model considers that a single type of item is produced in a multistage manufacturing process consisting of n number of stages.
2. In today's competitive market environment, product demand information cannot be estimated with accuracy. Therefore, this research work considers highly uncertain product demand, which is taken as a TFN (for instance, see Chang [14]).
3. Manufacturing companies keep their production rates higher than the demand rate to avoid shortages in the system. Therefore, this study considers that the production rate of each manufacturing stage is constant with the condition $P > \lambda$ and shortages are not allowed.
4. In the long run production process, the system moves to an "out-of-control" state and produces defective products, along with good quality products. This defective proportion may not be constant throughout nor follow specific probability distribution. Thus, this research work considers an uncertain defective proportion at each stage of the multistage production system and treats it as a TFN.
5. Production cost of the manufacturing system consists of various components. This study considers costs of energy, labor, and inspection costs at each manufacturing stage as the components of production cost.
6. Inter-stage material transfer cost and time are trivial in the multistage production systems. Therefore, this study considers that the inter-stage material transfer cost and time are negligible.

2.3. Mathematical Model

This section develops a multi-objective model for the proposed lean-green manufacturing system with aforementioned notation and assumptions. Figure 2 presents the inventory behavior of stage-j.

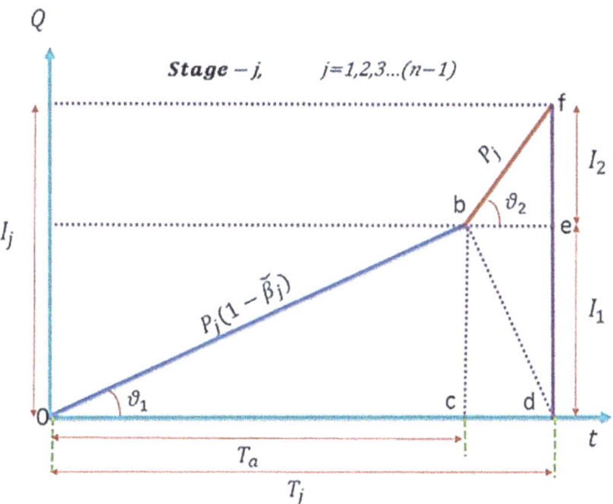

Figure 2. Inventory behavior of stage-j (reference: Tayyab and Sarkar [12]).

Manufacturing stage-j starts with zero inventory and achieves maximum inventory I_j in time $T_j = (1 + \beta_j) \times \frac{Q}{P_j}$ where $\tilde{\beta}_j$ is uncertain defective proportion at stage-j.

Regular production is carried out during $(0, T_a)$, and reworking of defective items is performed during the interval (T_a, T_j). Average inventory of stage-j is computed as below.

$$\overline{I_j} = \frac{1}{T_j} \times \left[\int_0^{T_a} Q(t).dt + \int_{T_a}^{T_j} Q(t).dt \right],$$

which provides

$$\overline{I}_j = \frac{Q}{2} \times \left[1 - \frac{\widetilde{\beta}_j^2}{1 + \widetilde{\beta}_j} \right].$$

(1)

Figure 3 shows inventory behavior of the final production stage (stage-n). Regular production, along with demand fulfilment, is carried out during $(0, T_{1n})$, reworking of defective items is performed from (T_{1n}, T_{2n}), and pure consumption occurs during the interval (T_{2n}, T_n). Cycle time of stage-n is obtained as $T_n = \frac{Q}{\widetilde{\lambda}}$, and the total time of the complete manufacturing system is simply computed as

$$T = \frac{Q}{\widetilde{\lambda}} \times \left[1 + \widetilde{\lambda} \times \sum_{j=1}^{n-1} \left(\frac{1 + \widetilde{\beta}_j}{P_j} \right) \right].$$

(2)

From Figure 3, total inventory of manufacturing stage-n is obtained as

$$I_n = \int_0^{T_{1n}} Q(t).dt + \int_{T_{1n}}^{T_{2n}} Q(t).dt + \int_{T_{2n}}^{T_n} Q(t).dt = \frac{Q^2}{2\widetilde{\lambda}} \times \left[1 - \frac{\widetilde{\lambda}}{P_n} \times \left(1 + \widetilde{\beta}_n + \widetilde{\beta}_n^2 \right) \right].$$

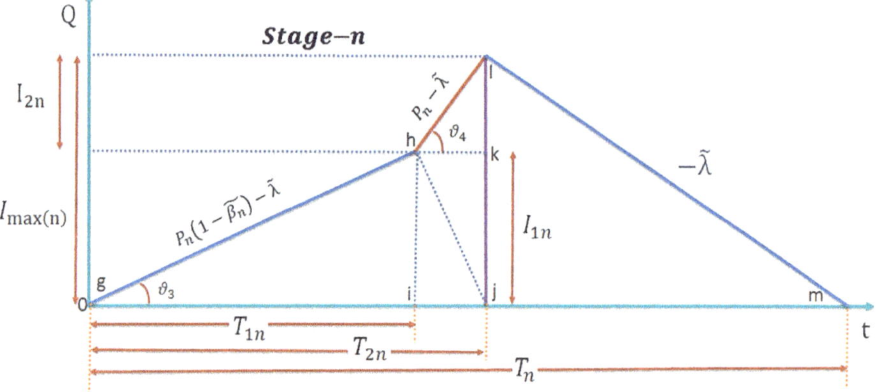

Figure 3. Inventory behavior of stage-n (reference: Tayyab and Sarkar [12]).

From above, the average inventory of finished items in the system is obtained as

$$\overline{I} = \frac{Q}{2P_n} \times \left[\frac{P_n - \widetilde{\lambda} \times \left(1 + \widetilde{\beta}_n + \widetilde{\beta}_n^2 \right)}{1 + \widetilde{\lambda} \times \sum_{j=1}^{n-1} \left(\frac{1 + \widetilde{\beta}_j}{P_j} \right)} \right].$$

(3)

2.3.1. Objective Function Formulation with Constant Demand and Imperfect Proportion

To obtain a sustainable lot size, this paper proposes a multi-objective production model for an imperfect multistage lean manufacturing system. Multi-objectives contain economic objective for cost minimization, and an environmental objective for CO_2 emissions reduction to develop a lean-green manufacturing system.

Cost Minimization Objective (f_{Cost})

The cost minimization objective of the proposed multi-objective production model with constant demand rate and defective proportion is formulated in Equation (4). The first part of the objective function depicts the setup cost of the system. The second term shows the order processing cost and reworking cost of the system. The final term in the objective function represents the inventory carrying cost of finished items in the system.

$$
\begin{aligned}
TC(Q) \quad &= \frac{1}{T} \times \sum_{i=1}^{n} A_{ci} + \frac{1}{T}\left[Q \times \sum_{i=1}^{n} C_{ci} \times (1+\beta_i) \right] + h_c \times \frac{Q}{2P_n} \times \left[\frac{P_n - \lambda \times \left(1 + \beta_n + \beta_n{}^2\right)}{1 + \lambda \times \sum_{j=1}^{n-1}\left(\frac{1+\beta_j}{P_j}\right)} \right] \\[2mm]
&= \frac{\sum_{i=1}^{n} A_{ci} + Q \times \sum_{i=1}^{n} C_{ci} \times (1+\beta_i)}{\frac{Q}{\lambda} \times \left[1 + \lambda \times \sum_{j=1}^{n-1}\left(\frac{1+\beta_j}{P_j}\right) \right]} + h_c \times \frac{Q}{2P_n} \times \left[\frac{P_n - \lambda \times \left(1 + \beta_n + \beta_n{}^2\right)}{1 + \lambda \times \sum_{j=1}^{n-1}\left(\frac{1+\beta_j}{P_j}\right)} \right].
\end{aligned}
\tag{4}
$$

CO_2 Emissions Minimization Objective $\left(f_{CO_2}\right)$

CO_2 minimization objective of the proposed multi-objective production model in fuzzy form is formulated in Equation (5). The first part of the objective function depicts CO_2 emissions during setup of the system. The second term shows emissions during the order processing and the reworking of the defective items in the system. The final term in the objective function represents emissions during inventory carrying of finished items in the system.

$$
\begin{aligned}
TE(Q) \quad &= \frac{1}{T} \times \sum_{i=1}^{n} A_{ei} + \frac{1}{T}\left[Q \times \sum_{i=1}^{n} C_{ei} \times (1+\beta_i) \right] + h_e \times \frac{Q}{2P_n} \times \left[\frac{P_n - \lambda \times \left(1 + \beta_n + \beta_n{}^2\right)}{1 + \lambda \times \sum_{j=1}^{n-1}\left(\frac{1+\beta_j}{P_j}\right)} \right], \\[2mm]
&= \frac{\sum_{i=1}^{n} A_{ei} + Q \times \sum_{i=1}^{n} C_{ei} \times (1+\beta_i)}{\frac{Q}{\lambda} \times \left[1 + \lambda \times \sum_{j=1}^{n-1}\left(\frac{1+\beta_j}{P_j}\right) \right]} + h_e \times \frac{Q}{2P_n} \times \left[\frac{P_n - \lambda \times \left(1 + \beta_n + \beta_n{}^2\right)}{1 + \lambda \times \sum_{j=1}^{n-1}\left(\frac{1+\beta_j}{P_j}\right)} \right].
\end{aligned}
\tag{5}
$$

2.3.2. Objective Function Formulation with Fuzzy Demand and Imperfect Proportion

This model considers defective proportion at each stage of the manufacturing process as highly uncertain. Thus, an uncertainty control approach is required to handle this vagueness in system information. Fuzzy theory is most suitable for such type of operating conditions under uncertain environment [14]. Therefore, defective proportion is taken as a TFN, and the decomposition principle, along with the signed distance approach, are applied to convert the fuzzy objective functions into their equivalent crisp forms. For the fuzzy set $\widetilde{\kappa} = (\Delta_1, \Delta_2, \Delta_3)$, defined on R with $\Delta_1 < \Delta_2 < \Delta_3$ is a TFN with a membership function

$$
\varphi_{\widetilde{\kappa}}(x) = \begin{cases}
\frac{x - \Delta_1}{\Delta_2 - \Delta_1} & if \quad \Delta_1 \le x \le \Delta_2, \\[2mm]
\frac{\Delta_3 - x}{\Delta_3 - \Delta_2} & if \quad \Delta_2 \le x \le \Delta_3, \\[2mm]
0 & otherwise.
\end{cases}
$$

According to decomposition principle, $\alpha - cut$ of $\widetilde{\kappa} = (\Delta_1, \Delta_2, \Delta_3)$ is $\kappa(\alpha) = \left[\kappa_L(\alpha), \kappa_U(\alpha)\right]$ for $\alpha \in [0, 1]$ where $\kappa_L(\alpha) = \Delta_1 + (\Delta_2 - \Delta_1)\alpha$ and $\kappa_U(\alpha) = \Delta_3 - (\Delta_3 - \Delta_2)\alpha$. Then the signed distance of $\widetilde{\kappa}$ to $\widetilde{0_1}$ is determined as $d\left(\widetilde{\kappa}, \widetilde{0_1}\right) = \frac{1}{4}(\Delta_1 + 2\Delta_2 + \Delta_3)$.

Using the aforementioned uncertainty control approach, uncertainty in product demand and the defective proportion data for both objective functions is handled as below.

Fuzzy Membership Function Development

For the product demand of the manufacturing process represented by a fuzzy set $\tilde{\lambda} = (\lambda_1, \lambda_2, \lambda_3)$, defined on R with $\lambda_1 < \lambda_2 < \lambda_3$ is a TFN with a membership function

$$\omega_{\tilde{\lambda}}(\rho) = \begin{cases} \frac{\rho - \lambda_1}{\lambda_2 - \lambda_1} & if & \lambda_1 \le \rho \le \lambda_2, \\ \frac{\lambda_3 - \rho}{\lambda_3 - \lambda_2} & if & \lambda_2 \le \rho \le \lambda_3, \\ 0 & otherwise. \end{cases}$$

On the similar scale, for the defective proportion at each stage of the manufacturing process represented by a fuzzy set $\tilde{\beta} = (\delta_1, \delta_2, \delta_3)$, defined on R with $\delta_1 < \delta_2 < \delta_3$ is a TFN with a membership function

$$\varphi_{\tilde{\beta}}(x) = \begin{cases} \frac{x - \delta_1}{\delta_2 - \delta_1} & if & \delta_1 \le x \le \delta_2, \\ \frac{\delta_3 - x}{\delta_3 - \delta_2} & if & \delta_2 \le x \le \delta_3, \\ 0 & otherwise. \end{cases}$$

Fuzzification

Fuzzified form of cost minimization objective (f_{Cost}) and CO_2 minimization objective (f_{CO_2}) are developed as below.

$$\widetilde{TC}(Q) = \frac{\sum\limits_{i=1}^{n} A_{ci} + Q \times \sum\limits_{i=1}^{n} C_{ci} \times \left(1 + d\left(\tilde{\beta}_i, \tilde{0}_1\right)\right)}{\frac{Q}{d(\tilde{\lambda}, \tilde{0}_1)} \times \left[1 + d\left(\tilde{\lambda}, \tilde{0}_1\right) \times \sum\limits_{j=1}^{n-1} \left(\frac{1 + d\left(\tilde{\beta}_j, \tilde{0}_1\right)}{P_j}\right)\right]} + h_c \times \frac{Q}{2P_n} \times \left[\frac{P_n - d\left(\tilde{\lambda}, \tilde{0}_1\right) \times \left(1 + d\left(\tilde{\beta}_n, \tilde{0}_1\right) + d\left(\tilde{\beta}_n, \tilde{0}_1\right)^2\right)}{1 + d\left(\tilde{\lambda}, \tilde{0}_1\right) \times \sum\limits_{j=1}^{n-1} \left(\frac{1 + d\left(\tilde{\beta}_j, \tilde{0}_1\right)}{P_j}\right)}\right],$$

and

$$\widetilde{TE}(Q) = \frac{\sum\limits_{i=1}^{n} A_{ei} + Q \times \sum\limits_{i=1}^{n} C_{ei} \times \left(1 + d\left(\tilde{\beta}_i, \tilde{0}_1\right)\right)}{\frac{Q}{d(\tilde{\lambda}, \tilde{0}_1)} \times \left[1 + d\left(\tilde{\lambda}, \tilde{0}_1\right) \times \sum\limits_{j=1}^{n-1} \left(\frac{1 + d\left(\tilde{\beta}_j, \tilde{0}_1\right)}{P_j}\right)\right]} + h_e \times \frac{Q}{2P_n} \times \left[\frac{P_n - d\left(\tilde{\lambda}, \tilde{0}_1\right) \times \left(1 + d\left(\tilde{\beta}_n, \tilde{0}_1\right) + d\left(\tilde{\beta}_n, \tilde{0}_1\right)^2\right)}{1 + d\left(\tilde{\lambda}, \tilde{0}_1\right) \times \sum\limits_{j=1}^{n-1} \left(\frac{1 + d\left(\tilde{\beta}_j, \tilde{0}_1\right)}{P_j}\right)}\right].$$

Defuzzification

Defuzzification is the conversion of the fuzzy linguistic variable to its equivalent crisp form for further computations of the objective functions. This research work has used the signed distance method of defuzzification for uncertain objective functions of the model.

For TFN, the signed distance of $\tilde{\lambda}$ to $\tilde{0}_1$ is determined as

$$d\left(\tilde{\lambda}, \tilde{0}_1\right) = \frac{1}{4}(\lambda_1 + 2\lambda_2 + \lambda_3),$$

and the signed distance of $\tilde{\beta}$ to $\tilde{0}_1$ is obtained as

$$d\left(\tilde{\beta}, \tilde{0}_1\right) = \frac{1}{4}(\delta_1 + 2\delta_2 + \delta_3).$$

Putting this in the fuzzy objective functions defuzzify them in their equivalent crisp form as below.

$$[TC]^{Equivalent\ crisp} = \frac{\left(\frac{1}{4}(\lambda_1 + 2\lambda_2 + \lambda_3)\right) \times \left(\sum\limits_{i=1}^{n} A_{ci} + Q \times \sum\limits_{i=1}^{n} C_{ci} \times \left(1 + \left(\frac{1}{4}(\delta_{1i} + 2\delta_{2i} + \delta_{3i})\right)\right)\right)}{Q \times \left[1 + \left(\frac{1}{4}(\lambda_1 + 2\lambda_2 + \lambda_3)\right) \times \sum\limits_{j=1}^{n-1} \left(\frac{1 + \left(\frac{1}{4}(\delta_{1j} + 2\delta_{2j} + \delta_{3j})\right)}{P_j}\right)\right]}$$

$$+ h_c \times \frac{Q}{2P_n} \times \left[\frac{P_n - \left(\frac{1}{4}(\lambda_1 + 2\lambda_2 + \lambda_3)\right) \times \left(1 + \left(\frac{1}{4}(\delta_{1n} + 2\delta_{2n} + \delta_{3n})\right) + \left(\frac{1}{4}(\delta_{1n} + 2\delta_{2n} + \delta_{3n})\right)^2\right)}{1 + \left(\frac{1}{4}(\lambda_1 + 2\lambda_2 + \lambda_3)\right) \times \sum\limits_{j=1}^{n-1} \left(\frac{1 + \left(\frac{1}{4}(\delta_{1j} + 2\delta_{2j} + \delta_{3j})\right)}{P_j}\right)}\right],$$

and

$$[TE]^{Equivalent\ crisp} = \frac{\left(\frac{1}{4}(\lambda_1+2\lambda_2+\lambda_3)\right)\times\left(\sum_{i=1}^{n}A_{ei}+Q\times\sum_{i=1}^{n}C_{ei}\times\left(1+\left(\frac{1}{4}(\delta_{1i}+2\delta_{2i}+\delta_{3i})\right)\right)\right)}{Q\times\left[1+\left(\frac{1}{4}(\lambda_1+2\lambda_2+\lambda_3)\right)\times\sum_{j=1}^{n-1}\left(\frac{1+\left(\frac{1}{4}\left(\delta_{1j}+2\delta_{2j}+\delta_{3j}\right)\right)}{P_j}\right)\right]}$$

$$+h_e\times\frac{Q}{2P_n}\times\left[\frac{P_n-\left(\frac{1}{4}(\lambda_1+2\lambda_2+\lambda_3)\right)\times\left(1+\left(\frac{1}{4}(\delta_{1n}+2\delta_{2n}+\delta_{3n})\right)+\left(\frac{1}{4}(\delta_{1n}+2\delta_{2n}+\delta_{3n})\right)^2\right)}{1+\left(\frac{1}{4}(\lambda_1+2\lambda_2+\lambda_3)\right)\times\sum_{j=1}^{n-1}\left(\frac{1+\left(\frac{1}{4}\left(\delta_{1j}+2\delta_{2j}+\delta_{3j}\right)\right)}{P_j}\right)}\right].$$

In order to verify that the $[TC]^{Crisp}$ attends minimum value in the interval $(0,T)$, one can simply apply analytical technique. As $[TC]^{Crisp}$ is a minimization objective with production quantity (Q_{Cost}) as a decision variable, thus, principle minor must be positive. From necessary conditions, $\frac{d[TC]^{Crisp}}{dQ_{Cost}}=0$.

From sufficient conditions, it is verified that

$$\frac{d^2[TC]^{Crisp}}{dQ^2_{Cost}}=\frac{2\lambda\sum_{i=1}^{n}A_{ci}}{Q^3\left(1+\frac{1}{4}(\lambda_1+2\lambda_2+\lambda_3)\times\sum_{j=1}^{n-1}\frac{1+\frac{1}{4}\left(\delta_{1j}+2\delta_{2j}+\delta_{3j}\right)}{P_j}\right)}>0,$$

which proves that $[TC]^{Crisp}$ is strictly convex in Q^*_{Cost}. Hence, $[TC]^{Crisp}$ has a minimum value in the interval $(0,T)$.

Similarly, it can be proved that $[TE]^{Crisp}$ attends the minimum value in the interval $(0,T)$ as the following: $[TE]^{Crisp}$ is a CO_2 minimization objective with production quantity (Q_{CO_2}) as a decision variable. Thus, the principle minor must be positive. From necessary conditions, $\frac{d[TE]^{Crisp}}{dQ_{CO_2}}=0$.

From sufficient conditions, it is verified that

$$\frac{d^2[TE]^{Crisp}}{dQ^2_{CO_2}}=\frac{2\lambda\sum_{i=1}^{n}A_{ei}}{Q^3\left(1+\frac{1}{4}(\lambda_1+2\lambda_2+\lambda_3)\sum_{j=1}^{n-1}\frac{1+\frac{1}{4}\left(\delta_{1j}+2\delta_{2j}+\delta_{3j}\right)}{P_j}\right)}>0,$$

which proves that $[TE]^{Crisp}$ is strictly convex in $Q^*_{CO_2}$. Hence, $[TE]^{Crisp}$ has a minimum value in the interval $(0,T)$.

3. Solution Procedure

Goal programming (GP) is a well-known method of solving multi-objective optimization problems. There are several variants of GP used under different conditions. WFGP is proven to outperform in the scenario of an uncertain environment [12,20]. In this method, decision-makers decide the aspiration level and maximum acceptable level of each objective, which is then converted into a fuzzy membership function and the satisfaction level of each objective is determined. Then, the satisfaction level of each objective is maximized according to the predetermined importance criteria provided by the decision-makers. This research work applies WFGP to solve the proposed multi-objective nonlinear programing model.

Solution steps for the proposed multi-objective multi-stage lean manufacturing model are provided below.

1. Determine $\alpha-$ extreme solutions

Each objective function is solved separately to obtain target (aspiration) level of the function. Then maximum acceptable level of each objective function is determined by keeping aspiration level of other objective as an equality constraint, and a pay-off table (POF) is developed.

2. Develop Fuzzy membership function (FMF)

Fuzzy membership function (μ_k) for satisfaction level (TFN) of each objective is developed in this step by using aspiration level and maximum acceptable level of each objective determined in the previous step.

$$\mu_k = \begin{cases} 0 & if \quad f_k \geq f_k^{\alpha-accept}, \\ \dfrac{f_k^{\alpha-accept}-f_k}{f_k^{\alpha-accept}-f_k^{\alpha-aspiration}} & if \quad f_k^{\alpha-aspiration} < f_k < f_k^{\alpha-accept}, \\ 1 & if \quad f_k \leq f_k^{\alpha-aspiratio}, \end{cases}$$

where $f_k^{\alpha-aspiration}$ and $f_k^{\alpha-accept}$ are the $\alpha-$extreme solutions of objective function k.

3. Develop WFGP model

WFGP is an interactive technique which requires expert human intervention to assign priority weights to each objective function in accordance with the decision-maker's choice. Thus, the decision-makers determine importance weight (Z_k) for satisfaction level of each objective function k. WFGP model is then developed as below, which is in turn solved through a metaheuristic approach.

$$Maximize \quad \sum_{k=1}^{K} Z_k \times \mu_k$$

$$Subject\ to \quad \dfrac{f_k^{\alpha-accept}-f_k}{f_k^{\alpha-accept}-f_k^{\alpha-aspiration}} = \mu_k, \quad \forall k$$

$$\sum_{k=1}^{K} Z_k = 1, \quad \forall k$$

$$0 \leq \mu_k \leq 1. \quad \forall k$$

4. Experimental Analysis

To illustrate the practical implication of the developed multi-objective production model, a numerical experiment is carried out for a five-stage lean-green manufacturing process with the appropriate data set.

4.1. Numerical Data

Table 1 shows the numerical data (see Notation for units of parameters), modified from Tayyab and Sarkar [12] and Sarkar and Mahapatra [24].

Table 1. Data for numerical experiment.

Parameter	Value	Parameter	Value
n	5	$(\lambda_1, \lambda_2, \lambda_3)$	(460 490 610)
P_i	[810 795 780 765 750]	A_{ic}	[50 35 65 50 20]
C_{ic}	[5 2.5 3 4.2 7]	$(\delta_{1i}, \delta_{2i}, \delta_{3i})$	[0.05 0.10 0.17]
A_{ie}	[100 70 130 100 40]	C_{ie}	[10 5 6 8.4 14]
h_c	25	h_e	135
Z_{Cost}	0.60	Z_{CO_2}	0.40

4.2. Numerical Solution Procedure

Multi-objective WFGP model is solved with the given data set (Table 1) in the following steps:

Step 1. Find $\alpha-$ positive extreme solution of each objective function by solving them separately and record the values (Table 2) as their respective aspiration levels (where $\lambda_1 = \lambda_2 = \lambda_3$ and $\delta_{1i} = \delta_{2i} = \delta_{3i}$ for the model with constant demand and imperfect proportion).

Table 2. Aspiration levels of model objectives.

Model Consideration	Objective	Aspiration Level
Constant demand and imperfect proportion	f_{Cost} (\$/unit time) f_{CO_2} (units/ unit time)	3453.67 7326.03
Fuzzy demand and imperfect proportion	f_{Cost} (\$/unit time) f_{CO_2} (units/ unit time)	3467.737 7318.47

Step 2. Find $\alpha-$ negative extreme solution of each objective function by taking the aspiration level of each objective as an equality constraint for second objective and record the values (Table 3) as the form of a pay-off table (POT).

Table 3. Payoff table for multi-objectives.

Model Consideration	POT	f_{Cost} (\$/unit time)	f_{CO_2} (units/unit time)
Constant demand and imperfect proportion	f_{Cost} (\$/unit time) f_{CO_2} (units/unit time)	3453.67 7460.49	3494.71 7326.03
Fuzzy demand and imperfect proportion	f_{Cost} (\$/unit time) f_{CO_2} (units/ unit time)	3467.37 7441.69	3504.92 7318.47

Step 3. Develop fuzzy membership functions of the satisfaction level (μ_k) for each objective as below (for instance, regarding model with fuzzy parameters).

$$\mu_{Cost} = \begin{cases} 0 & if \quad f_{Cost} \geq 3504.92, \\ \frac{3504.92 - f_{Cost}}{3504.92 - 3467.37} & if \quad 3467.37 < f_{Cost} < 3504.92, \\ 1 & if \quad f_k \leq 3467.37, \end{cases}$$

and

$$\mu_{CO_2} = \begin{cases} 0 & if \quad f_{CO_2} \geq 7441.69, \\ \frac{7441.69 - f_{CO_2}}{7441.69 - 7318.47} & if \quad 7318.47 < f_{CO_2} < 7441.69, \\ 1 & if \quad f_{CO_2} \leq 7318.47. \end{cases}$$

Step 4. Decide the importance weight (Z_k) for each objective function and develop a WFGP model as below.

$$\begin{aligned} Maximize \quad & Z_{Cost} \times \mu_{Cost} + Z_{CO_2} \times \mu_{CO_2} \\ Subject\ to \quad & \frac{3504.92 - f_{Cost}}{3504.92 - 3,467.37} = \mu_{Cost}, \\ & \frac{7441.69 - f_{CO_2}}{7441.69 - 7318.47} = \mu_{CO_2}, \\ & \sum_{k=1}^{K} Z_k = 1, \qquad \forall k \\ & 0 \leq \mu_k \leq 1. \qquad \forall k \end{aligned}$$

Step 5. Obtain optimal tradeoff solutions of the single-objective model developed in Step 4 through a heuristic approach.

4.3. Computational Results

As the developed WFGP model is a nonlinear maximization problem, a heuristic approach is required to solve the model. Therefore, interior point optimization is applied using MATLAB R2015a with system specifications of 4GB RAM and 1.80 GHz processor speed, and the optimal solution for the model with uncertain information is achieved in 11.54 s. Table 4 and Figure 4 show the optimal tradeoff values of the conflicting objective functions. Table 5 shows the optimal values of decision variables in the multi-objective optimization model.

Table 4. Optimal results of multi-objectives.

Model Consideration	Attributes	f_{Cost}^*	$f_{CO_2}^*$
Constant demand and imperfect proportion	Objective value	3460.20 ($/unit time)	7373.45 (units/unit time)
	Satisfaction level	84.09%	64.73%
Fuzzy demand and imperfect proportion	Objective value	3473.35 ($/unit time)	7361.92 (units/unit time)
	Satisfaction level	84.07%	64.74%

Table 5. Optimal decision variable values.

Model Consideration	Decision Variable	Optimal Value
Constant demand and imperfect proportion	Q_{Cost}^* (items)	177.15
	$Q_{CO_2}^*$ (items)	107.81
	$Q_{sustainable}^*$ (items)	145.04
Fuzzy demand and imperfect proportion	Q_{Cost}^* (items)	194.93
	$Q_{CO_2}^*$ (items)	118.63
	$Q_{sustainable}^*$ (items)	159.59

4.4. Result Analysis and Discussion

Results indicate a considerate tradeoff between conflicting objectives of the proposed model. Table 4 verifies that the optimal values of the objective functions lie within the acceptable limits, as defined in the POT. One can observe that the optimal cost of the model with constant demand and imperfect proportion is less than the model with fuzzy demand and imperfect proportion. This variation is due to the fact that the model with fuzzy parameters is handling a wide range of scattered information by the addition of a certain cost.

4.4.1. Satisfaction Level Achievement

Results indicate that the proposed fuzzy model has successfully achieved 84.07% satisfaction level for cost objective and 64.74% satisfaction level for CO_2 emissions objective as computed below.

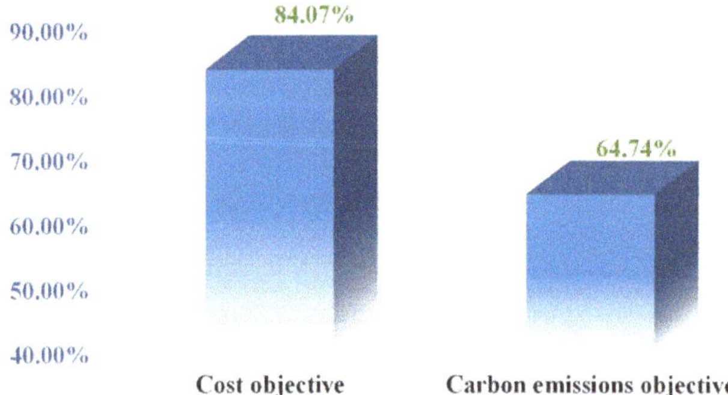

Figure 4. Optimal satisfaction levels for the model with uncertain information.

$$\mu_{Cost} = \frac{3504.92 - 3473.35}{3504.92 - 3467.37} = 84.07\%,$$

and

$$\mu_{CO_2} = \frac{7441.69 - 7361.92}{7441.69 - 7318.47} = 64.74\%.$$

One can compute the contribution of an achieved satisfaction level of each conflicting objective in the proposed multi-objective lean-green multistage manufacturing model as $contribution_k = \frac{\mu_k}{\sum_k \mu_k}$, which indicates that the contribution of μ_{Cost} and μ_{CO_2} are 56.49% and 43.51%, respectively (Figure 5) in the overall performance of the proposed multi-objective multistage lean-green manufacturing system.

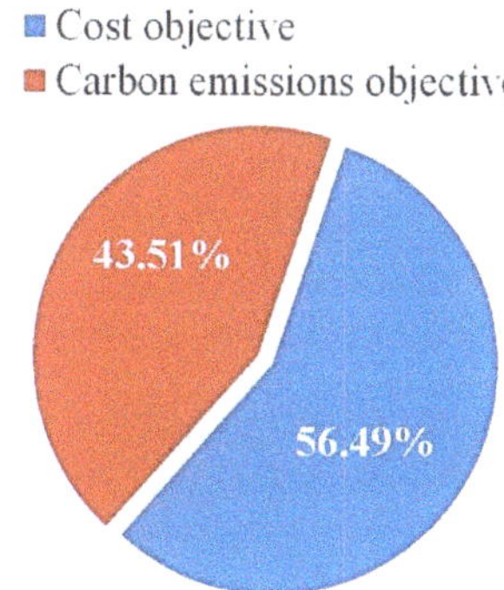

Figure 5. Contribution of satisfaction levels for the model with uncertain information.

4.4.2. Effect of Uncertain Environment

Proposed model is analyzed for different instances of uncertain process information by varying values of fuzzy deviational variables and their effect is studied. Table 6 shows the optimal tradeoff among model objectives for below levels of uncertainty in the imperfect production proportion of the manufacturing system.

$$case\ 1: \ (\delta_{1i}, \delta_{2i}, \delta_{3i}) = (0.05 \quad 0.10 \quad 0.25),$$
$$case\ 2: \ (\delta_{1i}, \delta_{2i}, \delta_{3i}) = (0.10 \quad 0.10 \quad 0.10),$$
$$case\ 3: \ (\delta_{1i}, \delta_{2i}, \delta_{3i}) = (0.00 \quad 0.10 \quad 0.17).$$

Table 6. Effect of uncertain information on optimal results.

Case	$Q^*_{sustainable}$	f_{Cost}	f_{CO_2}
Case 1	165.58	3473.08	7341.14
Case 2	158.23	3473.31	7366.82
Case 3	156.26	3473.15	7373.96

It can be observed from Table 6 that the uncertainty in information bears a significant degree of impact on the optimal trade-off value of each objective function in the model.

4.4.3. Sensitivity Analysis

The effects of various cost and CO_2 emission parameters on optimal results of the multi-objective function of the model with uncertain demand and imperfect proportion are studied by varying their values from -50% to $+50\%$, as shown in Table 7.

Sensitivity analysis of the model shows that the production cost and corresponding CO_2 emissions of the manufacturing system create the highest impact on the total cost and emissions of the system, respectively. A 50% decrease in the production cost of the system reduces 45.29% of the system cost, and vice versa. Similarly, a 50% decrease in CO_2 emissions of the system during production reduces 42.64% emissions of the system, and vice versa. Variations in the setup cost and inventory carrying cost have a trivial impact on the system cost. Similar behavior is shown by the variations in CO_2 emissions of the system during the order setup and inventory carrying process. Figures 6 and 7 show the effect of variations in various key parameters on the cost and CO_2 emissions minimization objective as a whole.

Table 7. Sensitivity analysis for important parameters of numerical experiment.

Change in Parameter(s) (%)		Change in TC^* (%)	Change in Parameter(s) (%)		Change in TE^* (%)
	-50	-2.78		-50	-4.34
A_{ci}	-25	-1.29	A_{ei}	-25	-1.96
	$+25$	$+1.17$		$+25$	$+1.71$
	$+50$	$+2.26$		$+50$	$+3.34$
	-50	-45.29		-50	-42.64
C_{ci}	-25	-22.64	C_{ei}	-25	-21.32
	$+25$	$+22.64$		$+25$	$+21.32$
	$+50$	$+45.29$		$+50$	$+42.64$
	-50	-2.78		-50	-4.34
h_c	-25	-1.29	h_e	-25	-1.96
	$+25$	$+1.17$		$+25$	$+1.71$
	$+50$	$+2.26$		$+50$	$+3.34$

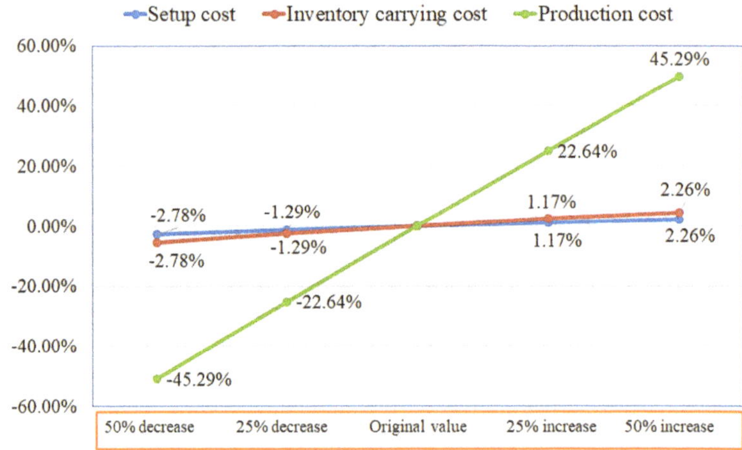

Figure 6. Sensitivity analysis of cost objective for the model with uncertain information.

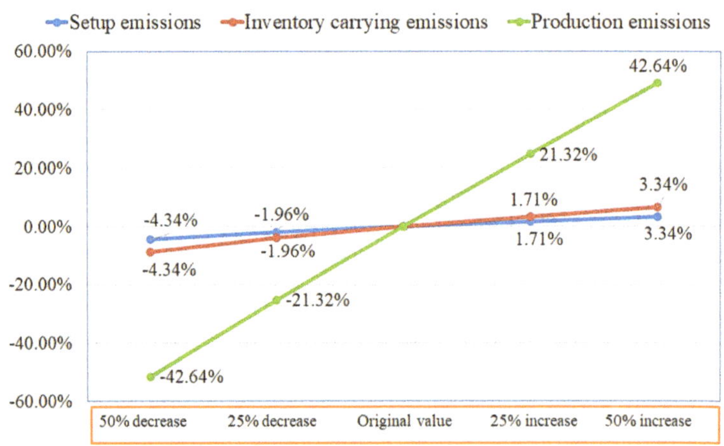

Figure 7. Sensitivity analysis of CO_2 emissions objective for the model with uncertain information.

4.5. Managerial Insights

Analysis of the results verify that the proposed model has provided a practical approach to achieve optimal production policy for imperfect multistage lean manufacturing system. Below managerial insights are inferred from the model analysis:

Insight 1. The proposed study provides a handy tool to the managers and decision-makers of the imperfect manufacturing process to transform it into lean-green manufacturing system by reducing excessive inventories and CO_2 emissions of the complete multistage manufacturing process. This improves the effectiveness of the system decisions to attain economic as well as environmental sustainability.

Insight 2. Uncertainty control analysis indicates that the proposed model is significantly capable of handling high uncertainty in the parametric information by implementing WFGP approach. Model has achieved reasonable satisfaction levels of the conflicting objective functions under high uncertainty. Thus, managers can conclude best possible tradeoffs among corporate objectives through implementation of this study and can compute sustainable lot size for a sustainable manufacturing process.

Insight 3. Sensitivity analysis of the system confirms that the impact of setup cost and inventory carrying cost is same on the total cost of the system. Thus, managers must wisely decide about making efforts to reduce setup cost or carrying cost first. This can be done by implementing lean strategies including SMED (single minute exchange of dyes) and FIFO (first-in-first-out) approaches for cutting down setup cost and inventory carrying costs, respectively.

5. Conclusions

This research work provides an optimal production policy for uncertain process information in a multistage lean manufacturing process. Market demand information of the product is uncertain in nature. Imperfect manufacturing process produces defective products at an uncertain rate, which are reworked to convert them into perfect quality products and reduce wastages. To control this uncertainty in the market demand and manufacturing process, the decomposition principle and the signed distance method of fuzzy theory are applied. The manufacturing process is analyzed for its environmental concerns, and a sustainable lot size is obtained through metaheuristic approach for simultaneous achievement of economic as well as environmental sustainability. The WFGP approach is applied to handle the multi-objective model with conflicting objectives for cost and CO_2 reductions of the system. An experimental study is performed to verify the practical implication of the model, and results are evaluated through an uncertainty control analysis and sensitivity analysis of the important parameters.

Experimental results verify that the proposed model provides a real-world approach to achieve the optimal production policy for an imperfect multistage lean manufacturing system by achieving an 84.07% satisfaction level for the cost-minimization objective and 64.74% satisfaction level for the CO_2 emissions objective function. Imperative managerial insights are inferred from the analysis of the optimal results, which indicate the robustness of the proposed model and its applicability in the real-world lean-green manufacturing processes. The proposed study can be extended in several directions. Consideration of controllable production rate of the manufacturing system [25–27] can be the immediate possible extensions. Other potential extensions may include planned backlogging [28,29], partial backlogging [30–32], trade credit policies [33,34], random defective rate [35,36], product deterioration [37], and constrained manufacturing environment [38] in the production system.

Author Contributions: Conceptualization, M.T. and B.S.; Methodology, M.T.; Software, M.T.; Validation, B.S.; Formal analysis, M.T.; Investigation, M.T.; Resources, M.T. and B.S.; Data curation, M.T.; Writing—original draft preparation, M.T.; Writing—review and editing, M.U. and B.S.; Visualization, M.T., M.U. and B.S.; Supervision, B.S.

Funding: This research received no external funding.

Conflicts of Interest: The authors declare no conflict of interest.

Nomenclature

Indices

i	number of production stages, $i = 1, 2, 3 \ldots n$
j	number of production stages, $j = 1, 2, 3 \ldots (n-1)$
k	number of model objectives, $k = 1, 2, 3 \ldots K$

Decision variables

Q^*_{Cost}	optimal lot size to achieve minimum system cost (items)
$Q^*_{CO_2}$	optimal lot size to achieve minimum carbon emissions (items)
$Q^*_{sustainable}$	optimal lot size to achieve economic and environmental sustainability (items)

Parameters

P_i	production rate of manufacturing stage-i (items/year)
$\widetilde{\lambda}$	customer demand (items/year) (fuzzy number)
$\widetilde{\beta}_i$	imperfect production proportion at stage-i (fuzzy number)
A_{ic}	setup cost of manufacturing stage-i (\$/setup)
C_{ic}	production cost at manufacturing stage-i (\$/item)
h_c	inventory carrying cost of finished items (\$/item/year)
A_{ie}	CO_2 emissions during setup at stage-i (units/setup)
C_{ie}	CO_2 emissions during product manufacturing at stage-i (units/item)
h_e	CO_2 emissions during inventory carrying of finished items (units/item/year)
$\overline{I}_i$	average inventory of stage-i (items per unit time)
TC	total cost of the system (\$/unit time)
TE	total CO_2 emissions of the system (units/unit time)
Z_k	importance weight of objective-k for WFGP model
μ_k	satisfaction level of objective-k
δ_i, λ_i	fuzzy deviational variables

Abbreviations

Below abbreviations are used for the development of this model.

WFGP	Weighted fuzzy goal programming
EPQ	Economic production quantity
TFN	Triangular fuzzy number
TC	Total cost
TE	Total Emissions
POF	Pay-off table
FMF	Fuzzy membership function

References

1. Hodge, G.L.; Goforth Ross, K.; Joines, J.A.; Thoney, K. Adapting lean manufacturing principles to the textile industry. *Prod. Plan. Control* **2011**, *22*, 237–247. [CrossRef]
2. Fisher, M. Process improvement by poka-yoke. *Work Study* **1999**, *48*, 264–266. [CrossRef]
3. McNamara, P. Psychological factors affecting the sustainability of 5S lean. *Int. J. Lean Enterp. Res.* **2014**, *1*, 94–111. [CrossRef]
4. Bilalis, N.; Scroubelos, G.; Antoniadis, A.; Emiris, D.; Koulouriotis, D. Visual factory: Basic principles and the 'zoning' approach. *Int. J. Prod. Res.* **2002**, *40*, 3575–3588. [CrossRef]
5. Sugimori, Y.; Kusunoki, K.; Cho, F.; Uchikawa, S. Toyota production system and kanban system materialization of just-in-time and respect-for-human system. *Int. J. Prod. Res.* **1977**, *15*, 553–564. [CrossRef]
6. Fresner, J. Starting continuous improvement with a cleaner production assessment in an Austrian textile mill. *J. Clean. Prod.* **1998**, *6*, 85–91. [CrossRef]
7. Ozturk, E.; Koseoglu, H.; Karaboyaci, M.; Yigit, N.O.; Yetis, U.; Kitis, M. Sustainable textile production: Cleaner production assessment/eco-efficiency analysis study in a textile mill. *J. Clean. Prod.* **2016**, *138*, 248–263. [CrossRef]

8.   Taft, E.W. The most economical production lot. *Iron Age* **1918**, *101*, 1410–1412.

9.   Sarkar, B.; Sana, S.S.; Chaudhuri, K. An imperfect production process for time varying demand with inflation and time value of money—An EMQ model. *Expert Syst. Appl.* **2011**, *38*, 13543–13548. [CrossRef]

10.   Chiu, S.W.; Chiu, Y.S.; Yang, J.C. Combining an alternative multi-delivery policy into economic production lot size problem with partial rework. *Expert Syst. Appl.* **2012**, *39*, 2578–2583. [CrossRef]

11.   Sarkar, B.; Chaudhuri, K.; Moon, I. Manufacturing setup cost reduction and quality improvement for the distribution free continuous-review inventory model with a service level constraint. *J. Manuf. Syst.* **2015**, *34*, 74–82. [CrossRef]

12.   Tayyab, M.; Sarkar, B. Optimal batch quantity in a cleaner multi-stage lean production system with random defective rate. *J. Clean. Prod.* **2016**, *139*, 922–934. [CrossRef]

13.   Kim, M.S.; Sarkar, B. Multi-stage cleaner production process with quality improvement and lead time dependent ordering cost. *J. Clean. Prod.* **2017**, *144*, 572–590. [CrossRef]

14.   Chang, H.C. An application of fuzzy sets theory to the EOQ model with imperfect quality items. *Comput. Oper. Res.* **2004**, *31*, 2079–2092. [CrossRef]

15.   Priyan, S.; Manivannan, P. Optimal inventory modeling of supply chain system involving quality inspection errors and fuzzy defective rate. *OPSEARCH* **2017**, *54*, 21–43. [CrossRef]

16.   Zeballos, L.J.; Méndez, C.A.; Barbosa-Povoa, A.P.; Novais, A.Q. Multi-period design and planning of closed-loop supply chains with uncertain supply and demand. *Comput. Chem. Eng.* **2014**, *66*, 151–164. [CrossRef]

17.   Sarkar, B.; Ganguly, B.; Sarkar, M.; Pareek, S. Effect of variable transportation and carbon emission in a three-echelon supply chain model. *Transport. Res. Part E Logist. Transp. Rev.* **2016**, *91*, 112–128. [CrossRef]

18.   Xu, Z.; Pokharel, S.; Elomri, A.; Mutlu, F. Emission policies and their analysis for the design of hybrid and dedicated closed-loop supply chains. *J. Clean. Prod.* **2017**, *142*, 4152–4168. [CrossRef]

19.   Gandhi, N.S.; Thanki, S.J.; Thakkar, J.J. Ranking of drivers for integrated lean-green manufacturing for Indian manufacturing SMEs. *J. Clean. Prod.* **2018**, *171*, 675–689. [CrossRef]

20.   Galeazzo, A.; Furlan, A.; Vinelli, A. Lean and green in action: Interdependencies and performance of pollution prevention projects. *J. Clean. Prod.* **2014**, *85*, 191–200. [CrossRef]

21.   Diaz-Elsayed, N.; Jondral, A.; Greinacher, S.; Dornfeld, D.; Lanza, G. Assessment of lean and green strategies by simulation of manufacturing systems in discrete production environments. *CIRP Ann. Manuf. Technol.* **2013**, *62*, 475–478. [CrossRef]

22.   Thanki, S.; Thakkar, J. Value-value load diagram: A graphical tool for lean-green performance assessment. *Prod. Plan. Control* **2016**, *27*, 1280–1297. [CrossRef]

23.   Jaber, M.Y.; Khan, M. Managing yield by lot splitting in a serial production line with learning, rework and scrap. *Int. J. Prod. Econ.* **2010**, *124*, 32–39. [CrossRef]

24.   Sarkar, B.; Mahapatra, A.S. Periodic review fuzzy inventory model with variable lead time and fuzzy demand. *Int. Trans. Oper. Res.* **2017**, *24*, 1197–1227. [CrossRef]

25.   Sarkar, B. An inventory model with reliability in an imperfect production process. *Appl. Math. Comput.* **2012**, *218*, 4881–4891. [CrossRef]

26.   Sarkar, B.; Sana, S.S.; Chaudhuri, K. Optimal reliability, production lotsize and safety stock: An economic manufacturing quantity model. *Int. J. Manag. Sci. Eng. Manag.* **2010**, *5*, 192–202. [CrossRef]

27.   Shin, D.; Guchhait, R.; Sarkar, B.; Mittal, M. Controllable lead time, service level constraint, and transportation discounts in a continuous review inventory model. *RAIRO Oper. Res.* **2016**, *50*, 921–934. [CrossRef]

28.   Sarkar, B.; Moon, I. Improved quality, setup cost reduction, and variable backorder costs in an imperfect production process. *Int. J. Prod. Econ.* **2014**, *155*, 204–213. [CrossRef]

29.   Sarkar, B.; Ahmed, W.; Kim, N. Joint effects of variable carbon emission cost and multi-delay-in-payments under single-setup-multiple-delivery policy in a global sustainable supply chain. *J. Clean. Prod.* **2018**, *185*, 421–445. [CrossRef]

30.   Sarkar, B.; Cárdenas-Barrón, L.E.; Sarkar, M.; Singgih, M.L. An economic production quantity model with random defective rate, rework process and backorders for a single stage production system. *J. Manuf. Syst.* **2014**, *33*, 423–435. [CrossRef]

31.   Sarkar, B.; Sarkar, S. An improved inventory model with partial backlogging, time varying deterioration and stock-dependent demand. *Econ. Model.* **2013**, *30*, 924–932. [CrossRef]

32.   Sarkar, B. A production-inventory model with probabilistic deterioration in two-echelon supply chain management. *Appl. Math. Model.* **2013**, *37*, 3138–3151. [CrossRef]

33. Sarkar, B.; Sana, S.S.; Chaudhuri, K. An inventory model with finite replenishment rate, trade credit policy and price-discount offer. *J. Ind. Eng.* **2013**, 1–18. [CrossRef]

34. Cárdenas-barrón, L.E.; Sarkar, B.; Treviño-Garza, G. Easy and improved algorithms to joint determination of the replenishment lot size and number of shipments for an EPQ model with rework. *Math. Comput. Appl.* **2013**, *18*, 132–138. [CrossRef]

35. Kang, C.W.; Ullah, M.; Sarkar, B.; Hussain, I.; Akhtar, R. Impact of random defective rate on lot size focusing work-in-process inventory in manufacturing system. *Int. J. Prod. Res.* **2017**, *55*, 1748–1766. [CrossRef]

36. Ahmed, W.; Sarkar, B. Impact of carbon emissions in a sustainable supply chain management for a second generation biofuel. *J. Clean. Prod.* **2018**, *186*, 807–820. [CrossRef]

37. Sarkar, B. An EOQ model with delay in payments and time varying deterioration rate. *Math. Comput. Model.* **2012**, *55*, 367–377. [CrossRef]

38. Sarkar, M.; Sarkar, B.; Iqbal, M. Effect of Energy and Failure Rate in a Multi-Item Smart Production System. *Energies* **2018**, *11*, 2958. [CrossRef]

© 2018 by the authors. Licensee MDPI, Basel, Switzerland. This article is an open access article distributed under the terms and conditions of the Creative Commons Attribution (CC BY) license (http://creativecommons.org/licenses/by/4.0/).

Article

Effect of Time-Varying Factors on Optimal Combination of Quality Inspectors for Offline Inspection Station

Muhammad Babar Ramzan [1], Shehreyar Mohsin Qureshi [2], Sonia Irshad Mari [3], Muhammad Saad Memon [3], Mandeep Mittal [4,*], Muhammad Imran [5] and Muhammad Waqas Iqbal [6]

[1] Department of Garment Manufacturing, National Textile University, Faisalabad 37610, Pakistan; babar_ramzan@yahoo.com

[2] Department of Industrial and Manufacturing Engineering, NED University of Engineering and Technology, Karachi 75270, Pakistan; sheheryar@neduet.edu.pk

[3] Department of Industrial Engineering and Management, Mehran University of Engineering and Technology, Jamshoro 76062, Pakistan; sonia.irshad@faculty.muet.edu.pk (S.I.M.); saad.memon@faculty.muet.edu.pk (M.S.M.)

[4] Department of Mathematics, Amity Institute of Applied Science, Amity University, Noida 201313, India

[5] Department of Industrial and Management Engineering, Hanyang University, Seoul 15588, Korea; Imran.ime13@gmail.com

[6] Department of Industrial Engineering, Hongik University, Seoul 04066, Korea; Waqastextilion@gmail.com

* Correspondence: mittal_mandeep@yahoo.com; Tel.: +91-989-140-2516

Received: 29 November 2018; Accepted: 29 December 2018; Published: 7 January 2019

Abstract: With advanced manufacturing technology, organizations like to cut their operational cost and improve product quality, yet the importance of human labor is still alive in some manufacturing industries. The performance of human-based systems depends much on the skill of labor that varies person to person within available manpower. Much work has been done on human resource and management, however, allocation of manpower based on their skill yet not investigated. For this purpose, this study considered offline inspection system where inspection is performed by the human labor of varying skill levels. A multi-objective optimization model is proposed based on Time-Varying factors; inspection skill, operation time and learning behavior. Min-max goal programming technique was used to determine the efficient combination of inspectors of each skill level at different time intervals of a running order. The optimized results ensured the achievement of all objectives of inspection station: the cost associated with inspectors, outgoing quality and inspected quantity. The obtained results proved that inspection performance of inspectors improves significantly with learning and revision of allocation of inspectors with the proposed model ensure better utilization of available manpower, maintain good quality and reduce cost as well.

Keywords: human-based production system; offline inspection; optimization; inspection cost; outgoing quality; learning behavior

---

1. Introduction

Human labor and factors associated are one of the main engineering disciplines and play a vital role in maintaining good quality, manufacturing cost and productivity. Although researchers are working on automation and hybrid systems, however, the importance of human labor is not limited yet. One of the important advantage, of human labor over automation, is their decision-making ability. These type of abilities in human labor improves as skill level and experience increases. Organizations always want to utilize their manpower efficiently according to their capacity, yet there is a lack of studies that evaluate the human labor based on their skills. The present study has been conducted to fill this study gap by considering the skill level, operation time and learning behavior of human labor. For this purpose, a human-based quality control system is considered where most of the processes are carried out by human labor. Quality Control (QC) is an important part of quality management system that consists of monitoring activities along with quality planning, quality assurance and quality improvement [1]. The main objective of QC is to maintain the good level of quality by mitigating the root causes of defective products [2]. Inspection is the main activity of QC that is performed to decide the product's conformance and non-conformance at different stages of manufacturing [3]. The process of inspection is investigated here to highlight the importance of inspection skill and inspection time in a manufacturing environment where learning affects significantly. Inspection can either be online or offline where online inspect the product during the process and offline inspect the product after the completion of the process [4,5]. Although online inspection has been considered as an economical method, however sometimes it is not feasible. Thus inspection process has to be done offline on finished or semi-finished products [6]. This paper also considered offline station where 100% inspection is done by human labor.

In human based manufacturing setups, learning behavior imparts significant enhancements in the performance of labor and their skill improves with the passage of time. Researchers believed that any organization that learns faster will have a competitive advantage in the future. However, this learning varies person to person within an organization and help to classify the available manpower into their respective skill levels. Six types of learning have been identified and one of them is learning by doing like inspection process performed by human labor. While planning for the new order, allocation of manpower is done once, that is, before the start of order and same labor is used until the completion of the job. However, in the actual scenario, human labor learns from their experience and improve their performance with the passage of time. Thus, they will be able to do more work with better efficiency because of their improved skills. In this situation, the organization must revise the allocation of labor that may bring advantages like better utilization of available manpower, achievement of inspection targets, reduce cost and maintain good quality. Recently, the optimal number of quality inspectors have been determined to minimize cost [7,8]. However, the effect of Time-Varying factors, like learning behavior, on an efficient combination of inspectors is not considered yet. This study has kept this factor in contact and the process of inspection is investigated here to determine the optimal number of quality inspectors for inspection station over different time periods.

2. Literature Review

In past, plenty of work has been done on offline inspection to reduce the overall cost, increase company profit and improve product quality. These objectives have been achieved by giving due consideration to inspection policies, inspection systems and optimization of process target values. One of the pioneering work in developing inspection policy was done by Herer and Raz [9] to reduce the inspection cost using dynamic programming. The similar objective was also achieved by calculating optimal lot size and expected number of inspections [10]. After that plenty of work has been conducted to improve the effectiveness of inspection policy. Anily and Grosfeld-Nir [11] determined inspection policy and lot size for a single production run. Further investigation was done with two-time parameters and multiple productions run with rigid demand. Wang and Meng [12] developed a joint optimization model to determine the total cost function. Their model was compared with three policies like no inspection, full inspection and disregard the first *s* (DTF-*s*) items policies by a numerical example. Avinadav and Perlman [13] studied such process to minimize the cost by determining the optimal inspection interval. Sarkar and Saren [14] developed an inspection policy for an imperfect manufacturing system that has inspection error and warranty cost to reduce the inspection cost.

Other inspection strategies have also been developed that includes inspection disposition (ID) policy and inspection disposition and rework (IDR) policy. Raz, et al. [15] developed the first ID policy to minimize the cost function by solving the problem of economic optimization. After that their ID policy was extensively studied by other researchers with the consideration of different assumptions [16–21].

Continues sampling plan (CSP) is also a pioneering method of inspection in which 100% inspection and sampling inspection is alternatively conducted [22]. The basic sampling plan known as CSP-1 was developed by Dodge [23] to monitor the average outgoing quality level (AOQL). After that many modifications have been incorporated in the procedure of original CSP-1 by considering different assumptions [24–30].

The studies have also been conducted to optimize the process parameters by many investigators of quality control. After the pioneering work done by Springer [31], number of studies have been conducted to minimize the expected cost. Earlier a process target model (PTM) was proposed to optimize a single objective for three different types of screening problems [32]. Their aim was to cancel out the effect of error in inspection through the conception of cut off points that helped to divide the products into grade one, grade two and scrap. Duffuaa, et al. [33] proposed another PTM to increase the profit by assuming the independent characteristics of quality for a two-stage process. This PTM was also modified using acceptance sampling by Duffuaa, et al. [34] to achieve the same objectives. Recently, Multi-Objective Optimization (MOO) problem has been explored to find out the value of process parameters: income, profit and product uniformity [35–37]. The pioneer work on MOO was done considering 100% inspection policy to optimize the objective functions and Pareto optimal points were ranked by proposing an algorithm [36]. Their MOO model further reviewed by considering the sampling inspection however similar results were attained [37]. A further improvement was done considering two types of inspection errors because inspection system may be error-prone. After comparing the results of revised and previous models, it was concluded that inspection errors have a major effect on profit and uniformity. This study also worked on MOO and considered three important parameters to measure the performance of human labor while performing inspection process.

The philosophy of learning behavior is not only to improve the productivity but also look for other aspects that support the process of learning. That's why a number of studies have found a relationship with quality control techniques and learning which was summarized by Jaber [38]. This combination of learning and quality control was first suggested by Koulamas [39] to evaluate the effect of product design on quality and cost. Teng and Thompson [40] worked on the learning behavior of workers and assessed that how it affects the quality and cost of the product. Similarly, Franceschini and Galetto [41] reduced the non-confirming quantity in production plant by improving the skill of workers. Jaber and Guiffrida [42] worked on wright's learning curve [43] and proposed a quality learning curve (QLC) for a process that generates defects and required rework also. Further, this QLC was investigated by relaxing its different assumptions. Like Jaber and Guiffrida [44] assumed that an imperfect process can be interrupted to maintain quality and improved system's performance. Similarly, Jaber and Khan [45] further relaxed two assumptions and considered scrap after production along with a rework of a lot. They concluded that optimal performance improves with learning and deteriorates when learning in rework becomes faster. It is observed that quantity and cost of production have a direct link with quality and this subject will have particular interest when combined with learning behavior. A number of researchers have been investigating errors in screening, however, the relationship between quality and screening need to be studied further Jaber [38]. This study has considered learning behavior in the proposed model and its effect on inspection performance of inspectors.

Despite the above-mentioned literature, the researchers have investigated this research area with respect to the application of new trends, techniques and methodologies in the human labor selection and job assignment. It includes artificial intelligence, genetic algorithm, goal programming, fuzzy logic, data mining and data envelopment analysis [46–49]. In a human based production environment, assigning the job to workers according to their competence is an important step to keep overall cost in control and maintain production efficiency. A fuzzy logic interface method has been proposed to assign and verify production jobs to human labor. The proposed method has been applied to a discrete manufacturing system to reduce the cost due to human errors [46]. Similarly, a synchronized job assignment model has been proposed to overcome the problem of human performance due to deviation in skill level and fluctuation of cycle time. A multi-objective simulation integrated hybrid genetic algorithm was used as a job assignment model such that it promotes teamwork and overcome the effect of varying skill level [49]. However, there is a lack of studies in which such new trends have been applied in human-based inspection system. This study has also addressed this gap by applying multi-objective goal programming to the offline inspection station.

Table 1 has summarized the span of work done on offline inspection. It indicates that how different researchers have contributed to the field under study and compare them with the present study. Despite much work, human-based inspection system has not been studied yet considering the effect of learning on the performance of labor with different skill levels. The present study has focused this gap to contribute to the current literature and investigated that how learning behavior of different inspectors affect the inspection performance and total manpower required for inspection station. The MOO model has been presented here that can determine the group of inspectors having different skill levels such that all objectives of inspection station are achieved.

This study also incorporates the effect of learning behavior on inspection skill of human labor in terms of quality, cost and quantity. The proposed model is able to determine the optimal values of inspectors at different time periods and compares that how the requirement of manpower varies from time to time due to learning.

Table 1. Summary of contribution made by previous studies.

Authors	Inspection					Learning Behavior	Study Objective
	Strategy	Error	Cost	Time	Skill		
Jaber and Guiffrida [42]		✓				✓	Proposed quality learning curve
Finkelshtein, Herer, Raz and Ben-Gal [16]	Sampling	✓	✓				Optimal ID policy
Duffuaa and Khan [50]	100%	✓	✓				Repeat inspection plan to measures the performance
Anily and Grosfeld-Nir [11]	Both		✓				Optimal inspection policy
Elshafei, et al. [51]	100%	✓	✓				Optimal inspection sequence for repeat inspection plan
Wang [17]	Sampling	✓	✓				Optimal ID policy
Duffuaa and Khan [52]	Both	✓	✓				Optimal inspection cycles
Wang and Hung [18]	Sampling	✓	✓				Optimal ID policy
Jaber and Guiffrida [44]						✓	Proposed QLC by relaxing assumptions
Wang and Meng [12]	Both	✓	✓				Optimal inspection policy
Colledani and Tolio [53]	Sampling	✓					Analytical method of evaluation
Tzimerman and Herer [6]	Sampling	✓	✓				Optimal inspection policy
Bendavid and Herer [19]	Sampling	✓	✓				Optimal ID policy
Vaghefi and Sarhangian [54]	Sampling	✓	✓				Optimal inspection policy
Wang, Sheu, Chen and Horng [20]	Sampling	✓	✓				Optimal ID policy
Yu, Yu and Wu [28]	Both	✓	✓				Optimal inspection policy
Khan, et al. [55]	100%	✓					Economic order quantity with learning in production
Jaber and Khan [45]						✓	Proposed QLC by relaxing assumptions
Yang [56]	Both		✓				Optimization of K- stage inspection system
Khan, et al. [57]	100%	✓	✓				Economic order quantity
Tsai and Wang [21]	Sampling	✓	✓				Optimal IDR Policy
Yu and Yu [29]	Both	✓	✓				Optimal inspection policy
Khan, et al. [58]	100%	✓	✓				Effect on human factors on cost of supply chain
Avinadav and Sarne [59]	Both	✓	✓				Selection of inspection systems
Aviradav and Perlman [13]	Sampling	✓	✓				Optimal inspection interval
Duffuaa and El-Ga'aly [36]	100%		✓				Maximization of income, profit, product uniformity
Duffuaa and El-Ga'aly [37]	Sampling		✓				Maximization of income, profit, product uniformity
Bouslah, et al. [60]	Sampling	✓					Joint production control and economic sampling plan
Khan, et al. [61]	100%	✓	✓		✓	✓	Integrated supply chain model
Liu and Liu [62]	Sampling	✓					Resubmitted sampling scheme
Aslam, et al. [63]	Sampling	✓					Mixed acceptance sampling plan
Yang and Cho [64]	100%		✓				Optimal inspection cycles
Mohammadi, et al. [65]	Sampling	✓	✓				Effective robust inspection planning
Duffuaa and El-Ga'aly [35]	Sampling	✓	✓				Maximization of income, profit, product uniformity
Sarkar and Saren [14]	Sampling	✓	✓				Product inspection policy
Ramzan and Kang [7]	Both	✓	✓		✓		MOO model to determine inspectors of different skills
Jaber [38]		✓	✓			✓	A review of studies linking quality with learning
Kang, Ramzan, Sarkar and Imran [8]	Both	✓	✓		✓		MOO model to determine inspectors for different products
This paper	Both	✓	✓	✓	✓		MOO model to determine optimal quality inspectors

3. Model Formulation

3.1. Definition of Research Problem

Figure 1 indicates the movement of input material through production unit and inspection stations to complete the manufacturing process that is assumed to be an imperfect process. The output products transfer to an offline inspection station for two-step inspection processes: 100% inspection followed by sampling inspection. The first inspection station has J number of inspectors that perform the process of 100% inspection. They have varying skill levels which are defined based on their quantity inspected per day and errors in inspection per day. Each inspector classifies the output products either confirming or non-conforming. A batch of conforming products, having fixed quantity N, is then presented for sampling inspection. On the other hand, non-conforming products can either be reworked or rejected.

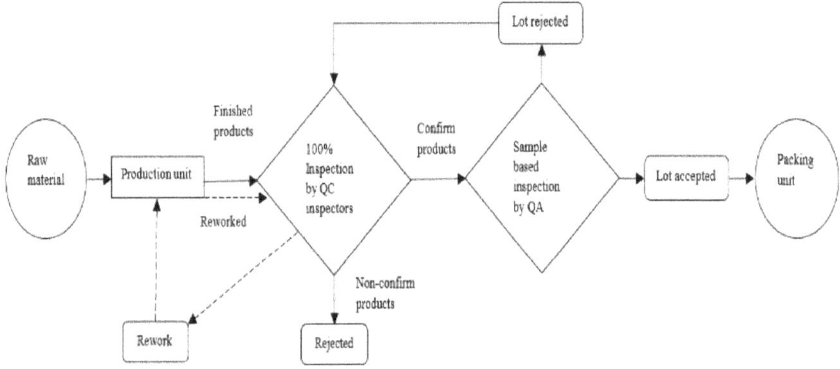

Figure 1. Flowchart of production unit and inspection stations of a manufacturing system.

A person with high inspection skill performs the Sampling inspection. The quantity n is randomly selected as a sample size from the presented batch/lot and the number of defective items d are separated. The value of d is compared with the threshold value c to make the final decision of lot acceptance or lot rejection. The decision will be to accept the lot if $d \leq c$ however lot will be rejected if $d > c$. The accepted lot is moved to the next process yet the rejected lot is returned to the same inspector for re-inspect. Defective items are separated from the rejected lot and exchanged with the conforming items to complete the batch for second sampling inspection. In order to calculate the quantity inspected per day and inspection cost per day of each inspector, number of accepted lots are used. Similarly, the value of outgoing quality is determined by the number of defectives items found. The value of Outgoing quality, accepted quantity and inspection cost depend on the number of inspectors and their skill levels that vary with the passage of time depending upon their learning behavior and experience.

3.2. Model Notations and Abbreviations

Model Notations

<u>Sets</u>

j	type of inspectors where, j = Low skill, Medium skill and High skill
l	low skill $l = 1, 2, 3, \ldots , L$
m	medium skill $m = 1, 2, 3, \ldots , M$
h	high skill $h = 1, 2, 3, \ldots , H$

Parameters

N	lot/batch size
n	sample size
IT_j	inspection time per unit taken by jth inspector
V	cost of inspection (\$/min)
MI	maximum allowable quality inspector
ST	standard time of inspection of particular product
TVC_T	target of total cost for inspection station
AOQ_T	target of outgoing quality for inspection station
TIQ_T	target of accepted quantity for inspection station

Input variables

d_j	number of defective items present in sample size n inspected of jth inspector
b_j	learning rate of jth inspector
$d_1^\pm$	deviational variables for cost of inspectors
$d_2^\pm$	deviational variables for outgoing quality
$d_3^\pm$	deviational variables for accepted quantity
OQ_j	average outgoing quality of jth quality inspectors
IQ_j	inspected quantity by jth quality inspectors
VC_j	variable cost of jth quality inspectors

Decision variables

NI_j	number of jth type of skilled labor

3.3. Outgoing Quality

Maintaining an acceptable level of Outgoing Quality (OQ) is one of the main objectives and its dependents on the skill of quality inspectors. The value of OQ can be measured by sample-based inspection process by determining the number of defectives present in inspected quantity. Let quantity Q of finished products move from the manufacturing unit to the offline station where 100% inspection is done. If Q_j is the total quantity inspected by each inspector j and p_j is the probability of separating the non-conforming (NC) products from confirming (C) products than the value of NC and C can be calculated as:

$$NC_j = p_j \times Q_j$$
$$C_j = (1 - p_j) \times Q_j$$

The quantity NC_j can either be sent for rework or rejected. Rework quantity (RW_j) and rejected quantity (RE_j) can be calculated by the following equations:

$$RW_j = \alpha_j \times NC_j = \alpha_j \times (p_j \times Q_j)$$

$$RE_j = (1 - \alpha_j) \times NC_j = (1 - \alpha_j) \times (p_j \times Q_j)$$

where α_j is the probability of rework-able quantity. Similarly, the conforming quantity is moved for the process of sampling inspection as a lot/batch of size N. The following equation can be used to determine the value of OQ,

$$OQ = \frac{no\ of\ defective\ items}{Sample\ size}$$

$$OQ_j = \frac{d_j}{n_j} \ \forall j$$

where d shows number of defects present in sample size n. Since, this study has focused on learning behavior of quality inspectors along with their skill and inspection time. Thus, with the passage of time, quality inspectors will make less inspection error that will improve their individual OQ as well as of inspection station. A similar concept of reduction in defective percentage was also addressed by Jaber and Guiffrida [44], keeping in view the wright's learning curve. Their suggested formula is used here:

$$OQ(w) = OQ_s \times w^{-b}$$

where OQ_s is the initial value, b is the learning rate and $OQ(w)$ is the value of outgoing quality level at wth week. Similarly, the value of OQ for any inspector j can be calculated as:

$$OQ_j(w) = OQ_s \times w^{-b_j} \; \forall j$$

where b_j is the learning rate of inspector j. This study is investigating the human labor of J types of skill levels, thus Average Outgoing Quality (AOQ) can be calculated as:

$$AOQ(w) = \frac{\sum_{j=1}^{J} OQ_j(w)}{\sum_j NI_j} \; \forall j$$

$$AOQ(w) = \frac{OQ_l(w) + OQ_m(w) + OQ_h(w)}{\sum_j NI_j}$$

$$AOQ(w) = \frac{NI_L\left(OQ_{s,\,l} \times w^{-b_l}\right) + NI_M\left(OQ_{s,\,m} \times w^{-b_m}\right) + NI_H\left(OQ_{s,\,h} \times w^{-b_h}\right)}{\sum_j NI_j}$$

where b_l, b_m, b_h indicate the learning rate of inspector with low, medium and high inspection skill respectively. Finally, the value of AOQ for j type of quality inspectors is calculated by Equation.

$$AOQ = \frac{\sum_{j=1}^{J} NI_j\left(OQ_{s,\,j} \times w^{-b_j}\right)}{\sum_j NI_j} \; \forall j$$

3.4. Inspection Quantity

The second objective is to achieve the target of total inspection quantity to avoid any bottleneck and skill of quality inspector play a key role in this regard. Inspected quantity IQ is a quantity accepted by lot sampling process and is calculated for each inspector according to his inspection time. Quantity inspected by a jth inspector can be calculated as:

$$IQ = \frac{Time\ available}{Inspection\ time}$$

$$IQ_j = \frac{TA}{IT_j} \; \forall j$$

where IT_j is the inspection time taken by the jth quality inspector to inspect one item. With the passage of time, the efficiency of each quality inspector improves and inspected quantity increases because of reduction in inspection time due to learning. To calculate the improvement in inspection time, concept of wright's learning curve [43] is used that suggests the exponential relationship between man hour and cumulative production.

$$IT(w) = IT_s \times w^{-b}$$

where IT_s is the initial value of inspection time, b is the learning rate and $IT(w)$ is the inspection time at wth week. Similarly, the value of IT_j and IQ_j for a jth inspector can be calculated as:

$$IT_j(w) = IT_s \times w^{-b_j} \; \forall j$$

$$IQ_j(w) = \frac{TA}{IT_j(w)} = \frac{TA}{IT_{s,j} \times w^{-b_j}} \; \forall j$$

where b_j is the learning rate of the jth quality inspector. Total inspected quantity TIQ of the offline station will include all J types of inspectors.

$$TIQ(w) = \sum_{j=1}^{J} IQ_j(w) \; \forall j$$

$$TIQ(w) = IQ_l(w) + IQ_m(w) + IQ_h(w)$$

$$TIQ(w) = NI_l \left(\frac{TA}{IT_{s,l} \times w^{-b_l}} \right) + NI_m \left(\frac{TA}{IT_{s,m} \times w^{-b_m}} \right) + NI_h \left(\frac{TA}{IT_{s,h} \times w^{-b_h}} \right)$$

Total inspected quantity by J type of quality inspectors can be calculated by the following equation:

$$TIQ(w) = \sum_{j=1}^{J} NI_j \left(\frac{TA}{IT_{s,j} \times w^{-b_j}} \right)$$

3.5. Inspection Cost

Total inspection cost consists fixed cost (setup, inspection material, salaried workers etc.) and variable cost that is related to human labor. Since this study is investigating the skill of inspectors that vary with experience and learning of human labor, thus change in inspected quantity of inspector will also vary the related cost. This study is focusing more on the cost of quality inspectors that will be further used to optimize the total variable cost associated with all quality inspectors. By using inspected quantity IQ_j, the VC_j can be calculated on the basis of time earned TE_j for the jth quality inspector.

$$VC_j = TE_j \times V$$

$$TE_j = IQ_j \times ST$$

Thus

$$VC_j = (IQ_j \times ST) \times V \; \forall j$$

where ST is the standard time of inspection of a particular product and V is the cost of inspection ($\$/min$). As described earlier, a decrease in IT is observed based on the learning behavior of quality inspector that increase the IQ. Thus, the value of VC_j for jth inspector at any stage w will be calculated as:

$$VC_j(w) = \left(\frac{TA}{IT_j(w)} \right) \times ST \times V$$

$$VC_j(w) = \left(\frac{TA}{IT_{s,j} \times w^{-b_j}} \right) \times ST \times V$$

Total Variable Cost (TVC) of offline station that has J type of inspectors, will be calculated by as:

$$TVC(w) = \sum_{j=1}^{J} VC_j(w) \; \forall j$$

$$TVC(w) = \sum_{j=1}^{J} \left(\frac{TA}{IT_{s,\,j} \times w^{-b_j}} \right) \times ST \times V$$

$$TVC(w) = VC_l(w) + VC_m(w) + VC_h(w)$$

$$TVC(w) = \left\{ NI_l \left(\frac{TA}{IT_{s,\,l} \times w^{-b_l}} \right) \times ST \times V \right\} + \left\{ NI_m \left(\frac{TA}{IT_{s,m} \times w^{-b_m}} \right) \times ST \times V \right\}$$
$$+ \left\{ NI_h \left(\frac{TA}{IT_{s,h} \times w^{-b_h}} \right) \times ST \times V \right\}$$

$$TVC(w) = \left\{ NI_l \left(\frac{TA}{IT_{s,\,l} \times w^{-b_l}} \right) + NI_m \left(\frac{TA}{IT_{s,m} \times w^{-b_m}} \right) + NI_h \left(\frac{TA}{IT_{s,h} \times w^{-b_h}} \right) \right\} \times ST \times V$$

Finally, the total variable cost of all J type of inspectors can be calculated by:

$$TVC = \left\{ \sum_{j=1}^{J} NI_j \left(\frac{TA}{IT_{s,\,j} \times w^{-b_j}} \right) \right\} \times ST \times V$$

3.6. Objective Functions

The optimal values of decision variables are obtained using goal programming (GP) which is a type of multi-objective decision making and is widely used by many authors where more than one objective has to be achieved. The proposed model is also multi-objective that optimizes three objectives as mentioned below.

To keep the total cost per day TVC of all inspectors less than the target value of cost TVC_T.

$$Z_1 = \left\{ \sum_{j=1}^{J} NI_j \left(\frac{TA}{IT_{s,\,j} \times w^{-b_j}} \right) \right\} \times ST \times V$$

To keep the quality level of inspection.

$$Z_2 = \sum_{j=1}^{J} NI_j \left(\frac{TA}{IT_{s,j} \times w^{-b_j}} \right)$$

To meet the daily target of inspection quantity.

$$Z_3 = \frac{\sum_{j=1}^{J} NI_j \left(OQ_{s,\,j} \times w^{-b_j} \right)}{\sum_j NI_j}$$

In order to achieve these three objectives, the best combination of inspectors with respect to their skill levels need to be determined. GP variant, Min-max or Chebyshev GP, is used to determine the optimum decision variables by satisfying all the objective functions. In Chebyshev GP, the unwanted deviation for three goals was normalized. Percentage normalization was used in which each deviation is divided by target value of its respective objective. The objective function minimizes the worst or maximal deviation (λ) from amongst the set of three goals [66]. The GP formulation can be presented as follows:

$$Minimize\ Z = \lambda$$

Subject to

$$\frac{d_1^+}{TVC_T} \le \lambda$$

$$\frac{d_2^+}{AOQ_T} \le \lambda$$

$$\frac{d_3^-}{TIQ_T} \leq \lambda$$

$$\left\{ \sum_{j=1}^{J} NI_j \left(\frac{TA}{IT_{s,j} \times w^{-b_j}} \right) \right\} \times ST \times V + d_1^- - d_1^+ = TVC_T$$

$$\frac{\sum_{j=1}^{J} NI_j \left(OQ_{s,j}\, w^{-b_j} \right)}{\sum_j NI_j} + d_2^- - d_2^+ = AOQ_T$$

$$\sum_{j=1}^{J} NI_j \left(\frac{TA}{IT_{s,j} \times w^{-b_j}} \right) + d_3^- - d_3^+ = TIQ_T$$

$$\sum_j NI_j \leq MI\ NI_j \geq 0$$

$$d_t^-,\ d_t^+ \geq 0\ \forall t \in \{1,2,3\}$$

4. Results and Discussion

The application of the proposed model is described here with the help of an example from garment manufacturing unit and an offline station is selected where the inspection of finished products is performed by human labor. The product selected for this study is a short sleeve polo shirt and the completion of the order will take 120 days. Since this study has incorporated the concept of learning and inspection skill of human labor will improve with the passage of time. This study has considered three skill levels, that is, low, medium and high, along with the three performance measures of inspection including cost, quality and quantity. Figure 2 indicates how these performance measures vary for three different skill levels.

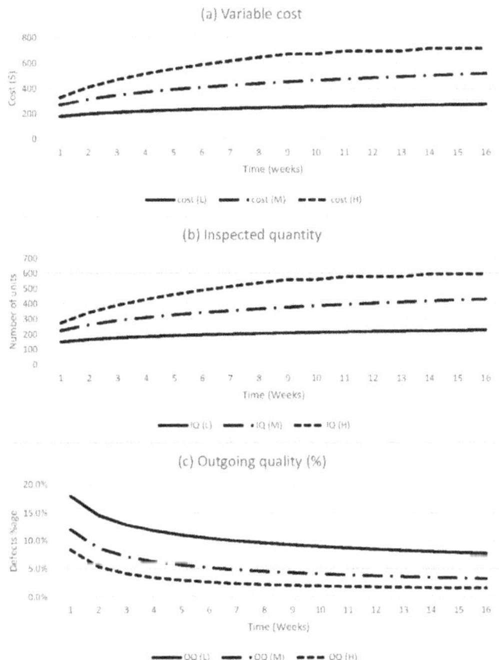

Figure 2. Influence of learning behavior on inspection performance of different inspectors.

At the start of order, the difference between three types of inspectors is not significant. However, the learning behavior varies person to person, that's why some inspectors learn quickly and improve their overall skill as compared to others. To achieve the objectives of the inspection station, an efficient combination of inspectors must be maintained that utilize the available manpower according to their skill levels. Thus, the requirement of manpower with respect to the skill levels will vary until we complete the order. That's why data were collected (Figure 2) for three different stages, that is, at the completion of 4th week (stage A), at the completion of 8th week (stage B) and at the completion of the 12th week (stage C). Keeping this scenario in view, data are collected for the selected product after the completion of each month and summarized in Table 2.

Table 2. Input data for short sleeve polo shirt.

Notation	At Stage A	At Stage B	At Stage C
ST (mins)	0.96	0.96	0.96
OQ_T	0.07	0.06	0.05
OQ	0.12	0.10	0.08
OQ_m	0.07	0.05	0.04
OQ_h	0.04	0.03	0.02
TC_T (\$)	3900	4690	5470
C_l (\$)	222	247	263
C_m (\$)	374	440	483
C_h (\$)	516	645	692
TIQ_T (Units)	3125	3750	4375
IQ_l (Units)	185	206	219
IQ_m (Units)	311	366	403
IQ_h (Units)	430	537	577
V (\$/min)	1.25	1.25	1.25
MI (Units)	12	12	12

In order to analyze the data in Table 2, an Optimization software, that is, Lingo 15.0, was applied by keeping the following system configuration: Intel® Core™ i7-7500U CPU @ 2.70 GHz Intel, 8.00 GB of RAM. Min-max GP method was used to calculate the optimal values of decision variables that also gave the optimized results of objective functions. The obtained results are summarized in Table 3 for all three stages and analysis can be divided into two parts: analysis of decision variable and analysis of objective function.

The decision variable analysis shows the optimum number of inspectors with their respective skill levels for each stage. These results ensure that all the objectives (cost, quantity and quality) have been achieved. Since the study incorporated learning behavior in this proposed model, thus the value of incoming quantity also vary along with the skill of inspectors as the order progress. At the early stage, when required targets of objective functions were low and the performance of inspectors of each skill level was also at the initial stage. The optimal combination that can achieve all targets of inspection station requires more inspectors with high inspection skill in comparison to low and medium. This is the confirmation of the fact that if an offline station consists of low skill inspectors mainly as compared to medium or high skill inspectors, the cost of inspection station may be low but the target of inspection quantity and quality level will be difficult to maintain for offline station. Therefore, the organization like to maintain an inspection station that consists of the best combination of inspectors to achieve all targets simultaneously.

As the order progress, the skill of each quality inspectors improved so as the incoming quantity which changed the targets of inspection station as well. Thus, at the second stage, the optimal results were obtained to satisfy the revised targets. However, this time, a combination of inspectors is changed and more medium skill inspectors are included. It is because of the fact that learning improves the performance of all inspectors and then the inspection station was able to achieve targets with less utilization of high skill labor. Similarly, at the last stage, the optimal combination consists of more low and medium skill quality inspectors to fulfill the revised targets. It is because of the fact that skill of both low and medium skill inspectors was improved enough meet the demands and less contribution was required from high skill inspectors.

Table 3. Optimized results of decision variables and objective functions for different stages.

Decision Variables	Values	Objectives	Target		Deviation Variables	
			Set	Achieved	d^+	d^-
Stage A						
Low skill	3	Inspection cost ($)	3900	3852	0	48
Medium skill	3	Outgoing quality	0.07	0.07	0	0
High skill	4	Inspected quantity (Units)	3125	3208	83	0
Stage B						
Low skill	3	Inspection cost ($)	4688	4671	0	17
Medium skill	6	Outgoing quality	0.06	0.06	0	0
High skill	2	Inspected quantity (Units)	3750	3888	138	0
Stage C						
Low skill	4	Inspection cost ($)	5469	5280	0	189
Medium skill	5	Outgoing quality	0.05	0.05	0	0
High skill	2	Inspected quantity (Units)	4375	4403	28	0

In a labor-intensive industrial setup like garment manufacturing, where the process of inspection is mainly performed by human labor, the presence of manpower with varying skill levels develops an environment that encourages the labor to compete with each other. Such a scenario provides them an opportunity to learn quickly that improve the level of their inspection skill at a faster rate. However, product type can significantly affect the rate of improvement in inspection skill. In case of a simple or basic garment, human labor can learn the things quickly because these products consist of fewer parts, a smaller number of operations or characteristics/features that a human inspector needs to inspect. On the other hand, this learning ability not only reduce but vary from person to person as the type of product moves to slightly complex, complicated or highly fashioned garments. This is because of the fact that these have more parts, increased number of operations and tough characteristics/features that an inspector need to inspect with concentration. Such products can affect the level of inspection skill of human labor that ultimately increases the inspection costs and decrease the outgoing quality. Similarly, the required manpower also changes with respect to product type to fulfill the requirements of the inspection station.

Analysis of the objective function, on the other hand, demonstrates the variance between the target values and the actual values of each goal at different stages, where underachieved values defined as d^- and overachieved value as d^+. Min-max GP method provided optimum results of decision variables such that the set target of each objective function is attained. Even though underachieved and overachieved values are also there for different objective functions but all these deviational values do not violate the given conditions. Like in Table 3, overachieved value of the inspection quantity (d^+) are 83, 138 and 28 for stage A, B and C respectively. However, it is still according to the constraints given in Section 3.6. Inspection quantity per day should not be less than the set target but presented results gave over achieved value, which is a positive side of the results. Similarly in Table 3, underachieved values of variable cost (d^-) are 48, 17 and 189 for stage A, B and C respectively. Since the constraint of the proposed model is to retain this variable cost low as much as possible so, these underachieved values also fulfill the already mentioned constraints.

In actual scenario, organization/managers allocate manpower only once, that is, at the start of the order and do not change till the completion of the order. This situation is not in favor of organizations as they are not using their labor according to their capacity. It may cause different problems like bottleneck and poor outgoing quality. Figure 3 demonstrates this fact, where the group of inspectors with their respective skill level is kept same for a full order. Variation in achieved values and targets values was evaluated. For this purpose optimal combination obtained at stage A (Table 3) is used here. It highlights not only the importance of skill of inspectors and learning but also explains that why it is important to revise the allocation of manpower at the periodical interval.

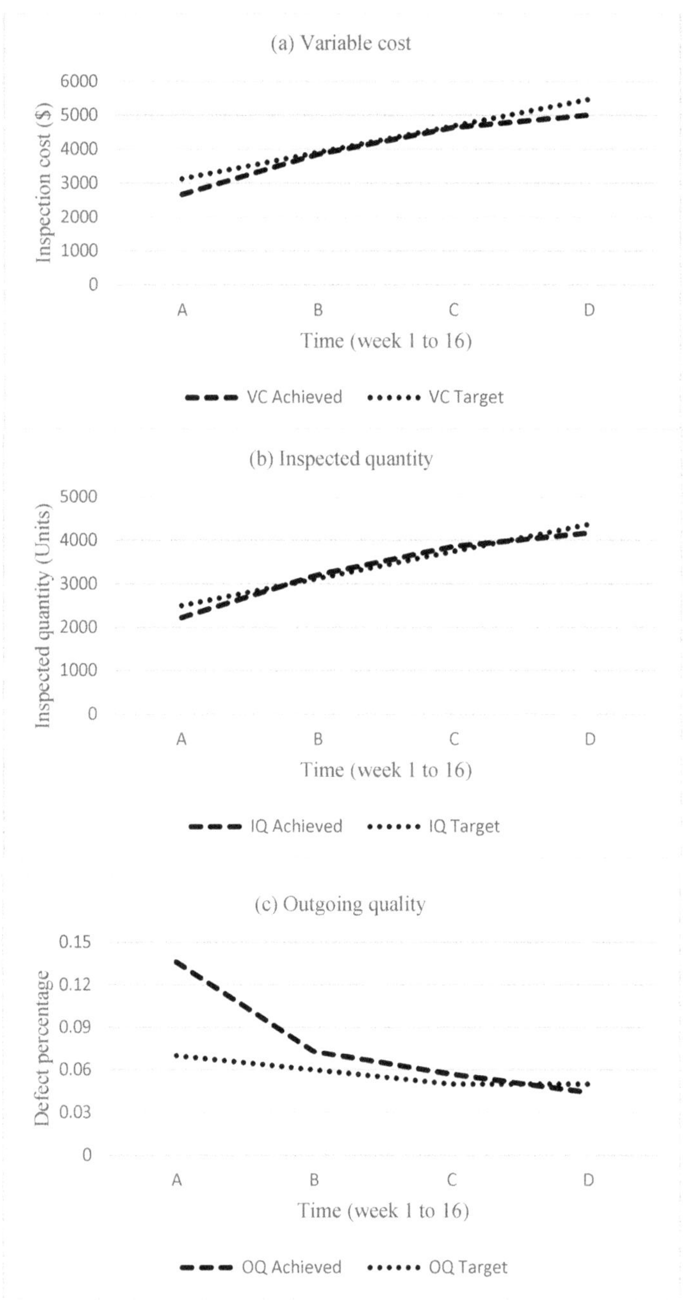

Figure 3. Evaluation of objective functions with the same combination of inspectors.

It is evident from the Figure 3 that the combination of inspectors of different skill levels could not achieve the targets throughout the order even though learning is also considered. Especially, outgoing quality is not kept under control due to the high percentage of inspection error. On the other hand, inspected quantity hit the targets on some stages but underachieved as well that creates a bottleneck. Such a situation will increase the workload on inspectors, increase the chance of overtime and affect the quality of product also due to work in progress. However, in the actual scenario, the performance of labor is not the same and they improve with time due to learning and experience. Thus, revision of optimal combination of inspectors at different intervals will not only save cost and improve quality but also avoid overtime and utilize the manpower of organization efficiently. In this way, the organization will be able to use their high skill labor for inspection of complex products where more skilled labor is required. This study also provides a way how to imply the available labor and get work from them according to their skill capacity. Also, such systems provide competitive environments that will help the employee to improve their skills. However, further work must be done to explore the ways that increase the learning process and also provide some bases to establish the pay/salary scale for employees based on their improvement with time.

5. Conclusions

In this study, multi-objective optimization model is developed to utilize the manpower efficiently and assign them job according to their capacity. For this purpose, offline inspection system is considered where inspection is performed by the manpower of varying skill levels. A multi-objective optimization model is proposed based on inspection skill, operation time, learning behavior. Min-max goal programming technique was used to determine the efficient combination of inspectors of each skill level at different time intervals of running order. The optimized results ensured that all the objectives of offline inspection station are attained that contains: the cost associated with inspectors, outgoing quality and inspected quantity. The results proved that the performance of inspectors improves significantly with learning and allocation of inspectors should be revised after regular interval keeping in view the improvement in their skill levels. For this purpose, the proposed model is helpful for organizations to ensure better utilization of available manpower, maintain good quality and reduce cost as well. Moreover, this study provides the basis to the researchers to further explore this research area for its practical application in the human-based manufacturing system. Such studies can be used to develop specific software for the assessment of human labor and job assignment. Also, this study has considered the offline inspection, while future work should also be done on online inspection.

Author Contributions: The contribution of the respective authors are as follows: M.B.R. conducted the complete study with main contribution in conceptualization and original draft preparation under the supervision of M.M. While S.I.M. and M.S.M. contributed in developing and defining the methodology, formal analysis, and validation. S.M.Q. conducted the process of revision and editing. Lastly, M.I. and M.W.I. jointly contributed in applying the optimization software and data collection from the relevant industry.

Funding: This research received no external funding.

Conflicts of Interest: The authors declare no conflict of interest.

References

1. Purushothama, B. *A Practical Guide on Quality Management in Spinning*; Woodhead Publishing: Delhi, India, 2011.
2. Babu, V.R. *Industrial Engineering in Apparel Production*; Woodhead Publisher: Delhi, India, 2012.
3. Ullah, M.; Kang, C.; Qureshi, A.S.M. A Study of Inventory Models for Imperfect Manufacturing setup Considering Work-in-Process Inventory. *J. Soc. Korea Ind. Syst. Eng.* **2014**, *37*, 231–238. [CrossRef]
4. Tirkel, I.; Rabinowitz, G. Modeling cost benefit analysis of inspection in a production line. *Int. J. Prod. Econ.* **2014**, *147*, 38–45. [CrossRef]
5. Tirkel, I.; Rabinowitz, G. The relationship between yield and flow time in a production system under inspection. *Int. J. Prod. Res.* **2012**, *50*, 3686–3697. [CrossRef]

6. Tzimerman, A.; Herer, Y.T. Off-line inspections under inspection errors. *IiE Trans.* **2009**, *41*, 626–641. [CrossRef]

7. Ramzan, M.B.; Kang, C.W. Minimization of inspection cost by determining the optimal number of quality inspectors in the garment industry. *Indian J. Fibre Text. Res.* **2016**, *41*, 346–350.

8. Kang, C.W.; Ramzan, M.B.; Sarkar, B.; Imran, M. Effect of inspection performance in smart manufacturing system based on human quality control system. *Int. J. Adv. Manuf. Technol.* **2018**, *94*, 4351–4364. [CrossRef]

9. Herer, Y.T.; Raz, T. Optimal parallel inspection for finding the first nonconforming unit in a batch—An information theoretic approach. *Manag. Sci.* **2000**, *46*, 845–857. [CrossRef]

10. Grosfeld-Nir, A.; Gerchak, Y.; He, Q.-M. Manufacturing to order with random yield and costly inspection. *Oper. Res.* **2000**, *48*, 761–767. [CrossRef]

11. Anily, S.; Grosfeld-Nir, A. An optimal lot-sizing and offline inspection policy in the case of nonrigid demand. *Oper. Res.* **2006**, *54*, 311–323. [CrossRef]

12. Wang, C.-H.; Meng, F.-C. Optimal lot size and offline inspection policy. *Comput. Math. Appl.* **2009**, *58*, 1921–1929. [CrossRef]

13. Avinadav, T.; Perlman, Y. Economic design of offline inspections for a batch production process. *Int. J. Prod. Res.* **2013**, *51*, 3372–3384. [CrossRef]

14. Sarkar, B.; Saren, S. Product inspection policy for an imperfect production system with inspection errors and warranty cost. *Eur. J. Oper. Res.* **2016**, *248*, 263–271. [CrossRef]

15. Raz, T.; Herer, Y.T.; Grosfeld-Nir, A. Economic optimization of off-line inspection. *IiE Trans.* **2000**, *32*, 205–217. [CrossRef]

16. Finkelshtein, A.; Herer, Y.T.; Raz, T.; Ben-Gal, I. Economic optimization of off-line inspection in a process subject to failure and recovery. *IiE Trans.* **2005**, *37*, 995–1009. [CrossRef]

17. Wang, C.-H. Economic off-line quality control strategy with two types of inspection errors. *Eur. J. Oper. Res.* **2007**, *179*, 132–147. [CrossRef]

18. Wang, C.-H.; Hung, C.-C. An offline inspection and disposition model incorporating discrete Weibull distribution and manufacturing variation. *J. Oper. Res. Soc. Japan* **2008**, *51*, 155–165. [CrossRef]

19. Bendavid, I.; Herer, Y.T. Economic optimization of off-line inspection in a process that also produces non-conforming units when in control and conforming units when out of control. *Eur. J. Oper. Res.* **2009**, *195*, 139–155. [CrossRef]

20. Wang, W.-Y.; Sheu, S.-H.; Chen, Y.-C.; Horng, D.-J. Economic optimization of off-line inspection with rework consideration. *Eur. J. Oper. Res.* **2009**, *194*, 807–813. [CrossRef]

21. Tsai, W.; Wang, C.-H. Economic optimization for an off-line inspection, disposition and rework model. *Comput. Ind. Eng.* **2011**, *61*, 891–896. [CrossRef]

22. Montgomery, D.C. *Introduction to Statistical Process Control*, 6th ed.; John Wiley & Sons, Inc.: Hoboken, NJ, USA, 2009.

23. Dodge, H.F. A sampling inspection plan for continuous production. *Ann. Math. Stat.* **1943**, *14*, 264–279. [CrossRef]

24. Chen, C.-H.; Chou, C.-Y. Design of a CSP-1 plan based on regret-balanced criterion. *J. Appl. Stat.* **2000**, *27*, 697–701. [CrossRef]

25. Govindaraju, K.; Kandasamy, C. Design of generalized CSP-C continuous sampling plan. *J. Appl. Stat.* **2000**, *27*, 829–841. [CrossRef]

26. Richard Cassady, C.; Maillart, L.M.; Rehmert, I.J.; Nachlas, J.A. Demonstrating Deming's kp rule using an economic model of the CSP-1. *Qual. Eng.* **2000**, *12*, 327–334. [CrossRef]

27. Chen, C.-H.; Chou, C.-Y. Economic design of continuous sampling plan under linear inspection cost. *J. Appl. Stat.* **2002**, *29*, 1003–1009. [CrossRef]

28. Yu, H.-F.; Yu, W.-C.; Wu, W.P. A mixed inspection policy for CSP-1 and precise inspection under inspection errors and return cost. *Comput. Ind. Eng.* **2009**, *57*, 652–659. [CrossRef]

29. Yu, H.-F.; Yu, W.-C. A Joint Inspection Policy between CSP-1 and Precise Inspection for Non-repairable Products under Inspection Errors and Return Cost. *J. Qual.* **2011**, *18*, 61–73.

30. Galindo-Pacheco, G.M.; Paternina-Arboleda, C.D.; Barbosa-Correa, R.A.; Llinás-Solano, H. Non-linear programming model for cost minimisation in a supply chain, including non-quality and inspection costs. *Int. J. Oper. Res.* **2012**, *14*, 301–323. [CrossRef]

31. Springer, C. A method for determining the most economic position of a process mean. *Ind. Qual. Control* **1951**, *8*, 36–39.

32. Duffuaa, S.O.; Siddiqui, A.W. Process targeting with multi-class screening and measurement error. *Int. J. Prod. Res.* **2003**, *41*, 1373–1391. [CrossRef]

33. Duffuaa, S.O.; Al-Turki, U.M.; Kolus, A.A. A process targeting model for a product with two dependent quality characteristics using 100% inspection. *Int. J. Prod. Res.* **2009**, *47*, 1039–1053. [CrossRef]

34. Duffuaa, S.; Al-Turki, U.; Kolus, A. Process-targeting model for a product with two dependent quality characteristics using acceptance sampling plans. *Int. J. Prod. Res.* **2009**, *47*, 4031–4046. [CrossRef]

35. Duffuaa, S.O.; El-Ga'aly, A. Impact of inspection errors on the formulation of a multi-objective optimization process targeting model under inspection sampling plan. *Comput. Ind. Eng.* **2015**, *80*, 254–260. [CrossRef]

36. Duffuaa, S.O.; El-Ga'aly, A. A multi-objective mathematical optimization model for process targeting using 100% inspection policy. *Appl. Math. Model.* **2013**, *37*, 1545–1552. [CrossRef]

37. Duffuaa, S.O.; El-Ga'aly, A. A multi-objective optimization model for process targeting using sampling plans. *Comput. Ind. Eng.* **2013**, *64*, 309–317. [CrossRef]

38. Jaber, M.Y. *Learning Curves: Theory, Models, and Applications*; CRC Press: Boca Raton, FL, USA, 2016.

39. Koulamas, C. Quality improvement through product redesign and the learning curve. *Omega* **1992**, *20*, 161–168. [CrossRef]

40. Teng, J.-T.; Thompson, G.L. Optimal strategies for general price-quality decision models of new products with learning production costs. *Eur. J. Oper. Res.* **1996**, *93*, 476–489. [CrossRef]

41. Franceschini, F.; Galetto, M. Asymptotic defectiveness of manufacturing plants: An estimate based on process learning curves. *Int. J. Prod. Res.* **2002**, *40*, 537–545. [CrossRef]

42. Jaber, M.Y.; Guiffrida, A.L. Learning curves for processes generating defects requiring reworks. *Eur. J. Oper. Res.* **2004**, *159*, 663–672. [CrossRef]

43. Wright, T.P. Factors affecting the cost of airplanes. *J. Aeronaut. sci.* **1936**, *3*, 122–128. [CrossRef]

44. Jaber, M.Y.; Guiffrida, A.L. Learning curves for imperfect production processes with reworks and process restoration interruptions. *Eur. J. Oper. Res.* **2008**, *189*, 93–104. [CrossRef]

45. Jaber, M.Y.; Khan, M. Managing yield by lot splitting in a serial production line with learning, rework and scrap. *Int. J. Prod. Econ.* **2010**, *124*, 32–39. [CrossRef]

46. Kłosowski, G.; Gola, A.; Świć, A. Application of fuzzy logic in assigning workers to production tasks. In Proceedings of the 13th International Conference Distributed Computing and Artificial Intelligence, Sevilla, Spain, 1–3 June 2016; pp. 505–513.

47. Kuo, Y.-H.; Kusiak, A. From data to big data in production research: The past and future trends. *Int. J. Prod. Res.* **2018**, 1–26. [CrossRef]

48. Jasiulewicz-Kaczmarek, M.; Saniuk, A. Human factor in sustainable manufacturing. In *International Conference on Universal Access in Human-Computer Interaction*; Springer: Cham, Switzerland, 2015; pp. 444–455.

49. Muhammad Imran, C.W.K. A Synchronized Job Assignment Model for Manual Assembly Lines Using Multi-Objective Simulation Integrated Hybrid Genetic Algorithm (MO-SHGA). *J. Korean Soc. Ind. Syst. Eng.* **2017**, *40*, 211–220. [CrossRef]

50. Duffuaa, S.; Khan, M. Impact of inspection errors on the performance measures of a general repeat inspection plan. *Int. J. Prod. Res.* **2005**, *43*, 4945–4967. [CrossRef]

51. Elshafei, M.; Khan, M.; Duffuaa, S. Repeat inspection planning using dynamic programming. *Int. J. Prod. Res.* **2006**, *44*, 257–270. [CrossRef]

52. Duffuaa, S.O.; Khan, M. A general repeat inspection plan for dependent multicharacteristic critical components. *Eur. J. Oper. Res.* **2008**, *191*, 374–385. [CrossRef]

53. Colledani, M.; Tolio, T. Performance evaluation of production systems monitored by statistical process control and off-line inspections. *Int. J. Prod. Econ.* **2009**, *120*, 348–367. [CrossRef]

54. Vaghefi, A.; Sarhangian, V. Contribution of simulation to the optimization of inspection plans for multi-stage manufacturing systems. *Comput. Ind. Eng.* **2009**, *57*, 1226–1234. [CrossRef]

55. Khan, M.; Jaber, M.; Wahab, M. Economic order quantity model for items with imperfect quality with learning in inspection. *Int. J. Prod. Econ.* **2010**, *124*, 87–96. [CrossRef]

56. Yang, M. Minimization of Inspection Cost in an Inspection System Considering the Effect of Lot Formation on AOQ. *Int. J. Manag. Sci.* **2010**, *16*, 119–135.

57. Khan, M.; Jaber, M.Y.; Bonney, M. An economic order quantity (EOQ) for items with imperfect quality and inspection errors. *Int. J. Prod. Econ.* **2011**, *133*, 113–118. [CrossRef]

58. Khan, M.; Jaber, M.; Guiffrida, A. The effect of human factors on the performance of a two level supply chain. *Int. J. Prod. Res.* **2012**, *50*, 517–533. [CrossRef]

59. Avinadav, T.; Sarne, D. Sequencing counts: A combined approach for sequencing and selecting costly unreliable off-line inspections. *Comput. Oper. Res.* **2012**, *39*, 2488–2499. [CrossRef]

60. Bouslah, B.; Gharbi, A.; Pellerin, R. Joint production and quality control of unreliable batch manufacturing systems with rectifying inspection. *Int. J. Prod. Res.* **2014**, *52*, 4103–4117. [CrossRef]

61. Khan, M.; Jaber, M.Y.; Ahmad, A.-R. An integrated supply chain model with errors in quality inspection and learning in production. *Omega* **2014**, *42*, 16–24. [CrossRef]

62. Liu, N.-C.; Liu, W.-C. The effects of quality management practices on employees' well-being. *Total Qual. Manag. Bus. Excell.* **2014**. [CrossRef]

63. Aslam, M.; Wu, C.-W.; Azam, M.; Jun, C.-H. Mixed acceptance sampling plans for product inspection using process capability index. *Qual. Eng.* **2014**, *26*, 450–459. [CrossRef]

64. Yang, M.H.; Cho, J.H. Minimisation of inspection and rework cost in a BLU factory considering imperfect inspection. *Int. J. Prod. Res.* **2014**, *52*, 384–396. [CrossRef]

65. Mohammadi, M.; Siadat, A.; Dantan, J.-Y.; Tavakkoli-Moghaddam, R. Mathematical modelling of a robust inspection process plan: Taguchi and Monte Carlo methods. *Int. J. Prod. Res.* **2015**, *53*, 2202–2224. [CrossRef]

66. Jones, D.; Tamiz, M. *Practical Goal Programming*; Springer: Berlin, Germany, 2010; Volume 141.

© 2019 by the authors. Licensee MDPI, Basel, Switzerland. This article is an open access article distributed under the terms and conditions of the Creative Commons Attribution (CC BY) license (http://creativecommons.org/licenses/by/4.0/).

 *mathematics* 

Article

A Single-Stage Manufacturing Model with Imperfect Items, Inspections, Rework, and Planned Backorders

Chang Wook Kang [1], Misbah Ullah [2], Mitali Sarkar [1], Muhammad Omair [3] and Biswajit Sarkar [1,*]

[1] Department of Industrial and Management Engineering, ERICA Campus, Hanyang University, Ansan, Gyeonggi-do 15588, Korea; cwkang57@hanyang.ac.kr (C.W.K.); mitalisarkar.ms@gmail.com (M.S.)
[2] Department of Industrial Engineering, University of Engineering and Technology, Peshawar 25000, Pakistan; misbah@uetpeshawar.edu.pk
[3] Department of Industrial Engineering, Jalozai Campus, University of Engineering and Technology, Peshawar 25000, Pakistan; muhamad.omair87@gmail.com
* Correspondence: bsbiswajitsarkar@gmail.com; Tel.: +82-107-498-1981

Received: 14 April 2019; Accepted: 16 May 2019; Published: 19 May 2019

Abstract: Each industry prefers to sell perfect products in order to maintain its brand image. However, due to a long-run single-stage production system, the industry generally obtains obstacles. To solve this issue, a single-stage manufacturing model is formulated to make a perfect production system without defective items. For this, the industry decides to stop selling any products until whole products are ready to fulfill the order quantity. Furthermore, manufacturing managers prefer product qualification from the inspection station especially when processes are imperfect. The purpose of the proposed manufacturing model considers that the customer demands are not fulfilled during the production phase due to imperfection in the process, however customers are satisfied either at the end of the inspection process or after reworking the imperfect products. Rework operation, inspection process, and planned backordering are incorporated in the proposed model. An analytical approach is utilized to optimize the lot size and planned backorder quantities based on the minimum average cost. Numerical examples are used to illustrate and compare the proposed model with previously developed models. The proposed model is considered more beneficial in comparison with the existing models as it incorporates imperfection, rework, inspection rate, and planned backorders.

Keywords: imperfect manufacturing system; backordering; defective products; rework; inspection

1. Introduction

Economic and production order quantity models have been extensively used in real industrial life for calculation of the optimum lot size. However, these models are based on the assumption that production processes result in perfect quality products always. There are various factors, which affect the manufacturing processes to make that assumption highly unrealistic. Engineers and technical experts reprocess these imperfect products and deliver them as per customer requirements after its qualification from the quality control section. For example, manufacturing processes involved in ultra-precision manufacturing industries including the automobile sector, ship manufacturing, defense and aerospace related products manufacturing industries, and tool manufacturing industries are a few of the common areas where the production of reworked and non-reworked products cannot be eliminated at all. Such products need qualification from the quality control department of the said organization in order to ensure that the product has been manufacturing as per customer requirements. In such setups, it is very rare that the product is delivered without its confirmation from inspection stations, i.e., demands fulfillment during the production stage. This proposed model re-considers the concept of Cárdenas-Barrón [1] to develop rework and backordering for the single-stage manufacturing setup.

Jamal et al. [2] introduced the rework operation to develop an optimal batch size for a single-stage production system. Two models were developed with the assumption that the rework operation can be performed either immediately or at the end of N-cycles. The optimal batch size was calculated based on the average minimum cost function. In the model, it is assumed that customer demands are met during the production process where demands are fulfilling during the production stage although the process is imperfect. Later, Cárdenas-Barrón [1] extended this model by incorporating the concept of planned backorders for a single-stage manufacturing setup. The rework operation was performed immediately at the end of the production process. Cárdenas-Barrón [1] obtained a closed-form solution for this model based on the average system cost. The optimum lot size and backorder quantity was calculated. Recently, Sarkar et al. [3] assumed that defective items may follow three kinds of distribution functions, i.e., uniform, triangular, and beta distribution. The analytical method has been used to calculate the optimal lot size and backorder quantity. These models assume imperfect processes during the production phase on one side and customer demands fulfillment on the other side. Therefore, inventory modeling in most of the models is based on $(p(1 - \gamma) - d)$. The production model based on $(p(1 - \gamma) - d)$ is based on the assumption that processes deviate from ideal conditions and are unable to produce quality products each and every time. This deviation from the ideal situation results in imperfection in processes that ultimately affect the products produced (p) by these processes. This imperfection results in the production of γ that need rectification / rework. Therefore, this number of items cannot be added to the total inventory buildup rather they need rectification. Hence, the net inventory is $(p(1 - \gamma) - d)$ instead of $(p - d)$, which is applicable in the case of the perfect manufacturing environment. Whereas simultaneous demands fulfillment during the production phase is relatively unrealistic. Products need qualification from an inspection station before they are delivered to the customer or warehouse especially when processes are imperfect. Optimum lot sizes were calculated for real life scenarios, resulting in extension to the basic economic order quantity (EOQ) and production quantity models. Lee and Rosenblatt [4] introduced an idea of larger investment in quality control approaches to obtain the benefit of lower defective rate, reduced setup, and holding costs. Ben-Daya and Hariga [5] developed an economic lot size model with an important realistic assumption that perfect production processes deviate at random rate and results in some non-conforming products. Salameh and Jaber [6] extended the EOQ model for electronic industry items in particular with the assumption that 100% screening is performed of all incoming units. They assumed that low quality products could be sold out at the reduced price in a single batch. Later, Goyal and Cárdenas-Barrón [7] used a simple algebraic approach for optimum lot size calculation. The results were almost similar when other approaches were utilized in an imperfect production environment.

Biswas and Sarker [8] developed an optimal lot size for a lean manufacturing system where scrap items were identified 'before', 'during' and 'after' the rework process. Shortages in the production process were met through a buffer station of the qualified finish products. Chiu et al. [9] initiated a random defective rate within an imperfect production system. Pal et al. [10] worked on the production model with a stochastic demand not a stochastic defective rate. Ojha et al. [11] considered Chiu et al.'s [9] model with a quality assurance and rework. Chiu et al. [12] made their own model [9] with service level constraint. Rahim and Al-Hajailan [13] worked on Ojha et al.'s [11] model with a time-varying defective rate. Sarkar [14] considers delay-in-payments and stock-dependent demand. Sarkar et al. [15] converted a single-stage production system into a multi-stage production system when the defective item's reworking would be after n cycles or in each cycle that would be decided, whereas they considered a constant defective rate. Sarkar et al. [16] incorporated lead-time demand in the integrated production model. Ullah and Kang [17] mathematically managed the rejections of defective products and inspection on work-in-process lot-size. Cárdenas-Barrón et al. [18–20] covered three production models with discrete deliveries, multiple shipments, and partial reworks. Wee et al. [21] proposed an alternative solution approach of the production model. Sarkar [22] worked on the improved production model with the concept of reliability without any defective products. Tayyab and Sarkar [23] extended Sarkar et al.'s [3] model with multi-stage but with constant defective products. Sarkar and Moon [24] introduced the model the under inflation and time value of money. Sarkar et al. [25] extended Sarkar's [22] model with time-dependent demand under the reliable production system. Sarkar et al. [26] developed an optimal reliable model with defective products. However, none of the researchers consider the demand after production and full inspection to fulfill the customer's demand. Wee et al. [27] used the renewal reward theorem (RRT) to calculate the average total profit for optimization of the economic production quantity lot size by considering the imperfect production environment and screening constraint. Researchers (Sana [28,29] and Chaudhuri [30]) concluded that imperfect quality items can be reworked for their practical utilization and significance.

The basic objective of the study is to convert and model the imperfect production system into a mathematical form. The purpose of the proposed production model considers the scenario where the customer demands are not fulfilled during the production phase due to imperfection in the process, and customers are satisfied either at the end of the inspection process or after reworking the imperfect products. The rework operation, inspection process, and planned backordering are incorporated in the proposed model. An analytical approach is utilized to optimize the lot size and planned backorder quantities based on the minimum average cost. The introduction has been given in this section. The rest of the paper is modeled as follows: The next section is related to the past and current research studies related to the imperfection in the production model. Section 3 describes the model development with notation, assumptions, and equations required for optimal solutions. In Section 4, numerical examples are used to evaluate and discuss the proposed model's results. Section 5 includes the results and discussion of the numerical experiments performed using the proposed methodology. The sensitivity analysis has been given in detailed form in Section 6. Conclusions and future directions of the research are presented in the last section.

2. Literature Review

Most of the research carried out in the past concentrated on the production process itself. Defective products are produced in an out-of-control status of the process. These products are either scrapped or reworked in most of the research literature. One major aspect of the production setup is the product qualification via the inspection process. Industries including aerospace, automobile, and pharmaceutical cannot deliver the product until and unless it is qualified by the inspection stage. Inspection is considered as a process that consumes resources (time, material, techniques) in a similar way as are consumed by other processes. Relatively less attention has been given to the inspection rate in conjunction with the manufacturing process. However, in most of the inventory related literature, it is assumed that inspection takes either minimal time or is performed in parallel to the manufacturing processes. Whereas inspection rate importance is much required in environments where production processes are imperfect as the probability of non-conforming products increases. Ben-Salem et al. [31,32] considered an environmental issue and sub-control issues within two alternative production maintenance systems where the production degradation is considered without stopping the demanded products for the market. Purohit et al. [33] extended a basic production model with maintenance planning within an integrated approach, even though they considered production and demand in parallel. Ullah et al. [34] explained the inspection errors within the inspection system within a tradition production system. Shin et al. [35] extended Ullah et al.'s [34] model with trade-credit financing. Kim and Sarkar [36] converted the basic single-stage production system to the multi-stage production system where they did not consider initial production and then demand. Diabat et al. [37] extended the basic production model with partial downstream down payment. Kang et al. [38] explained the effect of smart manufacturing technology based on the human quality control system. Sarkar et al. [39] extended the field of the imperfect production with a distribution free approach to calculate shortages during lead-time demand. Omair et al. [40] and Sarkar et al. [41] extended the production model by considering the imperfect production with the inspection process, rework, and rejected products. Kang et al. [42] extended Cárdenas-Barrón's [1] model with safety stock and planned backorders whereas they did not think about imperfect products.

Researchers highlighted the impact of the inspection process on the optimum lot size during the last few decades. Zhang and Gerchak [43] introduced a model based on the joint lot size in addition with the inspection policy. The single period manufacturing problem was considered with uncertainty in demand and inconstant yield. Inspection errors effect on total cost and lot size was highlighted by Ben-Daya and Rahim [44] in a multistage production setup. They assumed that inspection is performed at the end of every stage. Ben-Daya et al. [45] developed models for an integrated system highlighting the impact of different inspection policies including 'no inspection', 'sample based inspection', and 100 percent inspection. Khan et al. [46] presented an extensive review of EOQ models focusing on the model developed by the Salameh and Jaber model [6]. The review classified various extended models based on imperfect items, shortages backordering, quality, and supply chain. Razaei [47] incorporated backorder into the inventory model for an imperfect manufacturing setup. Analytical approaches were used to get the optimal lot size and shortages level. Wee et al. [48] extended Salameh and Jaber's model [6] for the imperfect production setup incorporating the backorders. Taleizadeh et al. [49] developed an economic production order quantity model with disturbance in processes that result in scrap and rework. They considered the backorder with cycle length and optimal backorder quantity as the decision variable. Hu et al. [50] introduced an inventory model that highlighted that the backorder cost per unit is increasing linearly with shortage time. Ganguly et al. [51] considered partial backorders in supply chain management to meet the demand in the market.

Table 1 highlights contributions made during the last several years in the field of inventory management focusing on imperfect production processes. It can be observed in Table 1 that the literature assumes imperfect production processes for a production quantity model along with rework operation, inspection process, and backorder process. Most of the considered research works focuses on the objective to minimize the cost of the production system under the optimization process except the work of Wee et al. [27] (maximizing total profit of the production system). All these extended models assume that customer demands are fulfilled in the production phase. However, products need qualification from an inspection station before they are delivered to the customer especially when the process is imperfect. Therefore, it is possible that demands fulfillment during the production phase is not appropriate as the probability of delivering imperfect quality products to the customer rises. Research gap exists in the available literature, to the best of our knowledge, to ensure qualified products delivery to the end user while working in random imperfect production setup. This paper is an approach towards fulfillment of this gap in the existing literature. From Table 1, it is found that several authors considered production and supply in parallel when imperfect products are in the system, which is impractical to maintain the brand image of the industry. The most benefitted way is that during production, the products supply should not be allowed to meet the demand as imperfect products are there. This is a major research gap within all studies in this direction. Thus, this proposed model is considered to solve this issue.

This model assumes that products can only be consumed at a demand rate when they are either qualified or processes are perfect. The management delivers the manufactured products to the inspection station where products are classified as either qualified, rejected or to be reworked. It is assumed that imperfect products are rewardable. However, the probability of rejected products is low and can be ignored. It may further be added that the backordering process has also been taken into consideration to calculate the optimum backorder quantity in an imperfect manufacturing environment. It is hoped that this model integrates all major aspects (rework, inspection rate and backordering) of the production setup to help managers in making decisions regarding the inventory level and backorder quantity for an imperfect production setup focusing on qualified products delivery to the customer.

Table 1. Author contribution in the production model on the basis of process, rework, inspection, backordering and demand.

Author (s)	Process			Rework Process			Inspection Process			Backordering Process			Demands Fulfillment (d)	
	Imperfect	Re-Workable	Scrap	Perfect	Not Allowed	Imperfect	No	Sampling	100%	No Back-Ordering	Lost Sale	Back-Ordering	During Production Phase	After Inspection
Cárdenas-Barrón [1]	√	√		√					√		√	√	√	
Salameh and Jaber [6]	√		√		√		√			√				
Ojha et al. [11]	√	√		√					√	√				
Chiu et al. [12]	√	√	√			√	√					√	√	
Sarkar et al. [15]	√	√		√			√				√	√	√	
Wee et al. [27]	√		√		√				√			√	√	
Sana [29]	√	√		√			√			√		√	√	
Ben-Daya and Rahim [44]	√		√	√				√					√	
Ben-Salem et al. [31]	√		√		√									
Ullah et al. [34]	√	√					√			√	√			
Kim and Sarkar [36]	√			√			√					√	√	
Kang et al. [38]	√			√			√			√			√	
Sarkar et al. [39]	√	√	√	√		√	√	√	√	√	√	√	√	
This Paper	√	√	√	√					√				√	√

3. Mathematical Model

The problem starts with the assumption that the optimal lot size and backorder quantity are derived for an imperfect manufacturing environment. Lot size Q arrives at the manufacturing station during each cycle. Due to process imperfection, the γ percent of total products needs rectification as the process goes out-of-control status during the production phase. All these imperfect products are reworkable. Due to imperfection in the process, manufactured products are not delivered to the customer directly, rather these products are processed through an inspection stage with a known inspection rate (M) in order to know about product qualification. On the other side, all reworkable products are re-processed after the inspection process. It is assumed that reworked products are qualified after reprocessing and no re-inspection is required.

Other assumptions are as follows:

1. Demand and production rates are known and constant. During production, no product is to be sold. After production completion, the demand is started as the production system contains the defective product.
2. Production rate is higher than the demand rate.
3. Reworkable products produced during the manufacturing phase are known in percentage.
4. Imperfect products produced during the production processes are re-processed on the same machine with 100% qualification.
5. Customer receives qualified perfect products only to maintain the good brand image of the industry.
6. Qualified products are delivered to the customer after the inspection process.
7. Planned backordering is allowed and the inventory storage space is unlimited.

The following notation has been used to develop the model.

Parameters

d	demand rate (units/unit time)
p	production rate (units/unit time)
h	inventory holding cost per unit time ($/unit/unit time)
t_1	time required for planned backorder units to be fulfilled when process starts again.
t_2	time required by machining station to manufacture lot size Q
t_3	time required to inspect manufactured units
t_4	time required to rework imperfect units
t_5	time required to consume on hand inventory
t_6	time required to build the backorder inventory
I_1	average inventory developed during machining of lot size Q (units/unit time)
I_2	average inventory produced during rework operation (units/unit time)
I_{max}	maximum average inventory produced during one cycle time (units/unit time)
γ	percentage of imperfect products produced during the production phase (%)
Z	backorder cost per unit product per unit time ($/unit backorder/ unit time)
c	cost of manufacturing per unit product ($/unit)
k	setup cost per lot size ($/setup)
C_i	inventory holding cost per unit of time ($/unit time)
C_b	backorder cost per unit time ($/unit time)
C_s	setup cost per unit time ($/unit time)
C_m	manufacturing cost per unit time ($/unit time)
$TC(Q,B)$	total cost per unit time ($/unit time)

Variables

Q	lot size (units)
B	backorder quantity (units)

The inventory over time for an economic order production quantity model for a complete cycle is shown below in Figure 1. The process considers planned backorders, manufacturing of lot size, inspection process, and rework process of imperfect products. As stated earlier, customer demands are not fulfilled during the production phase due to the process imperfection. The objective is to ensure qualified products for delivery to the customer. The inventory builds up during the production phase. Demands are not fulfilled in the production phase. Customer demands are fulfilled at the rate d during the rework process only. Some percentages of products are qualified at the rework stage and can be delivered to the customer in order to minimize inventory cost.

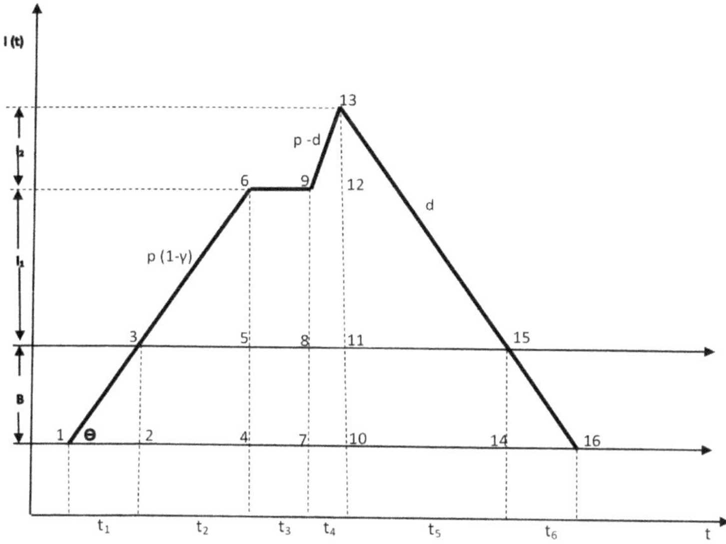

Figure 1. Inventory over time in an imperfect manufacturing environment with inspection and planned backordering.

Considering Figure 1, it is found that the time required to produce lot size Q at the production rate p is the sum of the time required by different processes to produce Q units. These processes include the time required for planned backorder units to be fulfilled when the process starts again (t_1), time required to manufacture lot size Q (t_2), time required to inspect the manufactured units (t_3), and time required to rework imperfect units (t_4), i.e.,

$$t_p = t_1 + t_2 + t_3 + t_4 \tag{1}$$

Obtaining these times in terms of the demand rate (d) and inventory lot size Q,

From Figure 1, the area of Δ 146 is the accumulated by production and backorder inventory, one can obtain

$$\tan \theta = \frac{I_1 + B}{t_1 + t_2}$$

$$p(1 - \gamma) = \frac{I_1 + B}{t_1 + t_2}$$

$$t_1 + t_2 = \frac{I_1 + B}{p(1 - \gamma)} \tag{2}$$

The time required for the inspection of manufactured products (t_3) is given by the following relation at the rate M,

$$t_3 = \frac{I_1 + B}{M} \tag{3}$$

Similarly, t_4 is the time required to rework imperfect products,

$$t_4 = \frac{\gamma Q}{p}$$

In addition, the area of Δ 91213 is the accumulated inventory during reworking, then the time can be obtained as follows:

$$t_4 = \frac{I_2}{(p-d)}$$

i.e.,

$$I_2 = \gamma Q\left(1 - \frac{d}{p}\right) \tag{4}$$

Therefore, (1) becomes

$$\frac{Q}{p} = \frac{I_1 + B}{p(1-\gamma)} + \frac{I_1 + B}{M} + \frac{\gamma Q}{p}$$

After some algebraic rearrangements and simplification, one can obtain

$$I_1 = \frac{MQ(1-\gamma)^2}{M + p(1-\gamma)} - B \tag{5}$$

Therefore, from (4) and (5), the maximum inventory is given by

$$I_{max} = I_1 + I_2 = \frac{QM\,(1-\gamma)^2}{M + p(1-\gamma)} - B + \gamma Q\left(1 - \frac{d}{p}\right) \tag{6}$$

Assuming

$$\frac{(1-\gamma)^2}{M + p(1-\gamma)} = \theta_1 \text{ and } \left(1 - \frac{d}{p}\right) = \theta_2, \text{ then}$$

$I_1 = MQ\theta_1 - B$
$I_2 = \gamma Q\theta_2$
$I_{max} = MQ\theta_1 + \gamma Q\theta_2 - B$

The average inventory can be found by taking the sum of the area of triangles and rectangles as shown in Figure 1 divided by the cycle time. The cycle time is defined as the time in which Q units inventory is consumed at the demand rate (d).

From the area Δ 356, the inventory for accruing Q quantity during inspection, area A_1 is given by

$$A_1 = \frac{I_1 \times t_2}{2}$$

where $t_2 = \frac{I_1}{p(1-\gamma)}$, putting in the above equation

$$A_1 = \frac{(MQ\theta_1 - B)^2}{2p(1-\gamma)} \tag{7}$$

The area of Δ 5698 is the area for holding inventory Q quantity during inspection, A_2 is as follows:

$$A_2 = I_1 \times t_3$$

where $t_3 = \frac{(I_1+B)}{M}$, thus

$$A_2 = \frac{(MQ\theta_1 - B)^2 + (B(MQ\theta_1 - B))}{M} \tag{8}$$

Similarly,

$$A_3 = I_1 \times t_4, \text{ where } t_4 = \frac{\gamma Q}{p}$$

$$A_3 = \left(\frac{MQ^2\theta_1\gamma}{p} - \frac{BQ\gamma}{p} \right) \tag{9}$$

The area of Δ 91213 is the area for reworking, which gives the building of inventory. One can have

$$A_4 = \frac{I_2 \times t_4}{2}$$

$$A_4 = \frac{Q^2\gamma^2\theta_2}{2p} \tag{10}$$

The area A_5 of $\Delta 111315$ is the area for demand of products without any production. Thus, the area can be found as follows:

$$A_5 = \frac{I_{max} \times t_5}{2}, \text{ where } t_5 = \frac{I_{max}}{d}$$

Therefore,

$$A_5 = \frac{(MQ\theta_1 + \gamma Q\theta_2 - B)^2}{2d} \tag{11}$$

Therefore, the average inventory $\left(I_{avg}\right)$ per unit cycle time is the summation of (7–10), and (11) divided by the cycle time (t), i.e.,

$$I_{avg} = \left\{ \frac{\left(\frac{(MQ\theta_1-B)^2}{2p(1-\gamma)}\right) + \left(\frac{(MQ\theta_1-B)^2+(B(MQ\theta_1-B))}{M}\right) + \left(\frac{MQ^2\theta_1\gamma}{p} - \frac{BQ\gamma}{p}\right) + \left(\frac{Q^2\gamma^2\theta_2}{2p}\right) + \left(\frac{(MQ\theta_1+\gamma Q\theta_2-B)^2}{2d}\right)}{\frac{Q}{d}} \right\} \tag{12}$$

Algebraic simplification results in the following expression:

$$I_{avg} = \{ \left(\frac{(M^2Q^2\theta_1^2+B^2-2\,BMQ\theta_1)}{2p(1-\gamma)} \right) + MQ^2\theta_1^2 - BQ\theta_1 + \frac{MQ^2\theta_1\gamma}{p} + \frac{Q^2\gamma^2\theta_2}{2p} - \frac{BQ\gamma}{p} + \frac{1}{2d}(M^2Q^2\theta_1^2 + B^2 + Q^2\theta_2^2\gamma^2 - 2BMQ\theta_1 - 2\,BQ\gamma\theta_2 + 2MQ^2\theta_1\theta_2\gamma))/(\frac{Q}{d}) \} \tag{13}$$

Hence, the total inventory holding cost per unit cycle time is given by

$$C_i = h\{ \left(\frac{(M^2Q^2\theta_1^2+B^2-2\,BMQ\theta_1)}{2p(1-\gamma)} \right) + MQ^2\theta_1^2 - BQ\theta_1 + \frac{MQ^2\theta_1\gamma}{p} + \frac{Q^2\theta_2\gamma^2}{2p} - \frac{BQ\gamma}{p} + \frac{1}{2d}(M^2Q^2\theta_1^2 + B^2 + Q^2\theta_2^2\gamma^2 - 2BMQ\theta_1 - 2\,BQ\gamma\theta_2 + 2MQ^2\theta_1\theta_2\gamma))/(\frac{Q}{d}) \} \tag{14}$$

The average backordering cost can be computed by calculating the area of Δ 123 and Δ 141516 which represent the average inventory in the form of backordered products per unit time (t).

Therefore,

$$\text{Backorder cost } (C_b) = (\text{Average backorder inventory}) \times (\text{Unit cos t per unit backorder product per unit of time})$$

$$= \frac{(\text{Area of } (\Delta\ 123) + \text{Area of}(\Delta\ 141516))}{t} \times (z)$$

$$= \frac{\left(\frac{B\times t_1}{2} + \frac{B\times t_6}{2}\right)}{\frac{Q}{d}}(z)$$

where $t_1 = \frac{B}{p(1-\gamma)}$, $t_6 = \frac{B}{d}$, which gives

$$C_b = \left(\frac{B^2(p(1-\gamma)+d)}{2p(1-\gamma)Q}\right)(z) \tag{15}$$

Other important costs taken into consideration are the setup cost (C_s) and the manufacturing cost (C_m).

$$C_s = \frac{kd}{Q} \tag{16}$$

$$C_m = cd(1+\gamma) \tag{17}$$

Therefore, following the relation for all associated costs including the inventory-holding cost, setup cost, backorder cost and manufacturing cost per unit of time.

$$TC\,(Q,B) = h\{(\frac{M^2Q^2\theta_1^2+B^2-2\,BMQ\theta_1}{2p(1-\gamma)} + MQ^2\theta_1^2 - BQ\theta_1 + \frac{MQ^2\theta_1\gamma}{p} + \frac{Q^2\theta_2\gamma^2}{2p} - \frac{BQ\gamma}{p} +$$
$$\tfrac{1}{2d}(M^2Q^2\theta_1^2 + B^2 + Q^2\theta_2^2\gamma^2 - 2BMQ\theta_1 - 2\,BQ\gamma\theta_2 + 2MQ^2\theta_1\theta_2\gamma))/(\tfrac{Q}{d})\} \tag{18}$$
$$+\,(\tfrac{B^2(p(1-\gamma)+d)}{2p(1-\gamma)Q})(z) + cd(1+\gamma) + \tfrac{kd}{Q}$$

After algebraic simplification, one obtains:

$$TC\,(Q,B) = \left(\frac{dhM^2\theta_1^2}{2p(1-\gamma)} + dhM\theta_1^2 + \frac{dh\theta_2\gamma^2}{2p} + \frac{dhM\theta_1\gamma}{p} + \frac{M^2h\theta_1{}^2}{2} + \frac{h\theta_2{}^2\gamma^2}{2} + hM\theta_1\theta_2\gamma\right)Q$$

$$\left(\frac{dh}{p(1-\gamma)} + h + \frac{(p(1-\gamma)+d)z}{p(1-\gamma)}\right)\frac{B^2}{2Q} - \left(\frac{dhM\theta_1}{p(1-\gamma)} + dh\theta_1 + \frac{dh\gamma}{p} + hM\theta_1 + h\theta_2\gamma\right)B + cd(1+\gamma) + \frac{kd}{Q}$$

The following symbols are used to simplify the above expression,

$$R_1 = \left(\frac{dhM^2\theta_1^2}{2p(1-\gamma)} + dhM\theta_1^2 + \frac{dh\theta_2\gamma^2}{2p} + \frac{dhM\theta_1\gamma}{p} + \frac{M^2h\theta_1{}^2}{2} + \frac{h\theta_2{}^2\gamma^2}{2} + hM\theta_1\theta_2\gamma\right)$$

$$R_2 = \left(\frac{dh}{p(1-\gamma)} + h + \frac{(p(1-\gamma)+d)z}{p(1-\gamma)}\right)$$

$$R_3 = \left(\frac{dhM\theta_1}{p(1-\gamma)} + dh\theta_1 + \frac{dh\gamma}{p} + hM\theta_1 + h\theta_2\gamma\right)$$

thus,

$$TC\,(Q,B) = Q(R_1) + \left(\frac{B^2}{2Q}\right)(R_2) - B(R_3) + \frac{kd}{Q} + cd(1+\gamma) \tag{19}$$

The optimum lot size (Q^*) and backorder quantity (B^*) can be found by minimization of the total cost function assuming Q and B as continues function.

Theorem 1. *$TC^*(Q,B)$ will have global maximum values at Q^* and B^* if $Q^* = \sqrt{\dfrac{2kd(R_2)}{2R_1R_2-(R_3)^2}}$ and $B^* = \dfrac{R_3}{R_2}$*
$$\left(\sqrt{\frac{2kd(R_2)}{2R_1R_2-(R_3)^2}}\right).$$

Proof. The following conditions need to be fulfilled in order to verify that (19) is a convex function.

$$\text{I)}\quad \frac{\partial^2 TC\,(Q,B)}{\partial Q^2} > 0,\quad \frac{\partial^2 TC\,(Q,B)}{\partial B^2} > 0$$

$$\text{II)}\quad \left(\frac{\partial^2 TC\,(Q,B)}{\partial Q^2}\right)\left(\frac{\partial^2 TC\,(Q,B)}{\partial B^2}\right) - \left(\frac{\partial^2 TC\,(Q,B)}{\partial Q\partial B}\right)^2 > 0 \tag{20}$$

□

Assuming Q and B as continues, first partial derivatives with respect to Q and B of (19) are shown in (21) and (22), respectively.

$$\frac{\partial TC\,(Q,\,B)}{\partial Q} = (R_1) - \left(\frac{B^2}{2Q^2}\right)(R_2) - \frac{kd}{Q^2} \tag{21}$$

$$\frac{\partial TC\,(Q,\,B)}{\partial B} = \left(\frac{B}{Q}\right)(R_2) - (R_3) \tag{22}$$

Similarly,

$$\frac{\partial^2 TC\,(Q,\,B)}{\partial Q^2} = \left(\frac{B^2}{Q^3}\right)(R_2) + \frac{2kd}{Q^3} > 0 \tag{23}$$

$$\frac{\partial^2 TC\,(Q,\,B)}{\partial B^2} = \left(\frac{R_2}{Q}\right) > 0 \tag{24}$$

$$\left(\frac{\partial^2 TC\,(Q,\,B)}{\partial Q^2}\right)\left(\frac{\partial^2 TC\,(Q,\,B)}{\partial B^2}\right) - \left(\frac{\partial^2 TC\,(Q,\,B)}{\partial Q \partial B}\right)^2 = \frac{2kd}{Q^4}(R_2) \tag{25}$$

The necessary and sufficient conditions for (19) to be convex have been fulfilled as shown in (23), (24), and (25). Therefore, an optimal solution exists at which the total cost function will be minimum. Thus, (21) and (22) are simultaneously solved to obtain the optimal lot size (Q) and backorder quantity (B) which are given by:

$$Q^* = \sqrt{\frac{2kd(R_2)}{2R_1R_2 - (R_3)^2}} \tag{26}$$

$$B^* = \frac{R_3}{R_2}\left(\sqrt{\frac{2kd(R_2)}{2R_1R_2 - (R_3)^2}}\right) \tag{27}$$

The optimal total cost can be obtained by substituting (26) and (27) in (19) as follows:

$$TC^*(Q^*, B^*) = \sqrt{\frac{2kd\left(2R_1R_2 - R_3^2\right)}{R_2}} + cd(1 + \gamma)$$

Therefore, $TC^*(Q^*, B^*)$ is the global minimum total cost at Q^* and B^*.

This proposed model could be reduced to the basic economic production quantity model (EPQ) if processes are assumed perfect, demand is fulfilled during the production phase and shortages are not expected. In addition, the optimum lot size and optimum backorder quantity obtained by (26) and (27) can be reduced to the mathematical model developed by Cárdenas-Barrón [1] if the assumption of meeting customer demands during the production phase is made.

4. Numerical Examples

Two numerical examples have been used to understand the proposed model utilization for practical engineering problems and its comparison with previously developed models. Data has been taken from the Cárdenas-Barrón's model [1]. Numerical examples illustrate the effect of imperfect products and the inspection rate on the optimum lot size and backorder quantity. The total cost, optimal lot size and optimal backorder quantity at different values of the defect rate and inspection rate are calculated.

4.1. Example 1

The parametric data used for the mathematical experiment of the proposed model include the demand rate (d) = 300 units per year, inspection rate (M) = 550 units per year, production rate (p) = 550 units per year, holding cost (h) = $50 per unit per year, shortage cost per unit (z) = $10 per unit short per year, manufacturing cost (c) = $7 per unit, setup cost ($k$) = $50 per unit per year.

4.2. Example 2

The similar type of analysis is performed for another example whose data set is relatively for higher demand and production rate. In this example, the demand rate (d) = 4800 units per year, production rate (p) = 24,000 units per year, setup cost (k) = $120 per lot size, holding cost (h) = $0.6 per unit per year, shortages cost (z) = $14 per unit short per year and manufacturing cost (c) = $3 per unit. The inspection rate is assumed as 36,000 units per year and defective products produced during the process are varied gradually over an interval (0, 40%).

5. Results and Discussion

Table 2 shows the impact of the defective rate on the optimal lot size (Q) and optimal backorder quantity (B). It can be observed that lot size increases with an increase in the percentage of the defective rate. Similarly, the total cost variation is also significant. Total cost is increased with an increase in imperfection during the manufacturing process. However, an increase in the optimal backorder quantity level is relatively low with an increase in the defective rate. Comparing these results with the results obtained by Cárdenas-Barrón [1] (Example 1), it can be observed that the total cost of our proposed model is comparable for the range of imperfection (0%, 40%). However, the lot size has been increased significantly. The backorder quantity increases up to 35% of the defective rate and then decreases with increases in defective products.

Table 2. Change in optimal lot size and backorder quantity with variation in the defective rate (Example 1).

$\gamma(\%)$	θ_1	Total Cost ($/Year)	Q^*(Units)	B^*(Units)
0	0.000909091	2423.44	93	52
1	0.000895477	2437.49	95	53
5	0.000841492	2493.71	104	57
10	0.00077512	2564.18	118	62
15	0.000710074	2635.20	136	69
20	0.000646465	2707.40	160	79
25	0.000584416	2782.06	191	90
30	0.000524064	2861.69	228	104
35	0.000465565	2950.74	259	113
40	0.000409091	3054.67	262	109

It can be observed in Table 3 that the total cost and optimal lot size increase with an increase in the defective rate significantly at higher demand and production rate. The higher the defective products produced, the higher will be the lot size to be ordered. However, the change in backorder units is less significant for higher production rate and demand rate when the holding cost per unit product is comparatively low. Comparing these results with the results obtained by Cárdenas-Barrón [1] (Example 2), the total cost of our proposed model has not been increased significantly although the inspection process and carrying inventory over a longer period have been incorporated. Relatively larger lot sizes have been proposed by our model.

The change in optimal lot size with the change in the inspection rate has been shown in Figure 2 below. Example 2 data has been used for further analysis. It can be observed that the optimal lot size goes on decreasing with an increase in the inspection rate. It may be noted that the inspection rate is increased from 24,000 units (equal to the production rate) to 42,000 units per year (one and half time higher than the production rate). Defective products rate remain fixed at 20%. All other data remain the same. The total cost and backorder quantity level has been shown in Table 4. It may be noted that the backorder quantity level doesn't change with significant change in the inspection rate at the fixed level of defective products produced.

Table 3. Change in optimal lot size and backorder quantity with variation in the defective rate (Example 2).

$\gamma(\%)$	θ_1	Total Cost ($/Year)	Q^* (Units)	B^*(Units)
0	0.000016	14,991.78	1947	52
1	0.000016	15,133.44	1954	52
5	0.000015	15,700.49	1985	52
10	0.000014	16,410.33	2020	52
15	0.000012	17,121.43	2052	52
20	0.000011	17,833.88	2080	52
25	0.000010	18,547.80	2103	51
30	0.000009	19,263.32	2120	51
35	0.000008	19,980.57	2131	51
40	0.000007	20,699.69	2135	50

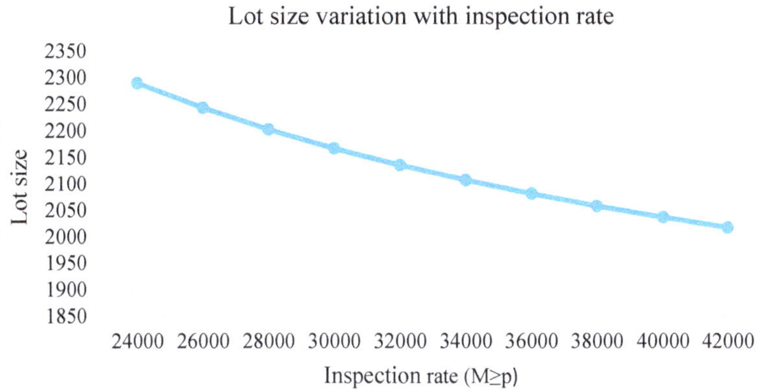

Lot size variation with inspection rate

Figure 2. Change in optimal lot size with change in the inspection rate.

Table 4. Change in optimal lot size and backorder quantity with variation in the inspection rate (Example 2).

M (Units Per Year)	θ_1	Total Cost ($/Year)	Q^*(Units)	B^* (Units)
24,000	0.0000148	17,783.27	2289	52
26,000	0.0000141	17,793.67	2243	52
28,000	0.0000135	17,803.15	2202	52
30,000	0.0000130	17,811.81	2166	52
32,000	0.0000125	17,819.77	2134	52
34,000	0.0000120	17,827.10	2106	52
36,000	0.0000115	17,833.88	2080	52
38,000	0.0000112	17,840.17	2057	52
40,000	0.0000108	17,846.02	2035	52
42,000	0.0000104	17,851.47	2016	52

At the end, both the inspection rate and defective products production rate are changed simultaneously with an ascending order and their impact on the lot size, backorder level, and total cost is evaluated. The data has been shown in Table 5. It can be observed that total cost increases with an increase in the inspection and defective rate. Moreover, it can be observed that for higher inspection rate relative to the production process (M>>>p), the optimal lot size calculated by our proposed model approaches the lot size calculated by using the economic order quantity model under ideal conditions.

Table 5. Change in optimal lot size and backorder quantity with variation in the defective rate and inspection rate (Example 2).

M (Units Per Year)	γ(%)	θ_1	Total Cost ($/Year)	Q^* (Units)	B^* (Units)
24,000	0	0.000020	14,914.97	2237	52
26,000	5	0.000018	15,644.61	2196	52
28,000	10	0.000016	16,371.63	2167	52
30,000	15	0.000014	17,096.52	2147	52
32,000	20	0.000125	17,819.77	2134	52
34,000	25	0.000011	18,541.88	2126	52
36,000	30	0.0000092	19,263.32	2120	51
38,000	35	0.0000078	19,984.59	2115	50
40,000	40	0.0000066	20,706.15	2109	50
42,000	45	0.0000054	21,428.45	2100	49

6. Sensitivity Analysis

In order to evaluate the impact of important parameters on the optimal lot size and optimal backorder quantity as proposed by the developed model, sensitive analysis is carried out. It is based on the data shown in Example 2.

Table 6 summarizes the effect of key parameters on the optimal lot size, optimal backorder quantity, and total cost function. Parameters' values have been changed from −50% to +50%. It can be observed that the lot size increases with the increase in values for parameters k, d, p, and γ. Furthermore, the lot size decreases for increased values for parameters h M, *and* z. Similarly, the optimal backorder quantity increases for increased values of parameters k, d, h, and p. The impact of parameters' M, γ, and z is reverse, i.e., optimal backorder quantity decreases for higher values of M and z. Furthermore, the impact of positive change in parameters k, d, h, M, γ, and z values increases the total cost function value. However, the total cost function value decreases with increased values for parameter p. The setup cost and holding cost are key costs for any production system. If all costs are fixed and these costs are increased, the total cost must be increased. Within these costs, it is found that the setup cost is more sensitive than the holding cost. It means that the industry can reduce further the total cost by reducing the setup cost using some initial investment. It is found that the production rate is increased whereas the total cost is reduced. In the proposed model, the production rate is constant and it's increasing value gives decreasing total cost. This implies that the industry can think about the controllable production rate within a certain limit of minimum production rate and maximum production rate. Then, the total cost of the production system can be reduced more. As it is a traditional production system, thus the inspection cost is controlled through human inspection. With the increasing value of this cost, the total cost increases; this implies that the industry manager should think about the smart production system where all inspections controlled through online by the machine and production rate is flexible. The rejection cost is the most sensitive cost as all types of efforts are there with all procedures but there are no products that can be sold to obtain revenue. Thus, the machinery system should be perfect always to produce perfect product always such that there should not be any rejection cost. The shortage cost should not be in the production system as the production rate is greater than the demand but as during production no demanded products are given to the market. Thus, this cost has to incorporate. Even though, the shortage cost is there, the total cost is increasing very less comparing to the other costs.

Table 6. Sensitivity analysis (Example 2).

Parameter	Changes (%)	Parameter Value	Q (Units)	B (Units)	Total Cost ($/Year)
Setup cost (k)	−50%	60	1471	37	17,671.65
	−25%	90	1801	45	17,759.68
	+25%	150	2325	58	17,899.26
	+50%	180	2547	63	17,958.36
Holding cost (h)	−50%	0.3	2910	37	17,675.81
	−25%	0.45	2389	45	17,762.20
	+25%	0.75	1870	58	17,896.04
	+50%	0.9	1716	63	17,951.37
Production rate (p)	−50%	12,000	1720	47	17,949.81
	−25%	18,000	1915	50	17,881.72
	+25%	30,000	2229	53	17,796.91
	+50%	36,000	2366	54	17,766.95
Inspection rate (M)	−50%	18,000	2479	53	17,744.79
	−25%	27,000	2222	52	17,798.52
	+25%	45,000	1990	51	17,858.99
	+50%	54,000	1927	51	17,877.76
Rejection (%)	−50%	0.1	2020	52	16,410.33
	−25%	0.15	2052	52	17,121.43
	+25%	0.25	2103	51	18,547.80
	+50%	0.3	2120	51	19,263.32
Shortage cost (z)	−50%	7.2	2123	102	17,822.63
	−25%	10.8	2094	69	17,830.06
	+25%	18	2071	42	17,836.21
	+50%	21.6	2065	35	17,837.78

7. Conclusions

In this paper, a single stage manufacturing model for the imperfect production setup with rework, inspection, and backordering has been developed. It is assumed that inventory is carried over a longer period of time in the production phase, as the process is imperfect. Due to process imperfection, products are not delivered directly to the customer as observed in most of the conventional production order quantity models. This model optimized the lot size and backorder quantity in an imperfect manufacturing setup when demands are fulfilled after the production process. This approach is against the conventional production order quantity models where processes are imperfect and demands are fulfilled simultaneously. Moreover, products were passed through an inspection station for their qualification. The results were compared via numerical examples to the previously developed model from literature. The numerical examples highlighted the importance of the proposed model as compared to the previous model by considering the quality of the product through the inspection process, however, the cost is slightly been increased. Furthermore, as the demand rate and production rate were increased, the total cost of this model has been reduced in a significant way. The model provides insights to manufacturing engineers working in an imperfect manufacturing setup. It can be best utilized when manufacturing industries deliver products after the inspection process. Total cost is increased when demand is not fulfilled during the production process due to process imperfection under given conditions. Furthermore, the proposed model approaches the production quantity model to calculate the optimal lot size, when the inspection rate is much higher in comparison to the production rate under ideal situations. The model can be extended to incorporate the partial backordering and multistage manufacturing environment to obtain an optimal lot size and backorder level. The consideration of the suppliers and buyers with the imperfection process and sample based inspection can be the extension of the study with the assumption that demands are fulfilled after the inspection process. This production can be incorporated within an integrated inventory model or within a supply chain model to extend it further. The production rate can be considered as variable to make it a smart production system to produce smart products.

Author Contributions: C.W.K. supervised the research work and presented an idea; B.S. conceived and designed the research plan by the support of mathematical modeling; M.U. performed the numerical experiments and wrote the original draft; M.S. analyzed the data; M.O. contributed by computer software tool to get numerical results. All authors worked equally in reporting revisions and updating the manuscript.

Conflicts of Interest: The authors declare no conflict of interest.

References

1. Cárdenas-Barrón, L.-E. Economic production quantity with rework process at a single-stage manufacturing system with planned backorders. *Comput. Ind. Eng.* **2009**, *57*, 1105–1113. [CrossRef]
2. Jamal, A.; Sarker, B.R.; Mondal, S. Optimal manufacturing batch size with rework process at a single-stage production system. *Comput. Ind. Eng.* **2004**, *47*, 77–89. [CrossRef]
3. Sarkar, B.; Cárdenas-Barrón, L.-E.; Sarkar, M.; Singgih, M.L. An economic production quantity model with random defective rate, rework process and backorders for a single stage production system. *J. Manuf. Syst.* **2014**, *33*, 423–435. [CrossRef]
4. Rosenblatt, M.-J.; Lee, H.-L. Economic production cycles with imperfect production processes. *IIE Trans.* **1986**, *18*, 48–55. [CrossRef]
5. Ben-Daya, M.; Hariga, M. Economic lot scheduling problem with imperfect production processes. *J. Oper. Res. Soc.* **2000**, *51*, 875–881. [CrossRef]
6. Salameh, M.; Jaber, M. Economic production quantity model for items with imperfect quality. *Int. J. Prod. Econ.* **2000**, *64*, 59–64. [CrossRef]
7. Goyal, S.-K.; Cárdenas-Barrón, L.-E. Note on: Economic production quantity model for items with imperfect quality—A practical approach. *Int. J. Prod. Econ.* **2002**, *77*, 85–87. [CrossRef]
8. Biswas, P.; Sarker, B.-R. Optimal batch quantity models for a lean production system with in-cycle rework and scrap. *Int. J. Prod. Res.* **2008**, *46*, 6585–6610. [CrossRef]
9. Chiu, S.-W.; Gong, D.-C.; Wee, H.-M. Effects of random defective rate and imperfect rework process on economic production quantity model. *Jpn. J. Ind. Appl. Math.* **2004**, *2*, 375–389. [CrossRef]
10. Pal, B.; Sana, S.-S.; Chaudhuri, K. A mathematical model on EPQ for stochastic demand in an imperfect production system. *J. Manuf. Syst.* **2013**, *32*, 260–270. [CrossRef]
11. Ojha, D.; Sarker, B.; Biswas, P. An optimal batch size for an imperfect production system with quality assurance and rework. *Int. J. Prod. Res.* **2007**, *45*, 3191–3214. [CrossRef]
12. Chiu, S.-W.; Ting, C.-K.; Chiu, Y.-S.-P. Optimal production lot sizing with rework, scrap rate, and service level constraint. *Math. Comput. Model.* **2007**, *46*, 535–549. [CrossRef]
13. Rahim, M.; Al-Hajailan, W. An optimal production run for an imperfect production process with allowable shortages and time-varying fraction defective rate. *Int. J. Adv. Manuf. Technol.* **2006**, *27*, 1170–1177. [CrossRef]
14. Sarkar, B. An EOQ model with delay in payments and stock dependent demand in the presence of imperfect production. *Appl. Math. Comput.* **2012**, *21*, 8295–8308. [CrossRef]
15. Sarker, B.-R.; Jamal, A.-M.-M.; Mondal, S. Optimal batch sizing in a multi-stage production system with rework consideration. *Eur. J. Oper. Res.* **2008**, *184*, 915–929. [CrossRef]
16. Sarkar, B.; Gupta, H.; Chaudhuri, K.; Goyal, S.-K. An integrated inventory model with variable lead time, defective units and delay in payments. *Appl. Math. Comput.* **2014**, *237*, 650–658. [CrossRef]
17. Ullah, M.; Kang, C.-W. Effect of rework, rejects and inspection on lot size with work-in-process inventory. *Int. J. Prod. Res.* **2014**, *5*, 2448–2460. [CrossRef]
18. Cárdenas-Barrón, L.E.; Taleizadeh, A.A.; Treviño-Garza, G. An improved solution to replenishment lot size problem with discontinuous issuing policy and rework, and the multi-delivery policy into economic production lot size problem with partial rework. *Expert Syst. Appl.* **2012**, *39*, 13540–13546. [CrossRef]
19. Cárdenas-Barrón, L.-E.; Sarkar, B.; Treviño-Garza, G. An improved solution to the replenishment policy for the EMQ model with rework and multiple shipments. *Appl. Math. Model.* **2013**, *37*, 5549–5554. [CrossRef]
20. Cárdenas-Barrón, L.-E.; Treviño-Garza, G.; Widyadana, G.A.; Wee, H.-M. A constrained multi-products EPQ inventory model with discrete delivery order and lot size. *Appl. Math. Comput.* **2014**, *230*, 359–370. [CrossRef]
21. Wee, H.-M.; Wang, W.-T.; Cárdenas-Barrón, L.-E. An alternative analysis and solution procedure for the EPQ model with rework process at a single-stage manufacturing system with planned backorders. *Comput. Ind. Eng.* **2013**, *64*, 748–755. [CrossRef]

22. Sarkar, B. An inventory model with reliability in an imperfect production process. *Appl. Math. Comput.* **2012**, *218*, 4881–4891. [CrossRef]

23. Tayyab, M.; Sarkar, B. Optimal batch quantity in a cleaner multi-stage lean production system with random defective rate. *J. Clean. Prod.* **2016**, *139*, 922–934. [CrossRef]

24. Sarkar, B.; Moon, I. An EPQ model with inflation in an imperfect production system. *Appl. Math. Comput.* **2011**, *217*, 6159–6167. [CrossRef]

25. Sarkar, B.; Mandal, P.; Sarkar, S. An EMQ model with price and time dependent demand under the effect of reliability and inflation. *Appl. Math. Comput.* **2014**, *231*, 414–421. [CrossRef]

26. Sarkar, B.; Sana, S.-S.; Chaudhuri, K. Optimal reliability, production lot size and safety stock in an imperfect production system. *Int. J. Math. Oper. Res.* **2010**, *2*, 467–490. [CrossRef]

27. Wee, H.-M.; Wang, W.-T.; Yang, P.-C. A production quantity model for imperfect quality items with shortage and screening constraint. *Int. J. Prod. Res.* **2012**, *51*, 1869–1884. [CrossRef]

28. Sana, S.-S. A production–inventory model in an imperfect production process. *Eur. J. Oper. Res.* **2010**, *200*, 451–464. [CrossRef]

29. Sana, S.-S. An economic production lot size model in an imperfect production system. *Eur. J. Oper. Res.* **2010**, *201*, 158–170. [CrossRef]

30. Sana, S.-S.; Chaudhuri, K. An EMQ model in an imperfect production process. *Int. J. Syst. Sci.* **2010**, *41*, 635–646. [CrossRef]

31. Ben-Salem, A.; Gharbi, A.; Hajji, A. Environmental issue in an alternative production-maintenance control for unreliable manufacturing system subject to degradation. *Int. J. Adv. Manuf. Technol.* **2015**, *77*, 383–398. [CrossRef]

32. Ben-Salem, A.; Gharbi, A.; Hajji, A. Production and uncertain green subcontracting control for an unreliable manufacturing system facing emissions. *Int. J. Adv. Manuf. Technol.* **2016**, *83*, 1787–1799. [CrossRef]

33. Purohit, B.-S.; Lad, B.K. Production and maintenance planning: An integrated approach under uncertainties. *Int. J. Adv. Manuf. Technol.* **2016**, *86*, 3179–3191. [CrossRef]

34. Ullah, M.; Kang, C.-W.; Sarkar, B. Human errors incorporation in work-in-process group manufacturing system. *Sci. Iran.* **2017**, *24*, 2050–2061.

35. Shin, D.; Mittal, M.; Sarkar, B. Effects of human errors and trade-credit financing in two-echelon supply chain model. *Eur. J. Ind. Eng.* **2018**, *12*, 465–503. [CrossRef]

36. Kim, M.; Sarkar, B. Multi-stage cleaner production process with quality improvement and lead time dependent ordering cost. *J. Clean. Prod.* **2017**, *144*, 572–590. [CrossRef]

37. Diabat, A.; Taleizadeh, A.-A.; Lashgari, M. A lot sizing model with partial downstream delayed payment, partial upstream advanced payment, and partial backordering for deteriorating items. *J. Manuf. Syst.* **2017**, *45*, 322–342. [CrossRef]

38. Kang, C.-W.; Babar, B.-M.; Sarkar, B.; Imran, M. Effect of inspection performance in smart manufacturing system based on human quality control system. *Int. J. Adv. Manuf. Technol.* **2018**, *94*, 4351–4364. [CrossRef]

39. Sarkar, B.; Kim, J.-S.; Kim, M.; Sarkar, M.; Iqbal, W. An application of distribution free approach through an integrated inventory model with imperfect production and backorders. *J. Manuf. Syst.* **2018**, *4*, 153–167.

40. Omair, M.; Sarkar, B.; Cárdenas-Barrón, L.E. Minimum quantity lubrication and carbon footprint: A step towards sustainability. *Sustainability* **2017**, *9*, 714. [CrossRef]

41. Sarkar, B.; Omair, M.; Choi, S.-B. A multi-objective optimization of energy, economic, and carbon emission in a production model under sustainable supply chain management. *Appl. Sci.* **2018**, *8*, 1744. [CrossRef]

42. Kang, C.-W.; Ullah, M.; Sarkar, B. Optimum ordering policy for an imperfect single-stage manufacturing system with safety stock and planned backorder. *Int. J. Adv. Manuf. Technol.* **2018**, *95*, 109–120. [CrossRef]

43. Zhang, X.; Gerchak, Y. Joint lot sizing and inspection policy in an EOQ model with random yield. *IIE Trans.* **1990**, *22*, 41–47. [CrossRef]

44. Ben-Daya, M.; Rahim, A. Optimal lot-sizing, quality improvement and inspection errors for multistage production systems. *Int. J. Prod. Res.* **2003**, *41*, 65–79. [CrossRef]

45. Ben-Daya, M.; Noman, S.M.; Hariga, M. Integrated inventory control and inspection policies with deterministic demand. *Comput. Oper. Res.* **2006**, *33*, 1625–1638. [CrossRef]

46. Khan, M.; Jaber, M.-Y.; Guiffrida, A.-L.; Zolfaghari, S. A review of the extensions of a modified EOQ model for imperfect quality items. *Int. J. Prod. Econ.* **2011**, *132*, 1–12. [CrossRef]

47. Rezaei, J. Economic order quantity model with backorder for imperfect quality items. In Proceedings of the 2005 IEEE International Engineering Management Conference, St. John's, NL, Canada, 11–14 September 2005; pp. 466–470.
48. Wee, H.-M.; Yu, J.; Chen, M.-C. Optimal inventory model for items with imperfect quality and shortage backordering. *Omega* **2007**, *35*, 7–11. [CrossRef]
49. Taleizadeh, A.-A.; Cárdenas-Barrón, L.-E.; Mohammadi, B. A deterministic multi product single machine EPQ model with backordering, scraped products, rework and interruption in manufacturing process. *Int. J. Prod. Econ.* **2014**, *150*, 9–27. [CrossRef]
50. Hu, W.-T.; Kim, S.-L.; Banerjee, A. An inventory model with partial backordering and unit backorder cost linearly increasing with the waiting time. *Eur. J. Oper. Res.* **2009**, *197*, 581–587. [CrossRef]
51. Ganguly, B.; Pareek, S.; Sarkar, B.; Sarkar, M.; Omair, M. Influence of controllable lead time, premium price, and unequal shipments under environmental effects in a supply chain management. *RAIRO-Oper. Res.* **2018**. [CrossRef]

 © 2019 by the authors. Licensee MDPI, Basel, Switzerland. This article is an open access article distributed under the terms and conditions of the Creative Commons Attribution (CC BY) license (http://creativecommons.org/licenses/by/4.0/).

Article

Multi-Product Production System with the Reduced Failure Rate and the Optimum Energy Consumption under Variable Demand

Shaktipada Bhuniya [1], Biswajit Sarkar [2],* and Sarla Pareek [1]

[1] Department of Mathematics and Statistics, Banasthali Vidyapith, Banasthali, Rajasthan 304 022, India; shakti13math@gmail.com (S.B.); spareek13@gmail.com (S.P.)

[2] Department of Industrial & Management Engineering, Hanyang University, Ansan Gyeonggi-do 15588, Korea

* Correspondence: bsbiswajitsarkar@gmail.com; Tel.: +82-31-400-5259 or +82-10-7498-1981; Fax: +82-31-400-5959

Received: 31 December 2018; Accepted: 26 February 2019; Published: 24 May 2019

Abstract: The advertising of any smart product is crucial in generating customer demand, along with reducing sale prices. Naturally, a decrease in price always increases the demand for any smart product. This study introduces a multi-product production process, taking into consideration the advertising- and price-dependent demands of products, where the failure rate of the production system is reduced under the optimum energy consumption. For long-run production systems, unusual energy consumption and machine failures occur frequently, which are reduced in this study. All costs related with the production system are included in the optimum energy costs. The unit production cost is dependent on the production rate of the machine and its failure rate. The aim of this study is to obtain the optimum profit with a reduced failure rate, under the optimum advertising costs and the optimum sale price. The total profit of the model becomes a complex, non-linear function, with respect to the decision variables. For this reason, the model is solved numerically by an iterative method. However, the global optimality is proved numerically, by using the Hessian matrix. The numerical results obtained show that for smart production, the maximum profit always occurs at the optimum values of the decision variables.

Keywords: inventory; smart production; variable demand; advertisement; energy; rework; system reliability

---

1. Introduction

Any production system may produce both perfect and imperfect products. During the long-run production process of a smart production system, there may be a chance of machine failure, due to machine breakdown, unskilled laborers, or interrupted energy suppliers, and so on. Due to these reasons, a defective product may be produced in two ways: at a constant rate or at a random rate. For a constant defect rate, the total number of defective items is constant and, for a random defect rate, the total number of defective products is a random variable. Many studies based on constant defect rates in single-item production systems have been carried out (see, for instance, [1]); however, very little research based on random defect rates (see, for instance, [2]) is available. Sarkar [3] proposed a model for a multi-item production system with a random defect rate and budget and space constraints. There has been no research, to our knowledge, on multi-item smart production systems with the optimum consumption of energy and a random defect rate for smart products, with advertisement, price-dependent demand patterns, and reduced failure rates. Therefore, this proposed model gives a new direction for production systems with budget and space constraints under the effect of energy.

For this type of (perfect and imperfect) production system, rework has a vital role in smart production, where the production system becomes out-of-control (from an in-control state) within a random time interval. It provides a reliable system by reworking the defective items. Some studies of imperfect production systems with deterioration are already available in the literature, such as Rossenblat and Lee [4], who discussed imperfect production systems; which was extended by Kim and Hog [5] by considering the deterioration of products in a production system to find the optimal production length. They indicated three types of deterioration methods—constant deterioration, linearly increasing deterioration, and exponentially increasing deterioration—for system moving from an in-control to an out-of-control state. Giri and Dohi [6] proposed a model which highlights random machine failure rates for single-item production systems, where the machine breakdown is stochastic and the preventive maintenance time for machine failure is also a random variable.

Sana et al. [7] gave an idea for a research model of an imperfect production system, by considering defective products with reduced prices, although this type of idea is now common in the market. Chiu et al. [8] proposed a model which assumes a rework policy of the defective products, using extra costs for the reworking of imperfect products. In the literature, there are few models based on energy in an imperfect smart production system. No study, to our knowledge, has considered the optimum energy consumption and its profit for a smart production system, under advertising- and price-dependent demands. Egea et al. [9] expressed, in a model, how to measure energy during the loading of a smart machine. González et al. [10] proposed a model for turbo-machinery components, using a total energy consideration.

Sarkar [11] considered an inventory model involving stock-dependent demand with delayed payments. This model indicated the replenishment policy for an imperfect production system with a finite replenish rate. A production-inventory model with deterioration and a finite replenishment rate was developed by Sarkar [12]. In this model, to maximize the profit, several discount offers for customers, to attract a large order size, were considered. Generally, imperfect product production depends on the production system reliability. Sarkar [13] developed an economic manufacturing quantity (EMQ) model with an investment in the production system for the development of a high system reliability with lower imperfect production. This model first considered an advertisement policy, where the demand depends on the price of products. An imperfect production model was considered by Chakraborty and Giri [14] with some imperfect products produced in an out-of-control system during preventive maintenance. An inspection was considered in this model to detect the defective items for the reworking process, although some defective products cannot be repaired.

Sarkar [15] investigated an economic production quantity (EPQ) model with imperfect products, where a back-ordering policy was included in the model, along with a reworking policy. To calculate the rate of defective items, three different distribution functions were used and the results are compared in this model. Sarkar and Saren [16] introduced a production model in an imperfect production system with an inspection policy. In their model, the production system becomes out-of-control in a random time interval. This model considered a quality inspector for choosing falsely an imperfect product and making a decision about quality, and vice versa. Over a fixed time period, the warranty policy also makes this model more realistic.

Pasandideh et al. [17] extended a production model for multiple products in a single-machine imperfect production system, where the imperfect products were classified by their nature, to consider whether to rework or scrap them. To make this model more realistic, they considered fully backlogging all shortages. An inventory model for a system with a non-stationary stochastic demand with a detailed analysis of the lot-size problem was developed by Purohit et al. [18]; a carbon-emissions mechanism was included with this model to make it a more generalized study. This model discussed labor issues with the training required involving work related to the machine. Sana [19] considered an EPQ lot-size model for imperfect production with defective items when the system becomes out-of-control. An optimal inventory for a repair model was initiated by Cárdenas-Barrón et al. [20].

Another two research models, based on deterioration and partial backlogging, were developed by Tiwari et al. [21,22].

Storage capacity for an inventory system plays an important role in any production house. Limited storage makes increasing production problematic. Due to this reason, shortages may occur. Huang et al. [23] developed an inventory model in which a rental warehouse was considered, with an associated cost, for fulfilling the required capacity in addition to the available warehouse. This inventory model investigated optimal retailer lot-size policies with delayed payments and space constraints. An inventory model for multiple products with limited space was proposed by Pasandideh and Niaki [24]. This model contained a non-linear integer programming problem and found an optimal solution for the available warehouse by adding space constraints. Through a genetic algorithm with a non-linear cost function and space constraint, a multi-stage inventory model was discussed by Hafshejani et al. [25]. An inventory model considering demand and limited space availability, where reliability depends on the unit production costs, was introduced by Mahapatra et al. [26].

In a production model, the budget for a production system is initially required for the manufacturer. Instead of a periodic budget, a limited budget is preferable. There is some on-going research into budget constraints, such as Taleizadeh et al. [27], who proposed a model, in a multi-item production system, for a reworking of defective items policy. They found the global minimum of the total cost by considering a service level and a budget constraint. Minimizing the total annual cost with a limited capital budget and calculating the optimal lot size and capital investment in a setup-costs model was explained by Hou and Lin [28]. Mohan et al. [29] introduced an inventory model, considering delayed payments, budget constraints, and permissible partial payment (with penalty) for a multi-item production system with a replenishment policy. Todde et al. [30] and Du et al. [31] published the basic energy models, based on energy consumption and energy analysis. Cárdenas-Barrón et al. [32] proposed an inventory model with an improved heuristic algorithm solving method for a just-in-time (JIT) system with the maximum available budget. Xu et al. [33] presented a bio-fuel model for the pyrolysis products of plants. Tomić and Schneider [34] discussed a method for recovering energy from waste using a closed-loop supply chain. Haraldsson and Johansson [35] developed an energy model, based on different energy efficiencies during production. Similarly, Dey et al. [36] and Sarkar et al. [37–39] developed their models based on energies, but did not consider advertising for smart products, reducing the sale price of products, or reducing the failure rate of a production system. Yao et al. [40] and Gola [41] put forth the valuable idea of formulating a model considering the reliability of a manufacturing system.

The world becomes smarter every day. Several researchers have discussed imperfect production processes, such as Tayyab et al. [42], Sarkar [43], Kim et al. [44], and Sarkar et al. [45], but the effect of a smart manufacturing system in any production model has not been discussed. The effects of energy and failure rate in a multi-item smart production system was first discussed by Sarkar et al. [3]. In reality, the demand for a particular product depends on various key factors, with advertisement of a particular product being one of them. Ideally, advertisement of a particular product increases the demand for that product. Thus, the total profit can be optimized when the demand depends on the advertisement of products. The relationship between product quality and advertising, in an analytical model, was developed by Chenavaz and Jasimuddin [46]; they also explained the positive and negative advertising–quality relationships. A two-level supply chain model was developed by Giri and Sharma [47], where the demand depends on the advertising cost. In this model, they considered a single-manufacturer, two-retailer system, where the retailers compete. In the same direction, Xiao et al. [48] formulated a two-echelon supply chain model for a single manufacturer and multiple retailers. In this model, they studied co-operative issues in advertising. Recently, Noh et al. [49] developed a two-echelon supply chain model, where the demand depends on the advertisement. They used the Stackelberge game policy to solve this model. Sale price, also, has a great impact on the demand for a product. The demand for a product gradually increases if the sale price is less, and vice versa. Constant demand is a business service that helps customers to find new consumers and penetrate new markets, which can optimize the inside-sales and marketing activities to achieve high-quality sales. Variable demand can predict and quantify changes that are caused by transportation conditions on the demand. As a high price negatively affects how likely clients are to buy products or services, assuming the demand to be price-dependent is more realistic. Karaoz et al. [50] considered an inventory model with price- and time-dependent demand, under the influence of complementary and substitute product sale prices. The finite replenishment inventory model was developed by considering the demand to be sensitive to changes in time and sale price. Sana [51] introduced the price-sensitive demand for perishable items in an inventory model. The demand for any inventory system is not always constant and may depend on time, sale price, and inventory. Pal et al. [52] developed a multi-item inventory model where the demand was sensitive to the sale price and price-break. Sarkar et al. [53] developed an EMQ model with price- and time-dependent demand, under the effects of reliability and inflation. Sarkar and Sarkar [54] established an inventory model, where the demand was inventory-dependent, and an algorithm was developed to maximize the profit. To maximize the vendor profit, the optimal ordering quantity and sale price were optimized by an analytical procedure. An integrated model, with development lead time and production rate, was discussed by Azadeh and Paknafs [55]. Several researchers have developed many models where the demand depends on the sale price of the products or advertising-dependent demand; however, price- and advertising-dependent demand in a smart production system has still not been considered, to our knowledge. Thus, this new direction is considered in this research. Table 1 shows the contribution of previous author(s).

Table 1. Author(s) contribution table.

Author(s)	Development Cost	Inspection	Demand	Advertisement	Energy	Rework	Reliability
Cárdenas-Barrón [1]			√			√	
Sana et al. [2]			√				
Sarkar et al. [3]	√	√	√		√	√	√
Rosenblatt and Lee [4]			√				
Sarkar [12]	√		√				√
Sarkar and Saren [16]			√				
Sana [19]	√		√				√
Taleizadeh et al. [27]			√				√
Mohan et al. [29]	√		√				
Cárdenas-Barrón [32]			√			√	√
This Paper	√	√	√	√	√	√	√

2. Problem Definition, Notation, Assumptions

2.1. Problem Definition

Smart production systems are modern production systems, in which the main aim is to produce smart products (such as mobile phones, computers, and so on) through some smart machines by some smart, skilled labors. The whole smart production process is controlled under the optimum energy consumptions. For example, a smart machine can produce more woolen clothes than an expert labor; in this case, quality and quantity will also be more developed than in simple production systems. An imperfect production system is a production system in which a defective product is produced in a long-run process, due to machine breakdown or any other problem. Energy consumption has an important role in imperfect production systems. In smart production systems, production of imperfect products can be controlled with sufficient consumption of energy. Our model is based on a smart production system for multiple products, where machine failure occurs in random time intervals, γ_i, in which there is an unusual amount of energy consumption and less reliable conditions and production completed in random time intervals, T_i . For this situation, the development cost used in this model is reliability-dependent, to reduce the machine failure rate and to minimize the consumption of energy. An inspection cost is used to detect defective items. For the popularity of the products in this model, advertisement costs and price-dependent demand patterns are used. To obtain more reliability, there are a several possibilities for the reworking of defective items before being sold to the customer. The aim of this model is to obtain the maximum profit for a multi-item production system with the conditions of energy consumption and a random defect rate, under the effects of advertisement and sale price. Process flow for multi-product production system are shown in the following Figure 1.

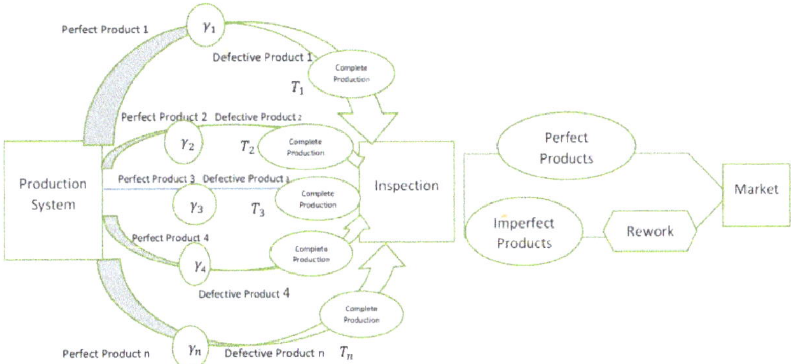

Figure 1. Complex production management with system reliability.

2.2. Assumptions

In developing the proposed model, the following assumptions are considered:

1. This is a multi-item production model, where defective products are produced after a random time. The system moves to an out-of-control state at a random time, γ_i, at a failure rate of α, and produces defective items. These items are then reworked to make them as new products.
2. This system is controlled, under energy consumption, for completely finished products only.
3. No shortages are allowed for this multi-item smart production system, as variable production rate is greater than the demand and the lead time is considered to be negligible.
4. The demand is assumed to be advertising- and price-dependent, to increase the demand pattern. It is taken as $D(S,y) = \frac{S_{max}-S}{S-S_{min}} + xy^{\mu}$, where x is the scaling parameter and μ is the shape parameter.
5. If the failure rate α decreases, then the system will be more reliable, and vice versa.
6. The system contains multiple items and, thus, there will be a possibility for space problems for the products, which affects the total budget. Thus, to make this model more realistic under the sufficient energy consumption, space, and budget constraints are considered in this model.

3. Model Formulation

The model studies about a smart production system for multi-item. During the time $t_i = 0$ to $t_i = t_{1i}$, the production continues with upward direction graph and for the time $[t_{1i}, T]$, there is no production, thus, the holding inventories positions are downstream direction with demand D. The production system becomes out-of-control from in-control in random time interval γ_i. See Figure 2 for the description of the production system. The governing differential equation of the inventory is

$$\frac{dI_{1i}(t_i)}{dt_i} = p_i - D, 0 \le t_i \le t_{1i} \tag{1}$$

with initial condition $I_{1i}(0) = 0$ and

$$\frac{dI_{2i}(t_i)}{dt_i} = -D, t_{1i} \leq t_i \leq T \tag{2}$$

with initial condition $I_{2i}(T) = 0$.

Solving the above differential equations, one can obtain

$$I_{1i} = (p_i - D)t_i, 0 \leq t_i \leq t_{1i} \tag{3}$$

$$I_{2i} = D(T - t_i), t_{1i} \leq t_i \leq T \tag{4}$$

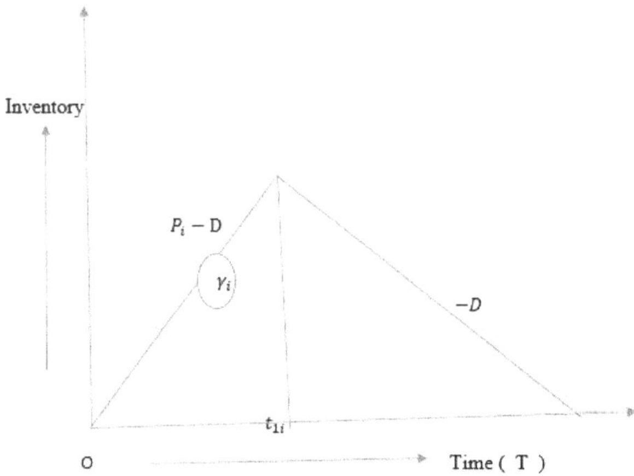

Figure 2. Economic Production quantity.

Considering the following costs in a multi-item smart production system, we must optimize for those which have a low failure rate and maximum profit.

Setup Cost (SC)

In this model, setup cost for product i is S_{ci} per setup and per setup energy consumption cost be S'_{ci}. Then, the average setup cost per cycle is

$$SC = \sum_{i=1}^{n}(S_{ci} + S'_{ci})\frac{D}{q_i}. \tag{5}$$

Holding Cost (HC)

For this production system, to calculate the holding cost it is necessary to find the total inventory by summation from $i = 1$ to $i = n$; further, to get the average inventory, the total inventory can be divided by cycle length. Thus, the total inventory can be calculated as

$$\begin{aligned}
\text{Total Inventory} &= \sum_{i=1}^{n}\left[\int_0^{t_{1i}} I_{1i}(t_i)dt_i + \int_{t_{1i}}^{T} I_{2i}(t_i)dt_i\right] \\
&= \sum_{i=1}^{n}\left[\int_0^{t_{1i}} (p_i - D)t_i dt_i + \int_{t_{1i}}^{T} D(T - t_i)dt_i\right]. \tag{6}
\end{aligned}$$

Hence, the total holding cost, with sufficient consumption of energy, can be calculated, as follows:

$$
\begin{aligned}
HC &= \sum_{i=1}^{n} \frac{(H_{ci} + H'_{ci})D}{2q_i} \left[\int_0^{t_{1i}} (p_i - D)t_i dt_i + \int_{t_{1i}}^{T} \left((p_i - D)\frac{q_i}{p_i} - Dt_i \right) dt_i \right] \\
&= \sum_{i=1}^{n} \frac{(H_{ci} + H'_{ci})q_i}{2} \left(1 - \frac{D}{p_i} \right).
\end{aligned}
\tag{7}
$$

Development Cost (DC)

The system becomes more reliable based on development costs. A high investment in the production system, in terms of development costs, makes for a low failure rate, and vice versa. The labor cost and energy resources cost are included in the development cost. Thus, the total development cost per unit time is

$$
DC = G + Ze^{r\frac{\alpha_{max}-\alpha}{\alpha-\alpha_{min}}}.
\tag{8}
$$

Inspection Cost (IC)

A multi-item smart production system is considered to be a long-run process, where the process is imperfect. Thus, an inspection cost plays an important role in detecting defective items, such that the defective products can be reworked easily. Thus, the inspection cost per unit cycle, under energy consumption, is

$$
\begin{aligned}
IC &= \sum_{i=1}^{n} (I_c + I'_c) q_i \times \frac{D}{q_i} \\
&= \sum_{i=1}^{n} (I_c + I'_c) D.
\end{aligned}
\tag{9}
$$

Rework Cost (RC)

For reworking the defective products, at first, inspection is needed for all defective products. To find the RC, it is important to know how many defective products there are, and how much the RC is for defective items. The rate of defective items is considered (see, for instance, [3]) to be $\beta P_i^{\delta}(t_i - \gamma_i)^{\tau}$, where $\delta \geq 0, \tau \geq 0$, and $t_i \geq \gamma_i$. Hence, the number of imperfect items produced by the machine is

$$
\begin{aligned}
E(K) &= \sum_{i=1}^{n} \left(\frac{\beta}{\tau+1} \right) p_i^{\delta+1} \int_0^{t_{1i}} (t_{1i} - \gamma_i)^{\tau+1} dH(\gamma_i) \\
&= \sum_{i=1}^{n} p_i^{\delta+1} \left(\frac{\beta}{\tau+1} \right) e^{\frac{-\alpha q_i}{p_i}} \chi \left(\alpha, \frac{q_i}{p_i} \right), \quad as \; t_{1i} = \frac{q_i}{p_i}.
\end{aligned}
\tag{10}
$$

For reworking the defective products to perfect state, a RC is needed, along with a consumption of energy cost. The rework cost (RC) per cycle can, thus, be calculated as

$$
RC = \sum_{i=1}^{n} (R_{ci} + R'_{ci}) \frac{D}{q_i} E(K).
\tag{11}
$$

Unit Production Cost (UPC)

The production cost depends on the raw material costs, the development cost, and the tool/die cost. It is proportional to the cost of the previous indicated costs. The quality of the raw materials affects the reliability of the products. In this model, the unit production cost is considered to be as follows:

$$
UPC = \sum_{i=1}^{n} [M_c + \frac{DC}{p_i} + \sigma p_i^{\xi}].
\tag{12}
$$

Advertisement Cost (AC)

The multi-item production system based on advertisement, and demand gradually increased due to huge amount investment on advertisement. This model consider advertisement cost to make popular of the smart products.

$$AC = \frac{hy^2}{2}.$$ (13)

Total Expected Profit (TEP)

The TEP per cycle is

$$
\begin{aligned}
\text{TEP}(q_i, p_i, S, \alpha, Y) &= \text{Revenue} - \text{HC} - \text{SC} - \text{IC} - \text{RC} - \text{AC} \\
&= \sum_{i=1}^{n} \left[D(S - \text{UPC}) - \frac{(H_{ci}+H'_{ci})q_i}{2}\left(1 - \frac{D}{p_i}\right) - (S_{ci} + S'_{ci})\frac{D}{q_i} - (I_c + I'_c)D - (R_{ci} + R'_{ci})\frac{D}{q_i}E(K) \right] - \frac{hy^2}{2} \\
&= \sum_{i=1}^{n} \left[\left(\frac{S_{max}-S}{S-S_{min}} + xy^{\mu}\right)(S - P_{ci}) - \frac{(H_{ci}+H'_{ci})q_i}{2}\left(1 - \frac{\left(\frac{S_{max}-S}{S-S_{min}}+xy^{\mu}\right)}{p_i}\right) \right. \\
&\quad - (S_{ci} + S'_{ci})\frac{\left(\frac{S_{max}-S}{S-S_{min}}+xy^{\mu}\right)}{q_i} - (I_c + I'_c)\left(\frac{S_{max}-S}{S-S_{min}} + xy^{\mu}\right) \\
&\quad \left. - (R_{ci} + R'_{ci})\frac{\left(\frac{S_{max}-S}{S-S_{min}}+xy^{\mu}\right)}{q_i}p_i^{\delta+1}\left(\frac{\beta}{\tau+1}\right)e^{\frac{-\alpha q_i}{p_i}}\chi\left(\alpha, \frac{q_i}{p_i}\right) \right] - \frac{hy^2}{2},
\end{aligned}
$$ (14)

where

$$
t_{2i} = \frac{\left(p_i - \left(\frac{S_{max}-S}{S-S_{min}} + xy^{\mu}\right)\right)q_i}{p_i\left(\frac{S_{max}-S}{S-S_{min}} + xy^{\mu}\right)},
$$ (15)

and

$$
\begin{aligned}
\chi\left(\alpha, \frac{q_i}{p_i}\right) &= \frac{t_1^{\tau+2}}{\tau+2} + \frac{\alpha t_1^{\tau+3}}{\tau+3} + \frac{\alpha^2 t_1^{\tau+4}}{\tau+4} + \frac{\alpha^3 t_1^{\tau+5}}{\tau+5} + \dots \dots \\
&= \sum_{i=1}^{\infty} \frac{t_1^{\tau+j+1}\alpha^{j+1}}{(j-1)!(\tau+j+1)} \\
&= \sum_{i=1}^{\infty} \frac{\left(\frac{q_i}{p_i}\right)^{\tau+j+1}\alpha^{j+1}}{(j-1)!(\tau+j+1)}.
\end{aligned}
$$ (16)

Constraints

Capital investment in a smart production system plays an important role, although the amount is limited. A sufficient amount of investment gives an opportunity to choose good-quality raw materials within a required time. Although in this model, defective items are produced and a rework facility is available, this may differ in other situations, with different investment budgets. This model considers a budget constraint and, to separate the imperfect products, managers define a specific quality level, which may or may not be chosen for reworking. Considering A for maximum space available for storing in square feet and B for maximum budget available. Sufficient spaces are allotted for storing good-quality products and for reworking imperfect products. Therefore, the profit function, including budget and space constraints, becomes

$$
\begin{aligned}
\text{TEP}(q_i, p_i, S, \alpha, Y) = \sum_{i=1}^{n} &\left[\left(\frac{S_{max}-S}{S-S_{min}} + xy^{\mu}\right)(S - P_{ci}) - \frac{(H_{ci}+H'_{ci})q_i}{2}\left(1 - \frac{\left(\frac{S_{max}-S}{S-S_{min}}+xy^{\mu}\right)}{p_i}\right) \right. \\
&- (S_{ci} + S'_{ci})\frac{\left(\frac{S_{max}-S}{S-S_{min}}+xy^{\mu}\right)}{q_i} - (I_c + I'_c)\left(\frac{S_{max}-S}{S-S_{min}} + xy^{\mu}\right) \\
&\left. - (R_{ci} + R'_{ci})\frac{\left(\frac{S_{max}-S}{S-S_{min}}+xy^{\mu}\right)}{q_i}p_i^{\delta+1}\left(\frac{\beta}{\tau+1}\right)e^{\frac{-\alpha q_i}{p_i}}\chi\left(\alpha, \frac{q_i}{p_i}\right) \right] - \frac{hy^2}{2},
\end{aligned}
$$ (17)

$\sum_{i=1}^{n} \phi_i q_i \le A$, and $\sum_{i=1}^{n} \psi_i q_i \le B$.

The model cannot be solved analytically. Thus, this model is solved through a numerical tool. The global maximum values of the decision variables are proved numerically.

4. Numerical Examples

We use three numerical examples to validate the model.

4.1. Example 1

In this section, numerical examples are provided to validate this model. In Table 2, for Example 1, the parametric values of the material, holding, rework, and inspection costs; maximum and minimum sale prices; maximum and minimum failure rate; energy costs due to different materials; and other shifting and scaling parameters are shown, which are used in the model and solved using the Mathematica 9.0 software (Wolfram Research, Champaign, IL, USA). The output values of the decision variables (production rate, production lot size, average sale price, advertising variable, and failure rate) are shown in Table 3.

Table 2. Input parameters of Example 1.

I_c ($/unit)	I_c' ($/unit)	H_{c1} ($/unit)	H_{c2} ($/unit)	M (Square Feet)
6	4	2	3.05	130
S_{ci} ($/unit)	S_{ci}' ($/unit)	R_{ci} ($/ defective items)	R_{ci}' ($/defective items)	β (unit)
1000	1100	110	120	5.9
S_{max} ($/unit)	S_{min} ($/unit)	α_{max} (unit)	α_{min} (unit)	δ (unit)
2000	100	0.90	0.10	0.8
ξ	h ($/year)	r	σ	G ($/item)
0.7	20,000	0.85	0.02	200
τ	x	μ	M_c ($/unit)	Z ($)
3	10	1.85	100	30

Table 3. Optimum results of Example 1.

p_1	p_2	q_1	q_2	S	y	α	Total Profit
598.50	530.29	60.17	51.70	565.31	0.25	0.50	2619.20
(units/year)	(units/year)	(units)	(units)	($/unit)	($/year)	(unit)	($/unit)

The TEP is maximum, as the values of the Hessian at the optimal values of the decision variables are $H_{11} = -0.0388796 < 0$; $H_{22} = 0.00254255 > 0$; $H_{33} = -1.01444 \times 10^{-7} < 0$; $H_{44} = 4.75811 \times 10^{-12} > 0$; $H_{55} = -1.5627 \times 10^{-10} < 0$; $H_{66} = 2.66536 \times 10^{-13} > 0$; $H_{77} = -1.87583 \times 10^{-8} < 0$.

4.2. Example 2

In Table 4, for Example 2, the parametric values of the material, holding, rework, and inspection costs; maximum and minimum sale prices; maximum and minimum failure rate; energy costs due to different materials; and other shifting and scaling parameters are shown, which are used in the model and solved using the Mathematica 9 software. The output values of the decision variables (production rate, production lot size, average sale price, advertising variable, and failure rate) are shown in Table 5.

Table 4. Input parameters of Example 2.

I_c ($/unit)	I_c' ($/unit)	H_{c1} ($/unit)	H_{c2} ($/unit)	M (Square feet)
8	5	3	5.05	120
S_{ci} ($/unit)	S_{ci}' ($/unit)	R_{ci} ($/defective items)	R_{ci}' ($/defective items)	β (unit)
1100	1300	110	120	3.8
S_{max} ($/unit)	S_{min} ($/unit)	α_{max} (unit)	α_{min} (unit)	δ (unit)
2000	250	0.50	0.10	0.65
ς	h ($/year)	r	σ	G ($/item)
0.7	5000	0.75	0.01	350
τ	x	μ	M_c ($/unit)	Z ($)
4	9	1.35	150	60

Table 5. Optimum results of Example 2.

p_1	p_2	q_1	q_2	S	y	α	**Total Profit**
707.48	596.59	125.82	106.539	379.61	1.04	0.50	6133.76
(units/year)	(units/year)	(units)	(units)	($/unit)	($/year)	(unit)	($/unit)

The TEP is maximum, as the values of the Hessian at the optimal values of the decision variables are $H_{11} = -0.0267288 < 0$; $H_{22} = 0.00137255 > 0$; $H_{33} = -1.37276 \times 10^{-7} < 0$; $H_{44} = 1.88938 \times 10^{-11} > 0$; $H_{55} = -8.46171 \times 10^{-10} < 0$; $H_{66} = 2.25542 \times 10^{-10} > 0$; $H_{77} = -1.21061 \times 10^{-6} < 0$.

4.3. Example 3

In Table 6, for Example 3, the parametric values of the material, holding, rework, and inspection costs; maximum and minimum sale prices; maximum and minimum failure rate; energy costs due to different materials; and other shifting and scaling parameters are shown, which are used in the model and solved using the Mathematica 9 software. The output values of the decision variables (production rate, production lot size, average sale price, advertising variable, and failure rate) are shown in Table 7.

Table 6. Input parameters of Example 3.

I_c ($/unit)	I_c' ($/unit)	H_{c1} ($/unit)	H_{c2} ($/unit)	M (Square Feet)
8	5	2	4.05	120
S_{ci} ($/unit)	S_{ci}' ($/unit)	R_{ci} ($/defective items)	R_{ci}' ($/defective items)	β (unit)
1100	1400	110	120	4.8
S_{max} ($/unit)	S_{min} ($/unit)	α_{max} (unit)	α_{min} (unit)	δ (unit)
1600	400	0.50	0.10	0.25
ς	h ($/year)	r	σ	G ($/item)
0.7	6000	0.65	0.01	350
τ	x	μ	M_c ($/unit)	Z ($)
4	10	1.35	150	60

Table 7. Optimum results of Example 3.

p_1	p_2	q_1	q_2	S	y	α	Total Profit
689.67	547.70	193.80	156.36	508.92	1.96	0.50	11914.6
(units/year)	(units/year)	(units)	(units)	($/unit)	($/year)	(unit)	($/unit)

The TEP is maximum, as the values of the Hessian at the optimal values of the decision variables are $H_{11} = -0.0119313 < 0$; $H_{22} = 0.000332921 > 0$; $H_{33} = -5.68341 \times 10^{-8} < 0$; $H_{44} = 1.57781 \times 10^{-11} > 0$; $H_{55} = -9.86172 \times 10^{-10} < 0$; $H_{66} = 8.62069 \times 10^{-10} > 0$; $H_{77} = -4.44002 \times 10^{-6} < 0$.

Table 8 compares the total expected profits (TEP) of the three given examples. There is are another scope to maximize the TEP—by increasing sale price, which depends on the total cost of making the products; although production rate and production lot size also give opportunity to maximize the TEP. Setup cost and energy consumption costs due to setup also have impacts on the TEP. Furthermore, advertising plays an important role; due to increased investment into advertisement, the examples show how the total profit can be extended.

Table 8. Comparative Study.

	Example 1	Example 2	Example 2
TEP	2619.20 $/unit	6133.76 $/unit	11914.6 $/unit

The following three dimension Figures 3–9 are total expected profit (TEP) versus different pair decision variables. The concavity of the figures indicated the profit of the model in different cases.

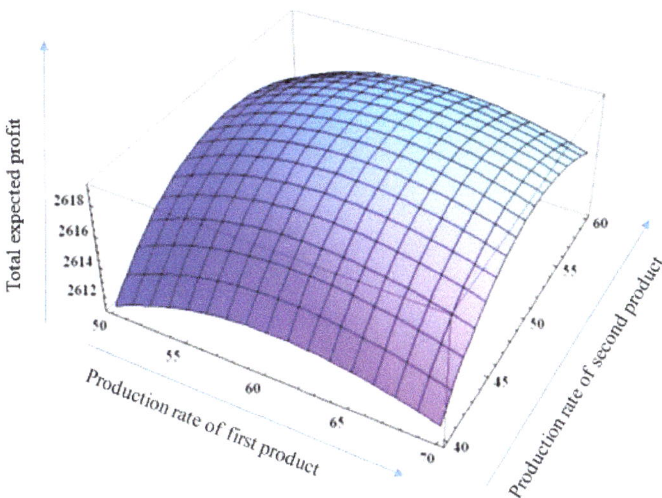

Figure 3. The total expected profit versus the production rate of two products (P_1, P_2).

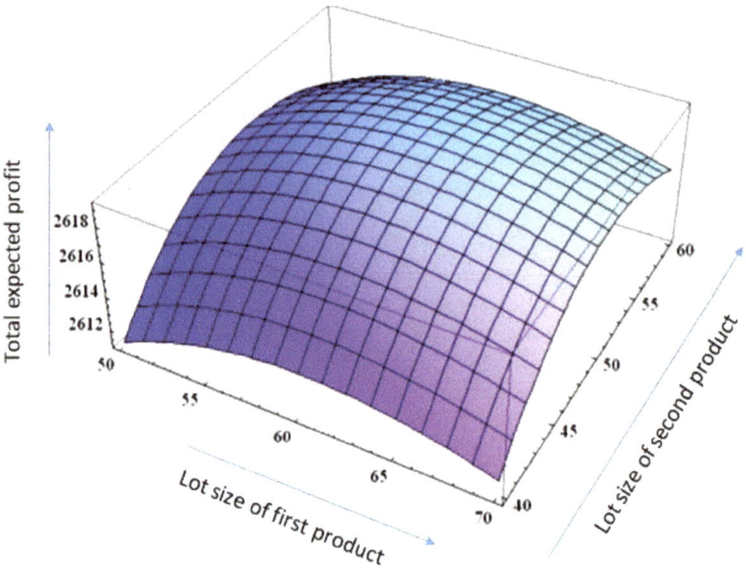

Figure 4. The total expected profit versus lot size of two products (q_1, q_2).

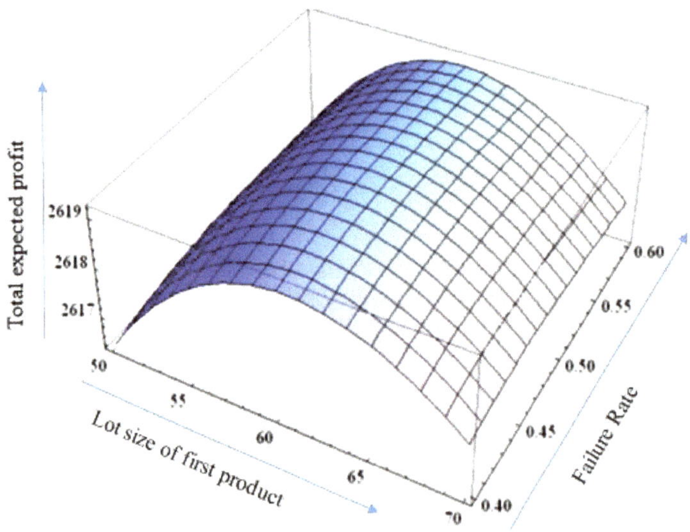

Figure 5. The total expected profit versus the production rate P_1 and the failure rate α.

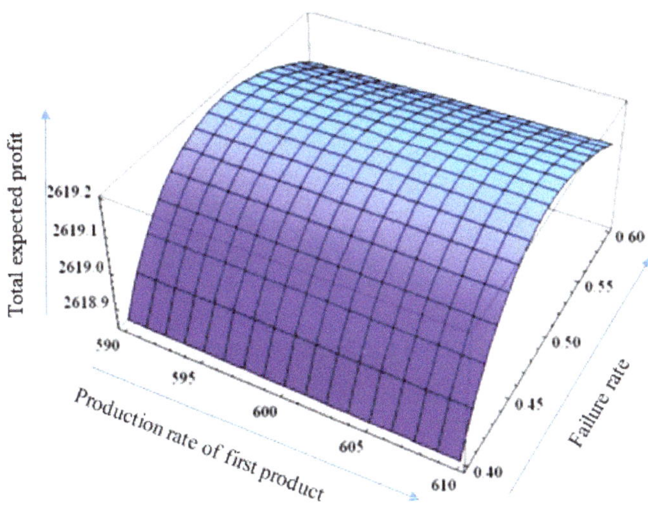

Figure 6. The total expected profit versus the production lot size q_1 and failure rate α.

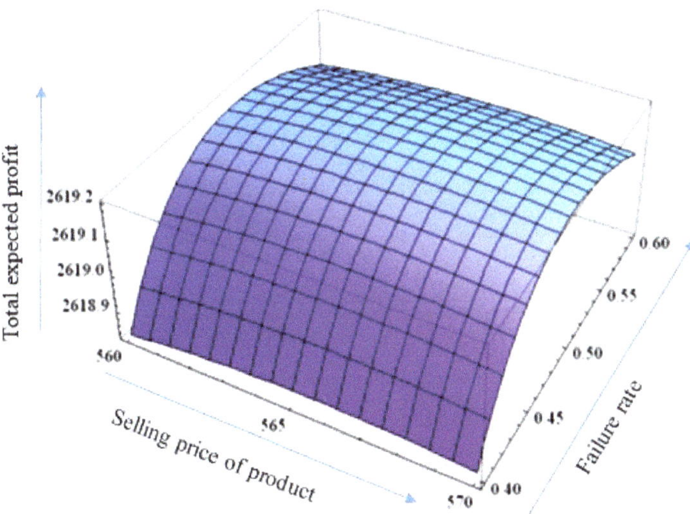

Figure 7. The total expected profit versus the advertising variable quantity and failure rate α of two products.

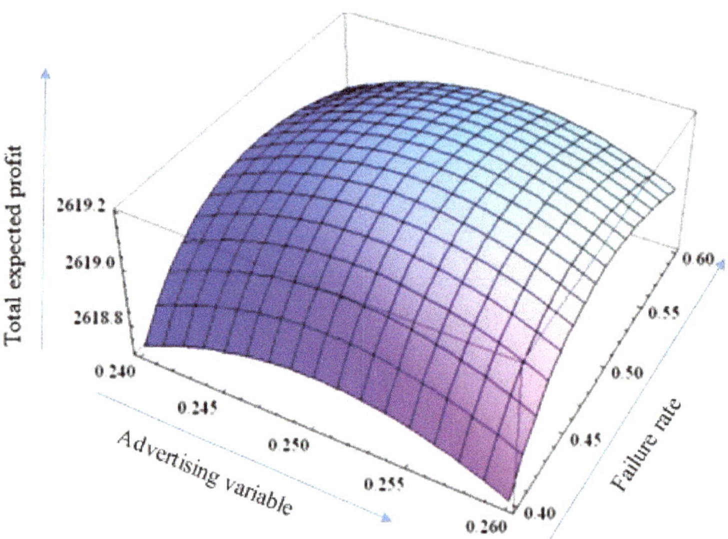

Figure 8. The total expected profit versus the selling price and the failure rate α.

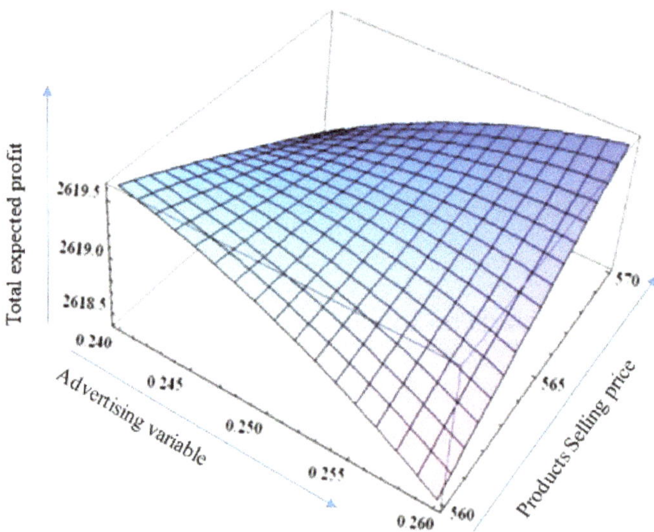

Figure 9. The total expected profit versus the advertising variable and the sale price.

5. Sensitivity Analysis

A sensitivity analysis of the cost and scaling parameters was conducted, and the major changes are summarized in Table 9 and Figure 10.

Table 9 shows how changes by certain percentages (-50%, -25%, $+25\%$, $+50\%$) in the cost and scaling parameters affect the total profit. Here NF indicates that the numerical result is not feasible. We can conclude the following from the sensitivity analysis.

Table 9. Sensitivity analysis of the key parameters.

Parameters	Changes (%)	TEP (%)	Parameters	Changes (%)	TEP (%)
I_c	-50	$+0.44$	I_c'	-50	$+0.29$
	-25	$+0.22$		-25	$+0.15$
	$+25$	-0.22		$+25$	-0.15
	$+50$	-0.44		$+50$	-0.29
H_{c1}	-50	$+1.32$	H_{c2}	-50	$+1.73$
	-25	$+0.61$		-25	$+0.18$
	$+25$	-0.54		$+25$	-0.70
	$+50$	-1.03		$+50$	-1.34
S_{ci}	-50	$+1.43$	S_{ci}'	-50	$+1.84$
	-25	$+0.66$		-25	$+0.84$
	$+25$	-0.58		$+25$	-0.74
	$+50$	-1.10		$+50$	-1.40
R_{ci}	-50	$+0.03$	R_{ci}'	-50	$+0.02$
	-25	$+0.01$		-25	$+0.01$
	$+25$	-0.01		$+25$	-0.01
	$+50$	-0.02		$+50$	-0.02
M_c	-50	$+16.18$	x	-50	-30.70
	-25	$+7.71$		-25	-9.36
	$+25$	-7.03		$+25$	$+4.74$
	$+50$	-13.49		$+50$	$+7.24$
σ	-50	$+0.13$	h	-50	NF
	-25	$+0.27$		-25	$+5.44$
	$+25$	-0.23		$+25$	-6.30
	$+50$	-0.12		$+50$	-13.06

1. That changes in energy cost due to inspection and energy cost per setup has low effects on the total optimal profit as, in a smart production system, there is an optimum consumption of energy for different purposes. Some industries do not consider inspection on the final product, due to the value-added process. Due to an increase of inspection, the reworking cost and other costs also increase, which has a large impact on the total optimal profit. On the other hand, inspection maintains the quality and correctness of the products produced in a smart production system.

2. Changes in holding cost had little impact on the total profit of the smart production system. This result can be justified, as the products are not held for a long time. Similarly, variation of the setup cost also has a lesser, but significant, effect on the total optimal profit. Increasing the value of the cost parameters decreases the total optimal profit, which is clearly shown in the sensitivity analysis table and agrees with reality.

3. On the other hand, the material costs in a multi-item smart production system has a great impact on the total profit. With an increase in material costs, all other production costs will increase and, thus, the profit decreases.

4. The value of the scaling parameters for advertising and the tool/die cost function also had effects on the total profit. These parameters behave similarly to other cost parameters for controlling the total optimal profit.

5. The effect of advertisement has a sensitive impact on the total profit. The advertisement of a product is very much important for generating a high popularity and increasing the market demand for the product. Due to increases in sales, the total optimal profit gradually increases. As in Table 9, it is found that eventually, an increasing value of advertising will reduce the total profit.

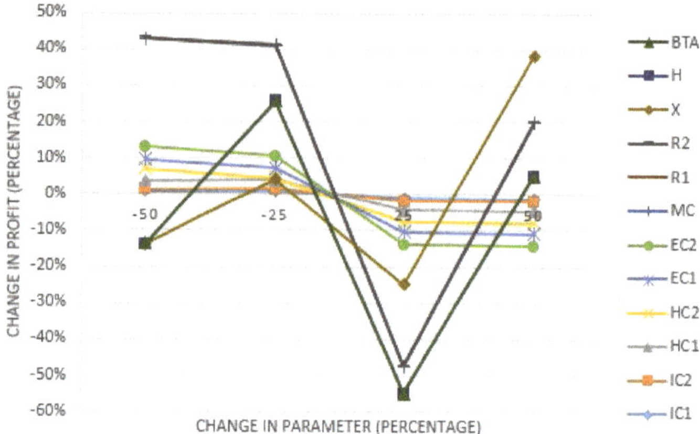

Figure 10. Change of total expected profit (TEP) in percentage versus parametric values in percentage.

6. Managerial Insights

Some recommendations for the industry are as follows:

1. The industry manager must concentrate on the matter of advertisement. This paper clearly shows the impact of advertisement on the total profit. To increase the popularity of products and, in turn, the demand for products, the industry manager should invest some costs into advertisement. In the competitive modern market, this matter plays a significant role for a smart production system. However, no research has considered the advertisement concept, until now, in the field of smart production system. Thus, the result of this study will help the industry manager to increase their profits.

2. Another support that industry managers of smart production systems can use are the strategies, obtained by this research, regarding energy consumption. If a smart production system uses a machine, instead of labor, energy consumption will be present and there will, subsequently, be energy costs to bear. Thus, this model will help the industry to maximize the profit.

3. This model considers random machine breakdowns. Using this idea, the industry manager can pay attention to the previous data of machine failures. This way, the failure rate can be reduced by using smart machines with optimum energy consumptions, and by optimizing development-cost investments.

7. Conclusions

The advertising of smart products has a large impact on the market demand, in addition to the reduction of sale prices, which has an important influence on the profit made from smart products. This study used these two ideas, along with the effects of the optimum use of energy consumption, random machine failures, and optimization of profit, in order to design a model for investigating the optimal decision variables for smart production systems, which gave strong benefits regarding the quality of the smart products. The profit equation of the model was a non-linear equation, and thus could not be solved with an analytical approach. A numerical tool was used to obtain the results, and the numerical findings gave a global optimum solution. The main limitation of the model is that the setup cost in the model is considered to be constant, which can be reduced by an initial investment. In this direction, the model can be extended (refer to the references Malik and Sarkar, [56]; Sarkar and Moon, [57]; Sarkar et al. [58]; Sarkar and Majumdar, [59]; Sarkar et al. [60]; and Majumdar et al. [61]). This model may also be extended by incorporating inspection errors, as the inspection may be done by human beings.

Author Contributions: Conceptualization, B.S. and S.B.; methodology, B.S. and S.B.; software, S.B.; validation, B.S., S.B. and S.P.; formal analysis, S.B.; investigation, B.S.; resources, B.S. and S.P.; data curation, S.B., B.S. and S.P.;

writing—original draft preparation, S.B.; writing—review and editing, B.S., S.B. and S.P.; visualization, B.S. and S.B.; supervision, B.S. and S.P.

Funding: This research received no external funding.

Acknowledgments: The authors are happy to acknowledge the support from two reviewers for revising the earlier versions of the paper.

Conflicts of Interest: The authors declare no conflict of interest.

Notations

Index

i	number of product ($i = 1, 2, 3, ..., n$)

Model Decision Variables

p_i	production rate for item i per cycle (unit/year)
q_i	production lot size for item i per cycle (units)
S	average selling price of the product (\$/unit)
y	advertising variable (\$/year)
α	failure rate, indicates reliability, known as system design variable

Input Parameters

N	available maximum budget (\$/planning period)
M	available maximum space for sorting in square feet
α_{max}	maximum system failure rate
α_{min}	minimum system failure rate
T	production length per cycle (year)
t_{1i}	required time for maximum inventory (year)
D_c	development cost per unit item (\$/unit)
P_c	production cost per unit item (\$/unit)
S_{max}	maximum selling price (\$/unit)
S_{min}	minimum selling price (\$/unit)
ψ_i	consumed budget of item i (\$/unit)
ϕ_i	occupied space for the unit item i (square feet per item)
K	number of defective items per cycle
I_c	inspection cost per item (\$/item)
D	demand for item i
$E(K)$	per unit production system expected number of defective items
H_{ci}	holding cost for item i (\$/unit/unit time)
S_{ci}	setup cost for item i per setup (\$/setup)
M_C	material cost for item i (\$/unit)
R_{ci}	rework cost for item i (\$/defective item)
R'_{ci}	energy consumption cost due to rework for item i (\$/item)
x	scaling parameter for the products
μ	shape parameter for the products
Z	resources fixed cost (\$)
G	fixed energy and labor cost, independent of reliability α (\$/item)
TEP	total expected profit per unit time (\$/year)
I'_c	energy consumption due to Inspection per unit cycle (\$/cycle)
S'_{ci}	energy consumption cost per setup (\$/setup)
r	shape parameter for development cost function
β	scaling parameter for defective item function
σ	scaling parameter for tool/die cost function
δ	shape parameter for defective item function
h	investment cost for advertisement
ξ	shape parameter for tool/die cost function

Random Variable

γ_i	system random time to move to an out-of-control state from an in-control situation, for item i.

References

1. Cárdenas-Barrón, L.E. Economic production quantity with rework process at a single-stage manufacturing system with planned backorders. *Comput. Ind. Eng.* **2009**, *57*, 1105–1113. [CrossRef]
2. Sana, S.S. A production-inventory model in an imperfect production process. *Eur. J. Oper. Res.* **2010**, *200*, 451–464. [CrossRef]
3. Sarkar, M.; Sarkar, B.; Iqbal, M.W. Effect of energy and failure rate in a multi-item smart production system. *Energies* **2018**, *11*, 2958. [CrossRef]
4. Rosenblatt, M.J.; Lee, H.L. Economic production cycles with imperfect production processes. *IIE Trans.* **1986**, *18*, 48–55. [CrossRef]
5. Kim, C.H.; Hong, Y. An optimal production run length in deteriorating production processes. *Int. J. Prod. Econ.* **1999**, *58*, 183–189.
6. Giri, B.; Dohi, T. Exact formulation of stochastic EMQ model for an unreliable production system. *J. Oper. Res. Soc.* **2005**, *56*, 563–575. [CrossRef]
7. Sana, S.S.; Goyal, S.K.; Chaudhuri, K. An imperfect production process in a volume flexible inventory model. *Int. J. Prod. Econ.* **2007**, *105*, 548–559. [CrossRef]
8. Chiu, Y.-S.P.; Chen, K.-K.; Cheng, F.-T.; Wu, M.-F. Optimization of the finite production rate model with scrap, rework and stochastic machine breakdown. *Comput. Math. Appl.* **2010**, *59*, 919–932. [CrossRef]
9. Egea, A.J.S.; Deferrari, N.; Abate, G.; Martínez Krahmer, D.; Lopez de Lacalle, L.N. Short-cut method to assess a gross available energy in a medium-load screw friction press. *Metals* **2018**, *8*, 173. [CrossRef]
10. González, H.; Calleja, A.; Pereira, O.; Ortega, N.; Lopez de Lacalle, L.N.; Barton, M. Super abrasive machining of integral rotary components using grinding flank tools. *Metals* **2018**, *8*, 24. [CrossRef]
11. Sarkar, B. An EOQ model with delay in payments and stock dependent demand in the presence of imperfect production. *Appl. Math. Comput.* **2012**, *218*, 8295–8308. [CrossRef]
12. Sarkar, B. An EOQ model with delay in payments and time varying deterioration rate. *Math. Comput. Model.* **2012**, *55*, 367–377. [CrossRef]
13. Sarkar, B. An inventory model with reliability in an imperfect production process. *Appl. Math. Comput.* **2012**, *218*, 4881–4891. [CrossRef]
14. Chakraborty, T.; Giri, B. Lot sizing in a deteriorating production system under inspections, imperfect maintenance and reworks. *Oper. Res.* **2014**, *14*, 29–50. [CrossRef]
15. Sarkar, B.; Cárdenas-Barrán, L.E.; Sarkar, M.; Singgih, M.L. An economic production quantity model with random defective rate, rework process and backorders for a single stage production system. *J. Manuf. Syst.* **2014**, *33*, 423–435. [CrossRef]
16. Sarkar, B.; Saren, S. Product inspection policy for an imperfect production system with inspection errors and warranty cost. *Eur. J. Oper. Res.* **2016**, *248*, 263–271. [CrossRef]
17. Pasandideh, S.H.R.; Niaki, S.T.A.; Nobil, A.H.; Cárdenas-Barrán, L.E. A multi-product single machine economic production quantity model for an imperfect production system under warehouse construction cost. *Int. J. Prod. Econ.* **2015**, *169*, 203–214. [CrossRef]
18. Purohit, A.K.; Shankar, R.; Dey, P.K.; Choudhary, A. Non-stationary stochastic inventory lot-sizing with emission and service level constraints in a carbon cap-and-trade system. *J. Clean. Prod.* **2016**, *113*, 654–661. [CrossRef]
19. Sana, S.S. An economic production lot size model in an imperfect production system. *Eur. J. Oper. Res.* **2010**, *201*, 158–170. [CrossRef]
20. Cárdenas-Barrán, L.E.; Chung, K.-J.; Kazemi, N.; Shekarian, E. Optimal inventory system with two backlog costs in response to a discount offer: Corrections and complements. *Oper. Res.* **2018**, *18*, 97–104. [CrossRef]
21. Tiwari, S.; Jaggi, C.K.; Gupta, M.; Cárdenas-Barrón, L.E. Optimal pricing and lot-sizing policy for supply chain system with deteriorating items under limited storage capacity. *Int. J. Prod. Econ.* **2018**, *200*, 278–290. [CrossRef]
22. Tiwari, S.; Cárdenas-Barrón, L.E.; Goh, M.; Shaikh, A.A. Joint pricing and inventory model for deteriorating items with expiration dates and partial backlogging under two-level partial trade credits in supply chain. *Int. J. Prod. Econ.* **2018**, *200*, 16–36. [CrossRef]
23. Huang, Y.-F.; Lai, C.-S.; Shyu, M.-L. Retailer's EOQ model with limited storage space under partially permissible delay in payments. *Math. Prob. Eng.* **2007**, *2007*, 90873. [CrossRef]

24. Pasandideh, S.H.R.; Niaki, S.T.A. A genetic algorithm approach to optimize a multi-products EPQ model with discrete delivery orders and constrained space. *Appl. Math. Comput.* **2008**, *195*, 506–514. [CrossRef]

25. Hafshejani, K.F.; Valmohammadi, C.; Khakpoor, A. Retracted: Using genetic algorithm approach to solve a multi-product EPQ model with defective items, rework, and constrained space. *J. Ind. Eng. Int.* **2012**, *8*, 1–8. [CrossRef]

26. Mahapatra, G.; Mandal, T.; Samanta, G. An EPQ model with imprecise space constraint based on intuitionists fuzzy optimization technique. *J. Mult.-Valued Log. Soft Comput.* **2012**, *19*, 409–423.

27. Taleizadeh, A.; Jalali-Naini, S.G.; Wee, H.-M.; Kuo, T.-C. An imperfect multi-product production system with rework. *Sci. Iran.* **2013**, *20*, 811–823.

28. Hou, K.-L.; Lin, L.-C. Investing in setup reduction in the EOQ model with random yields under a limited capital budget. *J. Inf. Optim. Sci.* **2011**, *32*, 75–83. [CrossRef]

29. Mohan, S.; Mohan, G.; Chandrasekhar, A. Multi-item, economic order quantity model with permissible delay in payments and a budget constraint. *Eur. J. Ind. Eng.* **2008**, *2*, 446–460. [CrossRef]

30. Todde, G.; Murgia, L.; Caria, M.; Pazzona, A. A comprehensive energy analysis and related carbon footprint of dairy farms, Part 2: Investigation and modeling of indirect energy requirements. *Energies* **2018**, *11*, 463. [CrossRef]

31. Du, Y.; Hu, G.; Xiang, S.; Zhang, K.; Liu, H.; Guo, F. Estimation of the diesel particulate filter soot load based on an equivalent circuit model. *Energies* **2018**, *11*, 472. [CrossRef]

32. Cárdenas-Barrón, L.E.; Treviño-Garza, G.; Widyadana, G.A.; Wee, H.-M. A constrained multi-products EPQ inventory model with discrete delivery order and lot size. *Appl. Math. Comput.* **2014**, *230*, 359–370. [CrossRef]

33. Xua, L.; Cheng, J.-H.; Liu, P.; Wang, Q.; Xu, Z.-X.; Liu, Q.; Shen, J.-Y.; Wang, L.-J. Production of bio-fuel oil from pyrolysis of plant acidified oil. *Renew. Energy* **2018**, *130*, 910–919. [CrossRef]

34. Tomic, T.; Schneider, D.R. The role of energy from waste in circular economy and closing the loop concept-Energy analysis approach. *Renew. Sustain. Energy Rev.* **2018**, *98*, 268–287. [CrossRef]

35. Haraldsson, J.; Johansson, M.T. Review of measures for improved energy efficiency in production-related processes in the aluminium industry-from electrolysis to recycling. *Renew. Sustain. Energy Rev.* **2018**, *93*, 525–548. [CrossRef]

36. Dey, B.; Sarkar, B.; Sarkar, M.; Pareek, S. An integrated inventory model involving discrete setup cost reduction, variable safety factor, selling-price dependent demand, and investment. *Rairo-Oper. Res.* **2019**, in press. [CrossRef]

37. Sarkar, B.; Sarkar, S.; Yun, W.Y. Retailer's optimal strategy for fixed lifetime products. *Int. J. Mach. Learn. Cybern.* **2016**, *7*, 121–133. [CrossRef]

38. Sarkar, B.; Majumder, A.; Sarkar, M.; Dey, B.K.; Roy, G. Two-echelon supply chain model with manufacturing quality improvement and setup cost reduction. *J. Ind. Manag. Opt.* **2017**, *13*, 1085–1104. [CrossRef]

39. Sarkar, B.; Chaudhuri, K.S.; Moon, I. Manufacturing setup cost reduction and quality improvement for the distribution free continuous-review inventory model with a service level constraint. *J. Manuf. Syst.* **2015**, *34*, 74–82. [CrossRef]

40. Yao, X.; Zhou, J.; Zhang, J.; Boer, C.R. From intelligence manufacturing to smart manufacturing for industry 4.0 driven by next generation artificial intelligence and further on. In Proceedings of the 2017 5th International Conference on Enterprise Systems (ES), Beijing, China, 22–24 September 2017. [CrossRef]

41. Gola, A. Reliability analysis of reconfigurable manufacturing system structures using computer simulation methods. *Eksploatacja I Niezawodnosc—Maintaience Reliab.* **2019**, *21*, 90–102. [CrossRef]

42. Tayyab, M.; Sarkar, B.; Yahya, B. Imperfect Multi-Stage Lean Manufacturing System with Rework under Fuzzy Demand. *Mathematics* **2019**, *7*, 13. [CrossRef]

43. Sarkar, B. Mathematical and analytical approach for the management of defective items in a multi-stage production system. *J. Clean. Prod.* **2019**, in press. [CrossRef]

44. Kim, M.S.; Kim, J.S.; Sarkar, B.; Sarakr, M.; Iqbal, M.W. An improved way to calculate imperfect items during long-run production in an integrated inventory model with backorders. *J. Manuf. Syst.* **2018**, *47*, 153–167. [CrossRef]

45. Sarkar, B.; Sett, B.K.; Sarkar, S. Optimal production run time and inspection errors in an imperfect production system with warranty. *J. Ind. Manag. Opt.* **2018**, *14*, 267–282. [CrossRef]

46. Chenavaz, R.Y.; Jasimuddin, S.M. An analytical model of the relationship between product quality and advertising. *Eur. J. Oper. Res.* **2017**, *263*, 295–307. [CrossRef]

47. Giri, B.C.; Sharma, S. Manufacturer's pricing strategy in a two-level supply chain with competing retailers and advertising cost dependent demand. *Econ. Model.* **2014**, *38*, 102–111. [CrossRef]

48. Xiao, D.; Zhou, Y.W.; Zhong, Y.; Xie, W. Optimal cooperative advertising and ordering policies for a two-echelon supply chain. *Comp. Ind. Eng.* **2018**, in press. [CrossRef]

49. Noh, J.S.; Kim, J.S.; Sarkar, B. Two-echelon supply chain coordination with advertising demand under Stackelberg game policy. *Eur. J. Ind. Eng.* **2019**, in press. [CrossRef]

50. Karaöz, M.; Erçlu, A.; Sütçü, A. An EOQ model with price and time dependent demand under the influence of complement and substitute product's selling-prices. *J. Alanya Fac. Bus.—Alanya Isletme Fakültesi Dergisi* **2011**, *3*, 21–32.

51. Sana, S.S. Price-sensitive demand for perishable items—An EOQ model. *App. Math. Mod.* **2011**, *217*, 6248–6259.

52. Pal, B.; Sana, S.S.; Chaudhuri, K. Multi-item EOQ model while demand is sales price and price break sensitive. *Eco. Mod.* **2012**, *29*, 2283–2288. [CrossRef]

53. Sarkar, B.; Mandal, P.; Sarkar, S. An EMQ model with price and time dependent demand under the effect of reliability and inflation. *Appl. Math. Comput.* **2014**, *231*, 414–421. [CrossRef]

54. Sarkar, B.; Sarkar, S. Variable deterioration and demand-An inventory model. *Econ. Mod.* **2013**, *31*, 548–556. [CrossRef]

55. Azadeh, A.; Paknafs, B. Integrated simulation modeling of business, maintenance and production systems for concurrent improvement of lead time, cost and production rate. *Ind. Eng. Manag. Syst.* **2017**, *15*, 403–431.

56. Malik, I.A.; Sarkar, B. Optimizing a Multi-Product Continuous—Review Inventory Model With Uncertain Demand, Quality Improvement, Setup Cost Reduction, and Variation Control in Lead Time. *IEEE Access* **2018**, *6*, 36176–36187. [CrossRef]

57. Sarkar, I.; Moon, I. Improved quality, setup cost reduction, and variable backorder costs in an imperfect production process. *Int. J. Prod. Econ.* **2014**, *155*, 204–213. [CrossRef]

58. Sarkar, B.; Mondal, B.; Sarkar, S. Quality improvement and backorder price discount under controllable lead time in an inventory model. *J. Manuf. Syst.* **2015**, *35*, 26–36. [CrossRef]

59. Sarkar, B.; Majumdar, A. Integrated vendor–buyer supply chain model with vendor's setup cost reduction. *Appl. Math. Comput.* **2013**, *224*, 362–371. [CrossRef]

60. Sarkar, B.; Saren, S.; Sinha, D.; Hur, S. Effect of unequal lot sizes, variable setup cost, and carbon emission cost in a supply chain model. *Math. Probl. Eng.* **2015**, *2015*, 1–13. [CrossRef]

61. Majumdar, A.; Guchhait, R.; Sarkar, B. Manufacturing quality improvement and setup cost reduction in an integrated vendor-buyer supply chain model. *Eur. J. Ind. Eng.* **2017**, *13*, 588–612. [CrossRef]

 © 2019 by the authors. Licensee MDPI, Basel, Switzerland. This article is an open access article distributed under the terms and conditions of the Creative Commons Attribution (CC BY) license (http://creativecommons.org/licenses/by/4.0/).

 *mathematics*

Article

Dynamic Pricing in a Multi-Period Newsvendor Under Stochastic Price-Dependent Demand

Mehran Ullah, Irfanullah Khan and Biswajit Sarkar *

Department of Industrial & Management Engineering, Hanyang University, Ansan, Gyeonggi-do 155 88, Korea; mehrandirvi@gmail.com (M.U.); irfanullah13@outlook.com (I.K.)
* Correspondence: bsbiswajitsarkar@gmail.com; Tel.: +82-107-498-1981

Received: 13 May 2019; Accepted: 2 June 2019; Published: 6 June 2019

Abstract: The faster growth of technology stipulates the rapid development of new products; with the spread of new technologies old ones are outdated and their market demand declines sharply. The combined impact of demand uncertainty and short life-cycles complicate ordering and pricing decision of retailers that leads to a decrease in the profit. This study deals with the joint inventory and dynamic pricing policy for such products considering stochastic price-dependent demand. The aim is to develop a discount policy that enables the retailer to order more at the start of the selling season thus increase the profit and market share of the retailer. A multi-period newsvendor model is developed under the distribution-free approach and the optimal stocking quantities, unit selling price, and the discount percentage are obtained. The results show that the proposed discount policy increases the expected profit of the system. Additionally, the stocking quantity and the unit selling price also increases in the proposed discount policy. The robustness of the proposed model is illustrated with numerical examples and sensitivity analysis. Managerial insights are given to extract significant insights for the newsvendor model with discount policy.

Keywords: price discounts; stochastic-price dependent demand; newsvendor; pricing policy

---

1. Introduction

In today's competitive economic environment, fluctuation in market demand is causing problems for businesses, especially those who are dealing with products having short life cycles. The example includes apparel products, fashion goods, seasonal clothing, personal computers, and other electronic goods. Demand uncertainty either leads to stock outs or overstocking; the short life cycles, further, aggravates the severity of the problem. In such situations, an important question arises: How to deal with these uncertain items? The newsvendor model offers a solution to such problems, in which the decision maker faces stochastic and exogenous demand. The newsvendor has to decide the order quantity ahead of the selling season by using the available information. The aim is to find the trade-off between overstocking and under-stocking conditions. In the classical newsvendor problem, the common practice is to dispose of the product at a salvage price considering no discount during the selling season. However, the past 30 years has reshaped the landscape of the retail industry; retailers prefer to avoid the stock outs by ordering larger lot sizes, and a discount is offered on the remaining stock at the end of the selling season. This increases the market share of the retailer because a strategic consumer waits patiently and buys the product when the price drops. This paper deals with such a business problem, in which the consumer can switch to alternative options with discounted prices; the retailer decides the lot size, finished products price, and the discount percentage on the price of the finished product at the end of the selling season.

As demand is selling price dependent, reducing the price, at the end of the selling season, increases the net demand. Moreover, in these strategic situations, the firms make decisions considering pricing and stocking. The retailer who deals in short-life-cycle/seasonal items faces stochastic demand and information on demand distribution is limited; however, the newsvendor has only an educated guess of the mean and variance of the demand. To find the optimal inventory level normal distribution of demand is employed; although, this does not provide the best performance compared to other distributions with the same mean and variance. Scarf [1] first addressed the distribution-free newsboy problem where mean and variance of demand were given, whereas the demand distribution assumption was not available. Considering the distribution-free approach, Scarf [1] developed a model for the optimal ordering quantity to get the maximum expected profit with the worst possible distribution of the demand. The min–max distribution-free approach developed by Scarf is good but is complex and difficult to understand. This problem was solved by Gallego and Moon [2], where they simplified Scarf's ordering rule for the newsboy problem.

This paper extends the classical newsvendor distribution-free model; the seller offers a progressive discount to generate more revenue. This model considers a retailer who is selling seasonal items and facing a stochastic demand. The retailer has only one opportunity to place the order before the start of the selling season. Reorders are not possible during the season; because the lead time is longer than the selling season. The newsboy takes the ordering decision based on historical data, which may possess high variance to the demand. In the classical newsvendor model, the selling price is assumed constant throughout the selling season, which prevents the retailer from generating revenue by price adjustment. This model is more realistic because it provides a price adjustment policy, which helps in generating extra revenue. This model specifically answers the following questions. What selling price should the retailer choose? How much to order considering the discounted policy later? And how much discount percent must the retailer offer at the end of the season?

The remainder of this paper is assembled as follows. In Section 2, this paper reviews the relevant literature. Section 3 provides notation and develops the mathematical model. Section 4 tests the developed model with numerical examples and provides a sensitivity analysis. In Section 5, the paper is concluded.

2. Literature Review

In this section we provide a summary of the relevant literature. This paper directly contributes to three streams of research, namely newsvendor models, price dependent stochastic demand, and price discounts in newsvendor. Therefore, relevant papers were reviewed in this section.

The classical newsboy problem is designed to find the optimal quantity of the items in a single period, probabilistic framework. The aim of calculating order quantity is to minimize the expected cost of overestimating and underestimating the probabilistic demand in the selling season. The newsboy problem has gained considerable attention since the pioneering paper of Arrow et al. [2]. Scarf [1] developed the min–max distribution-free procedure to calculate the optimal order quantity with only the mean and variance. Scarf proved the worst distribution of demand would be positive at two points. Gallego and Moon [3] simplified the proof of Scarf [1], the Scarf rule consists of a lengthy mathematical argument and is difficult to understand for researchers and managers. Anvari [4] maximizes the market value of the firm using the capital asset pricing model. An extensive literature review on the newsvendor problem can be found in the literature, for example in Khouja [5]. Bitran and Mondschein [6] dealt with seasonal items, they also developed a pricing policy based on time and inventory. As most of the newsvendor models deal with profit maximization, however, a few models deal with the probability of exceeding a specified minimum profit, as in Lau [7] and Parlar and Kevin Weng [8]. Sarkar [9] considered a service level constraint and variable lead time with the min–max distribution free approach, Sarkar [10] considered distribution-free approach with buyback contracts.

There exists an enormous amount of literature concerning the extension of the classical newsvendor model [11,12]. The comparison of applying the normal distribution and the scenario of the distribution-free approach (only mean and variance of the demand is available) with discounts is given by Moon et al. [13]; however, they did not consider price dependent demand, furthermore, their model is of a single period while this paper considered a multi period newsvendor problem. The consignment policy is considered in Sarkar et al. [14] in a single-period newsvendor problem with the distribution-free approach and retailer cost is reduced by the sustainable consignment contract. The classical inventory problem is extended by Kogan and Lou [15] to the multi-stage newsboy problem; further, they divided product flow into sequential stages for the reduction in underage or overage costs at the end-of-season. Matsuyama [16] considers a multi-period inventory problem; in their model, the ordering quantity is divided into many cycles, and if the inventory of the present cycle did not match that will affect the next cycle. The effect of coordination on stocking and pricing for two consecutive periods was studied by Lee [17], he analyzed a normal sales period and a leftover markdown sales period. Chen et al. [18] considers a multi-period supply chain model where the supplier is selling products to a multi-period newsvendor and calculated the optimal pricing for the supplier.

Petruzzi and Dada [19] argue that demand is price dependent in the newsvendor model, Khouja [20] developed the price-dependent demand policy and proved the concavity of the newsvendor problem. Arcelus et al. [21] considered the ordering policies in a newsvendor framework with a stochastic price-dependent demand. The classical single-period newsvendor model is extended in Chung et al. [22] with a price adjustment for the retailer in-season after receiving the demand; however, they assumed demand follows a normal distribution while this paper did not assume any distribution for the demand. Banerjee and Meitei [23] studied the effect of the selling price reduction and analyzed the optimal ordering policy for a single period inventory model. The demand is a stochastic function of the selling price in Abad [24] and he used the service level constraint to avoid the economic consequences of a stock-out situation. The relationships of the purchasing cost, price-dependent stochastic demand, and salvage value are shown by Ma et al. [25], they studied impacts of discounted schemes on the expected profit in a single-period newsvendor framework. Arcelus et al. [26] considered risk tolerance in the newsvendor problem with a price-dependent demand, and Hu and Su [27] optimized newsvendor expected profit with joint procurement planning and a pricing procedure.

The additive and multiplicative demand cases under the behavior of the strategic consumer are studied in Ye and Sun [28] in the newsvendor problem. Arcelus et al. [29] evaluated the pricing and ordering policies of a newsvendor facing risk-averse, risk-neutral, and risk-seeking situations with price-dependent stochastic demand. In their paper, they offer three types of sales policy: Pricing, rebates, and advertising for additive and multiplicative demand structure. He et al. [30] investigated the channel coordination issue in the newsvendor framework with the stochastic price and sales effort dependent demand. A coordination contract is studied by Chen et al. [31] in a stochastic demand case for fashionable products concerning the supplier–retailer channel. In their model, the supplier allows a fixed capacity of production, and retailers are not allowed to change the demand after the demand is realized. A decentralized supply chain is investigated in Chen and Bell [32], in this study the retailer determines the price and order quantity while facing the stochastic price-dependent demand and returns from the customer. Jadidi et al. [33] considered a single period model with quantity discounts, a transportation capacity problem, and price-dependent stochastic demand. Modak and Kelle [34] examined the dual-channel supply chain considering a stochastic demand dependent on retailer price and delivery-time.

The uncertain demand in the newsboy problem creates the need for price setting, which can lead to revenue maximization that is accomplished by increasing total sales. The newsboy deals with the overstocking situation by introducing discounts at the end of the season, he must clear the maximum leftover inventory by offering discounts on sales. Khouja [20] introduces the concept of multiple discounts in the newsvendor model, he further elaborated that discounts are applicable on the products until these items remain on the shelf. Khouja [35] extended Khouja [20] by offering quantity discounts with multiple discounts in the single period inventory problem. Cachon and Kök [36] develop the newsvendor model with a clearance price, they also showed how much discount should be offered if the inventory is remaining at the end of the season. The clearance sales theory is expressed by Nocke and Peitz [37], in their model the selling price was originally high; however, at the end-of selling period, they reduced the price to clear the remaining inventory. The effect of discounts on gift cards is analyzed by Khouja et al. [38] in the newsvendor framework. A discrete-time model is developed by Gupta et al. [39] for deciding clearance prices in the single period inventory problem by setting a bound on prices and the expected revenue. Cachon and Kök [36] considered a newsvendor problem with clearance pricing: Before the start of the selling season demand is stochastic and exogenous, during the season demand is endogenous and deterministic in their model. The optimal values for order quantity are calculated in Ullah and Sarkar [40]. Jammernegg and Kischka [41] developed a procedure to calculate the optimal values of the order quantity, expected profit, and selling price in the newsvendor framework. Mandal et al. [42] elaborates on the consumers switch to the alternative option as soon as they realize the lower price; likewise, firms make their pricing and stocking decisions. Hu and Su [27] eliminated the assumption on salvage and considered a clearance price as a decision variable. The revision of the interval environment is considered in Ruidas et al. [43].

3. Mathematical Model

The newsboy deals in short lifecycle items, which become obsolete after some time, the demand reduces after the selling season and other products with better performance replace them. This paper develops a joint pricing and inventory model for a multi-period newsboy with price-dependent stochastic demand. The newsboy also offers a discount at the end of the period to increase its revenue and market share. No assumption on demand is considered except that demand belongs to a class φ of the probability distribution functions with a known mean and variance; where a distribution-free approach is applied to solve the model.

3.1. Proposed Model

A multi-period newsvendor model was developed based on joint stocking and pricing decisions under demand uncertainty. The decision maker decides the order quantity at the start of the selling season. The newsvendor offers a discount on leftover inventory at the end of the season to enhance the sales; inventory still left after the discounted sales will be salvage in the next selling season. The combine multi-period and discount policy provide multiple selling opportunities, which reduce the risk considering the overstocking and understocking situations in the newsvendor model. The newsvendor is facing a price-dependent stochastic demand of the form $d_i(p, X) = a(p) + X_i$. The demand comprises of two components; the price dependent deterministic demand $(a(p))$ and random error (X_i), which is independent of the price; such type of demand considerations are common in the literature, for example see [19,41,44,45]. The deterministic price dependent demand is $a(p) = y - z * p$ that is linearly decreasing the function of the price p and it follows increasing the price elasticity (see [12,45,46]). Here y represents the market share and the z is the price sensitivity. The price dependent deterministic demand $a(p)$ is assumed to be positive. The stochastic price independent demand factor X_i, a random variable, is additive in nature because of the additive modeling framework. Further, the assumptions are considered that X_i follows a probability distribution function $f(X_i)$ and a cumulative probability distribution function $F(X_i)$ with mean μ_i and standard deviation σ_i.

The objective was to determine the stocking quantity (Q_i), selling price (p), and discount percentage (β) to maximize the expected profit of the newsvendor. The profit of the proposed system can be determined by calculating the revenue minus the total cost. In the proposed system, the revenue is obtained from three sources; selling finished products with price p, selling with discounted price $p(1 - \beta)$, and selling the leftover inventory from the previous period $((1 - \alpha(\beta))E(Q_{i-1} - d_{i-1})^+)$ with salvage value s. If the demand in the selling period does not exceed Q_i units, then the revenue is $pd_i(p, X)$, and the remaining units will be sold with a discount percentage β of the selling price (p). Otherwise the revenue is pQ_i, therefore, the total expected revenue, from selling with price p can be expressed as $pE(\min(Q_i, d_i))$. The cost consists of a purchasing cost cQ_i for Q_i units. The leftover after the discounted sales are held at a cost h per unit after the selling season that are salvaged in the next period with a salvage value s. Similarly, if the demand exceeds Q_i units, and the revenue is pQ_i, the shortage may arise and a penalty cost (b per unit) will be considered for the observed shortage. The profit for the multi-period newsvendor is the difference between the revenue generated $pE(\min(Q_i, d_i))$ and the cost incurred per item. The multi-period newsvendor expected profit function can be expressed as:

$$\pi = pE(\min(Q_i, d_i)) - cQ_i - bE(d_i - Q_i)^+ + \alpha(\beta)p(1-\beta)E(Q_i - d_i)^+ - (1-\alpha(\beta))hE(Q_i - d_i)^+ + s(1-\alpha(\beta))E(Q_{i-1} - d_{i-1})^+. \tag{1}$$

In Equation (1) the first term $pE(\min(Q_i, d_i))$ is the expected profit generated from selling products with the full price. The second term is the purchase cost, the third term is the shortage cost, when the demand exceeds the order quantity, and the fourth term shows the expected revenue from discounted sales. As $E(Q_i - d_i)^+$ is the expected leftover after the end of season when the full price sales are completed, $\alpha(\beta)$ is the percentage of expected sales when the discount is offered. Where, $\alpha(\beta)$ is an increasing function of the discount percentage (β). $p(1-\beta)$ shows the discounted price, thus, the total revenue from the discounted sales can be formulated as $(\alpha(\beta)p(1-\beta)E(Q_i - d_i)^+)$. From the initial leftover $(E(Q_i - d_i)^+)$, $\alpha(\beta)$ percent is sold with a discount then $(1-\alpha(\beta))$ percent remains after the discounted sales, for which the expected holding cost is $(1-\alpha(\beta))hE(Q_i - d_i)^+$, and revenue from salvaging the expected leftover from the previous period is $s(1-\alpha(\beta))E(Q_{i-1} - d_{i-1})^+$.

Where the discounted sales percentage $\alpha(\beta)$ is an increasing function of the discount percentage β (price), such that $\alpha(\beta) = \left(1 - e^{-\frac{\zeta * \beta}{\varrho}}\right)$. Where ζ and ϱ are the parameters; furthermore, $\alpha(\beta)$ shows a declining increment to increase in β and its behaviour is illustrated in Figure 1, when $\zeta = 0.2$ and $\varrho = 0.05$. Where, $\alpha(\beta)$ is a monotonically increasing function over β, as $\frac{\partial \alpha(\beta)}{\partial \beta} > 0$ and $\frac{\partial^2 \alpha(\beta)}{\partial \beta^2} < 0$, $\forall \beta$.

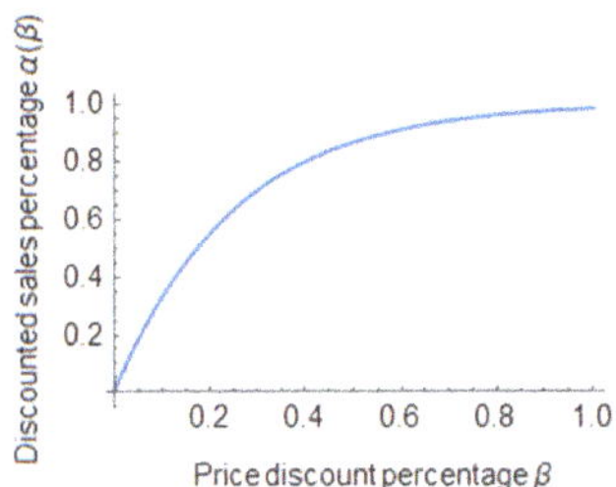

Figure 1. Discounted sales percentage plotted against the price discount percentage.

Observing that $\min(Q_i, d_i) = d_i - (d_i - Q_i)^+$, the expected profit can be written as,

$$\pi = pE(d_i) - cQ_i - (b+p)E(d_i - Q_i)^+ + (\alpha(\beta)p(1-\beta) - (1-\alpha(\beta))h)E(Q_i - d_i)^+ + (1-\alpha(\beta))sE(Q_{i-1} - d_{i-1})^+. \tag{2}$$

As the newsvendor has partial (no distribution knowledge) information on demand, considering this situation, the expected leftover stock is $(Q_i - d_i)^+$ can be formulated as:

$$(Q_i - d_i)^+ = \frac{E|Q_i - d_i| + E(Q_i - d_i)}{2}. \tag{3}$$

By Cauchy Schwartz inequality:

$$E|Q_i - d_i| \leq \sqrt{E(Q_i - d_i)^2} = \sqrt{E(Q_i^2 - 2Q_i d_i + d_i^2)};$$

$$E|Q_i - d_i| \leq \sqrt{E(Q_i^2) - E(2Q_i d_i) + E(d_i^2)}. \tag{4}$$

The price dependent stochastic demand for period i is:

$$d_i = d_i(p, X).$$

The expected value of the price dependent stochastic price-dependent demand is equal to the expected value of random error plus the deterministic price dependent demand, which is greater than zero as:

$$E(d_i) = \mu_i + (a(p))^+.$$

The price dependent deterministic demand during the season is the maximum perceived cumulative deterministic (riskless) demand, i.e., market share plus the price sensitivity for the cumulative deterministic demand multiplied by price, such that:

$$a(p) = y - z * p.$$

Notice that:

$$E\left(Q_i^2\right) = (Q_i)^2,$$

and

$$E\left(d_i^2\right) = (\mu_i + a)^2 + \sigma_i^2.$$

Putting the values in inequality (4):

$$E|Q_i - d_i| \leq \sqrt{(Q_i)^2 - 2(Q_i)(\mu_i + a) + (\mu_i)^2 + \sigma_i^2}$$

$$= \sqrt{\sigma_i^2 + (Q_i)^2 - 2(Q_i)(\mu_i + a) + (\mu_i + a)^2}.$$

After simplification it can be written as:

$$E|Q_i - d_i| \leq \sqrt{\sigma_i^2 + \sigma_r^2 + (Q_i - \mu_i - a)^2}.$$

Putting the $E|Q_i - d_i|$ in Equation (3),

$$E(Q_i - d_i)^+ \leq \frac{1}{2}\left(\sqrt{\sigma_i^2 + (Q_i - \mu_i - a)^2} + (Q_i - \mu_i - a)\right).$$

Similarly, one can easily prove that,

$$E(d_i - Q_i)^+ \leq \frac{1}{2}\left(\sqrt{\sigma_i^2 + (\mu_i + a - Q_i)^2} - (Q_i - \mu_i - a)\right),$$

and

$$E(Q_{i-1} - d_{i-1})^+ \leq \frac{1}{2}\left(\sqrt{\sigma_{i-1}^2 + (Q_{i-1} - \mu_{i-1} - a)^2} + (Q_{i-1} - \mu_{i-1} - a)\right).$$

Utilizing these inequalities, the expected profit can be written as:

$$\pi = \sum_{i=1}^{n} \left\{ \begin{array}{l} p(\mu_i + a) - cQ_i - \frac{(b+p)}{2}\left(\sqrt{\sigma_i^2 + (\mu_i + a - Q_i)^2} - (Q_i - \mu_i - a)\right) \\ +\left(\left(1 - e^{-\frac{\zeta\beta}{\varrho}}\right)p(1 - \beta) - h\left(e^{-\frac{\zeta\beta}{\varrho}}\right)\right)\frac{1}{2}\left(\sqrt{\sigma_i^2 + (Q_i - \mu_i - a)^2} + (Q_i - \mu_i - a)\right) \\ +\left(e^{-\frac{\zeta\beta}{\varrho}}\right)\frac{s}{2}\left(\sqrt{\sigma_{i-1}^2 + (Q_{i-1} - \mu_{i-1} - a)^2} + (Q_{i-1} - \mu_{i-1} - a)\right) \end{array} \right\}. \tag{5}$$

3.2. Optimal Policies

Taking the first derivative of (5) with respect to Q_i, p, and β, and equating to zero the optimal values of Q_i^*, p^*, and β^* can be obtained, such that,

$$\frac{\partial \pi}{\partial Q_i} = \Sigma_{i=1}^n \left(-c + \frac{1}{2}\left(-e^{-\frac{\beta\zeta}{\varrho}}h + \left(1 - e^{-\frac{\beta\zeta}{\varrho}}\right)p(1-\beta)\right)\left(1 + \frac{-y+pz+Q_i-\mu_i}{\sqrt{(-y+pz+Q_i-\mu_i)^2+\sigma_i^2}}\right) - \frac{1}{2}(b+p)\left(-1 - \frac{y-pz-Q_i+\mu_i}{\sqrt{(y-pz-Q_i+\mu_i)^2+\sigma_i^2}}\right)\right) = 0;$$

$$\frac{\partial \pi}{\partial p} = \Sigma_{i=1}^n \left(y - 2pz + \mu_i + \frac{1}{2}e^{-\frac{\beta\zeta}{\varrho}}s\left(z + \frac{z(-y+pz+Q_{-1+i}-\mu_{-1+i})}{\sqrt{(-y+pz+Q_{-1+i}-\mu_{-1+i})^2+\sigma_{-1+i}^2}}\right) + \frac{1}{2}\left(-e^{-\frac{\beta\zeta}{\varrho}}h + (1-e^{-\frac{\beta\zeta}{\varrho}})p(1-\beta))(z + \frac{z(-y+pz+Q_i-\mu_i)}{\sqrt{(-y+pz+Q_i-\mu_i)^2+\sigma_i^2}}\right) + \frac{1}{2}(1 - e^{-\frac{\beta\zeta}{\varrho}})(1-\beta)(-y + pz + Q_i - \mu_i + \sqrt{(-y+pz+Q_i-\mu_i)^2+\sigma_i^2}) - \frac{1}{2}(b+p)(-z - \frac{z(y-pz-Q_i+\mu_i)}{\sqrt{(y-pz-Q_i+\mu_i)^2+\sigma_i^2}}) + \frac{1}{2}(-y + pz + Q_i - \mu_i - \sqrt{(y-pz-Q_i+\mu_i)^2+\sigma_i^2})\right) = 0;$$

$$\frac{\partial \pi}{\partial \beta} = \Sigma_{i=1}^n \left(-\frac{e^{-\frac{\beta\zeta}{\varrho}}s\zeta\left(-y+pz+Q_{-1+i}-\mu_{-1+i}+\sqrt{(-y+pz+Q_{-1+i}-\mu_{-1+i})^2+\sigma_{-1+i}^2}\right)}{2\varrho} + \frac{1}{2}(-(1 - e^{-\frac{\beta\zeta}{\varrho}})p + \frac{e^{-\frac{\beta\zeta}{\varrho}}h\zeta}{\varrho} + \frac{e^{-\frac{\beta\zeta}{\varrho}}p(1-\beta)\zeta}{\varrho})(-y + pz + Q_i - \mu_i + \sqrt{(-y+pz+Q_i-\mu_i)^2+\sigma_i^2})\right) = 0.$$

Q_i^*, p^*, and β^* obtained from the above first order conditions are the global optimal if the hessian matrix of π is negative semidefinite.

See Appendix A for proof.

4. Numerical Example

The developed model is tested with a numerical experiment and sensitivity analysis of the input parameters. The numerical experiment considers a newsvendor problem with two periods. The data for the given example was taken from Alfares and Elmorra [47]. Here $c = 35.1$ \$/unit, $b = 14$ \$/unit, $h = 14$ \$/unit/period, $s = 10$ \$/unit, $\mu_1 = 100$, $\sigma_1 = 15$, $\mu_2 = 100$, $\sigma_2 = 15$, $y = 500$ units/period, $z = 5$, $\zeta = 0.05$, and $\rho = 0.08$.

Case 1. Case 1 is the proposed model with the discount policy. The newsvendor decides on the stocking quantity and price based on the available information. Initially, the product is offered to the consumer with the full price, and at the end of the season, the newsvendor offers a discount on the leftover inventory. The inventory left after the discounted sales are salvaged in the next season.

Case 2. This case is considered to study the impacts of a discount on the expected profit and optimal policies of the newsvendor. This case examines the traditional newsvendor that does not offer a discount. In this case, the newsvendor decides on the stocking quantity and price based on the available information. No discount is offered and, thus, the discount percentage is zero. All the leftover inventory after the selling season is salvaged. Results of both cases are summarised in Table 1.

Table 1. Optimal results of Case 1 and Case 2.

Decision Variables	Case 1	Case 2
Q_1 (Units)	219.77	218.25
Q_2 (Units)	217.95	216.54
P ($/product)	77.12	76.88
β (%)	51	0
Expected Profit π ($)	16,763.5	16,530

From the results, it is clear that the expected profit of the discount policy was higher compared to the non-discount policy. Furthermore, both the stocking quantities and the optimal price were higher in the discount policy compared to the non-discount policy. The higher stocking quantity decreases the risk of stock outs, and the overstocking risk was neutralized by the discount policy. Thus, the discount policy was more flexible compared to the traditional non-discount policy.

4.1. Sensitivity Analysis

The sensitivity analysis was performed for all the key parameters and results were compiled in Tables 2–4. The percentage of variation for all parameters is in the range of −50% to +50%. The sensitivity analysis results from Table 2 revealed the following insights on model parameters:

- Considering the profit of the system, the most effective parameter was purchasing cost; decreasing it by 50% increases the expected profit by 50.99%. However, on the positive side, the effect was a little lesser compared to the negative side. Increasing purchasing cost by 50% decreases the profit by 40.85%. Increasing purchasing cost decreases order quantity, increases the optimal price, and the discount percentage is almost unaffected. This shows that the discount percentage applies to both expensive and inexpensive products.
- The impacts of shortage cost, on the profit of the system, were almost symmetrical towards both negative and positive changes. Order quantity increased with an increase in the shortage cost, whereas, the effect on other variables was negligible.
- The holding cost directly influenced the discount percentage, increasing the holding cost increases the discount percentage. This provides interesting results for newsvendors with higher holding costs. They can increase their profits by discounted sales policy. We can see that the profit was more sensitive towards negative changes compared to the positive changes in the holding cost. The order quantity decreased whereas the price was unaffected.

Compared to the holding cost that had a direct relation with the discount percentage; the salvage value had an indirect relation with the discount percentage. Increasing the salvage value decreases discount percentage; the expected profit behaves in an almost symmetric way, with little high changes on the positive side. The sensitivity of the holding cost and salvage value provides instructing results for the decision maker. For retailers where holding was higher, a high discount policy was better; and for retailers with high salvage value, a low discount policy was the optimal one.

Table 2. Sensitivity analysis for the key operational parameter.

Parameter	Percent Change in Value	Decision Variables				Percent Change in Expected Profit
		Q_1	Q_2	p	β	
c	−50	271.7	266.8	68.6	0.51	50.99
	−25	244.6	241.9	72.8	0.51	24.18
	+25	195.7	194.3	81.3	0.50	−21.66
	+50	172.1	170.9	85.6	0.50	−40.85
b	−50	218.9	217.1	77.0	0.51	0.483
	−25	219.3	217.5	77.1	0.51	0.237
	+25	220.1	218.3	77.1	0.51	−0.230
	+50	220.5	218.6	77.1	0.51	−0. 453
h	−50	221.1	218.8	77.1	0.47	0.647
	−25	220.4	218.3	77.1	0.49	0.314
	+25	219.2	217.5	77.0	0.53	−0.297
	+50	218.2	217.2	77.0	0.54	−0.581
s	−50	218.9	218.0	77.1	0.52	−0.222
	−25	219.3	218.0	77.1	0.51	−0.113
	+25	220.2	217.8	77.1	0.50	0.117
	+50	220.8	217.8	77.1	0.49	0.240

This paper assumed a stochastic price-dependent demand that composed of a deterministic price dependent part and a random error. Table 3 provides a sensitivity analysis of the demand parameters; the following insights are obtained from the results:

- Although changing the mean of the random errors in the demand affects the profit, the results were symmetric in both the direction. Increasing μ_1 or μ_2 by 50% increased the profit by 12.36%, because, the expected value of demand $d_i(p, X_i)$ increased with increasing μ_1 and μ_2. Price of the finished product increased with increases in demand; however, the discount percentage remained the same. This means the discount policy applied to both low and higher demands newsvendors.
- Compared to the random error μ_1 and μ_2 the standard deviations had much less impact on the profit of the system. However, the changes, in profit, were symmetric to both positive and negative changes in standard deviations of the demand. The stocking quantity increased with an increase in standard deviation. The result was clear because increasing standard deviation increased the uncertainty; therefore, the stocking quantity was increased. Furthermore, the price remained unaffected; however, the discount percentage decreased with increasing σ_1. This means, for a less uncertain demand, the newsvendor should increase the discount percentage to increase its profit and market share. However, as the uncertainty increased the discount rate was reduced to avoid extra loses from high-expected salvage quantity.
- The deterministic price-dependent demand had two parameters, which were y and z. Where y is the potential market size for the deterministic demand and z is the price sensitivity of the consumer. Increasing or decreasing y directly increased or decreased the expected demand; therefore, the profit of the system was affected accordingly. However, the results were asymmetric and the effect grew as y increased.
- On the other hand, increasing z reduced the profit; however, the impacts were asymmetric and the decrease in profit declined as z increased. For a higher value of z, the consumer was more sensitive to price; therefore, the optimal price decreased with increasing z. The discount percentage, on the other hand, continuously increased with an increase in z. This means a higher discount percentage applied to a consumer having high price sensitivity and vice versa. Stocking quantity decreased with an increase in z because the realized deterministic demand decreased with increasing z.

Table 3. Sensitivity analysis for the demand parameter.

Parameter	Percent Change in Value	Decision Variables				Percent Change Expected Profit
		Q_1	Q_2	p	β	
μ_1	−50	181.9	230.1	74.6	0.51	−12.15
	−25	200.8	224.0	75.86	0.51	−6.17
	+25	238.9	211.8	78.3	0.51	6.361
	+50	527.6	205.7	79.6	0.50	12.91
μ_2	−50	231.9	180.1	74.6	0.51	−12.15
	−25	225.8	199.0	75.86	0.51	−6.17
	+25	213.6	236.8	78.3	0.51	6.361
	+50	207.6	255.7	79.64	0.50	12.91
σ_1	−50	216.5	217.4	77.2	0.52	1.763
	−25	218.1	217.7	77.1	0.51	0.881
	+25	221.3	218.1	77.0	0.50	−0.880
	+50	222.9	218.41	77.0	0.50	−1.761
σ_2	−50	219.2	215.5	77.2	0.50	1.98
	−25	219.4	216.7	77.1	0.50	0.99
	+25	220.0	219.1	77.0	0.51	−0.989
	+50	220.3	220.2	77.0	0.51	−1.979
z	−50	217.8	268.6	137.3	0.48	203.7
	−25	244.4	242.2	97.15	0.50	66.14
	+25	196.0	194.3	65.1	0.51	−37.67
	+50	172.7	171.1	57.1	0.52	−61.17
y	−50	81.4	79.8	53.7	0.52	−87.0
	−25	155.3	153.7	64.5	0.52	−53.2
	+25	284.0	281.9	89.7	0.50	72.0
	+50	348.1	345.9	102.2	0.49	162.8

The discounted portion of sales $(\alpha(\beta))$ is a function of the discount percentage with parameters ζ and ϱ. Increasing the discount percentage reduces the price and discounted sales increases. A sensitivity analysis of the parameters of the discounted portion is given in Table 4; the results showed that:

- With the increase in the parameter (ζ) value, the expected profit of the system showed an asymmetric increase. For a 25% decrease, the expected profit decreased by −0.32%; however, with further decreases the profit remains unaffected. On the positive side, the profit increase by 0.31% for a 25% increase, and 0.628% for a 50% increase in the value of ζ. The discount percent decreased with an increase in ζ, price remains unaffected, and the stocking quantity increased with increasing ζ.

- By increasing the value of the parameter (ϱ), the asymmetric negative change occurred in the profit, the order quantity decreased, the price almost remained the same, and the discount percentage increased.

Table 4. Sensitivity analysis for the key discounted sales parameter.

Parameter	Percent Change in Value	Decision Variables				Percent Change in Expected Profit
		Q_1	Q_2	p	β	
ζ	−50	218.2	217.2	77.0	0.53	−0.329
	−25	219.3	217.0	77.0	0.52	−0.329
	+25	220.1	218.3	77.1	0.50	0.318
	+50	220.6	218.7	77.2	0.49	0.628
ϱ	−50	221.6	219.5	77.3	0.47	1.226
	−25	220.3	218.4	77.2	0.49	0.423
	+25	219.4	217.6	77.0	0.51	−0.262
	+50	219.2	217.4	77.0	0.52	−0.441

4.2. Managerial Insights

Based on the obtained results, the following recommendations were suggested to managers:

- The proposed discount policy increases both the price and order quantities, thus, managers can order high quantity compared to non-discount policy. Higher ordering quantities decrease the risk of shortage cost, whereas, the discount percentage decreases the risk of overstocking and leftovers. Therefore, in the proposed policy, both the risks are minimized.
- Another important insight is that the discount percentage increases with a decrease in market uncertainty, this means, a retailer having low variable demand can order more with a higher discount percentage. However, retailers with a highly variable demand, orders low quantity with a low discount percentage.
- The optimal discount percentage increases as the consumer price sensitivity increases, therefore, managers that are dealing with markets having higher consumer sensitivity are advised to offer higher discount percentage. This increases the discounted percentage and higher profit and market shares can be achieved.
- For the single period problem, the optimal ordering quantity and price decreases, where the discount percentage increases; the discounts on the successive periods leads to an increase in order quantity compared to the single period newsvendor problem. The increase in the ordered quantity and price is the result of discount percentage that reduces salvage quantity for the subsequent period; however, the selling price of the items is still more than the salvaged value of the product.
- The associated risk with salvaging leftover and shortage is high in the stochastic environment and it can be reduced by implying the distribution free approach with discount offering—discounts are helpful for mangers in increasing the profit generation. The major objective of the manager is to maximize the expected profit, which can be achieved by the proposed discounted policy for leftover items in successive periods.

5. Conclusions

This paper studied joint pricing and inventory policies for the newsvendor model with discounted sales. The classical multi-period distribution-free newsvendor model was extended with a discount policy to increase the sales and profit of the retailer. Stochastic-price dependent demand was considered, and a distribution-free approach was applied to solve the model. No specific assumptions on the distribution of the random error in the demand were considered, except that it had a known mean and variance. Two numerical examples were considered, and the results showed that the proposed discount policy increased the sales and profit of the system. Furthermore, the discount policy provided more flexibility to the newsvendor in deciding the optimal price. With this policy, the newsvendor, initially, decided a higher price compared to the one without the discount policy, and later on, the leftover is discounted with a lower price. Thus, the newsvendor can catch both the strategic and non-strategic consumer at the same time. The sensitivity results showed that the discount policy was applicable to both expensive and inexpensive products; retailers with a higher holding cost can use this policy to increase their profit. This study considered the inventory and pricing policy for only one player, however, in practice, every business directly deals with upstream and downstream linkages. Therefore, this study can be extended by considering more than one player, such as the models developed by Sarkar [9,10]; in this case, two different discount policies can be considered, discount for the final consumer and discount for the newsvendor. Another limitation of this study was that we considered only the single discount per period, considering multiple discounts in one period is a more practical extension of this study. A third possible extension is to consider deteriorating products, such as the study done by Ullah et al. [48]; in this case, the salvage value of the deteriorated product must be zero.

Author Contributions: Conceptualization, M.U.; Methodology, B.S. and M.U.; Software, M.U.; Validation, B.S.; Formal analysis, M.U. and I.K.; Investigation, B.S. and M.U.; Resources, M.U., B.S., and I.K.; Data curation, B.S. and M.U.; Writing—original draft preparation, I.K. and M.U.; Writing—review and editing, I.K. and M.U.; Visualization, M.U.; Supervision, B.S.

Funding: This research received no external funding.

Conflicts of Interest: The authors declare no conflict of interest.

Notation

The following notation are used to establish the mathematical model:

Decision variables

p	Price of finished product per unit (\$/unit)
β	Discount percentage (percent of finished product price)
Q_i	Ordering quantity ith period (units)

Parameters

i	Index for selling period, where $i = 1 \ldots \ldots n$
b	Shortage cost per unit (\$/unit)
c	Purchasing cost per unit (\$/unit)
h	Holding cost (\$/unit/period)
s	Salvage value (\$/unit)
$d_i(p, X_i)$	Price dependent stochastic demand of period i
X_i	Random error in demand
μ_i	The expected value of random error
σ_i	The standard deviation of demand
$a(p) = y - z * p$	Deterministic price dependent demand in-season
y	Maximum perceived cumulative deterministic (riskless) demand i.e., market share (units/unit time)
z	Price sensitivity for cumulative deterministic demand
$E(d_i) = \mu_i + a(p)$	The expected value of price dependent stochastic demand
X^+	Max [X,0]
π	Expected profit (\$)

Appendix A

If,

$$a_{1,1} = \frac{\partial}{\partial Q_1}\frac{\partial \pi}{\partial Q_1}; a_{1,2} = \frac{\partial}{\partial p}\frac{\partial \pi}{\partial Q_1}; a_{1,3} = \frac{\partial}{\partial \beta}\frac{\partial \pi}{\partial Q_1}; a_{1,4} = \frac{\partial}{\partial Q_2}\frac{\partial \pi}{\partial Q_1}; a_{2,1} = \frac{\partial}{\partial Q_1}\frac{\partial \pi}{\partial p}; a_{2,2} =$$

$$\frac{\partial}{\partial p}\frac{\partial \pi}{\partial p}; a_{2,3} = \frac{\partial}{\partial \beta}\frac{\partial \pi}{\partial p}; a_{2,4} = \frac{\partial}{\partial Q_2}\frac{\partial \pi}{\partial p} a_{3,1} = \frac{\partial}{\partial Q_1}\frac{\partial \pi}{\partial \beta}; a_{3,2} = \frac{\partial}{\partial p}\frac{\partial \pi}{\partial \beta}; a_{3,3} = \frac{\partial}{\partial \beta}\frac{\partial \pi}{\partial \beta}; a_{3,4} = \frac{\partial}{\partial Q_2}\frac{\partial \pi}{\partial \beta};$$

$$a_{4,1} = \frac{\partial}{\partial Q_1}\frac{\partial \pi}{\partial Q_2}; a_{4,2} = \frac{\partial}{\partial p}\frac{\partial \pi}{\partial Q_2}; a_{4,3} = \frac{\partial}{\partial \beta}\frac{\partial \pi}{\partial Q_2}; a_{4,4} = \frac{\partial}{\partial Q_2}\frac{\partial \pi}{\partial Q_2},$$

then the Hessian matrix of (4) can be expressed as:

$$H = \begin{pmatrix} a_{1,1} & a_{1,2} & a_{1,3} & a_{1,4} \\ a_{2,1} & a_{2,2} & a_{2,3} & a_{2,4} \\ a_{3,1} & a_{3,2} & a_{3,3} & a_{3,4} \\ a_{4,1} & a_{4,2} & a_{4,3} & a_{4,4} \end{pmatrix}.$$

At the optimal solutions $Q_1{}^* = 219.76$, $Q_2{}^* = 217.94$, $p^* = 77.12$, and $\beta^* = 0.51$ the Hessian matrix can be written as,

$$H = \begin{pmatrix} -2.32 & -11.20 & -1.43 & 0.00 \\ -11.20 & -139.07 & -0.79 & 0.00 \\ -1.43 & -0.79 & -1677.99 & 0.00 \\ 0.00 & -13.49 & 1.47 & -2.79 \end{pmatrix}.$$

The first four principle minors are -2.32, $+197.81$, $-331,668.04$, and $+926,141.46$. Hence total profit is strictly concave at $Q_1{}^* = 219.76$, $Q_2{}^* = 217.94$, $p^* = 77.12$, and $\beta^* = 0.51$.

References

1. Scarf, H. A min-max solution of an inventory problem. In *Studies in the Mathematical Theory of Inventory and Production*; Stanford University Press: Palo Alto, CA, USA, 1958.
2. Arrow, K.J.; Harris, T.; Marschak, J. Optimal inventory policy. *Econometrica* **1951**, 250–272. [CrossRef]
3. Gallego, G.; Moon, I. The distribution free newsboy problem: Review and extensions. *J. Oper. Res. Soc.* **1993**, *44*, 825–834. [CrossRef]
4. Anvari, M. Optimality criteria and risk in inventory models: The case of the newsboy problem. *J. Oper. Res. Soc.* **1987**, *38*, 625–632. [CrossRef]
5. Khouja, M. The single-period (news-vendor) problem: Literature review and suggestions for future research. *Omega* **1999**, *27*, 537–553. [CrossRef]
6. Bitran, G.R.; Mondschein, S.V. Periodic pricing of seasonal products in retailing. *Manag. Sci.* **1997**, *43*, 64–79. [CrossRef]
7. Lau, H.-S. The newsboy problem under alternative optimization objectives. *J. Oper. Res. Soc.* **1980**, *31*, 525–535. [CrossRef]
8. Parlar, M.; Kevin Weng, Z. Balancing desirable but conflicting objectives in the newsvendor problem. *IIE Trans.* **2003**, *35*, 131–142. [CrossRef]
9. Sarkar, B.; Majumder, A.; Sarkar, M.; Kim, N.; Ullah, M. Effects of variable production rate on quality of products in a single-vendor multi-buyer supply chain management. *Int. J. Adv. Manuf. Technol.* **2018**, *99*, 567–581. [CrossRef]
10. Sarkar, B.; Ullah, M.; Kim, N. Environmental and economic assessment of closed-loop supply chain with remanufacturing and returnable transport items. *Comput. Ind. Eng.* **2017**, *111*, 148–163. [CrossRef]
11. Qin, Y.; Wang, R.; Vakharia, A.J.; Chen, Y.; Seref, M.M. The newsvendor problem: Review and directions for future research. *Eur. J. Oper. Res.* **2011**, *213*, 361–374. [CrossRef]
12. Porteus, E.L. Stochastic inventory theory. *Hdbk. Oper. Res. Manag. Sci.* **1990**, *2*, 605–652.
13. Moon, I.; Yoo, D.K.; Saha, S. The distribution-free newsboy problem with multiple discounts and upgrades. *Math. Probl. Eng.* **2016**, *2016*, 2017253. [CrossRef]
14. Sarkar, B.; Zhang, C.; Majumder, A.; Sarkar, M.; Seo, Y.W. A distribution free newsvendor model with consignment policy and retailer's royalty reduction. *Int. J. Prod. Res.* **2018**, *56*, 5025–5044. [CrossRef]
15. Kogan, K.; Lou, S. Multi-stage newsboy problem: A dynamic model. *Eur. J. Oper. Res.* **2003**, *149*, 448–458. [CrossRef]
16. Matsuyama, K. The multi-period newsboy problem. *Eur. J. Oper. Res.* **2006**, *171*, 170–188. [CrossRef]
17. Lee, C.H. Coordination on stocking and progressive pricing policies for a supply chain. *Int. J. Prod. Econ.* **2007**, *106*, 307–319. [CrossRef]
18. Chen, X.A.; Wang, Z.; Yuan, H. Optimal pricing for selling to a static multi-period newsvendor. *Oper. Res. Lett.* **2017**, *45*, 415–420. [CrossRef]
19. Petruzzi, N.C.; Dada, M. Pricing and the newsvendor problem: A review with extensions. *Oper. Res.* **1999**, *47*, 183–194. [CrossRef]
20. Khouja, M. The newsboy problem under progressive multiple discounts. *Eur. J. Oper. Res.* **1995**, *84*, 458–466. [CrossRef]
21. Arcelus, F.J.; Kumar, S.; Srinivasan, G. Retailer's response to alternate manufacturer's incentives under a single-period, price-dependent, stochastic-demand framework. *Decis. Sci.* **2005**, *36*, 599–626. [CrossRef]

22. Chung, C.-S.; Flynn, J.; Zhu, J. The newsvendor problem with an in-season price adjustment. *Eur. J. Oper. Res.* **2009**, *198*, 148–156. [CrossRef]
23. Banerjee, S.; Meitei, N.S. Effect of declining selling price: Profit analysis for a single period inventory model with stochastic demand and lead time. *J. Oper. Res. Soc.* **2010**, *61*, 696–704. [CrossRef]
24. Abad, P. Determining optimal price and order size for a price setting newsvendor under cycle service level. *Int. J. Prod. Econ.* **2014**, *158*, 106–113. [CrossRef]
25. Ma, S.; Jemai, Z.; Sahin, E.; Dallery, Y. Analysis of the Newsboy problem subject to price dependent demand and multiple discounts. *Am. Inst. Math. Sci.* **2018**, *14*, 931–951. [CrossRef]
26. Arcelus, F.J.; Kumar, S.; Srinivasan, G. Risk tolerance and a retailer's pricing and ordering policies within a newsvendor framework. *Omega* **2012**, *40*, 188–198. [CrossRef]
27. Hu, X.; Su, P. The newsvendor's joint procurement and pricing problem under price-sensitive stochastic demand and purchase price uncertainty. *Omega* **2018**, *79*, 81–90. [CrossRef]
28. Ye, T.; Sun, H. Price-setting newsvendor with strategic consumers. *Omega* **2016**, *63*, 103–110. [CrossRef]
29. Arcelus, F.J.; Kumar, S.; Srinivasan, G. Pricing, rebate, advertising and ordering policies of a retailer facing price-dependent stochastic demand in newsvendor framework under different risk preferences. *Int. Trans. Oper. Res.* **2006**, *13*, 209–227. [CrossRef]
30. He, Y.; Zhao, X.; Zhao, L.; He, J. Coordinating a supply chain with effort and price dependent stochastic demand. *Appl. Math. Model.* **2009**, *33*, 2777–2790. [CrossRef]
31. Chen, H.; Chen, Y.F.; Chiu, C.-H.; Choi, T.-M.; Sethi, S. Coordination mechanism for the supply chain with leadtime consideration and price-dependent demand. *Eur. J. Oper. Res.* **2010**, *203*, 70–80. [CrossRef]
32. Chen, J.; Bell, P.C. Coordinating a decentralized supply chain with customer returns and price-dependent stochastic demand using a buyback policy. *Eur. J. Oper. Res.* **2011**, *212*, 293–300. [CrossRef]
33. Jadidi, O.; Jaber, M.Y.; Zolfaghari, S. Joint pricing and inventory problem with price dependent stochastic demand and price discounts. *Comput. Ind. Eng.* **2017**, *114*, 45–53. [CrossRef]
34. Modak, N.M.; Kelle, P. Managing a dual-channel supply chain under price and delivery-time dependent stochastic demand. *Eur. J. Oper. Res.* **2019**, *272*, 147–161. [CrossRef]
35. Khouja, M. The newsboy problem with multiple discounts offered by suppliers and retailers. *Decis. Sci.* **1996**, *27*, 589–599. [CrossRef]
36. Cachon, G.P.; Kök, A.G. Implementation of the newsvendor model with clearance pricing: How to (and how not to) estimate a salvage value. *Manuf. Serv. Oper. Manag.* **2007**, *9*, 276–290. [CrossRef]
37. Nocke, V.; Peitz, M. A theory of clearance sales. *Econ. J.* **2007**, *117*, 964–990. [CrossRef]
38. Khouja, M.; Pan, J.; Zhou, J. Effects of gift cards on optimal order and discount of seasonal products. *Eur. J. Oper. Res.* **2016**, *248*, 159–173. [CrossRef]
39. Gupta, D.; Hill, A.V.; Bouzdine-Chameeva, T. A pricing model for clearing end-of-season retail inventory. *Eur. J. Oper. Res.* **2006**, *170*, 518–540. [CrossRef]
40. Ullah, M.; Sarkar, B. Smart and sustainable supply chain management: A proposal to use rfid to improve electronic waste management. In Proceedings of the International Conference on Computers and Industrial Engineering, Auckland, New Zealand, 2–5 December 2018.
41. Jammernegg, W.; Kischka, P. The price-setting newsvendor with service and loss constraints. *Omega* **2013**, *41*, 326–335. [CrossRef]
42. Mandal, P.; Kaul, R.; Jain, T. Stocking and pricing decisions under endogenous demand and reference point effects. *Eur. J. Oper. Res.* **2018**, *264*, 181–199. [CrossRef]
43. Ruidas, S.; Seikh, M.R.; Nayak, P.K.; Sarkar, B. A single period production inventory model in interval environment with price revision. *Int. J. Appl. Comput. Math.* **2018**, *5*, 7. [CrossRef]
44. Yao, L.; Chen, Y.F.; Yan, H. The newsvendor problem with pricing: Extensions. *Int. J. Manag. Sci. Eng. Manag.* **2006**, *1*, 3–16. [CrossRef]
45. Raza, S.A. Supply chain coordination under a revenue-sharing contract with corporate social responsibility and partial demand information. *Int. J. Prod. Econ.* **2018**, *205*, 1–14. [CrossRef]
46. Hsueh, C.-F. A bilevel programming model for corporate social responsibility collaboration in sustainable supply chain management. *Trans. Res. E Log.* **2015**, *73*, 84–95. [CrossRef]

47. Alfares, H.K.; Elmorra, H.H. The distribution-free newsboy problem: Extensions to the shortage penalty case. *Int. J. Prod. Econ.* **2005**, *93*, 465–477. [CrossRef]

48. Ullah, M.; Sarkar, B.; Asghar, I. Effects of preservation technology investment on waste generation in a two-echelon supply chain model. *Mathematics* **2019**, *7*, 189. [CrossRef]

© 2019 by the authors. Licensee MDPI, Basel, Switzerland. This article is an open access article distributed under the terms and conditions of the Creative Commons Attribution (CC BY) license (http://creativecommons.org/licenses/by/4.0/).

Article

Pricing Decision within an Inventory Model for Complementary and Substitutable Products

Ata Allah Taleizadeh [1], Masoumeh Sadat Babaei [2], Shib Sankar Sana [3] and Biswajit Sarkar [4],*

[1] School of Industrial Engineering, College of Engineering, University of Tehran, Tehran 456311155, Iran; taleizadeh@ut.ac.ir
[2] Department of Industrial Engineering, Islamic Azad University, South Tehran Branch, Tehran 1584743311, Iran; st_m_babaei@azad.ac.ir
[3] Principal, Kishore Bharati Bhagini Nivedita College, Ramkrishna Sarani, Behala, Kolkata 700060, India; shib_sankar@yahoo.com
[4] Department of Industrial & Management Engineering, Hanyang University, Ansan, Gyeonggi-do 15588, Korea
*   Correspondence: bsbiswajitsarkar@gmail.com; Tel.: +82-10-7498-1981

Received: 30 March 2019; Accepted: 31 May 2019; Published: 26 June 2019

Abstract: A combination of substitutable and complementary products is very important for any business industry to make all-round profit from different aspects. How deterioration affects complementary products or substitutable products is discussed in this study. This study investigates the pricing and inventory decisions for complementary and substitutable items which are deteriorating in nature. Four models are analyzed where the demand of one product is dependent upon the selling price and the price of another product. This paper tries to compute the optimum prices and order quantities to optimize the total profit, which is the main aim. Theoretically, this model is solved by a classical optimization method. Numerical examples demonstrate the applicability of this model. Results conclude that the total profit is dependent on the degree of substitutability and complementarity. A sensitivity analysis of optimal solutions is given to test the stability of the proposed model.

Keywords: inventory; pricing; complementary products; substitutable products; deterioration

---

1. Introduction

Everyday life starts with the use of a toothbrush and toothpaste, which are complementary products. Then, the energy for the whole day is derived from the breakfast where at least one seasonal fruit is mandatory for good health. The seasonal fruit can be replaced by another fruit based on the availability and price and is substitutable product. Whenever the increasing price of one product indicates the increasing sales rate of another product, then those products substitute each other [1]. Increasing the price of one item can help to increase the sale of another substitutable item. For example, Coca-Cola and Pepsi are two substitutable products as both can satisfy the purpose. One customer can buy Pepsi instead of Coca-Cola in any circumstances. Complementary products are used in combination with another product [2]. Single use of any part of complementary products has limited usage. The overall utility is increased whenever both products are used together. The market demand of complementary products depends upon both products [3]. Car and fuel, camera and memory card, toothpaste and brush—these are a few examples of complementary products. A car cannot move without proper fuel, which implies that the car is useless without fuel, and fuel is no longer important if the car is not there. To earn in all aspects, the manufacturer introduces substitutable as well as complementary products in the business. From the perspective of the inventory, complementary products need more storage and inventory, as both products are important. In reality, substitutable products give the manufacturer tough competition, as it is important to choose alternative products very wisely such that it can satisfy the demand as well as the quality of the original product. The pricing factor is the next important thing for substitutable products. The price of the alternative should be less than or equal to the original product. The situation may differ in case of the reverse situation when the alternative's price is greater than the original price. The situation gets more complicated when the product deteriorates in nature.

There are several researches who focus on substitutable and complementary products, while none of them have investigated both inventory and pricing decisions of the products simultaneously. Moreover, this investigation is done for deteriorating products. In this paper, four models are discussed for pricing and inventory decisions of two types of items, which may be complementary or substitutable, and deteriorating or non-deteriorating. The rest of the present paper is organized as follows: Section 2 presents a literature review, whereas problem definition, notation, and assumptions are defined in Section 3, and the mathematical modeling is presented in Section 4. Numerical examples, a sensitivity analysis, and managerial insights are provided in Section 5. Section 6 concludes the achievements of this proposed model. The last section provides the References of the study.

2. Literature Review

Pricing policy is an important part in the field of supply chain management [4–7]. Several research articles have been studied on inventory and pricing decisions for single or multi-products inventory models [8–10]. However, few of them studied pricing decisions for complementary or substitutable products. In such cases, the price, quality, and durability of the products play a crucial role to attract customers [5,9]. On the other hand, retailers face the cross-selling phenomenon for complementary products. McGillivray and Silver [11] investigated the full substitutability and its effect on total cost within an inventory model. Zhang et al. [12] illustrated an inventory system when the demand of a minor and a major item is correlated with cross-selling. Zhang [13] discussed an inventory problem for multiple products with the inventory replenishment. They proposed a search method which gives a global optimum solution. Yue et al. [14] investigated a duopoly market dealing with complementary products.

An inventory model was discussed by Liu and Yuan [15] for two types of products where the consumption rates of those products are correlated, and the replenishment rates are coordinated. Wei et al. [9] analyzed a non-coordination supply chain management consisting of two manufactures with one retailer for two complementary type products with the help of five different game theories: manufacturer-leader Bertrand, manufacturer-leader Stackelberg, retailer-leader Bertrand, retailer-leader Stackelberg, and Nash game (NG) models. Guchhait et al. [16] studied an advertisement-dependent demand scenario within supply chain management. They used a Stackelberg game policy to solve the model. Shavandi et al. [17] extended the pricing and lot sizing model of Abad [18] for multiple products. They proposed several pricing strategies of multi-products for deteriorating items which may be complementary, substitutable, or independent. The aim of the research was to maximize the profit optimization, along with optimum prices and production quantities.

An inventory model was studied by Balkhi and Benkherouf [19] for deteriorating products. The market demand was stock dependent and time dependent. A procedure was proposed by them to achieve the optimum replenishment schedule. Anjos et al. [20] considered perishable products and presented a continuous pricing strategy where the optimal pricing strategy can be explicitly characterized and easily implemented. Dye [21] extended an inventory model for perishable items with a time-dependent backlog by assuming a price-sensitive demand and variable deterioration with time. In that article, an improved algorithm was proposed by them to find out the optimal selling price of a product and replenishment schedule. Panda et al. [22] considered a single-item economic order quantity (EOQ) model with discounted sale as products were perishable in nature, and the demand was stock dependent. Pang [23] studied optimal dynamic pricing policies and inventory control policies for deteriorating products within a periodic type inventory model. Sarkar and Saren [24] investigated the effect of the trade-credit policy for exponentially deteriorating products. Sarkar et al. [25] developed a two-echelon supply chain model for deteriorating items, where single-setup-multi-delivery (SSMD) transportation policy was adopted.

Sett et al. [26] and Sarkar et al. [27] studied fixed lifetime products and their replenishment policy. The deterioration rate was variable, and the backlog was time dependent. Sarkar et al. [28] applied preservation technology for seasonal deteriorating products while Ullah et al. [29] used the preservation for waste generation. Recycling of the deteriorated products was investigated by Iqbal and Sarkar [30]. The market demand can be affected by the deterioration and a backlog scenario may appear in that case. The service level can help to get rid of this situation [31]. Gürler and Yilmaz [32] studied a supply chain management with two substitutable products in a newsboy problem. Demands for both products were independent, and the aim of the research was to maximize the retailer's and manufacturer's profits jointly. Kim and Bell [33] proposed that the production system of a firm sold their products to multiple market segments for a specific period. A price-driven scenario was examined by them, and the effect on the pricing strategy and production capacity of the firm were discussed.

Stavrulaki [34] investigated an inventory model for substitutable products. They explained the combined effect of the demand stimulation for stochastic demand. Karakul and Chan [35] studied the analytical implications of product substitutability and procurement decisions in a newsboy problem with two products. Thereafter, they established the unimodality of the expected profit of their previous work [36] with respect to the procurement and price of quantities. Birge et al. [37] developed an inventory model for substitutable products where market demand follows a uniform distribution. Netessine et al. [38] determined the optimum procurement for multi-substitutable items under exogenous prices of an integrated inventory-pricing problem. Parlar and Goyal [39] developed two integrated production–inventory models of two substitutable products under different prices and assumptions [40]. Tang and Yin [41] discussed an improved procedure to find the optimum selling price within an inventory framework where demand is price sensitive for substitutable products. Xia [42] suggested a two-echelon supply chain model including competitive multiple suppliers and multiple buyers, where each supplier offers one type of substitutable product to the buyers. Netessine et al. [43] discussed about centralized inventory policy where market demand is substitutable. Table 1 gives the contribution from different authors associated with the area of research.

Table 1. Author's contributions table.

Author(s)	Inventory	Pricing	Deterioration	Complementary Products	Substitutable Products
Sarkar and Lee [2]		√		√	
Sana [4]	√	√			
Smith et al. [7]	√	√			
Wei at al. [9]		√		√	
Mcgillivray and Silver [11]	√				√
Yue et al. [14]		√		√	
Balkhi and Benkherouf [19]	√		√		
Dye [21]	√	√	√		
Karakul and Chan [35]	√	√			√
Tang and Yin [41]	√	√			√
This study	√	√	√	√	√

3. Problem Description, Notation, and Assumptions

3.1. Problem Definition

Consider a market where the retailer intends to determine the order quantities of different products. The products may be complementary or substitutable, for which two different scenarios exist.

In case of complementary products, the retailer expects a higher consumption rate of a product when the consumption rate of its complementary product increases and vice versa. Moreover, the demand of every complementary product depends upon the price of the other such that the high price of one of them decreases the demand of the complementary product and vice versa. Obviously, the degree of complementarity between two products, which is between 0 and 1, has the main effect on the changes between their respective demands and prices. In this situation, the retailer determines the optimal order quantity and selling price of each product in such a way that the retailer can maximize the total profit.

In the second scenario, each product can be used instead of another one in the case of substitutable products. In this situation, the retailer faces the situation that the decreasing demand of a substitutable product increases the other product's demand. The demand of each substitutable product depends upon its own price and the price of another one such that the higher price of one of them increases the demand of its substitutable product and vice versa because the customer intends to use the substitutable product, which has a lower price compared to the other. Generally, the degree of substitutability between two products, which is between 0 and 1, has the main effects on the changes between their demands and prices. Therefore, the retailer should determine the optimum order quantity such that the total profit should maximize.

3.2. Notation

The notation related to this study is given in the Nomenclature section at the end of the study.

3.3. Assumptions

Assumptions are used to formulate the model are as follows:

1. A retailer maximizes the profit based on an economic order quantity (EOQ) model. The demand is dependent upon selling prices and lot sizes of two types of products. The demand of two complementary products is dependent upon each product's selling price and the degree of complementarity.
2. After a certain time, products start to deteriorate with a constant deterioration rate. As deterioration is considered in this model, deterioration cost per time unit for the product m is η'_m (\$ per unit). The deterioration rate of product 1 and product 2 is D_r.
3. The degree of complementarity between two products is given by θ, where $0 \leq \varphi \leq 1$, and the degree of substitutability between two products is given by d, where $0 \leq \omega \leq 1$.
4. The lead time is zero, and there are no shortages for both types of products.
5. The time horizon is infinite.

4. Mathematical Model

The scenarios for both non-deteriorating (Figure 1) and deteriorating (Figure 2) products are discussed.

In all cases, the aim is to find the optimal values of the order quantity, the cycle length, and selling prices for profit maximization.

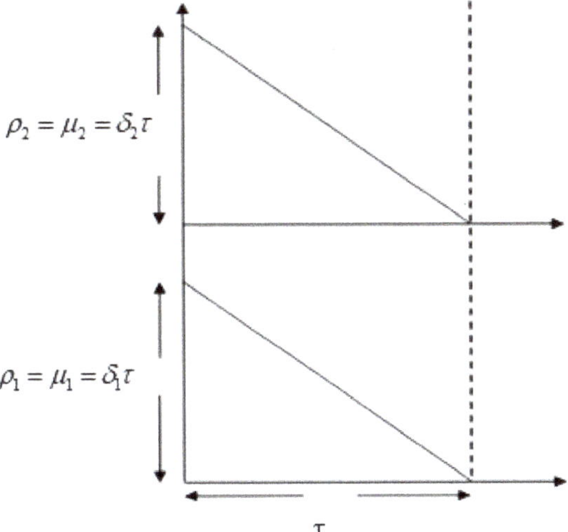

Figure 1. Inventory level of two products (complementary or substitutable) without shortage under EOQ policy.

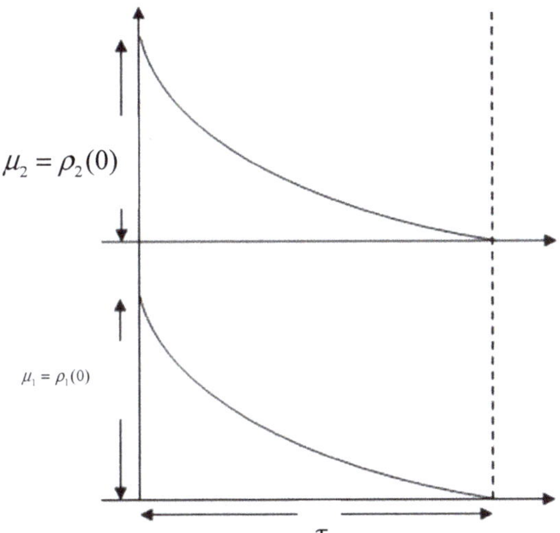

Figure 2. Inventory level of two deteriorating products (complementary or substitutable) without shortage under EOQ policy.

4.1. Non-Deteriorating Complementary Products

This section discusses the complementary products when products are of a non-deteriorating type. The demand [35,44] of two complementary products, product 1 and product 2, are

$$\delta_1(\gamma_1, \gamma_2) = \alpha - \beta\gamma_1 - \beta\varphi\gamma_2 \tag{1}$$

$$\delta_2(\gamma_1, \gamma_2) = \alpha - \beta\gamma_2 - \beta\varphi\gamma_1 \tag{2}$$

respectively. Here, all parameters (α, β, φ) are non-negative. The retailer is trying to obtain the optimal values of decision variables. Therefore, the average profit is

$$Maximize$$

$$\pi_C(\tau, \gamma_1, \gamma_2) = (\gamma_1 - \eta_1)\delta_1(\gamma_1, \gamma_2) + (\gamma_2 - \eta_2)\delta_2(\gamma_1, \gamma_2) - \left(\frac{G_1 + G_2}{\tau} + \frac{[\lambda_1\delta_1(\gamma_1,\gamma_2) + \lambda_2\delta_2(\gamma_1,\gamma_2)]}{2}\tau \right) \tag{3}$$

Expression (1) gives the revenue from product 1. Expression (2) explains the revenue from product 2, and Expression (3) defines the summation of ordering costs and holding costs. The purpose is to maximize the profit, using the following theorem:

Theorem 1. $\pi_C(\tau, \gamma_1, \gamma_2)$ *is strictly concave if*

$$G_1 + G_2 > \frac{\tau\beta}{2}\left[\tau(\lambda_1(\gamma_1 + \gamma_2\varphi) + \lambda_2(\gamma_2 + \gamma_1\varphi)) - 4\varphi\gamma_1\gamma_2 - 2(\gamma_1^2 + \gamma_2^2)\right].$$

Proof. See Appendix A.

Substituting the values of $\delta_1(\gamma_1, \gamma_2)$ and $\delta_2(\gamma_1, \gamma_2)$ within the first order derivative of the Equation (3) with respect to the decision variable τ, one can obtain

$$\frac{\partial \pi_C(\tau, \gamma_1, \gamma_2)}{\partial \tau} = \frac{G_1 + G_2}{\tau^2} + \frac{\lambda_1(\beta\gamma_1 - \alpha + \beta\varphi\gamma_2) + \lambda_2(\beta\gamma_2 - \alpha + \beta\varphi\gamma_1)}{2} \tag{4}$$

The first order derivatives of Equation (3) with respect to the variables γ_1 and γ_2 are given by

$$\frac{\partial \pi_C(\tau, \gamma_1, \gamma_2)}{\partial \gamma_1} = \frac{\tau\beta(\lambda_1 + \varphi\lambda_2)}{2} + \alpha + \beta\eta_1 - 2\beta\gamma_1 + \beta\varphi\eta_2 - 2\beta\varphi\gamma_2 \tag{5}$$

$$\frac{\partial \pi_C(\tau, \gamma_1, \gamma_2)}{\partial \gamma_2} = \frac{\tau\beta(\lambda_2 + \varphi\lambda_1)}{2} + \alpha + \beta\eta_2 - 2\beta\gamma_2 + \beta\varphi\eta_1 - 2\beta\varphi\gamma_1 \tag{6}$$

The required optimum values of γ_1 and γ_2 are as given by the following expressions as:

$$\gamma_1^* = \frac{2\alpha + \beta(1 + \varphi)(\lambda_1\tau + 2\eta_1)}{4\beta(1 + \varphi)} \tag{7}$$

$$\gamma_2^* = \frac{2\alpha + \beta(1 + \varphi)(\lambda_2\tau + 2\eta_2)}{4\beta(1 + \varphi)} \tag{8}$$

Using Equations (7) and (8) on the right-hand side of Equation (4) and equating them to zero, one can obtain

$$\alpha_1\tau^3 + \alpha_2\tau^2 + \alpha_4 = 0 \tag{9}$$

where

$$\alpha_1 = \beta(\lambda_1^2 + \lambda_2^2 + 2\varphi\lambda_1\lambda_2) \tag{10}$$

$$\alpha_2 = 2[\beta(\lambda_1\eta_1 + \lambda_2\eta_2) + \beta\varphi(\lambda_1\eta_2 + \lambda_2\eta_1) - \alpha(\lambda_1 + \lambda_2)] \tag{11}$$

$$\alpha_4 = 8(G_1 + G_2) \tag{12}$$

Equation (9) is a cubic polynomial and the discriminant is

$$\Delta = -4\alpha_2^3\alpha_4 - 27\alpha_1^2\alpha_4^2 \tag{13}$$

According to Δ, the discriminant has three types of values. $\Delta > 0$ gives three real roots of distinct quality, $\Delta = 0$ gives multiple real roots, and $\Delta < 0$ gives one real root and two complex roots. The following improved solution procedure has been developed to solve the problem numerically.

Solution algorithm

Step 1. Using Equations (10)–(12), calculate the coefficients of polynomial shown in Equation (9).

Step 2. All positive real roots of Equation (9) are found by using MATLAB software. Go to the Step (3).

Step 3. Determine the values (γ_1, γ_2) from Step (2).

Step 4. For all combinations of $(\tau, \gamma_1, \gamma_2)$, obtain the profit and check the concavity clause (Theorem 1). Select the maximum value as an optimal value of the profit. Then the related optimum decision variables are τ^*, γ_1^* and γ_2^*.

Step 5. Determine the optimal values of the order quantities using $\mu_1^* = \delta_1(\gamma_1^*, \gamma_2^*)\tau^*$ and $\mu_2^* = \delta_2(\gamma_1^*, \gamma_2^*)\tau^*$.

4.2. Non-Deteriorating Substitutable Products

This section discusses about non-deteriorating substitutable products. The demand of two substitutable products, product 1 and product 2, is defined as follows:

$$\delta_1(\gamma_1, \gamma_2) = \alpha - \beta\gamma_1 + \beta\omega\gamma_2 \tag{14}$$

$$\delta_2(\gamma_1, \gamma_2) = \alpha - \beta\gamma_2 + \beta\omega\gamma_1 \tag{15}$$

where all parameters are non-negative. The average profit is

$$\pi_S(\tau, \gamma_1, \gamma_2) = (\gamma_1 - \eta_1)\delta_1(\gamma_1, \gamma_2) + (\gamma_2 - \eta_2)\delta_2(\gamma_1, \gamma_2) - \left(\frac{G_1 + G_2}{\tau} + \frac{[\lambda_1\delta_1(\gamma_1, \gamma_2) + \lambda_2\delta_2(\gamma_1, \gamma_2)]}{2}\tau\right) \tag{16}$$

The following theorem is used to maximize the above profit.

Theorem 2. $\pi_S(\tau, \gamma_1, \gamma_2)$ *is strictly concave*

$$G_1 + G_2 > \frac{\tau\beta}{2}\left[\tau\lambda_1(\gamma_1 - \gamma_2\omega) + \tau\lambda_2(\gamma_2 - \gamma_1\omega) + 4\omega\gamma_1\gamma_2 - 2(\gamma_1^2 + \gamma_2^2)\right].$$

Proof. See Appendix B.

Now, Equation (16) gives the following expression with respect to the variables τ, γ_1, and γ_2.

$$\frac{\partial\pi_S(\tau, \gamma_1, \gamma_2)}{\partial\tau} = \frac{G_1 + G_2}{\tau^2} + \frac{\lambda_1(\beta\gamma_1 - \alpha - \beta\omega\gamma_2) + \lambda_2(\beta\gamma_2 - \alpha - \beta\omega\gamma_1)}{2} \tag{17}$$

$$\frac{\partial\pi_S(\tau, \gamma_1, \gamma_2)}{\partial\gamma_1} = \frac{\tau\beta(\lambda_1 - \omega\lambda_2)}{2} + \alpha - 2\beta\gamma_1 + \beta\eta_1 + 2\beta\omega\gamma_2 - \beta\omega\eta_2 \tag{18}$$

$$\frac{\partial\pi_S(\tau, \gamma_1, \gamma_2)}{\partial\gamma_2} = \frac{\tau\beta(\lambda_2 - \omega\lambda_1)}{2} + \alpha - 2\beta\gamma_2 + \beta\eta_2 + 2\beta\omega\gamma_1 - \beta\omega\eta_1 \tag{19}$$

Equations (18) and (19) give the optimum values of γ_1 and γ_2 as

$$\gamma_1^* = \frac{2\alpha + \beta(1 + \omega)(\lambda_1\tau + 2\eta_1)}{4\beta(1 + \omega)}, \text{ where } 0 \le \omega < 1 \tag{20}$$

$$\gamma_2^* = \frac{2\alpha + \beta(1 + \omega)(\lambda_2\tau + 2\eta_2)}{4\beta(1 + \omega)}, \text{ where } 0 \le \omega < 1 \tag{21}$$

Using Equations (20) and (21) on the right-hand side of Equation (17) and equating them to zero, one can obtain

$$\alpha_1 \tau^3 + \alpha_2 \tau^2 + \alpha_4 = 0 \tag{22}$$

where

$$\alpha_1 = \beta(\lambda_1{}^2 - 2\omega\lambda_1\lambda_2 + \lambda_2{}^2) \tag{23}$$

$$\alpha_2 = 2[-\alpha(\lambda_1 + \lambda_2) + \beta(\lambda_1\eta_1 + \lambda_2\eta_2) - \beta\omega(\lambda_1\eta_2 + \lambda_2\eta_1)] \tag{24}$$

$$\alpha_4 = 8(G_1 + G_2) \tag{25}$$

According to the discussion about the discriminant of cubic polynomial above, the following solution procedure is suggested for the substitutable products case.

Solution algorithm

Step 1. Using Equations (23)–(25), calculate the coefficients of the polynomial in Equation (22).

Step 2. After finding all possible roots of Equation (22) with the help of MATLAB software, go to the Step (3).

Step 3. From Step 2, determine the values of (γ_1, γ_2).

Step 4. For all values of $(\tau, \gamma_1, \gamma_2)$, determine the total profit and check the concavity clause (Theorem 2). Then select the maximum value as the optimum one. The related decision variables associated with the maximum profit are τ^*, γ_1^*, and γ_2^*.

Step 5. Determine optimal values of order quantities using $\mu_1^* = \delta_1(\gamma_1^*, \gamma_2^*)\tau^*$ and $\mu_2^* = \delta_2(\gamma_1^*, \gamma_2^*)\tau^*$.

4.3. Deteriorating Complementary Products

In this case, it is assumed that the retailer is doing the business with two deteriorating complementary products. For both products, the deterioration rates are constant. The differential equation $\frac{dp(t)}{dt} = -D_r p(t) - \delta(\gamma)$ shows that the inventory level changes over time, where $\rho(\tau) = 0$. Solving this differential equation yields $\rho(t) = \frac{\delta(\gamma)}{D_r}(e^{D_r(\tau-t)} - 1)$, and according to Figure 2, the order quantity of each product is equal to $\mu = \rho(0) = \frac{\delta(\gamma)}{D_r}(e^{D_r\tau} - 1)$. The holding cost is given by

$$\lambda \int_0^\tau \rho(t) = \lambda \int_0^\tau \frac{\delta(\gamma)}{D_r}(e^{D_r(\tau-t)} - 1)dt = \frac{\lambda\delta(\gamma)}{D_r{}^2}(e^{D_r\tau} - D_r\tau - 1) \tag{26}$$

Moreover, the deterioration cost is

$$\eta'\left(\underbrace{\rho(0) - \delta(\gamma)\tau}_{Deterioration\ Quantity}\right) = \eta'\left(\frac{\delta(\gamma)}{D_r}(e^{D_r\tau} - 1) - \delta(\gamma)\tau\right) \tag{27}$$

The total cost after utilizing the approximation of the Taylor series expansion, $e^{D_r\tau} = 1 + D_r\tau + \frac{(D_r\tau)^2}{2}$, is

$$TC = \frac{1}{\tau}\left[G + \overbrace{\frac{\lambda\delta(\gamma)}{D_r{}^2}(e^{D_r\tau} - D_r\tau - 1)}^{Holding\ Cost} + \underbrace{\frac{\eta'\delta(\gamma)}{D_r}(e^{D_r\tau} - 1) - \eta'\delta(\gamma)\tau}_{Deterioration\ Cost}\right]$$

$$e^{D_r \tau} = 1 + D_r \tau + \frac{(D_r \tau)^2}{2}$$

$$TC = \frac{1}{\tau}\left[G + \frac{\lambda \delta(\gamma)}{2}\tau^2 + \frac{\eta' \delta(\gamma) D_r \tau^2}{2} \right] = \frac{G}{\tau} + \frac{(\lambda + \eta' D_r)\delta(\gamma)}{2}\tau^2 \tag{28}$$

The demand of two deteriorating complementary products, product 1 and product 2, is as follows (as defined in Section 4.1):

$$\delta_1(\gamma_1, \gamma_2) = \alpha - \beta\gamma_1 + \beta\varphi\gamma_2 \tag{29}$$

$$\delta_2(\gamma_1, \gamma_2) = \alpha - \beta\gamma_2 + \beta\varphi\gamma_1 \tag{30}$$

Finally, the average profit is

$$\pi_C(\tau, \gamma_1, \gamma_2) = \begin{array}{l} (\gamma_1 - \eta_1)\delta_1(\gamma_1, \gamma_2) + (\gamma_2 - \eta_2)\delta_2(\gamma_1, \gamma_2) \\ -\left[\frac{G_1 + G_2}{\tau} + \frac{\tau(\lambda_1 + \eta_1' D_r)\delta_1(\gamma_1, \gamma_2)}{2} + \frac{\tau(\lambda_2 + \eta_2' D_r)\delta_2(\gamma_1, \gamma_2)}{2}\right] \end{array} \tag{31}$$

To optimize the profit, the following theorem is used.

Theorem 3. *$\pi_C(\tau, \gamma_1, \gamma_2)$ is strictly concave if*

$$G_1 + G_2 > \frac{\tau\beta}{2}[-4\varphi\gamma_1\gamma_2 - 2(\gamma_1^2 + \gamma_2^2) + \tau\gamma_1(\lambda_1 + \varphi\lambda_2) + \tau\gamma_2(\lambda_2 + \varphi\lambda_1) + \tau\eta_1' D_r(\gamma_1 + \varphi\gamma_2) + \tau\eta_2' D_r(\gamma_2 + \varphi\gamma_1)]$$

Proof. See Appendix C.

Here, the first order derivative of Equation (31) with respect to T yields

$$\frac{\partial \pi_c(\tau, \gamma_1, \gamma_2)}{\partial \tau} = \frac{G_1 + G_2}{\tau^2} + \frac{\left(\lambda_1 + \eta_1' D_r\right)(\beta\gamma_1 - \alpha + \beta\varphi\gamma_2) + \left(\lambda_2 + \eta_2' D_r\right)(\beta\gamma_2 - \alpha + \beta\varphi\gamma_1)}{2} \tag{32}$$

The first order derivatives of the Equation (31) for variables p_1 and p_2 are given by

$$\frac{\partial \pi_C(\tau, \gamma_1, \gamma_2)}{\partial \gamma_1} = \frac{\tau\beta((\lambda_1 + \eta_1' D_r) + \varphi(\lambda_2 + \eta_2' D_r))}{2} + \alpha - 2\beta(\gamma_1 + \varphi\gamma_2) + \beta(\eta_1 + \varphi\eta_2) \tag{33}$$

$$\frac{\partial \pi_C(\tau, \gamma_1, \gamma_2)}{\partial \gamma_2} = \frac{\tau\beta((\lambda_2 + \eta_2' D_r) + \varphi(\lambda_1 + \eta_1' D_r))}{2} + \alpha - 2\beta(\gamma_2 + \varphi\gamma_1) + \beta(\eta_2 + \varphi\eta_1) \tag{34}$$

Optimum values of the decision variables p_1 and p_2 are

$$\gamma_1^* = \frac{2\alpha + \beta(1 + \varphi)(2\eta_1 + \tau\lambda_1 + \tau\eta_1' D_r)}{4\beta(1 + \varphi)} \tag{35}$$

$$\gamma_2^* = \frac{2\alpha + \beta(1 + \varphi)(2\eta_2 + \tau\lambda_2 + \tau\eta_2' D_r)}{4\beta(1 + \varphi)} \tag{36}$$

Using Equations (35) and (36) in Equation (32) and equating them to zero for optimality, one can obtain

$$\alpha_1 \tau^3 + \alpha_2 \tau^2 + \alpha_4 = 0 \tag{37}$$

where

$$\alpha_1 = \beta\left[(\gamma_1^2 + \gamma_2^2) + D_r^2(\eta_1'^2 + \eta_2'^2) + 2\eta_1' D_r(\gamma_1 + \varphi\lambda_2) + 2\eta_2' D_r(\lambda_2 + \varphi\lambda_1) + 2\varphi(\lambda_1\lambda_2 + \eta_1'\eta_2' D_r^2)\right] \tag{38}$$

$$\alpha_2 = -2\left[(\gamma_1 + \eta_1' D_r)(\alpha - \beta(\eta_1 + \varphi\eta_2)) + (\gamma_2 + \eta_2' D_r)(\alpha - \beta(\eta_2 + \varphi\eta_1))\right] \tag{39}$$

$$\alpha_4 = 8(G_1 + G_2) \tag{40}$$

According to the discussion about the discriminant of cubic polynomial provided above (Section 4.1.), the following improved solution procedure is suggested for finding optimum solutions.

Solution algorithm

Step 1. Using Equations (38)–(40), calculate coefficients of polynomial shown in the Equation (37).

Step 2. Equation (37) gives all roots by using MATLAB software and then go to the Step (3).

Step 3. Determine all values of (γ_1, γ_2) for period from the Step (2).

Step 4. For all possible combinations of $(\tau, \gamma_1, \gamma_2)$, determine the total profit and check the concavity clause (Theorem 3). Select the maximum value as the optimal value of the profit. Then optimal values of the decision variables are τ^*, γ_1^*, and γ_2^*.

4.4. Deteriorating Substitutable Products

This case discusses deteriorating substitutable products. The demand of two deteriorating substitutable products, product 1 and product 2, is

$$\delta_1(\gamma_1, \gamma_2) = \alpha - \beta\gamma_1 + \beta\omega\gamma_2 \tag{41}$$

$$\delta_2(\gamma_1, \gamma_2) = \alpha - \beta\gamma_2 + \beta\omega\gamma_1 \tag{42}$$

where all parameters are non-negative. The average profit function is calculated as

$$\pi_S(\tau, \gamma_1, \gamma_2) = \begin{aligned}&(\gamma_1 - \eta_1)\delta_1(\gamma_1, \gamma_2) + (\gamma_2 - \eta_2)\delta_2(\gamma_1, \gamma_2)\\ &-[\tfrac{G_1+G_2}{\tau} + \tfrac{\tau(\lambda_1+\eta_1'D_r)\delta_1(\gamma_1,\gamma_2)}{2} + \tfrac{\tau(\lambda_2+\eta_2'D_r)\delta_2(\gamma_1,\gamma_2)}{2}]\end{aligned} \tag{43}$$

With the help of the following theorem, this profit can be optimized.

Theorem 4. $\pi_S(T, p_1, p_2)$ *is strictly concave if*

$$G_1 + G_2 > \tfrac{\tau\beta}{2}[4\omega\gamma_1\gamma_2 - 2(\gamma_1^2 + \gamma_2^2) + \tau\gamma_1(\lambda_1 - \omega\lambda_2) + \tau\gamma_2(\lambda_2 - \omega\lambda_1) + \tau\eta_1'D_r(\gamma_1 - \omega\gamma_2) + \tau\eta_2'D_r(\gamma_2 - \omega\gamma_1)]$$

holds.

Proof. See Appendix D.

From the necessary condition, Equation (43) gives the following expressions

$$\frac{\partial \pi_s(\tau, \gamma_1, \gamma_2)}{\partial \tau} = \frac{G_1 + G_2}{\tau^2} + \frac{\left(\lambda_1 + \eta_1'D_r\right)(\beta\gamma_1 - \alpha - \beta\omega\gamma_2) + \left(\lambda_2 + \eta_2'D_r\right)(\beta\gamma_2 - \alpha - \beta\omega\gamma_1)}{2} \tag{44}$$

$$\frac{\partial \pi_s(\tau, \gamma_1, \gamma_2)}{\partial \gamma_1} = \frac{\tau\beta((\lambda_1 + \eta_1'D_r) - \omega(\lambda_2 + \eta_2'D_r))}{2} + \alpha - 2\beta(\gamma_1 - \omega\gamma_2) + \beta(\eta_1 - \omega\eta_2) \tag{45}$$

$$\frac{\partial \pi_s(\tau, \gamma_1, \gamma_2)}{\partial \gamma_2} = \frac{\tau\beta((\lambda_2 + \eta_2'D_r) - \omega(\lambda_1 + \eta_1'D_r))}{2} + \alpha - 2\beta(\gamma_2 - \omega\gamma_1) + \beta(\eta_2 - \omega\eta_1) \tag{46}$$

Setting Equations (45) and (46) equal to zero, p_1 and p_2 are given as follows:

$$\gamma_1^* = \frac{2\alpha + \beta(1 - \omega)(2\eta_1 + \tau\lambda_1 + \tau\eta_1'D_r)}{4\beta(1 - \omega)}, \text{ where } 0 \le d < 1 \tag{47}$$

$$\gamma_2^* = \frac{2\alpha + \beta(1 - \omega)(2\eta_2 + \tau\lambda_2 + \tau\eta_2'D_r)}{4\beta(1 - \omega)}, \text{ where } 0 \le d < 1 \tag{48}$$

Using Equations (47) and (48) on the right-hand side of the Equation (44) and equating them to zero, one can obtain

$$\alpha_1 \tau^3 + \alpha_2 \tau^2 + \alpha_4 = 0 \tag{49}$$

where

$$\alpha_1 = \beta\left[(\gamma_1^2 + \gamma_2^2) + D_r^2(\eta_1'^2 + \eta_2'^2) + 2\eta_1' D_r(\gamma_1 - \omega\lambda_2) + 2\eta_2' D_r(\lambda_2 - \omega\lambda_1) - 2\omega(\lambda_1\lambda_2 + \eta_1'\eta_2' D_r^2)\right] \tag{50}$$

$$\alpha_2 = -2\left[(\gamma_1 + \eta_1' D_r)(\alpha - \beta(\eta_1 - \omega\eta_2)) + (\gamma_2 + \eta_2' D_r)(\alpha - \beta(\eta_2 - \omega\eta_1))\right] \tag{51}$$

$$\alpha_4 = 8(G_1 + G_2) \tag{52}$$

According to the discussion about discriminant of cubic polynomial provided above (Section 4.1), the following solution procedure can help to find the numerical solution for substitutable products.

Solution algorithm

Step 1. Calculate the coefficients of polynomial from the Equation (49) by using Equations (50) to (52).

Step 2. Equation (49) gives the roots of the equation using MATLAB software. Move to the Step 3.

Step 3. Find (γ_1, γ_2) using Step 2.

Step 4. For $(\tau, \gamma_1, \gamma_2)$, obtain the profit and check the concavity clause (Theorem 4). Choose the maximum value as the optimal value of the profit. Then related decision variables for the maximum profit are given by τ^*, γ_1^*, and γ_2^*.

Step 5. Determine optimal values of order quantities using $\mu_1^* = \delta_1(\gamma_1^*, \gamma_2^*)\tau^*$ and $\mu_2^* = \delta_2(\gamma_1^*, \gamma_2^*)\tau^*$.

5. Numerical Examples and Sensitivity Analysis

Four numerical examples are given to justify the mathematical model. All examples explain effects of changes of the degree of complementarity or substitutability.

5.1. Example 1: Non-Deteriorating Complementary Products

The values of all parameters are $G_1 = 120$, $G_2 = 100$, $\lambda_1 = 6$,, $\lambda_2 = 3$, $\alpha = 100$, $\beta = 0.4$, $\eta_1 = 20$, and $\eta_2 = 10$. By using Steps (1) to (5) of the proposed solution algorithm in Section 4.1, optimal values of τ, γ_1, γ_2, μ_1, and μ_2 are obtained. These values with some variation of φ are depicted in the Table 2. '*' demonstrates optimum results. Table 2 reveals the increasing value of φ gives decreasing values of $\gamma_1^*, \gamma_2^*, \mu_1^*, \mu_2^*$, and $\pi_C(\tau^*, \gamma_1^*, \gamma_2^*)$, but gives an increasing value of τ^*. More precisely, these changes on γ_1^* and γ_2^* are shown in Figure 3. Bold font indicates the optimum results.

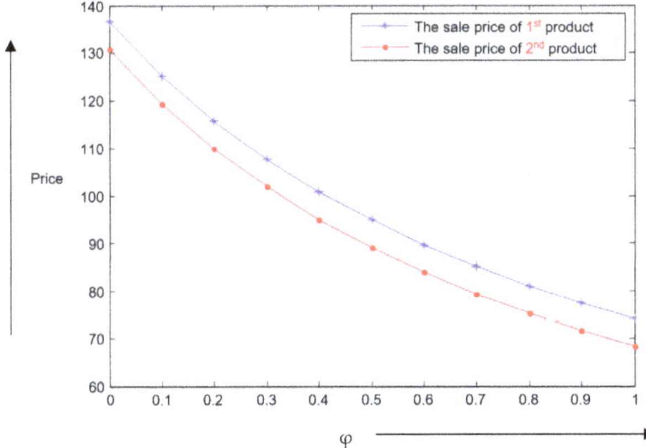

Figure 3. Changes of the complementarity degree upon selling prices.

Table 2. Optimum results for some variations of φ.

φ	Concavity	τ	γ_1	γ_2	μ_1	μ_2	Π_C
	True	93.3221	274.9832	199.9916	−932.5924	1866.8	1247.6
0	True	1.0290 *	136.5438 *	130.7719 *	46.7087 *	49.0849 *	10621 *
	False	−1.0180	-	-	-	-	-
	True	85.9137	252.5069	183.0716	−715.2833	1432.3	944.4092
0.1	True	1.0327 *	125.1854 *	119.4109 *	46.6255 *	48.7723 *	8548.6 *
	False	−1.0229	-	-	-	-	-
	True	79.5269	233.4570	168.8118	−547.7619	1097.4	715.6247
0.2	True	1.0362 *	115.7210 *	109.9438 *	46.4588 *	48.1415 *	7752.7 *
	False	−1.0229	-	-	-	-	-
	True	73.9641	217.1000	156.6269	-416.8036	835.5901	539.8298
0.3	True	1.0398 *	107.7135 *	101.9337 *	46.4588 *	48.1415 *	7752.7 *
	False	−1.0253	-	-	-	-	-
	True	69.0754	202.8988	146.0923	−313.2069	628.5371	402.6596
0.4	True	1.0433 *	100.8607 *	95.0682 *	46.3753 *	47.5031 *	6481.3 *
	False	−1.0278	-	-	-	-	-
	True	64.7452	190.4512	136.8923	−230.4240	463.1133	294.2216
0.5	True	1.0470 *	94.9038 *	89.1186 *	46.2917 *	47.5031 *	6481.3 *
	False	−1.0303	-	-	-	-	-
	True	60.8831	179.4496	128.7873	−163.7025	329.8139	207.5357
0.6	True	1.0606 *	89.7010 *	83.9130 *	46.2079 *	47.1809 *	5965.9 *
	False	−1.0328	-	-	-	-	-
	True	57.4169	169.6547	121.5921	−109.5329	221.6201	137.5717
0.7	True	1.0644 *	85.1110 *	79.3202 *	46.0401 *	46.5305 *	5108.5 *
	False	−1.0353	-	-	-	-	-
	True	54.2887	160.8775	115.1610	−65.2832	133.2681	80.6367
0.8	True	1.0581 *	81.0316 *	75.2380 *	46.0401 *	46.5305 *	5108.5 *
	False	−1.0379	-	-	-	-	-
	True	51.4514	152.9666	109.3780	−28.9524	60.7553	33.9747
0.9	True	1.0619 *	77.3824 *	71.5859 *	45.9560 *	46.2022 *	4748.4 *
	False	−1.0404	-	-	-	-	-
	True	48.8661	145.7992	104.1496	1.0005	1.0005	−4.5010
1.0	True	1.0658 *	74.0987 *	68.2993 *	45.8717 *	45.8717 *	4424.9 *
	False	−1.0430	-	-	-	-	-

"-" means infeasible result; "*" means optimum solution.

5.2. Example 2: Non-Deteriorating Substitutable Products

Values of parameters are $G_1 = 150$, $G_2 = 155$, $\lambda_1 = 4.5$, $\lambda_2 = 4$, $\alpha = 100$, $\beta = 0.3$, $\eta_1 = 15$, and $\eta_2 = 13$. Steps (1) to (5) of the solution algorithm proposed in the Section 4.2 help to obtain optimum values of τ, γ_1, γ_2, μ_1, and μ_2 for different values of d. Table 3 reveals that when ω increases, τ^*, γ_1^*, and γ_2^* decrease, but μ_1^*, μ_2^*, and $\pi_S(\tau^*, \gamma_1^*, \gamma_2^*)$ increase. The changes on γ_1^* and γ_2^* are shown in Figure 4. Figure 5 reveals that the profit decreases whenever the degree of complementarity increases and vice-versa. Therefore, the retailer who intends to work on complementary products should select two products with a smaller degree of complementary to make more profit. But if he/she intends to work on substitutable products, he/she should select two products with a larger degree of substitutability.

Table 3. Optimum results for some variations of ω.

ω	Concavity	τ	γ_1	γ_2	μ_1	μ_2	Π_C
	True	149.7187	342.6002	322.8854	−416.2286	469.2757	56.4733
0	True	1.2292 *	175.5495	174.3959 *	58.1837 *	58.6091 *	14,799 *
	False	−1.2170	-	-	-	-	-
	True		380.5094	358.6400	−566.5791	638.3149	77.4294
0.1	True		194.0644 *	192.9112 *	58.3172 *	58.7838 *	16,647 *
	False	−1.2170	-	-	-	-	-
	True	188.4626	427.8537	403.2959	−783.6958	882.4670	107.5095
0.2	True	1.2228 *	217.2090 *	216.0561 *	58.4503 *	58.9578 *	18,957 *
	False	−1.2149	-	-	-	-	-
	True	216.0547	488.6568	460.6500	−1110.2	1249.7	152.5353
0.3	True		246.9673 *	245.8149 *	58.5830 *	59.1312 *	21,929 *
	False	−1.2107	-	-	-	-	-
	True	252.7386	569.6088	537.0164	−1627.8	1831.9	223.6789
0.4	True	1.2165 *	286.6464 *	285.4943 *	58.7153 *	59.3039 *	25,894 *
	False	−1.2107	-	-	-	-	-
	True	303.8947	682.7149	643.7281	−2508.7	2822.8	344.5513
0.5	True	1.2134 *	342.1984 *	341.0467 *	58.8472 *	59.4761 *	31,445 *
	False	−1.2086	-	-	-	-	-
	True	380.1918	851.8825	803.3585	−4167.0	4688.3	572.1289
0.6	True	1.2103 *	425.5283 *	424.3770 *	58.9787 *	59.6475 *	39,774 *
	False	−1.2065	-	-	-	-	-
	True	506.2033	1132.5	1068.3	−7808.6	8784.9	1073.6
0.7	True	1.2073 *	564.4137 *	563.2628 *	59.1098 *	59.8184 *	53,659 *
	False	−1.2044	-	-	-	-	-
	True	753.9936	1689.1	1593.8	−18,250	20,531	2525.4
0.8	True	1.2043 *	842.1881 *	841.0376 *	59.2405 *	59.9887 *	81,433 *
	False	−1.2013	-	-	-	-	-
	True	1465.2	3322.5	3138.4	−72,373	81,420	10,300
0.9	True	1.2013 *	1675.5 *	1674.4 *	59.3708 *	60.1583 *	164,760 *
	False	-	-	-	-	-	-
	Undefined	22659	∞	∞	∞	∞	Undefined
1.0	Undefined	1.1983	$-\infty$	∞	∞	$-\infty$	Undefined
	Undefined	−1.1982	$-\infty$	∞	$-\infty$	∞	Undefined

"-" means infeasible result; "*" means optimum solution.

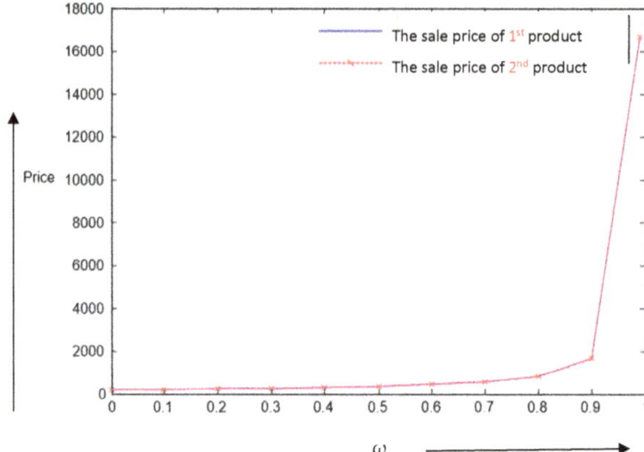

Figure 4. Changes in the degree of substitutability upon selling prices.

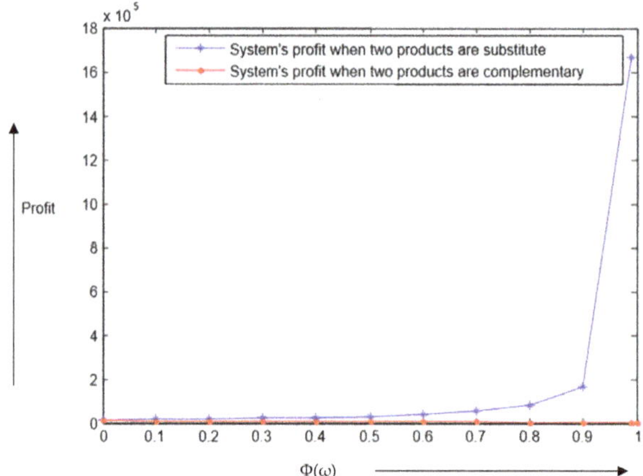

Figure 5. Effects of the changes of complementarity and substitutability degrees on profits.

5.3. Example 3: Deteriorating Complementary Products

The values of parameters for this example are $G_1 = 120$, $G_2 = 100$, $\lambda_1 = 6$, $\lambda_2 = 3$, $\alpha = 100$, $\beta = 0.4$, $\eta_1 = 20$, $\eta_2 = 10$, $\eta_1' = 10$, $\eta_2' = 5$ and $D_r = 0.01$. Using Steps (1) to (5) of the proposed solution algorithm in Section 4.3, the optimal values of τ, $\gamma_1, \gamma_2, \mu_1$, and μ_2 are obtained. Table 4 shows that $\gamma_1^*, \gamma_2^*, \mu_1^*, \mu_2^*$, and $\pi_C(\tau^*, \gamma_1^*, \gamma_2^*)$ decrease when φ increases, but at that time, τ^* increases (similar to Section 5.1). The changes on γ_1^* and γ_2^* are shown in Figure 6 precisely.

Table 4. Optimum results for some variations of φ (deterioration case).

φ	Concavity	τ	γ_1	γ_2	μ_1	μ_2	Π_C
	True	91.7921	274.9829	199.9914	−1503.0	3008.7	1247.6
0	True	1.0208 *	136.5567 *	130.7784 *	46.5584 *	48.9299 *	10,618 *
	False	−1.0096	-	-	-	-	-
	True	84.5050	252.5065	183.0715	−1105.7	2214.1	944.3665
0.1	True	1.0242 *	125.1983 *	119.4174 *	46.4762 *	48.6187 *	9486.7 *
	False	−1.0120	-	-	-	-	-

Table 4. *Cont.*

φ	Concavity	τ	γ_1	γ_2	μ_1	μ_2	Π_C
	True	78.2230	233.4567	168.8117	−817.1057	1637.0	715.5786
0.2	True	**1.0277 ***	**115.7340 ***	**109.9503 ***	**46.3938 ***	**48.3057 ***	**8545.1 ***
	False	−1.0144	-	-	-	-	-
	True	72.7513	217.0996	156.6267	−602.9090	1208.7	539.7803
0.3	True	**1.0313 ***	**107.7265 ***	**101.9402 ***	**46.3114 ***	**47.9908 ***	**7749.2 ***
	False	−1.0168	-	-	-	-	-
	True	67.9427	202.8984	146.0921	−441.0527	885.1454	402.6065
0.4	True	**1.0348 ***	**100.8638 ***	**95.0748 ***	**46.2288 ***	**47.6741 ***	**7067.7 ***
	False	−1.0193	-	-	-	-	-
	True	63.6836	190.4508	136.8921	−316.9006	636.9689	294.1649
0.5	True	**1.0384 ***	**94.9169 ***	**89.1251 ***	**46.1462 ***	**47.3553 ***	**6477.9 ***
	False	−1.0218	-	-	-	-	-
	True	59.8847	179.4492	128.7871	−220.4678	444.2343	207.4755
0.6	True	**1.0421 ***	**89.7142 ***	**83.9196 ***	**46.0634 ***	**47.0346 ***	**5962.4 ***
	False	−1.0243	-	-	-	-	-
	True	56.4753	169.6543	121.5918	−144.7758	292.9846	137.5079
0.7	True	**1.0458 ***	**85.1242 ***	**79.3268 ***	**45.9805 ***	**46.7119 ***	**5508.2 ***
	False	−1.0268	-	-	-	-	-
	True	53.3984	160.8770	115.1607	−84.8435	173.2576	80.5692
0.8	True	**1.0495 ***	**81.0449 ***	**75.2447 ***	**45.8975 ***	**46.3871 ***	**5105.1 ***
	False	−1.0293	-	-	-	-	-
	True	50.6076	152.9660	109.3778	−37.0492	77.8092	33.9034
0.9	True	**1.0533 ***	**77.3957 ***	**71.5926 ***	**45.8144 ***	**46.0602 ***	**4745.0 ***
	False	−1.0318	-	-	-	-	-
	True	48.0647	145.7986	104.1493	1.2845	1.2845	−4.5761
1.0	True	**1.0571 ***	**74.1121 ***	**68.3060 ***	**45.7312 ***	**45.7312 ***	**4421.4 ***
	False	−1.0344	-	-	-	-	-

"-" means infeasible result; "*" means optimum solution.

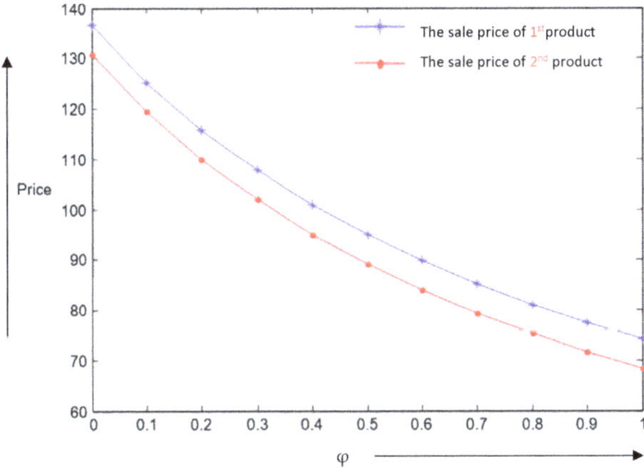

Figure 6. Effects of changes of complementarity degree on selling prices of deteriorating products.

5.4. Example 4: Deteriorating Substitutable Products

Fixed parametric values are $G_1 = 150$, $G_2 = 155$, $\lambda_1 = 4.5$, $\lambda_2 = 4$, $\alpha = 100$, $\beta = 0.3$, $\eta_1 = 15$, $\eta_2 = 13$, $\eta_1' = 7$, $\eta_2' = 6$, and $D_r = 0.01$. For different values of ω, one can use Steps (1) to (5) of proposed solution algorithm in Section 4.4. From the Table 5, it can be concluded that τ^*, γ_1^*, and γ_2^* decrease with increased values of ω, but μ_1^*, μ_2^*, and $\pi_S(\tau^*, \gamma_1^*, \gamma_2^*)$ increase (similar Section 5.2). The changes on γ_1^* and γ_2^* are presented in the Figure 7. Figure 8 describes that the profit decreases when the complementarity degree increases, but profit increases while the degree of substitutability increases. As a result, the retailer who intends to work on complementary products should select two products with a smaller degree of complementary in order to make more profit. If he/she intends to work on substitutable products, he/she should select two products with a larger degree of substitutability.

Table 5. Optimum results for some variations of ω (deterioration case).

ω	Concavity	τ	γ_1	γ_2	μ_1	μ_2	Π_S
	True	147.4584	342.6379	322.8369	−940.472	1060.9	56.9563
0	True	1.2199 *	175.5604 *	174.4049 *	58.0953 *	58.5208 *	14,795 *
	False	−1.2099	-	-	-	-	-
	True	164.4331	380.5500	358.5848	−1423.5	1604.6	78.1010
0.1	True	1.2167 *	194.0753 *	192.9202 *	58.2279 *	58.6945 *	16,643 *
	False	−1.2078	-	-	-	-	-
	True	185.6144	427.8978	403.2320	−2254.3	2539.8	108.4521
0.2	True	1.2136 *	217.2198 *	216.0651 *	58.3600 *	58.8676 *	18,954 *
	False	−1.2057	-	-	-	-	-
	True	212.7872	488.7046	460.5742	−3816.6	4298.5	153.8840
0.3	True	1.2104 *	246.9782 *	245.8238 *	56.4918 *	59.0400 *	21,926 *
	False	−1.2036	-	-	-	-	-
	True	248.9126	569.6604	536.9240	−7146.8	8047.2	225.6693
0.4	True	1.2073 *	286.6572 *	285.5032 *	58.6231 *	59.2118 *	25,890 *
	False	−1.2015	-	-	-	-	-
	True	299.2879	682.7698	643.6106	−15,703	17,678	347.6315
0.5	True	1.2043 *	342.2092 *	341.0556 *	58.7540 *	59.3829 *	31441 *
	False	−1.1994	-	-	-	-	-
	True	374.4167	851.9377	803.1996	−45,423	51134	577.2581
0.6	True	1.2012 *	425.5390 *	424.3859 *	58.8845 *	59.5534 *	39,770 *
	False	−1.1974	-	-	-	-	-
0.7	True	498.4871	1132.6	1068.0	−224,880	253,130	1083.2

"-" means infeasible result; "*" means optimum solution.

Managerial insights

Numerical studies justify the theoretical implications of this research in each case separately. The degree of complementarity of a product has an inverse impact on the selling price, i.e., the increasing complementarity gives less profit. This implies that if one product is more dependent or complement to the other product, the selling of the one product is dependent upon the other product. In the case that one of the complementary products is not available at the right time, the chances to sale of the other product are smaller. To handle this situation, industry managers need to take more care of the inventory and safety stock of the complementary products.

For the case of substitutable products, the increasing rate of the degree of substitutability has a direct impact on the profit. This implies that if customers can get any similar substitutable product, the customer is satisfied with the service. The absence of the original product does not have any impact on the profit. This study suggests to industry managers that they should introduce similar types of alternative products to the shops or their centers such that they can survive facing losses from original products. It helps to keep the brand image of the industry as well as the retailer. The scenario is the same for the case of deteriorated substitutable and complementary products. Therefore, the industry needs to create their strategies carefully for complementary products rather than substitutable products.

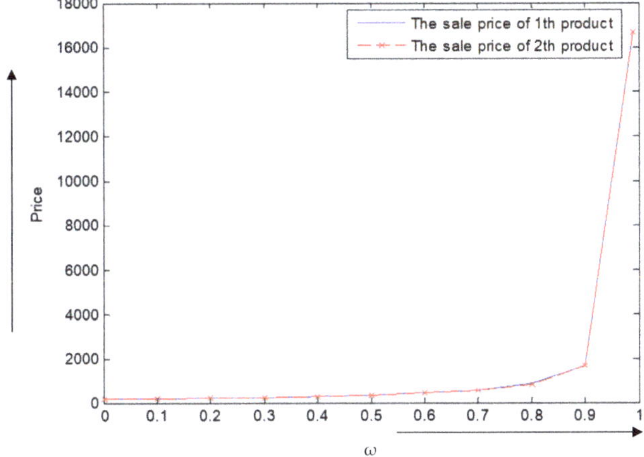

Figure 7. Changes in the degree of substitutability upon selling prices of deteriorating products.

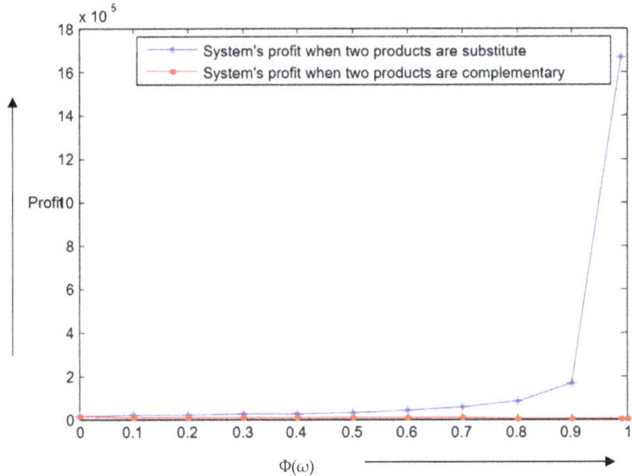

Figure 8. Changes of the complementarily and substitutability degrees upon profits of deteriorating products.

6. Conclusions

Four pricing models for two types of products which may be complementary or substitutable and non-deteriorating or deteriorating were studied analytically as well as numerically. For all cases, it was observed that selling prices and lot sizes of both products as well as the average profit were decreased, and the period length was increased while the degree of complementarity was increased and vice versa. The selling prices and order quantities of both products as well as average profit were increased, and the cycle length was decreased with increased values of the degree of substitutability. This study concluded that sell of complementary products should be with some smaller degree of complementarity and substitutable products with bigger degree of substitutability in order to make more profit. The present model can help a manager to fix optimal selling prices, ordered quantities, and cycle length for different scenarios so that the joint profit is maximized. In the existing literature, some research articles are focused on substitutable and complementary products but none of them have investigated both inventory and pricing decisions of products simultaneously. Therefore, the present article is quite new compared to existing literature. Lead time is considered as zero in this study, but in reality it is not possible. This study can be extended with random lead time and backorder. The business can be spread through multiple number of retailers. Extra service facilities can be included within the study [45–48] for deteriorating products and non-zero random lead time. The quality improvement [49,50] and preservation technology [28,29] for deteriorating products can improve the study in a realistic way.

Author Contributions: Methodology, data curation, writing—original draft preparation, M.S.B.; formal analysis, conceptualization, visualization S.S.S.; investigation, S.S.S. and B.S.; resources, validation, supervision, project administration, A.A.T.; writing—review and editing, funding acquisition, B.S.

Funding: This research was supported by the Basic Science Research Program through the National Research Foundation of Korea (NRF) funded by the Ministry of Education, Science and Technology (Project Number: 2017R1D1A1B03033846).

Conflicts of Interest: The authors declare no conflict of interest.

Nomenclature

Index	Description
m	1, 2
Decision variables	
γ_m	product m's selling price ($/unit)
μ_m	order quantity of product m (units)
τ	time interval between two replenishments
Parameters	
$\delta_1(\gamma_1, \gamma_2)$	market demand for product 1 (units)
$\delta_2(\gamma_1, \gamma_2)$	demand for product 2 (units)
ρ_m	inventory level at time τ of product m
$\propto$	basic potential demand (units)
β	self-price sensitivity
λ_m	holding cost of product m ($/unit/unit time)
η_m	purchasing cost of product m ($/unit)
η'_m	deterioration cost of product m ($/unit)
D_r	deterioration rate of product 1 and product 2
φ	degree of complementarity of two products, $0 \le \varphi \le 1$
ω	degree of substitutability of two products, $0 \le \omega \le 1$

Appendix A

Proof of Theorem 1. $\pi_C(\tau, \gamma_1, \gamma_2)$ is strictly concave, if $X.H.X^T = [\ \tau \quad \gamma_1 \quad \gamma_2\ ] \times H \times [\ \tau \quad \gamma_1 \quad \gamma_2\ ]^T < 0$ where

$$H = \begin{bmatrix} \frac{\partial^2 \pi_C}{\partial \tau^2} & \frac{\partial^2 \pi_C}{\partial \tau \partial \gamma_1} & \frac{\partial^2 \pi_C}{\partial \tau \partial \gamma_1} \\ \frac{\partial^2 \pi_C}{\partial \gamma_1 \partial \tau} & \frac{\partial^2 \pi_C}{\partial \gamma_1^2} & \frac{\partial^2 \pi_C}{\partial \gamma_1 \partial \gamma_2} \\ \frac{\partial^2 \pi_C}{\partial \gamma_2 \partial \tau} & \frac{\partial^2 \pi_C}{\partial \gamma_2 \partial \gamma_1} & \frac{\partial^2 \pi_C}{\partial \gamma_2^2} \end{bmatrix} = \begin{bmatrix} -\frac{2(G_1+G_2)}{\tau^3} & \frac{\beta(\lambda_1+\lambda_2\varphi)}{2} & \frac{\beta(\lambda_1\varphi+\lambda_2)}{2} \\ \frac{\beta(\lambda_1+\lambda_2\varphi)}{2} & -2\beta & -2\beta\varphi \\ \frac{\beta(\lambda_1\varphi+\lambda_2)}{2} & -2\beta\varphi & -2\beta \end{bmatrix} \tag{A1}$$

Now,

$$X \times H \times X^T = -\frac{2(G_1 + G_2)}{\tau} + \tau\beta\lambda_1(\gamma_1 + \gamma_2\varphi) + \tau\beta\lambda_2(\gamma_2 + \gamma_1\varphi) - 4\beta\varphi\gamma_1\gamma_2 - 2\beta(\gamma_1^2 + \gamma_2^2) \tag{A2}$$

It has to show that the expression in Equation (A2) is negative. Therefore, the profit function is strictly concave if

$$G_1 + G_2 > \frac{\tau\beta}{2}\left[\tau(\lambda_1(\gamma_1 + \gamma_2\varphi) + \lambda_2(\gamma_2 + \gamma_1\varphi)) - 4\varphi\gamma_1\gamma_2 - 2(\gamma_1^2 + \gamma_2^2)\right]$$

holds; hence the proof.

Appendix B

Proof of Theorem 2. $\pi_S(\tau, \gamma_1, \gamma_2)$ is a concave whenever $X.H.X^T = \begin{bmatrix} \tau & \gamma_1 & \gamma_2 \end{bmatrix} \times H \times \begin{bmatrix} \tau & \gamma_1 & \gamma_2 \end{bmatrix}^T < 0$ where

$$H = \begin{bmatrix} \frac{\partial^2 \pi_S}{\partial \tau^2} & \frac{\partial^2 \pi_S}{\partial \tau \partial \gamma_1} & \frac{\partial^2 \pi_S}{\partial \tau \partial \gamma_1} \\ \frac{\partial^2 \pi_S}{\partial \gamma_1 \partial \tau} & \frac{\partial^2 \pi_S}{\partial \gamma_1^2} & \frac{\partial^2 \pi_S}{\partial \gamma_1 \partial \gamma_2} \\ \frac{\partial^2 \pi_S}{\partial \gamma_2 \partial \tau} & \frac{\partial^2 \pi_S}{\partial \gamma_2 \partial \gamma_1} & \frac{\partial^2 \pi_S}{\partial \gamma_2^2} \end{bmatrix} = \begin{bmatrix} -\frac{2(G_1+G_2)}{\tau^3} & \frac{\beta(\lambda_1-\lambda_2\omega)}{2} & \frac{\beta(-\lambda_1\omega+\lambda_2)}{2} \\ \frac{\beta(\lambda_1-\lambda_2\omega)}{2} & -2\beta & 2\beta\omega \\ \frac{\beta(-\lambda_1\omega+\lambda_2)}{2} & 2\beta\omega & -2\beta \end{bmatrix} \tag{A3}$$

Thus, one can have

$$X \times H \times X^T = -\frac{2(G_1 + G_2)}{\tau} + \tau\beta\lambda_1(\gamma_1 - \gamma_2\omega) + \tau\beta\lambda_2(\gamma_2 - \gamma_1\omega) + 4\beta\omega\gamma_1\gamma_2 - 2\beta(\gamma_1^2 + \gamma_2^2) \tag{A4}$$

Then, it has to show that $X \times H \times X^T < 0$.
Hence, the profit function $\pi_S(\tau, \gamma_1, \gamma_2)$ is concave if

$$G_1 + G_2 > \frac{\tau\beta}{2}\left[\tau(\lambda_1(\gamma_1 - \gamma_2\omega) + \lambda_2(\gamma_2 - \gamma_1\omega)) + 4\omega\gamma_1\gamma_2 - 2(\gamma_1^2 + \gamma_2^2)\right]$$

holds. The proof is completed here.

Appendix C

Proof of Theorem 3. $\pi_C(\tau, \gamma_1, \gamma_2)$ is strictly concave if $X.H.X^T = \begin{bmatrix} \tau & \gamma_1 & \gamma_2 \end{bmatrix} \times H \times \begin{bmatrix} \tau & \gamma_1 & \gamma_2 \end{bmatrix}^T < 0$ where

$$H = \begin{bmatrix} \frac{\partial^2 \pi_C}{\partial \tau^2} & \frac{\partial^2 \pi_C}{\partial \tau \partial \gamma_1} & \frac{\partial^2 \pi_C}{\partial \tau \partial \gamma_1} \\ \frac{\partial^2 \pi_C}{\partial \gamma_1 \partial \tau} & \frac{\partial^2 \pi_C}{\partial \gamma_1^2} & \frac{\partial^2 \pi_C}{\partial \gamma_1 \partial \gamma_2} \\ \frac{\partial^2 \pi_C}{\partial \gamma_2 \partial \tau} & \frac{\partial^2 \pi_C}{\partial \gamma_2 \partial \gamma_1} & \frac{\partial^2 \pi_C}{\partial \gamma_2^2} \end{bmatrix}$$

$$= \begin{bmatrix} -\frac{2(G_1+G_2)}{\tau^3} & \frac{\beta(\gamma_1+\eta_1'D_r)}{2} + \frac{\beta\varphi(\gamma_2+\eta_2'D_r)}{2} & \frac{\beta(\gamma_2+\eta_2'D_r)}{2} + \frac{\beta\varphi(\gamma_1+\eta_1'D_r)}{2} \\ \frac{\beta(\gamma_1+\eta_1'D_r)}{2} + \frac{\beta\varphi(\gamma_2+\eta_2'D_r)}{2} & -2\beta & -2\beta\varphi \\ \frac{\beta(\gamma_2+\eta_2'D_r)}{2} + \frac{\beta\varphi(\gamma_1+\eta_1'D_r)}{2} & -2\beta\varphi & -2\beta \end{bmatrix} \tag{A5}$$

Now,

$$\begin{aligned} X \times H \times X^T = \; & -\frac{2(G_1+G_2)}{\tau} - 4\beta\varphi\gamma_1\gamma_2 - 2\beta(\gamma_1^2 + \gamma_2^2) + \tau\beta\gamma_1(h_1 + \varphi h_2) + \tau\beta\gamma_2(h_2 + \varphi h_1) \\ & + \tau\beta\eta_1'D_r(\gamma_1 + \varphi\gamma_2) + \tau\beta\eta_2'D_r(\gamma_2 + \varphi\gamma_1) \end{aligned} \tag{A6}$$

Now, it is needed to show that $X \times H \times X^T < 0$.
Therefore, the profit function $\pi_C(\tau, \gamma_1, \gamma_2)$ is concave if

$$G_1 + G_2 > \frac{\tau\beta}{2}[-4\varphi\gamma_1\gamma_2 - 2(\gamma_1^2 + \gamma_2^2) + \tau\gamma_1(\lambda_1 + \varphi\lambda_2) + \tau\gamma_2(\lambda_2 + \varphi\lambda_1) + \tau\eta_1'D_r(\gamma_1 + \varphi\gamma_2) + \tau\eta_2'D_r(\gamma_2 + \varphi\gamma_1)]$$

holds. The proof is completed here.

Appendix D

Proof of Theorem 4. $\pi_S(\tau, \gamma_1, \gamma_2)$ is a strictly concave function if $X.H.X^T = \begin{bmatrix} \tau & \gamma_1 & \gamma_2 \end{bmatrix} \times H \times \begin{bmatrix} \tau & \gamma_1 & \gamma_2 \end{bmatrix}^T < 0$ holds. where

$$
H = \begin{bmatrix}
\frac{\partial^2 \pi_C}{\partial \tau^2} & \frac{\partial^2 \pi_C}{\partial \tau \partial \gamma_1} & \frac{\partial^2 \pi_C}{\partial \tau \partial \gamma_1} \\
\frac{\partial^2 \pi_C}{\partial \gamma_1 \partial \tau} & \frac{\partial^2 \pi_C}{\partial \gamma_1^2} & \frac{\partial^2 \pi_C}{\partial \gamma_1 \partial \gamma_2} \\
\frac{\partial^2 \pi_C}{\partial \gamma_2 \partial \tau} & \frac{\partial^2 \pi_C}{\partial \gamma_2 \partial \gamma_1} & \frac{\partial^2 \pi_C}{\partial \gamma_2^2}
\end{bmatrix}
$$
$$
= \begin{bmatrix}
-\frac{2(G_1+G_2)}{\tau^3} & \frac{\beta(\lambda_1+\eta_1' D_r)}{2} - \frac{\beta\omega(\lambda_2+\eta_2' D_r)}{2} & \frac{\beta(\lambda_2+\eta_2' D_r)}{2} - \frac{\beta\omega(\lambda_1+\eta_1' D_r)}{2} \\
\frac{\beta(\lambda_1+\eta_1' D_r)}{2} - \frac{\beta\omega(\lambda_2+\eta_2' D_r)}{2} & -2\beta & 2\beta\omega \\
\frac{\beta(\lambda_2+\eta_2' D_r)}{2} - \frac{\beta\omega(\lambda_1+\eta_1' D_r)}{2} & 2\beta\omega & -2\beta
\end{bmatrix} \tag{A7}
$$

Thus,

$$
X \times H \times X^T = -\frac{2(G_1+G_2)}{\tau} + 4\beta\omega\gamma_1\gamma_2 - 2\beta(\gamma_1^2 + \gamma_2^2) + \tau\beta\gamma_1(\lambda_1 - \omega\lambda_2) + \tau\beta\gamma_2(\lambda_2 - \omega\lambda_1) \\
+ \tau\beta\eta_1' D_r(\gamma_1 - \omega\gamma_2) + \tau\beta\eta_2' D_r(\gamma_2 - \omega\gamma_1) \tag{A8}
$$

Then, it has to show that $X \times H \times X^T < 0$.
Then, the profit function $\pi_S(\tau, \gamma_1, \gamma_2)$ is concave if

$$
G_1 + G_2 > \frac{\tau\beta}{2}[4\omega\gamma_1\gamma_2 - 2(\gamma_1^2 + \gamma_2^2) + \tau\gamma_1(\lambda_1 - \omega\lambda_2) + \tau\gamma_2(\lambda_2 - \omega\lambda_1) + \tau\eta_1' D_r(\gamma_1 - \omega\gamma_2) + \tau\eta_2' D_r(\gamma_2 - \omega\gamma_1)]
$$

holds. The proof is completed here.

References

1. Shocker, A.D.; Bayus, B.L.; Kim, N. Product complments and substitutes in the real world: The relevence of "other products". *J. Market.* **2004**, *68*, 28–40. [CrossRef]
2. Sarkar, M.; Lee, Y.H. Optimum pricing strategy for complementary products with stochastic reservation price in a supply chain model. *J. Ind. Manag. Optim.* **2017**, *13*, 1553–1586.
3. Avgeropoulos, S.; Sammut-Bonnici, T.; McGee, J. Complementary products. *Encycl. Manag.* **2015**, *12*. [CrossRef]
4. Sana, S.S. Price-sensitive demand for perishable items—An EOQ model. *Appl. Math. Comput.* **2011**, *217*, 6248–6259.
5. Sarkar, B.; Saren, S.; Wee, H.M. An inventory model with variable demand, component cost and selling price for deteriorating items. *Econ. Model.* **2013**, *30*, 306–310. [CrossRef]
6. Salvietti, L.; Smith, N.R.; Cárdenas-Barrón, L.E. A stochastic profit-maximising economic lot scheduling problem with price optimisation. *Eur. J. Oper. Res.* **2014**, *8*, 193–221. [CrossRef]
7. Smith, N.R.; Robles, J.L.; Cárdenas-Barrón, L.E. Optimal pricing and production master planning in a multiperiod horizon considering capacity and inventory constraints. *Math. Probl. Eng.* **2009**, *932676*, 15. [CrossRef]
8. Soon, W. A review of multi-product pricing models. *Appl. Math. Comput.* **2011**, *217*, 8149–8165. [CrossRef]
9. Wei, J.; Zhao, J.; Li, Y. Pricing decisions for complementary products with firms' different market powers. *Eur. J. Oper. Res.* **2013**, *224*, 507–519. [CrossRef]
10. Dey, B.; Sarkar, B.; Sarkar, M.; Pareek, S. An integrated inventory model involving discrete setup cost reduction, variable safety factor, selling-price dependent demand, and investment. *RAIRO-Oper. Res.* **2019**, *53*, 39–57. [CrossRef]
11. Mcgillivray, R.; Silver, E. Some concepts for inventory control under substitutable demand. *Inf. Syst. Oper. Res.* **1978**, *16*, 47–63. [CrossRef]
12. Zhang, R.Q.; Kaku, I.; Xiao, Y.Y. Deterministic EOQ with partial backordering and correlated demand caused by cross-selling. *Eur. J. Oper. Res.* **2011**, *210*, 537–551. [CrossRef]
13. Zhang, R.Q. An extension of partial backordering EOQ with correlated demand caused by cross-selling considering multiple minor items. *Eur. J. Oper. Res.* **2012**, *220*, 876–881. [CrossRef]

14. Yue, X.; Mukhopadhyay, S.K.; Zhu, X. A Bertrand model of pricing of complementary goods under information asymmetry. *J. Bus. Res.* **2006**, *59*, 1182–1192. [CrossRef]
15. Liu, L.; Yuan, X.M. Coordinated replenishments in inventory systems with correlated demands. *Eur. J. Oper. Res.* **2000**, *123*, 490–503. [CrossRef]
16. Guchhait, R.; Sarkar, M.; Sarkar, B.; Pareek, S. Single-vendor multi-buyer game theoretic model under multi-factor dependent demand. *Int. J. Inventory Res.* **2017**, *4*, 303–332. [CrossRef]
17. Shavandi, H.; Mahlooji, H.; Nosratian, N.E. A constrained multi-product pricing and inventory control problem. *Appl. Soft Comput.* **2012**, *12*, 2454–2461. [CrossRef]
18. Abad, P.L. Optimal pricing and lot-sizing under conditions of perishability, finite production and partial backordering and lost sale. *Eur. J. Oper. Res.* **2003**, *144*, 677–685. [CrossRef]
19. Balkhi, Z.T.; Benkherouf, L. On an inventory model for deteriorating items with stock dependent and time-varying demand rates. *Comput. Oper. Res.* **2004**, *31*, 223–240. [CrossRef]
20. Anjos, M.F.; Cheng, R.C.; Currie, C.S. Optimal pricing policies for perishable products. *Eur. J. Oper. Res.* **2005**, *166*, 246–254. [CrossRef]
21. Dye, C.Y. Joint pricing and ordering policy for a deteriorating inventory with partial backlogging. *Omega* **2007**, *35*, 184–189. [CrossRef]
22. Panda, S.; Saha, S.; Basu, M. An EOQ model for perishable products with discounted selling price and stock dependent demand. *Cent. Eur. J. Oper. Res.* **2009**, *17*, 31. [CrossRef]
23. Pang, Z. Optimal dynamic pricing and inventory control with stock deterioration and partial backordering. *Oper. Res. Lett.* **2011**, *39*, 375–379. [CrossRef]
24. Sarkar, B.; Majumder, A.; Sarkar, M.; Dey, B.K.; Roy, G. Two-echelon supply chain model with manufacturing quality improvement and setup cost reduction. *J. Ind. Manag. Optim.* **2017**, *13*, 1085–1104. [CrossRef]
25. Sarkar, B.; Saren, S. Partial trade-credit policy of retailer with exponentially deteriorating items. *Int. J. Appl. Comput. Math.* **2015**, *1*, 343–368. [CrossRef]
26. Sett, B.K.; Sarkar, B.; Sarkar, B.; Yun, W.Y. Optimal replenishment policy with variable deterioration for fixed lifetime products. *Sci. Iran.* **2016**, *23*, 2318–2329.
27. Sarkar, B.; Sett, B.K.; Roy, G.; Goswami, A. Flexible setup cost and deterioration of products in a supply chain model. *Int. J. Appl. Comput. Math.* **2016**, *2*, 25–40. [CrossRef]
28. Sarkar, B.; Mandal, B.; Sarkar, S. Preservation of deteriorating seasonal products with stock-dependent consumption rate and shortages. *J. Ind. Manag. Optim.* **2017**, *13*, 187–206. [CrossRef]
29. Ullah, M.; Sarkar, B.; Asghar, I. Effects of preservation technology investment on waste generation in a two-echelon supply chain model. *Mathematics* **2019**, *7*, 20. [CrossRef]
30. Iqbal, M.W.; Sarkar, B. Recycling of lifetime dependent deteriorated products through different supply chains. *RAIRO-Operat. Res.* **2019**, *53*, 129–156. [CrossRef]
31. Dey, B.K.; Sarkar, B.; Pareek, S. A two-echelon supply chain management with setup time and cost reduction, quality improvement and variable production rate. *Mathematics* **2019**, *7*, 25. [CrossRef]
32. Gürler, Ü.; Yılmaz, A. Inventory and coordination issues with two substitutable products. *Appl. Math. Model.* **2010**, *34*, 539–551. [CrossRef]
33. Kim, S.W.; Bell, P.C. Optimal pricing and production decisions in the presence of symmetrical and asymmetrical substitution. *Omega* **2011**, *39*, 528–538. [CrossRef]
34. Stavrulaki, E. Inventory decisions for substitutable products with stock-dependent demand. *Int. J. Prod. Econ.* **2011**, *129*, 65–78. [CrossRef]
35. Karakul, M.; Chan, L.M.A. Analytical and managerial implications of integrating product substitutability in the joint pricing and procurement problem. *Eur. J. Oper. Res.* **2008**, *190*, 179–204. [CrossRef]
36. Karakul, M.; Chan, L.M.A. Joint pricing and procurement of substitutable products with random demands–a technical note. *Eur. J. Oper. Res.* **2010**, *201*, 324–328. [CrossRef]
37. Birge, J.R.; Drogosz, J.; Duenyas, I. Setting single-period optimal capacity levels and prices for substitutable products. *Int. J. Flex. Manuf. Syst.* **1998**, *10*, 407–430. [CrossRef]
38. Netessine, S.; Dobson, G.; Shumsky, R.A. Flexible service capacity: Optimal investment and the impact of demand correlation. *Oper. Res.* **2002**, *50*, 375–388. [CrossRef]
39. Parlar, M.; Goyal, S. Optimal ordering decisions for two substitutable products with stochastic demands. *Opsearch* **1984**, *21*, 1–15.

40. Pasternack, B.A.; Drezner, Z. Optimal inventory policies for substitutable commodities with stochastic demand. *Nav. Res. Logist.* **1991**, *38*, 221–240. [CrossRef]
41. Tang, C.S.; Yin, R. Joint ordering and pricing strategies for managing substitutable products. *Prod. Oper. Manag.* **2007**, *16*, 138–153. [CrossRef]
42. Xia, Y. Competitive strategies and market segmentation for suppliers with substitutable products. *Eur. J. Oper. Res.* **2011**, *210*, 194–203. [CrossRef]
43. Netessine, S.; Rudi, N. Centralized and competitive inventory models with demand substitution. *Oper. Res.* **2003**, *51*, 329–335. [CrossRef]
44. Gupta, S.; Loulou, R.J.M.S. Process innovation, product differentiation, and channel structure: Strategic incentives in a duopoly. *Market. Sci.* **1998**, *17*, 301–316. [CrossRef]
45. Sarkar, M.; Guchhait, R.; Sarkar, B. Modelling for service solution of a closed-loop supply chain with the presence of third party logistics. In *Advances in Production Management Systems. Production Management for Data-Driven, Intelligent, Collaborative, and Sustainable Manufacturing*; APMS 2018; IFIP Advances in Information and Communication Technology; Springer: Cham, Switzerland, 2018; Volume 535, pp. 320–327.
46. Khanna, A.; Kishore, A.; Sarkar, B.; Jaggi, C. Supply chain with customer-based two-level credit policies under an imperfect quality environment. *Mathematics* **2018**, *6*, 35. [CrossRef]
47. Guchhait, R.; Pareek, S.; Sarkar, B. Application of distribution-free approach in integrated and dual-channel supply chain under buyback contract. In *Handbook of Research on Promoting Business Process Improvement Through Inventory Control Techniques*; IGI Global: Hershey, PA, USA, 2018; pp. 388–426.
48. Guchhait, R.; Dey, B.K.; Bhuniya, S.; Mandal, B.; Bachar, R.; Chauduri, K.S.; Wee, H.M.; Sarkar, B. Investment for process quality improvement and setup cost reduction in an imperfect production process with warranty policy and shortages. *RAIRO-Oper. Res.* **2019**. [CrossRef]
49. Shin, D.; Guchhait, R.; Sarkar, B.; Mittal, M. Controllable lead time, service level constraint, and transportation discounts in a continuous review inventory model. *RAIRO-Oper. Res.* **2016**, *50*, 921–934. [CrossRef]
50. Majumder, A.; Guchhait, R.; Sarkar, B. Manufacturing quality improvement and setup cost reduction in a vendor-buyer supply chain model. *Eur. J. Ind. Eng.* **2017**, *11*, 588–612. [CrossRef]

 © 2019 by the authors. Licensee MDPI, Basel, Switzerland. This article is an open access article distributed under the terms and conditions of the Creative Commons Attribution (CC BY) license (http://creativecommons.org/licenses/by/4.0/).

 *mathematics*

Article

The Inventory Model for Deteriorating Items under Conditions Involving Cash Discount and Trade Credit

Kun-Jen Chung [1,2], Jui-Jung Liao [3], Shy-Der Lin [4] and Sheng-Tu Chuang [5] and Hari Mohan Srivastava [6,7,*]

[1] College of Business, Chung Yuan Christian University, Chung-Li 32023, Taiwan
[2] School of Business, National Taiwan University of Science and Technology, Taipei 10607, Taiwan
[3] Department of Business Administration, Chihlee University of Technology, Banqiao District, New Taipei 22050, Taiwan
[4] Department of Applied Mathematics and Business Administration, Chung Yuan Christian University, Chung-Li 32023, Taiwan
[5] Department of Applied Mathematics, Chung Yuan Christian University, Chung-Li 30323, Taiwan
[6] Department of Mathematics and Statistics, University of Victoria, Victoria, BC V8W 3R4, Canada
[7] Department of Medical Research, China Medical University Hospital, China Medical University, Taichung 40402, Taiwan
*   Correspondence: harimsri@math.uvic.ca; Tel.: +1-250-472-5313 or +1-250-477-6960

Received: 26 March 2019; Accepted: 27 June 2019; Published: 2 July 2019

Abstract: In the year 2004, Chang and Teng investigated an inventory model for deteriorating items in which the supplier not only provides a cash discount, but also allows a permissible delay in payments. The main purpose of the present investigation is three-fold, as follows. First, it is found herein that Theorem 1 of Chang and Teng (2004) has notable shortcomings in terms of their determination of the optimal solution of the annual total relevant cost $Z(T)$ by adopting the Taylor-series approximation method. Theorem 1 in this paper does not make use of the Taylor-series approximation method in order to overcome the shortcomings in Chang and Teng (2004) and alternatively derives all the optimal solutions of the annual total relevant cost $Z(T)$. Secondly, this paper systematically revisits the annual total relevant cost $Z(T)$ in Chang and Teng (2004) and presents in detail the mathematically correct ways for the derivations of $Z(T)$. Thirdly, this paper not only shows that Theorem 1 of Chang and Teng (2004) is not necessarily true for finding the optimal solution of the annual total relevant cost $Z(T)$, but it also demonstrates how Theorem 1 in this paper can locate all of the optimal solutions of $Z(T)$. The mathematical analytic investigation presented in this paper is believed to be useful for correct managerial considerations and managerial decisions.

Keywords: inventory modelling and optimization; Trade-credit financing; Cash discounts; permissible delays in payments; Supply chain management; economic order quantity (EOQ); mathematical solution procedure; deteriorating items; mathematical analytic tools and techniques; managerial considerations and managerial decisions

JEL: Primary 91B24; 93C15; Secondary 90B30

---

1. Introduction, Motivation and Preliminaries

Recently, Stokes [1] mentioned that since trade credit arises spontaneously with a firm's purchases, it is understood to be one of the most flexible sources of short-term financing which is available to firms. From the viewpoint of important managerial considerations and managerial decisions, we need to include the idea to offer trade credit and the determination of the firm's terms of sale. Additionally, the decision of the purchasing firm to take (or not to take) any advantage of a cash discount and the motivations behind such a decision are also important in managerial considerations. By assuming the increasing salience of a sales promotion tool, Arcelus et al. [2] analyzed the advantages and disadvantages of the two most common payment reduction schemes. These schemes include a cash discount and a delay in the payment of the merchandise. Obviously, a cash discount can encourage the customer to pay cash upon the delivery and to reduce thereby the default risk. One may consider the permissible delay in payments as a kind of price reduction. Therefore, clearly, the permissible delay in payments will not only attract new customers for the firm, but it will also considerably increase the firm's sales.

Considering the importance of trade credit and several other features of the firm's cash-discount problem, Hill and Riener [3] considered a model for determining the cash discount in the firm's credit policy. Huang and Chung [4], on the other hand, studied the optimal replenishment and payment polices in the EOQ model under cash discount and trade credit. Huang [5] made an attempt to adopt the payment rule which was already discussed by Chung and Huang [6] as well as the policy of cash discount which was investigated by Huang and Chung [4] with a view to developing the buyer's inventory model. In fact, Chung [7] mentioned several shortcomings in the investigation by Huang [8] and presented the correct solution procedure for it. In addition, Ouyang et al. [9] studied an inventory model with the policy of non-instantaneous receipt under the cash discount and trade credit. Huang and Hsu [10] extended the work of Ouyang et al. [9] to hold true in a more general situation. Sana and Chaudhuri [11] modelled the retailer's profit-maximizing strategy in the situation when the retailer is confronted with the supplier's offer of trade credit and price discount on the purchase of a given merchandise. Finally, it was Ho et al. [12] who investigated and determined the optimal pricing, ordering, shipping, and payment policy with a view to maximizing the joint expected total profit per unit time under a two-part strategy, i.e., cash discount and delayed payment. Feng et al. [13] explored the retailer's optimal replenishment and payment policies in the economic production quantity (EPQ) model under the policy of cash discount as well as a two-level trade-credit policy. Yang [14] considered the optimal order and payment policies for deteriorating items in the analysis of discount cash flows under such alternatives as (for example) conditionally permissible delay in payments and cash discount. Quite recently, in an interesting paper, Taleizadeh et al. [15] considered the payment-delay policy in multi-product single-machine EPQ model with repair failure and back-ordering. Our above-detailed literature review reveals the fact that research about the inventory model under the cash-discount and trade-credit conditions still constitutes a popular topic in the study of operations and inventory management. The impact of deterioration in any given system is remarkably important. It is, therefore, necessary to manage the product's deterioration as, for instance, in fruit and vegetables which are known to deteriorate over time. Thus, for example, a model for exponentially decaying inventory was considered by Chare and Schrader [16], and Philip [17] discussed an inventory model with a three-parameter Weibull distribution rate and with no shortages. Later, Philip's aforementioned model in [17] was extended by Shah [18] who introduced shortages as well. We refer the reader to the remarkable survey by Aggarwal and Jaggi [19] dealing with some ordering policies of deteriorating items under permissible delay in payments. An order-level lot size inventory model with the inventory-level dependent demand and deterioration was discussed by Sarker et al. [20]. Liao et al. [21], on the other hand, studied an inventory model for deteriorating items under inflation and permissible delay in payments. Chang et al. [22] investigated and determined the optimal cycle time for deteriorating items under the policy of trade credit. The study by Arcelus et al. [2] included the retailer's pricing and trade-credit policy, as well as the inventory policies for deteriorating items. The work of Manna and Chaudhuri [23] presented an EOQ model involving the ramp-type demand rate, the time-dependent deterioration rate, the unit production cost, as well as shortages. In the existing literature on the subject of our present investigation, one can find papers which are related to variable deterioration such as those by Sana and Chaudhary [11], Skouri et al. [24], Sett et al. [25], Sarkar [26], Sarkar and Sarkar [27] and Sarker et al. [20]. Other related recent developments on the subject-matter of this article can be found in the works by (for example) Chung et al. (see [6,7,28–31]),

Feng et al. [13], Hill and Riener [3], Ho et al. [12], Huang et al. (see [4,5,10]), Liao et al. (see [21,32,33]), and Srivastava et al. [34].

It should be remarked that Sarkar and Sarkar [27] studied a significantly improved inventory model involving what may be referred to probabilistic deterioration. Their solution of the model depended heavily upon Control Theory. On the other hand, Sarkar et al. [35] considered an EOQ model for deteriorating items together with time-dependent increasing demand. They found the component cost as well as the selling price as a continuous time-rate. Sarkar and Sarkar [27] also considered an inventory model involving infinite replenishment rate, stock-dependent demand, time-varying deterioration rates as well as partial backlogging. A production-inventory model involving probabilistic deterioration in two-echelon supply chain was discussed by Sarkar [26]. Derivation of the marketing policy for non-instantaneous deteriorating items involving a generalized-type deterioration and holding-cost rates can indeed be found in the investigation by Shah et al. [18]. Chung et al. [28] considered inventory modelling involving non-instantaneous receipt and exponentially deteriorating items in the case of an integrated three-layer supply chain system under two-level trade-credit. A discussion in the case of several lot-sizing policies for deterioration items under two-level trade credit with partial trade credit to credit-risk retailer and limited storage capacity can be seen in the work by Liao et al. [33]. Wu et al. [36] presented inventory-related policies for perishable products with expiration dates and advance-cash-credit payment schemes. By making use of the Stackelberg and the Nash equilibrium solution, Jaggi et al. [37] explored inventory and credit decisions in the case of deteriorating items with displayed stock-dependent demand involving the two-echelon supply chain. Many other remarkable studies about deteriorating items can be found in the article by Kawale and Sanas [38] (see also [39,40]).

By combining all elements of the trade credit, the cash discount and the deteriorating items, Chang and Teng [41] studied a certain inventory model for deteriorating items in the case when the supplier does not only provide a cash discount, but also allows a permissible delay in payments. We remark in passing that a cash discount can encourage the customer to pay on delivery and it can also reduce the default risk. In the past few years, marketing researchers and practitioners in the area of supply chain management appear to have recognized and understood the phenomenon that the supplier offers a permissible delay in payment to the retailer if the outstanding amount is paid within the permitted fixed settlement period, known as the trade-credit period. During the trade-credit period, the retailer is allowed to accumulate revenues received upon selling items and earning interests. Consequently, without any incentive to make early payments and with the possibility and prospect of earning interest by means of the accumulated revenue which is received during the credit period, the retailer chooses to postpone payment until the last moment of the permissible period which is allowed by the supplier. Thus, clearly, the offer of the trade credit does lead to delayed cash inflow and to thereby increase the risk of cash-flow shortage as well as bad debt. From the suppliers' viewpoint, it is always hoped that they will be able to find a trade-credit policy to increase sales and to decrease the risk of cash-flow shortage and bad debt. In reality, however, especially on the side of the operations management, a supplier is generally willing and ready to provide the retailer with either a cash discount or a permissible delay in payments or both.

The main purpose of the present investigation is three-fold as stated below.

1. To observe that Theorem 1 of Chang and Teng [41] has notable shortcomings in their determination of the optimal solution of the annual total relevant cost $Z(T)$ by adopting the Taylor-series approximation method. Theorem 1 in this paper does not make use of the Taylor-series approximation method in order to overcome the shortcomings in Chang and Teng [41] and thereby derive the optimal solutions of the annual total relevant cost $Z(T)$.

2. To systematically revisit the annual total relevant cost $Z(T)$ in Chang and Teng [41] and to present in detail the mathematically correct ways for the derivations of $Z(T)$.

3. To not only show that Theorem 1 of Chang and Teng [41] is not necessarily true for finding the optimal solution of the annual total relevant cost $Z(T)$, but to also demonstrate how Theorem 1 in this paper can locate all of the optimal solutions of $Z(T)$.

The mathematically correct analytic investigation of the model, which we have presented in this paper, is believed to be useful for *correct* managerial considerations and *right* managerial decisions (see also [41]).

2. The Mathematical Modelling of the Problem

This paper adopts the same assumptions and notations as described in Chang and Teng [41].

Assumptions

(1) The demand for the item is constant with time.
(2) Shortages are not allowed.
(3) Replenishment is instantaneous.
(4) During the time the account is not settled, generated sales revenue is deposited in an interest-bearing account. At the end of this period (that is, M_1 or M_2), the customer pays the supplier the total amount in the interest-bearing account, and then starts paying off the amount owed to the supplier whenever the customer has money obtained from sales.
(5) Time horizon is infinite.

Notations

$D =$ the demand rate per year.
$h =$ the unit holding cost per year excluding interest charges.
$p =$ the selling price per unit.
$c =$ the unit purchasing cost, with $c < p$.
$I_c =$ the interest charged per \$ in stocks per year by the supplier or a bank.
$I_d =$ the interest earned per \$ per year.
$S =$ the ordering cost per order.
$r =$ the cash discount rate, $0 < r < 1$.
$\theta =$ the constant deterioration rate, where $0 \leq \theta < 1$.
$M_1 =$ the period of cash discount.
$M_2 =$ the period of permissible delay in settling account, with $M_2 > M_1$.
$T =$ the replenishment time interval.
$I(t) = \dfrac{D}{\theta} \left[e^{\theta(T-t)} - 1 \right]$, where $0 \leq t \leq T$.
Policy I: The customer accepts the cash discount and makes the full payment at time M_1.
Policy II: The customer does not accept the cash discount and makes the full payment at time M_2.
$\mathbf{TVC}_1(T) =$ the annual total relevant cost when the customer adopts Policy (I).
$\mathbf{TVC}_2(T) =$ the annual total relevant cost when the customer adopts Policy (II).
$T_1^* =$ the optimal replenishment time of $\mathrm{TVC}_1(T)$.
$T_2^* =$ the optimal replenishment time of $\mathrm{TVC}_2(T)$.
$Z(T) =$ the annual total relevant cost

$$= \begin{cases} \mathrm{TVC}_1(T) & \text{if the customer adopts Policy I} \\ \mathrm{TVC}_2(T) & \text{if the customer adopts Policy II.} \end{cases}$$

$T^* =$ the optimal replenishment time of $Z(T)$.

$$\overline{W_1} = \frac{\ln\left(\dfrac{p\theta M_1 + p I_d \theta M_1^2}{c(1-r)} + 1 \right)}{\theta}$$

$$\overline{W_3} = \frac{\ln\left(\dfrac{p\theta M_2 + \dfrac{p I_d \theta M_2^2}{2}}{c} + 1 \right)}{\theta}$$

The annual total relevant cost $Z(T)$ consists of the following items:

(a) The cost of placing order;
(b) The cost of purchasing units;
(c) The cost of carrying inventory (excluding interest changes);
(d) The cost discount earned;
(e) The interest earned from sales revenue during the permissible period $[0, M_1]$ or $[0, M_2]$
(f) The cost of interest change for unsold items after the permissible delay M_1 or M_2.

Therefore, we have

$$\text{the annual total relevant cost} = (a) + (b) + (c) - (d) - (e) + (f) \tag{1}$$

Chang and Teng [41] reveal that

$$\text{(a) Cost of placing order} = \frac{S}{T}, \tag{2}$$

$$\text{(b) Cost of purchasing units} = \frac{cD}{\theta T}\left(e^{\theta T} - 1\right), \tag{3}$$

$$\text{(c) Cost of carrying inventory} = \frac{hD}{\theta^2 T}\left(e^{\theta T} - 1\right) - \frac{hD}{\theta}, \tag{4}$$

$$\text{(d) Cash discount earned} = \begin{cases} \frac{rcD}{\theta T}\left(e^{\theta T} - 1\right) & \text{if Policy I is adopted;} \tag{5a} \\ 0 & \text{if Policy II is adopted.} \tag{5b} \end{cases}$$

Regarding the interest charged and earned, the following four possible cases, which are based upon the customer's two choices (Policy I and Policy II), occur:

Case I. The customers adopt Policy (I) and $\overline{W_1} > M_1$

(A) $T > M_1$

Chang and Teng [41] showed that

(e)

$$\text{The interest earned per year} = \frac{pI_d DM_1^2}{2T} \tag{6}$$

(f) The internet payable per year

The customer buys $I(0)$ units at time 0, and owes $c(1-r)I(0)$ to the supplier. At the time M_1, the customer sells DM_1 units in total, and has pDM_1 plus the interest earned $\frac{pI_d DM_1^2}{2}$ to pay the supplier. From the difference between the total purchase cost $c(1-r)I(0)$ and the total amount of money in the account, i.e.,

$$pDM_1 + \frac{pI_d DM_1^2}{2},$$

the following two cases to occur:

(i) If

$$pDM_1 + \frac{pI_d DM_1^2}{2} > c(1-r)I(0), \tag{7}$$

Equation (3) in Chang and Teng [41] implies that

$$\overline{W_1} = \frac{\ln\left(\frac{p\theta M_1 + pI_d\theta M_1^2}{c(1-r)} + 1\right)}{\theta} > T \geq M_1 \tag{8}$$

and

$$\text{the interest payable per year} = 0 \tag{9}$$

From Equations (2) to (9), the annual total relevant cost $Z_5(T)$ is given by

$$Z_5(T) = \frac{S}{T} + \frac{D\left[h + c\theta(1-r)\right]}{\theta^2 T}\left(e^{\theta T} - 1\right)$$
$$-\frac{hD}{\theta} - \frac{pI_d DM_1^2}{2T} \quad \text{if} \quad M_1 \leq T < \overline{W_1}. \tag{10}$$

(ii) If

$$pDM_1 + \frac{pI_d DM_1^2}{2} \leq c(1-r)I(0), \tag{11}$$

we have $T \geq \overline{W_1}$. Chang and Teng [41] demonstrated that

the interest payable per year

$$= \frac{I_c}{2pDT}\left[\frac{c(1-r)D}{\theta}\left(e^{\theta T} - 1\right) - pDM_1\left(1 + \frac{I_d M_1}{2}\right)\right]^2. \tag{12}$$

From Equations (2) to (6) and (10) to (12), we find the annual total relevant cost $Z_1(T)$ given by

$$Z_1(T) = \frac{S}{T} + \frac{D\left[h + c\theta(1-r)\right]}{\theta^2 T}\left(e^{\theta T} - 1\right) - \frac{hD}{\theta} - \frac{pI_d DM_1^2}{2T}$$
$$+ \frac{I_c}{2pDT}\left[\frac{c(1-r)D}{\theta}\left(e^{\theta T} - 1\right) - pDM_1\left(1 + \frac{I_d M_1}{2}\right)\right]^2$$
$$\text{if} \quad T \geq \overline{W_1}. \tag{13}$$

(B) $T \leq M_1$

Under Case I (B), the customer sells DT units in total at time T, and has cDT to pay the supplier in full at time M_1. Therefore, Chang and Teng [41] derived the annual total relevant cost $Z_2(T)$ as follows:

$$Z_2(T) = \frac{S}{T} + \frac{D\left[h + c\theta(1-r)\right]}{\theta^2 T}\left(e^{\theta T} - 1\right)$$
$$-\frac{hD}{\theta} - pI_d D\left(M_1 - \frac{T}{2}\right) \quad \text{if} \quad 0 < T < M_1 \tag{14}$$

Combining Case I (A) and Case I (B), we have

$$\text{TVC}_1(T) = \begin{cases} Z_2(T) & \text{if} \quad 0 < T \leq M_1 & \text{(15a)} \\ Z_5(T) & \text{if} \quad M_1 < T < \overline{W_1} & \text{(15b)} \\ Z_1(T) & \text{if} \quad T \geq \overline{W_1} & \text{(15c)} \end{cases}$$

Since

$$Z_2(M_1) = Z_5(M_1) \quad \text{and} \quad Z_5(\overline{W_1}) = Z_1(\overline{W_1}),$$

the function $\text{TVC}_1(T)$ is continuous when $T > 0$ if $\overline{W_1} > M_1$. For convenience, all the items $Z_2(T)$, $Z_5(T)$ and $Z_1(T)$ are defined on $T > 0$. Case 1 in Chang and Teng [41] only discussed Equation (11) and ignored Equation (7) of this paper. When Equation (11) of this paper holds true, then Equation (3) in Chang and Teng [41] implies that

$$T \geq \overline{W_1} > M_1 \tag{16}$$

Case I (B) and Equation (16) reveal that the domain of the function $\text{TVC}_1(T)$ is the set given by $(0, M_1] \cup [\overline{W_1}, \infty)$, but not the set $(0, \infty)$. The interval $(M_1, \overline{W_1})$ is not contained in the domain of the function $\text{TVC}_1(T)$. Therefore, the annual total relevant cost of Cases 1 and 2 in Chang and Teng [41] can be expressed as follows:

$$\text{TVC}_1(T) = \begin{cases} Z_2(T) & \text{if } 0 < T \le M_1 \\ Z_1(T) & \text{if } T \ge \overline{W_1}. \end{cases}$$

The functional behavior of $\text{TVC}_1(T)$ on the interval $(M_1, \overline{W_1})$ was not discussed in Chang and Teng [41]. This does not make sense, since the domain of $\text{TVC}_1(T)$ should be $(0, \infty)$. The correct derivation of $\text{TVC}_1(T)$ should consider Equations (7) and (11) together. Consequently, it is the shortcoming of the modelling of Chang and Teng [41]. Equations (15a), (15b) and (15c) correct the claims made by Chang and Teng [41].

Case II. The customer adopts Policy I and $\overline{W_1} \le M_1$

(C) $T > M_1$

If $T \ge M_1 \ge \overline{W_1}$, then Equation (11) holds true. Following the same arguments as in Case I (A) (ii), we find that

the interest payable per year

$$= \frac{I_c}{2pDT} \left[\frac{c(1-r)D}{\theta} \left(e^{\theta T} - 1 \right) - pDM_1 \left(1 + \frac{I_d M_1}{2} \right) \right]^2. \tag{17}$$

According to Equations (2) to (6) and (17), the annual total relevant cost $\text{TVC}_1(T)$ can be expressed as follows:

$$\text{TVC}_1(T) = \begin{cases} Z_2(T) & \text{if } 0 < T \le M_1 & \text{(18a)} \\ Z_1(T) & \text{if } M_1 < T & \text{(18b)} \end{cases}$$

Since $Z_2(M_1) < Z_1(M_1)$, the function $\text{TVC}_1(T)$ is continuous except when $T = M_1$ if $M_1 \ge \overline{W_1}$.

Case III. The customer adopts Policy II and $\overline{W_3} > M_2$

Following the same arguments as in Case I, the annual total relevant cost $\text{TVC}_2(T)$ can be expressed as follows:

$$\text{TVC}_2(T) = \begin{cases} Z_4(T) & \text{if } 0 < T \le M_2, & \text{(19a)} \\ Z_6(T) & \text{if } M_2 < T < \overline{W_3}, & \text{(19b)} \\ Z_3(T) & \text{if } T \ge \overline{W_3}, & \text{(19c)} \end{cases}$$

where

$$Z_4(T) = \frac{S}{T} + \frac{D(h+c\theta)}{\theta^2 T} \left(e^{\theta T} - 1 \right) - \frac{hD}{\theta} - pI_d D \left(M_2 - \frac{T}{2} \right), \tag{20}$$

$$Z_6(T) = \frac{S}{T} + \frac{D(h+c\theta)}{\theta^2 T} \left(e^{\theta T} - 1 \right) - \frac{hD}{\theta} - \frac{pI_d DM_2^2}{2T}, \tag{21}$$

$$Z_3(T) = \frac{S}{T} + \frac{D(h+c\theta)}{\theta^2 T} \left(e^{\theta T} - 1 \right) - \frac{hD}{\theta} - \frac{pI_d DM_2^2}{2T} + \frac{I_c}{2pDT} \left[\frac{cD}{\theta} \left(e^{\theta T} - 1 \right) - pDM_2 \left(1 + \frac{I_d M_2}{2} \right) \right]^2 \tag{22}$$

and

$$\overline{W_3} = \frac{\ln \left(\dfrac{p\theta M_2 + \dfrac{pI_d \theta M_2^2}{2}}{c} + 1 \right)}{\theta}. \tag{23}$$

Since

$$Z_4(M_2) = Z_6(M_2) \quad \text{and} \quad Z_6(\overline{W_3}) = Z_3(\overline{W_3}),$$

the function $TVC_2(T)$ is continuous when $T > 0$ if $M_2 < \overline{W_3}$. For convenience, all the items $Z_4(T)$, $Z_6(T)$ and $Z_3(T)$ are defined on $T > 0$. Let us now use the inequalities given by Equations (24) and (25) as follows:

$$pDM_2 + \frac{pI_dDM_2^2}{2} > cI(0) \tag{24}$$

and

$$pDM_2 + \frac{pI_dDM_2^2}{2} \leq cI(0), \tag{25}$$

respectively. Case 3 in Chang and Teng [41] only discussed Equation (25) and ignored Equation (24) above. When Equation (25) in this paper holds true, Equation (3) in Chang and Teng [41] implies that

$$T \geq \overline{W_3} > M_2. \tag{26}$$

Case II (D) and Equation (26) reveal that the domain of the function $TVC_2(T)$ is the set $(0, M_2] \cup [\overline{W_3}, \infty)$, but not the set $(0, \infty)$. The interval $(M_2, \overline{W_3})$ is not contained in the domain of the function $TVC_2(T)$. Therefore, the annual total relevant cost of Cases 3 and 4 in Chang and Teng [41] can be expressed as follows:

$$TVC_2(T) = \begin{cases} Z_4(T) & \text{if} \ \ 0 < T \leq M_2; \\ Z_3(T) & \text{if} \ \ T > \overline{W_3}. \end{cases}$$

The functional behavior of $TVC_2(T)$ on $(M_2, \overline{W_3})$ was not discussed in Chang and Teng [41]. This exclusion does not make sense, since the domain of $TVC_2(T)$ should be $(0, \infty)$. The correct derivation of $TVC_2(T)$ should, in fact, consider the equations (24) and (25) together. Consequently, it is another shortcoming of the modelling by Chang and Teng [41]. Equations (19a), (19b) and (19c) correct the claims by Chang and Teng [41].

Case IV. The customer adopts Policy II and $\overline{W_3} \leq M_2$.

Following the same arguments as in Case II, the annual total relevant cost $TVC_2(T)$ can be expressed as follows:

$$TVC_2(T) = \begin{cases} Z_4(T) & \text{if} \ \ 0 < T \leq M_2; & \text{(27a)} \\ Z_3(T) & \text{if} \ \ M_2 < T. & \text{(27b)} \end{cases}$$

Since $Z_4(M_2) < Z_3(M_2)$, the function $TVC_2(T)$ is continuous except when $T = M_2$ if $M_2 \geq \overline{W_3}$. Upon combining Cases I to IV, the annual total relevant cost $Z(T)$ can be expressed as follows:

$$Z(T) = \begin{cases} TVC_1(T) & \text{if the customer adopts Policy I;} & \text{(28a)} \\ TVC_2(T) & \text{if the customer adopts Policy II.} & \text{(28b)} \end{cases}$$

The objective here is to determine which policy $[T_1^*$ (Policy I) or T_2^* (Policy II)] satisfies the following condition:

$$Z(T^*) = \min\{TVC_1(T_1^*), TVC_2(T_2^*)\}. \tag{29}$$

3. The Convexity of $Z_i(T)$

By making use of Equations (10), (13), (14), (20), (21) and (22), we find for the first and the second derivatives of the annual total relevant cost $Z_2(T)$ that

$$Z_2'(T) = -\frac{S}{T^2} + \frac{D\left[h + c\theta(1-r)\right]}{\theta^2 T^2}\left(\theta T e^{\theta T} - e^{\theta T} + 1\right) + \frac{pI_d D}{2}, \tag{30}$$

$$Z_2''(T) = \frac{2S}{T^3} + \frac{2D\left[h + c\theta(1-r)\right]}{\theta^2 T^3}\left(e^{\theta T} - 1 - \theta T e^{\theta T} + \frac{1}{2}\theta^2 T^2 e^{\theta T}\right), \tag{31}$$

$$Z_5'(T) = -\frac{S}{T^2} + \frac{D\left[h + c\theta(1-r)\right]}{\theta^2 T^2}\left(\theta T e^{\theta T} - e^{\theta T} + 1\right) + \frac{pI_d D M_1^2}{2T^2}, \tag{32}$$

$$Z_5''(T) = \frac{2S - pI_d D M_1^2}{T^3} + \frac{D\left[h + c\theta(1-r)\right]}{\theta^2 T^3}\left(e^{\theta T} - 1 - \theta T e^{\theta T} + \frac{1}{2}\theta^2 T^2 e^{\theta T}\right), \tag{33}$$

$$Z_1'(T) = -\frac{S}{T^2} + \frac{Dh + B_1\theta^2}{\theta^2 T^2}\left(\theta T e^{\theta T} - e^{\theta T} + 1\right) + \frac{pI_d D M_1^2}{2T^2}$$
$$+ \frac{I_c}{2pDT^2}\left(2B_1^2\theta T e^{2\theta T} - B_1^2 e^{2\theta T} - 2B_1 W_1 \theta T e^{\theta T} + 2B_1 W_1 e^{\theta T} - W_1^2\right), \tag{34}$$

$$Z_1''(T) = \frac{2S}{T^3} + \frac{2(Dh + B_1\theta^2)}{\theta^2 T^3}\left(e^{\theta T} - 1 - \theta T e^{\theta T} + \frac{1}{2}\theta^2 T^2 e^{\theta T}\right) - \frac{pI_d D M_1^2}{T^3}$$
$$+ \frac{I_c}{2pDT^3}\left[4B_1^2(\theta T)^2 e^{2\theta T} - 4B_1^2\theta T e^{2\theta T} + 2B_1^2 e^{2\theta T} - 2B_1 W_1(\theta T)^2 e^{\theta T}\right.$$
$$\left. + 4B_1 W_1 \theta T e^{\theta T} - 4B_1 W_1 e^{\theta T} + 2W_1^2\right], \tag{35}$$

$$Z_4'(T) = -\frac{S}{T^2} + \frac{D\left[h + c\theta\right]}{\theta^2 T^2}\left(\theta T e^{\theta T} - e^{\theta T} + 1\right) + \frac{pI_d D}{2}, \tag{36}$$

$$Z_4''(T) = \frac{2S}{T^3} + \frac{2D\left[h + c\theta\right]}{\theta^2 T^3}\left(e^{\theta T} - 1 - \theta T e^{\theta T} + \frac{1}{2}\theta^2 T^2 e^{\theta T}\right), \tag{37}$$

$$Z_6'(T) = -\frac{S}{T^2} + \frac{D\left[h + c\theta\right]}{\theta^2 T^2}\left(\theta T e^{\theta T} - e^{\theta T} + 1\right) + \frac{pI_d D M_2^2}{2T^2}, \tag{38}$$

$$Z_6''(T) = \frac{2S - pI_d D M_2^2}{T^3} + \frac{D\left[h + c\theta\right]}{\theta^2 T^3}\left(e^{\theta T} - 1 - \theta T e^{\theta T} + \frac{1}{2}\theta^2 T^2 e^{\theta T}\right), \tag{39}$$

$$Z_3'(T) = -\frac{S}{T^2} + \frac{Dh + B_3\theta^2}{\theta^2 T^2}\left(\theta T e^{\theta T} - e^{\theta T} + 1\right) + \frac{pI_d D M_2^2}{2T^2}$$
$$+ \frac{I_c}{2pDT^2}\left(2B_3^2\theta T e^{2\theta T} - B_3^2 e^{2\theta T} - 2B_3 W_3 \theta T e^{\theta T} + 2B_3 W_3 e^{\theta T} - W_3^2\right) \tag{40}$$

and

$$Z_3''(T) = \frac{2S}{T^3} + \frac{2(Dh + B_3\theta^2)}{\theta^2 T^3}\left(e^{\theta T} - 1 - \theta T e^{\theta T} + \frac{1}{2}\theta^2 T^2 e^{\theta T}\right) - \frac{pI_d D M_2^2}{T^3}$$
$$+ \frac{I_c}{2pDT^3}\left[4B_3^2(\theta T)^2 e^{2\theta T} - 4B_3^2\theta T e^{2\theta T} + 2B_3^2 e^{2\theta T} - 2B_3 W_3(\theta T)^2 e^{\theta T}\right.$$
$$\left. + 4B_3 W_3 \theta T e^{\theta T} - 4B_3 W_3 e^{\theta T} + 2W_3^2\right], \tag{41}$$

where

$$A_1 = pDM_1 \left(1 + \frac{I_d M_1}{2}\right) \quad \text{and} \quad B_1 = \frac{c(1-r)D}{\theta},$$

$$A_3 = pDM_2 \left(1 + \frac{I_d M_2}{2}\right) \quad \text{and} \quad B_3 = \frac{cD}{\theta}$$

and

$$W_1 = A_1 + B_1 \quad \text{and} \quad W_3 = A_3 + B_3.$$

We now suppose that

$$G = 2S - pI_d DM_2^2 \tag{42}$$

Then, clearly, we have the results asserted by Theorem 1 below.

Theorem 1. *Each of the following assertions holds true*:

(A) $e^{\theta T} - 1 - \theta T e^{\theta T} + \frac{1}{2}\theta^2 T^2 e^{\theta T} > 0$ *if* $T > 0$.

(B)   (i) $\theta T e^{\theta T} - e^{\theta T} + 1$ *is increasing if* $T > 0$.

    (ii) $\theta T e^{\theta T} - e^{\theta T} + 1 > \dfrac{\theta^2 T^2}{2}$ *if* $T > 0$.

(C) $2B_1^2 \theta T e^{2\theta T} - B_1^2 e^{2\theta T} - 2B_1 W_1 \theta T e^{\theta T} + 2B_1 W_1 e^{\theta T} - W_1^2 \geq 0$ *if* $T \geq \overline{W_1}$

(D) $2B_3^2 \theta T e^{2\theta T} - B_3^2 e^{2\theta T} - 2B_3 W_3 \theta T e^{\theta T} + 2B_3 W_3 e^{\theta T} - W_3^2 \geq 0$ *if* $T \geq \overline{W_3}$

(E) *Both $Z_2(T)$ and $Z_4(T)$ are convex on $T > 0$.*

(F) *If $G > 0$, then both $Z_5(T)$ and $Z_6(T)$ are convex on $T > 0$.*

(G) *If $3B_1 > A_1$, then $Z_1(T)$ are convex on $T > 0$.*

(H) *If $3B_3 > A_3$, then $Z_3(T)$ are convex on $T > 0$.*

Proof.

(A) See Lemma 1 in Huang and Liao [8].

(B) Let

$$f(x) = xe^x - e^x + 1 \tag{43}$$

and

$$k(x) = xe^x - e^x + 1 - \frac{x^2}{2}. \tag{44}$$

Equations (43) and (44) yield

$$f'(x) = xe^x > 0 \quad \text{if} \quad x > 0 \tag{45}$$

and

$$k'(x) = x(e^x - 1) > 0 \quad \text{if} \quad x > 0. \tag{46}$$

Equations (45) and (46) imply that both functions $f(x)$ and $k(x)$ are increasing when $x > 0$. Therefore, we get

$$f(x) > f(0) = 0 \quad \text{and} \quad k(x) > k(0) = 0.$$

We now set $x = \theta T$. Consequently, we have

  (i) $\theta T e^{\theta T} - e^{\theta T} + 1$ is increasing if $T > 0$.

  (ii) $\theta T e^{\theta T} - e^{\theta T} + 1 > \dfrac{\theta^2 T^2}{2}$ if $T > 0$.

(C) Let

$$g(T) = 2B_1^2 \theta T e^{2\theta T} - B_1^2 e^{2\theta T} - 2B_1 W_1 \theta T e^{\theta T} + 2B_1 W_1 e^{\theta T} - W_1^2. \tag{47}$$

Then

$$g'(T) = 2\theta^2 T B_1^2 e^{\theta T} \left(2e^{\theta T} - \frac{W_1}{B_1}\right). \tag{48}$$

So, we have

$$g'(T) > 0 \ \text{ if } \ T \geq \overline{W_1}. \tag{49}$$

Equation (47) implies that the function $g(T)$ is increasing if $T \geq \overline{W_1}$. Therefore, we have

$$g(T) > g(\overline{W_1}) = 0 \ \text{ if } \ T \geq \overline{W_1} \tag{50}$$

Consequently, we obtain

$$2B_1^2 \theta T e^{2\theta T} - B_1^2 e^{2\theta T} - 2B_1 W_1 \theta T e^{\theta T} + 2B_1 W_1 e^{\theta T} - W_1^2 \geq 0$$

if $T \geq \overline{W_1}$.

(D) The proof is similar to that of Theorem 1 (C).

(E) Equations (31), (37) and Theorem 1 (A) imply that $Z_2''(T) > 0$ and $Z_4''(T) > 0$. Therefore, both functions $Z_2(T)$ and $Z_4(T)$ are convex on $T > 0$.

(F) If $G > 0$, then $2S - pI_dDM_1^2 \geq G > 0$. Equations (33) and (39) imply that $Z_5''(T) > 0$ and $Z_6''(T) > 0$. Therefore, both functions $Z_5(T)$ and $Z_6(T)$ are convex when $T > 0$ if $G > 0$.

(G) See, for details, Lemma 2 in the paper by Huang and Liao [8].

(H) See, for details, Lemma 2 in the paper by Huang and Liao [8].

□

4. Theorems for the Optimal Cycle T^* of $Z(T)$

Equations (30), (32), (34), (36), (38) and (40) yield

$$Z_2'(M_1) = Z_5'(M_1)$$
$$= \frac{1}{\theta^2 M_1^2} \left\{-S\theta^2 + D\left[h + c\theta(1-r)\right]\left(\theta M_1 e^{\theta M_1} - e^{\theta M_1} + 1\right) + \frac{pI_dD\theta^2 M_1^2}{2}\right\}, \tag{51}$$

$$Z_5'(\overline{W_1}) = Z_1'(\overline{W_1})$$
$$= \frac{1}{\theta^2 \overline{W_1}^2} \left\{-S\theta^2 + D\left[h + c\theta(1-r)\right]\left(\theta\overline{W_1} e^{\theta\overline{W_1}} - e^{\theta\overline{W_1}} + 1\right) + \frac{pI_dD\theta^2 M_1^2}{2}\right\}, \tag{52}$$

$$Z_1'(M_1) = \frac{1}{\theta^2 M_1^2} \left\{-S\theta^2 + D\left[h + c\theta(1-r)\right]\left(\theta M_1 e^{\theta M_1} - e^{\theta M_1} + 1\right) + \frac{pI_dD\theta^2 M_1^2}{2}\right.$$
$$\left. + \frac{I_c\theta^2}{2pD}\left[2B_1^2\theta M_1 e^{2\theta M_1} - B_1^2 e^{2\theta M_1} - 2B_1 W_1 \theta M_1 e^{\theta M_1} + 2B_1 W_1 e^{\theta M_1} - W_1^2\right]\right\}, \tag{53}$$

$$
\begin{aligned}
Z_4'(M_2) &= Z_6'(M_2) \\
&= \frac{1}{\theta^2 M_2^2} \left\{ -S\theta^2 + D(h + c\theta)\left(\theta M_2 e^{\theta M_2} - e^{\theta M_2} + 1 \right) + \frac{p I_d D \theta^2 M_2^2}{2} \right\},
\end{aligned}
\tag{54}
$$

$$
\begin{aligned}
Z_6'(\overline{W_3}) &= Z_3'(\overline{W_3}) \\
&= \frac{1}{\theta^2 \overline{W_3}^2} \left\{ -S\theta^2 + D(h + c\theta)\left(\theta \overline{W_3} e^{\theta \overline{W_3}} - e^{\theta \overline{W_3}} + 1 \right) + \frac{p I_d D \theta^2 M_2^2}{2} \right\}
\end{aligned}
\tag{55}
$$

and

$$
\begin{aligned}
Z_3'(M_2) &= \frac{1}{\theta^2 M_2^2} \left\{ -S\theta^2 + D(h + c\theta)\left(\theta M_2 e^{\theta M_2} - e^{\theta M_2} + 1 \right) + \frac{p I_d D \theta^2 M_2^2}{2} \right. \\
&\left. \quad + \frac{I_c \theta^2}{2pD} \left[2 B_3^2 \theta M_2 e^{2\theta M_2} - B_3^2 e^{2\theta M_2} - 2 B_3 W_3 \theta M_2 e^{\theta M_2} + 2 B_3 W_3 e^{\theta M_2} - W_3^2 \right] \right\}.
\end{aligned}
\tag{56}
$$

Case I. Policy I is adopted and $M_1 < \overline{W_1}$.
Let

$$
\begin{aligned}
\triangle_{25} &= \theta^2 M_1^2 Z_2'(M_1) - \theta^2 M_1^2 Z_5'(M_1) \\
&= -S\theta^2 + D\left[h + c\theta(1 - r) \right]\left(\theta M_1 e^{\theta M_1} - e^{\theta M_1} + 1 \right) + \frac{p I_d D \theta^2 M_1^2}{2}
\end{aligned}
\tag{57}
$$

and

$$
\begin{aligned}
\triangle_{51} &= \theta^2 \left(\overline{W_1} \right)^2 Z_5'(\overline{W_1}) = \theta^2 \left(\overline{W_1} \right)^2 Z_1'(\overline{W_1}) \\
&= -S\theta^2 + D\left[h + c\theta(1 - r) \right]\left(\theta \overline{W_1} e^{\theta \overline{W_1}} - e^{\theta \overline{W_1}} + 1 \right) + \frac{p I_d D \theta^2 M_1^2}{2}.
\end{aligned}
\tag{58}
$$

Since $M_1 < \overline{W_1}$, Theorem 1 (B) implies that

$$
\triangle_{51} > \triangle_{25}
\tag{59}
$$

Case II. Policy I is adopted and $M_1 \geq \overline{W_1}$.
Let

$$
\begin{aligned}
\triangle_2 &= \theta^2 M_1^2 Z_2'(M_1) \\
&= -S\theta^2 + D\left[h + c\theta(1 - r) \right]\left(\theta M_1 e^{\theta M_1} - e^{\theta M_1} + 1 \right) + \frac{p I_d D \theta^2 M_1^2}{2}
\end{aligned}
\tag{60}
$$

and

$$\triangle_1 = \theta^2 M_1^2 Z_1'(M_1)$$
$$= -S\theta^2 + D\left[h + c\theta(1-r)\right]\left(\theta M_1 e^{\theta M_1} - e^{\theta M_1} + 1\right) + \frac{pI_d D\theta^2 M_1^2}{2}$$
$$+ \frac{I_c \theta^2}{2pD}\left[2B_1^2 \theta M_1 e^{2\theta M_1} - B_1^2 e^{2\theta M_1} - 2B_1 W_1 \theta M_1 e^{\theta M_1} + 2B_1 W_1 e^{\theta M_1} - W_1^2\right].$$

(61)

Now, since $M_1 \geq \overline{W_1}$, Theorem 1 (C) implies that

$$\triangle_1 > \triangle_2.$$

(62)

Case III. Policy II is adopted and $M_2 < \overline{W_3}$.

Let

$$\triangle_{46} = \theta^2 M_2^2 Z_4'(M_2) = \theta^2 M_2^2 Z_6'(M_2)$$
$$= -S\theta^2 + D(h + c\theta)\left(\theta M_2 e^{\theta M_2} - e^{\theta M_2} + 1\right) + \frac{pI_d D\theta^2 M_2^2}{2}$$
$$> \triangle_{25} = \triangle_2$$

(63)

and

$$\triangle_{63} = \theta^2 \left(\overline{W_3}\right)^2 Z_4'(\overline{W_3}) = \theta^2 \left(\overline{W_3}\right)^2 Z_3'(\overline{W_3})$$
$$= -S\theta^2 + D(h + c\theta)\left(\theta \overline{W_3} e^{\theta \overline{W_3}} - e^{\theta \overline{W_3}} + 1\right) + \frac{pI_d D\theta^2 M_2^2}{2}$$

(64)

Since $M_2 < \overline{W_3}$, Theorem 1 (B) implies that

$$\triangle_{63} > \triangle_{46} > \triangle_{25} = \triangle_2$$

(65)

Case IV. Policy II is adopted and $M_2 \geq \overline{W_3}$.

Let

$$\triangle_4 = \theta^2 M_2^2 Z_4'(M_2)$$
$$= -S\theta^2 + D(h + c\theta)\left(\theta M_2 e^{\theta M_2} - e^{\theta M_2} + 1\right) + \frac{pI_d D\theta^2 M_2^2}{2}$$
$$= \triangle_{46} > \triangle_{25} = \triangle_2$$

(66)

and

$$\triangle_3 = \theta^2 M_2^2 Z_3'(M_2)$$
$$= -S\theta^2 + D(h + c\theta)\left(\theta M_2 e^{\theta M_2} - e^{\theta M_2} + 1\right) + \frac{pI_d D\theta^2 M_2^2}{2}$$
$$+ \frac{I_c \theta^2}{2pD}\left[2B_3^2 \theta M_2 e^{2\theta M_2} - B_3^2 e^{2\theta M_2} - 2B_3 W_3 \theta M_2 e^{\theta M_2} + 2B_3 W_3 e^{\theta M_2} - W_3^2\right].$$

(67)

Thus, since $M_2 \geq \overline{W_3}$, Theorem 1 (D) implies that

$$\triangle_3 \geq \triangle_4.$$

(68)

Henceforth, in our investigation, we assume that

$$G > 0, \tag{69}$$

$$3B_1 > A_1 \tag{70}$$

and

$$3B_3 > A_3. \tag{71}$$

Theorem 1 (E) to Theorem 1 (H), together, imply that the function $Z_i(T)$ is convex when $T > 0$ for all $i = 1, 2, 3, 4, 5, 6$. Let T_i denote the root of the following equation:

$$Z_i(T) = 0 \qquad (i = 1, 2, 3, 4, 5, 6) \tag{72}$$

From the convexity of the function $Z_i(T)$ when $T > 0$, we conclude that

$$Z_i'(T) = \begin{cases} < 0 & \text{if } 0 < T < T_i; \tag{73a} \\ = 0 & \text{if } T = T_i; \tag{73b} \\ > 0 & \text{if } T > T_i. \tag{73c} \end{cases}$$

Therefore, clearly, the function $Z_i(T)$ is decreasing on $(0, T_i]$ and increasing on $[T_i, \infty)$ for $i = 1, 2, 3, 4, 5, 6$.

Proposition 1. *Suppose that Policy I is adopted and $M_1 < \overline{W_1}$. Then the following assertions hold true:*

(A) If $\triangle_{25} > 0$, then $T_1^* = T_2$.
(B) If $\triangle_{25} \leq 0 < \triangle_{51}$, then $T_1^* = T_5$.
(C) If $\triangle_{51} \leq 0$, then $T_1^* = T_1$.

Proof.

(A) If $\triangle_{25} > 0$, then $\triangle_{51} > \triangle_{25} > 0$. From Equations (73a), (73b) and (73c), we thus find that

    (i) $Z_2(T)$ is decreasing on $(0, T_2]$ and increasing on $[T_2, M_1]$.
    (ii) $Z_5(T)$ is increasing on $[M_1, \overline{W_1}]$.
    (iii) $Z_1(T)$ is increasing on $[\overline{W_1}, \infty)$.

Since the function $TVC_1(T)$ is continuous when $T > 0$ if $M_1 < \overline{W_1}$, Equations (15a), (15b) and (15c) and the above observations (i) to (iii) imply that $T_1^* = T_2$.

(B) If $\triangle_{25} \leq 0 < \triangle_{51}$, from Equations (73a), (73b) and (73c), we find that

    (iv) $Z_2(T)$ is decreasing on $(0, M_1]$.
    (v) $Z_5(T)$ is decreasing on $[M_1, T_5]$ and increasing on $[T_5, \overline{W_1}]$.
    (vi) $Z_1(T)$ is increasing on $[\overline{W_1}, \infty)$.

Since the function $TVC_1(T)$ is continuous when $T > 0$ if $M_1 < \overline{W_1}$, Equations (15a), (15b) and (15c), together with the above observations (iv) to (vi), imply that $T_1^* = T_5$.

(C) If $\triangle_{51} \leq 0$, then $\triangle_{25} < \triangle_{51} \leq 0$. From Equations (73a), (73b) and (73c), we obtain

    (vii) $Z_2(T)$ is decreasing on $(0, M_1]$.
    (viii) $Z_5(T)$ is decreasing on $[M_1, \overline{W_1}]$.
    (ix) $Z_1(T)$ is decreasing on $[\overline{W_1}, T_1]$ and increasing on $[T_1, \infty)$.

Since $TVC_1(T)$ is continuous on $T > 0$ if $M_1 < \overline{W_1}$, Equations (15a), (15b) and (15c), and the above observations (vii) to (ix), imply that $T_1^* = T_1$.

$\square$

Proposition 2. *Suppose that Policy I is adopted and $M_1 \geq \overline{W_1}$. Then the following assertions hold true:*

(A) *If $\triangle_2 > 0$, then $T_1^* = T_2$.*
(B) *If $\triangle_2 \leq 0 < \triangle_1$, then $T_1^* = M_1$.*
(C) *If $\triangle_1 \leq 0$, then $T_1^* = M_1$ or T_1 is associated with the least cost.*

Proof.

(A) If $\triangle_2 > 0$, then $\triangle_1 \geq \triangle_2 > 0$. From Equations (73a), (73b) and (73c), we have

    (i) $Z_2(T)$ is decreasing on $(0, T_2]$ and increasing on $[T_2, M_1]$.
    (ii) $Z_1(T)$ is increasing on (M_1, ∞).

Since $Z_1(M_1) > Z_2(M_1)$, Equations (18a) and (18b), together with the above observations (i) and (ii), imply that $T_1^* = T_2$.

(B) If $\triangle_2 \leq 0 < \triangle_1$, from Equations (73a), (73b) and (73c), we have

    (iii) $Z_2(T)$ is decreasing on $(0, M_1]$.
    (iv) $Z_1(T)$ is increasing on (M_1, ∞).

Since $Z_1(M_1) > Z_2(M_1)$, Equations (18a) and (18b) and the above observations (iii) and (iv) imply that $T_1^* = M_1$.

(C) If $\triangle_1 \leq 0$, then $\triangle_2 \leq \triangle_1 \leq 0$. Thus, from Equations (73a), (73b) and (73c), we get

    (v) $Z_2(T)$ is decreasing on $(0, M_1]$.
    (vi) $Z_1(T)$ is decreasing on $(M_1, T_1]$ and increasing on $[T_1, \infty)$.

Since $Z_1(M_1) > Z_2(M_1)$, Equations (18a) and (18b), together with the above observations (v) and (vi), imply that $T_1^* = M_1$ or T_1 is associated with the least cost.

$\square$

Proposition 3. *Suppose that Policy II is adopted and $M_2 < \overline{W_3}$. Then the following assertions hold true:*

(A) *If $\triangle_{46} > 0$, then $T_2^* = T_4$.*
(B) *If $\triangle_{46} \leq 0 < \triangle_{63}$, then $T_2^* = T_6$.*
(C) *If $\triangle_{63} \leq 0$, then $T_2^* = T_3$.*

Proof. The proof of Proposition 3 is similar to that of Proposition 1. We, therefore, choose to skip the details involved. $\square$

Proposition 4. *Suppose that Policy II is adopted and $M_2 \geq \overline{W_3}$. Then the following assertions hold true:*

(A) *If $\triangle_4 > 0$, then $T_2^* = T_4$.*
(B) *If $\triangle_4 \leq 0 < \triangle_3$, then $T_2^* = M_2$.*
(C) *If $\triangle_3 \leq 0$, then $T_2^* = M_2$ or T_3 is associated with the least cost.*

Proof. The proof of Proposition 4 would run parallel to that of Proposition 2. The details involved are, therefore, omitted. $\square$

Theorem 2. *Suppose that*

$$M_1 < \overline{W_1} \quad and \quad M_2 < \overline{W_3}.$$

Then each of the following assertions holds true:

(i)   If $\triangle_{25} > 0$, then $T^* = T_2$ or T_4 is associated with the least cost.

(ii) If $\triangle_{25} \leq 0 < \triangle_{51}$ and $\triangle_{46} > 0$, then $T^* = T_5$ or T_4 is associated with the least cost.

(iii) If $\triangle_{25} \leq 0 < \triangle_{51}$ and $\triangle_{46} \leq 0 < \triangle_{63,}$, $T^* = T_5$ or T_6 is associated with the least cost.

(iv) If $\triangle_{25} \leq 0 < \triangle_{51}$ and $\triangle_{63} \leq 0$, then $T^* = T_5$ or T_3 is associated with the least cost.

(v)   If $\triangle_{51} \leq 0$ and $\triangle_{46} > 0$, then $T^* = T_1$ or T_4 is associated with the least cost.

(vi) If $\triangle_{51} \leq 0$ and $\triangle_{46} \leq 0 < \triangle_{63}$, then $T^* = T_1$ or T_6 is associated with the least cost.

(vii) If $\triangle_{51} \leq 0$ and $\triangle_{63} \leq 0$, then $T^* = T_1$ or T_3 is associated with the least cost.

Proof. The demonstration of Theorem 2 would make use of Propositions 1 and 3. $\square$

Theorem 3. *Suppose that* $M_1 < \overline{W_1}$ *and* $M_2 \geq \overline{W_3}$. *Then each of the following assertions holds true:*

(i)   If $\triangle_{25} > 0$, then $T^* = T_2$ or T_4 is associated with the least cost.

(ii) If $\triangle_{25} \leq 0 < \triangle_{51}$ and $\triangle_4 > 0$, then $T^* = T_5$ or T_4 is associated with the least cost.

(iii) If $\triangle_{25} \leq 0 < \triangle_{51}$ and $\triangle_4 \leq 0 < \triangle_3$, then $T^* = T_5$ or M_2 is associated with the least cost.

(iv) If $\triangle_{25} \leq 0 < \triangle_{51}$ and $\triangle_3 \leq 0$, then $T^* = T_5$, then M_2 or T_3 is associated with the least cost.

(v)   If $\triangle_{51} \leq 0$ and $\triangle_4 > 0$, then $T^* = T_1$ or T_4 is associated with the least cost.

(vi) If $\triangle_{51} \leq 0$ and $\triangle_4 \leq 0 < \triangle_3$, then $T^* = T_1$ or M_2 is associated with the least cost.

(vii) If $\triangle_{51} \leq 0$ and $\triangle_3 \leq 0$, then $T^* = T_1$, M_2 or T_3 is associated with the least cost.

Proof. The proof of Theorem 3 follows from Propositions 1 and 4. $\square$

Theorem 4. *Suppose that* $M_1 \geq \overline{W_1}$ *and* $M_2 < \overline{W_3}$. *Then each of the following assertions holds true:*

(i)   If $\triangle_2 > 0$, then $T^* = T_2$ or T_4 is associated with the least cost.

(ii) If $\triangle_2 \leq 0 < \triangle_1$ and $\triangle_{46} > 0$, then $T^* = M_1$ or T_4 is associated with the least cost.

(iii) If $\triangle_2 \leq 0 < \triangle_1$ and $\triangle_{46} \leq 0 < \triangle_{63}$, then $T^* = M_1$ or T_6 is associated with the least cost.

(iv) If $\triangle_2 \leq 0 < \triangle_1$ and $\triangle_{63} \leq 0$, then $T^* = M_1$ or T_3 is associated with the least cost.

(v)   If $\triangle_1 \leq 0$ and $\triangle_{46} > 0,$, then $T^* = M_1$, T_1 or T_4 is associated with the least cost.

(vi) If $\triangle_1 \leq 0$ and $\triangle_{46} \leq 0 < \triangle_{63}$, then $T^* = M_1$, T_1 or T_6 is associated with the least cost.

(vii) If $\triangle_1 \leq 0$ and $\triangle_{63} \leq 0$, then $T^* = M_1$, T_1 or T_3 is associated with the least cost.

Proof. Theorem 4 can be proven by applying Propositions 2 and 3. $\square$

Theorem 5. *Suppose that* $M_1 \geq \overline{W_1}$ *and* $M_2 \geq \overline{W_3}$. *Then each of the following assertions holds true:*

(i)   If $\triangle_2 > 0$, then $T^* = T_2$ or T_4 is associated with the least cost.

(ii) If $\triangle_2 \leq 0 < \triangle_1$ and $\triangle_4 > 0$, then $T^* = M_1$ or T_4 is associated with the least cost.

(iii) If $\triangle_2 \leq 0 < \triangle_1$ and $\triangle_4 \leq 0 < \triangle_3$, then $T^* = M_1$ or M_2 is associated with the least cost.

(iv) If $\triangle_2 \leq 0 < \triangle_1$ and $\triangle_3 \leq 0$, then $T^* = M_1$, M_2 or T_3 is associated with the least cost.

(v)   If $\triangle_1 \leq 0$ and $\triangle_4 > 0$, then $T^* = M_1$, T_1 or T_4 is associated with the least cost.

(vi) If $\triangle_1 \leq 0$ and $\triangle_4 \leq 0 < \triangle_3$, then $T^* = M_1$, T_1 or M_2 is associated with the least cost.

(vii) If $\triangle_1 \leq 0$ and $\triangle_3 \leq 0$, then $T^* = M_1$, T_1, M_2 or T_3 is associated with the least cost.

Proof. It is easy to derive Theorem 5 by making use of Propositions 2 and 4. $\square$

5. Discussions Concerning Theorem 1 of Chang and Teng [41]

(A)   About Theorem 1 (1) in Chang and Teng [41]:
Equation (29) reveals that

$$Z(T^*) = \min\{\mathrm{TVC}_1(T_1^*), \mathrm{TVC}_2(T_2^*)\}$$

Therefore, clearly, we have $T^* = T_1^*$ (Policy I) or T_2^* (Policy II) associated with the least cost. Chang and Teng [41] do not make the comparison between $TVC_1(T_1^*)$ and $TVC_2(T_2^*)$. Consequently, in general, the claimed assertion of Theorem 1 (1) in Chang and Teng [41] is not necessarily true.

(B) About Theorem 1 (2) in the paper by Chang and Teng [41]:

If

$$2S = [h + c\theta(1 - r) + pI_d] DM_1^2,$$

our Theorem 1 (B) (ii) implies that

$$Z_2'(M_1) = \frac{1}{\theta^2 M_1^2} \left\{ -S\theta^2 + D\left[h + c\theta(1 - r)\right] \left(\theta M_1 e^{\theta M_1} - e^{\theta M_1} + 1 \right) + \frac{pI_d D\theta^2 M_1^2}{2} \right\}$$

$$> \frac{1}{\theta^2 M_1^2} \left\{ -S\theta^2 + D\left[h + c\theta(1 - r)\right] \frac{\theta^2 M_1^2}{2} + \frac{pI_d D\theta^2 M_1^2}{2} \right\}$$

$$= \frac{1}{2M_1^2} \left\{ -2S + [h + c\theta(1 - r) + pI_d] DM_1^2 \right\}$$

$$= 0.$$

Since $Z_2'(M_1) > 0$, M_1 is not the optimal solution of $TVC_1(T)$. Therefore, in general, the result claimed in Theorem 1 (2) of Chang and Teng [41] is not true.

(C) About Theorem 1 (3) in the paper by Chang and Teng [41]:

Let T_{CT}^* denote the optimal solution obtained by Theorem 1 in Chang and Teng [41]. In this case, we consider the following example.

Example 1. *Given $D = 500$ units/year, h = \$4/unit/year, $I_c = 0.09$/year, $I_d = 0.06$/year, $c = \$30$ per unit, $p = \$35$ per unit, $r = 0.02$, $Q = 0.07$, $M_1 = 30$ days = 30/365 years, $M_2 = 56$ days = 56/365 years and $S = \$13.85$ per order. Then*

$$(h + c\theta + pI_d)DM_2^2 > 2S > [h + c\theta(1 - r) + pI_d] DM_1^2,$$

$G = 2.9839 > 0$, $A_1 = 1441.9028$, $B_1 = 210000$, $A_3 = 2697.2895$, $B_3 = 214285.7143$, $\overline{W_1} = 0.09775$, $\overline{W_3} = 0.178696983$, $M_1 = 0.08219$ years and $M_2 = 0.1787$ years. We thus observe that $M_1 < \overline{W_1}$, $M_2 < \overline{W_3}$, $3B_1 > A_1$, $3B_1 > A_1$ and $3B_3 > A_3$. Furthermore, we get $\triangle_{25} = -1.59779 \times 10^{-4} < 0$, $\triangle_{46} = 0.1698 > 0$ and $\triangle_{51} = 0.02795 > 0$. Then, by applying Theorem 2 (ii) of this paper, we have $T^* = T_5$ or T_4. The familiar Intermediate Value Theorem (see, for example, Varberg et al. [42]) can now be used to locate T_5 and T_4. We thus find that $T_5 = 0.08231$, $T_4 = 0.08207$, $TVC_1(T_5) = 14950.0759$ and $TVC_2(T_4) = 15176.1460$. Since

$$TVC_1(T_5) < TVC_2(T_4),$$

we have $T^* = T_5$. Moreover, by applying the Intermediate Value Theorem, we conclude that

$$0 < T_1 < \overline{W_1},$$

since $\triangle_{51} > 0$, $G > 0$ and

$$\lim_{T \to 0^+} Z_5'(T) = -\infty.$$

Therefore, T_1 does not satisfy Equation (24) in Chang and Teng [41]. Consequently, Theorem 1 (3) in Chang and Teng [41] can be used. We then get $T_{CT}^* = T_4$. However, the accurate optimum solution of the above Example should be $T^* = T_5$. Therefore, by contradiction, Theorem 1 (3) in Chang and Teng [41] is not necessarily true.

(D) About Theorem 1 (4) in the paper by Chang and Teng [41]:
The proof in this case is similar to that in (B) above. Therefore, Theorem 1 (4) in Chang and Teng [41] is not true.

(E) About Theorem 1 (5) in the paper by Chang and Teng [41]:
Our reasoning here is the same as that of (A) above. Therefore, Theorem 1 (5) in the work of Chang and Teng [41] is not necessarily true.

By incorporating (A) to (E) above, it is concluded that in general, Theorem 1 in Chang and Teng [41] is not necessarily true.

6. Concluding Remarks and Observations

In our present investigation, we have successfully divided all our mathematical analytic derivations of the annual total relevant cost $Z(T)$ into the following four cases:

(1) $M_1 < \overline{W}_1$
(2) $M_1 \geq \overline{W}_1$
(3) $M_2 < \overline{W}_3$
(4) $M_2 \geq \overline{W}_3$

When the above Case 2 and Case 4 hold true, the annual total relevant costs in this paper are seen to be consistent with those of Chang and Teng [41]. However, if the above Case 1 and Case 3 hold true, then the annual total relevant costs in the work by Chang and Teng [41] are observed to be incorrect. Furthermore, this paper has also indicates that Theorem 1 in Chang and Teng [41] is based on the assumption that θT is small. However, our present investigation does not include this assumption. On the other hand, in general, Theorem 1 in the work by Chang and Teng [41] is not necessarily true. Theorems 2 to 5 in this paper have been fruitfully used to characterize the optimal solutions and to demonstrate the fact that they can locate all optimal solutions of $Z(t)$. By incorporating the above arguments. we conclude that our present investigation has not only removed all those shortcomings in the paper by Chang and Teng [41], but it has also presented solvable ways for the problem considered by Chang and Teng [41]. Consequently, in this paper, we have corrected and substantially improved the work of Chang and Teng [41]. Therefore, it can significantly reduce the cost of the inventory model.

The mathematically correct analytic investigation of the model, which we have presented in this paper, is believed to be useful for *correct* managerial considerations and *right* managerial decisions.

The proposed model, for which we have presented a mathematical analytic investigation in this article, is capable of being extended in several different directions. Among other such possibilities of extension and generalization of our study here, it may be worthwhile to extend the constant demand rate to hold true in the case of a more realistic situation when the time-varying demand rate is a function of the time, the selling price, the advertisement of the product quality, and sundry other considerations. Yet another direction for future research on the subject-matter of our present investigation is the possibility of generalization and extension of the model with a view to allowing for shortages, quantity discounts, inflation rates, and other business-related considerations.

Author Contributions: Conceptualization, K.-J.C., J.-J.L. and S.-D.L.; methodology, K.-J.C., J.-J.L., S.-D.L. and H.M.S.; software, S.-D.L. and S.-T.C.; validation, S.-D.L. and H.M.S.; formal analysis, K.-J.C., J.-J.L., S.-D.L. and H.M.S.; investigation, K.-J.C., J.-J.L. and S.-D.L.; resources, K.-J.C., J.-J.L. and S.-D.L.; data curation, S.-D. Lin and S.-T.C.; writing–original draft preparation, K.-J.C., J.-J.L., S.-D.L. and S.-T.C.; writing–review and editing, K.-J.C., J.-J.L., S.-D.L. and H.M.S.; visualization, J.-J.L. and S.-T.C.; supervision, K.-J.C., S.-D.L. and H.M.S.; project administration, K.-J.C., J.-J.L., S.-D.L. and H.M.S.; funding acquisition, Not Applicable.

Funding: This research received no external funding.

Conflicts of Interest: The authors declare no conflicts of interest.

References

1. Stokes, J.R. Dynamic cash discounts when sales volume is stochastic. *Quart. Rev. Econ. Financ.* **2005**, *45*, 144–160. [CrossRef]
2. Arcelus, F.J.; Shah, N.H.; Srinivasan, G. Retailer's pricing, credit and inventory policies for deteriorating items in response to temporary price/credit incentives. *Int. J. Prod. Econ.* **2003**, *81–82*, 153–162. [CrossRef]
3. Hill, N.C.; Riener, K.D. Determining the cash discount in the firm's credit policy. *Financ. Manag.* **1979**, *8*, 68–73. [CrossRef]
4. Huang, Y.-F.; Chung, K.-J. Optimal replenishment and payment policies in the EOQ model under cash discount and trade credit. *Asia-Pac. J. Oper. Res.* **2003**, *20*, 177–190.
5. Huang, Y.-F. Optimal retailer's replenishment decisions in the EPQ model under two levels of trade credit policy. *Eur. J. Oper. Res.* **2007**, *176*, 1577–1591. [CrossRef]
6. Chung, K.-J.; Huang, T.-S. The optimal retailer's ordering policies for deteriorating items with limited storage capacity under trade credit financing. *Int. J. Prod. Econ.* **2007**, *106*, 127–145. [CrossRef]
7. Chung, K.-J. The complete proof on the optimal ordering policy under cash discount and trade credit. *Int. J. Syst. Sci.* **2010**, *41*, 467–475. [CrossRef]
8. Huang, K.-N.; Liao, J.-J. A simple method to locate the optimal solution for exponentially deteriorating items under trade credit financing. *Comput. Math. Appl.* **2008**, *56*, 965–977. [CrossRef]
9. Ouyang, L.-Y.; Yang, C.-T.; Chan, Y.-L.; Cárdenas-Barrón, L.E. A comprehensive extension of the optimal replenishment decisions under two levels of trade credit policy depending on the order quantity. *Appl. Math. Comput.* **2013**, *224*, 268–277.
10. Huang, Y.-F.; Hsu, K.-H. An EOQ model under retailer partial trade credit policy in supply chain. *Int. J. Prod. Econ.* **2008**, *112*, 655–664. [CrossRef]
11. Sana, S.S.; Chaudhuri, K.S. A deterministic EOQ model with delays in payments and price discount offers. *Eur. J. Oper. Res.* **2008**, *184*, 509–533. [CrossRef]
12. Ho, C.-H.; Ouyang, L.-Y.; Su, C.-H. Optimal pricing, shipment and payment policy for an integrated supplier–buyer inventory model with two-part trade credit. *Eur. J. Oper. Res.* **2008**, *187*, 496–510. [CrossRef]
13. Feng, H.-R.; Li, J.; Zhao, D. Retailer's optimal replenishment and payment policies in the EPQ model under cash discount and two-level trade credit policy. *Appl. Math. Model.* **2013**, *37*, 3322–3339. [CrossRef]
14. Yang, C.-T. The optimal order and payment policies for deteriorating items in discount cash flows analysis under the alternatives of conditionally permissible delay in payments and cash discount. *Top* **2010**, *18*, 429–443. [CrossRef]
15. Taleizadeh, A.A.; Sarkar, B.; Hasani, M. Delayed payment policy in multi-product single-machine economic production quantity model with repair failure and partial backordering. *J. Ind. Manag. Optim.* **2019**, 1684–1688. [CrossRef]
16. Ghare, P.M.; Scharder, G.P. A model for exponentially decaying inventory. *J. Ind. Engrg.* **1963**, *14*, 238–243.
17. Philip, G.C. A generalized EOQ model for items with Weibull distribution deterioration. *AIIE Trans.* **1974**, *6*, 159–162. [CrossRef]
18. Shah, Y.K. An order-level lot-size inventory model for deteriorating items. *AIIE Trans.* **1977**, *9*, 108–112. [CrossRef]
19. Aggarwal, S.P.; Jaggi, C.K. Ordering policies of deteriorating items under permissible delay in payments. *J. Oper. Res. Soc.* **1995**, *46*, 658–662. [CrossRef]
20. Sarkar, B.; Saren, S.; Cárdenas-Barrón, L.E. An inventory model with trade-credit policy and variable deterioration for fixed lifetime products. *Ann. Oper. Res.* **2015**, *229*, 677–702. [CrossRef]
21. Liao, J.-J.; Huang, K.-N.; Chung, K.-J.; Ting, P.-S.; Lin, S.-D.; Srivastava, H.M. Some mathematical analytic arguments for determining valid optimal lot size for deteriorating items with limited storage capacity under permissible delay in payments. *Appl. Math. Inform. Sci.* **2016**, *10*, 915–925. [CrossRef]
22. Chang, H.-J.; Hung, C.-H.; Dye, C.-Y. An inventory model for deteriorating items with linear trend demand under the condition of permissible delay in payments. *Prod. Plan. Control.* **2001**, *12*, 274–282. [CrossRef]
23. Manna, S.K.; Chaudhuri, K.S. An EOQ model with ramp type demand rate, time dependent deterioration rate, unit production cost and shortages. *Eur. J. Oper. Res.* **2006**, *171*, 557–566. [CrossRef]

24. Skouri, K.; Konstantaras, I.; Manna, S.K. Chaudhuri, K.S.; Inventory models with ramp type demand rate, time dependent deterioration rate, unit production cost and shortages. *Ann. Oper. Res.* **2011**, *191*, 73–95. [CrossRef]

25. Sett, B.K.; Sarkar, B.; Goswami, A. A two-warehouse inventory model with increasing demand and time varying deterioration. *Sci. Iran.* **2012**, *19*, 1969–1977. [CrossRef]

26. Sarkar, B. An EOQ model with delay in payments and time varying deterioration rate. *Math. Comput. Model.* **2012**, *55*, 367–377. [CrossRef]

27. Sarkar, B.; Sarkar, S. An improved inventory model with partial backlogging, time varying deterioration and stock-dependent demand. *Econ. Model.* **2013**, *30*, 924–932. [CrossRef]

28. Chung, K.-J.; Cárdenas-Barrón, L.E.; Ting, P.-S. An inventory model with non-instantaneous receipt and exponentially deteriorating items for an integrated three layer supply chain system under two levels of trade credit. *Int. J. Prod. Econ.* **2014**, *155*, 310–317. [CrossRef]

29. Chung, K.-J.; Liao, J.-J.; Ting, P.-S.; Lin, S.-D.; Srivastava, H.M. A unified presentation of inventory models under quantity discounts, trade credits and cash discounts in the supply chain management. *Revista de la Real Academia de Ciencias Exactas, Físicas y Naturales Serie A Matemáticas (RACSAM)* **2018**, *112*, 509–538. [CrossRef]

30. Chung, K.-J.; Lin, S.-D.; Srivastava, H.M. The complete solution procedures for the mathematical analysis of some families of optimal inventory models with order-size dependent trade credit and deterministic and constant demand. *Appl. Math. Comput.* **2012**, *219*, 141–156. [CrossRef]

31. Chung, K.-J.; Lin, S.-D.; Srivastava, H.M. The inventory models under conditional trade credit in a supply chain system. *Appl. Math. Model.* **2013**, *37*, 10036–10052. [CrossRef]

32. Liao, J.-J.; Huang, K.-N.; Chung, K.-J.; Lin, S.-D.; Ting, P.-S.; Srivastava, H.M. Mathematical analytic techniques for determining the optimal ordering strategy for the retailer under the permitted trade-credit policy of two levels in a supply chain system. *Filomat* **2018**, *32*, 4195–4207. [CrossRef]

33. Liao, J.-J.; Huang, K.-N.; Chung, K.-J.; Ting, P.-S.; Lin, S.-D.; Srivastava, H.M. Lot-sizing policies for deterioration items under two-level trade credit with partial trade credit to credit-risk retailer and limited storage capacity. *Math. Methods Appl. Sci.* **2017**, *40*, 2122–2139. [CrossRef]

34. Srivastava, H.M.; Chung, K.-J.; Liao, J.-J.; Lin, S.-D. and Chuang, S.-T. Some modified mathematical analytic derivations of the annual total relevant cost of the inventory model with two levels of trade credit in the supply chain system. *Math. Methods Appl. Sci.* **2019**, *42*, 3967–3977. [CrossRef]

35. Sarker, B.R.; Mukherjee, S.; Balan, C.V. An order-level lot size inventory model with inventory-level dependent demand and deterioration. *Int. J. Prod. Econ.* **1997**, *48*, 227–236. [CrossRef]

36. Wu, J.-A.; Teng, J.-T.; Chan, Y.-L. Inventory policies for perishable products with expiration dates and advance-cash-credit payment schemes. *Int. J. Syst. Sci. Oper. Logist.* **2018**, *5*, 310–326. [CrossRef]

37. Jaggi, C.K.; Gupta, M.; Kausar, A.; Tiwari, S. Inventory and credit decisions for deteriorating items with displayed stock dependent demand in two-echelon supply chain using Stackelberg and Nash equilibrium solution. *Ann. Oper. Res.* **2019**, *274*, 309–329. [CrossRef]

38. Kawale, S.; Sanas, Y. A review on inventory models under trade credit. *Int. J. Math. Oper. Res.* **2017**, *11*, 520–543. [CrossRef]

39. Sarker, B.R.; Jamal, A.M.M.; Wang, S.-J. Supply chain models for perishable products under inflation and permissible delay in payment. *Comput. Oper. Res.* **2000**, *27*, 59–75. [CrossRef]

40. Shah, N.H.; Soni, H.N.; Patel, K.A. Optimizing inventory and marketing policy for non-instantaneous deteriorating items with generalized type deterioration and holding cost rates. *Omega* **2013**, *41*, 421–430. [CrossRef]

41. Chang, C.-T.; Teng, J.-T. Retailer's optimal ordering policy under supplier credits. *Math. Methods Oper. Res.* **2004**, *60*, 471–483. [CrossRef]

42. Varberg, D.; Purcell, E.J.; Rigdon, S.E. *Calculus*, 9th ed.; Pearson Education Incorporated: Upper Saddle River, NJ, USA, 2007.

 © 2019 by the authors. Licensee MDPI, Basel, Switzerland. This article is an open access article distributed under the terms and conditions of the Creative Commons Attribution (CC BY) license (http://creativecommons.org/licenses/by/4.0/).

Article

The Quantitative Analysis of Workers' Stress Due to Working Environment in the Production System of the Automobile Part Manufacturing Industry

Muhammad Omair [1], Misbah Ullah [2], Baishakhi Ganguly [3], Sahar Noor [2], Shahid Maqsood [2] and Biswajit Sarkar [4,*]

1 Department of Industrial Engineering, Jalozai Campus, University of Engineering and Technology, Peshawar 25000, Pakistan
2 Department of Industrial Engineering, University of Engineering and Technology, Peshawar 25000, Pakistan
3 Department of Mathematics & Statistics, Banasthali University, Rajasthan 304022, India
4 Department of Industrial and Management Engineering, Hanyang University, Ansan 15588, Korea
* Correspondence: bsbiswajitsarkar@gmail.com; Tel.: +82-1074981981

Received: 28 April 2019; Accepted: 8 July 2019; Published: 15 July 2019

Abstract: Production now requires the management of production processes and operations on the basis of customers' demand to ensure the best combination of technology and humans in the system. The role of the humans in the production process is very significant for the production and quality of the product. The production system depends upon technology and human factors and is highly influenced by the working conditions of the workers, that is, work load, physical, dealings, job timings and so forth. In the current global economy, minimizing production costs is a serious priority for the industries. However, the costs of bad working conditions increase the intensity of the average stress among employees to cause extra costs by affecting the workers' efficiency and products' quality, which is invisible in the eyes of decision makers. This research identifies the cost of workers' stress by developing a linkage between the economic benefits of the firms and the social upgrading of the workers. A numerical example of a production based system is performed to represent the real-time application of the proposed model. A sensitivity analysis is also carried out to quantify the impact of average stress among workers on the production system. Sequential quadratic programming is used to optimize the given nonlinear model for production planning. The optimal results influence ergonomics awareness and the relationship with the safety culture among managers in a firm. It is concluded that efficient and effective production cannot be possible without considering the working conditions of humans in the firm. Managerial insights are also generated from the implications of the results and sensitivity analysis.

Keywords: production; imperfect production; defective rate; workplace stress; workers' efficiency

1. Introduction

The production industry is considered one of the key indicators for the development of a nation. Human resources are a significant approach to a firm's performance and it is believed that the most important assets of the firm are its people. The workforce has been an important variable for management to maximize the revenues [1]. Despite these assessments, managers are giving a relatively low priority to workers. That is the reason why, when a firm needs to cut costs, they first look to the investments in the worker, that is, wages, training and firing [2]. Human workers require social development and deserve a good working environment to avoid workplace stress. It is the responsibility of the firm to provide a safe and healthy environment for the workers for better performance and productivity.

Work-related stress and workplace violence are widely recognized as major challenges to occupational health and safety [3–5]. Most injuries at work occur from physical stress and the strain to perform repetitive and overused tasks, over a long time, which results in damaged joints, muscles, and tendons [6]. Job stress can be defined as the occurrence of harmful physical and emotional response when the requirement of the work does not fulfil the capability of worker [7]. It is considered as a major challenge to the individual on the basis of mental and physical health, and it also damages the organizational health [8]. Job insecurity and physical exertion at the workplace also cause stress. Stressed workers are more likely to be unhealthy, less motivated, poor productivity and a less safe working environment, which ultimately produces a bad impact on the success factor of the firm in a competitive market [9]. It is estimated that job stress costs the state economy, health care and lost productivity. The latest figures state that the estimated cost of work-related stress costs the UK economy are £7 billion a year in sick pay and lost production [10].

Even in advanced high-technology based industries, the physical demands of work are still high, which is directed to produce the environment of physical hazards and work injuries [11]. However, limited work has been done on the impact of stresses on work performance. Similarly, many factors can reduce the impact of work stress during work, but also a very little work has been done to cover these individual and organizational factors. An important source of stress is job strain, which is faced by workers at the workplace. Job insecurity and physical exertion at workplace also cause stress. Even in advanced high-technology based industries, the physical demands of work are still high, which is directed to produce an environment of physical hazards and work injuries [12].

Traditionally decision makers consider common costs, that is, manufacturing, labor cost, holding cost and maintenance cost to take decisions in inventory and production systems. Despite the significant importance of non-ergonomic working conditions and work injuries, these are ignored. Managers cannot justify the investments in any project regarding the working environment unless it is economically feasible and beneficial to the firm. Accordingly, to manage the workplace, it is extremely significant to understand the costs and economic benefits of breakeven time [13]. This study represents the significance of controlling workplace stress due to the working environment through a mathematical model. The inclusion of stress level among workers is anticipated in considering effective and efficient production. The mathematical analysis of average stress is a big contribution, which is inversely related to the worker's efficiency and product's quality. The contribution of this research is extended to quantify the production loss and required labor due to the stress. The analysis provides a platform for production managers to make investments in favor of workers and can be utilized to support the research with the objective to determine the factors that bring awareness and a safety culture in the firm.

This paper is structured as follows: Section 2 present a literature survey regarding stress level, and imperfect production. Furthermore, Section 3 is related to the verbal problem statement to discuss the imperfect production system and drawbacks in the form of workplace stress. The formulation of a mathematical model considering notation, assumptions along with solution methodology is given in Section 4. Section 5 depict the numerical experiment for the practical implication of the mathematical model including data collection and data analysis. Section 6 is related to the numerical results of the experiment performed in the automobile part manufacturing firm. Section 7 presents the sensitivity analysis of the model to determine the effect of workers' stress on the production system. The directions and recommendations for the support of management firms are also given. Finally, Section 8 finds the conclusions of this study.

2. Literature Survey

Effective and efficient production systems rely on the working conditions of the worker. The objective is to improve the working environment by optimising the effort of workers to enhance the firm's performance and promote human well-being. Most of the ergonomic-related work in developing countries is based in the industrial sector. Researchers also worked to create a link between working environment and cost. But the problem arises of how to quantify the invisible average costs of work-related injuries [14]. There is no specific and unique methodology in the literature, which could understand and calculate the cost of displeasure due to bad work conditions and the average cost of pain due to work-related injuries.

Most of the researchers have worked on the economic and environmental aspect of the production system i.e., [15–18]. However, now a work has been started to encourage the social dimension of the system, and in this direction several authors contributed to measure the cost associated in such cases. The cost of work injuries due to bad working environment is calculated by using the friction method, capital method, and willingness to pay method [19]. Furthermore, it also observed that working hours have a huge impact on work-related injuries because longer shift timings increase the probability of an accident [20]. Ruhm (2000) observes a good relationship between macroeconomic conditions and mortality, which he attributes to bad working conditions, the physical exertion of the worker, and work stress when job hours are extended [21]. Work stressors are environmental factors at work that lead to individual strain, that is, potentially harmful reactions of the individual [22].

The most common job stressors considered by researchers are chronic, for example, job conflict. Few researchers also focus on the shorter-term stressors, also called acute stressors. Examples of acute stressors might include something as annoying as a research assistant encountering computer shutdowns [23]. Chronic stressors are usually conceptualized and measured generically (i.e., the same for all jobs), while measures of events or acute stressors tend to be more job-specific, both conceptually and operationally [24]. This study considers the chronic stressors. The questionnaire and the data collection are based on the chronic stressors among the workers. The consequences of high stress among workers are production loss, bad quality work, work injuries, hiring new workers, time lost, legislation, legal expenses, lost jobs and training.

Knauth (1998) [25] discussed the effect of different attributes of working hours on fatigue. Early morning shifts, night shifts, extended working hours, and short recesses are the significant factors, which may cause accidents at work and reduce productivity. The concept behind the risk of fatigue among workers due to shift schedule is to keep it simple [26]. Inconvenient working environments cause stress, which does not allow the maximum utilization of personal ability to perform well [27]. A global survey estimated that due to high stress levels, 90% of the workers were disengaged and among 57% were absolutely disconnected from their work [28]. Another issue caused by increase stress level among worker is production loss. If a worker is normal, then he can work efficiently and effectively to achieve his target production, on the other hand if he feel any stress or pressure then definitely it will cause production loss [29]. Abraham Maslow imagined the categorized employee behaviour patterns in five levels of survival, security, social, stability and satisfaction for the knowledge-based global economy [30].

The effect of intangible cost on production and inventory can be analysed in various production and supply chain models. Most of the researchers ([31–36]) worked on imperfect production systems to help the managers in planning and controlling the bad quality items. However, very few researchers analyzed the cause to reduce the imperfect production in the system. There are many factors affecting the production flow to produce imperfection in the form of reworks, rejections and scraps. Errors can be generated by man, machine or material. This research covers the imperfect production occurring due to the average stress among unskilled workers caused by non-ergonomic working conditions. Previous studies analyzed the effect of stress on the efficiency of the workers and still its effect on the production system was missing. Mansour (2016) [26] only related the stress among workers due to bad working conditions and injuries to the efficiency of the systems. However, this research paper considers the stress level among workers that not only affects the efficiency of the system but also the defective rate. The contribution of this study is to incorporate the workplace depending efficiency and defective rate in the production model as an extension. The impact of stress on the production system is valid theoretically but still there is a gap to find it in tangible dimensions. The production model is developed to analyze the quantitative effect of stress on the total cost and required workers in the production system. The mathematical analysis of the proposed production model is evident to provide the importance of the good working environment, where there is less human stress to avoid the economic loss due to degraded efficiency and high defective rate.

3. Verbal Problem Statement

The research is performed by converting the theoretical idea into the mathematical model and be analysed by the nonlinear programming technique. The model considered an optimization problem to minimize the total cost of production system provided with the limitation of budget, production, and inventory space. The launch of a new product by the combination of human and technology is a big challenge for production planning and control. The developing countries are mainly concerned about manpower as compared to technology because the firms can easily acquire cheap labor against expensive high tech machines. That is the reason, in such scenarios the efficiency and the performance of unskilled labor is highly dependent upon ergonomic conditions and other moral support from the firm, which may result in a high stress level among workers in typical production firms. The issue can be clearly highlighted by considering a mass production system of automobile spare part industry, where there is a need to manage the resources in the form of unskilled workers and machines. The machines are performing the same operation of mass production and worker is working on it. The task might be repetitive or time consuming with an effect of bad working conditions. A single machine problem is considered to experiment the research idea represented in this paper. The flow diagram of imperfect quality production system of three automobile parts *A*, *B*, and *C* parts is illustrated in Figure 1.

The raw material from the inventory transported to the production department, where operators are working on machines for manufacturing of a automobile parts. A production planing and control decisions are looking to find the resources in terms of machines and labor required in production department, which depends upon the production rate and demand rate. The efficiency of operators is influenced by the average stress. The finished parts are shifted to quality department for inspection, where reworked parts are backtracked to machines while the scrapped items are recycled. After managing the imperfect products, the finished goods are transported to the final inventory. The defective items (bad quality items) are produced due to errors caused by man, machine or material, that is the reason the total defective rate is introduced to be the sum of the initial and variable defective rate. The former is the result of machine errors while the latter depends upon the average stress among workers. The main reasons for psychological stress among workers are injuries, moral degradation and so forth.

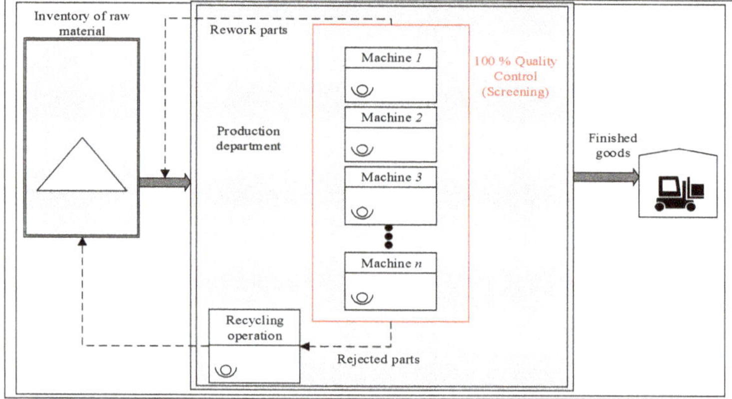

Figure 1. Representation of imperfect quality production system.

Let x_i represents a scaled input qualitative variable. The relationship between these qualitative input variables can be mapped into a single value to represent an average stress level among employees. Accordingly, the stress can be expressed as given in Equation (1).

$$s \;=\; \sum_{i=1}^{n}(w_i.x_i) \tag{1}$$

where s is stress level; x_i is the scaled elements of the work conditions that cause stress; w_i is the normalized weights.

3.1. Stress Level and Efficiency

The stressed worker cannot provide his best utilization according to his capabilities. The affiliation of the worker's stress level with efficiency can be formulated as given in Equation (2), where ρ is efficiency of the workers. The efficiency is directly proportional to the average stress (s) and the relationship is drawn by a curve as shown in Figure 2 [26].

$$\rho(s) \;=\; e^{-s/m} \tag{2}$$

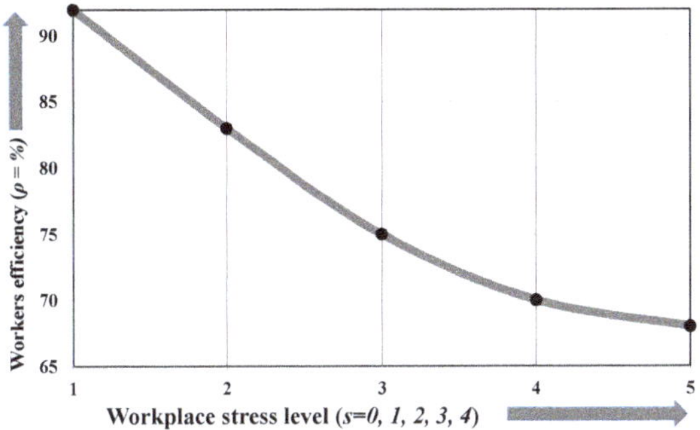

Figure 2. Relation between workers' stress and efficiency.

3.2. Stress Level and Defective Rate

The expression for defective rate includes variable defective rate depending upon the average stress with initial defective rate to cover the rest of factors, that is, machine, material and so forth, is given as in Equation (3). The Initial defective rate is considered to follow a uniform distribution. A direct relationship between stress and defective rate can be shown as in Figure 3, where μ is the total defective rate of the production system, μ_0 is the initial defective rate, τ and ϵ are the scaling factors and s is average stress among worker.

$$\mu(s) \;=\; \mu_0 + \tau \times s^{\epsilon} \tag{3}$$

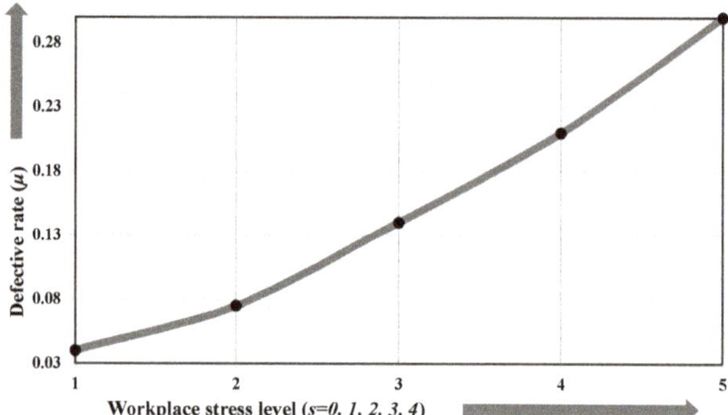

Figure 3. Relation between workers' stress and defective rate.

4. Formal Problem Statement

4.1. Assumptions

There are following assumptions for the proposed model.

1. The model is considered for multiple type items.
2. All products are screened and screening cost is incurred on each item [37].
3. Defective rate is stress level-dependent and the initial defective rate is considered as uniformly distributed.
4. Average efficiency of workers is also a function of stress level.
5. The defective items are reworked to make the perfect quality products.
6. Some parts are rejected after reworking the operation, which is recycled [38].

4.2. Notation

The research is based on the optimization of the decision variables to provide significant support to the managers and experts in production planing and control phase. Indeed this research provides a production resource planing for any production system with given data. There are three decision variables proposed in the mathematical modelling, that is, cycle time (T), number of machines required (K_j), and number of workers required (L_j). The cycle time provides data regarding the total time of the production system to process all the required parts with respect to the given demand. There are K_j number of machines working similar operations, the capacity of the system depends on the number of machines. Similarly, the number of workers L_j depend upon number of machines. The proposed research based on mathematical model will be effectively and efficiently providing information to the managers regarding exact number of machines and number of workers required to process parts by fulfilling the annual demand. The notations of the parameters and decision variables of the proposed imperfect production model are enlisted comprehensively as.

4.3. Mathematical Modelling

A mathematical model based on a single-stage production system with defective in the form of reworked and rejected items include fixed and variable costs, that is, capital cost, labor cost, setup cost, manufacturing cost, inventory carrying cost, reworking cost, energy cost and recycling cost. The capital cost is time-related, which consists of initial investment for purchasing and installing machine units. Cost of manufacturing in this model is associated with the machine/workstation used to manufacture the product. The total inventory of the production system is holding as given in Figure 4, where P is the production rate of the system, D is the demand, $I(t)$ is the total inventory of the system, the cycle time T is divided into small time fractions (i.e., t_1, t_2, and, t_3), h_1 and h_2 are the parameters used to represent the heights of the inventories during production and reworking process respectively. It is a continuous production scenario for one production cycle T. The model is based on pure production system where the production is going on and the demand D is also fulfilling simultaneously during t_1 and t_2. The reworking operations are done during time interval t_2. The production stopped at t_2, then there will be no production and inventory is going down to zero during t_3. Theocratically, the total cost of production is the sum of capital cost, setup cost, manufacturing cost, holding, backorder cost, energy, inspection, reworking and recycling cost. Mathematically, these fractions of the total cost can be mathematically calculated as given in Appendix A for the support of managers and industries.

By considering the interaction between the labors and machines, the total cost function is the sum of the fixed costs to represent the capital cost and variable costs in terms of labor cost. The objective function is to minimize the total cost of production as given in Equation (4).

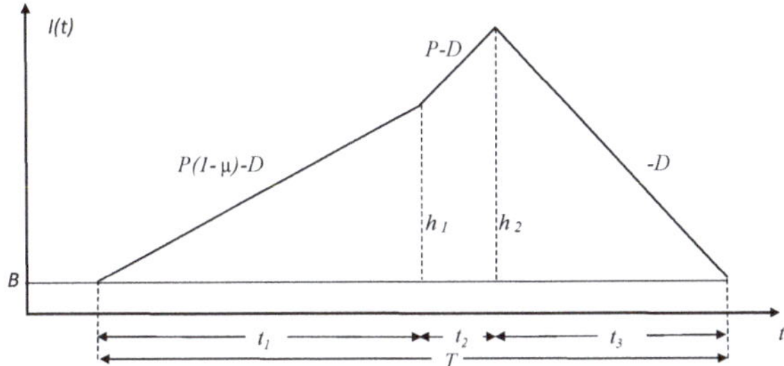

Figure 4. Economic production Quantity model with variable defective rate and backorders.

$$
\begin{aligned}
\text{Minimize cost} \quad = \quad & \text{Capital cost} + \text{Setup cost} + \text{Manufacturing cost} + \text{Labor cost} + \\
& \text{Holding cost} + \text{Backorder cost} + \text{Energy cost} + \text{Reworking cost} + \\
& \text{Screening cost} + \text{Recycling cost}
\end{aligned}
\tag{4}
$$

The mathematical form of the objective function to minimize the total cost of production can be given as in Equation (5). The model is limited by the budget, production, and capacity constraints.

$$
\begin{aligned}
Min\ TC \ = \ \sum_{j=1}^{J} \Bigg[& \frac{K_j V_j}{T} + \frac{S_j}{T} + M_j D_j + \frac{\rho(s)L_j W_j}{T} + \left(\frac{H_j D_j T\left(1-\left(1+\mu(s)+\mu(s)^2\right)\frac{D_j}{P_j}\right)}{2} + \frac{H_j B_j^2(1-\mu(s))}{2D_j T(1-\mu(s)-\frac{D_j}{P_j})} \right. \\
& \left. - H_j B_j \right) + \left(\frac{F_j B_j}{T} + \frac{y_j B_j^2(1-\mu(s))}{2D_j T(1-\mu(s)-\frac{D_j}{P_j})} \right) + \pi U_j D_j + \mu(s)M_j D_j + \frac{\theta_j}{T} + D_j \psi_j + \eta u_j \mu(s)D_j \Bigg]
\end{aligned}
\tag{5}
$$

Subject to

Budget constraint: The budget constraint is the limitation given by the management on the availability of the resources. There is a cost associated with the machines and number of workers. The addition or removal of the machine or worker in the production plan will effect the cost budget and balance. That is the reason the combination of the workers and machines should by utilized in this way to not exceed than the total budget as represented by Equation (6).

$$
K_j V_j + W_j L_j \ = M_j
\tag{6}
$$

Production function constraint: The production function provides a constraint to relate the production quantity with the number of workers and number of machines. Where there is a combine effect of availability factor A_f (%) for machines and workers, and an efficiency (ρ) of the workers is multiplied with the number of workers. The α and β are the respective shares of machines and workers in the production system depending upon the level of automation.

$$D_j T = A_f K_j^{\alpha} \left(\rho(s) L_j \right)^{\beta} \tag{7}$$

Space constraint: The inventory of the production system is also controlled depending upon the capacity of the storage system or warehouse. Therefore, the space constrain (volume based) is very important to limit the maximum quantity depending on the size of each part.

$$c_j D_j T \leq C_j \tag{8}$$

Non-negative constraint: All the decision variable i.e., T, K_j, and L_j are non negative.

$$T, K_j, L_j \geq 0 \tag{9}$$

5. NumericaL Experiment

Most of the data used for analysis of the proposed mode has been taken from the numerical experiment done by Sarkar et al., (2014) [39] except the data related to the average stress among workers in a production system.

5.1. Data Collection

The average stress is measured by the workplace stress scale on the basis of a questionnaire as given in Appendix B. The scale was designed and made by the Marlin Company, North Haven, CT, USA, and the American Institute of Stress (AIS) in 1978, Yonkers, NY, USA [40]. There are two basic type of stressors, that is, chronic and acute. Chronic stressors are usually conceptualized and measured generically (i.e., the same for all jobs), while measures of events or acute stressors tend to be more job-specific, both conceptually and operationally [23]. The chronic stressors are considered to collect the data from the workers because the research covered basic but not specific working conditions and the production system rely on long term data for making decisions. That is the reason a questionnaire is considered to reflect the data from the workers facing generic stresses throughout their job. The questionnaire is standard and general, which can be applied to any circumstances to find the average stress among workers. Each question indicates the specific performance factor of the worker chronically and in generic, that is, job satisfaction, safe working conditions, deadlines, job pressure, skills utilization and so forth. Responses of the workers are collected by using a five-point scale starting from one to five representing never, rarely, sometimes, often and very often, respectively [41]. This scale of questionnaires was utilized to evaluate the function of occupational stress among workers [42]. A survey was conducted among five automobile spare parts industries. A Total of 150 questionnaires were distributed among unskilled workers. The demographics of the participating workers are as follows.

1.   All workers are participated in the survey irrespective of the age and health.
2.   They are working on the production floor as an operator or helper.
3.   The respondent unskilled workers are performing manual activities, e.g., loading, unloading, helping, operating, heavy working, manual forging, cutting etc.

The survey was conducted during fall 2018. It was performed at an Industrial estate in Pakistan. The data was collected from the workers in two ways—those who could understand filled the questionnaire by themselves, otherwise most of them were interviewed for collecting the data. The workers were chosen randomly to avoid bias. Almost five general manufacturing industries related to the automobile part were selected for the collection of data for workplace stress to reflect the general conditions of workers. These industries are working as a separate firms to manufacture automobile spare parts, each consist of almost 400–500 workers. 150 workers were selected randomly among total 2200 workers from five industries (at the rate of 6%), where 130 workers returned the questionnaires and among which 12 were rejected due to incomplete and unreliable information. Therefore, a total 118 questionnaires were considered for the analysis of this research work and the response rate was about 78.7%. The sample size of 118 is enough to reflect the population of the workers among manufacturing industries.

5.2. Data Analysis

For the reliability and adequacy of the data, a Cronbach's reliability test was performed. The objective of the test is to check the average differences among each data. The test value for each question is calculated, which was more than 0.7 and is commonly recommended for the validation and reliability of the data. As all values result were above 0.70, a good reliability resulted for data sets, and these data sets were accepted for further analysis [43]. The value of stress among workers was found to be 2.6 as a normal average on a five-point Likert scale. The formula for central tendency (mean) was utilized to find the mean average of the workplace stress among workers. The value 2.6 value was considered as a general average stress existed among unskilled workers on the production floor of manufacturing system. The value of workplace stress above 2.6 will be assumed as "High Stressed" and lower value will be considered as a "Low stressed". The five-point Likert scale had qualitative levels from 1 to 5, tagged as never, rarely, sometimes, often, and very often respectively. The never and very often stress levels are the ideal situations for industry. The sensitivity analysis is performed on the basis of these mentioned stress levels to find the changing cost of production and number of labor required.

The data related to the production of each automobile part on the basis of demand, production rate, backorders and, energy utilization are given as in Table 1. The data related to the production rate, setup, holding, demand and manufacturing cost are taken from the work done by Sarkar et al., (2014) [39]. The data related to the energy, machine cost, defective rate, recycling cost, and labor costs are considered directly from the industry because these are depending on the industrial conditions and state regulations. The inspection data is collected from the research study of Sarkar (2016) [44]. The data related to stress are incorporated by the detailed survey using the questionnaire as discussed in the data collection section.

Table 1. Data related to the production and demand.

Item Type	Demand (Units)	Production Rate (Units/Year)	Backorders (Units)	Scrap (%)	Energy (KWh)
A	180	450	25	0.2	0.45
B	200	550	30	0.3	0.5
C	210	580	35	0.4	0.58

The data related to the cost of production are given in Table 2, that is, manufacturing, holding, setup, backorder, labor and machine costs. Manufacturing costs includes running costs incurred on each item. The backorder cost is applied for extra resources when shortages are occurred to fulfill the demand. When parts are transported from production department to quality department, imperfection is produced.

Table 2. Data related to the production costs.

Item Type	Manufacturing Cost ($/Unit)	Setup Cost ($)	Holding Cost ($/Unit/Year)	Fixed Backorder ($)	Variable Backorder ($/Unit)	Machine Cost ($)	Labor Cost ($/Unit)
A	6	45	47	9	0.9	450	1.7
B	7	50	50	10	1	500	2
C	8	55	56	10.5	1.25	580	2.2

The cost related to imperfect production is given as in Table 3. Inspection is carried out after production at quality department, where all parts are checked for defective and sorted the pass, rework and rejected parts. Inspection process acquire cost to perform some visual as well as lab testing. Recycling include process of converting the semi-finished item into raw material to reduce the waste and carry a cost.

Table 3. Data related to the imperfect production.

Item Type	Variable Inspection Cost ($/Unit)	Fixed Inspection Cost ($)	Recycling Cost ($/Unit)	Energy Cost ($/Unit)	Initial Defective Rate (%)
A	0.01	9.5	1.8	5	2.5
B	0.02	10	2	5	2.5
C	0.03	10.5	2.5	5	2.5

6. Results and Discussions

The model is nonlinear and complex enough to solve by using any analytical method. However, there are many iterative methods available to find the optimal solution of nonlinear optimization model with many decision variables and more constraints. The sequential quadratic programming (SQP) are the most effective methods to solve nonlinear equations [45]. The method of SQP is based on Newton's method in the best way to deal with the unconstrained optimizations [46]. The SQP deals with the quadratic programming problems, to find an optimal solution. SQP methods represent the state of the art in nonlinear programming methods. Schittkowski [47] has implemented and tested a version that outperforms every other tested method in terms of efficiency, accuracy, and percentage of successful solutions, over a large number of test problems. The method closely mimics Newton's method for unconstrained optimization. At each major iteration, an approximation is made of the Hessian using a quasi-Newton updating method. This is then used to generate a QP subproblem whose solution is used to form a search direction for a line search procedure.

The proposed production model in the form of nonlinear problem was coded in MATLAB for the analysis and by using the methodology of SQP, optimal solution and results are obtained [48]. The optimal solution of the production model is given as in Table 4, which is a complete production plan for the manufacturing of three items *A*, *B* and *C*. To meet the target in the form of demand rate and by the capability of the production system in term of production rate, the required machines and labor for each item is calculated. The cycle time of the complete production plan is almost three months (0.329 years). The optimal solution of the imperfect production model, that is, total cost (TC) considering the impact of workers' stress is found as $428,248.

Table 4. Optimal solution of the production model.

Sr. No.	Item Type	Decision Variable	Optimal Result	Objective Function (TC)
1		(T)	0.329 years	
2	A	K_1	56 machines	
3	B	K_2	60 machines	
4	C	K_3	60 machines	$428,248
5	A	L_1	111 workers	
6	B	L_2	123 workers	
7	C	L_3	131 workers	

The optimal results and solutions have been found to support the production planning phase of the manufacturing firm. Optimal requirements of resources in the form of workers and machines are quantified with the minimum expense of total cost (TC). This paper presents the impact of average stress on the production system with defective rate and workers' efficiency as a function of workers' stress. However, stress elements are generated among worker, whatever their cost, such stress factors have personal as well as economic consequences and can cause poor productivity and unavailability. Stress factors are required to be calculated economically for the benefit of decision makers. That is the reason, further analysis is required to find the sensitivity of the model by the effect of stress levels.

7. Sensitivity Analysis

Stresses can be divided into chronic and acute [23]. Given the differences between acute and chronic stressors, they may differ in their relations with individual strains and performance [24]. Stressors that are more job specific (whether chronic or acute) may have the greatest impact on individual strains and performance, because they are most salient to employees in a particular job. The chronic job stressors are considered. The stress factor is a variable function and it depends upon various aspects of the production system for example, repetitive and cyclic work type are common in mass production systems to make the job more boring and tedious, which is one of the causes of high stress. However, workers feel less stress in case of job production where the management invested on the training and development for handling a variety of tasks. Similarly, stress also depending upon the firm's policies regarding workers' health care and safety. Sometimes, there are a lot of medical facilities and insurances available for labors in low-income management firms and on the other hand, a very few incentives are available in high-quality firms.

There is a need to check the model on different levels of stress existed across various production firms. The stress factor negatively affects the workers' performance and efficiency, which is directed to lessen the workforce and increase the number of rejections. Hence there are two changes occur, firstly the actual number of labor will be reduced from the standard requirements due to the decreased performance level and secondly, the output production will be reduced to meet the target level. Therefore, to compensate the loss efficiency and maintain the output quantity at a certain level, the amount of labor should be increased. This can be done either by hiring extra labor in terms of cost or increasing the production schedule time. In both cases, the total cost (TC) will be increased. The detailed analysis and variations of TC and labors at each level of stress are given as in Table 5. These cases are given as following.

1. The scale can divided into five values of s levels from never, rarely, sometimes, often, and very often. An analysis is carried to quantify the exact amount of labor required by increasing the level of workers' stress ranging from $s = 0$ to $s = 4$ as given in Table 5. In first case, by considering L_1 it is observed that when stress is increasing then there is no changes occurs. The reason is the demand for first part A is almost 180, which can be easily fulfilled by the same number of workers, that is, 111. The change in average stress would not create any variation in the production system to disturb the manual workers.

2. By comparing the status of the labors required for part B and C, that is, L_2 and L_3 respectively at extreme levels of the stress, the labors required at $s = 0$ and $s = 4$ are 119 and 126 for item B while 124 and 137 for item C respectively. Indeed, the analysis is providing a quantitative impact of the stress level on the number of workers required in the production system.

3. Since the stress is a variable factor and there is also a need to analyze the sensitivity of the proposed model for the TC by changing the stress among workers. It is clear that the model is sensitive to the stress on the basis of TC and labors required for each part, that is, the total cost of production and labor requirements are increased by increasing the stress from level to level. However, the mathematical calculations are required. When comparing the extreme levels of the workplace stress, it is found that at $s = 0$, the total cost of production is around \$420,601, and at $s = 4$, the value increased to \$432,410.

4. The machines (K_1, K_2, and K_3) are not changing by increasing the average stress among the workers because the capital cost is enough that it cannot be effected by the average stress among workers. The efficiency $\rho(s)$ of the workers is inversely proportional whereas the defective rate $\mu(s)$ is directly proportional to the workplace stress.

Hence the mathematical analysis for the impact of the stress on the workers required and TC is providing an evidence to think and plan for the good working environment to the workers. Where, there is less chances to induce stress among workers due to which the efficiency of the workers is at a maximum level and there are less chances to create rejections due to workers. These meaningful results in different scenarios are beneficial to understand the economic loss (almost \$40,000) and consequences of the production system due to high-level stress among workers, which will pressurize the managers to improve the safety culture and working environment.

Table 5. Sensitivity analysis of the model with respect to the stress level of worker.

Parameter	Stress Level					
	$s = 0$	$s = 1$	$s = 2$	$s = 2.6$	$s = 3$	$s = 4$
				(This Paper)		
TC	420,601	422,469	426,464	428,248	429,438	432,410
L_1	111	111	111	111	111	111
L_2	119	120	123	123	123	126
L_3	124	126	131	131	132	137
K_1	56	56	56	56	56	56
K_2	60	60	60	60	60	60
K_3	60	60	60	60	60	60
$\rho(s)$	1	0.93	0.875	0.84	0.818	0.765
$\mu(s)$	0.025	0.035	0.053	0.066	0.076	0.105

8. Conclusions

This paper investigates the meaningful hidden costs, which are not considered by management in the planning phase of the production system. The injuries and the poor working conditions are the causes of the high workers' stress. The stress exists among workers in every environment, but the intensity and level of stress are different. Poor work conditions will amplify the stress among employees, which will significantly influence the production system. On the other hand, a more ergonomic workplace and safer practices will benefit the corporations. In-depth, the stress affects the efficiency of the workers and production rate causes the management to hire more workers and material for compensation causes more cost. The average stress among workers of traditional spare part industries is calculated by the detailed survey for practical application of the model. The incorporation of stress level in production enhances the quality of decision making to consider for the optimal solution. To get the optimal solution, the solution methodology of sequential quadratic programming (SQP) is selected, which uses Langrage multiplier directly and is based on the equations of Karush-Kuhn-Tucker (KKT). The sensitivity analysis provides a detailed analysis of the stress at each level in different scenarios for the sensitivity of the proposed imperfect production model. The results addressed the increased amount of invested labor and defective rate due to the increased stress level. These increments affect the total cost of production, which are not estimated and are ignored. The objective value of this research is to create awareness among production managers by calculating the tangible cost of workplace stress to control.

A set of recommendations are drawn from the optimal results obtained by the numerical analysis of production model. In case of high stress, almost 3, 11, and 21 extra workforce required for part A, B, and C respectively ($15,020) and overall economic loss is in the production cost is almost $47,000. The study motivates the decision makers to also include the quantitative factor of workplace stress in the production planing and control phase. The managers need to aware about the causing factors of stress among workers to minimize the economic loss (almost $47,000). The data collection and data analysis from the detailed questionnaire provide a better understanding of each performance factor related to the worker stress, which can be analysed for improving the average efficiency of the workers. It is one of the social responsibility of the production system to improve and maintain the workers' job satisfaction, safe working conditions, work deadlines, job pressure and skills utilization to avoid inducing stress. Furthermore, by this research the managers are required to calculate tangibly the average stress among workers quarterly or bimonthly for efficient and effective production with minimum rejections.

The understanding and the extensions of the proposed model can be extended into a socially responsible production model, one of the significant and major areas, but unfortunately it has less value in the eyes of managers. Future extensions of the models are also possible under certain conditions. Demand as a function of a sustainable product can be incorporated for better results. The purposed model concluded that stress among workers affects the production system due to the efficiency of workers and defective items. There is a need to make production more reliable by considering intangible costs, which seem invisible but ultimately affect the total cost of production under variable production quantity. A long-term strategic analysis is also necessary to represent the stress affecting capital units (machines). In addition, the detailed validation of the quantitative impact of workplace stress on the production is also required, which might include a survey based on questionnaires. The respondents will be the industrial managers, experts, and academicians. The objective is to validate the changes occurred in the production cost due to change in the workplace stress as an outcome of this research. Furthermore the proposed model can be compared against the research work done by Sarkar et al. (2014) [39] on the basis of the total cost by considering the same assumptions. The data collected from the manufacturing firms can be utilized to find the stress levels of the industries on the basis of the collected data. The analysis could be helpful for industries to identify the weak areas and the potential to improve their stress levels to achieve fewer rejections and higher efficiency in the production system. Overall, this research creates an awareness among managers to understand the economic value of workers' stress level in production.

Author Contributions: All the authors contributed equally at every stage of this research work. However, the individual contributions in the research study are as following i.e., writing–original draft preparation and methodology, M.O.; writing–review and editing, S.N.; resources and investigation, M.U.; software, S.M.; data curation and formal analysis, B.G.; supervision and conceptualization, B.S.

Funding: This research received no external funding.

Conflicts of Interest: The authors declare no conflict of interest.

Indices

j the index used to indicate the number of items, $j = 1, 2, ..., J$

Decision variables

n number of cycles (number)

T Cycle time of production (years)

L_j Number of labors work to manufacture jth item (numbers)

K_j Number of capital units utilized to manufacture jth item (numbers)

Parameters

W_j average wage of labor to manufacture jth item (\$/labor)

B_j backorder of jth item (\$/unit/year)

F_j fixed backorder cost to fulfill the shortages of jth item (\$/year)

y_j variable backorder cost to fulfill the shortages of item jth (\$/unit)

H_j holding cost of each item per cycle (\$/unit/year)

P_j production rate (units/year)

V_j cost of each machine unit require to manufacture jth item (\$/machine/year)

D_j demand rate, units per planning period (unit/year)

U_j units of energy utilised to manufacture jth item (kwh/unit)

s average stress among workers (number)

m scale element for stress level of worker (constant)

S_j setup cost for jth item (\$/year)

R_j reworking cost of jth item (\$/unit)

α labor share (%)

β capital share (%)

A_f availability factor (%)

θ_j fixed inspection cost of item jth (\$/year)

ψ_j variable inspection cost of item jth (\$/unit)

π_j cost of energy per unit required for jth item (\$/kWh/unit)

η_j scrap rate for recycling (%)

v_j recycling cost of item jth (\$/unit)

$\mu(s)$ total defective rate (percentage)

μ_0 initial defective rate (percentage)

τ scaling parameter (positive constant)

ϵ shape parameter (positive constant)

c capacity of each item (%)

C total capacity of inventory (%)

$\rho(s)$ the average efficiency of workers (%)

$I(t)$ total inventory of the production system (units)

TC   total cost of production model(\$/cycle)

Appendix A

Generally, managers are facing complexities to calculate the total cost of production. Therefore, a mathematical expressions of all the costs i.e., setup, manufacturing, labor, energy, inspection, reworking, recycling, and holding cost as a part of production system are represented from Equations (A1)–(A8).

Capital cost (CC)

This cost is an independent of the production quantity. It is time related, which consists of initial investment and setup cost of production as given in Equation. Cost of capital in this model is associated with the machine/workstation used to manufacture product.

$$CC = \sum_{j=1}^{J} \frac{K_j V_j}{T} \tag{A1}$$

Setup cost (SC)

The fixed cost including initial cost for each setup of the production system.

$$SC = \sum_{j=1}^{J} \frac{S_j}{T} \tag{A2}$$

Manufacturing cost (MC)

The basic cost depending on the production quantity of the system to meet the targeted demand.

$$MC = \sum_{j=1}^{J} M_j D_j \tag{A3}$$

Labor cost (LC)

The labor cost includes the wages and salaries of the workers depending on the production cycle time.

$$LC = \sum_{j=1}^{J} \frac{\rho(s) L_j W_j}{T} \tag{A4}$$

Energy cost (EC)

Energy cost incurred on all the production quantity due to the energy consumed by the machines, equipments, and utilities etc.

$$EC = \sum_{j=1}^{J} \pi U_j D_j \tag{A5}$$

Inspection cost (IC)

The insepction cost is incurred on all the items to check whether the parts are correct, rejected, or should move towards reworking operations.

$$IC = \sum_{j=1}^{J} \frac{\theta_j}{T} + D_j \psi_j \tag{A6}$$

Reworking cost (RWC)

The reworking cost also affects the total cost for reworking operations on a fraction of defective items.

$$RWC = \sum_{j=1}^{J} \mu(s) M_j D_j \tag{A7}$$

Recycling cost (RC)

The rejected items are move towards recycling process for the regeneration of the raw material.

$$RC = \sum_{j=1}^{J} \eta u_j \mu(s) D_j \tag{A8}$$

Holding cost and Backorder cost

The holding cost is referred to the cost of carrying inventory in the production house, which includes rents, salaries, insurance etc. It also depends on the time during which the final product will be held in inventory. The backorder cost is also incurred on the items produced to fulfill the shortages. Both the holding and backorder costs are taken from the work of Sarkar et al., (2014) [39].

Appendix B

Table A1. Questionnaire from the American Institute of Stress.

Sr.	Questions	Never	Rarely	Sometimes	Often	Very Often
1	In general, I am not particularly proud or satisfied with my job.	1	2	3	4	5
2	Conditions at work are unpleasant or sometimes even unsafe.	1	2	3	4	5
3	I feel that my job is negatively affecting my physical or emotional well-being.	1	2	3	4	5
4	I have too much work to do and/or too many unreasonable deadlines.	1	2	3	4	5
5	I find it difficult to express my opinions or feelings about my job conditions to my superiors.	1	2	3	4	5
6	I feel that job pressures interfere with my family or personal life.	1	2	3	4	5
7	I don't have adequate control or input over my work duties.	1	2	3	4	5
8	I do not receive appropriate appreciation or rewards for good performance.	1	2	3	4	5
9	I cannot utilize my skills and talents fully at work.	1	2	3	4	5
10	I tend to have frequent arguments with superiors, co-workers, or customers.	1	2	3	4	5

References

1. Perla, A.; Nikolaev, A.; Pasiliao, E. Workforce management under social link based corruption. *Omega* **2018**, *78*, 222–236. [CrossRef]
2. Barney, J.B. *Gaining and Sustaining Competitive Advantage*; Addison-Wesley Pub. Co.: Boston, MA, USA, 1997.
3. Choudhry, R.-M.; Fang, D. Why operatives engage in unsafe work behavior: Investigating factors on construction sites. *Saf. Sci.* **2008**, *46*, 566–584. [CrossRef]
4. Leka, S.; Jain, A.; Zwetsloot, G.; Cox, T. Policy-level interventions and work-related psychosocial risk management in the European Union. *Work Stress* **2010**, *24*, 298–307. [CrossRef]
5. EU-OSHA–European Agency for Safety and Health at Work. *Expert Forecast on Emerging Psychosocial Risks Related to Occupational Safety and Health*; European Agency for Safety and Health at Work: Bilbao, Spain, 2007.
6. Kumar, S. Theories of musculoskeletal injury causation. *Ergonomics* **2001**, *44*, 17–47. [CrossRef] [PubMed]
7. NIOSH. *Stress...at Work*; National Institute for Occupational Safety and Health (NIOSH): Washington, DC, USA, 1999; Volume 20, pp. 99–101.
8. Karasek, R. Demand/control model: A social-emotional, and psychological approach to stress risk and active behavior development. In *ILO Encyclopedia of Occupational Health and Safety*; ILO: Paris, France, 1998.
9. Leka, S.; Griffiths, A.; Cox, T.; World Health Organization (WHO). *Work Organisation and Stress: Systematic Problem Approaches for Employers, Managers and Trade Union Representatives*; WHO: Geneva, Switzerland, 2003.

10. Palmer, S.; Cooper, C.; Thomas, K. A Model of Work Stress. Counselling at Work-Winter. Available online: https://s3.amazonaws.com/academia.edu.documents/31468591/acw_winter04_a.pdf?response-content-disposition=inline%3B%20filename%3DPalmer_S._Cooper_C._and_Thomas_K._2004_..pdf&X-Amz-Algorithm=AWS4-HMAC-SHA256&X-Amz-Credential=AKIAIWOWYYGZ2Y53UL3A%2F20190714%2Fus-east-1%2Fs3%2Faws4_request&X-Amz-Date=20190714T064506Z&X-Amz-Expires=3600&X-Amz-Signed-Headers=host&X-Amz-Signature=99276df961605f827720de490055880ce6bf5eefbf248336ac81476d42ddf3d9 (accessed on 6 April 2019).

11. Dollard, M.-F.; Metzer, J.-C. Psychological research, practice, and production: The occupational stress problem. *Int. J. Stress Manag.* **1999**, *6*, 241–253. [CrossRef]

12. Park, J. Work stress and job performance. *Perspect. Labour Income* **2008**, *20*, 7.

13. Hendrick, H.-W. Determining the cost–benefits of ergonomics projects and factors that lead to their success. *Appl. Ergon.* **2003**, *34*, 419–427. [CrossRef]

14. Mrozek, J.-R.; Taylor, L.-O. What determines the value of life? A meta-analysis. *J. Policy Anal. Manag.* **2002**, *21*, 253–270. [CrossRef]

15. Habib, M.-S.; Sarkar, B.; Tayyab, M.; Saleem, M.-W.; Hussain, A.; Ullah, M.; Omair, M.; Iqbal, M.-W. Large-scale disaster waste management under uncertain environment. *J. Clean. Prod.* **2019**, *212*, 200–222. [CrossRef]

16. Omair, M.; Noor, S.; Hussain, I.; Maqsood, S.; Khattak, S.-B.; Akhtar, R.; Haq, I.-U. Sustainable development tool for Khyber Pakhtunkhwa's dimension stone industry. *Technol. J.* **2015**, *20*, 160–165.

17. Omair, M.; Noor, S.; Maqsood, S.; Nawaz, R. Assessment of Sustainability in Marble Quarry of Khyber Pakhtunkhwa Province Pakistan. *Int. J. Eng. Technol.* **2014**, *14*, 84–89.

18. Kang, C.-W.; Imran, M.; Omair, M.; Ahmed, W.; Ullah, M.; Sarkar, B. Stochastic-Petri Net Modeling and Optimization for Outdoor Patients in Building Sustainable Healthcare System Considering Staff Absenteeism. *Mathematics* **2019**, *7*, 499. [CrossRef]

19. Amador-Rodezno, R. An overview to cersss self evaluation of the cost-benefit on the investment in occupational safety and health in the textile factories: A step by step methodology. *J. Saf. Res.* **2005**, *36*, 215–229.

20. Boone, J.; van Ours, J.-C. Are recessions good for workplace safety? *J. Health Econ.* **2006**, *25*, 1069–1093. [CrossRef] [PubMed]

21. Ruhm, C.-J. Are recessions good for your health? *Q. J. Econ.* **2000**, *115*, 617–650. [CrossRef]

22. Beehr, T.A.; Johnson, L.B.; Nieva, R. Occupational stress: Coping of police and their spouses. *J. Organ. Behav.* **1995**, *16*, 3–25. [CrossRef]

23. Caplan, R.D.; Jones, K.-W. Effects of work load, role ambiguity, and type A personality on anxiety, depression, and heart rate. *J. Appl. Psychol.* **1975**, *60*, 713–719. [PubMed]

24. Motowidlo, S.-J.; Packard, J.-S.; Manning, M.-R. Occupational stress: Its causes and consequences for job performance. *J. Appl. Psychol.* **1986**, *71*, 618–629. [CrossRef]

25. Knauth, P. Innovative worktime arrangements. *Scand. J. Work Environ. Health* **1998**, *24*, 13–17.

26. Mansour, M. Quantifying the intangible costs related to non-ergonomic work conditions and work injuries based on the stress level among employees. *Saf. Sci.* **2016**, *82*, 283–288. [CrossRef]

27. Wilkinson, J. Shift work and fatigue-and the possible consequences. *Occup. Health Wellbeing* **2013**, *65*, 27–30.

28. Dyble, J. Workplace stress linked to lower productivity. *Empl. Benefits* **2014**, *9*, 2014.

29. Locke, E.-A. Toward a theory of task motivation and incentives. *Organ. Behav. Hum. Perform.* **1968**, *3*, 157–189. [CrossRef]

30. Harvard, P.-S. Maslow, mazes, minotaurs; updating employee needs and behavior patterns in a knowledge-based global economy. *J. Knowl. Econ.* **2010**, *1*, 117–127. [CrossRef]

31. Wee, H.-M.; Yu, J.; Chen, M.-C. Optimal inventory model for items with imperfect quality and shortage backordering. *Omega* **2007**, *35*, 7–11. [CrossRef]

32. Ganguly, B.; Pareek, S.; Sarkar, B.; Sarkar, M.; Omair, M. Influence of controllable lead time, premium price, and unequal shipments under environmental effects in a supply chain management. *RAIRO-Operations Research.* **2018**.://doi.org/10.1051/ro/2018041. [CrossRef]

33. Goyal, S.-K.; Cárdenas-Barrón, L.-E. Note on: Economic production quantity model for items with imperfect quality–a practical approach. *Int. J. Prod. Econ.* **2002**, *77*, 85–87. [CrossRef]

34. Wook Kang, C.; Ullah, M.; Sarkar, M.; Omair, M.; Sarkar, B. A Single-Stage Manufacturing Model with Imperfect Items, Inspections, Rework, and Planned Backorders. *Mathematics* **2019**, *7*, 446. [CrossRef]

35. Bouslah, B.; Gharbi, A.; Pellerin, R. Integrated production, sampling quality control and maintenance of deteriorating production systems with AOQL constraint. *Omega* **2016**, *61*, 110–126, doi:10.1016/j.omega.2015.07.012. [CrossRef]

36. Jeang, A. Simultaneous determination of production lot size and process parameters under process deterioration and process breakdown. *Omega* **2012**, *40*, 774–781, doi:10.1016/ j.omega.2011.12.005. [CrossRef]

37. Sarkar, B.; Omair, M.; Choi, S.-B. A multi-objective optimization of energy, economic, and carbon emission in a production model under sustainable supply chain management. *Appl. Sci.* **2018**, *8*, 1744. [CrossRef]

38. Omair, M.; Sarkar, B.; Cárdenas-Barrón, L.-E. Minimum quantity lubrication and carbon footprint: A step towards sustainability. *Sustainability* **2017**, *9*, 714. [CrossRef]

39. Sarkar, B.; Cárdenas-Barrón, L.-E.; Sarkar, M.; Singgih, M.-L. An economic production quantity model with random defective rate, rework process and backorders for a single stage production system. *J. Manuf. Syst.* **2014**, *33*, 423–435. [CrossRef]

40. The Marlin Company and American Institute of Stress (AIS). *The Workplace Stress Scale*; The Marlin Company and American Institute of Stress: North Haven, CT, USA and Yonkers, NY, USA, 1978.

41. Aghilinejad, M.; Zargham Sadeghi, A.-A.; Sarebanha, S.; Bahrami-Ahmadi, A. Role of occupational stress and burnout in prevalence of musculoskeletal disorders among embassy personnel of foreign countries in Iran. *Iran. Red Crescent Med. J.* **2014**, *16*, e9066. [CrossRef] [PubMed]

42. McCalister, K.-T.; Dolbier, C.-L.; Webster, J.-A.; Mallon, M.-W.; Steinhardt, M.-A. Hardiness and support at work as predictors of work stress and job satisfaction. *Am. J. Health Promot.* **2006**, *20*, 183–191. [CrossRef] [PubMed]

43. Idrees, M.; Hafeez, M.; Kim, J.-Y. Workers' age and the impact of psychological factors on the perception of safety at construction sites. *Sustainability* **2017**, *9*, 745. [CrossRef]

44. Sarkar, B. Supply chain coordination with variable backorder, inspections, and discount policy for fixed lifetime products. *Math. Probl. Eng.* **2016**, *2016*. [CrossRef]

45. Birgin, E.-G.; Haeser, G.; Ramos, A. Augmented lagrangians with constrained subproblems and convergence to second-order stationary points. *Comput. Optim. Appl.* **2018**, *69*, 51–75. [CrossRef]

46. Mostafa, N.; Khajavi, M. Optimization of welding parameters for weld penetration in FCAW. *J. Achiev. Mater. Manuf. Eng.* **2006**, *16*, 132–138.

47. Schittkowski, K. NLPQL: A fortran subroutine solving constrained nonlinear programming problems. *Ann. Oper. Res.* **1986**, *5*, 485–500. [CrossRef]

48. Theodorakatos, N.-P.; Manousakis, N.-M.; Korres, G.-N. A sequential quadratic programming method for contingency-constrained phasor measurement unit placement. *Int. Trans. Electr. Energy Syst.* **2015**, *25*, 3185–3211. [CrossRef]

© 2019 by the authors. Licensee MDPI, Basel, Switzerland. This article is an open access article distributed under the terms and conditions of the Creative Commons Attribution (CC BY) license (http://creativecommons.org/licenses/by/4.0/).

Article

A Generalized Process Targeting Model and an Application Involving a Production Process with Multiple Products

Mohammad A. M. Abdel-Aal * and Shokri Z. Selim

Systems Engineering Department, King Fahd University of Petroleum and Minerals, P.O. Box 5063, Dhahran 31261, Saudi Arabia; selim@kfupm.edu.sa

* Correspondence: mabdelaal@kfupm.edu.sa or m.abdelaal82@gmail.com; Tel.: +966-583-569-338

Received: 6 July 2019; Accepted: 1 August 2019; Published: 3 August 2019

Abstract: This paper presents a generalized targeting model that subsumes most known targeting problems. In this paper, a recurrent state is defined as a condition that requires reprocessing or rework. The generalized model can accommodate one or two specifications limits and can be used for the following quality characteristics: The nominal-the-better, the larger-the-better, and the smaller-the-better. This model can be used to find the optimal mean of a quality characteristic, as well as the optimal specification limits. In addition, the paper studies the conditions under which the solution to the proposed model can provide a global solution. The paper shows that, for some of the special cases and under very general conditions, the optimal lower limit should be zero and the optimal upper limit should be infinity. This paper proves that the expected profits improve for the case where only a lower limit on the quality characteristic is used, if a recurrent state is included by adding an optimized upper limit. A special case of the model is used to study the problem of determining a common mean for multiple products, as well as the optimal upper specification limits for each product. A solution procedure for maximizing the expected profits and obtaining the optimal solution is introduced. A numerical example is presented.

Keywords: quality control; targeting model; process mean; strict quasi-concavity; recurrent states; optimized limits

---

1. Introduction

Targeting models are used to find the optimal process parameters that will result in the least amount of losses, due to reworked or scrapped items, while satisfying the specification limit(s). Achieving minimum costs and high quality are prerequisites for attaining higher market shares and profit. Taguchi introduced three types of loss functions; nominal-the-better quality characteristic, N-type, smaller-the-better quality characteristic, S-type, and larger-the-better quality characteristic, L-type [1]. This paper has three main goals. The first goal is to introduce a general model for determining the optimal process parameters for targeting models, where the manufacturing process involves rework and/or scrapping of non-conforming items. The second goal is to show which of the model special cases can yield a global optimal solution. To this end, the paper investigates the strict quasi-concavity of the expected profit. The third goal is to show that adding a fictitious upper limit to a model that involves only a lower limit will result in a higher expected profits.

Over the last four decades, several papers have presented models, developed for determining the optimal process mean under various manufacturing settings. Table 1 summarizes some of the studies in the literature. The third column in Table 1 specifies which specification limits are considered in the paper, where L and U stand for the lower, and upper specification limits, respectively. The fourth column in the table indicates whether the mentioned paper discusses rework when non-conforming items are produced. The last column in Table 1 specifies the specification limit(s), whose violation will result in rework. In the following paragraphs, a discussion of some of the most important models in the literature is provided.

Table 1. Summary of some of the targeting models.

Reference	Spec. Limits L	Spec. Limits U	Rework Included	Rework Conditions x < L	Rework Conditions x > U
Springer (1951) [2]	✓	✓			
Bettes (1962) [3]	✓	✓	✓	✓	✓
Hunter and Kartha (1977) [4]	✓				
Arcelus and. Banerjee (1985) [5]	✓				
Golhar and Pollack (1988) [6]	✓	✓	✓	✓	✓
Pugh (1988) [7]	✓	✓			
Rahim and Banerjee (1988) [8]		✓			
Bai and Lee (1993) [9]	✓		✓	✓	
Dodson (1993) [10]	✓	✓			
Lee and Kim (1994) [11]	✓	✓	✓	✓	✓
Chen and Chung (1996) [12]	✓				
Pulak and Al-Sultan (1996) [13]	✓				
Al-Sultan and Al-Fawzan (1997) [14]	✓	✓			
Lee and Jang (1997) [15]	✓	✓			
Liu and Raghavachari (1997) [16]	✓	✓	✓	✓	✓
Roan et al. (1997) [17]	✓				
Wen and Mergen (1999) [18]	✓	✓			
Kim et al. (2000) [19]	✓		✓	✓	
Lee et al. (2000) [20]	✓	✓	✓	✓	
Rahim and Al-Sultan (2000) [21]	✓	✓			
Roan et al. (2000) [22]	✓				
Williams et al. (2000) [23]	✓				
Chen and Chou (2002) [24]	✓	✓			
Lee and Elsayed (2002) [25]	✓	✓	✓	✓	
Duffuaa and Siddiqui (2003) [26]	✓				
Bowling et al. (2004) [27]	✓	✓	✓	✓	✓
Chen (2004) [28]	✓	✓			
Rahim and Tuffaha (2004) [29]	✓	✓			
Chen and Chou (2005) [30]	✓	✓			
Chen (2006) [31]	✓	✓	✓		✓
Hong et al. (2006) [32]	✓		✓	✓	
Lee et al. (2007) [33]	✓				
Khasawneh et al. (2008) [34]	✓	✓	✓		✓
Chen and Kao (2009) [35]	✓	✓	✓		✓
Darwish (2009) [36]	✓		✓	✓	
Duffuaa et al. (2009) [37]	✓		✓	✓	
Chen and Khoo (2010) [38]	✓	✓	✓		✓
	✓	✓			

Table 1. *Cont.*

Reference	Spec. Limits		Rework Included	Rework Conditions	
	L	U		$x < L$	$x > U$
Park et al. (2011) [39]	✓				
Selim and Al-Zu'bi (2011) [40]	✓	✓	✓		✓
Darwish et al. (2013) [41]	✓				
Duffuaa and El-Ga'aly (2013a) [42]	✓	✓	✓	✓	
Duffuaa and El-Ga'aly (2013b) [43]	✓		✓	✓	
Peng and Khasawneh (2014) [44]	✓	✓	✓		✓
Chen et al. (2015a) [45]	✓				
Chen et al. (2015b) [46]	✓	✓	✓		✓
Chen and Kan (2015) [47]	✓	✓	✓		✓
Dodd et al. (2015) [48]	✓	✓	✓		✓
Duffuaa and El-Ga'aly (2015) [49]	✓		✓	✓	
Raza and Turiac (2016) [50]	✓	✓	✓	✓	
Raza et al. (2016) [51]	✓	✓	✓	✓	
This paper (General model)	✓				
	✓		✓	✓	
		✓			
	✓		✓		✓
	✓	✓			
	✓	✓	✓	✓	
	✓	✓	✓		✓
	✓	✓	✓	✓	✓

To the best of the authors' knowledge, Springer [2] was the first study that developed a targeting problem with two-sided specification limits. In that paper, a product is accepted if its quality characteristic falls within pre-defined specification limits; otherwise it is rejected. Bettes [3] presented a study for setting the optimum values for the process mean and upper specification limit. Golhar and Pollack [6] presented a model for determining the optimal process mean and upper specification limit for a process where the item is accepted if the considered quality characteristic falls between the specification limits, and is reworked if it does not. Rahim and Al-Sultan [21] studied the problem of jointly determining the optimum mean and variance of a process. Rahim and Tuffaha [29] considered a targeting problem, where the product is sold in a primary market if its quality characteristic falls within the specification limits; otherwise, it is sold in a secondary market.

In order to determine the probability distribution of the quality characteristic of the production process, a sample of data on the quality characteristic should be collected from the production line. The observations are to be tested for the best probability distribution fit. The adequacy of the fit should be assessed by goodness-of-fit tests, i.e., chi-square test. Interested readers are referred to [37]. Another technique is to apply Burr's density function to fit the set of collected data of the quality characteristic, given the first four moments of the data of the process characteristic; i.e., the mean, standard deviation, skewness coefficient, and kurtosis coefficient, can be reasonably accurately estimated. Then, Burr distribution can be easily transferred to any normal/non-normal distribution. Interested readers are referred to [52,53].

In some cases, it would be impractical or even impossible to measure the quality characteristic. In these cases, a variable that is correlated with the quality characteristic is used to estimate the quality of the product. This correlated variable is known as a surrogate variable, and it is known to be relatively easier to measure. Many studies have investigated models with surrogate variables [9,11,15,25,35,39].

Bai and Lee [9] presented the problem of setting the optimum target value of the process mean and the lower specification limit of a cement bag-filling process. Measuring the weight of a bag is difficult, due to the high-speed packing; therefore, the milliampere reading of a load cell is used as a surrogate.

Lee and Kim [11] extended the work of Bai and Lee [9], and considered a filling process with a pre-defined lower specification limit on the quality characteristic of interest. The material in a container is inspected based on a surrogate variable. Under-filled and over-filled containers are emptied and refilled. The optimal process mean, upper and lower specification limits are determined simultaneously by maximizing the profit function.

Lee and Jang [15] considered a production process with three-class screening. They constructed two profit models; one model is based on the performance variable, while the other is based on a surrogate variable.

The preceding studies addressed the case where a single product type is considered. However, situations arise where multiple product types are produced through a common production process. Lee et al. [33] developed a targeting model for a production process with multiple products. They determined the optimal common process mean; where a product is accepted if its quality characteristic is above a predefined lower limit, otherwise it is scrapped. Park et al. [39] extended the work of Lee et al. [33] and discussed a model that contains surrogate variables. The model enabled the estimation of the common process mean and the optimum lower limit for the performance variables.

In Section 2 of this paper, a generalized targeting model, that subsumes most of the known N-type targeting models, is developed. Special cases of the model cover the S-type and the L-type. This model can be used to find the optimal mean of a quality characteristic and the optimal specification limits. Section 3 provides a discussion on the quasi-concavity of the model and its special cases. This section identifies cases where an optimal lower limit is zero and an optimal upper limit is infinity. The section also identifies those cases where the upper limit is a finite value that can be easily obtained using line search methods. In Section 4, a case for introducing an upper limit on the quality characteristic for a process targeting model, that aims to improve expected profits, is presented. Section 5 provides an application for the proposed model by considering the model presented in [33], where several products have a common quality characteristic, while holding different cost parameters. The objective is to find the optimal common mean of the quality characteristic. Each product has its own specification limits. The solution procedure of the multiple products model is presented in Section 6. An illustrative numerical example and discussion of the results are given in Section 7. Finally, Section 8 concludes the paper.

The main contributions and novel merits of the manuscript can be summarized as:

1. Developing a generalized targeting model for process quality control. This generalized model subsumes most of the known N-type targeting models. Special cases of the model cover the S-type and the L-type.
2. Providing a solution procedure to find the optimal mean of a quality characteristic and optimal specification limits for the generalized targeting model.
3. Very useful managerial insights have been extracted from the numerical experiments.

The following notation is used throughout the paper:

X	the random variable of the quality characteristic under consideration
x	the realized value of X
L	the lower specification limit of X
U	the upper specification limit of X
$g_0(x)$	the production cost function
$g_a(x)$	the net profit function for an accepted product
$g_L(x)$	the cost incurred when $x < L$
$g_U(x)$	the cost incurred when $x > U$
TP	the total profit function
$E(TP)$	the expected total profit
$f(\cdot)$	the probability density function of X with mean μ and variance σ_i^2
$F(\cdot)$	the cumulative distribution function of X

In addition to the above notation, the following notation is used in the multiple product model:

X_i	the random value of the quality characteristic of the ith product type, $i = 1, 2, \dots, n$
x_i	the realized value of X_i
μ^*	the optimal common process mean
L_i	the lower specification limit of X_i
U_i	the upper specification limit of X_i
A_i	the selling price of one unit of the ith product
B	the fixed production cost per unit
C	the variable production cost proportional to X_i
I_c	the inspection cost per unit
S	the scrapping cost per unit
R	the rework cost per unit
α_i	the proportion of the ith product type produced, $\sum_{i=1}^{n} \alpha_i = 1$

2. Generalized Targeting Model

In this section, a general targeting model that incorporates most of the known targeting models is developed. This model may be used for finding the optimal mean and/or variance of a quality characteristic. It may also be used to determine the optimal "specification" limits for the quality characteristic. It is assumed that there is a single quality characteristic, x, with a probability density function $f(x)$. Without loss of generality, it is assumed that $\infty > x \geq 0$.

The cost structure is described as follows: The production cost, $g_0(x)$ is a function of the realized value of the quality characteristic. This includes raw material cost, manufacturing cost, and quality control and/or inspection costs.

It is assumed that there is a two-sided specification limit on the quality characteristic, a lower and an upper specification limit, L, and U, respectively. If the quality characteristic is less than the threshold L, i.e., $x < L$, an extra cost $g_L(x)$ is incurred. This could be the cost of scrapping the item, or reworking it. It is assumed that this cost is dependent on the value of the quality characteristic. This cost includes all expenses associated with scrapping or reworking the product. The expected cost in this case is defined as $G_L(L) = \int_{x=0}^{L} g_L(x) f(x) dx$. If the quality characteristic is between L and U, then the item is accepted. In this case, revenue may be collected, but some additional costs may be incurred, such as finishing, shipping, or packaging. The notation, $g_a(x)$, is used to denote the difference between the revenue and those additional costs. The expected value of this function is given by $G_a(L, U) = \int_{x=L}^{U} g_a(x) f(x) dx$. If the quality characteristic exceeds U, i.e., $x > U$, a cost is incurred. This could be the cost of scrapping or reworking the item. As in the case where $x < L$, this cost may be dependent on x. This cost is denoted by $g_U(x)$. The expected cost in this case is given by $G_U(U) = \int_{x=U}^{\infty} g_U(x) f(x) dx$. The functions g_0, g_a, g_L, and g_U may also reflect any of Taguchi's loss functions and in this case the model may have one or two limits dependent on the type of the quality characteristic.

In addition, the following assumptions are considered:

(a) The quality characteristic assumes non-negative values,
(b) The cost structure and parameters remain the same after rework,
(c) The probability density function of the quality characteristic does not change after rework is performed, and
(d) There is no limit on the number of rework attempts.

Next, the concepts of terminal and recurrent states of an item are introduced. An item is in a terminal state if it does not require any further processing, i.e., the item is scrapped or accepted. An item is in a recurrent state if it requires rework.

The literature of process targeting models discuss the case where a product is in a terminal state once it is produced. These models appear in Table 1 with "No" in the rework column. For some other models, the item is in a terminal state only if the quality characteristic is acceptable, i.e., rework is performed if $x < L$, or $x > U$, models in [3,6,11,16,27]. Other models assume that the terminal state is $x \leq U$, i.e., rework is needed if $x > U$, models [31,34,35,38,40,44,46–48]. Finally, there are models where a recurrent state is limited to the case where $x < L$, models [9,19,20,25,32,36,37,42,43,49–51].

Before introducing the new model, let's turn to the "specification" limits on the quality characteristic. Usually, specification limits are set by outside agents, such as downstream manufacturing processes, regulatory agencies, or clients/consumers. However, one may envision situations where this may not be the case. Consider a filling operation that has a lower specification limit on the fill volume. The lower limit is set in the interest of the consumer. However, the producer will incur a loss for overfilling. In this case, the producer may wish to introduce an upper limit for the fill volume to have a cap on lost material. A similar situation applies to coating processes, where a lower limit on the thickness of the coating is specified by standards. However, the manufacturer may want to set an upper limit on the thickness to avoid the cost of additional coating material. On the other hand, thick coating may have a significant effect on the product quality. It may result in uneven texture finish and may initiate cracking and delamination of paints, due to external forces, which will definitely affect the product life and performance. In these scenarios, the decision maker has to set a "suitable" upper limit.

Consider a quality characteristic that has to satisfy both the lower and upper limits. If the quality characteristic is outside the limits further processing is performed, i.e., a recurrent state is reached. In this case, the total profit, TP is given by:

$$TP = \begin{cases} E(TP) - g_L(x) & L > x \\ -g_0(x) + g_a(x) & L \leq x \leq U. \\ E(TP) - g_U(x) & x > U \end{cases} \tag{1}$$

The expected value of the total profit is given by:

$$E(TP) = \frac{G_a(L,U) - G_0(L,U) - G_L(L) - G_U(U)}{F(U) - F(L)}, \tag{2}$$

where $G_0(L,U) = \int_{x=L}^{U} g_0(x)f(x)dx$ is the expected cost of producing a unit within the limits. One can interpret the expression in Equation (2) as the expected profit, given that the quality characteristic is within the limits. The denominator is the probability of not being in a recurrent state. Now, the following three special cases of the model shown in Equation (2) are considered:

I.   Consider the case where the recurrent state is reached when $x > U$. In this case, the random variable, TP, will be the same as in Equation (1) above with the exception that if $x < L$, then $TP = -g_0(x) - g_L(x)$, thus:

$$TP = \begin{cases} -g_0(x) - g_L(x) & L > x \\ -g_0(x) + g_a(x) & L \leq x \leq U. \\ E(TP) - g_U(x) & x > U \end{cases}$$

The probability of not being in a recurrent state is $F(U)$. Hence Equation (2) becomes,

$$E(TP) = \frac{G_a(L,U) - G_{0U}(U) - G_L(L) - G_U(U)}{F(U)}, \tag{3}$$

where $G_{0U}(U) = \int_{x=0}^{U} g_0(x)f(x)dx$. If there is no lower limit; i.e., $L = 0$, then the model shown in Equation (1) becomes,

$$TP = \begin{cases} -g_0(x) + g_a(x) & x \leq U \\ E(TP) - g_U(x) & x > U \end{cases}.$$

Therefore, the model shown in Equation (2) reduces to,

$$E(TP) = \frac{G_a(U) - G_{0U}(U) - G_U(U)}{F(U)},$$

where $G_a(U) = \int_{x=0}^{U} g_a(x)f(x)dx$. This is typically the known model for S-type quality characteristic with recurrent state when $x > U$.

II. Similarly, if the recurrent state is reached when $x < L$, then the model shown in Equation (1) reduces to

$$TP = \begin{cases} E(TP) - g_L(x) & L > x \\ -g_0(x) + g_a(x) & L \le x \le U. \\ -g_0(x) - g_U(x) & x > U \end{cases}$$

Hence, the model shown in Equation (2) becomes,

$$E(TP) = \frac{G_a(L, U) - G_{0L}(L) - G_L(L) - G_U(U)}{1 - F(L)}, \tag{4}$$

where $G_{0L}(L) = \int_{x=L}^{\infty} g_0(x)f(x)dx$. For this special case, if there is no upper limit; i.e., $U = \infty$, then the model shown in Equation (1) becomes:

$$TP = \begin{cases} E(TP) - g_L(x) & L > x \\ -g_0(x) + g_a(x) & L \le x \end{cases}.$$

Hence, the model shown in Equation (2) reduces to,

$$E(TP) = \frac{G_a(L) - G_{0L}(L) - G_L(L)}{1 - F(L)}.$$

This represents the known model for L-type quality characteristic with recurrent state when $L > x$.

III. Finally, if there is no recurrent state; i.e., the item is scrapped if not accepted, then the model shown in Equation (1) becomes,

$$TP = \begin{cases} -g_0(x) - g_L(x) & L > x \\ -g_0(x) + g_a(x) & L \le x \le U. \\ -g_0(x) - g_U(x) & x > U \end{cases}$$

Thus, the model shown in Equation (2) reduces to,

$$E(TP) = G_a(L, U) - G_{00} - G_L(L) - G_U(U), \tag{5}$$

where $G_{00} = \int_{x=0}^{\infty} g_0(x)f(x)dx$. Similar to special cases I and II, if there is only a single limit U, and the acceptance condition is $x \le U$, then one has to substitute 0 for L in any integral involving L, and the model shown in Equation (5) becomes,

$$E(TP) = G_a(U) - G_{0U} - G_U(U).$$

This represents the known model for S-type quality characteristic without recurrent state. Similarly, as in special case II, if the acceptance condition is $x \geq L$, then one has to substitute ∞ for U in any integral involving U, hence the model shown in Equation (5) becomes,

$$E(TP) = G_a(L) - G_{00} - G_L(L).$$

This represents the known model for L-type quality characteristic without recurrent state.

The conclusion of this section is as follows, the model shown in Equation (2) is a general model and S-type quality characteristic and L-type quality characteristic models are special cases of the model shown in Equation (2). This general model and its special cases subsume most of the known targeting models that appear in the literature. In addition, this generalized model can be used for setting the optimum mean, variance, upper or lower limits, as will be shown in next sections.

It is important to understand whether the solution obtained is a local or global optimum. In the case of the maximizing a function, a global solution is guaranteed if the function involved is concave. However, concavity of a function is a strict condition. A less restrictive one is that the function is strictly quasi-concave. In the next section, the quasi-concavity of the models given by Equations (2)–(5) will be discussed.

3. Quasi-Concavity of the Generalized Targeting Model

Strictly quasi-concave functions have the property that a local maximum point is also global. A function, $h(x)$, is quasi-concave if the set $S_\alpha = \{x | h(x) \geq \alpha\}$ is convex for any scalar α and $S_\alpha \neq \Phi$. Also, $h(x)$ is quasi-concave if, for any x_1 and x_2 in its domain and $0 \leq \lambda \leq 1$, $h(\lambda x_1 + (1 - \lambda)x_2) \geq \min\{h(x_1), h(x_2)\}$ [54]. There is an extra method to characterize the strict quasi-concavity of a function of a single variable. A function of one variable is strictly quasi-concave if there is a point, p, in its domain where the function is increasing at points before p, and decreasing at points after p. This implies that the slope before p is positive, while it becomes negative at points after p. This property will be used in this paper.

The next subsections will attempt to prove the obtained results for general differentiable functions f, g_a, g_0, g_U, and g_L. In some cases no result can be obtained for general functions, for those cases, this paper adopts a cost structure that is widely used in most papers in the literature, where the cost of violating the limits, and that of accepting the quality characteristic, are constants, while that of production is a linear function, i.e., $g_L(x) = S$, $g_U(x) = R$, $g_a(x) = A$, and $g_0(x) = B + Cx$.

3.1. Quasi-Concavity of Equation (2) with Respect to $\mu, L,$ and U

In this subsection, the quasi-concavity of the function in Equation (2) with respect to the mean, and the limits, L and U is studied.

Assume that the limits L and U are known and the mean, μ, is to be determined. No result can be obtained about the quasi-convexity of Equation (2) as a function of μ. If the above cost structure is adopted, the expected profit function given by Equation (2) becomes:

$$E(TP) = A - B - \frac{C \int_{x=L}^{U} x f(x) dx + SF(L) + R[1 - F(U)]}{F(U) - F(L)}. \tag{6}$$

Counter-examples can be constructed to show that the function given by Equation (6) is not quasi-concave if $f(x)$ is a normal density function.

Next, consider the case where μ and U are known and the lower limit, L, is to be determined. The partial derivative of the profit function in Equation (6) with respect to L is given by:

$$\frac{\partial E(TP)}{\partial L} = f(L) \frac{CL[F(U) - F(L)] - C \int_{x=L}^{U} x f(x) dx - SF(U) - R[1 - F(U)]}{[F(U) - F(L)]^2}.$$

Note that the second term in the numerator exceeds the first term, hence, the partial derivative is negative and the function achieves its maximum value at $L = 0$. This result is valid for any differentiable density function, $f(x)$.

Next, consider the case where μ and L are known and the upper limit, U, is to be determined. Assume that $f(x)$ is differentiable. Using the same cost structure presented earlier, then the partial derivative of $E(TP)$ shown in Equation (6) with respect to U is given by:

$$\frac{\partial E(TP)}{\partial U} = f(U)\frac{-CU[F(U) - F(L)] + C\int_{x=L}^{U} xf(x)dx + SF(L) + R(1 - F(L))}{[F(U) - F(L)]^2}.$$

The sum of the first two terms in the numerator is negative and is a strictly decreasing function of U. Eventually, this sum goes to $-\infty$ as U goes to ∞.

Since this part of the discussion is interested in $U \geq \mu$, then there are two cases:

a. If the numerator at $U = \mu$ is negative then it remains negative for $U > \mu$ and the maximum of $E(TP)$ is achieved at $U = \mu$.

b. If the numerator at $U = \mu$ is positive, then as U increasers, the numerator decreases and eventually becomes negative. Hence $E(TP)$ changes sign from positive to negative, and is strictly quasi-concave.

3.2. Quasi-Concavity of Equation (3) with Respect to μ, L, and U

Counter examples can be constructed to show that Equation (3) is not quasi-concave with respect to μ for the general functions and even for the special cost structure introduced at the beginning of this section.

The partial derivative of Equation (3) with respect to L is given by:

$$\frac{\partial E(TP)}{\partial L} = -f(L)\frac{g_a(L) + g_L(L)}{F(U)} < 0.$$

Hence, the maximum of Equation (3) is achieved if L is set to 0.

The partial derivative of Equation (3) with respect to U, adopting the above cost-structure, is given by:

$$\frac{\partial E(TP)}{\partial U} = f(U)\frac{-CUF(U) + C\int_{x=0}^{U} xf(x)dx + (A + S)F(L) + R}{F(U)^2}.$$

The proof in this case is similar to that in Section 3.1. If $-C\mu.F(\mu) + C\int_{x=0}^{\mu} xf(x)dx + (A + S)F(L) + R \leq 0$, then the maximum profit is attained when $U = \mu$. Otherwise, the function $E(TP)$ is strictly quasi-concave function in U.

3.3. Quasi-Concavity of Equation (4) with Respect to μ, L and U

Counter examples can be constructed to show that Equation (4) is not quasi-concave with respect to μ for the general functions and even for the special cost structure introduced at the beginning of this section. Using the same cost structure presented earlier, then Equation (4) simplifies to:

$$E(TP) = \frac{(A + R)F(U) - (A - B + S)F(L) - C\int_{x=L}^{\infty} xf(x)dx - R - B}{1 - \Gamma(L)}. \tag{7}$$

The partial derivative of Equation (7) with respect to L is given by:

$$\frac{\partial E(TP)}{\partial L} = f(L)\frac{CL[1 - F(L)] - C\int_{x=L}^{\infty} xf(x)dx - (A + R)(1 - F(U)) - S}{[1 - F(L)]^2}.$$

Note that the second term in the numerator exceeds the first term, hence the numerator is negative and the maximum of the function is achieved at $L = 0$.

Finally, the partial derivative of Equation (4) with respect to U is given by:

$$\frac{\partial E(TP)}{\partial U} = f(U)\frac{g_a(U) + g_U(U)}{1 - F(L)} > 0.$$

Hence, the maximum is achieved by setting U at its largest possible value.

3.4. Quasi-Concavity of Equation (5) with Respect to μ, L, and U

If the cost structure described above is used, then Equation (5) simplifies to:

$$E(TP) = (A + R)F(U) - (A + S)F(L) - C\mu - B - R. \tag{8}$$

If L and U are known and $f(x)$ is the density function of the normal distribution, then $F(U)$ is concave for $U > \mu$ and $F(L)$ is convex for $L < \mu$. Hence, Equation (8) is a concave function in μ.

For the general function shown in Equation (5), it is straightforward to show that:

$$\frac{\partial E(TP)}{\partial L} = -f(L)[g_a(L) + g_L(L)] < 0,$$

$$\frac{\partial E(TP)}{\partial U} = f(U)[g_a(U) + g_U(U)] > 0.$$

Hence, the maximum of the function is achieved if $L = 0$ and if U is set at the largest possible value.

Table 2 summarizes the above results. In order to show that two parameters are fixed and the third one is a decision variable, that third parameter is written as an argument of the function TP. For example, $TP(\mu)$ means that L and U are fixed and μ is a decision variable.

4. Optimizing the Limits of a Quality Characteristic

Table 2 shows that the optimal lower limit is zero or the least acceptable value. Only the cases modeled by Equations (2) and (3) provide a chance to optimize the upper limit. Usually, specification limits are determined based on technical or contractual conditions. Hence, optimizing the upper limit does not seem natural in this case. However, there is room for improving the expected profit by including an upper limit on the quality characteristic to reduce the cost of raw material usage, as shown in the next sections. Consider a case modeled by Equation (4) with $U = \infty$. Here, the product is accepted as long as the quality characteristic exceeds the lower specification limit. If an upper limit is introduced along with an associated recurrent state, then the model described by Equation (2) is on hand. A similar case is obtained when the case is modeled by Equation (5) with $U = \infty$. Then, introducing an upper limit and a recurrent state will result in a model given by Equation (3). In Section 5, this approach is implemented. This implementation shows that the expected profit has improved.

Table 2. Quasi-concavity of TP as a function of one parameter.

Model	$TP(\mu)$	$TP(U)$	$TP(L)$
Equation (2)		Strictly quasi-concave [2]	Optimal $L = 0$ [2]
Equation (3)	Not strictly quasi-concave [2]	Optimal $U = \infty$ or the function is strictly quasi-concave [2]	Optimal $L = 0$ [1]
Equation (4)		Optimal $U = \infty$ [1]	Optimal $L = 0$ [2]
Equation (5)	Concave [2]		Optimal $L = 0$ [1]

[1] The is true for general functions. [2] This is true for the special cost structure.

5. Multiple Products Model

In this section, the generalized model that is presented in Section 2 is applied and an application, where a model given by Equation (5), is modified as described in the previous section, resulting in an improvement in the objective function. Lee et al. [33] considered a facility that manufactures multiple products (these products can be considered as different grades of the same product). The products are sold at different prices. Consider a quality characteristic that is common to all of these products, where the mean of the characteristic is the same among all the products. A product is scrapped if $x < L$, otherwise it is accepted. The current paper extends their work in different directions. To limit the losses of the manufacturer, an upper limit is added for the quality characteristic of each product. The upper limit assigned to each product is to be determined. If the upper limit is exceeded, a penalty is incurred. This penalty can be viewed as a rework cost. The resulting model involves the products' selling prices, production costs, inspection costs, scrap costs, and rework costs.

Consider a case of n products, and assume that the quality characteristic of product i is normally distributed and given by X_i with variance σ_i^2. X_1, X_2, ..., X_n are assumed to be independent normal random variables with a common mean μ. Every item is inspected for conformance to the specification limits L_i and U_i. The inspection operation incurs a fixed inspection cost I_c, independent of product type. If $U_i \geq x_i \geq L_i$, the item is accepted and sold at a price A_i, i.e., $g_a(x) = A_i - I_c$. On the other hand, if $x_i < L_i$, the item fails to meet the lower specification limit; thus it is scrapped at a cost S, which is a fixed scrapping cost for all product types, i.e., $g_L(x) = S + I_c$. Moreover, if $x_i > U_i$, the item is reworked where a penalty or rework cost, R, is incurred, and hence, $g_U(x) = R + I_c$. The reworked item is inspected again and may be accepted, scrapped or reworked again. The per unit production cost is assumed to be a linear function of x_i, hence, $g_0(x) = B + Cx_i$, where B and C are fixed, and variable costs per unit, respectively. The scenario described above fits the model given by Equation (3), therefore, the expected profit is given by:

$$E(TP) = \sum_{i=1}^{n} \alpha_i \frac{G_a(U_i) - G_{0U}(U_i) - G_L(L_i) - G_U(U_i)}{F(U_i)}, \tag{9}$$

where:

α_i is the proportion of product i of the total production,

$$G_a(U_i) = \int_{x_i = L_i}^{U_i} (A_i - I_c)f(x_i)dx_i ,$$

$$G_{0U}(U_i) = \int_{x_i = 0}^{U_i} (B + Cx_i)f(x_i)dx_i,$$

$$G_L(L_i) = \int_{x_i = 0}^{L_i} (S + I_c)f(x_i)dx_i,$$

$$G_U(U_i) = \int_{x_i = U_i}^{\infty} (R + I_c)f(x_i)dx_i.$$

By substitution for the above equations into Equation (9), the final form of the total expected profit is given by:

$$E(TP) = H(\mu, U) = R - B - C\mu + \sum_{i=1}^{n} \alpha_i H_i(\mu, U_i), \tag{10}$$

where $U = (U_1, U_2, ..., U_n)$ and

$$H_i(\mu, U_i) = \frac{C\sigma_i f(U_i) - (R + I_c) - (A_i + S)F(L_i)}{F(U_i)} + A_i. \tag{11}$$

The problem is to maximize $H(\mu, U)$ by determining the optimum common mean, μ, and the upper limits, U_i for $i = 1, 2, \ldots, n$. Note that if $U_i = \infty$, and $R = 0$, the model of [33] is retrieved.

6. Solution Procedure

The function $H(\mu, U)$ is neither concave nor quasi-concave in μ. Hence, it may have local optima. Figure 1 shows the function behavior versus changes in μ.

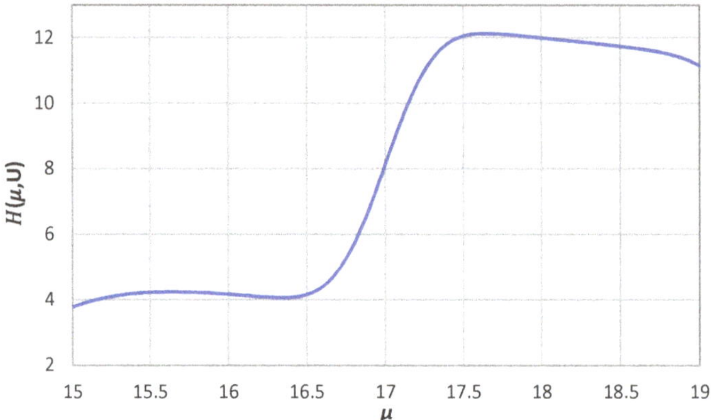

Figure 1. An illustrative example of $H(\mu, U)$ as a function of μ.

The proposed solution procedure includes an exhaustive search of μ. μ is searched in the range $[\mu_{\min}, \mu_{\max}]$. For each value of μ, the optimum values of $U_1, U_2, \ldots, U_n$ are obtained. Section 3 proved that the single variable function $H(U_i)$ is strictly quasi-concave in U_i. Therefore, the following solution procedure is applied:

1. Set $\mu_{\min} = \max\limits_{i = 1, 2, \ldots, n} \{L_i\}$, $\mu_{\max} = \mu_{\min} + \lambda \max\limits_{i = 1, 2, \ldots, n} \{\sigma_i\}$, where λ is a positive scalar. For normally distributed quality characteristics, it is recommended to take $\lambda \in [4, 7]$, this will guarantee covering more than 99.99% of the expected output of the production process. Also, set $H^* = 0$, and $\mu^* = \mu_1 = \mu_{\min}$.
2. Solve the problems $\max_{U_i > \mu_k} H_i(\mu_k, U_i)$ for $i = 1, \ldots, n$ using any line search method. Let U_{ik} be the solution for $i = 1, \ldots, n$.
3. Calculate $H(\mu_k, U_k)$ where $U_k = (U_{1k}, U_{2k}, \ldots, U_{nk})$. If $H(\mu_k, U_k) > H^*$, set $H^* = H(\mu_k, U_k)$, $U^* = U_k$ and $\mu^* = \mu_k$. Go to Step 4.
4. Update the value of the process mean; set $\mu_{k+1} = \mu_k + \delta$, where δ is a small positive scalar. If $\mu_{k+1} > \mu_{\max}$, stop; otherwise go to Step 2.

In the next section, the solution approach discussed above is implemented and a discussion of the effect of including an upper bound on the quality characteristic is presented.

7. Results and Discussion

This section presents a numerical example to find the optimum common mean for three products using the solution procedure presented in Section 6. The same data of the example presented in [33] is utilized. Subsequently, the impact on the total expected profit as a result of setting an upper limit to the quality characteristic of each product is investigated.

7.1. An Illustrative Example

This example considers a process where three types of electronic devices are processed through a common process of copper plating. The input data of the parameters are identical to the data in [33] and are shown below:

$$B = \$0.6, \ C = \$0.1, \ S = \$0.25, \ I_c = \$0.008,$$

$$A_1 = \$3.05, \ \alpha_1 = 0.4, \ \sigma_1^2 = 1.11 \ \mu m^2, \ L_1 = 13 \ \mu m,$$

$$A_2 = \$3.25, \ \alpha_2 = 0.3, \ \sigma_2^2 = 1.22 \ \mu m^2, \ L_2 = 14 \ \mu m,$$

$$A_3 = \$3.45, \ \alpha_3 = 0.3, \ \sigma_3^2 = 1.25 \ \mu m^2, \ L_3 = 15 \ \mu m.$$

In addition to the above parameters, a rework cost is introduced; $R = \$0.2$.

In the exhaustive search of μ, an increment $\delta = 0.001$, and $\lambda = 3$ are used.

The optimal solution is given by: $\mu^* = 17.1$, $U_1^* = 19.4052$, $U_2^* = 19.8687$, $U_3^* = 21.8496$, and $E(TP) = 0.8549$. The optimum common process mean μ^* is the same as that of [33]. However, the expected total profit is greater than that of [33] with a 9.25% enhancement. This is due to the upper limits being set.

The following section discusses the impact of the rework cost value, R, on the profit improvement associated with including an upper limit on the value of the quality characteristic.

7.2. The Impact of Setting Upper Specification Limits

The following section discusses the impact of the rework cost value, R, on the profit improvement associated with including an upper limit on the value of the quality characteristic.

Specification limits are considered in targeting models because quality control issues require them. However, this paper proposes including them as a *stop loss* measure. In the model presented in [33], a product is accepted once the quality characteristic exceeds the lower limit, regardless of how large it may be. This means that material loss could be high. It is proposed to introduce an upper limit on the quality characteristic. The introduction of the upper limit is associated with a penalty or rework cost. This subsection investigates the impact of the value of the rework cost on the improvement in the expected profit, given by Equation (10) and that of [33]. The improvement is computed using Equation (12) below, where $E(TP)^*$ is the maximum expected profit as obtained by the current model in Equation (10) and $E(TP)^{\text{Lee}}$ is the maximum profit obtained by Lee et al. [33]:

$$\% \text{ Improvement in } E(TP) = \frac{E(TP)^* - E(TP)^{\text{Lee}}}{E(TP)^{\text{Lee}}} * 100(\%). \tag{12}$$

Figure 2 shows that as the rework cost, R, increases, the percentage of improvement decreases. The probability of rework becomes almost nil for $R \geq 0.4$.

It is clear from Figure 2 that setting upper specification limits has a major impact on $E(TP)$, and there is a significant improvement in its value. It is also worth noting that, regardless of the value of the rework cost, R, there will be an improvement in $E(TP)$. As R increases, the percentage of improvement approaches 9.12%. If the model does not include rework, then it will consider the upper limits at infinity; i.e., the model described in [33], and this will result in losses from the excessive use of raw material. Therefore, rework in the analysis of the model is needed if one is interested in reducing the loss from the excess raw material.

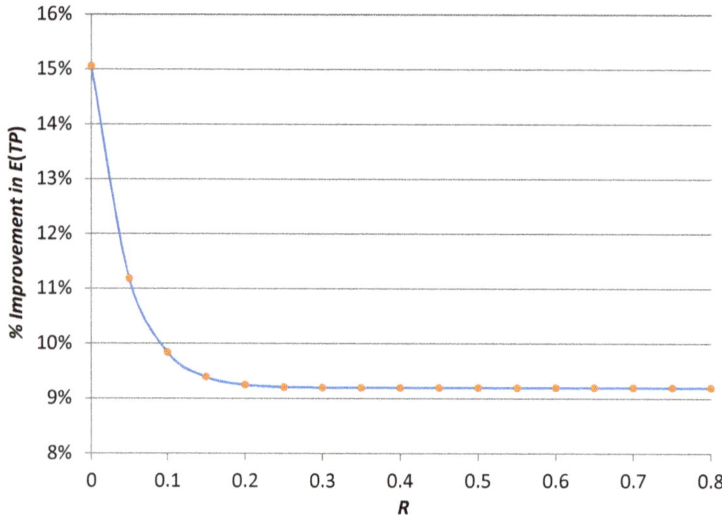

Figure 2. The impact on $E(TP)$ by setting upper limits with different values of R.

8. Conclusions

This paper proposed a generalized targeting model that subsumes most of the known targeting models of that category. Firstly, a generalized model for the N-type targeting models is introduced. The concept of a terminal and recurrent state of an item is introduced. An item reaches a terminal state if no further processing is required; otherwise, it is in a recurrent state. A recurrent state implies further work will be performed on the item. This differentiation of states gave rise to several special cases of the general model. These special cases can be used to model L-type and S-type quality characteristics. An important issue, discussed in this paper, is the conditions under which the obtained solution can be a global solution. To accomplish this task, the strict quasi-concavity of the developed models is examined. Strict quasi-concavity implies that a local maximum is also global. It is shown that the profit function is strictly quasi-concave if one or both specification limits are to be found, assuming that the mean and the other limit are fixed. If one is interested in finding the optimal mean of the quality characteristic, while the specification limits are fixed, one will face the disappointing fact that the profit function is not quasi-concave for all cases, except for the model where there are no recurrent states. This paper used a special case of the proposed model to obtain the optimal common mean of a quality characteristic and the upper specification limits for multiple products. The quality characteristics of the products were assumed to follow independent normal distributions with known variances. A procedure for determining the optimal solution is introduced and implemented on an example having three products. The results show that, incorporating upper specification limits on the quality characteristic has a significant improvement in the value of the optimal expected profit.

A practical case that needs further research is when there is a limit on the number of times an item is allowed to be in a recurrent state. This could be due to excessive loss of material or deterioration of some physical properties. In this paper, it is assumed that the cost of rework is independent of the number of rework cycles the item goes through. This assumption may not be applicable in all cases. Finally, the assumption that the quality characteristic keeps the same probability distribution regardless of number of rework cycles may be relaxed and further research will be needed in this case.

Author Contributions: Modeling, M.A.M.A. and S.Z.S.; formal analysis, M.A.M.A. and S.Z.S.; methodology, M.A.M.A. and S.Z.S.; supervision, S.Z.S.; writing—original draft, M.A.M.A.

Funding: This research received no external funding.

Acknowledgments: The authors wish to acknowledge the support of King Fahd University of Petroleum and Minerals.

Conflicts of Interest: The authors declare no conflict of interest.

References

1. Taguchi, G. *Introduction to Quality Engineering: Designing Quality into Products and Processes*; Quality Resources: Tokyo, Japan, 1986.
2. Springer, H.C. A method for determining the most economic position of a process mean. *Ind. Qual. Control* **1951**, *8*, 36–39.
3. Bettes, D.C. Finding an Optimum Target Value in Relation to a Fixed Lower Limit and an Arbitrary Upper Limit. *J. R. Stat. Soc. Ser. C Appl. Stat.* **1962**, *11*, 202–210. [CrossRef]
4. Hunter, W.G.; Kartha, C.P. Determining the Most Profitable Target Value for a Production Process. *J. Qual. Technol.* **1977**, *9*, 176–181. [CrossRef]
5. Arcelus, F.J.; Banerjee, P.K. Selection of the Most Economical Production Plan in a Tool-Wear Process. *Technometrics* **1985**, *27*, 433–437. [CrossRef]
6. Golhar, D.Y.; Pollock, S.M. Determination of the Optimal Process Mean and the Upper Limit for a Canning Problem. *J. Qual. Technol.* **1988**, *20*, 188–192. [CrossRef]
7. Pugh, A.G. An algorithm for economically setting a uniformly-shifting process. *Comput. Ind. Eng.* **1988**, *14*, 237–240. [CrossRef]
8. Rahim, M.A.; Banerjee, P.K. Optimal production run for a process with random linear drift. *Omega* **1988**, *16*, 347–351. [CrossRef]
9. Bai, D.S.; Lee, M.K. Optimal target values for a filling process when inspection is based on a correlated variable. *Int. J. Prod. Econ.* **1993**, *32*, 327–334. [CrossRef]
10. Dodson, B.L. Determining the Optimal Target Value for a Process with Upper and Lower Specifications. *Qual. Eng.* **1993**, *5*, 393–402. [CrossRef]
11. Lee, M.K.; Kim, G.S. Determination of the optimal target values for a filling process when inspection is based on a correlated variable. *Int. J. Prod. Econ.* **1994**, *37*, 205–213. [CrossRef]
12. Chen, S.L.; Chung, K.J. Determination of the optimal production run and the most profitable process mean for a production process. *Int. J. Prod. Res.* **1996**, *34*, 2051–2058. [CrossRef]
13. Pulak, M.F.S.; Al-Sultan, K.S. The optimum targeting for a single filling operation with rectifying inspection. *Omega* **1996**, *24*, 727–733. [CrossRef]
14. Al-Sultan, K.S.; Al-Fawzan, M.A. An extension of Rahim and Banerjee's model for a process with upper and lower specification limits. *Int. J. Prod. Econ.* **1997**, *53*, 265–280. [CrossRef]
15. Lee, K.; Jang, J.S. The optimum target values for a production process with three-class screening. *Int. J. Prod. Econ.* **1997**, *49*, 91–99. [CrossRef]
16. Liu, W.; Raghavachari, M. The target mean problem for an arbitrary quality characteristic distribution. *Int. J. Prod. Res.* **1997**, *35*, 1713–1728. [CrossRef]
17. Roan, J.; Gong, L.; Tang, K. Process mean determination under constant raw material supply. *Eur. J. Oper. Res.* **1997**, *99*, 353–365. [CrossRef]
18. Wen, D.; Mergen, A.E. Running a Process with Poor Capability. *Qual. Eng.* **1999**, *11*, 505–509. [CrossRef]
19. Kim, Y.J.; Cho, B.R.; Phillips, M.D. Determination of the Optimal Process Mean with the Consideration of Variance Reduction and Process Capability. *Qual. Eng.* **2000**, *13*, 251–260. [CrossRef]
20. Lee, M.K.; Hong, S.H.; Kwon, H.M.; Kim, S.B. Optimum process mean and screening limits for a production process with three-class screening. *Int. J. Reliab. Qual. Saf. Eng.* **2000**, *7*, 179–190. [CrossRef]
21. Rahim, M.A.; Al-Sultan, K.S. Joint determination of the optimum target mean and variance of a process. *J. Qual. Maint. Eng.* **2000**, *6*, 192–199. [CrossRef]
22. Roan, J.; Gong, L.; Tang, K. Joint determination of process mean, production run size and material order quantity for a container-filling process. *Int. J. Prod. Econ.* **2000**, *63*, 303–317. [CrossRef]
23. Williams, W.W.; Tang, K.; Gong, L. Process improvement for a container-filling process with random shifts. *Int. J. Prod. Econ.* **2000**, *66*, 23–31. [CrossRef]
24. Chen, C.H.; Chou, C.Y. Determining the Optimum Process Mean of a One-Sided Specification Limit. *Int. J. Adv. Manuf. Technol.* **2002**, *20*, 439–441. [CrossRef]

25. Lee, M.K.; Elsayed, E.A. Process mean and screening limits for filling processes under two-stage screening procedure. *Eur. J. Oper. Res.* **2002**, *138*, 118–126. [CrossRef]

26. Duffuaa, S.O.; Siddiqui, A.W. Process targeting with multi-class screening and measurement error. *Int. J. Prod. Res.* **2003**, *41*, 1373–1391. [CrossRef]

27. Bowling, S.R.; Khasawneh, M.T.; Kaewkuekool, S.; Cho, B.R. A Markovian approach to determining optimum process target levels for a multi-stage serial production system. *Eur. J. Oper. Res.* **2004**, *159*, 636–650. [CrossRef]

28. Chen, C.H. Determining the Optimum Process Mean of a One-sided Specification Limit with the Linear Quality Loss Function of Product. *J. Appl. Stat.* **2004**, *31*, 693–703. [CrossRef]

29. Rahim, M.A.; Tuffaha, F. Integrated model for determining the optimal initial settings of the process mean and the optimal production run assuming quadratic loss functions. *Int. J. Prod. Res.* **2004**, *42*, 3281–3300. [CrossRef]

30. Chen, C.H.; Chou, C.Y. Determining the Optimum Process Mean under a Log-normal Distribution. *Qual. Quant.* **2005**, *39*, 119–124. [CrossRef]

31. Chen, C.H. The Optimum Selection of Imperfect Quality Economic Manufacturing Quantity and Process Mean by Considering Quadratic Quality Loss Function. *J. Chin. Inst. Ind. Eng.* **2006**, *23*, 12–19. [CrossRef]

32. Hong, S.H.; Kwon, H.-M.; Lee, M.K.; Cho, B.R. Joint optimization in process target and tolerance limit for L-type quality characteristics. *Int. J. Prod. Res.* **2006**, *44*, 3051–3060. [CrossRef]

33. Lee, M.K.; Kwon, H.M.; Hong, S.H.; Kim, Y.J. Determination of the optimum target value for a production process with multiple products. *Int. J. Prod. Econ.* **2007**, *107*, 173–178. [CrossRef]

34. Khasawneh, M.T.; Bowling, S.R.; Cho, B.R. A Markovian approach to determining process means with dual quality characteristics. *J. Syst. Sci. Syst. Eng.* **2008**, *17*, 66–85. [CrossRef]

35. Chen, C.H.; Kao, H.S. The determination of optimum process mean and screening limits based on quality loss function. *Expert Syst. Appl.* **2009**, *36*, 7332–7335. [CrossRef]

36. Darwish, M.A. Economic selection of process mean for single-vendor single-buyer supply chain. *Eur. J. Oper. Res.* **2009**, *199*, 162–169. [CrossRef]

37. Duffuaa, S.O.; Al-Turki, U.M.; Kolus, A.A. A process targeting model for a product with two dependent quality characteristics using 100% inspection. *Int. J. Prod. Res.* **2009**, *47*, 1039–1053. [CrossRef]

38. Chen, C.-H.; Khoo, M.B.C. Optimum Process Mean Setting for Product with Rework Process. *Tamkang J. Sci. Eng.* **2010**, *13*, 375–384.

39. Park, T.; Kwon, H.M.; Hong, S.-H.; Lee, M.K. The optimum common process mean and screening limits for a production process with multiple products. *Comput. Ind. Eng.* **2011**, *60*, 158–163. [CrossRef]

40. Selim, S.Z.; Al-Zu'bi, W.K. Optimal means for continuous processes in series. *Eur. J. Oper. Res.* **2011**, *210*, 618–623. [CrossRef]

41. Darwish, M.A.; Abdulmalek, F.; Alkhedher, M. Optimal selection of process mean for a stochastic inventory model. *Eur. J. Oper. Res.* **2013**, *226*, 481–490. [CrossRef]

42. Duffuaa, S.O.; El-Ga'aly, A. A multi-objective mathematical optimization model for process targeting using 100% inspection policy. *Appl. Math. Model.* **2013**, *37*, 1545–1552. [CrossRef]

43. Duffuaa, S.O.; El-Ga'aly, A. A multi-objective optimization model for process targeting using sampling plans. *Comput. Ind. Eng.* **2013**, *64*, 309–317. [CrossRef]

44. Peng, C.-Y.; Khasawneh, M.T. A Markovian approach to determining optimum process means with inspection sampling plan in serial production systems. *Int. J. Adv. Manuf. Technol.* **2014**, *72*, 1299–1323. [CrossRef]

45. Chen, C.H.; Chou, C.Y.; Kan, C.C. Modified economic production and raw material model with quality loss for conforming product. *J. Ind. Prod. Eng.* **2015**, *32*, 196–203. [CrossRef]

46. Chen, C.H.; Khoo, M.B.C.; Chou, C.Y.; Kan, C.C. Joint determination of process quality level and production run time for imperfect production process. *J. Ind. Prod. Eng.* **2015**, *32*, 219–224. [CrossRef]

47. Chen, C.H.; Kan, C.C. Tolerance design based on specified process capability value. *Int. J. Manag. Sci. Eng. Manag.* **2015**, *10*, 210–214. [CrossRef]

48. Dodd, C.; Scanlan, J.; Marsh, R.; Wiseall, S. Improving profitability of optimal mean setting with multiple feature means for dual quality characteristics. *Int. J. Adv. Manuf. Technol.* **2015**, *81*, 1767–1780. [CrossRef]

49. Duffuaa, S.O.; El-Ga'aly, A. Impact of inspection errors on the formulation of a multi-objective optimization process targeting model under inspection sampling plan. *Comput. Ind. Eng.* **2015**, *80*, 254–260. [CrossRef]

50. Raza, S.A.; Turiac, M. Joint optimal determination of process mean, production quantity, pricing, and market segmentation with demand leakage. *Eur. J. Oper. Res.* **2016**, *249*, 312–326. [CrossRef]
51. Raza, S.A.; Abdullakutty, F.C.; Rathinam, S. Joint determination of process mean, price differentiation, and production decisions with demand leakage: A multi-objective approach. *Appl. Math. Model.* **2016**, *40*, 8446–8463. [CrossRef]
52. Lin, Y.C.; Chou, C.Y. On the design of variable sample size and sampling intervals $\overline{X}$ charts under non-normality. *Int. J. Prod. Econ.* **2005**, *96*, 249–261. [CrossRef]
53. Chen, C.H.; Chou, C.Y. Optimum process mean, standard deviation and specification limits settings under the Burr distribution. *Eng. Comput.* **2017**. [CrossRef]
54. Bazaraa, M.S.; Sherali, H.D.; Shetty, C.M. *Nonlinear Programming: Theory and Algorithms*, 3th ed.; Wiley-Interscience: Hoboken, NJ, USA, 2006.

© 2019 by the authors. Licensee MDPI, Basel, Switzerland. This article is an open access article distributed under the terms and conditions of the Creative Commons Attribution (CC BY) license (http://creativecommons.org/licenses/by/4.0/).

Article

Supply Chain with Customer-Based Two-Level Credit Policies under an Imperfect Quality Environment

Aditi Khanna [1], Aakanksha Kishore [1], Biswajit Sarkar [2,*] and Chandra K. Jaggi [1]

[1] Department of Operational Research, Faculty of Mathematical Sciences, New Academic Block, University of Delhi, Delhi 110007, India; dr.aditikhanna.or@gmail.com (A.K.); kishore.aakanksha@gmail.com (A.K.); ckjaggi@yahoo.com (C.K.J.)

[2] Department of Industrial & Management Engineering, Hanyang University, Ansan, Gyeonggi-do 15588, South Korea

* Correspondence: bsbiswajitsarkar@gmail.com; Tel.: +82-10-7498-1981

Received: 8 November 2018; Accepted: 25 November 2018; Published: 3 December 2018

Abstract: The present model develops a three-echelon supply chain, in which the manufacturer offers full permissible delay to the whole seller, while the latter, in turn, adopts distinct trade credit policies for his subsequent downstream retailers. The type of credit policy being offered to the retailers is decided on the basis of their past profiles. Hence, the whole seller puts forth full and partial permissible delays to his old and new retailers respectively. This study considers bad debts from the portion of new retailers who fail to make up for the delayed part of the partial payment. The analysis shows that it is beneficial for the whole seller to make shorter contracts, particularly with new retailers, along with the fetching of a higher fraction of initial purchase cost from them. In addition to the above-described scenario, the lot received by the whole seller from the manufacturer is not perfect, and it contains some defects for which he employs an inspection process before selling the items to the retailers. In order to make the study more realistic, Type-I, as well as Type-II misclassification errors, and the case of out-of-stock are considered. The impact of Type-I error has been found to be crucial in the study. The present paper determines the optimal policy for the whole seller by maximizing the expected total profit per unit time. For the optimality of the solution, theoretical results are provided. Finally, a numerical example and a sensitivity analysis are done to validate the model.

Keywords: inventory; defectives; inspection errors; full trade credit; partial trade credit

---

1. Introduction and Literature Overview

In recent years, credit financing has become one of the most desirable/important parts for most organizations. In order to sustain oneself in the competitive market, everyone has to look for the possibility of strengthening their financial assets so that the organization does not face a financial crisis while making/delivering goods to their customers. Most of the big firms/organizations have adopted a culture of trade credit or a bank loan to overcome the lack of financial assets. Nowadays, it is often seen that both manufacturers and whole sellers are offering partial trade credit policies to their end-users. The majority of the recent studies have looked into the downstream partial trade credit financing as a strategy to reduce risks by depositing a collateral amount at the beginning of the purchase, and then providing a complete permissible delay on the remaining amount. With changing market trends and increasing competition, it is indeed profitable for any whole seller to adopt a blend of both full and partial trade credit strategies, instead of treating all retailers as equivalent or providing partially permissible delay to them. However, the nature of policy may change with the type of retailers arriving at the whole seller's doorstep, depending upon the loyalty of end user. Generally, it is observed in real life that retailers act differently towards the terms and conditions of various payment options, discounts, sales, promotion strategies, etc. With regard to this, the whole seller can face some non-payment risks on the remaining amount of partial permissible delays by some retailers. For the financial benefits, as well as for the reduction of non-payment risks, this study provides a balanced promotional strategy to the whole seller, with the bifurcation of all the retailers, first into old and new types, by looking at the past records at his end. Furthermore, it turns out that sometimes all of the new retailers may not be capable/ interested in completing their respective partial payments, due to the whole seller. Thus, it is practical enough to further categorize new retailers into good and bad types. Good retailers are those who pay both immediate and delayed parts of the partially permissible delay, while bad ones are those who pay only the immediate part and fail to return the delayed part payment to the whole seller. The failure of making delayed partial payments from a proportion of new retailers is referred to as bad debts in the model. So, in total, the retailers have been bifurcated twice in the present research. Such a dual bifurcation of retailers into an old and new category and then new retailers into the good and bad category has not gained much attention in past research. Furthermore, the firms have to take appropriate quality control measures, in order to satisfy today's discerning consumers. Conventionally, it was presumed that the lots received by any whole seller/retailer were all perfect, and so there was no inspection processes involved. However, the production process may be faulty many times at the manufacturer's end, and produce some defectives. Therefore, it becomes inevitable for the whole seller to inspect the entire lot before selling it to the retailers. Nevertheless, it is found that because of unavoidable human errors, the inspection process is imperfect and leads to Type-I as well as Type-II misclassification errors. Type-I misclassification error is the scrapping of perfect items, while Type-II misclassification error is the sale of defectives by mistake. As a result of Type-II misclassification errors, the wrongly inspected defectives, when passed on to the retailers, lead to sales return. As the demand is satisfied through perfect items only, if the count of inspected perfect items becomes less than the demand during the inspection process, then shortages are bound to occur. Hence, the model considers fully backlogged shortages. This rational combination of problems features defective items, faulty inspection procedures, Type-I as well as Type-II errors, sales returns, and distinct trade credit policy through the dual bifurcation of customers, along with shortages that are encountered by numerous retail sectors that face such kind of problem sets, such as the textile industry, multi-brand retail stores like Shoppers Stop, Lifestyle, and so on. Therefore, the main objective of this study is to provide a simple formulation for firms who have been targeting vital spheres such as supreme standards of quality, promotional tools, effective inventory management, team building, etc.

Literature Overview

This section discusses the previous work done in the research area of inventory, considering both trade credit and the presence of imperfect quality items. This helps in throwing light upon the research contribution of the study and the existing literature.

Full and partial trade credit policy: In today's ever-evolving business world, it is crucial for the supply chain players to adopt the best policies for maximizing their overall profit. Trade credit is nowadays a widely accepted strategy to stimulate the manufacturer's/whole seller's demand, and boost their respective sales. It not only gives the whole seller an opportunity to delay the payment process until the credit limit, but it benefits him in terms of earning interest on the revenue fetched from sales. In the literature of this subject, Haley and Higgins [1] were the first to develop the concept of trade credit in the economic order quantity (EOQ) model. Further, Goyal [2], Davis, and Gaither [3], and Aggarwal and Jaggi [4], and their references, made significant contributions in elaborating the concept of one-stage trade credit. Soon after, it was found to be equally favorable for the lower members of the supply chain, to offer the permissible delay, to some extent. Many vital contributors related to two-stage trade credit financing were made, namely Teng et al. [5], Su et al. [6], Jaggi et al. [7], and their references. All of the above-mentioned work dealt with optimal order quantity with the permissible delay in payments, assuming that the supplier offered the retailer only a fully permissible delay in payments. However, it is more viable to assume that the supplier would offer the retailer a partially permissible delay in payments to avoid financial risks. Considering the two-stage partial downstream trade credit policy, Huang and Hsu [8] expanded Huang's [9] model where the retailer received a full trade credit by the supplier, but offered only a partial credit period to his subsequent downstream customers. To bring the existing research of trade credit policy closer to pragmatism, Teng [10], and Jaggi and Verma [11] investigated the retailer's optimal ordering policies by offering distinct trade credits to their respective customers. The former offered full and partial trade credit to his good and bad customers, while the later differentiated their customers on the basis of the old and the new. However, they did not regard any bad debts as a loss to the retailer that may occurr due to the default risks attached in offering partial trade credit policy. Furthermore, Jaggi et al. [12], Jaggi [13], Taleizadeh et al. [14], and Giri and Sharma [15] contributed by formulating an economic ordering model under two-stage trade credit financing. Besides these, some recent works related to credit/default risks in permissible delay policy contributed a lot in the literature, those of Shi and Zhang [16], Tiwari et al. [17], Wu and Chan [18], Chen and Teng [19], Shah [20], Sarkar and Saren [21], Wu et al. [22], Mahata and De [23], and Wu et al. [24]. Although the use of downstream partial trade credit policy was gaining attention among academia to reduce the threats related to credit-risk customers, none of them had classified the end retailers into four categories, firstly as old and new and later as good and bad types, to further look upon bad debts as a direct loss to the whole seller, which is the main focus of this study.

Imperfect quality and screening errors: In the traditional EOQ or economic production quantity (EPQ) models, the assumption of perfect-quality items in the lot received by the whole seller/retailer was taken as being permanent. Soon, this hypothesis was violated, as it did not seem to be realistic enough for carrying out future research. In view of this, several researchers devoted a great amount of effort to develop EOQ/EPQ models for defective items, viz., Porteus [25], Cárdenas-Barrón et al. [26], Lee and Rosenblatt [27], and Zhang and Gerchak [28]. Soon after, Salameh and Jaber [29] extended the traditional EOQ/EPQ model by accounting for imperfect quality items when using the conventional formulae. That paper considered the issue that poor-quality items were sold as a single batch by the end of the 100% screening process. Further extensions of Salameh and Jaber's [29] paper were given by Cárdenas-Barrón [30], Goyal and Cárdenas-Barrón [31], Papachristos and Konstantaras [32], and many others. Moreover, Maddah and Jaber [33] rectified a flaw in an EOQ model characterized by a random fraction of imperfect quality and a screening process. Due to unavoidable human errors in the inspection process, it becomes valid to presume Type-I and Type-II errors in the inspection process. In this direction, the first few researchers who came up with human errors in the inspection planning were Rauf et al. [34], Sarkar and Saren [35], and Duffua and Khan [36]. In recent years, Khan et al. [37] and Sett et al. [38] threw light on the inspection errors, and constructed models by considering imperfect quality without and with the occurrence of shortages, respectively. Recently, Hsu and Hsu [39] developed an EOQ model for imperfect quality items, along with screening errors and fully backlogged shortages and sales returns.

Imperfect quality along with trade credit:It is economical to incorporate permissible delays when considering the defectives in the study, as it helps to elevate the profit margins. This has been realized by many researchers, as they have collaborated imperfect quality, the screening process, and one/two-stage trade credit policies in their work. Some of the important contributions in this field are those of Zhou et al. [40], Khanra et al. [41], Khanna et al. [42], Palanivel and Uthayakumar [43], Taleizadeh et al. [44], and Khanna et al. [45]. None of the previous studies that worked on trade credit along with imperfect quality, imperfect inspection, and shortages had presumed the dual bifurcation of customers, i.e., firstly into old and new types (on the basis of past records), and secondly, the new ones into good and bad types (on the basis of partial payment fulfillment) at the whole seller's end, in a three-echelon supply chain. Table 1 demonstrates the important research gap, and highlights the contributions to the existing literature.

Table 1. Comparison of major attributes of some previous work with current model.

Author(s)	Imperfect Quality	Screening Errors	Credit Policy		Bifurcation of Customer		Bad Debts	Shortages	Optimality
			Upstream	Downstream	Old and New	Good and Bad			
Teng [10]	No	No	Yes (Full)	Yes (Distinct)	No	Yes	No	No	Closed-form
Jaggiand Verma [11]	No	No	Yes (Partial)	Yes (Partial)	No	No	No	No	Mathematical
Jaggi et al. [12]	No	No	Yes (Full)	Yes (Distinct)	Yes	No	No	No	Closed-form
Jaggi et al. [13]	No	No	Yes (Partial)	Yes (Partial)	No	No	No	No	Algebraic
Hsu and Hsu [39]	Yes	Yes	No	No	No	No	No	Yes	Closed-form
Taleizadeh et. al. [14]	No	No	Yes (Partial)	No	No	No	No	Yes	Closed-form
Wu and Chan [18]	No	No	Yes (Full)	Yes (Partial)	No	No	No	No	Mathematical
Chang et. al. [41]	Yes	Yes	No	Yes (Full)	No	No	No	No	Mathematical
Palanivel and Uthayakumar [43]	Yes	No	Yes (Full)	No	No	No	No	No	Graphical
Taleizadeh et. al. [44]	Yes	No	Yes (Full)	Yes (Full)	No	No	No	Yes	Mathematical
Sarkar and Saren [21]	No	No	Yes (Full)	Yes (Partial)	No	No	No	No	Mathematical
Zhou et. al. [40]	Yes	Yes	Yes (Full)	No	No	No	No	Yes	Closed-form
Mahata and De [23]	No	No	Yes (Full)	Yes (Partial)	No	No	No	No	Mathematical
Khanna et al. [45]	Yes	Yes	Yes (Full)	Yes (Full)	No	No	No	No	Mathematical
This paper	Yes	Yes	Yes (Full)	Yes (Partial)	Yes	Yes	Yes	Yes	Closed-form

Contribution: The proposed model can be viewed as a better reflection of today's business scenario for many retail industries. The single type of customer bifurcation (viz., either old and new, or good and bad) has gained some attention by a few researchers, but the present paper highlights the importance of dual customer bifurcation in a two-stage trade credit scenario under imperfect quality and imperfect inspection circumstances. In the current study, both upstream full and downstream, both full and partial types of trade credit policies, are incorporated in a three-level supply chain (manufacturer–whole seller–retailers). The type of trade credit policy is decided on the basis of the past profile of the retailers by the whole seller. Henceforth, the whole seller puts forth full and partial trade credit policies to his old and new retailers, respectively. It is assumed that all of the end retailers are not credit-worthy, and thus, one can incorporate bad debts at this juncture, a concept that has been left unaccounted by many researchers. The model tries to highlight the practical losses that occur many times due to bad debts, by bringing undue charges on the whole seller. Bad debts arise from the portion of new retailers that fail to make up for the remaining part of the partial permissible delay, and are defined as the uncollectible accounts that occur as a result of the customer being unable to fulfill their obligation to pay an outstanding debt, due to bankruptcy or other financial problems. This expense is the cost of a business having the inability to collect its debts. These are the lost customers, and it is reasonable to assume that the same bad customers/debtors never return to the cycle, and they are predicted to have changed their supplier (for fear of legal punishments and heavy penalties). Henceforth, it is intuitive to assume that bad debtors always arise as fresh faces to the whole seller, and this is why it is inevitable for him to face such losses in every cycle. Each time the cycle is repeated, there is a whole lot of new customers, out of whom some fail out to settle the delayed payment part, and their rough percentage is known by using past data. As a result, the customers have been practically bifurcated dually in the model. Another important aspect of this study is derived by assuming that the lot received from the manufacturer contains a certain percentage of defective items, which are dealt by a careful inspection process at the whole seller's end. Errors in inspection are usually ignored by many researchers in their study. Hence, Type-I and Type-II errors are further added into the system. For maintaining quality standards, demand is satisfied by perfect items only. In order to avoid shortages during the inspection process, the rate of inspection is set as being greater than the demand rate. The main purpose of combining this kind of scenario involving lot-sizing modeling with defective items, and inspection with a trade credit model, is to give the benefit of the doubt to the whole seller under the presence of imperfect quality items received from him. Some common causes of the presence of defectives in the received lot can be wear and tear in transit when large quantities are under consideration, mishandling of products, manufacturing defects, etc. In today's competitive environment, as the supply chain gets bigger, it becomes difficult for the players to work for their individual benefits. Hence, sellers who deal with quantities in bulk tend to offer permissible delays to their subsequent members, in order to facilitate demand. To compensate the losses incurred due to actual defectives and inspection errors, trade credit is looked upon as an opportunity to raise revenue. Though these two areas are distinct, they are definitely among the practical challenges faced by many non-manufacturing firms, which has motivated the move converge these two parameters in the model. Finally, the model presumes the occurrence of all the aforementioned factors, along with planned backorders, to fill the gap between EOQ and realism.

The remaining paper is planned in the following manner: in Section 2, construction of the analytical model is elaborated with the help of assumptions, notation, and mathematical formulation; In Section 3, a solution methodology is provided to reach the optimal solution, with the help of two lemmas; Section 4 explains the solution procedure to search the optimal case out of all situations of the trade credit policy; Section 5 validates the theoretical results with the help of a numerical process, and presents thorough managerial insights with the help of a sensitivity analysis; finally, Section 6 concludes the paper by providing a conclusion and model limitations, along with future research directions.

2. Analytical Model

2.1. Assumptions

The proposed model is based on the following assumptions.

(1)   The replenishment rate is finite. The lot received by the whole seller contains a certain proportion of defectives. The inspection process is imperfect and leads to Type-I and Type-II errors. The inspection rate is higher than the demand rate.

(2)   The demand rate is constant, uniform, and deterministic. The demand is satisfied with perfect items only. The time period is infinite, and the lead time is negligible.

(3)   In the upstream supply chain, a full trade credit policy is offered to the whole seller by the manufacturer. In the downstream supply chain, full trade credit is provided to all old retailers, while partial trade credit is given to all new retailers by the whole seller.

(4)   New retailers are categorized into good and bad types, based on their paying/not paying of complete dues on time, respectively. Bad debts are obtained by the fraction of new retailers who fail to pay for the delayed part of the partial trade credit.

(5)   Shortages are allowed, and they are fully backlogged.

2.2. Problem Description

The present study deals with a three-layer supply chain where there is a presence of defectives ($\alpha\%$) in the whole seller's lot (y) received from the manufacturer. The defect percentage (α) is a random variable with a known probability distribution function, taken as $f(\alpha)$. To deal with the existing imperfect quality items, and to provide customers with perfect items only, the whole seller conducts an inspection process of the entire lot, with the rate of inspection being greater than the demand rate, to avoid shortages during the inspection process. In order to compensate for some of the costs/losses incurred during quality control, the whole seller intends to gear up for some promotional techniques, and hence prefers to offer distinct trade credit to his subsequent retailers, according to past records/data. The current model is analyzed by offering full trade credit via upstream, i.e., the whole seller receives full trade credit from the manufacturer with the credit limit (M). However, in downstream, the whole seller offers full and partial trade credit policies to the proportion of old retailers (K) and new retailers ($1 - K$), respectively, with a credit limit (N). Amongst the new retailers ($1 - K$), good and bad types of new retailers arise, with proportions (R) and ($1 - R$) respectively, where (R) is estimated by using past data. All of the new retailers have to make a payment on (δ) units to the whole seller at the time of purchase of the product, and rest of the payment is supposed to be made at (N), where (δ) is the fraction of the selling price (s). However, in practice, it turns out that some new retailers ($1 - R$)($1 - K$) are unable to return for the settlement of the account at (N) with the whole seller. As a result, the whole seller earns interest at the rate of (I_e) on (δ) units of every purchase made by all the new retailers. However, the interest on ($1 - \delta$) units is earned by the whole seller at the rate of (I_e), through the proportions of ($1 - \delta$)(R)($1 - K$) only. Resultantly, there is an additional interest charge at the rate of (I_p) on the whole seller because of bad debts, i.e., ($1 - \delta$)($1 - R$)($1 - K$), as the inventory sold to them remains unpaid until the time inventory becomes zero (T'). At the time of the settlement of his account with the manufacturer at (M), the whole seller pays for all of his purchases at price (c), and incurs a cost at the rate of (I_p) for the items in stock, and for the sold items that are as yet unpaid by the proportions of both old and new retailers until (M).

The above-described sequence of trade credit in the present inventory cycle is shown in Figure 1.

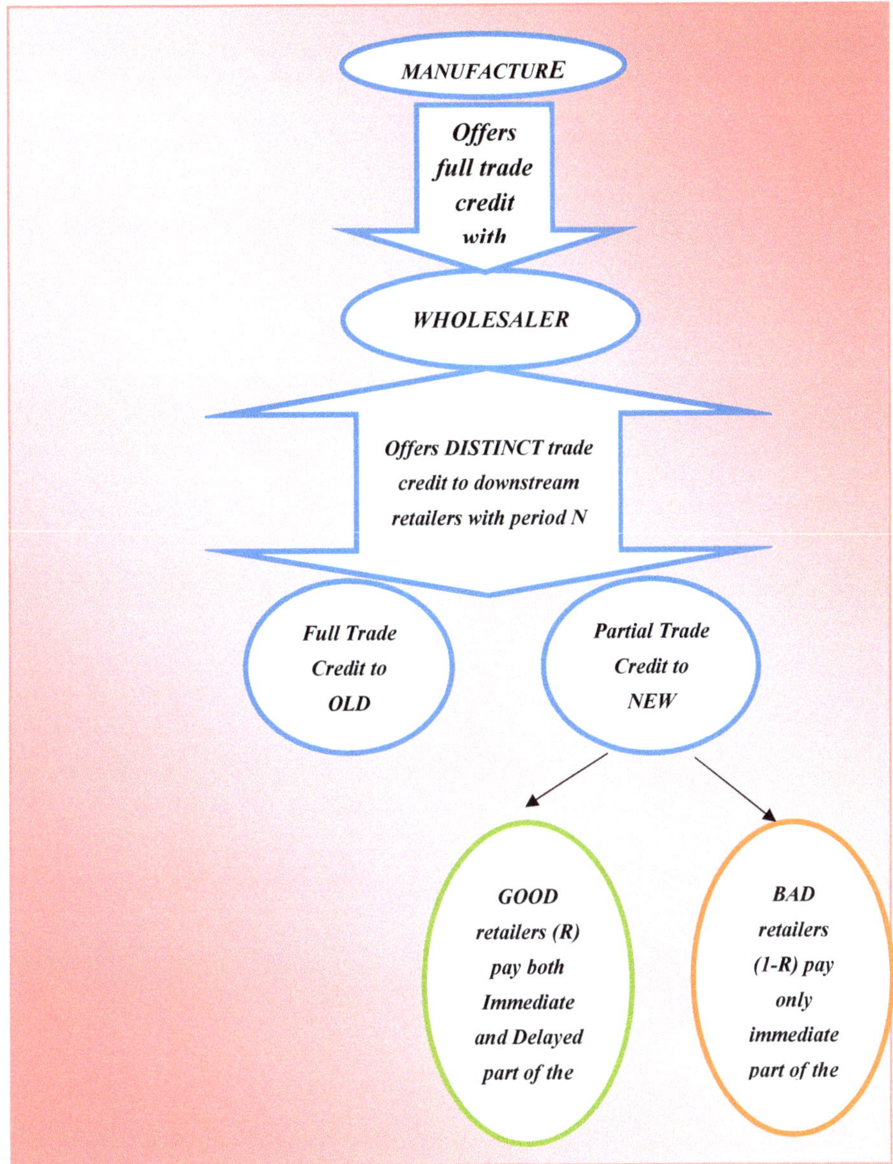

Figure 1. Sequence of permissible delay in the inventory cycle.

2.3. Mathematical Modeling

In this segment, a mathematical model befitting the above-described problem and its assumptions has been formulated. In this inventory model, the whole seller receives a batch of size y from the manufacturer, comprising of some defectives. The inventory at the whole seller's place begins to deplete through perfect items only, whereas the defectives keep accumulating until the end of the screening procedure, as shown in Figure 2.

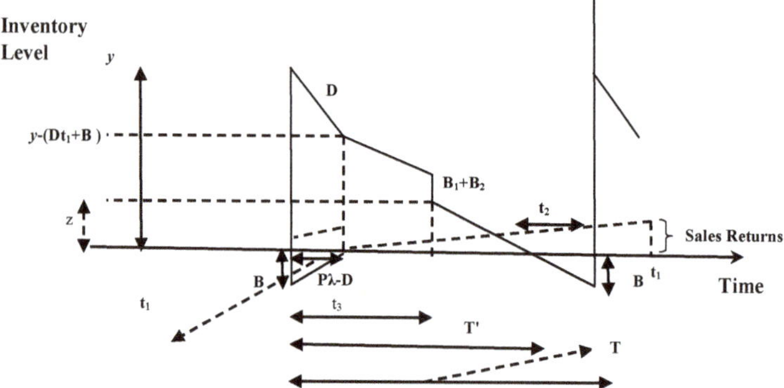

Figure 2. Inventory behavior of the system.

At the beginning of the cycle, the inspection and sale of perfect items start simultaneously at the whole seller's side. However, due to some errors committed by the inspector, there is the occurrence of Type-I and Type-II errors, with their respective proportions being $q_1 = P_r$(items screened as defects | non-defective items) and $q_2 = P_r$(items not screened as defects | defective items)$(0 < q_1 < q_2 < 1)$, following the probability density function of $f(q_1)$ and $f(q_2)$, respectively. It is assumed that (q_1) and (q_2) are independent of defect proportions (α). All of the items involving inspection errors are estimated inter-dependently by (q_1), (q_2), and (y). The flow of events is illustrated in Figure 3.

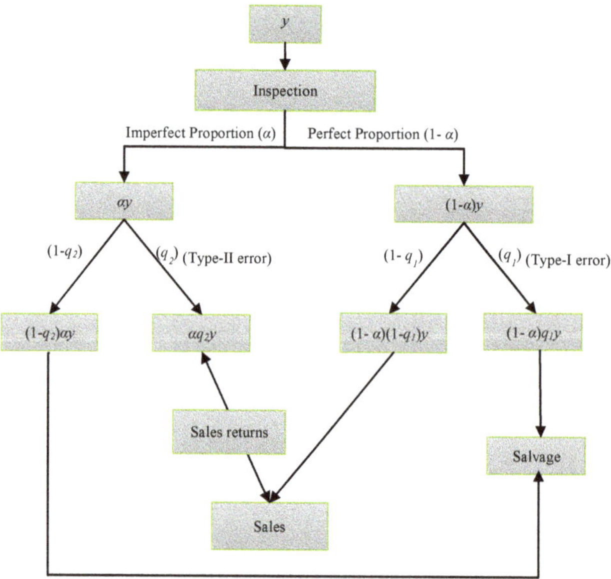

Figure 3. Flowchart of events in the inventory cycle.

As explained from Figure 3, the demand D is satisfied through perfect items only, i.e., $\alpha q_2 y + (1-\alpha)(1-q_1)y$. The B items that are intended to eliminate the backorders in each cycle are filled at the rate of $[\alpha q_2 + (1-\alpha)(1-q_1)]\lambda - D$. Let the proportion of perfect items be denoted as P.

$$\text{Therefore, } P = \alpha q_2 + (1-\alpha)(1-q_1). \tag{1}$$

Since P is a combination of random variables viz. α, q_1, q_2, P is a random variable.

$$E[P] = E[\alpha]E[q_2] + (1 - E[\alpha])(1 - [q_1]),\tag{2}$$

where $E[.]$ is the expected value operator.

Cycle length (T) = total number of perfect items sold/demand rate, i.e.:

$$T = \frac{\alpha q_2 y + (1 - \alpha)(1 - q_1)y}{D} = \frac{Py}{D},\tag{3}$$

$$E[T] = \frac{E[P]y}{D}.\tag{4}$$

i.e.:

$$E[T] = \frac{\{E[\alpha]E[q_2] + (1 - E[\alpha])(1 - E[q_1])\}y}{D}.\tag{5}$$

On account of a Type-I error, the whole seller rejects a fraction $(1 - \alpha)q_1 y$ of perfect items $(1 - \alpha)y$, while as a result of a Type-II error, he accepts a fraction $(\alpha q_2 y)$ of the total defectives (αy) by mistake. Due to quality dissatisfaction by the sale of defective items, the fraction $(\alpha q_2 y)$ re-enters the system continuously, similar to demand, until the end of inspection process (t_3), and it is held in the inventory for a complete cycle length (T). It is assumed that all of the sales returns $(\alpha q_2 y)$, referred as (B_2), are not replaced with perfect items, but are refunded fully. Thus, the total items to be salvaged immediately after the inspection process at a reduced price (v) are $((1 - q_2)\alpha y + (1 - \alpha)q_1 y)$, and these are referred to as (B_1) in the study. In total, the imperfect inspection process leaves $(1 - q_2)\alpha y$ and $(1 - \alpha)(1 - q_1)\alpha y$ as the actual defectives and non-defectives, respectively. The final outcome of the perfect items ready for sale after the assimilation of inspection errors is $((1 - \alpha)(1 - q_1)y + \alpha q_2 y)$.

$$\text{So, the total items salvaged are } B_1 = \alpha(1 - q_2)y + (1 - \alpha)q_1 y.\tag{6}$$

$$\text{And, the total sales returns are } B_2 = \alpha q_2 y.\tag{7}$$

The perfect items coming out of the inspection process are first used to satisfy the demand at the rate of $((1 - \alpha)(1 - q_1) + \alpha q_2)\lambda$. Then, the remaining perfect items are used to satisfy the shortages. During the time t_1, all the shortages are completely eliminated from the inventory cycle, as detailed below:

$$t_1 = \frac{B}{[\alpha q_2 + (1 - \alpha)(1 - q_1)]\lambda - D} = \frac{B}{P\lambda - D}.\tag{8}$$

Since, it is beneficial for the whole seller to allow a certain amount of shortage as these to help with cutting down inventory-holding costs to a significant level, time period t_2 is assigned to gather shortages, which are assumed to be fully backlogged in the model in the following manner:

$$t_2 = \frac{B}{D}.\tag{9}$$

In order to compete in the imperfect quality environment, it becomes essential to inspect the entire lot before reaching out in the mainstream. Thus, the inspection time is calculated as below:

$$t_3 = \frac{y}{\lambda}.\tag{10}$$

After the end of the inspection process, the inventory level gradually decreases only due to demand (D), and reaches zero at time (T_1).

$$\text{Thus, } T = T_1 + t_2.\tag{11}$$

$$\text{i.e., } T_1 = T - t_2 = \frac{Py - B}{D}, \tag{12}$$

$$\text{Therefore, } T_1 = \frac{z}{D} + t_3. \tag{13}$$

The value of z can be obtained by equating Equations (10), (12), and (13) as follows:

$$z = y\left(P - \frac{D}{\lambda}\right) - B. \tag{14}$$

2.3.1. Relevant Costs

Various cost components taken into account are:

(i)   Setup cost, which includes the fixed cost per cycle, i.e.:

$$SC = A. \tag{15}$$

(ii)   Purchase cost which includes variable cost per cycle, i.e.:

$$PC = cy. \tag{16}$$

(iii)   Inspection cost, which includes the cost of inspections per cycle, i.e.:

$$IC = iy. \tag{17}$$

(iv)   Type-I error cost, which includes the cost of rejecting a perfect item, i.e.:

$$EC_1 = c_r(1 - \alpha)q_1 y. \tag{18}$$

(v)   Type-II error cost, which includes the cost of accepting an imperfect item, i.e.:

$$EC_2 = c_a \alpha q_2 y. \tag{19}$$

(vi)   The inventory holding cost is the cost of carrying all non-defectives and defective items, plus the items returned from the market, i.e.:

$$HC = h\left\{\frac{1}{2}(2y - (B + Dt_1))t_1 + \frac{1}{2}(y - (B + Dt_1) + z + B_1)(t_3 - t_1) + \frac{1}{2}z(T' - t_3) + \frac{1}{2}B_2 T\right\}. \tag{20}$$

(vii)   Backordering cost is the cost of the total shortages that have occurred, i.e.:

$$BC = \frac{1}{2}c_B B(t_1 + t_2) \tag{21}$$

Therefore, by using Equations (15)–(21), the total cost per cycle is given by:

$$\text{Total cost } (T.C.) = SC + PC + IC + EC_1 + EC_2 + HC + BC \tag{22}$$

$$
\begin{aligned}
= A &+ y[c + i + c_r(1 - \alpha)q_1 + c_a \alpha q_2] + y\left\{
\begin{array}{l}
h\frac{B}{P\lambda - D} - h\frac{B}{2\lambda}\left(1 + \frac{D}{P\lambda - D}\right) - h\frac{B}{2\lambda} - h\frac{B}{2(P\lambda - D)} \\
- h\frac{B}{2(P\lambda - D)}\left(P - \frac{D}{\lambda}\right) - h\frac{B[\alpha(1 - q_2) + (1 - \alpha)q_1]}{2(P\lambda - D)} - h\frac{B}{D}\left(P - \frac{D}{\lambda}\right)
\end{array}
\right\} \\
&+ y^2\left\{h\frac{1}{2\lambda} + h\frac{1}{2\lambda}\left(P - \frac{D}{\lambda}\right) + h\frac{1}{2\lambda}[\alpha(1 - q_2) + (1 - \alpha)q_1] + h\frac{1}{2D}\left(P - \frac{D}{\lambda}\right)^2 + h\frac{\alpha q_2 P}{2D}\right\} \\
&+ B^2\left\{-h\frac{1}{2(P\lambda - D)}\left(1 + \frac{D}{P\lambda - D}\right) + h\frac{1}{P\lambda - D} + h\frac{D}{2(P\lambda - D)^2} + h\frac{1}{2D}\right\} + \frac{1}{2}c_B B^2\left(\frac{1}{P\lambda - D} + \frac{1}{D}\right)
\end{aligned} \tag{23}
$$

2.3.2. Sales Revenue

The total sales revenue consists of four parts:

(i)   The sales from sorted perfect items is:

$$R_1 = s[\alpha q_2 y + (1-\alpha)(1-q_1)y]. \tag{24}$$

(ii)   Revenue loss from sales return:

$$R_2 = -s\alpha q_2 y. \tag{25}$$

(iii)   Revenue loss from bad debts:

$$R_3 = -[s(1-\delta)][(1-\alpha)(1-q_1)][(1-R)(1-K)]y. \tag{26}$$

(iv)   Sales from total scrap items:

$$R_4 = v(B_1 + B_2) = v[\alpha(1-q_2)y + (1-\alpha)q_1 y + \alpha q_2 y]. \tag{27}$$

Therefore, by using Equations (24)–(27), the total sales revenue per cycle is given by:
Total revenue $(T.R.) = R_1 + R_2 + R_3 + R_4$

$$= s(1-\alpha)(1-q_1)y - s(1-\delta)(1-R)(1-K)(1-\alpha)(1-q_1)y + v[\alpha(1-q_2)y + (1-\alpha)q_1 y + \alpha q_2 y]. \tag{28}$$

2.3.3. Expected Total Profit per Unit Time

In the present model, two-stage trade credit policies are used differently via the upstream and downstream supply chains. The whole seller receives a credit period M from the manufacturer, and offers a credit period N to his subsequent downstream retailers. By offering full trade credit to all his old retailers, the whole seller starts earning interest from N up to M. Since, partial trade credit is given to all new retailers, therefore, the whole seller fetches a fraction of the purchase amount initially (i.e., immediate payment), and earns interest from time 0 to M. The remaining part (i.e., delayed payment) is paid at time N by good retailers only. From the bad credit retailers, the whole seller does not receive any delayed payment, and funds the amount from his own pocket. Hence, in addition to the unsold inventory, the whole seller finances a portion of the sold inventory (i.e., bad debts) from M to T.

Thus, depending upon the value of M, N, and T, eight different cases arise. The interest earned and paid is calculated distinctly for old and new retailers. For new retailers, immediate and delayed payment parts have been presented:

Case (i) $M \leq N \leq T' \leq T$

Case (ii) $T \leq M \leq N$

Case (iii) $N \leq M \leq t_1 \leq t_1 + N \leq T + N$

Case (iv) $N \leq t_1 + N \leq M \leq t_3 + N \leq T + N$

Case (v) $N \leq t_3 + N \leq M \leq T + N$

Case (vi) $N \leq T \leq M \leq T + N \leq T + N$

Case (vii) $N \leq T + N \leq M \leq T + N$

Case (viii) $N \leq T' + N \leq T + N \leq M$

Case (i) $M \leq N \leq T' \leq T$

This is the case of the smallest credit period when the first payment is still due from the retailers towards the whole seller. However, there is little interest that is earned from the initial payments made by all of the new retailers. Hence, the whole seller arranges the finances for the whole inventory to pay to the manufacturer at the time of settlement of his account (See Table 2).

Subcase 1: Interest earned and interest paid from the portion of old retailers (K)

The credit period M lies before the first payment N, so that there is no interest earned, while interest is paid on the entire inventory for the portion of old retailers.

$$\text{Interest earned } (I_{e11}) = 0 \tag{29}$$

$$\text{Interest payable } (I_{p11}) = \tfrac{1}{2}cI_pK[Py - DM - (P\lambda - D)M](T' - M) + \tfrac{1}{2}cI_pKDT'^2 + cI_pKDT'(N - M) \\ + \tfrac{1}{2}cI_pK(P\lambda - D)t_1^2 + cI_pKB(N - M) \tag{30}$$

Part 1: Immediate payment (δ)

The whole seller starts accumulating revenue from time 0 by the initial payment that is made by all of the new retailers, and he earns interest till M. The interest is paid by him into the inventory for the fraction of the initial price unsold to the portion of new retailers.

$$\text{Interest earned } (I_{e12}) = \frac{1}{2}\delta pI_e(1 - K)DM^2 + \frac{1}{2}\delta pI_e(1 - K)(P\lambda - D)M^2. \tag{31}$$

$$\text{Interest payable } (I_{p12}) = \frac{1}{2}\delta cI_p(1 - K)[Py - DM - (P\lambda - D)M](T' - M) \tag{32}$$

Part 2: Delayed payment ($1 - \delta$)

Since the first payment remains due by the new retailers, the whole seller funds the full inventory on the remaining fraction of the selling price. Here, the whole seller is not earning any interest:

$$\text{Interest earned from good retailers } (I_{e13}) = 0 \tag{33}$$

$$\text{Interest payable due to bad debts } (I_{p13}) \tag{34}$$

$$\text{Interest payable on the unsold inventory}(I_{p14}) = \tfrac{1}{2}(1 - \delta)cI_p(1 - K)[Py - DM - (P\lambda - D)M](T' - M) + \tfrac{1}{2}(1 - \delta)cI_p(1 - K)DT'^2 \\ + (1 - \delta)cI_p(1 - K)DT'(N - M) + \tfrac{1}{2}(1 - \delta)cI_p(1 - K)(P\lambda - D)t_1^2 + (1 - \delta)cI_p(1 - K)B(N - M) \tag{35}$$

Furthermore, there is additional interest paid on the salvage items in the present **Case (i)**, for the period $(M, t_3 + N)$.

$$\text{Interest payable on salvage items } (I_{p15}) = cI_p[\alpha y + (1 - \alpha)q_1 y](t_3 + N - M) \tag{36}$$

$$\text{Total Profit } T.P._{\cdot1} = T.R. - T.C. + I_{e11} + I_{e12} + I_{e13} - I_{p11} - I_{p12} - I_{p13} - I_{p14} - I_{p15} \tag{37}$$

$$i.e., T.P._{\cdot1} = yG_1 - y^2G_2 + BG_3 - B^2G_4 + yBG_5 + G_6 \tag{38}$$

where $G_1, \ldots, G_6$ is expanded in Appendix A.

As $G_1, G_2, \ldots, G_6$ comprise of terms of random variables, viz., α, q_1, q_2, therefore, by using the Maddah and Jaber [33] approach, one can obtain the expected value of the total profit per unit time:

$$E[Z_1(y, B)] = \left\{ \frac{E[G_1]}{E[P]}D - \frac{E[G_2]}{E[P]}Dy + \frac{E[G_3]}{E[P]}\frac{DB}{y} - \frac{E[G_4]}{E[P]}\frac{DB^2}{y} + \frac{E[G_5]}{E[P]}DB + \frac{E[G_6]}{E[P]}\frac{D}{y} \right\} \tag{39}$$

where $E[G_1]$ is elaborated in Appendix B, and likewise other expectations can be derived.

Table 2. Different cases of trade credit based on different values of M, N and T.

Cases	Immediate Payment Part	Delayed Payment Part
Case (i) $M \leq N \leq T' \leq T$	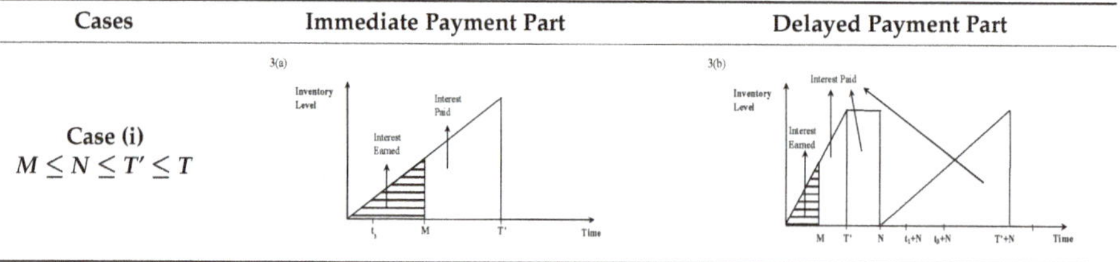	

Table 2. *Cont.*

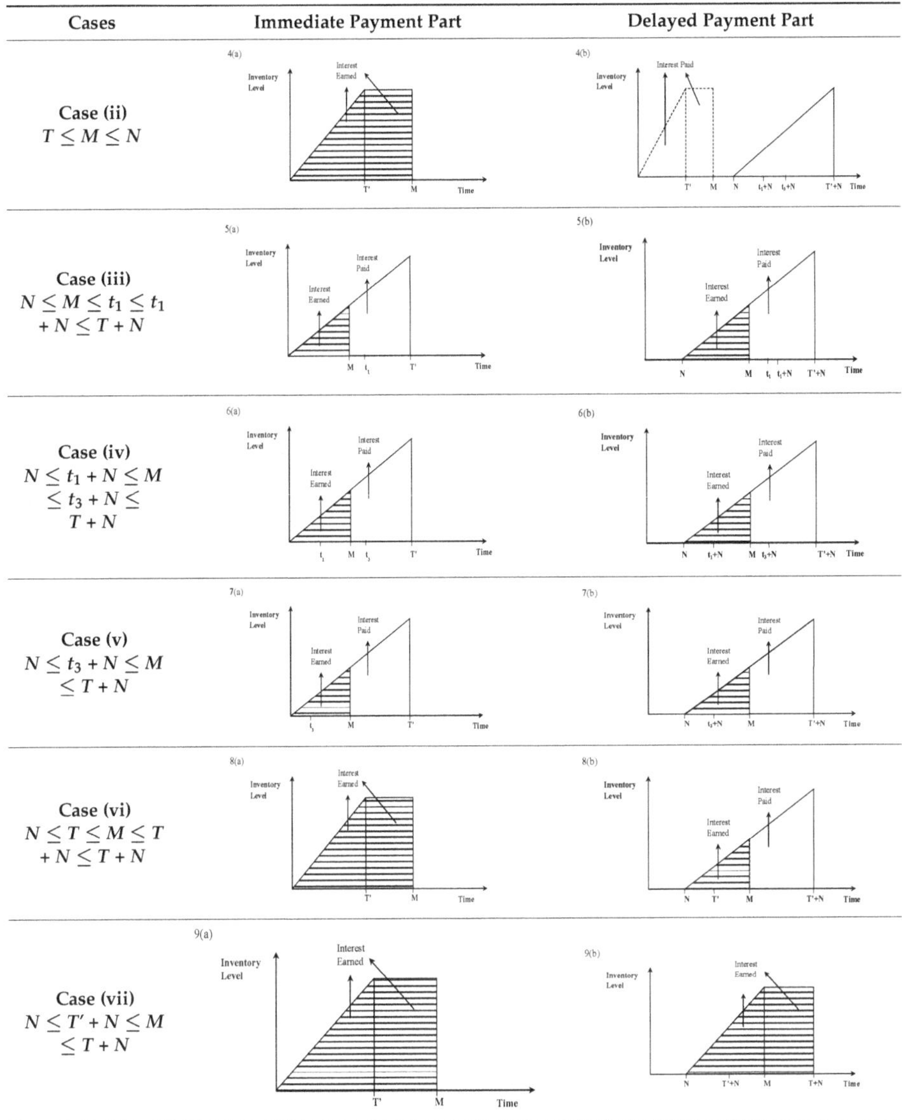

Case (ii) $T \leq M \leq N$

In this case, interest is earned from the immediate part of the partial payment only. The remaining inventory, which includes both defectives and perfect items, is sold to old and new retailers. However, he takes a loan to finance this stock (see Table 2).

Subcase 2: Interest earned and interest paid from the portion of old retailers (K)

From the portion of old retailers, the whole seller finances the complete stock. Since M lies before the first payment time N, so there is additional interest that is paid on both, viz., the entire inventory, and shortages for the period (M, N) and $(M, t_3 + N)$, respectively.

$$\text{Interest earned } (I_{e21}) = 0 \tag{40}$$

$$\text{Interest payable } (I_{p21}) = \frac{1}{2}cI_pKDT'^2 + cI_pKDT'(N-M) + \frac{1}{2}cI_pK(P\lambda - D)t_1^2 + cI_pKB(t_1 + N - M) \tag{41}$$

Subcase 3: Interest earned and interest paid from the portion of new retailers $(1 - K)$

From the proportion of new retailers, there is interest earned on the immediate payment part only. Funds are arranged to make the payment for the unsold inventory for the fraction of remaining selling price from the new retailers.

Part 1: Immediate payment (δ)

From the initial partial payment made by all the new retailers, the whole seller earns interest on the revenue generated from sales up to M. There is extra interest that is gained on an entire lot of inventory and shortages for lengths $(M - T')$ and $(M - t_1)$, respectively.

$$\text{Interest earned } (I_{e22}) = \frac{1}{2}\delta sI_e(1-K)DT'^2 + \delta sI_e(1-K)DT'(M-T') + \frac{1}{2}\delta sI_e(1-K)(P\lambda - D)t_1^2 + \delta sI_e(1-K)B(M - t_1) \tag{42}$$

$$\text{Interest payable } (I_{p22}) = 0 \tag{43}$$

Part 2: Delayed payment $(1 - \delta)$

From the delayed payment, there is no interest earned, since the first payment has not been made by any retailer until M. The whole seller has to finance the complete lot by the time of settlement of his account with the manufacturer.

$$\text{Interest earned from good retailers } (I_{e23}) = 0 \tag{44}$$

$$\text{Interest payable due to bad debts } (I_{p23}) = 0 \tag{45}$$

$$\text{Interest payable on the unsold inventory} (I_{p24}) = \frac{1}{2}(1-\delta)cI_p(1-K)DT'^2 + (1-\delta)cI_p(1-K)DT'(N-M)$$
$$+ \frac{1}{2}(1-\delta)cI_p(1-K)(P\lambda - D)t_1^2 + (1-\delta)cI_p(1-K)B(t_1 + N - M) \tag{46}$$

Furthermore, there is additional interest that is paid on the salvage items in the present **Case (ii)**, for the period $(M, t_3 + N)$.

$$\text{Interest payable on salvage items} (I_{p25}) = cI_p[\alpha y + (1 - \alpha)q_1 y](t_3 + N - M) \tag{47}$$

$$\text{Total Profit } T.P._2 = T.R. - T.C. + I_{e21} + I_{e22} + I_{e23} - I_{p21} - I_{p22} - I_{p23} - I_{p24} - I_{p25} \tag{48}$$

$$\text{i.e., } T.P._2 = yG_7 - y^2G_8 + BG_9 - B^2G_{10} + yBG_{11} + G_{12}. \tag{49}$$

where $G_7, \ldots, G_{12}$ is expanded in Appendix A.

As $G_7, \ldots, G_{12}$ comprise of terms of random variables, viz., α, q_1, q_2, therefore, by using the Maddah and Jaber [33] approach, one can obtain the expected value of the total profit per unit time:

$$E[Z_2(y, B)] = \left\{ \frac{E[G_7]}{E[P]}D - \frac{E[G_8]}{E[P]}Dy + \frac{E[G_9]}{E[P]}\frac{DB}{y} - \frac{E[G_{10}]}{E[P]}\frac{DB^2}{y} + \frac{E[G_{11}]}{E[P]}DB + \frac{E[G_{12}]}{E[P]}\frac{D}{y} \right\} \tag{50}$$

Case (iii) $\quad N \le M \le t_1 \le t_1 + N \le T + N$

In this case, the whole seller earns interest from the sales that are drawn from both old and new (good only) retailers, whose payments commence from N onwards. Thus, there is interest earned from all the new retailers by the initial payment that is made for the period $(0, M)$. There is additional interest paid by the whole seller, due to the portion of bad debts apart from the unsold inventory and salvage items (see Table 2).

Subcase 4: Interest earned and interest paid from the portion of old retailers (K)

The whole seller earns interest from old retailers as per demand consumption and shortage backordering for the time period N to M. Since all the credit sales are not realized until M, the whole seller pays interest on the unsold inventory for the fraction of old retailers.

$$\text{Interest earned }(I_{e31}) = \frac{1}{2}sI_eKD(M-N)^2 + \frac{1}{2}sI_eK(P\lambda - D)(M-N)^2. \tag{51}$$

$$\text{Interest payable}(I_{p31}) = \frac{1}{2}cI_pK[Py - D(M-N) - (P\lambda - D)(M-N)](T'+N-M). \tag{52}$$

Subcase 5: Interest earned and interest paid from the portion of new retailers $(1-K)$

In contrast to the previous sub case, the whole seller pays interest in three subparts from the fraction of new retailers. However, he earns interest from both the immediate and delayed parts of the partial payments.

Part 1: Immediate payment (δ)

From the initial partial payment made by all the new retailers, the whole seller earns interest for the period $(0, M)$, as per demand and shortage fulfillment. He pays interest on the unsold inventory for the fraction of the initial price paid by all the new retailers from time M to T'.

$$\text{Interest earned }(I_{e32}) = \frac{1}{2}\delta sI_e(1-K)DM^2 + \frac{1}{2}\delta sI_e(1-K)(P\lambda - D)M^2. \tag{53}$$

$$\text{Interest payable }(I_{p32}) = \frac{1}{2}\delta cI_p(1-K)[Py - DM - (P\lambda - D)M](T'-M) \tag{54}$$

Part 2: Delayed payment $(1-\delta)$

From the delayed payment, the interest-earning period for the whole seller starts from N up to M, as he begins to receive his first payment at N by the good retailers only. Interest is payable on the unsold items in stock for the remaining fraction of the price by the portion of new retailers for the time M to $T'+N$, and for the inventory sold to bad retailers.

$$\text{Interest earned from good retailers}(I_{e33}) = \frac{1}{2}(1-\delta)sI_eR(1-K)D(M-N)^2 \\ +\frac{1}{2}(1-\delta)sI_eR(1-K)(P\lambda - D)(M-N)^2 \tag{55}$$

$$\text{Interest payable due to bad debts}(I_{p33}) = \frac{1}{2}(1-\delta)cI_p(1-R)(1-K)D(M-N)^2 \\ +\frac{1}{2}(1-\delta)cI_p(1-R)(1-K)(P\lambda - D)(M-N)^2 \tag{56}$$

$$\text{Interest payable on the unsold inventory}(I_{p34}) = \frac{1}{2}(1-\delta)cI_p(1-K) \\ [Py - D(M-N) - (P\lambda - D)(M-N)](T'+N-M) \tag{57}$$

Furthermore, there is additional interest paid on the salvage items in **Case (iii)** for the period $(M, t_3 + N)$.

$$\text{Interest payable on salvage items }(I_{p35}) = cI_p[\alpha y + (1-\alpha)q_1 y](t_3 + N - M). \tag{58}$$

$$\text{Total Profit } T.P._3 = T.R. - T.C. + I_{e31} + I_{e32} + I_{e33} - I_{p31} - I_{p32} - I_{p33} - I_{p34} - I_{p35} \tag{59}$$

$$\text{i.e., } T.P._3 = yG_{13} - y^2G_{14} + BG_{15} - B^2G_{16} + yBG_{17} + G_{18}. \tag{60}$$

where $G_{13}, \ldots, G_{18}$ is expanded in Appendix A.

As $G_{13}, \ldots, G_{18}$ comprise of terms of random variables viz. α, q_1, q_2, therefore, by using the Maddah and Jabber [33] approach, one can obtain the expected value of the total profit per unit time:

$$E[Z_3(y,B)] = \left\{ \frac{E[G_{13}]}{E[P]} D - \frac{E[G_{14}]}{E[P]} Dy + \frac{E[G_{15}]}{E[P]} \frac{DB}{y} - \frac{E[G_{16}]}{E[P]} \frac{DB^2}{y} + \frac{E[G_{17}]}{E[P]} DB + \frac{E[G_{18}]}{E[P]} \frac{D}{y} \right\}. \quad (61)$$

Case (iv) $\quad N \leq t_1 + N \leq M \leq t_3 + N \leq T + N$

In addition to the interest earned and paid as in the previous **Case (iii)**, the whole seller not only earns interest on the demand and shortage backorder, but thus on a complete lot of shortages, since it is fully eliminated by the time of settlement of the account. Interest is paid on the unsold inventory and salvage items, and thus on the inventory sold to bad credit retailers (see Table 2).

Subcase 6: Interest earned and interest paid from the portion of old retailers (K)

The whole seller earns interest from old retailers as per the sales, through demand and shortage satisfaction for the time period N to M, and on a complete lot of shortages from $t_1 + N$ to M. Since all of the items are not sold until M, the whole seller pays interest on the unsold inventory for the fraction of old retailers.

$$\text{Interest earned } (I_{e41}) = \frac{1}{2} s I_e KD(M-N)^2 + \frac{1}{2} s I_e K (P\lambda - D) t_1^2 + s I_e KB(M - t_1 - N). \quad (62)$$

$$\text{Interest payable } (I_{p41}) = \frac{1}{2} c I_p K [Py - D(M-N) - B](T' + N - M). \quad (63)$$

Subcase 7: Interest earned and interest paid from the portion of new retailers $(1 - K)$

In this sub case, the whole seller pays interest in three subparts from the fraction of new retailers. He also earns interest from both the immediate and delayed parts of the partial payments made.

Part 1: Immediate payment (δ)

From the initial partial payment made by all of the new retailers, the whole seller earns interest on the revenue generated from sales through demand and shortage fulfillment, up to M. Further, he pays interest on the unsold inventory for the fraction of the initial price that is made by all the new retailers from period (M, T'), excluding the shortages, as these are completely eliminated and compensated by $t_1 + N$.

$$\text{Interest earned } (I_{e42}) = \frac{1}{2} \delta s I_e (1-K) DM^2 + \frac{1}{2} \delta s I_e (1-K)(P\lambda - D) t_1^2 + \delta s I_e (1-K) B(M - t_1). \quad (64)$$

$$\text{Interest payable } (I_{p42}) = \frac{1}{2} \delta c I_p (1-K)(Py - DM - B)(T' - M). \quad (65)$$

Part 2: Delayed payment $(1 - \delta)$

From the delayed payment, the whole seller starts earning interest on the demand satisfied by good retailers, only from time N to M. Since shortages have been met at $t_1 + N$, there is additional interest that is generated from them. Finances are arranged for the unsold inventory, and thus for the items sold to bad credit customers.

$$\text{Interest earned from good retailers} (I_{e43}) = \tfrac{1}{2}(1-\delta) s I_e R(1-K) D(M-N)^2 + \tfrac{1}{2}(1-\delta) s I_e R(1-K)(P\lambda - D) t_1^2$$
$$+ (1-\delta) s I_e R(1-K) B(M - t_1 - N) \quad (66)$$

$$\text{Interest payable due to bad debts} (I_{p43}) = \tfrac{1}{2}(1-\delta) c I_p (1-R)(1-K) D(M-N)^2$$
$$+ \tfrac{1}{2}(1-\delta) c I_p (1-R)(1-K)(P\lambda - D) t_1^2 + (1-\delta) c I_p (1-R)(1-K) B(M - t_1 - N) \quad (67)$$

$$\text{Interest payable on the unsold inventory} (I_{p14}) (I_{p44}) = \tfrac{1}{2}(1-\delta) c I_p (1-K)[Py - D(M-N) - B](T' + N - M) \quad (68)$$

Furthermore, there is additional interest paid on the salvage items in this **Case (iv)** for the period $(M, t_3 + N)$.

$$\text{Interest payable on salvage items}(I_{p45}) = cI_p[\alpha y + (1 - \alpha)q_1 y](t_3 + N - M). \tag{69}$$

$$\text{Total Profit } T.P._4 = T.R. - T.C. + I_{e41} + I_{e42} + I_{e43} - I_{p41} - I_{p42} - I_{p43} - I_{p44} - I_{p45} \tag{70}$$

$$\text{Therefore, } T.P._4 = yG_{19} - y^2 G_{20} + BG_{21} - B^2 G_{22} + yBG_{23} + G_{24}. \tag{71}$$

where $G_{19}, \ldots, G_{24}$ is expanded in Appendix A.

As $G_{19}, \ldots, G_{24}$ comprise of terms of random variables, viz., α, q_1, q_2, therefore, by using the Maddah and Jaber [33] approach, we obtain the expected value of the total profit per unit time:

$$E[Z_4(y, B)] = \left\{ \frac{E[G_{19}]}{E[P]} D - \frac{E[G_{20}]}{E[P]} Dy + \frac{E[G_{21}]}{E[P]} \frac{DB}{y} - \frac{E[G_{22}]}{E[P]} \frac{DB^2}{y} + \frac{E[G_{23}]}{E[P]} DB + \frac{E[G_{24}]}{E[P]} \frac{D}{y} \right\}. \tag{72}$$

Case (v) $N \leq t_3 + N \leq M \leq T' + N$

In this case, all of the expressions coincide with that of the previous case, except for the value of the salvage items. Here, interest is earned on a lot of scrap items instead of paying (see Table 2).

Subcase 8: Interest earned and interest paid from the portion of old retailers (K)

Here, the whole seller earns interest from the sales acquired from old retailers as per demand as well as shortages till M. Interest is paid on the unsold inventory after time point M. So, the results are similar to those discussed in subcase 6.

$$\text{Interest earned } (I_{e51}) = I_{e41} \tag{73}$$

$$\text{Interest payable } (I_{p51}) = I_{p41} \tag{74}$$

Subcase 9: Interest earned and interest paid from the portion of new retailers $(1 - K)$

Here, the whole seller earns interest from the portion of good retailers only while pays undue interest because of the bad debtors assumed in the model. There is also interest charged on the unsold items in the inventory as discussed in subcase 7.

Part 1: Immediate payment (δ)

$$\text{Interest earned } (I_{e52}) = I_{e42} \tag{75}$$

$$\text{Interest payable } (I_{p52}) = I_{p42} \tag{76}$$

Part 2: Delayed payment $(1 - \delta)$

$$\text{Interest earned from good retailers}(I_{e53}) = I_{e43} \tag{77}$$

$$\text{Interest payable due to bad debts}(I_{p53}) = I_{p43} \tag{78}$$

$$\text{Interest payable on the unsold inventory}(I_{p54}) = I_{p44} \tag{79}$$

In the present **Case (v)**, there is interest earned on the salvage items for the period $(t_3 + N, M)$.

$$\text{Interest earned on salvage items}(I_{e54}) = vI_e[\alpha y + (1 - \alpha)q_1 y](M - t_3 - N) \tag{80}$$

$$\text{Total Profit } T.P._5 = T.R. - T.C. + I_{e51} + I_{e52} + I_{e53} - I_{p51} - I_{p52} - I_{p53} - I_{p54} - I_{p55} \tag{81}$$

$$\text{Thus, } T.P._5 = yG_{25} - y^2 G_{26} + BG_{27} - B^2 G_{28} + yBG_{29} + G_{30}. \tag{82}$$

where $G_{24}, \ldots, G_{30}$ is expanded in Appendix A.

As $G_{24}, \dots, G_{30}$ comprise of the terms of random variables, viz., α, q_1, q_2, therefore, by using the Maddah and Jaber [33] approach, the expected value of the total profit per unit time is as follows:

$$E[Z_5(y, B)] = \left\{ \frac{E[G_{25}]}{E[P]} D - \frac{E[G_{26}]}{E[P]} Dy + \frac{E[G_{27}]}{E[P]} \frac{DB}{y} - \frac{E[G_{28}]}{E[P]} \frac{DB^2}{y} + \frac{E[G_{29}]}{E[P]} DB + \frac{E[G_{30}]}{E[P]} \frac{D}{y} \right\} \quad (83)$$

Case (vi) $\quad N \leq T' \leq M \leq T' + N \leq T + N$

In this case, the whole seller earns interest, not only from the current demand, but thus by the demand that is backordered from both old and new retailers. Interest is earned on the whole amount of shortages and salvage items. On the other hand, interest is paid on the unsold inventory, and thus on a fraction of the sold inventory due to bad debts. The difference from the previous cases lies in the calculations of the immediate part (see Table 2).

Subcase 10: Interest earned and interest paid from the portion of old retailers (K)

The whole seller earns interest on the average sales revenue that is received during the period (N, M) through demand and shortage satisfaction. After time $t_1 + N$, interest is earned on the full shortage level up to M. Since all of the items are not sold till M, the whole seller finances the unsold inventory for the fraction of the old retailers.

$$\text{Interest earned}(I_{e61}) = \frac{1}{2} s I_e K D (M - N)^2 + \frac{1}{2} s I_e K (P\lambda - D) t_1^2 + s I_e K B (M - t_1 - N). \quad (84)$$

$$\text{Interest payable}(I_{p61}) = \frac{1}{2} c I_p K [Py - D(M - N) - B](T' + N - M). \quad (85)$$

Subcase 11: Interest earned and interest paid from the portion of new retailers $(1 - K)$

In this sub case, apart from the unsold inventory, the whole seller pays interest only in the delayed part, and not in the immediate part. However, he earns interest on both the shortages and demand in the immediate and delayed parts of the partial payments made.

Part 1: Immediate payment (δ)

There is no item remaining in the stock in this part of the partially permissible delay. So, no interest is paid, while interest is earned on the full inventory, along with the shortages on the fraction of the initial price that is paid by all of the new retailers until credit limit M.

$$\text{Interest earned}(I_{e62}) = \frac{1}{2} \delta I_e s (1 - K) DT'^2 + \delta s I_e (1 - K) DT'(M - T') + \frac{1}{2} \delta s I_e (1 - K)(P\lambda - D) t_1^2 + \delta s I_e (1 - K) B(M - t_1). \quad (86)$$

$$\text{Interest payable}(I_{p62}) = 0. \quad (87)$$

Part 2: Delayed payment $(1 - \delta)$

The whole seller earns interest on the demand that is satisfied by good retailers, only from time N to M from the delayed payment. Again, there is additional interest that is generated from shortages, since these have been fully met by $t_1 + N$. The unsold inventory is financed, along with the items sold to bad credit customers from the whole seller's own pocket.

$$\text{Interest earned from good retailers}(I_{e63}) = \frac{1}{2}(1 - \delta) s I_e R (1 - K) D (M - N)^2 + \frac{1}{2}(1 - \delta) s I_e R (1 - K)(P\lambda - D) t_1^2 + (1 - \delta) s I_e R (1 - K) B (M - t_1 - N) \quad (88)$$

$$\text{Interest payable due to bad debts}(I_{p63}) = \frac{1}{2}(1 - \delta) c I_p (1 - R)(1 - K) D (M - N)^2 + \frac{1}{2}(1 - \delta) c I_p (1 - R)(1 - K)(P\lambda - D) t_1^2$$
$$+ (1 - \delta) c I_p (1 - R)(1 - K) B (M - t_1 - N) \quad (89)$$

$$\text{Interest payable on the unsold inventory}(I_{p64}) = \frac{1}{2}(1 - \delta) c I_p (1 - K)[Py - D(M - N) - B](T' + N - M). \quad (90)$$

Furthermore, there is additional interest earned on the salvage items in **Case (vi)** for the period $(t_3 + N, M)$.

$$\text{Interest earned on salvage items}(I_{e64}) = vI_e[\alpha y + (1 - \alpha)q_1 y](M - t_3 - N). \tag{91}$$

$$\text{Total Profit } T.P._6 = T.R. - T.C. + I_{e61} + I_{e62} + I_{e63} + I_{e64} - I_{p61} - I_{p62} - I_{p63} - I_{p64} \tag{92}$$

$$\text{Hence, } T.P.6 = yG_{31} - y^2G_{32} + BG_{33} - B^2G_{34} + yBG_{35} + G_{36}. \tag{93}$$

where $G_{31}, \ldots, G_{36}$ is expanded in Appendix A.

As $G_{31}, \ldots, G_{36}$ comprise of terms of random variables, viz., α, q_1, q_2, therefore, by using Maddah and Jaber [33] approach, one can get the expected value of total profit per unit time:

$$E[Z_6(y, B)] = \left\{ \frac{E[G_{31}]}{E[P]}D - \frac{E[G_{32}]}{E[P]}Dy + \frac{E[G_{33}]}{E[P]}\frac{DB}{y} - \frac{E[G_{34}]}{E[P]}\frac{DB^2}{y} + \frac{E[G_{35}]}{E[P]}DB + \frac{E[G_{36}]}{E[P]}\frac{D}{y} \right\}. \tag{94}$$

Case (vii) $N \leq T' + N \leq M \leq T + N$

This is the case of the largest credit period, where the entire inventory level becomes zero before the credit limit M. Thus, in this scenario, there is no interest that is paid by the whole seller, except for some bad debts. Interest is earned on the sales that are consumed by demand and shortages. Further, there is additional interest earned on the backordered lot of shortages for $(M - t_3 - N)$ length of time, and on the complete lot for the time period $(T' + N, M)$ (see Table 2).

Subcase 12: Interest earned and interest paid from the portion of old retailers (K)

The whole seller earns interest on average sales revenue for the period $(0, T')$, and on full sales revenue for a length of $(M - T' - N)$ from the portion of the old retailers. Since all of the items are exhausted by M, there is no interest paid.

$$\text{Interest earned } (I_{e71}) = \frac{1}{2}sI_eKDT'^2 + sI_eKDT'(M - T' - N) + \frac{1}{2}sI_eK(P\lambda - D)t_1^2 + sI_eKB(M - t_1 - N). \tag{95}$$

$$\text{Interest payable}(I_{p71}) = 0 \tag{96}$$

Subcase 13: Interest earned and interest paid from the portion of new retailers $(1 - K)$

Interest is not paid because of zero leftover stock in both parts of the partially permissible delay. However, some interest is still paid on the sold and unfinanced inventory in the delayed part, due to bad retailers.

Part 1: Immediate payment (δ)

Interest is earned on the whole lot and on shortages by the fraction of instant payment made by all of the new retailers. Since there is no inventory left in stock, no interest is paid by the whole seller.

$$\text{Interest earned } (I_{e72}) = \frac{1}{2}\delta sI_e(1 - K)DT'^2 + \delta sI_e(1 - K)DT'(M - T') + \frac{1}{2}\delta sI_e(1 - K)(P\lambda - D)t_1^2 + \delta sI_e(1 - K)B(M - t_1). \tag{97}$$

$$\text{Interest payable}(I_{p72}) = 0 \tag{98}$$

Part 2: Delayed payment $(1 - \delta)$

From the delayed payment, interest is earned on revenue generated from sales up to M on the remaining partial payment made by good retailers only. There is additional interest earned on the full inventory and on all of the shortages. However, there is some interest paid on the remaining payment from the proportion of bad credit retailers.

$$\text{Interest earned from good retailers}(I_{e73}) = \tfrac{1}{2}(1 - \delta)sI_eR(1 - K)DT'^2 + (1 - \delta)sI_eR(1 - K)DT'(M - T' - N) + \tfrac{1}{2}(1 - \delta)sI_eR(1 - K)(P\lambda - D)t_1^2 \tag{99}$$
$$+ (1 - \delta)sI_eR(1 - K)B(M - t_1 - N)$$

Interest payable due to bad debts $(I_{p73}) = \frac{1}{2}(1-\delta)cI_p(1-R)(1-K)DT'^2 + (1-\delta)cI_p(1-R)(1-K)DT'(M-T'-N)$
$+ \frac{1}{2}(1-\delta)cI_p(1-R)(1-K)(P\lambda-D)t_1^2 + (1-\delta)cI_p(1-R)(1-K)B(M-t_1-N)$
$$(100)$$

$$\text{Interest payable on the unsold inventory}\,(I_{p74}) = 0 \tag{101}$$

Furthermore, there is additional interest that is earned on the salvage items in **Case (vii)** for the period $(t_3 + N, M)$.

$$\text{Interest earned on salvage items}\,(I_{e74}) = vI_e[\alpha y + (1-\alpha)q_1 y](M - t_3 - N). \tag{102}$$

$$\text{Total Profit }T.P._7 = T.R. - T.C. + I_{e71} + I_{e72} + I_{e73} + I_{e74} - I_{p71} - I_{p72} - I_{p73} - I_{p74} \tag{103}$$

$$\text{Thus, }T.P._7 = yG_{37} - y^2 G_{38} + BG_{39} - B^2 G_{40} + yBG_{41} + G_{42}. \tag{104}$$

where $G_{37}, \ldots, G_{42}$ is expanded in Appendix A.

As $G_{37}, \ldots, G_{42}$ comprise of terms of random variables, viz., α, q_1, q_2, therefore, by using the Maddah and Jaber [33] approach, one can obtain the expected value of the total profit per unit time:

$$E[Z_7(y,B)] = \left\{ \frac{E[G_{37}]}{E[P]}D - \frac{E[G_{38}]}{E[P]}Dy + \frac{E[G_{39}]}{E[P]}\frac{DB}{y} - \frac{E[G_{40}]}{E[P]}\frac{DB^2}{y} + \frac{E[G_{41}]}{E[P]}DB + \frac{E[G_{42}]}{E[P]}\frac{D}{y} \right\}. \tag{105}$$

Case (viii) $\quad N \leq T' + N \leq T + N \leq M$

The present case completely overlaps with the previous **Case (vii)**. This case does not count. Thus, in total, there exist even distinct cases for obtaining the whole seller's expected total profit per unit time, viz.:

$$E\big[Z_j(y,B)\big] = \begin{cases} E[Z_1(y,B)] \text{ when } M \leq N \leq T' \leq T \\ E[Z_2(y,B)] \text{ when } T \leq M \leq N \\ E[Z_3(y,B)] \text{ when } N \leq M \leq t_1 \leq t_1 + N \leq T + N \\ E[Z_4(y,B)] \text{ when } N \leq t_1 + N \leq M \leq t_3 + N \leq T + N \\ E[Z_5(y,B)] \text{ when } N \leq t_3 + N \leq M \leq T' + N \leq T + N \\ E[Z_6(y,B)] \text{ when } N \leq T' \leq M \leq T' + N \leq T + N \\ E[Z_7(y,B)] \text{ when } N \leq T' + N \leq M \leq T + N \end{cases}$$

3. Theoretical Results for Optimality

The whole seller intends to maximize his expected total profit per unit time by jointly optimizing the replenishment quantity and the backorder level. In this section, the optimality of the objective function is established in the form of two lemmas. Here, optimality is shown for **Case (i)** only, and likewise, the optimality for other six cases can be derived by substituting the values of G_i's in respective cases.

Lemma 1. *The function of whole seller's expected total profit per unit time is concave.*

Proof. To prove the global concavity of the expected profit function, the following two second-order sufficient conditions of global optimality must be satisfied for all seven cases:

$$\left(\frac{\partial^2 E[Z_j(y,B)]}{\partial y^2} \right) \leq 0; \left(\frac{\partial^2 E[Z_j(y,B)]}{\partial B^2} \right) \leq 0$$

and $\qquad\qquad\qquad\qquad\qquad\qquad\qquad$ for j.

$$\left(\frac{\partial^2 E[Z_j(y,B)]}{\partial y \partial B} \right)^2 - \left(\frac{\partial^2 E[Z_j(y,B)]}{\partial y^2} \right)\left(\frac{\partial^2 E[Z_j(y,B)]}{\partial B^2} \right) \leq 0$$

$\square$

Case (i) $M \leq N \leq T' \leq T$

By taking a first-order partial derivative of $E[Z_1(y, B)]$ with respect to y and B, one can obtain:

$$\frac{\partial}{\partial y} E[Z_1(y, B)] = -\frac{E[G_2]}{E[P]} D - \frac{E[G_3]}{E[P]} \frac{DB}{y^2} + \frac{E[G_4]}{E[P]} \frac{DB^2}{y^2} - \frac{E[G_6]}{E[P]} \frac{D}{y^2}. \tag{106}$$

$$\frac{\partial}{\partial B} E[Z_1(y, B)] = \frac{E[G_3]}{E[P]} \frac{D}{y} - \frac{2E[G_4]}{E[P]} \frac{DB}{y} + \frac{E[G_5]}{E[P]} D. \tag{107}$$

By taking a second-order partial derivative of $E[Z_1(y, B)]$ with respect to y and B, one can obtain:

$$\frac{\partial^2}{\partial y^2} E[Z_1(y, B)] = \frac{E[G_3]}{E[P]} \frac{2DB}{y^3} - \frac{E[G_4]}{E[P]} \frac{2DB^2}{y^3} + \frac{E[G_6]}{E[P]} \frac{2D}{y^3}. \tag{108}$$

$$\frac{\partial^2}{\partial B^2} E[Z_1(y, B)] = -\frac{2E[G_4]}{E[P]} \frac{D}{y}. \tag{109}$$

$$\text{Again, } \frac{\partial^2}{\partial y \partial B} E[Z_1(y, B)] = -\frac{E[G_3]}{E[P]} \frac{D}{y^2} + \frac{E[G_4]}{E[P]} \frac{2DB}{y^2}. \tag{110}$$

Therefore, by using Equations (108)–(110) it is obtained that:

$$\left(\frac{\partial^2}{\partial y \partial B} E[Z_1(y, B)] \right)^2 - \left(\frac{\partial^2}{\partial y^2} E[Z_1(y, B)] \right) \left(\frac{\partial^2}{\partial B^2} E[Z_1(y, B)] \right) = \frac{E^2[G_3]}{E^2[P]} \frac{D^2}{y^4} - \frac{E[G_6]E[G_4]}{E^2[P]} \frac{4D^2}{y^4}. \tag{111}$$

The sufficient conditions of concavity are derived in Appendix C:

Lemma 2. *The optimal solution (y^*, B^*) that maximizes the whole seller's expected total profit per unit time for Case (i) is written as:*

$$y^* = \frac{\left\{ \begin{array}{l} \left\{ \begin{array}{l} -\frac{1}{2}(h + c_B)\left(\frac{1}{\{E[\alpha]E[q_2]+(1-E[\alpha])(1-E[q_1])\}\lambda - D} + \frac{1}{D} \right) + \frac{1}{2D}cI_p[1 - \delta(1-K)] \\ +\frac{1}{2}\frac{1}{\{E[\alpha]E[q_2]+(1-E[\alpha])(1-E[q_1])\}\lambda - D}cI_p[1 - \delta(1-K)] \end{array} \right\} B^2 \\[2mm] -\{cI_p[1 - \delta(1-K)](N - M) - \frac{1}{2D}cI_p\{E[\alpha]E[q_2] + (1-E[\alpha])(1-E[q_1])\}\lambda M\}B \\[2mm] -\left\{ \begin{array}{l} -\{E[\alpha]E[q_2] + (1-E[\alpha])(1-E[q_1])\}(1-\delta)cI_pK\lambda M^2 + \frac{1}{2}\{E[\alpha]E[q_2] + (1-E[\alpha])(1-E[q_1])\}\delta pI_e(1-K)\lambda M^2 \\[1mm] -cI_p[1 - \delta(1-K)]B(N-M) + \frac{1}{2}\{E[\alpha]E[q_2] + (1-E[\alpha])(1-E[q_1])\}\delta cI_p(1-K)yM \\[1mm] +(\frac{1}{2} - \delta)\{E[\alpha]E[q_2] + (1-E[\alpha])(1-E[q_1])\}cI_p\lambda M^2 - A \end{array} \right\} \end{array} \right\}}{\left\{ \begin{array}{l} -\left\{ \begin{array}{l} h\frac{1}{2\lambda} + h\frac{1}{2\lambda}(\{E[\alpha]E[q_2] + (1-E[\alpha])(1-E[q_1])\} - \frac{D}{\lambda}) + h\frac{1}{2\lambda}[E[\alpha](1-E[q_2]) + (1-E[\alpha])E[q_1]] \\[1mm] +h\frac{1}{2D}(\{E[\alpha]E[q_2] + (1-E[\alpha])(1-E[q_1])\} - \frac{D}{\lambda})^2 + h\frac{E[\alpha]E[q_2]}{2D}\{E[\alpha]E[q_2] + (1-E[\alpha])(1-E[q_1])\} \end{array} \right\} \\[2mm] +\frac{1}{D}cI_p\{E[\alpha]E[q_2] + (1-E[\alpha])(1-E[q_1])\}^2 - \frac{1}{2D}\delta cI_p(1-K)\{E[\alpha]E[q_2] + (1-E[\alpha])(1-E[q_1])\}^2 \\[2mm] +cI_p\frac{1}{\lambda}\{E[\alpha] + (1-E[\alpha])E[q_1]\} \end{array} \right\}} \tag{112}$$

$$B^* = \frac{\begin{array}{l} \{cI_p[1 - \delta(1-K)](N - M) - \frac{1}{2D}cI_p\{E[\alpha]E[q_2] + (1-E[\alpha])(1-E[q_1])\}\lambda M\} \\[2mm] +\left\{ \begin{array}{l} \frac{h}{2}\frac{(1-\{E[\alpha]E[q_2]+(1-E[\alpha])(1-E[q_1])\})}{\{E[\alpha]E[q_2]+(1-E[\alpha])(1-E[q_1])\}\lambda - D} - \frac{h}{2}\frac{\{E[\alpha](1-E[q_2])+(1-E[\alpha])E[q_1]\}}{\{E[\alpha]E[q_2]+(1-E[\alpha])(1-E[q_1])\}\lambda - D} \\[1mm] -h\frac{1}{D}(\{E[\alpha]E[q_2] + (1-E[\alpha])(1-E[q_1])\} - \frac{D}{\lambda}) + \frac{1}{2D}cI_p\{E[\alpha]E[q_2] + (1-E[\alpha])(1-E[q_1])\} + \\[1mm] (1-\delta)cI_p(1-K)\frac{\{E[\alpha]E[q_2]+(1-E[\alpha])(1-E[q_1])\}}{D} \end{array} \right\} y \end{array}}{\left\{ \begin{array}{l} -(h + c_B)\left(\frac{1}{\{E[\alpha]E[q_2]+(1-E[\alpha])(1-E[q_1])\}\lambda - D} + \frac{1}{D} \right) + \frac{1}{D}cI_p[1 - \delta(1-K)] \\[1mm] +\frac{1}{\{E[\alpha]E[q_2]+(1-E[\alpha])(1-E[q_1])\}\lambda - D}cI_p[1 - \delta(1-K)] \end{array} \right\}} \tag{113}$$

Proof. To determine the optimal values of y and B, say y^* and B^*, which maximize the function of $E[Z_1(y, B)]$, the first-order necessary condition of optimality must be satisfied:

$$\frac{\partial}{\partial y}E[Z_1(y, B)] = 0 \text{ and } \frac{\partial}{\partial B}E[Z_1(y, B)] = 0.$$

On setting Equation (106) to be equal to zero, one can get:

$$y^* = \frac{\left\{\begin{array}{l} -\frac{1}{2}(h + c_B)\left(\frac{1}{\{E[\alpha]E[q_2]+(1-E[\alpha])(1-E[q_1])\}\lambda-D} + \frac{1}{D}\right) + \frac{1}{2D}cI_p[1 - \delta(1 - K)] \\[4pt] +\frac{1}{2}\frac{1}{\{E[\alpha]E[q_2]+(1-E[\alpha])(1-E[q_1])\}\lambda-D}cI_p[1 - \delta(1 - K)] \end{array}\right\}B^2 - \left\{cI_p[1 - \delta(1 - K)](N - M) - \frac{1}{2D}cI_p\{E[\alpha]E[q_2] + (1 - E[\alpha])(1 - E[q_1])\}\lambda M\right\}B - \left\{\begin{array}{l} -\{E[\alpha]E[q_2] + (1 - E[\alpha])(1 - E[q_1])\}(1 - \delta)cI_pK\lambda M^2 + \frac{1}{2}\{E[\alpha]E[q_2] + (1 - E[\alpha])(1 - E[q_1])\}\delta pI_e(1 - K)\lambda M^2 \\[4pt] -cI_p[1 - \delta(1 - K)]B(N - M) + \frac{1}{2}\{E[\alpha]E[q_2] + (1 - E[\alpha])(1 - E[q_1])\}\delta cI_p(1 - K)yM \\[4pt] +(\frac{1}{2} - \delta)\{E[\alpha]E[q_2] + (1 - E[\alpha])(1 - E[q_1])\}cI_p\lambda M^2 - A \end{array}\right\}}{\left\{\begin{array}{l} -\left\{\begin{array}{l} h\frac{1}{2\lambda} + h\frac{1}{2\lambda}(\{E[\alpha]E[q_2] + (1 - E[\alpha])(1 - E[q_1])\} - \frac{D}{\lambda}) + h\frac{1}{2\lambda}[E[\alpha](1 - E[q_2]) + (1 - E[\alpha])E[q_1]] \\[4pt] +h\frac{1}{2D}(\{E[\alpha]E[q_2] + (1 - E[\alpha])(1 - E[q_1])\} - \frac{D}{\lambda})^2 + h\frac{E[\alpha]E[q_2]}{2D}\{E[\alpha]E[q_2] + (1 - E[\alpha])(1 - E[q_1])\} \end{array}\right\} \\[4pt] +\frac{1}{D}cI_p\{E[\alpha]E[q_2] + (1 - E[\alpha])(1 - E[q_1])\}^2 - \frac{1}{2D}\delta cI_p(1 - K)\{E[\alpha]E[q_2] + (1 - E[\alpha])(1 - E[q_1])\}^2 \\[4pt] +cI_p\frac{1}{\lambda}\{E[\alpha] + (1 - E[\alpha])E[q_1]\} \end{array}\right\}} \tag{114}$$

On setting Equation (107) to be equal to zero, one can find:

$$B^* = \frac{\left\{cI_p[1 - \delta(1 - K)](N - M) - \frac{1}{2D}cI_p\{E[\alpha]E[q_2] + (1 - E[\alpha])(1 - E[q_1])\}\lambda M\right\} + \left\{\begin{array}{l} \frac{h}{2}\frac{(1-\{E[\alpha]E[q_2]+(1-E[\alpha])(1-E[q_1])\})}{\{E[\alpha]E[q_2]+(1-E[\alpha])(1-E[q_1])\}\lambda-D} - \frac{h}{2}\frac{\{E[\alpha](1-E[q_2])+(1-E[\alpha])E[q_1]\}}{\{E[\alpha]E[q_2]+(1-E[\alpha])(1-E[q_1])\}\lambda-D} \\[4pt] -h\frac{1}{D}\left(\{E[\alpha]E[q_2] + (1 - E[\alpha])(1 - E[q_1])\} - \frac{D}{\lambda}\right) + \frac{1}{2D}cI_p\{E[\alpha]E[q_2] + (1 - E[\alpha])(1 - E[q_1])\}+ \\[4pt] (1 - \delta)cI_p(1 - K)\frac{\{E[\alpha]E[q_2]+(1-E[\alpha])(1-E[q_1])\}}{D} \end{array}\right\}y}{\left\{\begin{array}{l} -(h + c_B)\left(\frac{1}{\{E[\alpha]E[q_2]+(1-E[\alpha])(1-E[q_1])\}\lambda-D} + \frac{1}{D}\right) + \frac{1}{D}cI_p[1 - \delta(1 - K)] \\[4pt] +\frac{1}{\{E[\alpha]E[q_2]+(1-E[\alpha])(1-E[q_1])\}\lambda-D}cI_p[1 - \delta(1 - K)] \end{array}\right\}} \tag{115}$$

Hence, y^* and B^*, are the optimal values of y and B. Therefore, the global optimality of y^*, B^* is achieved for **Case (i)**. $\square$

4. Solution Procedure

In order to find whole seller's optimal ordering policy that maximizes the expected total profit per unit time, the following procedure is proposed:

Step 0: Input all parameters.

Step 1: Determine the optimal values of y^*, B^* for **Case (i)** from Equations (114) and (115), respectively. Using y^*, calculate the corresponding value of t_3^* and T^* from Equations (3) and (10), respectively. If $M \leq N \leq T' \leq T$, then calculate $E[Z_1(y, B)]$ from Equation (39), and go to Step 8, otherwise go to Step 2.

Step 2: Determine the optimal values of y^*, B^* for **Case (ii)** by substituting the particular values of $G_i's$ ($i = 7, \ldots, 12$) in Equations (114) and (115), respectively. Using y^*, calculate the corresponding value of t_3^* and T^* from Equations (10) and (3), respectively. If $T \leq M \leq N$, then calculate $E[Z_2(y, B)]$ from Equation (50) and go to Step 8; else, go to Step 3.

Step 3: Determine the optimal values of y^*, B^* for **Case (iii)** by substituting the particular values of $G_i's$ ($i = 13, \ldots, 18$) in Equations (114) and (115), respectively. Using y^*, calculate the corresponding value of t_3^* and T^* from Equations (10) and (3), respectively. If $N \leq M \leq t_1 \leq t_1 + N \leq T + N$, then calculate $E[Z_3(y, B)]$ from Equation (61) and go to Step 8; else, go to Step 4.

Step 4: Determine the optimal values of y^*, B^* for **Case (iv)** by substituting the particular values of $G_i's$
($i = 19, \ldots, 24$) in Equations (114) and (115), respectively. Using y^*, calculate the corresponding
value of t_3^* and T^* from Equations (10) and (3), respectively. If $N \le t_1 + N \le M \le t_3 + N \le T + N$, then calculate $E[Z_4(y, B)]$ from Equation (72) and go to Step 8; else, go to Step 5.

Step 5: Determine the optimal values of y^*, B^* for **Case (v)** by substituting the particular values of $G_i's$
($i = 25, \ldots, 30$) in Equations (114) and (115), respectively. Using y^*, calculate the corresponding
value of t_3^* and T^* from Equations (10) and (3), respectively. If $N \le t_3 + N \le M \le T' + N \le T + N$, then calculate $E[Z_5(y, B)]$ from Equation (83) and go to Step 8; else, go to Step 6.

Step 6: Determine the optimal values of y^*, B^* for **Case (vi)** by substituting the particular values of $G_i's$
($i = 31, \ldots, 36$) in Equations (114) and (115), respectively. Using y^*, calculate the corresponding
value of t_3^* and T^* from Equations (10) and (3), respectively. If $N \le T' \le M \le T'+N \le T+N$, then calculate $E[Z_6(y, B)]$ from Equation (94) and go to Step 8; else, go to Step 7.

Step 7: Determine the optimal values of y^*, B^* for **Case (vii)** by substituting the particular values
of $G_i's$ ($i = 37, \ldots, 42$) in Equations (114) and (115), respectively. Using y^*, calculate the
corresponding value of t_3^* and T^* from Equations (10) and (3), respectively. If $N \le T'+N \le M \le T+N$, then calculate $E[Z_7(y, B)]$ from Equation (105) and go to Step 8.

Step 8: Procedure terminates.

5. Numerical Analysis

This subsection validates the developed model with the help of a numerical analysis. The optimal
order quantity (y^*), the optimal backorder quantity (B^*), and the expected profit per unit time $E^*[Z(y, B)]$
are found out for a given set of parameters as detailed out in Tables 3 and 4.

5.1. Numerical Experiments

Table 3. Numerical data from the Teng [10] model.

Description	Symbol	Value	Units
Set-up cost	A	12	\$/cycle
Purchase cost	C	0.5	\$/unit
Selling price	s	1	\$/unit
Holding cost	h	0.2	\$/unit/year
Interest earned	I_e	0.1	\$/year
Interest paid	I_p	0.08	\$/year

Table 4. Other parameters for the numerical example.

Description	Symbol	Value	Units
Defect proportion	α	U~(0.5,0.15)	-
Type-1 error proportion	q_1	U~(0.1,0.3)	-
Type-1 error proportion	q_2	U~(0.1,0.3)	-
Probability density function	$f(\alpha)$	1/(0.15–0.5)	-
Probability density function	$f(q_1)$	1/(0.3–0.1)	-
Probability density function	$f(q_2)$	1/(0.3–0.1)	-
Demand rate	D	5000	units/year
Inspection rate	λ	8500	units/year
Inspection cost	i	0.15	\$/unit
Salvage cost	v	0.35	\$/unit
Type-I error cost	c_r	0.05	\$/unit
Type-II error cost	c_a	0.1	\$/unit
Backorder cost	c_B	0.2	\$/unit/year
Fraction of the purchase amount	δ	20	%
Fraction of old retailers	K	40	%
Fraction of good retailers	R	70	%
Whole seller's credit limit	M	40	days
Retailer's credit limit	N	10	days

5.2. Sensitivity Analysis

To analyze the changes in the values of the parameters, sensitivity analysis has been performed to study the variations. For simplicity of analysis, deterministic values of α, q_1, and q_2 are used instead of the expected ones. The effect of changes in the main parameters α, q_1, q_2 and the parameters δ, K, R, M, and N are observed on the optimal order quantity y^*, optimal backorder quantity B^*, optimal cycle length T^*, optimal total profit per unit time $Z^*(y, B)$, optimal total revenue per unit time ($T.R.U.^*$), and optimal total cost per unit time ($T.C.U.^*$). Results have been summarized in Tables 5–8.

Table 5. Optimal values.

Description	Symbol	Value	Units
Order size	y^*	709.47	units/cycle
Backorder quantity	B^*	66.81	units/cycle
Expected total profit per unit time	$E[Z^*(y, B)]$	663.26	$/year
Cycle length	T^*	45.78	days

Table 6. Impact of α on an optimal replenishment policy.

q_1		0	0.01	0.02	0.03	0.04
y^*	↑	708.79	708.93	709.47	710.60	712.65
B^*	↓	90.01	79.28	66.81	51.84	33.03
T^*	↓	46.67	46.21	45.78	45.39	45.05
$Z^*(y,B)$	↓	705.98	684.81	663.26	641.35	619.11
TRU^*	↑	4464.52	4484.02	4503.90	4524.22	4544.94
$R1$	↓	639.32	633.08	627.17	621.77	617.15
$R2$	↓	−1.41	−1.42	−1.42	−1.42	−1.42
$R3$	↑	−91.85	−90.96	−90.10	−89.33	−88.66
$R4$	↑	24.80	27.05	29.30	31.58	33.92
TCU^*	↑	3773.67	3814.06	3855.10	3896.73	3938.81
SC	-	12.00	12.00	12.00	12.00	12.00
PC	↑	354.39	354.47	354.73	355.29	356.32
IC	↑	106.31	106.34	106.42	106.58	106.89
EC_1	↑	0	0.32	0.63	0.95	1.28
EC_2	-	0.14	0.14	0.14	0.14	0.14
HC	↑	9.20	9.28	9.35	9.42	9.46
BC	↓	0.46	0.37	0.26	0.16	0.06

Table 7. Impact of q_1 on an optimal replenishment policy.

q_1		0	0.01	0.02	0.03	0.04
y^*	↑	708.79	708.93	709.47	710.60	712.65
B^*	↓	90.01	79.28	66.81	51.84	33.03
T^*	↓	46.67	46.21	45.78	45.39	45.05
$Z^*(y,B)$	↓	705.98	684.81	663.26	641.35	619.11
TRU^*	↑	4464.52	4484.02	4503.90	4524.22	4544.94
$R1$	↓	639.32	633.08	627.17	621.77	617.15
$R2$	↓	−1.41	−1.42	−1.42	−1.42	−1.42
$R3$	↑	−91.85	−90.96	−90.10	−89.33	−88.66
$R4$	↑	24.80	27.05	29.30	31.58	33.92
TCU^*	↑	3773.67	3814.06	3855.10	3896.73	3938.81
SC	-	12.00	12.00	12.00	12.00	12.00
PC	↑	354.39	354.47	354.73	355.29	356.32
IC	↑	106.31	106.34	106.42	106.58	106.89
EC_1	↑	0	0.32	0.63	0.95	1.28
EC_2	-	0.14	0.14	0.14	0.14	0.14
HC	↑	9.20	9.28	9.35	9.42	9.46
BC	↓	0.46	0.37	0.26	0.16	0.06

Table 8. Impact of q_2 on an optimal replenishment policy.

q_2		0	0.02	0.03	0.04	0.05
y^*	↓	712.66	709.47	707.90	706.34	704.80
B^*	↑	63.82	66.81	68.25	69.65	71.02
T^*	↓	45.89	45.78	45.73	45.68	45.64
$Z^*(y,B)$	↓	666.57	663.26	661.61	659.97	658.33
TRU^*	↓	4514.13	4503.90	4498.82	4493.75	4488.68
$R1$	↓	628.56	627.17	626.49	625.82	625.16
$R2$	↓	0.00	−1.42	−2.12	−2.83	−3.52
$R3$	↑	−90.51	90.10	−89.91	−89.71	−89.52
$R4$	↓	29.43	29.30	29.24	29.17	29.11
TCU^*	↑	3861.81	3855.10	3851.76	3848.41	3845.08
SC	-	12.00	12.00	12.00	12.00	12.00
PC	↓	356.33	354.74	353.95	353.17	352.40
IC	↓	106.90	106.42	106.19	105.95	105.72
EC_1	↓	0.64	0.64	0.64	0.64	0.63
EC_2	↑	0.00	0.14	0.21	0.28	0.35
HC	↑	9.37	9.36	9.36	9.35	9.35
BC	↓	0.24	0.27	0.28	0.29	0.30

The following observations have been derived from Tables 5–8.

- As exhibited from Table 5, the optimal values of the order quantity (y^*), the backorder level (B^*), the cycle length (T^*), and the expected value of the total profit per unit time ($Z^*(y, B)$) show a decreasing trend, while the values of the total revenue per unit time and the total cost per unit time increase with the defect proportion (α). The only portion of the revenue being increased with a rise in the number of defectives is from the sale of the scrap items. However, the total cost majorly increases, due to the holding cost and the sales returns cost. With the increased count of the defectives in the system, the demand decreases, due to the frustration in retailers dealing with the faulty items, resulting in the loss of orders and shortages.

- From Table 6, it is observed that with an increase in the proportion of Type-I errors (q_1), the optimal values of backorder level (B^*), cycle length (T^*), and the expected value of total profit per unit time ($Z^*(y, B)$) show declining trends, while the order quantity (y^*) increases along with total revenue and total cost values. The Type-I error causes a direct financial loss to the whole seller as his inspection team discards some perfect items at a reduced price, leading to a fall in profit values and cycle length. The increase in the sale of salvage items majorly contributes to raising the revenue. In order to satisfy the demand with perfect items, the whole seller needs to order more; hence, the purchase cost, inspection cost, Type-1 error cost, and holding costs increases with the rise in the value of $(y)^*$. Further, with a higher number of orders, there is a lessening of shortages as the demand can now be satisfied in a better way.

- It is clear from Table 7 that with an increment in the proportion of Type-II errors (q_2), the optimal values of order quantity (y^*), cycle length (T^*), and expected value of the total profit per unit time ($Z^*(y, B)$), and the total revenue exhibit a declining nature, while the optimal backorder level (B^*), and total cost values show a rise. The Type-II error causes a penalty and goodwill loss to the whole seller by the sale of some defectives to the retailers, resulting in sales returns. Because of these defect returns, there is an addition to the inventory of the system, leading to increase in holding cost and Type-II error cost. Owing to frustration and quality dissatisfaction, the retailers may not be willing to purchase more, so a decrement is observed in the values of $(y)^*$, and thus in the total profit of the system. Due to the lack of purchases made, there is an increase in backorder level, i.e., $(B)^*$.

Observations from Table 8:

- With the increase in the proportion of initial payments (δ), there is a lowering of order quantity (y^*), while there is increase in the optimal values of the expected total profit per unit time ($Z^*(y, B)$), backorder level (B^*), cycle length (T^*), and total cost. An increase in the initial partial payment leads to a reduction in the number of retailers who are interested in bulk purchases, majorly due to the discount/credit offered. With lesser demand come fewer orders and a shortening of the cycle length. However, due to an elevation in the fraction of initial payments, it becomes more profitable for the whole seller, as he can now generate extra revenue by keeping the additional amounts in an interest-bearing account.

- With a growth in the proportion of old retailers (K), the optimal values of order quantity (y^*) and cycle length (T^*) decrease, while the optimal values of the expected total profit per unit time ($Z^*(y, B)$), and backorder level (B^*) show increasing behaviors. With the higher value of (K), the whole seller is able to offer full permissible delays to a greater number of retailers, resulting in his increased profit values. However, since the y^* values are found to decrease, this implies that the whole seller cannot rely on old retailers fully to achieve maximum possible sales, but should focus on promoting new retailers simultaneously.

- It is evident that an enhancement in the proportion of good retailers (R) increases the optimal values of the expected total profit per unit time ($Z^*(y, B)$), backorder level (B^*), and total cost, while reducing the optimal values of order quantity (y^*) and cycle length (T^*) slightly. With an increasing number of good retailers, who make complete partial payments, the whole seller is able to minimize his losses that directly occur due to bad debts, and hence, his profit values rise. The order quantity and cycle length reveal a marginal decreasing trend, owing to the fact that the whole seller is interested in ordering less but more frequently to achieve the target sales.

- Upon increasing the whole seller's credit period (M), the optimal values of the expected total profit per unit time ($Z^*(y, B)$), backorder level (B^*), and total cost increase, while the optimal values of order quantity (y^*) and cycle length (T^*) decrease. With the increase of the whole seller's credit limit, he is able to put the revenue generated through sales into an interest-bearing account for a longer duration, and hence boost up his profit values considerably. However, with a reduction in the order quantity, there will be higher number of customers waiting for delivery, i.e., there is an increase in backorder levels.

- The higher credit limit of retailers (N), the less are the optimal values of the expected total profit per unit time ($Z^*(y, B)$), backorder level (B^*), cycle length (T^*), order quantity (y^*), and total cost. When the first payment to the whole seller is delayed, he is unable to put revenue into his account for a longer period, thus affecting his profit values negatively. With larger N, the major source of revenue generation comes from the immediate payments by all the new retailers. The whole seller experiences a reduction in the order quantity to avoid a larger risk being attached to the bad credit retailers who do not pay the remaining part of the partially permissible delay at N.

6. Summary

6.1. Managerial Insights

The study helps the whole seller to make the best decision by providing some useful managerial insights on how to deal with different types of retailers arriving at his doorstep, in a cost-effective and less compromising manner. Firstly, the model is able to prove that a combination of full and partial trade credit is beneficial for managers, to elevate their profit margins, and to attract new retailers, respectively. This is quite intuitive, as the whole seller can start to accumulate revenue on sales and earn interest on them before the settlement of his account with the manufacturer. In view of this, the whole seller should choose such a manufacturer who agrees on the greater length of delayed periods. On the contrary, he should try to make an agreement with shorter credit limits with his fellow retailers, to maximize his expected profit units. To make the best use of the trade credit policy, the whole seller prefers to order less, but with higher frequency. This way, it becomes cost-effective for the whole seller, as the inventory holding cost is diminished by the fast replenishment of the items. In order to make better use of the partial trade credit policy via downstream ends, it is beneficial for him to raise the level of backorders, on account of the initial amount that is received from the partially permissible delay. However, the overall cost of the system is observed to increase, due to the presence of bad credit retailers in the study, as these bring extra interest charges on the whole seller on the sold yet unpaid inventory. Furthermore, the analysis of various defect-related factors indicates that a higher number of defectives has an adverse effect on the entire cost of the system, as it leads to an additional increase in the cost of misclassifications during the inspection process. The screening errors that are considered in the study are of great importance, and it is proven here that a Type-I error has a higher impact on the supply chain cost than a Type-II error. This difference comes from the fact the former leads to the erroneous scrapping of perfect items, resulting in an increase of the optimal order quantity, and a decrease of backorders, along with a huge decrease in the profit values, while the latter reflects the opposite trends on these optimal values, with a marginal decrease in profit values. On a wider perspective, managers should keep looking into the sources of imperfections, as their reduction/elimination can boost the demand and the profit values substantially and simultaneously reduce the proportion of misclassifications.

6.2. Concluding Remarks

The present paper analyzes the impact of the dual bifurcation of retailers in a three-layer supply chain (manufacturer-whole seller-retailer) under a two-stage trade-credit policy, along with quality control measures. Owing to a stable and trustworthy relationship with the manufacturer, the whole seller receives a full permissible delay by the manufacturer, while he in turn offers distinct types of trade credit policies to his fellow retailers. To be closer to reality, the whole seller first categorizes all of the retailers approaching him into old and new types, using past record/data. Then, according to the fulfillment of payments by new retailers, he calls them good or bad. Due to this dual bifurcation, the retailer is able to avoid various financial and reputation-related risks by giving partial trade credit policies to all of the new retailers, and secondly, he is able to raise the revenue by offering full trade credit to all of the old retailers. Bad debts play an extensive role in the model, as these bring undue interest charges on the whole seller. Such a practical scenario had not gained much importance in the past. Other common aspects that are integrated into the model are related to quality and its control measures. The lot received by the whole seller contains some defectives and goes through 100% inspection process before reaching the retailers. Type-I and Type-II errors are incorporated to make the model more realistic. By considering all of the aforementioned factors in the study, the present model provides the managers with a more pragmatic model that has wide applicability, especially in many retail industries. Based on the different situations that may occur in a trade credit scenario, closed-form solutions for all the seven cases have been obtained mathematically. An algorithm has been used to determine the optimal order quantity and the optimal backorder quantity. The paper concludes by providing an extensive sensitivity for various key parameters over the decision variables to derive many useful managerial insights.

6.3. Future Research and Limitations

The model is limited to a static inventory model, where cycles are eventually repeated, and bad debtors are not penalized in later periods. Hence, they receive partial trade credit policy in all successive cycles. However, there can be a more practical dynamic model where the bad debtors either do not receive any trade credit at all, or they receive only a reduced trade credit volume. In this way, the bad debtors would have a higher effect on the strategy, which will be closer to trade credit financing in practice. As the model majorly focuses on different forms of permissible delays at the whole seller's end, it will be more justifiable if the credit period is linked to the order quantity, or if the demand function is taken as credit-dependent rather than constant or both. The whole seller's problem with large orders and limited storage space can be solved by two ware housings. Further, to obtain better accuracy in the results, it will be relevant to incorporate inflation and the time-value of money in the study.

Author Contributions: Conceptualization, visualization, A.K. (Aditi Khanna).; methodology, formal analysis, data curation, software, writing—original draft preparation, A.K. (Aakanksha Kishore); investigation, writing—review and editing, project administration, funding acquisition, B.S.; validation, resources, supervision, C.K.J.

Funding: This research received no external funding.

Conflicts of Interest: The authors declare no conflict of interest.

Nomenclature

The following nomenclature is used throughout the paper development.

Index

j	number of cases (j=1, 2, 3, 4, 5, 6, 7)

Parameters

D	demand rate in units per unit time (units/time)
λ	inspection rate in units per unit time, $\lambda > D$
A	proportion of imperfect items (a random variable with known probability density function)
q_1	proportion of Type-I imperfection errors (a random variable with known probability density function)
q_2	proportion of Type-II imperfection errors (a random variable with known probability density function)
$E(.)$	expected value operator
$E(\theta)$	expected value of θ
A	setup cost for each cycle ($/setup)
c	purchase cost per item ($/item)
i	inspection cost per item ($/item)
s	selling price ($/item)
v	salvage cost ($<s$) ($/item)
c_r	cost of committing a Type-I error ($/item)
c_a	cost of committing a Type-II error ($/item)
c_B	backordering cost per unit per unit time ($/item)
H	holding cost per unit time per unit time ($/unit/unit time)
δ	fraction of the purchase cost made at the initial time of the credit period
K	percentage of old retailers (estimated from the past data)
R	percentage of good retailers (estimated from the past data)
M	credit period offered by the manufacturer to the whole seller to settle his accounts (time unit)
N	credit period offered by the whole seller to the retailers to settle his accounts (time unit)
I_e	interest earned per unit per unit time ($/unit/unit time)
I_p	interest paid per unit per unit time ($/unit/unit time)
$f(\alpha)$	probability density function of defective items
$f(q_1)$	probability density function of Type-I error
$f(q_2)$	probability density function of Type-II error

Independent Decision variables

y	whole seller's order lot size for each cycle (units)
B	whole seller's backorder size for each cycle (units)

Dependent Decision variables

T	cycle length
$T.C.U.$	whole seller's total cost per unit time
$T.R.U.$	whole seller's total revenue per unit time
$T.P._j$	whole seller's total profit for j
$Z_j(y, B)$	whole seller's total profit per unit time for j
$E[Z_j(y,B)]$	whole seller's expected total profit per unit time for j

Appendix A

$$G_1 = s(1-\alpha)(1-q_1) - s(1-\delta)(1-R)(1-K)(1-\alpha)(1-q_1) + v[\alpha(1-q_2) + (1-\alpha)q_1 + \alpha q_2]$$
$$-c - i - c_r(1-\alpha)q_1 - c_a\alpha q_2 - \tfrac{1}{2D}cI_pKP^2\lambda M - \tfrac{1}{2}cI_pKPM - cI_pKP(N-M)$$
$$+\tfrac{1}{2D}\delta cI_p(1-K)P^2\lambda M - (1-\delta)cI_p(1-K)P(N-M) + \tfrac{1}{2D}(1-\delta)cI_p(1-K)P^2\lambda M$$
$$+\tfrac{1}{2}(1-\delta)cI_p(1-K)PM - cI_p[\alpha + (1-\alpha)q_1](N-M) \tag{A1}$$

$$G_2 = -\left\{ h\tfrac{1}{2\lambda} + h\tfrac{1}{2\lambda}\left(P - \tfrac{D}{\lambda}\right) + h\tfrac{1}{2\lambda}[\alpha(1-q_2) + (1-\alpha)q_1] + h\tfrac{1}{2D}\left(P-\tfrac{D}{\lambda}\right)^2 + h\tfrac{\alpha q_2 P}{2D} \right\}$$
$$+\tfrac{1}{D}cI_pKP^2 + \tfrac{1}{2D}\delta cI_p(1-K)P^2 + \tfrac{1}{D}(1-\delta)cI_p(1-K)P^2 + cI_p\tfrac{1}{\lambda}\{\alpha + (1-\alpha)q_1\} \tag{A2}$$

$$G_3 = cI_pK(N-M) - \tfrac{1}{2D}cI_pKP\lambda M - \tfrac{1}{2D}\delta cI_p(1-K)P\lambda M$$
$$-\tfrac{1}{2D}(1-\delta)cI_p(1-K)P\lambda M + (1-\delta)cI_p(1-K)(N-M) \tag{A3}$$

$$G_4 = -\left\{ -h\tfrac{1}{2(P\lambda-D)}\left(1 + \tfrac{D}{P\lambda-D}\right) + h\tfrac{1}{P\lambda-D} + h\tfrac{D}{2(P\lambda-D)^2} + h\tfrac{1}{2D} \right\} - \tfrac{1}{2}c_B\left(\tfrac{1}{P\lambda-D} + \tfrac{1}{D}\right)$$
$$+\tfrac{1}{2D}cI_pK + \tfrac{1}{2(P\lambda-D)}cI_pK + \tfrac{1}{2D}(1-\delta)cI_p(1-K) + \tfrac{1}{2(P\lambda-D)}(1-\delta)cI_p(1-K) \tag{A4}$$

$$G_5 = h\tfrac{1}{P\lambda-D} - h\tfrac{1}{2\lambda}\left(1 + \tfrac{D}{P\lambda-D}\right) - h\tfrac{1}{2\lambda} - h\tfrac{1}{2(P\lambda-D)} - h\tfrac{1}{2(P\lambda-D)}\left(P - \tfrac{D}{\lambda}\right)$$
$$-h\tfrac{\{\alpha(1-q_2)+(1-\alpha)q_1\}}{2(P\lambda-D)} - h\tfrac{1}{D}\left(P-\tfrac{D}{\lambda}\right)cI_pK\tfrac{P}{D} + cI_pK\tfrac{P}{D} + \tfrac{1}{2D}cI_pKP$$
$$+\tfrac{1}{2D}\delta cI_p(1-K)P + \tfrac{1}{2D}(1-\delta)cI_p(1-K)P + (1-\delta)cI_p(1-K)\tfrac{P}{D} \tag{A5}$$

$$G_6 = \tfrac{1}{2}\delta pI_e(1-K)DM^2 - \tfrac{1}{2}cI_pKP\lambda M^2 + \tfrac{1}{2}\delta pI_e(1-K)(P\lambda - D)M^2 - cI_pKB(N-M)$$
$$+\tfrac{1}{2}\delta cI_p(1-K)(Py - P\lambda M)M + \tfrac{1}{2}(1-\delta)cI_p(1-K)P\lambda M^2 - (1-\delta)cI_p(1-K)B(N-M) - A \tag{A6}$$

$$G_7 = s(1-\alpha)(1-q_1) - s(1-\delta)(1-R)(1-K)(1-\alpha)(1-q_1) + v[\alpha(1-q_2) + (1-\alpha)q_1 + \alpha q_2]$$
$$-c - i - c_r(1-\alpha)q_1 - c_a\alpha q_2 + \delta sI_e(1-K)PM - cI_pKP(N-M)$$
$$-(1-\delta)cI_p(1-K)P(N-M) - cI_p\{\alpha + (1-\alpha)q_1\}(N-M) \tag{A7}$$

$$G_8 = -\left\{ h\tfrac{1}{2\lambda} + h\tfrac{1}{2\lambda}\left(P-\tfrac{D}{\lambda}\right) + h\tfrac{1}{2\lambda}[\alpha(1-q_2)+(1-\alpha)q_1] + h\tfrac{1}{2D}\left(P-\tfrac{D}{\lambda}\right)^2 + h\tfrac{\alpha q_2 P}{2D} \right\}$$
$$-\left\{ \tfrac{P^2}{2D}\delta sI_e(1-K) - \tfrac{P^2}{D}\delta sI_e(1-K) - \tfrac{P^2}{2D}cI_pK - \tfrac{1}{2D}(1-\delta)cI_p(1-K)P^2 - cI_p\tfrac{[\alpha+(1-\alpha)q_1]}{\lambda} \right\} \tag{A8}$$

$$G_9 = -\delta sI_e(1-K)M + \delta sI_e(1-K)M + cI_pK(N-M) - cI_pK(N-M)$$
$$+(1-\delta)cI_p(1-K)(N-M) - (1-\delta)cI_p(1-K)(N-M) \tag{A9}$$

$$G_{10} = -\left\{ -h\tfrac{1}{2(P\lambda-D)}\left(1 + \tfrac{D}{P\lambda-D}\right) + h\tfrac{1}{P\lambda-D} + h\tfrac{D}{2(P\lambda-D)^2} + h\tfrac{1}{2D} \right\} + \tfrac{1}{2}c_BB^2\left(\tfrac{1}{P\lambda-D} + \tfrac{1}{D}\right)$$
$$-\left[\begin{array}{l} \tfrac{1}{2D}\delta sI_e(1-K) - \tfrac{1}{D}\delta sI_e(1-K) + \tfrac{1}{2(P\lambda-D)}\delta sI_e(1-K) - \tfrac{1}{(P\lambda-D)}\delta sI_e(1-K) \\[4pt] -\tfrac{1}{2D}cI_pK - \tfrac{1}{2(P\lambda-D)}cI_pK - \tfrac{1}{(P\lambda-D)}cI_pK - \tfrac{1}{2D}(1-\delta)cI_p(1-K) - \tfrac{1}{2(P\lambda-D)}(1-\delta)cI_p(1-K) \end{array} \right] \tag{A10}$$

$$G_{11} = h\tfrac{1}{P\lambda-D} - h\tfrac{1}{2\lambda}\left(1 + \tfrac{D}{P\lambda-D}\right) - h\tfrac{1}{2\lambda} - h\tfrac{1}{2(P\lambda-D)} - h\tfrac{1}{2(P\lambda-D)}\left(P-\tfrac{D}{\lambda}\right)$$
$$-h\tfrac{\{\alpha(1-q_2)+(1-\alpha)q_1\}}{2(P\lambda-D)} - h\tfrac{1}{D}\left(P-\tfrac{D}{\lambda}\right) - \delta pI_e(1-K)\tfrac{P}{D} + \delta pI_e(1-K)\tfrac{2P}{D}$$
$$+cI_pK\tfrac{P}{D} + (1-\delta)cI_p(1-K)\tfrac{P}{D} \tag{A11}$$

$$G_{12} = -A. \tag{A12}$$

$$G_{13} = s(1-\alpha)(1-q_1) - s(1-\delta)(1-R)(1-K)(1-\alpha)(1-q_1) + v[\alpha(1-q_2) + (1-\alpha)q_1 + \alpha q_2]$$
$$-c - i - c_r(1-\alpha)q_1 - c_a\alpha q_2 + \tfrac{1}{2}cI_pKP(M-N) + \tfrac{1}{2D}cI_pP^2\lambda(M-N) + \tfrac{1}{2D}\delta cI_p(1-K)P^2\lambda M \tag{A13}$$

$$G_{14} = -\left\{ h\tfrac{1}{2\lambda} + h\tfrac{1}{2\lambda}\left(P-\tfrac{D}{\lambda}\right) + h\tfrac{1}{2\lambda}[\alpha(1-q_2)+(1-\alpha)q_1] + h\tfrac{1}{2D}\left(P-\tfrac{D}{\lambda}\right)^2 + h\tfrac{\alpha q_2 P}{2D} \right\}$$
$$-\left\{ -\tfrac{1}{2D}cI_pP^2 - \tfrac{1}{2D}\delta cI_p(1-K)P^2 - \tfrac{1}{2D}(1-\delta)cI_p(1-K)P^2 \right\} \tag{A14}$$

$$G_{15} = -\frac{1}{2D}cI_pP\lambda(M-N) - \frac{1}{2D}\delta cI_p(1-K)P\lambda M - \frac{1}{2D}(1-\delta)cI_p(1-K)P\lambda(M-N). \tag{A15}$$

$$G_{16} = -\left\{-h\frac{1}{2(P\lambda-D)}\left(1+\frac{D}{P\lambda-D}\right)+h\frac{1}{P\lambda-D}+h\frac{D}{2(P\lambda-D)^2}+h\frac{1}{2D}\right\}-\frac{1}{2}c_B\left(\frac{1}{P\lambda-D}+\frac{1}{D}\right). \tag{A16}$$

$$\begin{aligned}
G_{17} = {}& h\frac{1}{P\lambda-D} - h\frac{1}{2\lambda}\left(1+\frac{D}{P\lambda-D}\right) - h\frac{1}{2\lambda} - h\frac{1}{2(P\lambda-D)} - h\frac{1}{2(P\lambda-D)}\left(P-\frac{D}{\lambda}\right)\\
& -h\frac{\{\alpha(1-q_2)+(1-\alpha)q_1\}}{2(P\lambda-D)} - h\frac{1}{D}\left(P-\frac{D}{\lambda}\right) + \frac{1}{2D}cI_pP + \frac{1}{2D}\delta cI_p(1-K)P + \frac{1}{2D}(1-\delta)cI_p(1-K)P
\end{aligned} \tag{A17}$$

$$\begin{aligned}
G_{18} = {}& \tfrac{1}{2}sI_eKD(M-N)^2 + \tfrac{1}{2}sI_eK(P\lambda-D)(M-N)^2 + \tfrac{1}{2}\delta sI_e(1-K)DM^2\\
& +\tfrac{1}{2}\delta sI_e(1-K)(P\lambda-D)M^2 + \tfrac{1}{2}(1-\delta)sI_eR(1-K)D(M-N)^2\\
& +\tfrac{1}{2}(1-\delta)sI_eR(1-K)(P\lambda-D)(M-N)^2 - \tfrac{1}{2}(1-\delta)cI_p(1-R)(1-K)D(M-N)^2\\
& -\tfrac{1}{2}(1-\delta)cI_p(1-R)(1-K)(P\lambda-D)(M-N)^2 - \tfrac{1}{2}cI_pKP\lambda(M-N)^2\\
& -\tfrac{1}{2}\delta cI_p(1-K)P\lambda M^2 - \tfrac{1}{2}(1-\delta)cI_p(1-K)P\lambda(M-N)^2 - A
\end{aligned} \tag{A18}$$

$$\begin{aligned}
G_{19} = {}& s(1-\alpha)(1-q_1) - s(1-\delta)(1-R)(1-K)(1-\alpha)(1-q_1) + v[\alpha(1-q_2)+(1-\alpha)q_1+\alpha q_2]\\
& -c - i - c_r(1-\alpha)q_1 - c_a\alpha q_2 + \tfrac{1}{2}cI_pK(M-N)P + \tfrac{1}{2}cI_pKP(M-N) + \tfrac{1}{2}\delta cI_p(1-K)MP\\
& +\tfrac{1}{2}\delta cI_p(1-K)MP + \tfrac{1}{2}(1-\delta)cI_p(1-K)(M-N)P + \tfrac{1}{2}(1-\delta)cI_p(1-K)(M-N)P\\
& -cI_p[\alpha+(1-\alpha)q_1](N-M)
\end{aligned} \tag{A19}$$

$$\begin{aligned}
G_{20} = {}& -\left\{h\frac{1}{2\lambda} + h\frac{1}{2\lambda}\left(P-\frac{D}{\lambda}\right) + h\frac{1}{2\lambda}[\alpha(1-q_2)+(1-\alpha)q_1] + h\frac{1}{2D}\left(P-\frac{D}{\lambda}\right)^2 + h\frac{\alpha q_2 P}{2D}\right\}\\
& -\left\{-\frac{1}{2D}cI_pKP^2 - \frac{1}{2D}\delta cI_p(1-K)P^2 - \frac{1}{2D}(1-\delta)cI_p(1-K)P^2 - cI_p[\alpha+(1-\alpha)q_1]\frac{1}{\lambda}\right\}
\end{aligned} \tag{A20}$$

$$\begin{aligned}
G_{21} = {}& sI_eK(M-N) + \delta sI_e(1-K)M + (1-\delta)sI_eR(1-K)(M-N) - \tfrac{1}{2}cI_pK(M-N) - \tfrac{1}{2}cI_pK(M-N)\\
& +\tfrac{1}{2}\delta cI_p(1-K)M - \tfrac{1}{2}\delta cI_p(1-K)M - (1-\delta)cI_p(1-R)(1-K)(M-N)\\
& -\tfrac{1}{2}(1-\delta)cI_p(1-K)(M-N) - \tfrac{1}{2}(1-\delta)cI_p(1-K)(M-N)
\end{aligned} \tag{A21}$$

$$\begin{aligned}
G_{22} = {}& -\left\{-h\frac{1}{2(P\lambda-D)}\left(1+\frac{D}{P\lambda-D}\right)+h\frac{1}{P\lambda-D}+h\frac{D}{2(P\lambda-D)^2}+h\frac{1}{2D}\right\}-\frac{1}{2}c_B\left(\frac{1}{P\lambda-D}+\frac{1}{D}\right)\\
& -\left\{\begin{aligned}
&\frac{1}{2(P\lambda-D)}sI_eK - pI_eK\frac{1}{(P\lambda-D)} + \frac{1}{2(P\lambda-D)}\delta sI_e(1-K) - \delta sI_e(1-K)\frac{1}{(P\lambda-D)}\\
&+\frac{1}{2(P\lambda-D)}(1-\delta)sI_eR(1-K) + (1-\delta)sI_eR(1-K)\frac{1}{(P\lambda-D)} - \frac{1}{2D}cI_pK\\
&-\frac{1}{2D}\delta cI_p(1-K) - \frac{1}{2(P\lambda-D)}(1-\delta)cI_p(1-R)(1-K) - (1-\delta)cI_p(1-R)(1-K)\frac{1}{(P\lambda-D)}\\
&-\frac{1}{2D}(1-\delta)cI_p(1-K)
\end{aligned}\right\}
\end{aligned} \tag{A22}$$

$$\begin{aligned}
G_{23} = {}& h\frac{1}{P\lambda-D} - h\frac{1}{2\lambda}\left(1+\frac{D}{P\lambda-D}\right) - h\frac{1}{2\lambda} - h\frac{1}{2(P\lambda-D)} - h\frac{1}{2(P\lambda-D)}\left(P-\frac{D}{\lambda}\right)\\
& -h\frac{[\alpha(1-q_2)+(1-\alpha)q_1]}{2(P\lambda-D)} - h\frac{1}{D}\left(P-\frac{D}{\lambda}\right) + \frac{1}{2D}cI_pKP + \frac{1}{2D}cI_pKP - \frac{1}{2D}\delta cI_p(1-K)P\\
& +\frac{1}{2D}\delta cI_p(1-K)P + \frac{1}{2D}\delta cI_p(1-K)P - \frac{1}{2D}(1-\delta)cI_p(1-K)P + \frac{1}{2D}(1-\delta)cI_p(1-K)P\\
& -\left\{-\frac{1}{2D}cI_pKP^2 - \frac{1}{2D}\delta cI_p(1-K)P^2 - \frac{1}{2D}(1-\delta)cI_p(1-K)P^2 - cI_p[\alpha+(1-\alpha)q_1]\frac{1}{\lambda}\right\}
\end{aligned} \tag{A23}$$

$$\begin{aligned}
G_{24} = {}& \tfrac{1}{2}sI_eKD(M-N)^2 + \tfrac{1}{2}\delta sI_e(1-K)DM^2 + \tfrac{1}{2}(1-\delta)sI_eR(1-K)D(M-N)^2\\
& -\tfrac{1}{2}(1-\delta)cI_p(1-R)(1-K)D(M-N)^2 - \tfrac{1}{2}(1-\delta)cI_p(1-K)D(M-N)^2\\
& -\tfrac{1}{2}\delta cI_p(1-K)DM^2 - \tfrac{1}{2}cI_pKD(M-N)^2 - A
\end{aligned} \tag{A24}$$

$$\begin{aligned}
G_{25} = {}& s(1-\alpha)(1-q_1) - s(1-\delta)(1-R)(1-K)(1-\alpha)(1-q_1) + v[\alpha(1-q_2)+(1-\alpha)q_1+\alpha q_2]\\
& -c - i - c_r(1-\alpha)q_1 - c_a\alpha q_2 + \tfrac{1}{2}cI_pK(M-N)P + \tfrac{1}{2}cI_pKP(M-N) + \tfrac{1}{2}\delta cI_p(1-K)MP\\
& +\tfrac{1}{2}\delta cI_p(1-K)MP + \tfrac{1}{2}(1-\delta)cI_p(1-K)(M-N)P + \tfrac{1}{2}(1-\delta)cI_p(1-K)(M-N)P\\
& +vI_e[\alpha+(1-\alpha)q_1](N-M)
\end{aligned} \tag{A25}$$

$$G_{26} = \ -\left\{ h\frac{1}{2\lambda} + h\frac{1}{2\lambda}\left(P - \frac{D}{\lambda}\right) + h\frac{1}{2\lambda}[\alpha(1-q_2) + (1-\alpha)q_1] + h\frac{1}{2D}\left(P - \frac{D}{\lambda}\right)^2 + h\frac{\alpha q_2 P}{2D} \right\}$$
$$-\left\{ -\frac{1}{2D}cI_pKP^2 - \frac{1}{2D}\delta cI_p(1-K)P^2 - \frac{1}{2D}(1-\delta)cI_p(1-K)P^2 - cI_p[\alpha + (1-\alpha)q_1]\frac{1}{\lambda} \right\} \tag{A26}$$

$$G_{27} = \ sI_eK(M-N) + \delta sI_e(1-K)M + (1-\delta)sI_eR(1-K)(M-N) - \tfrac{1}{2}cI_pK(M-N) - \tfrac{1}{2}cI_pK(M-N)$$
$$+\tfrac{1}{2}\delta cI_p(1-K)M - \tfrac{1}{2}\delta cI_p(1-K)M - (1-\delta)cI_p(1-R)(1-K)(M-N)$$
$$-\tfrac{1}{2}(1-\delta)cI_p(1-K)(M-N) - \tfrac{1}{2}(1-\delta)cI_p(1-K)(M-N) \tag{A27}$$

$$G_{28} = \ -\left\{ -h\frac{1}{2(P\lambda - D)}\left(1 + \frac{D}{P\lambda - D}\right) + h\frac{1}{P\lambda - D} + h\frac{D}{2(P\lambda - D)^2} + h\frac{1}{2D} \right\} - \tfrac{1}{2}c_B\left(\frac{1}{P\lambda - D} + \frac{1}{D}\right)$$
$$-\left\{ \begin{array}{l} \frac{1}{2(P\lambda - D)}sI_eK - pI_eK\frac{1}{(P\lambda - D)} + \frac{1}{2(P\lambda - D)}\delta sI_e(1-K) - \delta sI_e(1-K)\frac{1}{(P\lambda - D)} \\[4pt] + \frac{1}{2(P\lambda - D)}(1-\delta)sI_eR(1-K) + (1-\delta)sI_eR(1-K)\frac{1}{(P\lambda - D)} - \frac{1}{2D}cI_pK \\[4pt] - \frac{1}{2D}\delta cI_p(1-K) - \frac{1}{2(P\lambda - D)}(1-\delta)cI_p(1-R)(1-K) - (1-\delta)cI_p(1-R)(1-K)\frac{1}{(P\lambda - D)} \\[4pt] - \frac{1}{2D}(1-\delta)cI_p(1-K) \end{array} \right\} \tag{A28}$$

$$G_{29} = \ h\frac{1}{P\lambda - D} - h\frac{1}{2\lambda}\left(1 + \frac{D}{P\lambda - D}\right) - h\frac{1}{2\lambda} - h\frac{1}{2(P\lambda - D)} - h\frac{1}{2(P\lambda - D)}\left(P - \frac{D}{\lambda}\right)$$
$$-h\frac{[\alpha(1-q_2) + (1-\alpha)q_1]}{2(P\lambda - D)} - h\frac{1}{D}\left(P - \frac{D}{\lambda}\right) + \frac{1}{2D}cI_pKP + \frac{1}{2D}cI_pKP - \frac{1}{2D}\delta cI_p(1-K)P$$
$$+\frac{1}{2D}\delta cI_p(1-K)P + \frac{1}{2D}\delta cI_p(1-K)P - \frac{1}{2D}(1-\delta)cI_p(1-K)P + \frac{1}{2D}(1-\delta)cI_p(1-K)P$$
$$-\left\{ -\frac{1}{2D}cI_pKP^2 - \frac{1}{2D}\delta cI_p(1-K)P^2 - \frac{1}{2D}(1-\delta)cI_p(1-K)P^2 - cI_p[\alpha + (1-\alpha)q_1]\frac{1}{\lambda} \right\} \tag{A29}$$

$$G_{30} = \ \tfrac{1}{2}sI_eKD(M-N)^2 + \tfrac{1}{2}\delta sI_e(1-K)DM^2 + \tfrac{1}{2}(1-\delta)sI_eR(1-K)D(M-N)^2$$
$$-\tfrac{1}{2}(1-\delta)cI_p(1-R)(1-K)D(M-N)^2 - \tfrac{1}{2}(1-\delta)cI_p(1-K)D(M-N)^2$$
$$-\tfrac{1}{2}\delta cI_p(1-K)DM^2 - \tfrac{1}{2}cI_pKD(M-N)^2 - A \tag{A30}$$

$$G_{31} = \ s(1-\alpha)(1-q_1) - s(1-\delta)(1-R)(1-K)(1-\alpha)(1-q_1) + v[\alpha(1-q_2) + (1-\alpha)q_1 + \alpha q_2]$$
$$-c - i - c_r(1-\alpha)q_1 - c_a\alpha q_2\delta sI_e(1-K)PM + vI_e[\alpha + (1-\alpha)q_1](M-N) + \tfrac{1}{2}cI_pKP(M-N)$$
$$-\tfrac{1}{2}cI_pK(M-N)P + \tfrac{1}{2}(1-\delta)cI_p(1-K)P(M-N) - \tfrac{1}{2}(1-\delta)cI_p(1-K)P(M-N) \tag{A31}$$

$$G_{32} = \ -\left\{ h\frac{1}{2\lambda} + h\frac{1}{2\lambda}\left(P - \frac{D}{\lambda}\right) + h\frac{1}{2\lambda}[\alpha(1-q_2) + (1-\alpha)q_1] + h\frac{1}{2D}\left(P - \frac{D}{\lambda}\right)^2 + h\frac{\alpha q_2 P}{2D} \right\}$$
$$-\left\{ \frac{1}{2D}\delta sI_e(1-K)P^2 - \frac{1}{D}\delta sI_e(1-K)MP^2 - vI_e\frac{[\alpha + (1-\alpha)q_1]}{\lambda} - \frac{1}{2D}cI_pKP^2 - \frac{1}{2D}(1-\delta)cI_p(1-K)P^2 \right\} \tag{A32}$$

$$G_{33} = \ sI_eK(M-N) - \delta sI_e(1-K)M + \delta sI_e(1-K)M + (1-\delta)sI_eR(1-K)(M-N)$$
$$-\tfrac{1}{2}cI_pK(M-N) + \tfrac{1}{2}cI_pK(M-N) - (1-\delta)cI_p(1-R)(1-K)(M-N)$$
$$-\tfrac{1}{2}(1-\delta)cI_p(1-K)(M-N) - \tfrac{1}{2}(1-\delta)cI_p(1-K)(M-N) \tag{A33}$$

$$G_{34} = \ -\left\{ -h\frac{1}{2(P\lambda - D)}\left(1 + \frac{D}{P\lambda - D}\right) + h\frac{1}{P\lambda - D} + h\frac{D}{2(P\lambda - D)^2} + h\frac{1}{2D} \right\} - \tfrac{1}{2}c_B\left(\frac{1}{P\lambda - D} + \frac{1}{D}\right)$$
$$-\left\{ \begin{array}{l} \tfrac{1}{2}sI_eK\frac{1}{(P\lambda - D)} - sI_eK\frac{1}{(P\lambda - D)} + \frac{1}{2D}\delta sI_e(1-K) - \frac{1}{D}\delta sI_e(1-K)M \\[4pt] + \frac{1}{2(P\lambda - D)}\delta sI_e(1-K) - \frac{1}{(P\lambda - D)}\delta sI_e(1-K) + \frac{1}{2(P\lambda - D)}(1-\delta)sI_eR(1-K) \\[4pt] - \frac{1}{(P\lambda - D)}(1-\delta)sI_eR(1-K) - \frac{1}{2D}cI_pK - \frac{1}{2(P\lambda - D)}(1-\delta)cI_p(1-R)(1-K) \\[4pt] - \frac{1}{(P\lambda - D)}(1-\delta)cI_p(1-R)(1-K) - \frac{1}{2D}(1-\delta)cI_p(1-K) \end{array} \right\} \tag{A34}$$

$$G_{35} = \ h\frac{1}{P\lambda - D} - h\frac{1}{2\lambda}\left(1 + \frac{D}{P\lambda - D}\right) - h\frac{1}{2\lambda} - h\frac{1}{2(P\lambda - D)} - h\frac{1}{2(P\lambda - D)}\left(P - \frac{D}{\lambda}\right)$$
$$-h\frac{[\alpha(1-q_2) + (1-\alpha)q_1]}{2(P\lambda - D)} - h\frac{1}{D}\left(P - \frac{D}{\lambda}\right) - \frac{1}{D}\delta sI_e(1-K) - \frac{2}{D}\delta sI_e(1-K)MP + \frac{1}{2D}cI_pKP$$
$$+\frac{1}{2D}cI_pKP + \frac{1}{2D}(1-\delta)cI_p(1-K)P + \frac{1}{2D}(1-\delta)cI_p(1-K)P \tag{A35}$$

$$G_{36} = \ \tfrac{1}{2}sI_eKD(M-N)^2 + \tfrac{1}{2}(1-\delta)sI_eR(1-K)D(M-N)^2 - \tfrac{1}{2}(1-\delta)cI_p(1-R)(1-K)D(M-N)^2$$
$$-\tfrac{1}{2}cI_pKD(M-N)^2 - \tfrac{1}{2}(1-\delta)cI_p(1-K)D(M-N)^2 - A \tag{A36}$$

$$G_{37} = \; s(1-\alpha)(1-q_1) - s(1-\delta)(1-R)(1-K)(1-\alpha)(1-q_1) + v[\alpha(1-q_2) + (1-\alpha)q_1 + \alpha q_2]$$
$$-c - i - c_r(1-\alpha)q_1 - c_a\alpha q_2 + sI_eKP(M-N) + \delta sI_e(1-K)PM + (1-\delta)sI_eR(1-K)P(M-N) \quad (A37)$$
$$+vI_e[\alpha + (1-\alpha)q_1](M-N) - (1-\delta)cI_p(1-R)(1-K)P(M-N)$$

$$G_{38} = \; -\left\{ h\tfrac{1}{2\lambda} + h\tfrac{1}{2\lambda}\left(P - \tfrac{D}{\lambda}\right) + h\tfrac{1}{2\lambda}[\alpha(1-q_2) + (1-\alpha)q_1] + h\tfrac{1}{2D}\left(P - \tfrac{D}{\lambda}\right)^2 + h\tfrac{\alpha q_2 P}{2D} \right\}$$
$$-\left\{ \begin{array}{l} -\tfrac{1}{2D}sI_eKP^2 - \tfrac{1}{2D}\delta sI_e(1-K)P^2 - \tfrac{1}{2D}(1-\delta)sI_eR(1-K)P^2 - vI_e\tfrac{[\alpha+(1-\alpha)q_1]}{\lambda} \\[4pt] +\tfrac{1}{2D}(1-\delta)cI_p(1-R)(1-K)P^2 \end{array} \right\} \quad (A38)$$

$$G_{39} = \; -sI_eK(M-N) + sI_eK(M-N) - \delta sI_e(1-K)M + \delta sI_e(1-K)M - (1-\delta)sI_eR(1-K)(M-N)$$
$$+(1-\delta)sI_eR(1-K)(M-N) + (1-\delta)cI_p(1-R)(1-K)(M-N) - (1-\delta)cI_p(1-R)(1-K)(M-N) \quad (A39)$$

$$G_{40} = \; -\left\{ -h\tfrac{1}{2(P\lambda-D)}\left(1 + \tfrac{D}{P\lambda-D}\right) + h\tfrac{1}{P\lambda-D} + h\tfrac{D}{(P\lambda-D)^2} + h\tfrac{1}{2D} \right\} - \tfrac{1}{2}c_B\left(\tfrac{1}{P\lambda-D} + \tfrac{1}{D}\right)$$
$$-\left\{ \begin{array}{l} -\tfrac{1}{2D}sI_eK - \tfrac{1}{2(P\lambda-D)}sI_eK - \tfrac{1}{2D}\delta sI_e(1-K) - \tfrac{1}{2(P\lambda-D)}\delta sI_e(1-K) \\[4pt] -\tfrac{1}{2D}(1-\delta)sI_eR(1-K) - \tfrac{1}{2(P\lambda-D)}(1-\delta)sI_eR(1-K) \\[4pt] +\tfrac{1}{2D}(1-\delta)cI_p(1-R)(1-K) + \tfrac{1}{2(P\lambda-D)}(1-\delta)cI_p(1-R)(1-K) \end{array} \right\} \quad (A40)$$

$$G_{41} = \; h\tfrac{1}{P\lambda-D} - h\tfrac{1}{2\lambda}\left(1 + \tfrac{D}{P\lambda-D}\right) - h\tfrac{1}{2\lambda} - h\tfrac{1}{2(P\lambda-D)} - h\tfrac{1}{2(P\lambda-D)}\left(P - \tfrac{D}{\lambda}\right)$$
$$-h\tfrac{[\alpha(1-q_2)+(1-\alpha)q_1]}{2(P\lambda-D)} - h\tfrac{1}{D}\left(P - \tfrac{D}{\lambda}\right)\tfrac{1}{D}sI_eKP + \tfrac{1}{D}\delta sI_e(1-K)P \quad (A41)$$
$$+\tfrac{1}{D}(1-\delta)sI_eR(1-K)P - \tfrac{1}{D}(1-\delta)cI_p(1-R)(1-K)P$$

$$G_{42} = -A. \quad (A42)$$

Appendix B

The expected value of all of the G_i's ($i = 1,2,3,\ldots,42$) are calculated in the following manner, e.g.:

$$E[G_1] = \; s(1-E[\alpha])(1-E[q_1]) - s(1-\delta)(1-R)(1-K)(1-E[\alpha])(1-E[q_1])$$
$$+v\{E[\alpha](1-E[q_2]) + (1-E[\alpha])E[q_1] + E[\alpha]E[q_2]\} - c - i - c_r(1-E[\alpha])E[q_1]$$
$$-c_aE[\alpha]E[q_2] - \tfrac{1}{2D}cI_pKP^2\lambda M - \tfrac{1}{2}cI_pKPM - cI_pKP(N-M) + \tfrac{1}{2D}\delta cI_p(1-K)P^2\lambda M \quad (A43)$$
$$-(1-\delta)cI_p(1-K)P(N-M) + \tfrac{1}{2D}(1-\delta)cI_p(1-K)P^2\lambda M + \tfrac{1}{2}(1-\delta)cI_p(1-K)PM$$
$$-cI_p\{E[\alpha] + (1-E[\alpha])E[q_1]\}(N-M)$$

Similarly, the expected values can be calculated for all of the remaining G_i's ($i = 2,3,4,\ldots,42$).

Appendix C

Case (i) $M \leq N \leq T' \leq T$

C1 : Proof of first sufficient condition of concavity.

To Prove : $\frac{\partial^2}{\partial y^2}E[Z_1(y,B)] \leq 0$

where $\frac{\partial^2}{\partial y^2}E[Z_1(y,B)] = \frac{2D}{\{E[\alpha]E[q_2]+(1-E[\alpha])(1-E[q_1])\}y^3} *$

$$\left\{ \{E[\alpha]E[q_2] + (1-E[\alpha])(1-E[q_1])\} * \left\{ \begin{array}{l} \left\{ \tfrac{1}{2}\delta pI_e(1-K)\lambda M^2 - cI_p\left\{ \begin{array}{l} \tfrac{\lambda MB}{2D} - \tfrac{1}{2}\delta(1-K)yM - \left((1-\delta)K + \tfrac{1}{2} - \delta\right)\lambda M^2 \\[4pt] +\tfrac{[1-\delta(1-K)]}{2} * \left(\tfrac{1}{\{E[\alpha]E[q_2]+(1-E[\alpha])(1-E[q_1])\}\lambda-D} + \tfrac{1}{D}\right) \end{array} \right\} \right\} \\[16pt] +\tfrac{1}{2}(h + c_B)B^2\left(\tfrac{1}{\{E[\alpha]E[q_2]+(1-E[\alpha])(1-E[q_1])\}\lambda-D} + \tfrac{1}{D}\right) - A \end{array} \right\} \right\} \leq 0$$

Proof : As $0 \leq K, R, \delta, E[\alpha], E[q_1], E[q_2] \leq 1$,

$\therefore$ the following inequalities hold true :

$(1 - K) \geq 0;\ (1 - R) \geq 0;\ (1 - \delta) \geq 0;\ (1 - E[\alpha]) \geq 0;\ (1 - E[q_1]) \geq 0; (1 - E[q_1]) \geq 0;$

$\{E[\alpha]E[q_2] + (1 - E[\alpha])(1 - E[q_1])\} \geq 0;$

The first condition of concavity holds true under the condition of :

$$\tfrac{1}{2}\delta p I_e (1 - K)\lambda M^2 + \tfrac{1}{2}(h + c_B)B^2 \left(\tfrac{1}{\{E[\alpha]E[q_2]+(1-E[\alpha])(1-E[q_1])\}\lambda - D} + \tfrac{1}{D} \right)$$

$$\leq c I_p \left\{ \begin{array}{l} \tfrac{\lambda MB}{2D} - \tfrac{1}{2}\delta(1 - K)yM - \left((1 - \delta)K + \tfrac{1}{2} - \delta\right)\lambda M^2 \\[2mm] + \tfrac{[1-\delta(1-K)]}{2} * \left(\tfrac{1}{\{E[\alpha]E[q_2]+(1-E[\alpha])(1-E[q_1])\}\lambda - D} + \tfrac{1}{D} \right) \end{array} \right\} + A$$

C2 : Proof of sec ond sufficient condition of concavity.

To Prove : $\frac{\partial^2}{\partial B^2} E[Z_1(y, B)] \leq 0$

where $\frac{\partial^2}{\partial B^2} E[Z_1(y, B)] = \frac{D}{\{E[\alpha]E[q_2]+(1-E[\alpha])(1-E[q_1])\}y} *$

$$\left\{ (h + c_B - c I_p [1 - \delta(1 - K)]) \left(\tfrac{1}{\{E[\alpha]E[q_2]+(1-E[\alpha])(1-E[q_1])\}\lambda - D} + \tfrac{1}{D} \right) \right\}$$

Proof : As $0 \leq K, \delta, E[\alpha], E[q_1], E[q_2] \leq 1$,

$\therefore$ the following inequalities hold true :

$(1 - K) \geq 0;\ (1 - \delta) \geq 0;\ (1 - E[\alpha]) \geq 0;\ (1 - E[q_1]) \geq 0; (1 - E[q_1]) \geq 0;$

$\{E[\alpha]E[q_2] + (1 - E[\alpha])(1 - E[q_1])\} \geq 0;$

The second condition of concavity holds true under the condition of :

$h + c_B \leq c I_p [1 - \delta(1 - K)]$

C3 : Proof of third sufficient condition of concavity.

To Prove : $\left(\frac{\partial^2}{\partial y \partial B} E[Z_1(y, B)] \right)^2 - \left(\frac{\partial^2}{\partial y^2} E[Z_1(y, B)] \right) \left(\frac{\partial^2}{\partial B^2} E[Z_1(y, B)] \right) \leq 0$

where $\left(\frac{\partial^2}{\partial y \partial B} E[Z_1(y, B)] \right)^2 - \left(\frac{\partial^2}{\partial y^2} E[Z_1(y, B)] \right) \left(\frac{\partial^2}{\partial B^2} E[Z_1(y, B)] \right) = \frac{1}{\{E[\alpha]E[q_2]+(1-E[\alpha])(1-E[q_1])\}^2} \frac{D^2}{y^4} *$

$$\left\{ \begin{array}{l} \left\{ c I_p [1 - \delta(1 - K)](N - M) - \tfrac{1}{2D} c I_p \{E[\alpha]E[q_2] + (1 - E[\alpha])(1 - E[q_1])\}\lambda M \right\}^2 - \\[3mm] \left\{ \begin{array}{l} (h + c_B)^2 \left(\tfrac{1}{\{E[\alpha]E[q_2]+(1-E[\alpha])(1-E[q_1])\}\lambda - D} + \tfrac{1}{D} \right)^2 + \left(\tfrac{c I_p}{D} \right)^2 [1 - \delta(1 - K)]^2 \\[3mm] + \tfrac{1}{\{\{E[\alpha]E[q_2]+(1-E[\alpha])(1-E[q_1])\}\lambda - D\}^2} (c I_p)^2 [1 - \delta(1 - K)]^2 \end{array} \right\} \end{array} \right\}$$

Proof : As $0 \leq K, \delta, E[\alpha], E[q_1], E[q_2] \leq 1$,

$\therefore$ the following inequalities hold true :

$(1 - K) \geq 0;\ (1 - \delta) \geq 0;\ (1 - E[\alpha]) \geq 0;\ (1 - E[q_1]) \geq 0; (1 - E[q_1]) \geq 0;$

$\{E[\alpha]E[q_2] + (1 - E[\alpha])(1 - E[q_1])\} \geq 0;$

Thus, the third condition of concavity holds true under the condition of :

$$\left\{ c I_p [1 - \delta(1 - K)](N - M) - \tfrac{1}{2D} c I_p \{E[\alpha]E[q_2] + (1 - E[\alpha])(1 - E[q_1])\}\lambda M \right\}^2 \leq$$

$$\left\{ \begin{array}{l} (h + c_B)^2 \left(\tfrac{1}{\{E[\alpha]E[q_2]+(1-E[\alpha])(1-E[q_1])\}\lambda - D} + \tfrac{1}{D} \right)^2 + \left(\tfrac{c I_p}{D} \right)^2 [1 - \delta(1 - K)]^2 \\[3mm] + \tfrac{1}{\{\{E[\alpha]E[q_2]+(1-E[\alpha])(1-E[q_1])\}\lambda - D\}^2} (c I_p)^2 [1 - \delta(1 - K)]^2 \end{array} \right\}$$

References

1. Haley, C.W.; Higgins, R.C. Inventory policy and trade credit financing. *Manag. Sci.* **1973**, *20*, 464–471. [CrossRef]
2. Goyal, S.K. Economic order quantity under conditions of permissible delay in payments. *J. Oper. Res. Soc.* **1985**, *36*, 335–338. [CrossRef]

3. Davis, R.A.; Gaither, N. Optimal ordering policies under conditions of extended payment privileges. *Manag. Sci.* **1985**, *31*, 499–509. [CrossRef]

4. Aggarwal, S.P.; Jaggi, C.K. Ordering policies of deteriorating items under permissible delay in payments. *J. Oper. Res. Soc.* **1995**, *46*, 658–662. [CrossRef]

5. Teng, J.T.; Chang, C.T.; Chern, M.S.; Chan, Y.L. Retailer's optimal ordering policies with trade credit financing. *Int. J. Syst. Sci.* **2007**, *38*, 269–278. [CrossRef]

6. Su, C.H.; Ouyang, L.Y.; Ho, C.H.; Chang, C.T. Retailer's inventory policy and supplier's delivery policy under two-level trade credit strategy. *Asia-Pac. J. Oper. Res.* **2007**, *24*, 613–630. [CrossRef]

7. Jaggi, C.K.; Goyal, S.K.; Goel, S.K. Retailer's optimal replenishment decisions with credit-linked demand under permissible delay in payments. *Eur. J. Oper. Res.* **2008**, *190*, 130–135. [CrossRef]

8. Huang, Y.F.; Hsu, K.H. An EOQ model under retailer partial trade credit policy in supply chain. *Int. J. Prod. Econ.* **2008**, *112*, 655–664. [CrossRef]

9. Huang, Y.F. Retailer's inventory policy under supplier's partial trade credit policy. *J. Oper. Res. Soc. Jpn.* **2005**, *48*, 173–182. [CrossRef]

10. Teng, J.T. Optimal ordering policies for a retailer who offers distinct trade credits to its good and bad credit customers. *Int. J. Prod. Econ.* **2009**, *119*, 415–423. [CrossRef]

11. Jaggi, C.K.; Verma, M. Ordering policies under supplier-retailer partial trade credit financing. *Opsearch* **2010**, *47*, 293–310. [CrossRef]

12. Jaggi, C.K.; Verma, M.; Kausar, A. Customer based two stage credit policies in a supply chain. In Proceedings of the 2011 International Conference on Industrial Engineering and Operations Management, Kuala Lumpur, Malaysia, 22–24 January 2011.

13. Jaggi, C.K.; Aggarwal, K.K.; Verma, M. Optimal retailer's ordering policies under two-stage partial trade credit financing in a supply chain. *Int. J. Ind. Syst. Eng.* **2012**, *10*, 277–299. [CrossRef]

14. Taleizadeh, A.A.; Pentico, D.W.; Jabalameli, M.S.; Aryanezhad, M. An EOQ model with partial delayed payment and partial back ordering. *Omega* **2013**, *41*, 354–368. [CrossRef]

15. Giri, B.C.; Sharma, S. Optimal ordering policy for an inventory system with linearly increasing demand and allowable shortages under two levels trade credit financing. *Oper. Res.* **2016**, *16*, 25–50. [CrossRef]

16. Shi, X.; Zhang, S. An incentive-compatible solution for trade credit term incorporating default risk. *Eur. J. Oper. Res.* **2010**, *206*, 178–196. [CrossRef]

17. Tiwari, S.; Ahmed, W.; Sarkar, B. Multi-item sustainable green production system under trade-credit and partial backordering. *J. Clean. Prod.* **2018**, *204*, 82–95. [CrossRef]

18. Wu, J.; Chan, Y.L. Lot-sizing policies for deteriorating items with expiration dates and partial trade credit to credit-risk customers. *Int. J. Prod. Econ.* **2014**, *155*, 292–301. [CrossRef]

19. Chen, S.C.; Teng, J.T. Inventory and credit decisions for time-varying deteriorating items with up-stream and down-stream trade credit financing by discounted cash flow analysis. *Eur. J. Oper. Res.* **2015**, *243*, 566–575. [CrossRef]

20. Shah, N.H. Retailer's replenishment and credit policies for deteriorating inventory under credit period-dependent demand and bad-debt loss. *TOP* **2015**, *23*, 298–312. [CrossRef]

21. Sarkar, B.; Saren, S. Partial trade-credit policy of retailer with exponentially deteriorating items. *Int. J. App. Comput. Math.* **2015**, *1*, 343–368. [CrossRef]

22. Wu, J.; Al-Khateeb, F.B.; Teng, J.T.; Cárdenas-Barrón, L.E. Inventory models for deteriorating items with maximum lifetime under downstream partial trade credits to credit-risk customers by discounted cash-flow analysis. *Int. J. Prod. Econ.* **2016**, *171*, 105–115. [CrossRef]

23. Mahata, G.C.; De, S.K. Supply chain inventory model for deteriorating items with maximum lifetime and partial trade credit to credit-risk customers. *Int. J. Manag. Sci. Eng. Manag.* **2017**, *12*, 21–32. [CrossRef]

24. Wu, C.; Zhao, Q.; Xi, M. A retailer-supplier supply chain model with trade credit default risk in a supplier-Stackelberg game. *Comput. Ind. Eng.* **2017**, *112*, 568–575. [CrossRef]

25. Porteus, E.L. Optimal lot sizing, process quality improvement and setup cost reduction. *Oper. Res.* **1986**, *34*, 137–144. [CrossRef]

26. Cárdenas-Barrón, L.E.; Sarkar, B.; Treviño-Garza, G. Easy and improved algorithms to joint determination of the replenishment lot size and number of shipments for an EPQ model with rework. *Math. Comput. Appl.* **2013**, *18*, 3138–3151. [CrossRef]

27. Lee, H.L.; Rosenblatt, M.J. Simultaneous determination of production cycle and inspection schedules in a production system. *Manag. Sci.* **1987**, *33*, 1125–1136. [CrossRef]
28. Zhang, X.; Gerchak, Y. Joint lot sizing and inspection policy in an EOQ model with random yield. *IIE Trans.* **1990**, *22*, 41–47. [CrossRef]
29. Salameh, M.K.; Jaber, M.Y. Economic production quantity model for items with imperfect quality. *Int. J. Prod. Econ.* **2000**, *64*, 59–64. [CrossRef]
30. Cárdenas-Barrón, L.E. Observation on: Economic production quantity model for items with imperfect quality. *Int. J. Prod. Econ.* **2000**, *67*, 201. [CrossRef]
31. Goyal, S.K.; Cárdenas-Barrón, L.E. Note on: Economic production quantity model for items with imperfect quality—A practical approach. *Int. J. Prod. Econ.* **2002**, *77*, 85–87. [CrossRef]
32. Papachristos, S.; Konstantaras, I. Economic ordering quantity models for items with imperfect quality. *Int. J. Prod. Econ.* **2006**, *100*, 148–154. [CrossRef]
33. Maddah, B.; Jaber, M.Y. Economic order quantity for items with imperfect quality: Revisited. *Int. J. Prod. Econ.* **2008**, *112*, 808–815. [CrossRef]
34. Raouf, A.; Jain, J.K.; Sathe, P.T. A cost-minimization model for multi characteristic component inspection. *AIIE Trans.* **1983**, *15*, 187–194.
35. Sarkar, B.; Saren, S. Product inspection policy for an imperfect production system with inspection errors and warranty cost. *Eur. J. Oper. Res.* **2016**, *248*, 263–271. [CrossRef]
36. Duffuaa, S.O.; Khan, M. Impact of inspection errors on the performance measures of a general repeat inspection plan. *Int. J. Prod. Res.* **2005**, *43*, 4945–4967. [CrossRef]
37. Khan, M.; Jaber, M.Y.; Bonney, M. An economic order quantity (EOQ) for items with imperfect quality and inspection errors. *Int. J. Prod. Econ.* **2011**, *133*, 113–118. [CrossRef]
38. Sett, B.; Sarkar, S.; Sarkar, B. Optimal buffer inventory and inspection errors in an imperfect production system with regular preventive maintenance. *Int. J. Adv. Manuf. Technol.* **2017**, *90*, 545–560. [CrossRef]
39. Hsu, J.T.; Hsu, L.F. An EOQ model with imperfect quality items, inspection errors, shortage back ordering, and sales returns. *Int. J. Prod. Econ.* **2013**, *143*, 162–170. [CrossRef]
40. Zhou, Y.; Chen, C.; Li, C.; Zhong, Y. A synergic economic order quantity model with trade credit, shortages, imperfect quality and inspection errors. *Appl. Math. Model.* **2016**, *40*, 1012–1028. [CrossRef]
41. Khanra, S.; Mandal, B.; Sarkar, B. An inventory model with time dependent demand and shortages under trade credit policy. *Econ. Mod.* **2013**, *35*, 349–355. [CrossRef]
42. Khanna, A.; Kishore, A.; Jaggi, C. Impact of inflation and trade credit policy in an inventory model for imperfect quality items with allowable shortages. *Control Cybern.* **2016**, *45*, 37–82.
43. Palanivel, M.; Uthayakumar, R. An inventory model with imperfect items, stock dependent demand and permissible delay in payments under inflation. *RAIRO-Oper. Res.* **2016**, *50*, 473–489. [CrossRef]
44. Taleizadeh, A.A.; Lashgari, M.; Akram, R.; Heydari, J. Imperfect economic production quantity model with upstream trade credit periods linked to raw material order quantity and downstream trade credit periods. *Appl. Math. Model.* **2016**, *40*, 8777–8793. [CrossRef]
45. Khanna, A.; Kishore, A.; Jaggi, C. Strategic production modeling for defective items with imperfect inspection process, rework, and sales return under two-level trade credit. *Int. J. Ind. Eng. Comput.* **2017**, *8*, 85–118. [CrossRef]

© 2018 by the authors. Licensee MDPI, Basel, Switzerland. This article is an open access article distributed under the terms and conditions of the Creative Commons Attribution (CC BY) license (http://creativecommons.org/licenses/by/4.0/).

Article

Product Channeling in an O2O Supply Chain Management as Power Transmission in Electric Power Distribution Systems

Biswajit Sarkar [1], Muhammad Tayyab [1] and Seok-Beom Choi [2,*]

[1] Department of Industrial & Management Engineering, Hanyang University, Ansan, Gyeonggi-do 155 88, Korea; bsbiswajitsarkar@gmail.com (B.S.); mtayyabntu@yahoo.com (M.T.)

[2] Department of International Business, Cheju Halla University, Jeju-do 63092, Korea

* Correspondence: sbchoi777@naver.com; Tel.: +82-10-3854-2765

Received: 14 November 2018; Accepted: 18 December 2018; Published: 20 December 2018

Abstract: With the aim of delivering goods and services to customers, optimal delivery channel selection is a significant part of supply chain management. Several heuristics have been developed to solve the variants of distribution center allocation and vehicle routing problems. In reality, small-scale suppliers cannot afford research and development departments to optimize their distribution networks. In this context, this research work develops a model for an online to offline (O2O) supply chain management network of a small-scale household electric components manufacturer for delivering goods to its distribution centers and retailers. Retailers are acquired by the company through investment in the O2O channel of e-commerce. Electric power transmission and distribution is considered as representative of the product distribution network. A model is developed using a combination of the supply chain management technique and power transmission terminologies. The constrained linear programming model is solved through the linear programming tool of the LINGO optimization software and the global optimum results for the proposed quantity allocation problem are achieved. A numerical experiment is provided to illustrate the practical applicability of the model and the optimal results are analyzed for model robustness.

Keywords: supply chain management; e-commerce; O2O channel; customer acquisition cost; transshipments; electric power distribution

---

1. Introduction

Logistics management is a major contributor to reduce operational costs as well as enhance the competitiveness and service level of the industries where the facilitation and acceleration of efficient logistics management depends on the design of the distribution network (DN) [1]. Location allocation, routing, and inventory are the three major decisions in a distribution network design problem (DNDP). Facility location and customer allocation to facilities are constituents of the location allocation. Meanwhile, vehicle routing problems (VRPs) and inventory control problems are considered sub-areas of routing and inventory, respectively.

VRPs are extensively studied by researchers to support the planning of city logistics patterns [2]. Most of the industries are located outside metro cities, from where line-haul vehicles need to transport huge amounts of commodities to the cities. Large-sized vehicles are used to serve this purpose, which are generally forbidden to deliver the goods inside cities due to the risk of accidents. For this purpose, the transshipment of cargo between line-haul vehicles and inside city delivery vehicles is mandatory [3], creating the necessity of supply chain modeling for the distribution of products in cities. At the first level, cargo is transported to specific points in the outskirts of cities, and from there it is transported to the distribution centers at various places within those cities.

A similar concept is used for electric power transmission within cities. Electric power transmission and distribution systems are applied for the supply of electric power to end users. The transmission of the electricity from power generation sites to consumers is conducted through transmission lines, which traverse long distances [4]. Figure 1 shows the general transmission and distribution flow of electric power from power generation stations to consumer locations. High-voltage electric power cables transmit electric power to grid stations which are located at the outskirts of cities, and from there the stepped-down electric power is distributed to different zones of those cities.

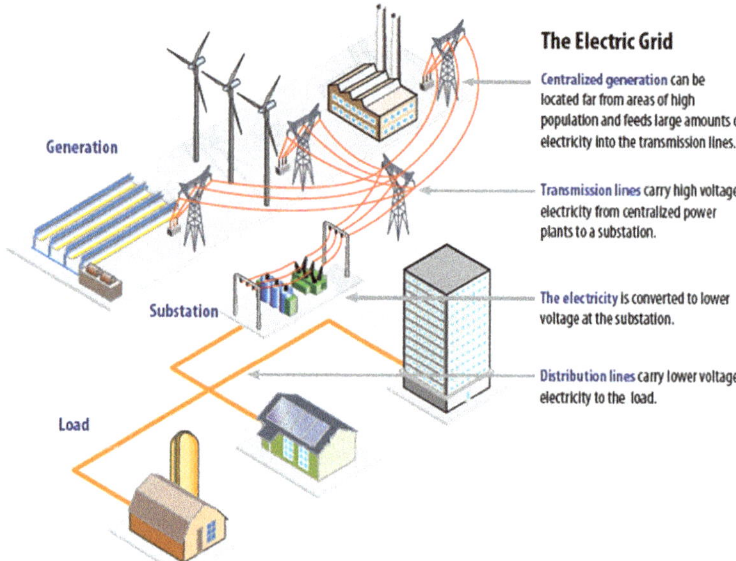

Figure 1. General flow of transmission and distribution network for electric power supply.

This phenomenon can be thought of as a two-tier distribution network, as heavy transportation vehicles deliver goods to the depots located outside cities, and light vehicles subsequently carry the products to the distribution centers or retailers located in different areas of those cities. Several heuristics to solve the variants of vehicle routing and distribution center location problems are available in the supply chain literature, most of which bear limited generic application. In addition, every manufacturing or distribution organization cannot necessarily afford research and development centers to design optimal routes and supply chain cost minimization procedures. For instance, some manufacturers of small electrical components (household bulbs and tube-lights) may not be able to afford a research and development segment in their hierarchical structure.

The accelerated development and extended usage of the internet and e-commerce has drawn the attention of businesses towards the integration of online and offline sales channels. The integration of offline and online channels is often termed as a multichannel context [5]. Traditional enterprises can deliver diverse types of products and services by the incorporation of online sales channels in their businesses. The number of companies taking advantage of this opportunity is growing rapidly. This enables potential customers to browse the product catalogs, price information, availability of the product, and even order the products prior to visiting the physical stores. Therefore, online sales channels can positively improve the sales of physical (offline) stores. This effective mode of sales is titled as online to offline (O2O) commerce [6]. Recently, a case study conducted by Chang et al. [7] provided factual proof of the effectiveness of the integration of offline and online sales channels to increase the sales order.

Further, for the O2O sales channel, the customer acquisition cost is required to convince the potential consumer or customer to purchase a specific product; this cost is considered a primary business metric. It determines the worth of the end customer to the business and, through it, the return on investment from the customers can be obtained. This enables businesses to evaluate their investment decisions on a single customer in order to improve profitability [8]. Organizations make investments for online marketing as well as offline marketing of their products. Then, to compute the customer acquisition cost, each customer who buys the product or service is asked about the channel of information through which he came to know about that product. A handful of data is collected for a specific period, and customer acquisition cost for both types of channels is obtained as follows:

$$\text{Customer acquisition cost for channel A} = \frac{\text{Total marketing investment in channel A}}{\text{Number of customers acquired through channel A}}.$$

Empirical evidence has proved that the customer acquisition cost plays a vital role for value creation in the business strategy. This cost has several components including online and offline marketing, subsidies, dealer commissions, administration costs, and bonuses. Most importantly, successful investment in the customer acquisition cost is directly associated with future corporate profits. It determines the customer retention, technology adoption, and improved market shares [9]. To incorporate an e-commerce strategy, this study considers a supply chain management model that acquires customers through the O2O channel and takes into account a certain customer acquisition cost for its O2O channel to enhance the future profitability of the system.

In this context, this paper develops a modified product distribution approach towards cost optimization through delivery network design for a small-scale electrical components manufacturer. The proposed model considers that the firm's target O2O market is available retailers in specific cities, and their manufacturing facilities are located at a distance from those cities. The company owns a fleet of large transport vehicles (trucks) for delivering goods to its main distribution center located near the outskirts of the cities, from where company-owned small-sized transportation vehicles distribute the commodities to sub-distribution centers (retailers) located inside various zones of those cities. This paper provides a solution to this supply chain model by considering it as an electric power transmission and distribution network. Terminologies of the electric power distribution are used to model the situation in order to provide ease in its applicability by the management of companies that have limited knowledge of supply chain management (SCM) techniques.

The structure of the paper is as follows: Section 2 presents a brief review of the relevant literature. The mathematical model is developed in Section 3, and Section 4 provides a numerical illustration of the model. Finally, Section 5 highlights the important concluding remarks and further research directions for the proposed model.

2. Literature Review

All the activities included in the flow of commodities from vendors/suppliers to the end users are part of the supply chain management (SCM). Developing an efficient flow of products in a supply chain with acceptable service level and minimum costs requires optimization plans falling in the category of SCM [10]. Various quantitative methods to manage multi-echelon SCM have been reported in the literature since the seminal work by Clark and Scarf [11]. Sabri and Beamon [12] developed an SCM model by the combination of planning decisions and strategic design using an iterative solution procedure. Nozick and Turnquist [13] used a linear function to determine the safety stock and then combined it with a facility location problem. Sarrafha et al. [14] provided a network design for a multi-echelon supply chain by integrating procurement, production, and deliveries in their model.

Transshipments in SCM helps to reduce risks due to shortages and stock-out conditions. Much of the literature takes transshipment into consideration. Most researchers have focused on optimal control strategies of transshipments without capacity constraints. Common supplier multi-retailer supply chain models with transshipments between retailers were considered by Rudi et al. [15],

Olsson [16], and Tang and Yen [17] with the consideration of infinite capacity of the central supplier. Axsäter et al. [18] developed a transshipment model for n retailers with stock out case, when the transshipments are made from the outside backup warehouse. Liao et al. [19] as well as Noham and Tzur [20] considered emergency orders and multiple items without capacity constraints, whereas Lee and Park [21] studied some incentives for each retailer to inflate the order in equilibrium in the presence of transshipments opportunity.

A major objective of VRPs found in the literature is the distance minimization, where the distance among each pair of retailers is kept constant in a classical VRP, and the total distance of the supply chain is calculated as the sum of all inter-customer distances on the route of a designated vehicle [22]. Crainic et al. [3] developed a VRP variant with time windows at the corresponding customers. Perboli et al. [23] presented three sets of two-echelon VRPs with four satellites and 50 customers. Their branch and cut algorithm can solve instances up to 21 customers to near-optimal solutions. Jepsen et al. [24] formulated a branch and cut procedure to solve VRPs with the constraint of a limited number of vehicles per satellite, and solved 47 out of 93 instances to their near-optimal solutions. Martínez-Salazar et al. [25] formulated a routing problem with the objective of minimizing customer waiting time. Breunig et al. [26] proposed a hybrid metaheuristic for two-echelon location routing and two-echelon vehicle routing problems.

Generally, distribution centers (DCs) function as inventory storage as well as transfer locations. Thus, the optimization of DC locations and their capacities under the influence of market demands is among the vital decisions of SCM [27]. Apte and Viswanathan [28] discussed some techniques to improve the overall efficiencies of logistics and distribution networks. Murali et al. [29] integrated the pre-positioning of the relief goods and facility location to determine the number of humanitarian aid centers needed, and their locations, after a disaster has occurred. Chaiwuttisak et al. [30] presented an integer programming model considering the allocation problem to improve the supply of blood products by reducing distribution centers and transportation cost.

The electric power system consists of a large-scale network of several electrical components to transmit and utilize electric power, where the optimal power flow (OPF) to obtain a steady-state operation point with minimum cost is the area of interest for researchers [31]. The classical method of OPF has the basics provided by power flow solutions in Newton's method [32] and the interior point method [33]. Several studies in recent years have used non-conventional approaches to obtain the OPF. Bose et al. [34] solved the convex dual OPF problem through tree networks. Sivasubramani and Swarup [35] presented a differential evolution algorithm and sequential quadratic programming for efficient solutions of OPF. Divshali et al. [36] used a particle swarm optimization technique to solve OPF with dynamic security constraints in deregulated power systems.

3. Mathematical Model

The following assumptions are considered in the model formulation.

3.1. Assumptions

1. The model considers a supply chain management network to deliver a single type of product from the manufacturing facilities of a company to the acquired multi-retailers.
2. For a robust supply chain management, flow of goods between various nodes should be smooth. Accordingly, this research work considers that all the connections between the sending and receiving nodes remain active throughout.
3. Line losses are assumed to be negligible, which indicates that the non-deteriorating product type is considered to design the proposed supply chain management model.
4. Different types of delivery vehicles are required for product transportation between the involved parties. This paper considers that the delivery vehicles capacity is sufficient.
5. This paper considers multi-retailers as the ultimate product-receiving parties in the supply chain management model and their annual demand of product is assumed to be known and constant.

3.2. Model Formulation

Keeping in view the abovementioned assumptions, a mathematical model is developed in this section. Consider a household electric switches manufacturing group having various manufacturing plants in industrial zones outside a specific city. From these manufacturing facilities, the commodities are transported to the company-owned warehouses in the outskirts of the city. Commodities are then delivered to various retail stores within the city area through small trucks. The model considers the manufacturing capacity of each manufacturing facility as the power generation capacity, the distance between nodes as the length of electric power transmission wires, the shipment cost as the resistance per unit length of the power transmission lines, the product quantity transferred from one node to another as the amount of power to be transmitted in energy packets, the fixed resistance at each customer as the customer acquisition cost for the O2O channel, and the retail store demand as the load of a consumer zone. Figure 2 illustrates the flow of power transmission among different nodes.

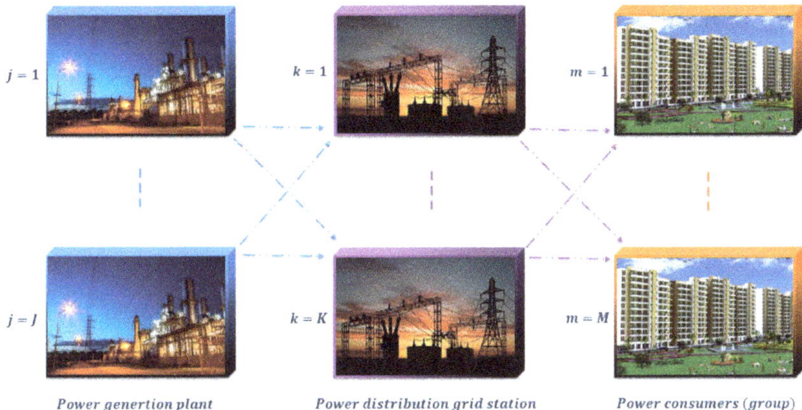

Figure 2. Flow of power transmission process.

The objective function of the proposed supply chain model is formulated as below.

$$Minimize \sum_j \sum_k R_{jk} d_{jk} w_{jk} x_k + \sum_k \sum_m R_{km} d_{km} w_{km} x_k + \sum_k \sum_n Z_{kn}, \tag{1}$$

subject to:

$$\sum_k w_{jk} \le W_j, \ \forall j \tag{2}$$

$$\sum_j w_{jk} \le W_k, \ \forall k \tag{3}$$

$$\sum_m w_{km} \le W_k, \ \forall k \tag{4}$$

$$\sum_k w_{km} = W_m, \ \forall m \tag{5}$$

$$\sum_j w_{jk} - \sum_m w_{km} = 0, \ \forall k \tag{6}$$

$$x_k \in \{0, 1\}, \ \forall k \tag{7}$$

$$w_{jk}, w_{km} \ge 0. \tag{8}$$

The objective function in Equation (1) is a resistance minimization objective of the supply chain. First term of the objective function indicates the resistance faced by the power flow from power plants to grid stations (representative of the shipment cost from a supplier to distribution centers), the second term indicates the resistance faced by the power flow from grid stations to consumer zones (representative of the shipment cost from the distribution centers to the consumer zones), and the third term shows the fixed resistance at each individual customer (indicating the customer acquisition cost for the O2O channel). Constraint (2) verifies that the total supply from plants does not exceed their supply capacity, and constraint (3) confirms that the total supply from plants to grid stations does not exceed the distribution center's capacity of grid stations. Similarly, constraints (4) and (5) are supply and demand constraints for the second part (grids to consumer zones) of the supply chain network. Constraint (6) is a transshipment constraint, and constraints (7) and (8) are binary and non-negativity constraints, respectively.

4. Numerical Experiment

An experimental study is performed to illustrate the practical applicability of the model with four power plants, three distribution centers, and five consumer zones with 50 consumers each. Resistance to the power flow from the power station to the grid stations and from the grid station to the consumer zones is considered as 0.05 ohm/MW/km. The fixed resistance at each customer is 1.5 ohm. Table 1 shows the demand of the consumer zones. Tables 2 and 3 show the capacities of the power generation plants and grid stations, respectively. Table 4 shows the length of the transmission wires between the power generation plants and the grid stations, and Table 5 summarizes the length of the transmission wires between the grid stations and consumer zones.

Table 1. Demand of consumer zones (MW).

Consumer Zone	Demand
1	250
2	350
3	300
4	450
5	350

Table 2. Power generation capacity of power plants (MW).

Power Plant	Capacity
1	500
2	400
3	350
4	450

Table 3. Capacity of grid stations (MW).

Grid Station	Capacity
1	700
2	450
3	550

Table 4. Length of power transmission lines between plants and grid stations (km).

	Grid Station 1	Grid Station 2	Grid Station 3
Power plant 1	160	240	180
Power plant 2	1480	1120	1220
Power plant 3	180	2360	180
Power plant 4	200	220	240

Table 5. Length of power transmission lines between grid stations and consumer zones (km).

	Grid Station 1	Grid Station 2	Grid Station 3
Consumer zone 1	260	360	60
Consumer zone 2	280	380	280
Consumer zone 3	300	200	500
Consumer zone 4	320	420	240
Consumer zone 5	340	440	340

4.1. Computational Results

The objective function of the proposed model is a mixed integer linear programming model (MILP) to minimize the resistance to the power flow of the power distribution system (shipment cost of the supply chain). The model was solved in LINGO 16.0 with computer specifications of 4 GB RAM and 2.30 GHz processor speed. The optimal results were achieved in 5.71 seconds, where the minimum resistance, i.e., the cost of the proposed supply chain network, was obtained as 56,044.71 ohm = $56,044.71. Figure 3 shows the optimal channel in the proposed supply chain. Tables 6 and 7 show the optimal values of binary variables and Tables 8 and 9 show the optimal quantity allocation to each selected channel of the supply chain.

Table 6. Optimal values of binary variables for the first part of the supply chain.

	Grid Station 1	Grid Station 2	Grid Station 3
Power plant 1	1	0	1
Power plant 2	0	1	0
Power plant 3	1	0	1
Power plant 4	1	1	0

Table 7. Optimal values of binary variables for the second part of the supply chain.

	Grid Station 1	Grid Station 2	Grid Station 3
Consumer zone 1	0	0	1
Consumer zone 2	1	1	0
Consumer zone 3	0	1	0
Consumer zone 4	1	0	1
Consumer zone 5	1	1	0

Table 8. Optimal transmission quantities from generation plants to grid stations (MW).

	Grid Station 1	Grid Station 2	Grid Station 3
Power plant 1	280	0	220
Power plant 2	0	400	0
Power plant 3	20	0	330
Power plant 4	400	50	0

Table 9. Optimal transmission quantities from grid stations to consumer zones (MW).

	Grid Station 1	Grid Station 2	Grid Station 3
Consumer zone 1	0	0	250
Consumer zone 2	275	75	0
Consumer zone 3	0	300	0
Consumer zone 4	150	0	300
Consumer zone 5	275	75	0

4.2. Results Analysis and Discussion

Results of the numerical experiment verify that the proposed model successfully achieved a wise selection and quantity transmission channel in the supply chain. Figure 3 shows a diagrammatic view of the optimal selection and allocation for the experimental analysis of the model.

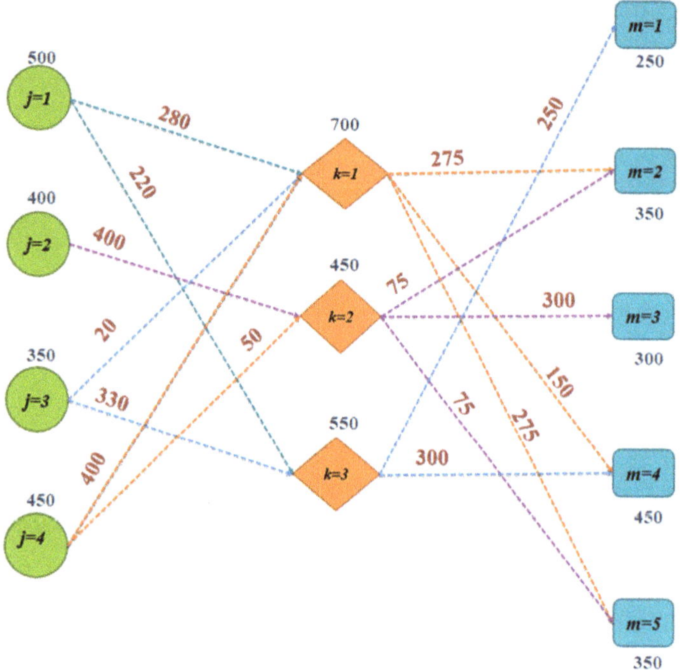

Figure 3. Optimal transmission policy.

Supply Chain Demand Satisfaction

The robustness of the supply chain model lies in its accurate demand satisfaction. Figure 3 confirms that the optimal transmission possibility obtained through this research work precisely satisfied the demand of all the consumer zones. For instance, the total demand of consumer zone 4 was 450 MW, which was fulfilled by transmitting 150 MW from grid station 1 and 300 MW from grid station 3 (Figure 4). One can observe the same for all other consumer zones from the optimal transmission policy of the proposed model.

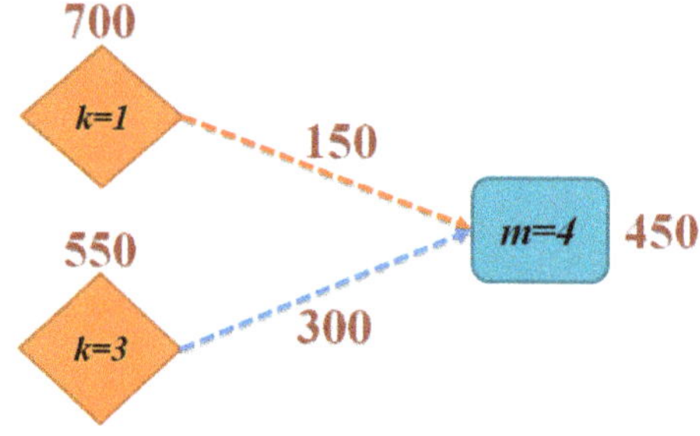

Figure 4. Demand satisfaction for consumer zone 4.

5. Concluding Remarks

This research work developed a model for the O2O supply chain management of a household electric components manufacturer for delivering goods to its distribution centers and retailers. In the proposed model, electric power transshipment and distribution is considered as product distribution network and is solved using power transmission terminologies. A supply chain management model for a household electric switches manufacturing group having various manufacturing plants in industrial zones outside a specific city was proposed. From these manufacturing facilities, the commodities are shipped to company-owned warehouses in the outskirts of the city. Commodities are then delivered to various retail stores within the city area through small trucks. These multi-retailers are the clients of the product that are acquired by making investments in O2O e-commerce. The model considers the manufacturing capacity of each facility as the power generation capacity, the distance between nodes as the length of the electric power transmission wires, the product quantity transferred from one node to another as the amount of power transmitted, and the retail store demand as the load of a consumer zone.

A numerical experiment for the mixed integer linear programming model was carried out using LINGO optimization software. Analysis of the results verified that the model successfully satisfied the retailer demand of each consumer zone while also minimizing the total cost of the supply chain. The model can be extended to several different scenarios, including the consideration of shortages that are backordered [37–40], controllable production rate [41–45], optimal trade credit policies [46–49], quantity allocation problems [50,51], different defective rework options [46], and variable deterioration [47]. An immediate possible extension to this paper could be the incorporation of an uncertain defective rate [52,53] in the supply chain.

Author Contributions: Conceptualization, B.S. and M.T.; Methodology, B.S. and M.T.; Software, B.S. and M.T.; Validation, B.S.; Formal analysis, B.S. and M.T.; Investigation, B.S. and M.T.; Resources, B.S. and M.T.; Data curation, B.S. and M.T.; Writing—original draft preparation, B.S. and M.T.; Writing—review and editing, B.S. and S.-B.C.; Visualization, B.S. and M.T.; Supervision, S.-B.C.

Funding: This work was supported by the National Research Foundation Grant funded by the Korean Government. (NRF-2017S1A2A2036758).

Conflicts of Interest: The authors declare no conflict of interest.

Nomenclature

Indices

j power generation facility ($j = 1, 2, 3 \ldots J$)
k grid station ($k = 1, 2, 3 \ldots K$)
m power consumer zone ($m = 1, 2, 3 \ldots M$)
n number of customers in consumer zone k ($n = 1, 2, 3 \ldots N_k$)

Decision variables

w_{jk} power transmission from power plant j to grid station k (MW)
w_{km} power transmission from grid station k to consumer zone m (MW)
x_k selection parameter for grid station k (binary variable)

Parameters

R_{jk} resistance to power flow between node j and node k (ohm)
R_{km} resistance to power flow between node k and node m (ohm)
d_{jk} length of power transmission wire between node j and node k (km)
d_{km} length of power transmission wire between node k and node m (km)
Z_{kn} fixed resistance to power flow at each customer (ohm)
W_j power generation capacity of plant j (MW)
W_k power transmittal capacity of grid station k (MW)
W_m load requirement of consumer group m (MW)

Abbreviations

The abbreviations listed below are used for the development of this model.

O2O	Online to offline commerce
DN	Distribution network
DNDP	Distribution network design problem
VRP	Vehicle routing problem
SCM	Supply chain management
DC	Distribution center
OPF	Optimal power flow

References

1. Ghorbani, A.; Jokar, M.R.A. A hybrid imperialist competitive-simulated annealing algorithm for a multisource multi-product location-routing-inventory problem. *Comput. Ind. Eng.* **2016**, *101*, 116–127. [CrossRef]
2. Van Duin, J.H.R.; Tavasszy, L.A.; Taniguchi, E. Real time simulation of auctioning and re-scheduling processes in hybrid freight markets. *Transport. Res. B Meth.* **2007**, *41*, 1050–1066. [CrossRef]
3. Crainic, T.G.; Ricciardi, N.; Storchi, G. Models for evaluating and planning city logistics systems. *Transp. Sci.* **2009**, *43*, 432–454. [CrossRef]
4. Rosa, G.; Costa, M.A. Robust functional analysis for fault detection in power transmission lines. *Appl. Math. Model.* **2016**, *40*, 9067–9978. [CrossRef]
5. Yang, S.; Lu, Y.; Chau, P.Y. Why do consumers adopt online channel? An empirical investigation of two channel extension mechanisms. *Decis. Support Syst.* **2013**, *54*, 858–869. [CrossRef]
6. Cassab, H.; MacLachlan, D.L. Interaction fluency: A customer performance measure of multichannel service. *Int. J. Prod. Perform. Manag.* **2006**, *55*, 555–568. [CrossRef]
7. Chang, Y.W.; Hsu, P.Y.; Yang, Q.M. Integration of online and offline channels: A view of O2O commerce. *Internet Res.* **2018**, in press. [CrossRef]
8. Chen, P.S.; Hitt, L.M. Switching cost and brand loyalty in electronic markets: Evidence from on-line retail brokers. In Proceedings of the Twenty First International Conference on Information Systems, Brisbane, Australia, 10–13 December 2000; pp. 134–144.
9. Livne, G.; Simpson, A.; Talmor, E. Do customer acquisition cost, retention and usage matter to firm performance and valuation? *J. Bus. Financ. Account.* **2011**, *38*, 334–363. [CrossRef]
10. Simchi-Levi, D.; Kaminsky, P.; Simchi-Levi, E. *Managing the Supply Chain: Definitive Guide*; Tata McGraw-Hill Education: New York, NY, USA, 2004.
11. Clark, A.J.; Scarf, H. Optimal policies for a multi-echelon inventory problem. *Manag. Sci.* **1960**, *6*, 475–490. [CrossRef]
12. Sabri, E.H.; Beamon, B.M. A multi-objective approach to simultaneous strategic and operational planning in supply chain design. *Omega* **2000**, *28*, 581–598. [CrossRef]
13. Nozick, L.K.; Turnquist, M.A. Inventory, transportation, service quality and the location of distribution centers. *Eur. J. Oper. Res.* **2001**, *129*, 362–371. [CrossRef]
14. Sarrafha, K.; Rahmati, S.H.; Niaki, S.T.; Zaretalab, A. A bi-objective integrated procurement, production, and distribution problem of a multi-echelon supply chain network design: A. new tuned MOEA. *Comput. Oper. Res.* **2015**, *54*, 35–51. [CrossRef]
15. Rudi, N.; Kapur, S.; Pyke, D.F. A two-location inventory model with transshipment and local decision making. *Manag. Sci.* **2001**, *47*, 1668–1680. [CrossRef]
16. Olsson, F. Optimal policies for inventory systems with lateral transshipments. *Int. J. Prod. Econ.* **2009**, *118*, 175–184. [CrossRef]
17. Tang, S.L.; Yan, H. Pre-distribution vs. post-distribution for cross-docking with transshipments. *Omega* **2010**, *38*, 192–202. [CrossRef]
18. Axsäter, S.; Howard, C.; Marklund, J. A distribution inventory model with transshipments from a support warehouse. *IIE Trans.* **2013**, *45*, 309–322. [CrossRef]

19. Liao, Y.; Shen, W.; Hu, X.; Yang, S. Optimal responses to stockouts: Lateral transshipment versus emergency order policies. *Omega* **2014**, *49*, 79–92. [CrossRef]

20. Noham, R.; Tzur, M. The single and multi-item transshipment problem with fixed transshipment costs. *Nav. Res. Log.* **2014**, *61*, 637–664. [CrossRef]

21. Lee, C.; Park, K.S. Inventory and transshipment decisions in the rationing game under capacity uncertainty. *Omega* **2016**, *65*, 82–97. [CrossRef]

22. Cinar, D.; Gakis, K.; Pardalos, P.M. A 2-phase constructive algorithm for cumulative vehicle routing problems with limited duration. *Expert. Syst. Appl.* **2016**, *56*, 48–58. [CrossRef]

23. Perboli, G.; Tadei, R.; Vigo, D. The two-echelon capacitated vehicle routing problem: Models and math-based heuristics. *Transport. Sci.* **2011**, *45*, 364–380. [CrossRef]

24. Jepsen, M.; Spoorendonk, S.; Ropke, S. A branch-and-cut algorithm for the symmetric two-echelon capacitated vehicle routing problem. *Transport. Sci.* **2013**, *47*, 23–37. [CrossRef]

25. Martínez-Salazar, I.; Angel-Bello, F.; Alvarez, A. A customer-centric routing problem with multiple trips of a single vehicle. *J. Oper. Res. Soc.* **2015**, *66*, 1312–1323. [CrossRef]

26. Breunig, U.; Schmid, V.; Hartl, R.F.; Vidal, T. A large neighbourhood based heuristic for two-echelon routing problems. *Comput. Oper. Res.* **2016**, *76*, 208–225. [CrossRef]

27. Zhuge, D.; Yu, S.; Zhen, L.; Wang, W. Multi-period distribution center location and scale decision in supply chain network. *Comput. Ind. Eng.* **2016**, *101*, 216–226. [CrossRef]

28. Apte, U.M.; Viswanathan, S. Effective cross docking for improving distribution efficiencies. *Int. J. Log.* **2000**, *3*, 291–302. [CrossRef]

29. Murali, P.; Ordóñez, F.; Dessouky, M.M. Facility location under demand uncertainty: Response to a large-scale bio-terror attack. *Soc. Econ. Plan. Sci.* **2012**, *46*, 78–87. [CrossRef]

30. Chaiwuttisak, P.; Smith, H.; Wu, Y.; Potts, C.; Sakuldamrongpanich, T.; Pathomsiri, S. Location of low-cost blood collection and distribution centres in Thailand. *Oper. Res. Health Care* **2016**, *9*, 7–15. [CrossRef]

31. Chen, M.J.; Hsu, Y.F.; Wu, Y.C. Modified penalty function method for optimal social welfare of electric power supply chain with transmission constraints. *Int. J. Electr. Power* **2014**, *57*, 90–96. [CrossRef]

32. Tinney, W.F.; Hart, C.E. Power flow solution by Newton's method. *IEEE Trans. Power. Appl. Syst.* **1967**, *PAS-86*, 1449–1460. [CrossRef]

33. Singh, B.; Mahanty, R.; Singh, S.P. Optimal rescheduling of generators for congestion management and benefit maximization in a decentralized bilateral multi-transactions power network. *Int. J. Emerg. Electr. Power Syst.* **2013**, *14*, 25–32. [CrossRef]

34. Bose, S.; Gayme, D.F.; Low, S.; Chandy, K.M. Optimal power flow over tree networks. In Proceedings of the IEEE 49th Annual Allerton Conference on Communication, Control, and Computing (Allerton), Monticello, IL, USA, 28–30 September 2011; pp. 1342–1348.

35. Sivasubramani, S.; Swarup, K.S. Sequential quadratic programming based differential evolution algorithm for optimal power flow problem. *IET Gener. Transm. Dis.* **2011**, *5*, 1149–1154. [CrossRef]

36. Divshali, P.H.; Hosseinian, S.H.; Azadani, E.N.; Abedi, M. Application of bifurcation theory in dynamic security constrained optimal dispatch in deregulated power system. *Elect. Eng.* **2011**, *93*, 157. [CrossRef]

37. Sarkar, B.; Moon, I. Improved quality, setup cost reduction, and variable backorder costs in an imperfect production process. *Int. J. Prod. Econ.* **2014**, *155*, 204–213. [CrossRef]

38. Sarkar, B. A production-inventory model with probabilistic deterioration in two-echelon supply chain management. *Appl. Math. Model.* **2013**, *37*, 3138–3151. [CrossRef]

39. Sarkar, B. An inventory model with reliability in an imperfect production process. *Appl. Math. Comput.* **2012**, *218*, 4881–4891. [CrossRef]

40. Sarkar, B.; Sana, S.S.; Chaudhuri, K. Optimal reliability, production lotsize and safety stock: An economic manufacturing quantity model. *Int. J. Manag. Sci. Eng. Manag.* **2010**, *5*, 192–202. [CrossRef]

41. Moon, I.; Shin, E.; Sarkar, B. Min–max distribution free continuous-review model with a service level constraint and variable lead time. *Appl. Math. Comput.* **2014**, *229*, 310–315. [CrossRef]

42. Shin, D.; Guchhait, R.; Sarkar, B.; Mittal, M. Controllable lead time, service level constraint, and transportation discounts in a continuous review inventory model. *RAIRO Oper. Res.* **2016**, *50*, 921–934. [CrossRef]

43. Sarkar, B.; Mahapatra, A.S. Periodic review fuzzy inventory model with variable lead time and fuzzy demand. *Int. Trans. Oper. Res.* **2017**, *24*, 1197–1227. [CrossRef]

44. Sarkar, B.; Cárdenas-Barrón, L.E.; Sarkar, M.; Singgih, M.L. An economic production quantity model with random defective rate, rework process and backorders for a single stage production system. *J. Manuf. Syst.* **2014**, *33*, 423–435. [CrossRef]

45. Sarkar, B.; Ahmed, W.; Kim, N. Joint effects of variable carbon emission cost and multi-delay-in-payments under single-setup-multiple-delivery policy in a global sustainable supply chain. *J. Clean. Prod.* **2018**, *185*, 421–445. [CrossRef]

46. Sarkar, B.; Sana, S.S.; Chaudhuri, K. An inventory model with finite replenishment rate, trade credit policy and price-discount offer. *J. Ind. Eng.* **2013**, *2013*. [CrossRef]

47. Sarkar, B.; Sarkar, S. An improved inventory model with partial backlogging, time varying deterioration and stock-dependent demand. *Econ. Model.* **2013**, *30*, 924–932. [CrossRef]

48. Kang, C.W.; Ullah, M.; Sarkar, B.; Hussain, I.; Akhtar, R. Impact of random defective rate on lot size focusing work-in-process inventory in manufacturing system. *Int. J. Prod. Res.* **2017**, *55*, 1748–1766. [CrossRef]

49. Ahmed, W.; Sarkar, B. Impact of carbon emissions in a sustainable supply chain management for a second generation biofuel. *J. Clean. Prod.* **2018**, *186*, 807–820. [CrossRef]

50. Cárdenas-barrón, L.E.; Sarkar, B.; Treviño-garza, G. Easy and improved algorithms to joint determination of the replenishment lot size and number of shipments for an EPQ model with rework. *Math. Comput. Appl.* **2013**, *18*, 132–138. [CrossRef]

51. Sarkar, B. An EOQ model with delay in payments and time varying deterioration rate. *Math. Comput. Model.* **2012**, *55*, 367–377. [CrossRef]

52. Tayyab, M.; Sarkar, B. Optimal batch quantity in a cleaner multi-stage lean production system with random defective rate. *J. Clean. Prod.* **2016**, *139*, 922–934. [CrossRef]

53. Kim, M.S.; Sarkar, B. Multi-stage cleaner production process with quality improvement and lead time dependent ordering cost. *J. Clean. Prod.* **2017**, *144*, 572–590. [CrossRef]

© 2018 by the authors. Licensee MDPI, Basel, Switzerland. This article is an open access article distributed under the terms and conditions of the Creative Commons Attribution (CC BY) license (http://creativecommons.org/licenses/by/4.0/).

 *mathematics*

Article

Interactive Fuzzy Multi Criteria Decision Making Approach for Supplier Selection and Order Allocation in a Resilient Supply Chain

Sonia Irshad Mari [1], **Muhammad Saad Memon** [1,*], **Muhammad Babar Ramzan** [2,*], **Sheheryar Mohsin Qureshi** [3] **and Muhammad Waqas Iqbal** [4]

[1]   Department of Industrial Engineering and Management, Mehran University of Engineering and Technology, Jamshoro, Sindh 76062, Pakistan; sonia.irshad@faculty.muet.edu.pk
[2]   Department of Garment Manufacturing, National Textile University, Faisalabad 37610, Pakistan
[3]   Department of Industrial and Manufacturing Engineering, NED University of Engineering and Technology, Karachi 75270, Pakistan; sheheryar@neduet.edu.pk
[4]   Department of Industrial Engineering, Hongik University, Seoul 04066, Korea; waqastextilion@gmail.com
*   Correspondence: saad.memon@faculty.muet.edu.pk (M.S.M.); babar_ramzan@yahoo.com (M.B.R.); Tel.: +923-332-888-606 (M.S.M.)

Received: 24 December 2018; Accepted: 25 January 2019; Published: 1 February 2019

Abstract: Modern supply chains are vulnerable to high impact, low probability disruption risks. A supply chain usually operates in such a network of entities where the resilience of one supplier is critical to overall supply chain resilience. Therefore, resilient planning is a key strategic requirement in supplier selection decisions for a competitive supply chain. The aim of this research is to develop quantitative resilient criteria for supplier selection and order allocation in a fuzzy environment. To serve the purpose, a possibilistic fuzzy multi-objective approach was proposed and an interactive fuzzy optimization solution methodology was developed. Using the proposed approach, organizations can tradeoff between cost and resilience in supply networks. The approach is illustrated using a supply chain case from a garments manufacturing company.

Keywords: resilient supply chain; supplier selection; fuzzy optimization; disruption risks

---

1. Introduction

Outsourcing is a competitive strategy in the global supply chain. Evaluating and selecting the best set of suppliers is a challenging decision in outsourcing and it plays a significant role in supply chain performance [1,2]. Traditionally the supplier selection and order allocation decision is made based on cost and quality criteria. However, modern supply chains are more prone to unexpected High Impact Low Probability (HILP) and Low Impact High Probability (LIHP) disruption events [3]. HILP disruption events are commonly known as random disruptions risks such as man-made and natural disasters, whereas LIHP disruptions are targeted disruptions such as day-to-day operational risks. Tang and Tomlin [4] proposed six disruption sources in the supply chain and among them, supplier performance is most frequent. The role of the supplier selection decision in these supply chain risk has only been partially explored in the literature [5]. Multi-sourcing strategies and are now common to many supply chains in order to minimize the supplier's disruption risks [6]. For example, during a fire in a plant of Philips Electronics in 2010 disrupt two of its major customers: Ericsson and Nokia. Ericsson lost about a month of production and suffered $200 million while Nokia recovered due to its multi-sourcing ability [7]. Toyota Motor Corp. lost billions of dollars in 2010 during product recall due to its part sourcing from one supplier for many car models. These examples show that multi-sourcing strategies work well. On the contrary, multi-sourcing strategies failed during some

HILP disruption events. For example, Japan earthquake disrupted many semiconductor supply chains. Chinese Firm ZTE Corp. faced shortages of batteries and LCD screens due to all of its suppliers in the affected region. Ford Motor Co. and General Motor Co. faced shortages of auto parts and stop production due to the shutdown of two Hitachi Ltd.'s plants. These historical events suggest that the supplier selection criteria should be extended to new resilience capabilities [8]. Thus it crucial to provide a reliable level of resilience to the supply side to protect such shortages especially during HILP events [9].

Several studies have been conducted to consider resilience in the supply chain [8,10,11]. The concept of resilience in specific to the supplier selection problem has also been discussed by several authors [12–25]. Most of these studies focused on multiple sourcing and operational performance of suppliers. However, to the best of authors knowledge, this is the first study which focuses on supply network by considers supply density, resilience score of supplier's locations, and transit time in addition to other operational criteria. This paper aims to develop a supplier selection and order allocation model to build a resilient supply chain in response to HILP disruptions. To do so, a possibilistic multi-objective fuzzy optimization-based model with a new resilience objective is proposed which consists of supply density, resilience score, and transit time. Furthermore, the proposed model is solved using Tiwari, et al. [26] weighted additive approach and Werners [27] fuzzy and operator methods. Fuzzy based multi-objective approaches are widely used in supplier selection problem to deal with uncertain information [28,29]. This research answers the following questions: (i) Which supplier is selected based on the importance given to each objective? (ii) How much to purchase from each selected supplier?

The remainder of the paper is organized as follows. Section 2 provides related literature. Section 3 includes problem description and mathematical model for supplier selection and order allocation with a new resilience objective. Section 4 comprises of the proposed possibilistic fuzzy based solution methodology. Section 5 presents a numerical example to show the application of proposed supplier selection and order allocation model. Sections 6 and 7 discussed the results of the proposed mathematical model and solution methodology. Finally, Section 8 presents some conclusions and future directions drawn from the study.

2. Literature Review

The word resilience first coined by Holling [30] in the context of ecology. According to Holling [30], the resilience is the ability of a system to absorb changes in state variables, driving variable and parameters, and still persist. Due to the increase in complexity and uncertainty in the business environment, several studies shown interest in the concept of resilience in a managerial perspective. Hamel and Valikangas [31] defined resilience as a capacity for continuous reconstruction. Sheffi [32] defined resilience in terms of enterprise resilience as the ability of an organization to successfully confront the unforeseen. Sutcliffe and Vogus [33] stated that resilience is the (1) ability to absorb strain and improve the functionality of organization despite the presence of difficulty or (2) ability to bounce back after disturbances. More recently, Woods [34] defined resilience in simple terms as system's ability to bounce back after disruptions and to bounce forward through learning from those disruption events and increase the system's adaptive capacity for handling uncertain events.

The concept of supply chain resilience gained prominent importance during recent years in supply chain risk management research [35]. Supply chain resilience is a relatively new concept to mitigate risks that can be defined as the ability to reduce the probability of a disruption, to reduce the impact of disruption, and to reduce the recovery time to normal performance [36]. Supply chain resilience has been defined by several authors in simpler and broader terms. Most of these studies described supply chain resilience as the ability to withstand disruptions and converge to the original state or to a new desirable state. Despite the increasing number of publication in supply chain resilience area, most of the researches provided qualitative insights and there is a limited number of quantitative modelling techniques available [37]. These qualitative models used different performance measures for designing resilient supply chains. Priya Datta, et al. [38] developed the framework to improve operational resilience. Falasca, Zobel and Cook [36] proposed three determinants (density, complexity, and node criticality) of supply chain resilience for supply chain design. Azevedo, et al. [39] proposed GResilient index to assess supply chain resilience using the Delphi technique. Miller-Hooks, et al. [40] proposed a transportation network resilience model using stochastic programming. They presented the expected fraction of demand fulfilment after disruption as resilience metric.

Supply chain usually functions in the system of parties where the resilience of one party (e.g. a supplier) is critical for overall supply chain resilience. As discuss earlier, many major disruptions break down supply networks and it takes a long time to recover. Whereas, the probability of disruption may be reduced by developing a resilient network and it takes considerably less recovery time [41]. Suppliers constitute the most important role in the performance of the supply chain, therefore, the resilience of the supply network is expected to contribute and increase overall supply chain resilience [42]. Despite its importance, there is very limited research conducted which consider resilience of supply network. The literature on resilient supplier selection is summarized in Table 1. Haldar, Ray, Banerjee and Ghosh [20] developed a resilient supplier selection model using AHP-QFD method to rate the suppliers. They used five resilience criteria (density, complexity, node criticality, responsiveness and re-engineering) for the supplier selection process. More recently, Haldar, Ray, Banerjee and Ghosh [21] proposed a fuzzy group decision-making approach for resilient supplier selection using triangular and trapezoidal fuzzy numbers. They considered investment, responsiveness, and capacity of holding inventory as resilient criteria. Sawik [22] proposed resilience strategies for supplier selection and order allocation problem under disruption risks. Recently, Sawik [23] proposed a stochastic mixed integer model for supplier selection and customer order scheduling using disruption risks for single and dual sourcing. Yilmaz-Börekçi, İşeri Say and Rofcanin [24] proposed a scale for measuring supplier resilience within supply networks. Mari, et al. [43] proposed a resilient and sustainable supply chain network model. They used expected disruption cost as a resilient metric. Torabi, Baghersad and Mansouri [9] developed a bi-objective stochastic model to trade-off the resilience level of the supply network and system cost. Memon, Lee and Mari [28] proposed supplier selection model using grey and uncertainty theories. Rajesh and Ravi [25] proposed a resilient supplier selection method using grey relational analysis. They consider supplier's performance, responsiveness, risk reduction, technical capabilities and sustainability as selection criteria. Sahu, Datta and Mahapatra [14] proposed fuzzy-VIKOR based supplier selection framework by combining the general selection strategy and resilience strategy. They considered investment capacity, responsiveness, and inventory capacity as resilience criteria. Hosseini and Al Khaled [13] developed a resilience score of suppliers based on eight criteria namely backup supplier contracting, surplus inventory, location separation, robustness, reliability, reorganization, rerouting, and restoration. They proposed a hybrid ensemble and AHP approach to select a suitable supplier. Pramanik, et al. [44] developed a resilient supplier selection model using AHP-TOPSIS and Sen, et al. [45] proposed g-resilient supplier selection model using dominance-based fuzzy decision making. Wang, Zhang, Chong and Wang [18] developed a resilient supplier selection model for construction supply chain by integrating building information modelling and geographical information system. Parkouhi and Ghadikolaei [12] proposed FANP and grey-VIKOR based resilient supplier selection model. They considered benefits, opportunities, costs, and risks as resilience criteria. López and Ishizaka [15] proposed a coupled method of FCM and AHP for supplier selection for a resilient supply chain. Recently, Jabbarzadeh, Fahimnia and Sabouhi [17] proposed stochastic bi-objective optimization model considering fuzzy c-means to developed sustainable supply chain network that performs resiliently. Sabouhi, Pishvaee and Jabalameli [16] presented integrated stochastic and fuzzy DEA based model for supplier selection considering resilience. They considered multi-sourcing, supplier fortification, and emergency inventory as resilience criteria. More recently, Parkouhi, et al. [46] proposed a resilient supplier selection framework using grey-DEMATEL approach.

Table 1. The classification of literature in resilient supplier selection.

Authors	Resilience Criteria	Types of Risks		Solution Methodology				
		Network Risks	Operational Risks	Possibilistic Programming	Stochastic Programming	Fuzzy Programming	Grey Programming	Other
Wang, Herty and Zhao [6]	Multi-supplier, production capacity, product quality, production cost	-	✓	-	-	-	-	Fluid-dynamic models
Torabi, Baghersad and Mansouri [9]	Multiple sourcing, fortifying supplier, pre-positioned inventories, backup supplier, and supplier's business continuity	-	✓	✓	✓	-	-	-
Parkouhi and Ghadikolaei [12]	Benefits, Opportunities, costs, and risks	-	✓	-	-	✓	✓	-
Hosseini and Al Khaled [13]	Absorptive capacity, adaptive capacity, and restorative capacity	✓	✓	-	-	-	-	Predictive analytics models
Sahu, Datta and Mahapatra [14]	Investment capacity, Responsiveness, and Inventory capacity	-	✓	-	-	✓	-	-
López and Ishizaka [15]	Flexibility, Visibility, Anticipation, Recovery, Security, Adaptability, Financial strength, Market position, and Collaboration	-	✓	-	-	✓	-	-
Sabouhi, Pishvaee and Jabalameli [16]	multi-sourcing, supplier fortification, and emergency inventory	-	✓	-	✓	✓	-	-
Jabbarzadeh, Fahimnia and Sabouhi [17]	Extra production capacities, Multiple sourcing, and Backup suppliers	-	✓	-	✓	✓	-	-
Hosseini and Barker [19]	Absorptive capacity, adaptive capacity, and restorative capacity	✓	✓	-	-	-	-	Bayesian network
Haldar, Ray, Banerjee and Ghosh [21]	Investment capacity, Responsiveness, and Emergency inventory holding capacity	-	✓	-	-	✓	-	-
Rajesh and Ravi [25]	Responsiveness, risk reduction, and Technical support	-	✓	-	-	-	✓	-
Parkouhi, Ghadikolaei and Lajimi [46]	Safety, Visibility, Environmental Controls, Trust, Flexibility, Support Services, Future Manufacturing Capabilities, and others	-	✓	-	-	-	✓	-
This paper	Supply density, Transit time, Resilience score of supplier's locations	✓	✓	✓	-	✓	-	-

Most of the above studies considered operational risks for resilient supplier selection problem. This is the first time that a possibilistic fuzzy multi-objective model is proposed for resilient supplier selection and order allocation problem. Furthermore, for the first time in the literature supply density, resilience index score, and transit time are considered as supply selection criteria for resilient supply network.

Falasca, Zobel and Cook [36] proposed three characteristics (node criticality, supply chain complexity, and supply chain density) for building a resilient supply chain. Among them, supply chain density is the most important resilient criteria when designing supply networks under HILP disruptions. This is due to fact that denser supply networks are vulnerable to HILP disruption risks [36]. For example, 1999s Taiwan earthquake ended up having a significant effect on the entire global PC supply chain, because of the high concentration of computer component manufacturers in Hsinchu, Taiwan [47]. This example shows that the selection of a large number of suppliers from each region is a vulnerable multi-sourcing strategy, hence, this paper proposed supply density-based approach to tackle this problem. Furthermore, every country or territory has different resilient capabilities FMGlobal [48] and it affects the performance of the supply chain. Therefore, this study also proposed resilience index score-based criteria to supplier selection. The resilience index score is proposed by FMGlobal [48] is a data-driven tool to rank the countries to supply chain disruption risks. Nine key drivers of supply chain risks are considered and grouped into three categories namely: economic, risk quality, and supply chain factors. These nine drivers include local supplier quality, quality of fire risk management, GDP per capita, oil intensity, quality of hazard risk management, exposure to natural hazards, corruption control, infrastructure, and political risks [49]. Transit time is the last resilient criteria for supplier selection considered in this study. Transit time is an important indicator of supply chain flexibility [50]. Transit time reduction is one of the wildly used criteria to mitigate supply risks [51].

3. Problem Formulation

In this study, a garment manufacturer is assumed which want to select a suitable set of suppliers for the required material. All the model parameters are considered as fuzzy parameters. Five objectives considered in this study, namely: a cost which includes purchase and transportation costs, the rejection rate of suppliers, transit time from suppliers, supply density, and supplier resilience score based on their locations.

3.1. Mathematical Model Notations

• Indices	
s *existing suppliers*	$s = 1, 2, \ldots S$
l *number of objective functions*	$l = 1, 2, \ldots L$
• Parameters	
$\tilde{d}$	Total demand of required material
$\tilde{u}_s$	Purchase cost of required material from supplier s
$\tilde{c}_s$	Transportation cost of material from supplier s
$\tilde{\rho}_s$	Percentage of the rejected material delivered by the supplier s
$\tilde{v}_s$	Capacity of supplier s
$\tilde{k}_s$	Minimum acceptable purchase quantity from supplier s
mx_s	Maximum number of suppliers selected for required material
$\tilde{t}_s$	Transit time from supplier s
ds^{ab}	Distance between selected supplier $a \in s$ and selected supplier $b \in s$ $(a \neq b)$
rs_s	Resilience index score of supplier s
• Decision Variables	
q_s	Purchase quantity from supplier s
$\Phi_s = \begin{cases} 1 \\ 0 \end{cases}$	1, If supplier s is selected, otherwise 0.
$\begin{matrix} \omega_{ab} \\ \omega'_{ab} \end{matrix} = \begin{cases} 1, \text{ If supplier } a \text{ and supplier } b \text{ are selected.} \\ 0, \text{otherwise} \end{cases}$	$\forall (a, b) \in s \text{ and } a \neq b$

3.2. Model Objectives

The Equation (1) represents the objective function for the cost. It is the sum of procurement cost and transportation cost from selected suppliers. The objective function of the total rate of rejection is estimated in Equation (2). The objective function for transit time from all selected supplier is calculated as shown in Equation (3). The Equation (4) shows the supply density for all selected suppliers. Total resilience index score objective is estimated in Equation (5). Where rs_s represents the resilience index score of supplier location obtained from FMGlobal [48].

$$\text{Minimize } f_{cost} = \sum_s (\tilde{u}_s + \tilde{c}_s\,)q_s \tag{1}$$

$$\text{Minimizes } f_{rej} = \sum_s \tilde{\rho}_s q_s \tag{2}$$

$$\text{Minimizes } f_{time} = \sum_s \tilde{t}_s \Phi_s \tag{3}$$

$$\text{Maximize } f_{den} = \frac{1}{\tilde{d}} \left(\sum_{a \in s} \sum_{\substack{b \in s \\ a \neq b}} ds^{ab} \Phi_s \right) \tag{4}$$

$$\text{Maximize } f_{res} = \sum_s \frac{rs_s q_s}{\tilde{d}} \tag{5}$$

3.3. Model Constraints

Constraint (6) ensures that total procured material should satisfy its demand.

$$\sum_s q_s = \tilde{d} \tag{6}$$

Constraint (7) is capacity restrictions on the supplier. Also, it controls the flow between the supplier and the buyer through a binary variable. Constraint (8) ensures that purchase quantity from the selected supplier will be more than its acceptable order quantity limit. Constraint (9) restricts the maximum allowable supplier to be selected for the required material.

$$q_s \leq \tilde{v}_s \Phi_s \tag{7}$$

$$q_s \geq \tilde{k}_s \Phi_s \tag{8}$$

$$\sum_s \Phi_s \leq mx_s \tag{9}$$

Constraints (10) and (11) determine the intra-stage flow between suppliers and buyer. If buyer received material from both supplier a and supplier b then $w_{ab} = \Phi_{a \in s} = \Phi_{b \in s} = 1$ and $w'_{ab} = 0$. On the contrary, if buyer not received material from both supplier a and supplier b then $w'_{ab} = 1$, $w_{ab} = 0$ and $\Phi_{a \in s} \neq \Phi_{b \in s}$.

$$2w_{ab} + w'_{ab} = \Phi_{a \in s} + \Phi_{b \in s}$$
$$\forall (a,b) \in s, \text{ and } a \neq b \tag{10}$$

$$w_{ab} + w'_{ab} \leq 1$$
$$\forall (a,b) \in s, \text{ and } a \neq b \tag{11}$$

4. Fuzzy Based Solution Methodology

Fuzzy based programming methods are highly used for multi-objective optimization because of their capability in measuring and adjusting the decision maker's satisfaction level of each objective function explicitly. In addition, the fuzzy theory is helpful to tackle the uncertain parameters related to supply chain optimization problem [52]. The main advantage of interactive fuzzy based approaches is that decision maker can efficiently achieve his/her preferences by controlling the search direction. The proposed solution methodology consists of the following steps.

Step 1: Convert uncertain mathematical model to equivalent auxiliary crisp

The proposed mathematical model is converted to an equivalent auxiliary crisp model. In this study, Jiménez, Arenas, Bilbao and Rodrı [51] approach is used which is based on an expected interval (EI) and expected value (EV) of fuzzy numbers. According to Jiménez, Arenas, Bilbao and Rodrı [51], the EI and EV of triangular fuzzy number (TFN) can be defined as in equation (12) and (13) respectively. Where ϑ^{pes} is the pessimistic value, ϑ^{mos} is the most likely value, and ϑ^{opt} is the optimimum value of triangular fuzzy number (ϑ). This research considered TFN because it is frequently used for a practical purpose [53].

$$EI(\widetilde{\vartheta}) = [E_1^\vartheta, E_2^\vartheta] = \left[\int_0^1 f_\vartheta^{-1}(x)dx, \int_0^1 g_\vartheta^{-1}(x)dx,\right] = \left[\frac{1}{2}(\vartheta^{pes} + \vartheta^{mos}), \frac{1}{2}(\vartheta^{mos} + \vartheta^{opt})\right] \tag{12}$$

$$EV(\widetilde{\vartheta}) = \frac{E_1^\vartheta + E_1^\vartheta}{2} = \frac{\vartheta^{pes} + 2\vartheta^{mos} + \vartheta^{opt}}{4} \tag{13}$$

Using the above Equations (12) and (13), the equivalent auxiliary crisp model can be formulated as follows.

$$\text{Minimize } f_{cost} = \sum_s \left(\frac{u_s^{pes} + 2u_s^{mos} + u_s^{opt} + c_s^{pes} + 2c_s^{mos} + c_s^{opt}}{4}\right) q_s \tag{14}$$

$$\text{Minimizes } f_{rej} = \sum_s \left(\frac{\rho_s^{pes} + 2\rho_s^{mos} + \rho_s^{opt}}{4}\right) q_s \tag{15}$$

$$\text{Minimizes } f_{time} = \sum_s \left(\frac{t_s^{pes} + 2t_s^{mos} + t_s^{opt}}{4}\right) \Phi_s \tag{16}$$

$$\text{Maximize } f_{den} = \frac{1}{\frac{d^{pes} + 2d^{mos} + d^{opt}}{4}} \left(\sum_{a \in s} \sum_{\substack{b \in s \\ a \neq b}} ds^{ab} \Phi_s\right) \tag{17}$$

$$\text{Maximize } f_{res} = \sum_s \frac{rs_s q_s}{\frac{d^{pes} + 2d^{mos} + d^{opt}}{4}} \tag{18}$$

Subject to

$$\sum_s q_s \geq \left[\frac{\alpha}{2}\left(\frac{d^{mos} + d^{opt}}{2}\right) + (1 - \frac{\alpha}{2})\left(\frac{d^{pes} + d^{mos}}{2}\right)\right] \tag{19}$$

$$\sum_s q_s \leq \left[\frac{\alpha}{2}\left(\frac{d^{pes} + d^{mos}}{2}\right) + (1 - \frac{\alpha}{2})\left(\frac{d^{mos} + d^{opt}}{2}\right)\right] \tag{20}$$

$$q_s \leq \Phi_s \left[\alpha\left(\frac{v_s^{pes} + v_s^{mos}}{2}\right) + (1 - \alpha)\left(\frac{v_s^{mos} + v_s^{opt}}{2}\right)\right] \tag{21}$$

$$q_s \geq \Phi_s \left[\alpha \left(\frac{k^{mos} + k^{opt}}{2} \right) + (1 - \alpha) \left(\frac{k^{pes} + k^{mos}}{2} \right) \right] \tag{22}$$

$$\sum_s \Phi_s \leq mx_s \tag{23}$$

$$2\omega_{ab} + \omega'_{ab} = \Phi_{a \in s} + \Phi_{b \in s} \tag{24}$$

$$\omega_{ab} + \omega'_{ab} \leq 1 \tag{25}$$

Step 2: Determine α- extreme solutions

To estimate the upper (α-UB) and lower (α-LB) bounds to each objective, the crisp model developed in Step 1 is solved for each objective along with its constraint.

Step 3: Determine fuzzy membership function

Develop the fuzzy membership function for each objective using lower (α-LB) and upper (α-UB) bound values. The linear memberships for fuzzy goals are given as follows. It is assumed that membership functions are linear based on preferences and satisfaction level.

$$\mu_{cost}(x) = \begin{cases} 1, & f_{cost}(x) \leq f_{cost}^{\alpha-LB} \\ \frac{f_{cost}^{\alpha-UB} - f_{cost}(x)}{f_{cost}^{\alpha-UB} - f_{cost}^{\alpha-LB}}, & f_{cost}^{\alpha-LB} < f_{cost}(x) \leq f_{cost}^{\alpha-UB} \\ 0, & f_{cost}(x) \geq f_{cost}^{\alpha-UB} \end{cases} \tag{26}$$

$$\mu_{rej}(x) = \begin{cases} 1, & f_{rej}(x) \leq f_{rej}^{\alpha-LB} \\ \frac{f_{rej}^{\alpha-UB} - f_{rej}(x)}{f_{rej}^{\alpha-UB} - f_{rej}^{\alpha-LB}}, & f_{rej}^{\alpha-LB} \leq f_{rej}(x) \leq f_{rej}^{\alpha-UB} \\ 0, & f_{rej}(x) \geq f_{rej}^{\alpha-UB} \end{cases} \tag{27}$$

$$\mu_{time}(x) = \begin{cases} 1, & f_{time}(x) \leq f_{time}^{\alpha-LB} \\ \frac{f_{time}^{\alpha-UB} - f_{time}(x)}{f_{time}^{\alpha-UB} - f_{time}^{\alpha-LB}}, & f_{time}^{\alpha-LB} \leq f_{time}(x) \leq f_{time}^{\alpha-UB} \\ 0, & f_{time}(x) \geq f_{time}^{\alpha-UB} \end{cases} \tag{28}$$

$$\mu_{den}(x) = \begin{cases} 1, & f_{den}(x) \geq f_{den}^{\alpha-UB} \\ \frac{f_{den}(x) - f_{den}^{\alpha-LB}}{f_{den}^{\alpha-UB} - f_{den}^{\alpha-LB}}, & f_{den}^{\alpha-UB} \leq f_{den}(x) \leq f_{den}^{\alpha-LB} \\ 0, & f_{den}(x) \leq f_{den}^{\alpha-LB} \end{cases} \tag{29}$$

$$\mu_{res}(x) = \begin{cases} 1, & f_{res}(x) \geq f_{res}^{\alpha-UB} \\ \frac{f_{res}(x) - f_{res}^{\alpha-LB}}{f_{res}^{\alpha-UB} - f_{res}^{\alpha-LB}}, & f_{res}^{\alpha-UB} \leq f_{res}(x) \leq f_{res}^{\alpha-LB} \\ 0, & f_{res}(x) \leq f_{res}^{\alpha-LB} \end{cases} \tag{30}$$

where $f_l^{\alpha-LB}$ is a minimum value of $f_l(x)$ and $f_l^{\alpha-UB}$ is a maximum value of $f_l(x)$ with predefined α. These values the lth objective depends on its nature. $f_l^{\alpha-LB}$ are set as the aspiration level of cost, rejection rate, and transit time objective. Whereas $f_l^{\alpha-UB}$ are set as the aspiration level of supply density and resilience index.

Step 4: Convert the multi-objective model into a single objective

The proposed model is converted into the single objective in this stage. In this paper, two most popular fuzzy based approaches i.e., Tiwari, Dharmar and Rao [26] *weighted additive approach* and Werners [27] *fuzzy and* operator are implemented.

The weighted additive approach allows the buyer to assign different weights to objectives in the simple additive fuzzy achievement function. Mathematical formulation of the weighted additive method is as follows.

$$\left.\begin{array}{ll} \text{maximize} & \sum_l w_l \mu_{zl}(x) \\ \text{subject to} & \mu_{zl}(x) \in [0,1], \quad \forall l \\ & x \geq 0 \end{array}\right\} \tag{31}$$

where w_l represents the weight of ith objective. Selection of weights are subjected choice of decision makers and some good techniques can be used to determine the weights such as FAHP and structural equation modelling.

Werners' fuzzy and operator method are widely used interactive method. The advantage of this method is that it is positively related to the compensation rate due to its strong monotonicity. Additionally, it is easy to handle and has generated reasonable consistent results in applications [54]. By adopting the Werner's' method following a single objective model can be formed.

$$\left.\begin{array}{ll} \text{maximize} & \gamma \zeta_0 + (1 - \gamma) \sum_l \zeta_1 \\ \text{subject to} & \mu_1(x) \geq \zeta_0 + \zeta_1, \quad \forall l \\ & \zeta_0, \zeta_1, \gamma \in [0,1] \end{array}\right\} \tag{32}$$

where, ζ_1 is the difference between satisfaction level of objectives their minimum satisfaction level ζ_0, That is, $\zeta_1 = \mu_1 - \zeta_0$. γ denotes the coefficient of compensation.

Step 5: Determine the solution method parameter

Determine the values of the relative importance of objectives (w_l) and coefficient of compensation (γ) to solve the mathematical model using both weighted additive approach and Werners' fuzzy and operator methods.

Step 6: Solve the model

In the last step, solve the model by using parameters of the mathematical model and solution methods. This process continues by varying the solution method parameters (i.e., γ and w_l) until decision makers are satisfied with the final solution. If decision makers want to modify the value of α, then restart the process from step 2.

5. An Illustration

The effectiveness of the proposed resilient supplier selection model and solution methodology is demonstrated in this section. The data relates to a realistic situation of a garment manufacturing sector as shown in Figure 1. The adopted situation can easily be extended to any other industry. Initially, the data are estimated as most likely values, these most likely values of the fuzzy parameter (fz^{mos}) are estimated using available information or hypothetically set based on realistic assumption. The pessimistic and optimistic values are estimated using $fz^{pes} = (1 - \eta_1)fz^{mos}$ and, where two random numbers η_1 and η_2 are assumed between 0.2 and 0.8 to estimate. Table 2 shows the data set of unit purchase cost, transportation cost, rejection rate, capacity, and resilience index score of potential suppliers. The unit purchase costs from each supplier are hypothetical set based on labour cost, land value, resource availability at supplier locations. Transportation cost and transit time (see Table 3) from each supplier to manufacturer are estimated from sea rates (https://www.searates.com/). The resilience index score of each supplier is estimated from FM, Global resilience index data-driven tool (https://www.fmglobal.com/) based on the location of suppliers. The distance (as the crow flies) between suppliers (see Table 4) are calculated using the Google maps (https://www.google.com/maps). The demand for raw materials is assumed as most likely 8000 units, minimum acceptable order quantity is assumed as 1000 units, and maximum three suppliers can be selected for required material.

Figure 1. Supply network of the problem under consideration.

Table 2. Model input data.

Potential Supplier Location	Purchase Cost of Material ($/unit)	Transportation Cost of Material ($/unit)	Percentage of the Rejected Material	Capacity of Suppliers (1000 units)	Resilience Score
Korea	(6,8,10)	(0.08,0.13,0.21)	(0.01,0.01,0.02)	(3.08,5,8.08)	42.1
China	(1,2,4)	(0.11,0.19,0.30)	(0.04,0.06,0.10)	(3.7,6,9.7)	45.3
Thailand	(4,6,8)	(0.05,0.09,0.14)	(0.02,0.03,0.05)	(3.08,5,8.08)	39
Bangladesh	(1,3,5)	(0.11,0.17,0.28)	(0.02,0.03,0.05)	(3.7,6,9.7)	29
India-(Calcutta)	(3,5,7)	(0.10,0.16,0.26)	(0.01,0.02,0.03)	(2.46,4,6.46)	27.1
India-(Hyderabad)	(3,4,5)	(0.01,0.02,0.03)	(0.01,0.02,0.03)	(4.62,7.5,12.12)	27.1
Pakistan	(3,5,7)	(0.06,0.10,0.16)	(0.01,0.02,0.03)	(3.08,5,8.08)	22.2
Turkey	(8,10,12)	(0.08,0.12,0.20)	(0.01,0.01,0.02)	(3.08,5,8.08)	38.4
Brazil	(6,8,10)	(0.09,0.14,0.22)	(0.02,0.04,0.06)	(3.7,6,9.7)	47.8
Mexico	(7,9,11)	(0.11,0.17,0.28)	(0.02,0.03,0.05)	(4.31,7,11.31)	44.8

Table 3. Transit time from suppliers.

Supplier Location	Transit Time (days)
Korea	(7.39,12,19.39)
China	(9.24,15,24.24)
Thailand	(1.23,2,3.23)
Bangladesh	(1.85,3,4.85)
India-(Calcutta)	(1.23,2,3.23)
India-(Hyderabad)	(0.62,1,1.62)
Pakistan	(3.08,5,8.08)
Turkey	(8.01,13,21.01)
Brazil	(14.78,24,38.78)
Mexico	(19.71,32,51.71)

Table 4. The distance between suppliers (kilometers).

Suppliers' location.	Korea	China	Thailand	Bangladesh	India (Calcutta)	India (Hyderabad)	Pakistan	Turkey	Brazil	Mexico
Korea	N/A	1028.16	3723.83	3783.61	4037.39	5212.7	5768.78	7641.12	16,739.92	12,029.52
China	1028.16	N/A	2743.18	3059.41	3304.68	4478.2	5273.64	7587.05	17,589.43	13,051.91
Thailand	3723.83	2743.18	N/A	1526.77	1616.16	2388.26	3709.84	6926.22	17,348.59	15,721.92
Bangladesh	3783.61	3059.41	1526.77	N/A	250.06	1432.18	2374.24	5408.55	16,067.89	15,092.77
India (Calcutta)	4037.39	3304.68	1616.16	250.06	N/A	1180.87	2186.91	5327.76	15,902.02	15,299.76
India (Hyderabad)	5212.7	4478.2	2388.26	1432.18	1180.87	N/A	1461.88	4788.23	14,979.22	15,878.77
Pakistan	5768.78	5273.64	3709.84	2374.24	2186.91	1461.88	N/A	3334.57	13,710.04	14,838.79
Turkey	7641.12	7587.05	6926.22	5408.55	5327.76	4788.23	3334.57	N/A	10,723.38	11,983.21
Brazil	16,739.92	17,589.43	17,348.59	16,067.89	15,902.02	14,979.22	13,710.04	10,723.38	N/A	5692.39
Mexico	12,029.52	13,051.91	15,721.92	15,092.77	15,299.76	15878.77	14,838.79	11,983.21	5692.39	N/A

6. Model Solution and Result Analysis

The proposed fuzzy multi-objective model is solved in Lingo 14.0. Lingo optimization software has been widely used in supply chain optimization problems [55,56]. According to the steps of the proposed methodology, the payoff values are estimated by solving the single objective model. Table 5 shows payoff values estimated from the single objective model. The aspiration level of cost, rejection, transit time, supply density, and resilience score are estimated as \$22,514.93, 0.025, 3.17 days, 0.286, and 47.89 respectively. Once the payoff values are estimated, the fuzzy membership functions of objectives are estimated as shown below.

$$\mu_{cost}(q_s) = \begin{cases} 1, f_{cost}(q_s) \leq 22514.93 \\ \frac{75796.12 - f_{cost}(q_s)}{75796.12 - 22514.93}, 22514.93 < f_{cost}(q_s) \leq 75796.12 \\ 0, f_{cost}(q_s) \geq 75796.12 \end{cases}$$

$$\mu_{rej}(q_s) = \begin{cases} 1, f_{rej}(q_s) \leq 0.025 \\ \frac{0.10 - f_{rej}(q_s)}{0.10 - 0.025}, 0.025 \leq f_{rej}(q_s) \leq 0.10 \\ 0, f_{rej}(q_s) \geq 0.10 \end{cases}$$

$$\mu_{time}(q_s) = \begin{cases} 1, f_{time}(q_s) \leq 3.17 \\ \frac{61.11 - f_{time}(q_s)}{61.11 - 3.17}, 3.17 \leq f_{time}(q_s) \leq 61.11 \\ 0, f_{time}(q_s) \geq 61.11 \end{cases}$$

$$\mu_{den}(q_s) = \begin{cases} 1, f_{den}(q_s) \geq 0.28 \\ \frac{f_{den}(q_s) - 0.017}{0.28 - 0.017}, 0.017 \leq f_{den}(q_s) \leq 0.28 \\ 0, f_{den}(q_s) \leq 0.017 \end{cases}$$

$$\mu_{res}(q_s) = \begin{cases} 1, f_{res}(q_s) \geq 47.89 \\ \frac{f_{res}(q_s) - 29.79}{47.89 - 29.79}, 29.79 \leq f_{res}(q_s) \leq 47.89 \\ 0, f_{res}(q_s) \leq 29.79 \end{cases}$$

Table 5. Payoff values.

Objective	Cost (\$)	Rejection (%)	Transit Time (days)	Supply Density	Resilience Score
Minimize Cost	22,514.93	0.097	19.04	0.022	38.2
Minimize Rejection	75,796.12	0.025	26.45	0.056	39.22
Minimize Transit time	38,143.47	0.052	3.17	0.017	29.79
Maximize Supply density	70,159.31	0.1	61.11	0.286	43.41
Maximize Resilience score	50,560.14	0.1	41	0.12	47.89

The proposed mathematical model is solved using both methods, that is, a weighted additive approach and Werners' 'fuzzy and' operator methods. A solution of illustrated case example using both methods is shown in Table 6. The result shows that both methods produce a comprehensive optimal solution. However, the weighted additive approach considers the importance of objectives based on the weight given to each objective. For example, when more importance is given to resilience score (i.e., $w_5 = 0.3$), it results in 78.6% achievement of aspiration level. On the other hand, when all objectives and equal importance (Werners' method) than resilience score is given, the objective results in 34% achievement of aspiration level. Comparative analysis of both methods is shown in Figure 2.

Table 6. Solution of a case example.

Model Objectives	Weighted Additive Approach [a]		Werners' 'Fuzzy and' Operator [b]	
	$\mu_l(x)$	$f_l(x)$	$\mu_l(x)$	$f_l(x)$
Cost	0.346	78,238.03	0.195	65,374.99
Rejection	0.051	0.12	0.366	0.072
Transit time	0.457	42.0	0.383	38.89
Supply density	0.975	0.27	0.972	0.272
Resilience score	0.786	75.5	0.803	44.33

$\alpha = 0.9$, [a] $w_1 = 0.2$, $w_2 = 0.1$, $w_3 = 0.2$, $w_4 = 0.2$, $w_5 = 0.3$, [b] $\gamma = 0.7$.

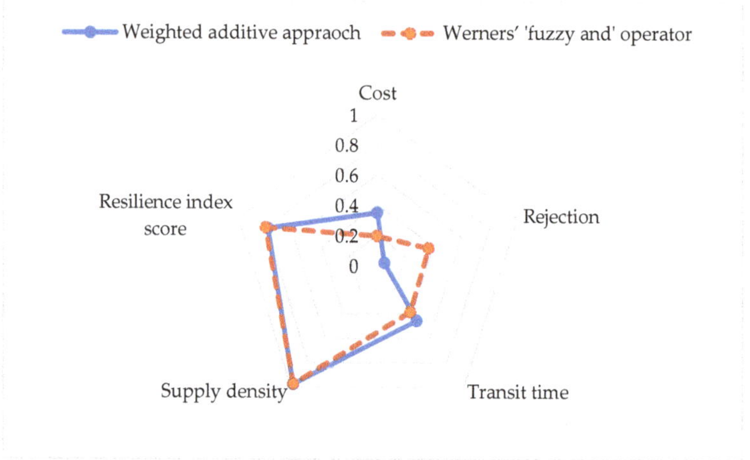

Figure 2. Achievement levels of objectives.

7. Sensitivity Analysis

Sensitivity analysis of both methods is carried out by varying methodology parameters (i.e., α, γ and w_l) as discussed in the final step of the proposed solution methodology. Tables 7 and 8 show the analysis result of the weighted additive approach and Werners' 'fuzzy and' operator method, respectively. Two cases are solved with the weighted additive approach: (1) more importance is given to resilience objective and (2) more importance given to cost objective.

The result shows that there is a tradeoff in economic objective and resilience objective. Hence, it can be said that the increase is supply chain resilience will tend to increase the total cost of the supply network. This is possible because economic supply networks are more dense networks in order to minimize transportation cost, hence any HILP disruption event such as earthquake or tsunami may disrupt more than one supplier. Therefore, it is suggested that companies should avoid denser supply networks to minimize risks from high impact low probability disruption event. Analysis result shows that 92% achievement of cost aspiration level will bring achievement of about only 6.1% supply density and 49.8% resilience index score aspiration levels respectively. On the other hand, 18.4% achievement of cost aspiration level will bring achievement of about 97% supply density and 88.1% resilience index score respectively.

Table 7. Sensitivity analysis of the weighted additive approach.

Alpha	Objective Weights	μ_{cost}	μ_{rej}	μ_{time}	μ_{den}	μ_{res}	f_{cost}	f_{rej}	f_{time}	f_{den}	f_{res}
0.9	$w_1 = 0.2, w_2 = 0.1, w_3 = 0.2, w_4 = 0.2, w_5 = 0.3$	0.184	0.200	0.365	0.97	0.881	65,941.28	0.085	39.9	0.27	45.7
	$w_1 = 0.3, w_2 = 0.2, w_3 = 0.2, w_4 = 0.1, w_5 = 0.2$	0.920	0.200	0.762	0.061	0.498	26,730.90	0.085	16.9	0.03	38.8
0.6	$w_1 = 0.2, w_2 = 0.1, w_3 = 0.2, w_4 = 0.2, w_5 = 0.3$	0.111	0.100	0.566	0.910	0.935	66,577.34	0.092	28.3	0.25	47.4
	$w_1 = 0.3, w_2 = 0.2, w_3 = 0.2, w_4 = 0.1, w_5 = 0.2$	0.851	0.200	0.762	0.061	0.634	27,727.65	0.085	16.9	0.03	42.6

Table 8. Sensitivity analysis of Werners' 'fuzzy and' operator.

α	γ	μ_{cost}	μ_{rej}	μ_{time}	μ_{den}	μ_{res}	f_{cost}	f_{rej}	f_{time}	f_{den}	f_{res}
	0.0–0.8	0.107	0.366	0.383	0.972	0.793	66,774.85	02.072	38.8	0.27	46.4
0.6	0.9	0.366	0.366	0.383	0.972	0.451	53,167.88	0.072	38.89	0.27	38.3
	1.0	0.407	0.566	0.543	0.407	0.406	51,054.51	0.057	29.62	0.12	37.3
	0.0–0.8	0.195	0.366	0.383	0.972	0.803	65,374.99	0.072	38.89	0.27	44.33
0.9	0.9	0.366	0.366	0.383	0.972	0.541	56,259.06	0.072	38.89	0.27	39.59
	1.0	0.392	0.566	0.543	0.407	0.392	54,899.34	0.057	29.6	0.12	36.88

Assume that decision makers required balance results and choose the best outcome at $\alpha = 0.9$, and $\gamma = 1.0$ (highlighted row in Table 8). Hence, the final decision of the illustrated example is shown in Figure 3. The optimal quantity of material purchase from suppliers are q (Korea) = 3751 units, q (Bangladesh) = 3655 units, and q (Turkey) = 1258 units.

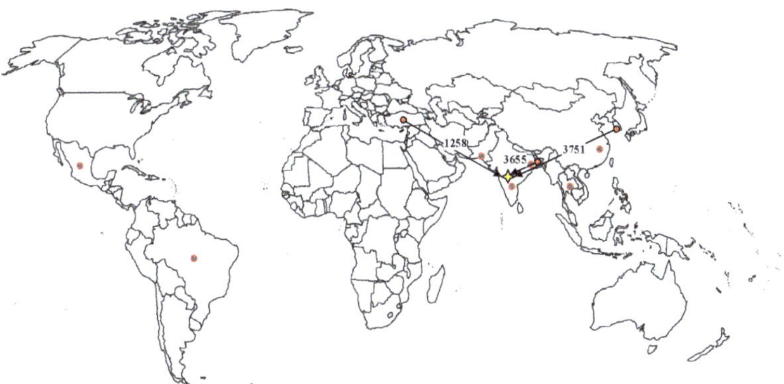

Figure 3. Final decision based on decision makers preferences.

8. Conclusion and Future Suggestions

This paper introduced a novel possibilistic fuzzy environment for resilient supplier selection and order allocation considering new resilience criteria (supply density, transit time and resilience index score) for the selection of suppliers. For this purpose, an interactive fuzzy multi-objective programming approach is introduced to reduce uncertainties inherent in the supplier selection decision.

A six-step solution methodology is designed to solve the proposed uncertain multi-objective model and a numerical case example is provided to show the applicability of the proposed model in a real situation.

This study significantly helps the practitioners who are trying to consider resilience in their supply network. The research results show the importance of supplier's location in order to minimize disruption risks. Furthermore, the proposed model will help the managers to effectively evaluate and select a suitable set of suppliers while considering cost and resilience simultaneously. Moreover, the proposed possibilistic fuzzy based solution methodology will be helpful for practitioners and academicians to tackle cognitive and stochastic uncertainties related to supplier evaluation and selection problem. This study also helps the academicians to analyze the importance of resilient supply networks under disruption risks. Additionally, the proposed multi-objective possibilistic fuzzy-based approach can be useful in another area of supply chain optimization.

Although this research gained important insights from the implementation of proposed resilience criteria and solution methodology, there are some limitations which may be considered in future research. This paper assumed triangular fuzzy numbers for model parameters, it will be interesting to compare the research results with other fuzzy numbers such as trapezoidal in future studies. Furthermore, this research only considered a disruption of the location where suppliers are located. However, it will be important to consider the disruption of transport links between suppliers and buyers. Another interesting possible direction for future research is to extend the proposed model to multi-commodity and multi-period planning horizon. Lastly, the analysis result indicates that economic networks are denser network and they may be vulnerable to disruption risks, it will be valuable to further investigate the relationship between the economic network (denser network) and disruption risks.

Author Contributions: Conceptualization, S.I.M.; Methodology, M.S.M.; Software, M.B.R.; Supervision, S.M.Q.; Validation, S.M.Q.; Writing—original draft, S.I.M.; Writing—review & editing, M.W.I.

Acknowledgments: This research was supported by the Higher Education Commission of Pakistan through the Startup Research Grant Program (SRPG#1299 and SRGP#1335).

Conflicts of Interest: The authors declare no conflict of interest.

References

1. Schoenherr, T.; Modi, S.B.; Benton, W.C.; Carter, C.R.; Choi, T.Y.; Larson, P.D.; Leenders, M.R.; Mabert, V.A.; Narasimhan, R.; Wagner, S.M. Research Opportunities in Purchasing and Supply Management. *Int. J. Prod. Res.* **2012**, *50*, 4556–4579. [CrossRef]

2. Kim, J.S.; Jeon, E.; Noh, J.; Park, J.H. A Model and an Algorithm for a Large-Scale Sustainable Supplier Selection and Order Allocation Problem. *Mathematics* **2018**, *6*, 325. [CrossRef]

3. Mari, I.S.; Lee, Y.H.; Memon, M.S.; Park, Y.S.; Kim, M. Adaptivity of Complex Network Topologies for Designing Resilient Supply Chain Networks. *Int. J. Ind. Eng.* **2015**, *22*, 102–116.

4. Tang, C.; Tomlin, B. The Power of Flexibility for Mitigating Supply Chain Risks. *Int. J. Prod. Econ.* **2008**, *116*, 12–27. [CrossRef]

5. Wang, C.-N.; Nguyen, V.T.; Thai, H.T.N.; Tran, N.N.; Tran, T.L.A. Sustainable Supplier Selection Process in Edible Oil Production by a Hybrid Fuzzy Analytical Hierarchy Process and Green Data Envelopment Analysis for the Smes Food Processing Industry. *Mathematics* **2018**, *6*, 302. [CrossRef]

6. Wang, X.; Herty, M.; Zhao, L. Contingent Rerouting for Enhancing Supply Chain Resilience from Supplier Behavior Perspective. *Int. Trans. Oper. Res.* **2016**, *23*, 775–796. [CrossRef]

7. Chopra, S.; Sodhi, M. Reducing the Risk of Supply Chain Disruptions. *Mit Sloan Manag. Rev.* **2014**, *55*, 72–80.

8. Barroso, A.P.; Machado, V.H.; Barros, A.R.; Machado, V.C. Toward a Resilient Supply Chain with Supply Disturbances. Presented at the 2010 IEEE International Conference on Industrial Engineering and Engineering Management (IEEM), Macao, China, 7–10 December 2010.

9. Torabi, S.A.; Baghersad, M.; Mansouri, S.A. Resilient Supplier Selection and Order Allocation under Operational and Disruption Risks. *Transp. Res. Part E: Logist. Transp. Rev.* **2015**, *79*, 22–48. [CrossRef]

10. Carvalho, H.; Azevedo, S.G.; Cruz-Machado, V. Agile and Resilient Approaches to Supply Chain Management: Influence on Performance and Competitiveness. *Logist. Res.* **2012**, *4*, 49–62. [CrossRef]
11. Ribeiro, P.J.; Barbosa-Povoa, A. Supply Chain Resilience: Definitions and Quantitative Modelling Approaches–a Literature Review. *Comput. Ind. Eng.* **2018**, *115*, 109–122. [CrossRef]
12. Parkouhi, V.S.; Ghadikolaei, A.S. A Resilience Approach for Supplier Selection: Using Fuzzy Analytic Network Process and Grey Vikor Techniques. *J. Clean. Prod.* **2017**, *161*, 431–451. [CrossRef]
13. Hosseini, S.; Al Khaled, A. A Hybrid Ensemble and Ahp Approach for Resilient Supplier Selection. *J. Intell. Manuf.* **2016**, 1–22. [CrossRef]
14. Sahu, A.K.; Datta, S.; Mahapatra, S.S. Evaluation and Selection of Resilient Suppliers in Fuzzy Environment: Exploration of Fuzzy-Vikor. *Benchmarking Int. J.* **2016**, *23*, 651–673. [CrossRef]
15. López, C.; Ishizaka, A. A Hybrid Fcm-Ahp Approach to Predict Impacts of Offshore Outsourcing Location Decisions on Supply Chain Resilience. *J. Bus. Res.* **2017**. [CrossRef]
16. Sabouhi, F.; Pishvaee, M.S.; Jabalameli, M.S. Resilient Supply Chain Design under Operational and Disruption Risks Considering Quantity Discount: A Case Study of Pharmaceutical Supply Chain. *Comput. Ind. Eng.* **2018**, *126*, 657–672. [CrossRef]
17. Jabbarzadeh, A.; Fahimnia, B.; Sabouhi, F. Resilient and Sustainable Supply Chain Design: Sustainability Analysis under Disruption Risks. *Int. J. Prod. Res.* **2018**, 1–24. [CrossRef]
18. Wang, T.-K.; Zhang, Q.; Chong, H.; Wang, X. Integrated Supplier Selection Framework in a Resilient Construction Supply Chain: An Approach Via Analytic Hierarchy Process (Ahp) and Grey Relational Analysis (Gra). *Sustainability* **2017**, *9*, 289. [CrossRef]
19. Hosseini, S.; Barker, K. A Bayesian Network Model for Resilience-Based Supplier Selection. *Int. J. Prod. Econ.* **2016**, *180*, 68–87. [CrossRef]
20. Haldar, A.; Ray, A.; Banerjee, D.; Ghosh, S. A Hybrid Mcdm Model for Resilient Supplier Selection. *Int. J. Manag. Sci. Eng. Manag.* **2012**, *7*, 284–292. [CrossRef]
21. Haldar, A.; Ray, A.; Banerjee, D.; Ghosh, S. Resilient Supplier Selection under a Fuzzy Environment. *Int. J. Manag. Sci. Eng. Manag.* **2014**, *9*, 147–156. [CrossRef]
22. Sawik, T. Selection of Resilient Supply Portfolio under Disruption Risks. *Omega* **2013**, *41*, 259–269. [CrossRef]
23. Sawik, T. Joint Supplier Selection and Scheduling of Customer Orders under Disruption Risks: Single vs. Dual Sourcing. *Omega* **2014**, *43*, 83–95. [CrossRef]
24. Yilmaz-Börekçi, D.; Say, A.İ.; Rofcanin, Y. Measuring Supplier Resilience in Supply Networks. *J. Chang. Manag.* **2015**, *15*, 64–82. [CrossRef]
25. Rajesh, R.; Ravi, V. Supplier Selection in Resilient Supply Chains: A Grey Relational Analysis Approach. *J. Clean. Prod.* **2015**, *86*, 343–359. [CrossRef]
26. Tiwari, R.N.; Dharmar, S.; Rao, J.R. Fuzzy Goal Programming—An Additive Model. *Fuzzy Sets Syst.* **1987**, *24*, 27–34. [CrossRef]
27. Werners, B.M. Aggregation Models in Mathematical Programming. In *Mathematical Models for Decision Support*; Springer: Berlin/Heidelberg, Germany, 1988; pp. 295–305.
28. Memon, M.S.; Lee, Y.H.; Mari, S.I. Group Multi-Criteria Supplier Selection Using Combined Grey Systems Theory and Uncertainty Theory. *Expert Syst. Appl.* **2015**, *42*, 7951–7959. [CrossRef]
29. Memon, M.S.; Mari, S.I.; Shaikh, F.; Shaikh, S.A. A Grey-Fuzzy Multiobjective Model for Supplier Selection and Production-Distribution Planning Considering Consumer Safety. *Math. Probl Eng.* **2018**, *2018*, 5259876. [CrossRef]
30. Holling, C.S. Resilience and Stability of Ecological Systems. *Annu. Rev. Ecol. Syst.* **1973**, *4*, 1–23. [CrossRef]
31. Hamel, G.; Valikangas, L. The Quest for Resilience. *Revista Icade Revista de las Facultades de Derecho y Ciencias Económicas y Empresariales* **2004**, *62*, 355–358.
32. Sheffi, Y. *The Resilient Enterprise: Overcoming Vulnerability for Competitive Advantage*; MIT Press: Cambridge, MA, USA, 2005.
33. Sutcliffe, M.K.; Vogus, T.J. Organizing for Resilience. *Posit. Organ. Scholarsh. Found. New Discip.* **2003**, *94*, 110.
34. Woods, D.D. Four Concepts for Resilience and the Implications for the Future of Resilience Engineering. *Reliab. Eng. Syst. Saf.* **2015**, *141*, 5–9. [CrossRef]
35. Zhalechian, M.; Torabi, S.A.; Mohammadi, M. Hub-and-Spoke Network Design under Operational and Disruption Risks. *Transp. Res. Part E Logist. Transp. Rev.* **2018**, *109*, 20–43. [CrossRef]

36. Falasca, M.; Zobel, C.W.; Cook, D. A Decision Support Framework to Assess Supply Chain Resilience. Presented at the 5th International ISCRAM Conference, Washington, DC, USA, 4–7 May 2008.
37. Mari, S.I.; Lee, Y.H.; Memon, M.S. Complex Network Theory-Based Approach for Designing Resilient Supply Chain Networks. *Int. J. Logist. Syst. Manag.* **2015**, *21*, 365–384. [CrossRef]
38. Priya Datta, P.; Christopher, M.; Allen, P. Agent-Based Modelling of Complex Production/Distribution Systems to Improve Resilience. *Int. J. Logist. Res. Appl.* **2007**, *10*, 187–203. [CrossRef]
39. Azevedo, S.G.; Govindan, K.; Carvalho, H.; Cruz-Machado, V. Gresilient Index to Assess the Greenness and Resilience of the Automotive Supply Chain. *Discuss. Pap. Bus. Econ.* **2011**, *8*. [CrossRef]
40. Miller-Hooks, E.; Zhang, X.; Faturechi, R. Measuring and Maximizing Resilience of Freight Transportation Networks. *Comput. Oper. Res.* **2012**, *39*, 1633–1643. [CrossRef]
41. Tang, C.S. Robust Strategies for Mitigating Supply Chain Disruptions. *Int. J. Logist. Res. Appl.* **2006**, *9*, 33–45. [CrossRef]
42. Sheffi, Y.; Rice, J.B., Jr. A Supply Chain View of the Resilient Enterprise. *Mit Sloan Manag. Rev.* **2005**, *47*, 41.
43. Mari, S.I.; Lee, Y.H.; Memon, M.S. Sustainable and Resilient Supply Chain Network Design under Disruption Risks. *Sustainability* **2014**, *6*, 6666–6686. [CrossRef]
44. Pramanik, D.; Haldar, A.; Mondal, S.C.; Naskar, S.K.; Ray, A. Resilient Supplier Selection Using Ahp-Topsis-Qfd under a Fuzzy Environment. *Int. J. Manag. Sci. Eng. Manag.* **2017**, *12*, 45–54. [CrossRef]
45. Sen, D.K.; Datta, S.; Mahapatra, S.S. Dominance Based Fuzzy Decision Support Framework for G-Resilient (Ecosilient) Supplier Selection: An Empirical Modelling. *Int. J. Sustain. Eng.* **2017**, *10*, 338–357. [CrossRef]
46. Parkouhi, S.V.; Ghadikolaei, A.S.; Lajimi, H.F. Resilient Supplier Selection and Segmentation in Grey Environment. *J. Clean. Prod.* **2019**, *207*, 1123–1137. [CrossRef]
47. Papadakis, I.S. Financial Performance of Supply Chains after Disruptions: An Event Study. *Supply Chain Manag. Int. J.* **2006**, *11*, 25–33. [CrossRef]
48. FMGlobal. The 2015 FM Global Resilience Index Annual Report. Available online: https://www.fmglobal.com/assets/pdf/Resilience_Methodology.pdf (accessed on 3 July 2018).
49. Mari, S.I.; Lee, Y.H.; Memon, M.S. Sustainable and Resilient Garment Supply Chain Network Design with Fuzzy Multi-Objectives under Uncertainty. *Sustainability* **2016**, *8*, 1038. [CrossRef]
50. Min, H.; Zhou, G. Supply Chain Modeling: Past, Present and Future. *Comput. Ind. Eng.* **2002**, *43*, 231–249. [CrossRef]
51. Jiménez, M.; Arenas, M.; Bilbao, A.; Rodrı, M.V. Linear Programming with Fuzzy Parameters: An Interactive Method Resolution. *Eur. J. Oper. Res.* **2007**, *177*, 1599–1609. [CrossRef]
52. Tayyab, M.; Sarkar, B.; Yahya, B. Imperfect Multi-Stage Lean Manufacturing System with Rework under Fuzzy Demand. *Mathematics* **2019**, *7*, 13. [CrossRef]
53. Li, X.; Chien, C.F.; Yang, L.; Gao, Z. The Train Fueling Cost Minimization Problem with Fuzzy Fuel Prices. *Flex. Serv. Manuf. J.* **2014**, *26*, 249–267. [CrossRef]
54. Selim, H.; Ozkarahan, I. A Supply Chain Distribution Network Design Model: An Interactive Fuzzy Goal Programming-Based Solution Approach. *Int. J. Adv. Manuf. Technol.* **2008**, *36*, 401–418. [CrossRef]
55. Babar Ramzan, M.; Qureshi, S.M.; Mari, S.I.; Memon, M.S.; Mittal, M.; Imran, M.; Iqbal, M.W. Effect of Time-Varying Factors on Optimal Combination of Quality Inspectors for Offline Inspection Station. *Mathematics* **2019**, *7*, 51. [CrossRef]
56. Sarkar, B.; Tayyab, M.; Choi, S.K. Product Channeling in an O2O Supply Chain Management as Power Transmission in Electric Power Distribution Systems. *Mathematics* **2019**, *7*, 4. [CrossRef]

© 2019 by the authors. Licensee MDPI, Basel, Switzerland. This article is an open access article distributed under the terms and conditions of the Creative Commons Attribution (CC BY) license (http://creativecommons.org/licenses/by/4.0/).

Article

Effects of Preservation Technology Investment on Waste Generation in a Two-Echelon Supply Chain Model

Mehran Ullah, Biswajit Sarkar * and Iqra Asghar

Department of Industrial & Management Engineering, Hanyang University, Ansan Gyeonggi-do 155 88, Korea; mehrandirvi@gmail.com or mehran@hanyang.ac.kr (M.U.); iqra_ntu60@yahoo.com or iqraasghar27@hanyang.ac.kr (I.A.)
* Correspondence: bsbiswajitsarkar@gmail.com; Tel.: +82-107-498-1981

Received: 27 December 2018; Accepted: 13 February 2019; Published: 17 February 2019

Abstract: This study develops an integrated production-inventory model for a two-echelon supply chain network with controllable probabilistic deterioration. The investment in preservation technology is considered a decision variable to control the deteriorated quantity of an integrated system. The objective of the study is to optimize preservation investment, the number of shipments and shipment quantity, so that the total cost per unit of time of the supply chain is minimized. The study proposes a solution method, and the results show that investment in preservation technology reduces the total supply chain cost by 13%. Additionally, preservation increases the lot size, thus increasing the production cycle length, which reduces the ordering cost of the system. Furthermore, this study shows that preservation leads to a reduction of solid waste from deteriorated products. Total deteriorated products reduced to 8 units from 235 units, hence, preservation generates positive environmental benefits along with economic impacts. The robustness of the proposed model is illustrated with a numerical example, sensitivity analysis, and graphical representations. Moreover, comparative study and managerial insights are given to extract significant insights from the model.

Keywords: preservation technology investment; refrigerated trucks; waste reduction; controllable probabilistic deterioration rate; supply chain management

---

1. Introduction

The phenomena of deterioration are referred to as spoilage, damage, vaporization or other changes in product quality or productivity because of environmental changes during storage. Products like semiconductor chips, battery, volatile liquids, seafood, and medicinal items like blood, deteriorate with time and lose their usefulness over their lifetime. In the existing literature, the deterioration is considered either a constant or random aspect of products. In reality, deterioration of products is a controllable factor which can be modified through investment in adequate preservation technology. Preservative technology contributes to the increased lifespan of materials by maximizing how efficiently they are processed. Similarly, preservation aids in lowering the amount of contamination generated in a commodity-to-product conversion or storage setup, and it also reduces the required number of material supplies too. For instance, the decay of the cement matrix under certain environmental factors is decreased using pozzolanic additives, which also reduces the calcium-leeching effect of cement substance in the environment [1]. Carbon emissions also reduce up to 12%, and energy efficiency is enhanced by utilizing 19% less natural reserves with this additive.

In multi-echelon supply chain structures, the impacts of deterioration are increased compared to the single player structure. Because the same product faces deterioration multiple times at every stage. A supply chain infrastructure usually comprises processing units and retail units. A product goes through several processes in a supply chain infrastructure; from supply to the conversion of raw materials, and, to the storage of finished products. The capacity of each unit to adjust production quantity and assure the adequately long shelf life is another important aspect of supply chain coordination. For instance, fruit or chemical processing units invest in decay preventive packaging or additives for their products. Then a retail unit must be equipped with appropriate preservation equipment (required environmental and temperature conditions) for reception and storage of such products. However, in the existing literature, simultaneous preservation efforts in multi-echelon supply chain management have been overlooked. Preservation investment is considered to be the personal preference of the individual players of the supply chain. However, in reality, multiple members of the supply chain can invest in their setups simultaneously to enhance the credibility and efficiency of preservation technology. There is a need to encourage production firms and retailers to organize, diversify and upgrade their commodities-inventory and business setup. Investments in infrastructure, production techniques, storage, packaging, and transportation are also required to reduce product deterioration and enhance supply chain system efficiency.

The deteriorated items waste is mainly associated with managerial and technical limitations in firms. Moreover, it is also related to a lack of coordination between the different players of a supply chain. As the supply chain is a dynamic system, the products are needed to be capable of readapting several processing environments as they face frequent changes from processing units to retail units. The lack of coordination between the supply chain increases the impact of deterioration from one stage to another. Consequently, waste due to the deterioration of products is generated in every stage of the supply chain. Inescapably, this also means that a large number of resources are drained away in the supply chain with no purpose fulfilled. Also, waste disposal costs of deteriorated items have not been considered in existing deteriorated inventory studies, where, practically supply chain members bear extra cost to deposit waste safely. The coordination of certain aspects, like costs and quality of products in a supply chain infrastructure, is a fundamental parameter in keeping product cost in-control and making a supply chain profitable. To overcome these limitations, this study highlights the waste conception due to deterioration in the supply chain system. Further, it develops an assessment study of the waste quantity management, and, provides a waste control strategy in terms of preservation investment.

The rest of the paper is organized as follows: A comprehensive literature review is provided in Section 2. Problem definition, notation and assumptions for the proposed model are presented in Section 3. The mathematical model formulation is addressed in Section 4. Numerical examples, the comparison between models with and without preservation investment and sensitivity analysis are provided in Section 5. Finally, the conclusion and future research are given in Section 6.

2. Literature Review

In this section, we thoroughly overview the related research areas of this paper. Three research areas are examined in this study, namely: controllable probabilistic deterioration rate; waste generation from deteriorated items; and preservation investment.

2.1. Controllable Probablistic Deterioration Rate

The consideration of deterioration has got significant attention in inventory-based research with constant deterioration rates, supply chain coordination and price discounts. For the first time, Ghare [2] analyzed the effect of the constant rate of decay on a classical EOQ model under constant demand and no shortages. However, practically, the life expectancy and failure rate of products are better observed with the variable deterioration rate. Covert and Philip [3] extended the case of product deterioration rate to a variable phenomenon. They consider a two-parameter Weibull distribution. Sarkar et al. [4] examines the time-varying deterioration rate for fixed lifetime products. They have also considered full trade and partial trade credit policy for deteriorated inventory. Philip [5], Mukhopadhyay et al. [6], and Hung [7] examined optimal inventory policy with the product loss because of deterioration, treating the deterioration rate as a time-varying function. Sarkar et al. [8] proposed an optimal cycle length model with time-varying deterioration and a partial backlogging rate. Shah et al. [9] discussed a time-varying deteriorated inventory system and introduced temporary price discounts for decayed products. Tadikamalla [10] studied the Gamma distribution based on the distribution pattern for the deterioration rate. Several authors examine the decay of products in terms of entropy [11,12]. The work relates entropy with the product's geometrical aspect and their intended physical performance. Sarkar [13] discuss a time-dependent demand and deterioration rate model. In this study, a trade-credit is offered to retailers by suppliers to buy extra deteriorated items with different discount rates. Ahmad et al. [14] reviewed the product deterioration and damage during transportation. Decaying inventory models for agricultural products were studied by Ning et al. [15], and the demand rate of products is considered as a function of product price and freshness. They examined a decrease in profit with increased cycle time for decaying products. Guariglia [16] examined the geometric configuration of products. The study discusses the performance of products based on their entropy characteristics.

Sarkar [17] examined the deteriorated inventory model with random rates of decay. The study examines different distributions for deterioration rates such as uniform, triangular and beta distributions. Iqbal and Sarkar [18] proposes a production model for an integrated imperfect and deteriorating production system. The study associates the total system cost with minimization regarding the production rate and cycle time. A deteriorating inventory model for the two-warehouse system was examined by Sett et al. [19]. The proposed model considers time-dependent deterioration for products with a quadratic demand function. An optimal replenishment policy model for deteriorating was developed by Sett et al. [20]. An economic manufacturing quantity model with deteriorating items and system unreliability under inflation and time value of money was studied by Sarkar and Sarkar [21]. They developed a reliable production system with investment in technology development to improve the reliability parameter. The study considers ramp-type demand for perishable products with a fixed life and deterioration as a linear function of time. A detailed review on deteriorating inventory models was provided by Bakker et al. [22]. The existing literature considers the phenomena of deterioration as a constant or variable parameter, whereas it is more realistic to recognize it as a controllable factor with making additional capital investments. To address the practical aspects of decay in the supply chain, we consider the controllable probabilistic deterioration of products in an integrated-inventory problem.

2.2. Waste Generation from Deteriorated Items

Product management after its life cycle has got much attention in literature. The need for repairable/recycled strategies for the used product inventory problem was acknowledged back in 1960. The recovered/repaired products could account for up to 50% of the original investment in inventory [23]. However, product waste generation due to deterioration before its use has not got much attention from researchers. Solid waste generation and resource wastage is predestined with deteriorating products. Some product's losses begin on production units even before reaching retail like foods, fruits and liquids. Deterioration cause changes in the quality, availability and wholesomeness of products that prevent them from being consumed and they end up as discarded material on both the manufacturer's and retailer's end. Deteriorated item's waste are actually well produced goods intended for consumption which are subsequently degraded or contaminated. These discarded deteriorated products have an adverse impact on the economic, social and environmental level. A multi-stage production system under uncertainties and process imperfection was studied in Tayyab et al. [24]. From an environmental point of view, disposing of deteriorated products contributes to additional greenhouse gas (GHG) emissions. It also leads to increased resource depletion and material wastage [25]. Financial impacts of deterioration on a firm are because of extra costs related to resource wastage and lost sales. Whereas, system quality of the firm refers to improved productivity and less wastage generation, where both rely on reduced system deterioration and maximized resource utilization.

The causes of decay in production systems are mainly associated with financial, technical and managerial limitations. These limitations affect production, storage, packaging techniques, transportation, and marketing systems. There is a need to improve factors which influence system resources to improve system productivity. The drivers for waste generation related to production, deterioration and handling were discussed by Thyberg and Tonjes [26]. Tayyab et al. [27] examined the manufacturing process to optimize a sustainable lot size to achieve simultaneous economic and environmental viability. The study also provides insights into waste prevention policies for a sustainable supply chain with deteriorating inventories. The pull approach and recycling concept is also considered in the literature to avoid potential waste from decayed and damaged products. Decay in raw material inventory, work-in-process inventory, and finished goods with proper disposal was discussed by Tibben-Lembke [28]. The efficient management of assets is considered a key parameter in overcoming the negative impacts of unsolicited costs on firm finances. Deterioration creates a waste of products in every supply chain, and this waste is generally considered a replaceable but not reusable commodity for a firm. For example, once deteriorated, then oils, resins, spare-part lubricators, chemicals, food, cosmetics, medicine, and biological matters are identified as non-reusable products. Inevitably, this also indicates that a large number of supplies are wasted away in the supply chain with no purpose fulfilled. Additionally, the deterioration increases the required quantity of raw material by contaminating non-deteriorated raw material inventory. In reality, supply chain members bear extra costs to deposit waste created from product deterioration. Hence, we introduced the waste disposal costs of deteriorated items to make this study more practical.

2.3. Preservation Tecniques

Economically, the deterioration of products has a direct and negative impact on the finances and brand image of a firm. High deterioration rates are entitled with higher annual costs, shortages and lost sales. On this account, business organizations are interested in comprehending the causes of deterioration and developing approaches to preserve their produced goods and increase profit. There is economic and environmental consciousness that pressurizes manufacturers to extend the useable lives of their products and reduce waste to conserve natural resources. Therefore, there is a need to initiate such product recovery systems in manufacturing units as well as at retail ends that reduce deterioration and wastage of resources. Practically, the deterioration rate of goods can be controlled and reduced through procedural/environmental changes and advanced equipment acquisition. These procedural changes entitled as preservation investment are meant to prevent, delay, or eventually reduce the deterioration/decay of goods. Many business entities invest in advanced equipment to extend product expiration dates. Murr and Morris [29] showed that a low storage temperature leads to a decrease in decay and increases the storage life of products. Investment in vacuum technology leads to a reduction in the deterioration rate of food and medicinal items. SO_2 fumigation was used to develop a color retention technique for fruits by Zauberman et al. [30]. Yang et al. [31] and Ho et al. [32] states that improved storage conditions reduce the total relevant inventory costs for deteriorating products. Waste management in the uncertain environment was examined by Habib et al. [33].

Nye et al. [34] investigates the effect of production batch size and investments in setup cost reduction. Lee [35] developed an economic assessment model for investment strategies in an imperfect production system and considered a preventive maintenance strategy to decrease system deterioration. Lin and Hou [36] examined an inventory system with random yields. This study considers added capital investment in the system to reduce yield variability and setup cost. An investment in supplier's management cost was added to reduce lead time by Hsu and Wee [37]. This study develops a deteriorating inventory replenishment model with an expiration date and uncertain lead time. Li et al. [38] developed a return-on-investment maximization model; the study examined the effect of the capital investment on setup and quality operations under budget constraints. The potential impact of investments in setup cost reduction and quality improvement was examined by Affisco et al. [39]. The study suggests that decreasing deterioration is not a linear function of investment. Uçkun et al. [40] developed an optimal investments level model to maximize system profit by reducing inaccuracies. The quality of products with a variable production rate in the supply chain was studied by Sarkar et al. [41]. Lee [42] proposes a cost/benefit model to asses and predict the impact of quality investments on returned profit in a multi-level production assembly. A non-linear mathematical model was proposed by Iqbal and Sarkar [43] to give an optimum amount of preservatives to increase the lifetime of products, while Wong et al. [44] studied waste reduction in pharmaceutical manufacturing. The study suggests that the preservation-based improved life length of products is associated with the price of the products, as the coordination of certain aspects like costs and quality of products in a supply chain infrastructure is a fundamental parameter in keeping product cost in-control and making a supply chain profitable. Therefore, in this study, the investment in preservation technology is considered a decision variable to control the deteriorated quantity and per unit total cost of an integrated system.

2.4. Research Gap

In the literature, at the retailer ends, different preservation investments are considered to decrease deterioration. As in supermarkets and retail shops, investment is made on refrigeration equipment to reduce the deterioration of foods, flowers and medicine. Preservation investments at the manufacturer's ends have got little attention in the literature, especially in the case of supply chain management, as a manufacturer and retailer jointly can invest to increase the productive life of their goods during and after production. For instance, by adding preservatives during production at the manufacturer's end and investment in vacuumed packaging, temperature control, and icing packs for storing and transporting goods by the retailer. Therefore, it is essential to examine the impact of shared deterioration control policies from multiple players in a supply chain system. To overcome these limitations, this study highlights the waste conception due to probabilistic deterioration in the supply chain system. Further, it develops an assessment of the generated waste quantity and also provides a waste control strategy in terms of preservation investment. Our objective is to examine the trade-off between additional investment cost to control probabilistic deterioration and resulting total cost change in a two-echelon supply chain. We also studied the impact of investment on solid waste generation in the supply chain due to product deterioration. A mathematical model is developed to determine the optimal level of investment on preservation technology, replenishment lot size and the number of deliveries per production cycle in a two-echelon supply chain model with a controllable probabilistic deterioration rate.

3. Problem Definition, Notation and Assumptions

3.1. Problem Definition

In this paper, a two-echelon supply chain is considered with a manufacturer and single retailer. The manufacturer produces products which deteriorate with a probabilistic deterioration rate λ. To reduce the net deterioration quantity, the manufacturer invests $\$(\aleph)$ in preservation equipment or technology. The reduction in deteriorated quantity is a function $w(\aleph)$ of the investment. For instance, after an investment $\aleph$, the net deterioration rate λ changes to $(\lambda - w(\aleph))$. The reduction rate is expressed as a function of the preservation technology cost $\aleph$, such that $w(\aleph) = \lambda(1 - e^{-g\aleph})$, where g is the shape parameter and it shows the effectiveness of the investment on the reduction of deterioration. The deterioration function $w(\aleph)$ is monotonically increasing in $\aleph$ and is at least twice differentiable. The relationship between preservation investment and reduction in deterioration rate is graphically illustrated in Figure 1. With a zero investment in preservation technology, the net deterioration rate remains the original deterioration rate λ. Whereas, with an increase in investment, the net deterioration rate reduces by a factor of $w(\aleph)$.

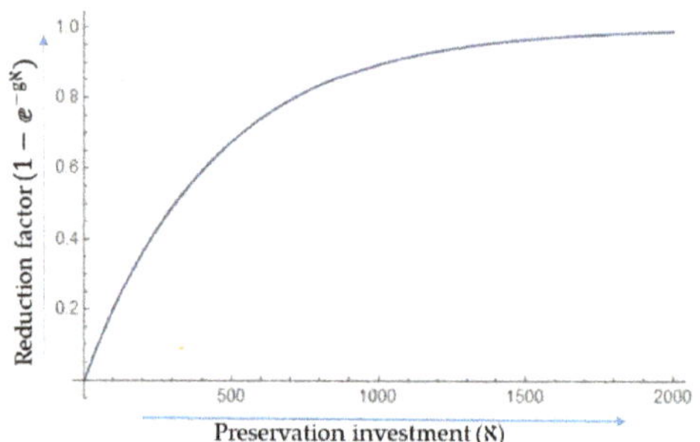

Figure 1. Investment factor plotted against reduction in deterioration rate.

We also consider preservation during transportation in this model for deteriorating items. Which means, the products are transported under special conditions, such as transportation by trucks installed with freezing units. The freezers installed on the truck are also operated by the energy obtained from the truck engine. Accordingly, we consider two types of variable transportation costs; which include the cost of transporting goods and preservation cost during transportation. The retailer's preservation investment is the cost of maintaining the required freezing conditions during transportation.

3.1.1. Notation

The following notation and assumptions are considered to formulate this model:

Constant demand (units/unit time)	x
Original deterioration rate, $\lambda > 0$ (unit/unit time)	λ
Cost of preservation technology, where $\aleph \geq 0$ (\$)	$\aleph$
Reduced deterioration rate, a function of $\aleph$ (unit/unit time)	$w(\aleph)$
Deterioration cost per unit (\$/unit)	C_λ
Waste disposal cost per deteriorated unit (\$/unit)	C_d
Inventory cycle length	T
The total cost of the integrated supply chain	TCSC
For manufacturer	
Manufacturer's production rate (unit/unit time)	P_m
Manufacturer's setup cost for a production batch (\$/batch)	S_m
Manufacturer's holding cost (\$/unit/unit time)	h_m
Manufacturer's production lot size per cycle (units)	Q
Area under the manufacturer's inventory level	Δ_m
Manufacturer's production time duration	T_1
Manufacturer's non-production time duration	T_2
For retailer	
Retailer's ordering cost (\$/order)	A_r
Retailer's holding cost (\$/unit/unit time)	h_r
Number of deliveries to retailer per production batch (units)	n
Retailer delivery lot size (units)	q
The duration between two successive deliveries to retailer	T_3
Freezing cost for transportation per unit item per km (\$/unit/km)	F_z
Fixed transportation cost per shipment (\$/delivery)	F
Truck capacity (units/truck)	K
Transportation cost per truck unit (\$/truck unit)	c_t
Distance covered (km)	l
Area under the Retailer's inventory level	Δ_r

3.1.2. Assumptions

An integrated inventory model is considered, and it is assumed that all information is shared; furthermore, both parties are interested in supply chain coordination. It is assumed that demand is constant and known, and the production rate of the manufacturer is greater than the demand rate. As, $p_m > x$. The finished product deteriorates at a random rate λ, which follows a uniform distribution. All of the deteriorated product is wasted, and supply chain members have to dispose of them properly. The original deterioration is reduced by investing in preservation technology, whereas, the reduction in deterioration is a function of the investment.

The retailer pays the cost of extra preservation during transportation.

4. Mathematical Modeling

In this section, we develop a single setup multiple delivery (SSMD) production model for deteriorating items. The manufacturer produces the production lot in one step but delivers in multiple shipments of a fixed quantity to the retailer at a constant shipment time. To control the deterioration rate of products, the manufacturer invests extra cost $\aleph$ in preservation technology. The total cycle time can be divided into two spans, T_1 and T_2 for the manufacturer. Where T_1 is the manufacturer's production uptime and T_2 is the non-production time, in which the manufacturer only makes deliveries to the retailer. The delivery times to the retailer are assigned as each delivery arrives at the time when all items from the previous delivery have been consumed. The supply chain logistics diagram is depicted in Figure 2.

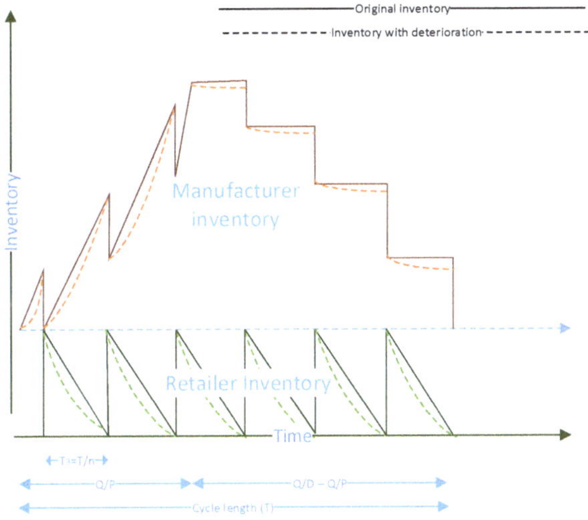

Figure 2. Inventory diagram for manufacturer and retailer.

4.1. The Retailer'S Cost

In the proposed system, the retailer needs $(Q = xT)$ quantity to satisfy market demand, for which the retailer incurs $\frac{A_r}{T}$ ordering cost per cycle. As an SSMD policy is considered, the manufacturer split the lot size into n equal shipments of size q, such that $(Q = nq)$. However, due to the deteriorating nature of products, the actual shipment size becomes,

$$q = T_3\left(x + \frac{(\lambda - w(\aleph)q)}{2}\right),$$

whereas, xT_3 is the market demand and $\left(\frac{(\lambda - w(\aleph)q)}{2}\right)T_3$ is the expected quantity that deteriorates during time T_3. As the retailer cycle time is $T_3 = T/n$, therfore,

$$q = \frac{T}{n}\left(x + \frac{(\lambda - w(\aleph)q)}{2}\right),$$

and,

$$\frac{n}{T} = \frac{q}{2}(\lambda - w(\aleph)q) + x \Rightarrow \frac{nq}{T(\lambda - w(\aleph)q)} = \frac{q}{2} + \frac{x}{(\lambda - w(\aleph))}.$$

Hence, the average inventory of the retailer can be given as,

$$\frac{q}{2} = \left(\frac{nq}{T(\lambda - w(\aleph))} - \frac{x}{(\lambda - w(\aleph))}\right). \tag{1}$$

Total deterioration at the retailer's end is the difference of total products received and the market demand fulfilled, which can be written as $(\lambda - w(\aleph))\Delta_r = (nq - xT)$. After simplifying, we get the time-weighted average inventory of the retailer, such that,

$$\frac{\Delta_r}{T} = \left(\frac{nq}{T(\lambda - w(\aleph))} - \frac{x}{(\lambda - w(\aleph))}\right). \tag{2}$$

From (1) and (2), $\Delta_r/T = \left(\frac{q}{2}\right)$, therefore, the retailer's holding cost can be written as $\frac{h_r \Delta_r}{T}$. The retailer average inventory is $\frac{\Delta_r}{T}$, for which the total deterioration cost is $\frac{C_\lambda(\lambda)\Delta_r}{T}$; however, after the manufacturer's investment in preservation technology, the net deterioration rate is reduced, for which the total deterioration cost at the retailer ends can be written as $\frac{C_\lambda(\lambda - w(\aleph))\Delta_r}{T}$. Each deteriorated item needs to be disposed of properly to avoid negative environmental impacts and pollution, for which, the total disposal cost is $\frac{C_d(\lambda - w(\aleph))\Delta_r}{T}$. It is assumed that transportation is the responsibility of the retailer, and therefore, transportation cost is added to the retailer's total cost function. The total transportation cost consists of a fixed transportation cost per shipment, the variable transportation cost, and cost of preservation during transportation in terms of freezing, which can be written as $\frac{1}{T}\left(nF + nq\left(\frac{l*c_t}{K} + F_z*l\right)\right)$. The total cost for the retailer consists of ordering cost, holding cost, deterioration cost, waste disposal cost, and transportation cost, which can be formulated as,

$$TC_r = \frac{1}{T}\left(A_r + h_r\Delta_r + \left(nF + nq\left(\frac{l*c_t}{K} + F_z*l\right)\right)\right) + (C_d + C_\lambda)\Delta_r(\lambda - w(\aleph)).$$

Therefore,

$$TC_r = \left(\frac{x}{nq} + \frac{\lambda - w(\aleph)}{2n}\right)\left(A_r + \left(nF + nq\left(\frac{l*c_t}{K} + F_z*l\right)\right)\right) + \frac{q}{2}(h_r + (C_d + C_\lambda)(\lambda - w(\aleph)). \tag{3}$$

4.2. The Manufacturer'S Cost

The manufacturer's total inventory is calculated with the help of Figure 2. It is clear that the production lot size for the manufacturer is the sum of the retailer demand and the expected number of deteriorated products r at the manufacturer's end, such that $(Q = nq + r)$. With probabilistic deterioration rate λ, the number of deteriorating items at the manufacturer's end is $(r = \lambda\Delta_m)$, which gives $\left(\frac{r}{\lambda} = \Delta_m\right)$. Therefore, the total number of deteriorated items for the whole supply chain is $\left(r + \frac{\lambda qT}{2}\right)$.

After simplification, we get the manufacturer's inventory as:

$$r + \frac{\lambda qT}{2} = \lambda T\left(\frac{Q\left(1 - \frac{x}{P}\right)}{P} + \frac{xq}{P}\right) \Rightarrow \Delta_m = qT\left(\frac{x}{P} + \frac{n-1}{2} - \frac{xn}{2p}\right).$$

For $(Q = nq + r)$ products, the production setup cost for the manufacturer is $\left(\frac{S_m}{T}\right)$ per cycle. Manufacturer's holding cost and deterioration cost for average inventory is $\left(\frac{h_m \Delta_m}{T}\right)$ and $\frac{C_\lambda(\lambda)\Delta_m}{T}$, respectively. With manufacturer's investment $\aleph$ in preservation technology, the deterioration rate decreases from λ to $(\lambda - w(\aleph))$. Accordingly, the deterioration cost for the manufacturer changes to $\frac{C_\lambda(\lambda - w(\aleph))\Delta_m}{T}$. The manufacturer also needs to dispose waste generated from net deteriorated items and pays an extra cost of $\frac{C_d(\lambda - w(\aleph))\Delta_m}{T}$. The manufacturer's cost includes setup cost, inventory holding cost, deterioration cost, preservation investment cost, and waste disposal cost, which can be derived as:

$$TC_m = \frac{1}{T}(S_m + h_m\Delta_m + \aleph + (\Delta_m(C_d + C_\lambda)(\lambda - w(\aleph)))). \tag{4}$$

Equation (4) can be simplified as given,

$$TC_m = \left(\frac{x}{nq} + \frac{(\lambda - w(\aleph))}{2n}\right)S_m + \aleph + q(h_m + (C_d + C_\lambda)(\lambda - w(\aleph)))\left(\frac{x}{p} + \frac{n-1}{2} - \frac{xn}{2p}\right). \tag{5}$$

4.3. The Total Supply Chain Cost

The total average cost of the supply chain is the sum of the manufacturer's and retailer's individual costs, as follows:

$$TCSC(q, n, \aleph) = TC_r + TC_m \Rightarrow \left(\frac{x}{nq} + \frac{(\lambda - w(\aleph))}{2n}\right)\left(A_r + S_m + \left(nF + nq\left(\frac{l*c_t}{K} + F_z*l\right)\right)\right) +$$
$$\frac{q}{2}\left((h_r + (C_d + C_\lambda)(\lambda - w(\aleph))) + \aleph + (h_m + (C_d + C_\lambda)(\lambda - w(\aleph)))\left(\frac{(2-n)x}{p} + n - 1\right)\right). \tag{6}$$

In this study, we have considered that the deterioration rate follows a continuous probability distribution function. As $\lambda = E[f(u)]$, with $f(u)$ following a uniform distribution as given below:

$$\lambda = E[f(u)] = \frac{a+b}{2}(a > 0, \, b > 0, \, a < b).$$

Now, the total cost function, Equation (6), can be written as:

$$TCSC(q, n, \aleph) = \left(\frac{x}{nq} + \left(\frac{(a+b)}{4n} - \frac{w(\aleph)}{2n}\right)\right)\left(A_r + S_m + \left(nF + nq\left(\frac{l*c_t}{K} + F_z*l\right)\right)\right) + (\aleph)$$
$$+ \frac{q}{2}\left(h_r + (C_d + C_\lambda)\left(\frac{a+b}{2} - w(\aleph)\right)\right) + \left(h_m + (C_d + C_\lambda)\left(\frac{a+b}{2} - w(\aleph)\right)\right)\left(\frac{x(2-n)}{p} + n - 1\right). \tag{7}$$

4.4. Solution Methodology

In this study, we have to find the optimal lot size q, preservation investment $\aleph$, and number of shipments to retailer n. Increasing the investment can control the deterioration rate up to some extent, but after the optimal point, investment contributes more to the total cost of the system compared to the reduction in deteriorated quantity. The objective is to examine the trade-off between the marginal cost of investment and the deteriorated quantities. Besides, a large order size q may increase the deteriorated quantity, yet, low order quantity would increase the ordering cost and transportation cost. A classical optimization technique is used to determine the optimal shipment quantity, number of shipments, and optimal investment in deterioration reduction. The developed problem consists of both continuous variables q and $\aleph$ and discrete variable n, which makes the model a MINLP problem. Due to the complexity of the objective function, it is impossible to solve it with ordinary optimization techniques. Therefore, this section provides a solution methodology to solve the problem and obtain the optimal policies for the proposed system.

Assuming that the lot size q and preservation investment $\aleph$ are any real nonzero numbers, then, at fixed n, there exists a unique q and $\aleph$, that minimizes the total cost and satisfies the following first-order conditions (FOC);

$$\frac{\partial(TCSC(q,n,\aleph))}{\partial(q)}$$
$$= \frac{\pi + (a+b)e^{-g\aleph}\left(p\left(\left(\frac{l*c_t}{K}+F_z*l\right)nq+nC_d\right)+C_\lambda n(p-x)+2C_\lambda x-C_d x(n-2)\right)}{4p} \tag{8}$$
$$= 0,$$

where, $\pi = 2h_r p - \dfrac{4(A_r+S_m+nF)px}{nq^2} - 2h_m(p-np+(n-2)x)$,
and,

$$\frac{\partial(TCSC(q,n,\aleph))}{\partial(\aleph)} = 1 - \frac{(a+b)S_m e^{-g\aleph}g}{4n} - \frac{(a+b)e^{-g\aleph}g\left(A_r+\left(nF+\left(\frac{l*c_t}{K}+F_z*l\right)nq\right)\right)}{4n}$$
$$-\tfrac{1}{4}(a+b)e^{-g\aleph}gq(C_d+C_\lambda) - \tfrac{1}{2}(a+b)e^{-g\aleph}gq(C_d+C_\lambda)\left(\tfrac{1}{2}(n-1)+\tfrac{x}{p}-\tfrac{nx}{2p}\right) = 0 \tag{9}$$

Furthermore, $TCSC(q,n,\aleph)$ is convex, if at q^* and $\aleph^*$, the Hessian of $TCSC(q,n,\aleph)$ is positive semidefinite. The convexity proof of the solution is provided in Appendix A.

As the number of shipments n is a discrete variable, the optimal number of shipments can be determined using the following necessary condition which n^* has to satisfy; for the optimal total cost TCSC at $q = q^*$ and $\aleph = \aleph^*$, the necessary condition is,

$$TCSC(q^*, \aleph^*, n^*-1) \geq TCSC(q^*, \aleph^*, n^*) \leq TCSC(q^*, \aleph^*, n^*+1)$$

And the optimal value of n^* is driven by Equation (10),

$$\frac{1}{2}(\gamma-1) \leq n \leq \frac{1}{2}(\gamma+1), \tag{10}$$

where,

$$\gamma = \frac{\left(\sqrt{q^2(h_m+\lambda(C_d+C_\lambda))}(p-x)+4A_r p(\lambda q+2x)+4S_m p(\lambda q+2x)-q(4A_r p+4S_m p+q(C_d+C_\lambda)(p-x))w(\aleph)\right)}{\left(q\sqrt{(p-x)}\sqrt{(h_m+\lambda(C_d+C_\lambda)-(C_d+C_\lambda)w(\aleph))}\right)}.$$

Solution algorithm

Step 1. Set n = 1 and $\aleph = 0$, find q^* from FOC (8)
Step 1.1. Using $q = q^*$, find $\aleph^*$ from FOC (9)
Step 1.2. Using $\aleph = \aleph^*$, find q^* from FOC (8)
Step 1.3. Repeat Step 1.1 and 1.2 until values of $\aleph^*$, find q^* stop changing, update $\aleph^*$, find q^* as optimal values at fixed n.
Step 1.4. Find the total cost $TCSC(q^*, \aleph^*, n)$ from (7). $TCSC(q^*, \aleph^*, n)$ is an optimal solution at fixed n.
Step 2. Set $n^* = n+1$, perform Step 1.1 to 1.4.
Step 2.1. If $TCSC(q^*, \aleph^*, n) \leq TCSC(q^*, \aleph^*, n-1)$ go to Step 2., else go to Step 3.
Step 3. Set $n^* = n-1$, $TCSC(q^*, \aleph^*, n)$ is the optimal solution.

5. Numerical Experiment

To get a deeper insight into the total costs and preservation technology investment relationship, we consider a numerical example with two cases in this section. The first case is the proposed model developed in the previous section. The second case considers a manufacturer that does not consider preservation investment, and consequently, the model reduces to two decision variables, with $\aleph = \$0$.

Case1:

The first case considers the preservation investment from the manufacturer's side. The data, taken from Sarkar [17], is given below with appropriate units.

A_r = \$25/order, S_m = \$800/batch, C_λ = \$40/unit, C_d = \$10/unit, F = \$50/shipment, h_r = \$7/unit/year, h_m = \$6/unit/year, p = 10,000 units/year, c_t = \$0.05/truck unit, K = 200 units/truck, F_z = \$0.00225/unit/truck, l = 400, x = 4800 units/year, g = 0.0075, a = 0.15, b = 0.25.

The optimal results with preservation investment are q $=$ 261.7 units, $\aleph$ $=$ \$492.5/Cycle, n $=$ 6/Cycle, with the total cost of the system TCSC = \$121,99.5/Cycle.

Case2:

In the second case of the given example, a manufacturer is considered who does not invest in preservation technology. The model is solved with the same data, and preservation investment is consider as $\aleph$ $=$ \$0. Then, the optimal solution without preservation investment is TCSC $=$ "\$"13,906.05/Cycle, q = 202.2 units, n = 6/Cycle. Results of both cases are compiled in Table 1.

Table 1. Optimal results of case 1 and case 2.

Model	q (Units)	n (Shipments)	$\aleph$ (\$)	TCSC (\$)	Generated Waste (Units)
With preservation investment	261.7	6	492.5	12,199.5	$\left(\lambda - \lambda\left(1 - e^{-g\aleph}\right)\right)nq = 7.76$
Without preservation investment	202.2	6	0	13,906.05	$\left(\lambda - \lambda\left(1 - e^{-g\aleph}\right)\right)nq = 235.46$

As shown in Table 1, the optimal lot size is q $=$ 261.7 units with preservation investment, and without preservation investment, the lot size reduces to q $=$ 202.2 units. The shipment size for both cases remains six shipments/ cycle. We also observe, that with an investment of ($\aleph$ $=$ \$492.5/cycle) the total cost of model TCSC reduces 13% from \$13,906.05 to \$12,199.5.

5.1. Sensitivity Analysis

To gain a deeper understanding of cost drivers in the proposed supply chain, sensitivity analysis is performed by varying the values of the input parameters within the range [−50% +50%]. The results of sensitivity analysis are compiled in Table 2.

It is clear from sensitivity results that the total cost of the system is highly sensitive to the variable cost of transportation c_t.

The freezing cost per unit item during transportation has the highest impact on total cost; up to 50% reduction in it, decreases the total cost by 18%. Even though this cost has a high impact on the total cost of the system, it does not affect the decision variables in any significant way.

The second most significant parameter regarding total cost variation is the manufacturer's holding cost. A 50% reduction in it decreases the total cost by 11.5%. Although, manufacturing holding cost does not change shipment size, however, its increment decreases production lot size by decreasing the number of shipments. Furthermore, it also reduces investment in preservation technology.

Although the retailer's holding cost affects the total cost, it has more influence on the number of shipments and shipment size. For a higher retailer's holding cost, the number of shipments increases and the shipment size decreases.

Constant transportation cost has the highest influence on the number of shipments. An increase in constant transportation cost decreases the shipment size considerably.

The most relevant parameter regarding investment in preservation, is the shape parameter g. Increasing the shape parameter increases the effectiveness of the technology and, hence, preservation investment is reduced.

A 50% increase in the manufacturer's setup cost increases the total cost by 10%. An increase in this cost leads to an increase in the number of shipments per cycle, which results in a decrease in the number of setups per year.

Although, deterioration per unit item does not affect total cost to a large extent, its effect on preservation investment is enormous. The decrease in deterioration cost decreases investment in preservation and vice versa. The effect of waste disposal cost is similar to deterioration cost, the more disposal cost per unit the deteriorated item has, the more investment is required.

Table 2. Sensitivity analysis of input parameters.

Parameter	% Change in Value	Decision Variables			% Change in Total Cost
		q	$\aleph$	n	
A_r	−50	261.70	491.70	6	−0.32
	−25	261.70	492.14	6	−0.16
	+25	261.70	492.90	6	0.16
	+50	261.70	493.30	7	0.31
S_m	−50	260.40	459.60	5	−11.77
	−25	261.20	478.40	5	−5.39
	+25	262.10	503.80	7	4.76
	+50	262.4	513.3	8	9.07
C_λ	−50	261.6	425.2	6	−0.55
	−25	261.7	463.1	6	−0.24
	+25	261.70	516.6	6	0.20
	+50	261.70	537.1	6	0.37
F	−50	184.80	482.40	8	−4.41
	−25	226.50	488.00	7	−2.02
	+25	292.70	496.30	5	1.77
	+50	320.8	499.7	5	3.38
h_r	−50	371.00	505.50	4	−4.40
	−25	302.40	497.50	5	−2.02
	+25	234.00	489.00	7	1.77
	+50	213.60	486.30	8	3.38
h_m	−50	260.50	527.50	9	−11.51
	−25	261.00	506.70	7	−5.28
	+25	262.40	481.80	6	4.66
	+50	263.20	473.30	5	8.87
c_t	−50	261.70	492.20	6	−1.97
	−25	261.74	492.95	6	−0.98
	+25	261.70	492.55	6	0.99
	+50	261.70	492.56	6	1.97
F_z	−50	261.77	492.24	6	−17.70
	−25	261.76	492.39	6	−8.85
	+25	261.72	492.68	6	8.86
	+50	261.70	492.83	6	17.71
g	−50	247.9	119.4	6	3.61
	−25	260.1	604.4	6	1.29
	+25	262.6	418.2	6	−0.83
	+50	263.3	364.9	6	−1.42

5.2. Comparative Study

The comparative study shows the relationship between decision variables with total cost and their inter-relationship. It shows how the optimal value of a decision variable and total cost changes with changing the value of other decision variables.

5.2.1. Relationship of Total Cost with Decision Variables

The impact of shipment size q on total cost is illustrated in Figure 3. It is clear that total cost increases with both an increase and decrease in the optimal value of q. Figure 4 shows the convexity of the objective function with preservation investment when other parameters are constant. Similarly, Figure 5 provides the impact of variation in shipment frequency on the total cost of the system.

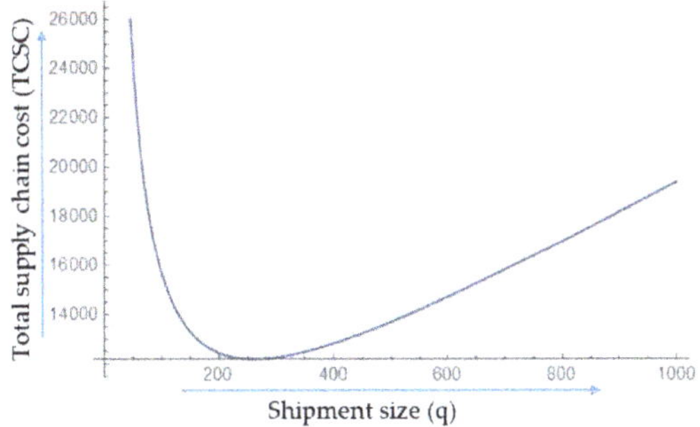

Figure 3. Relationship between total cost and shipment size.

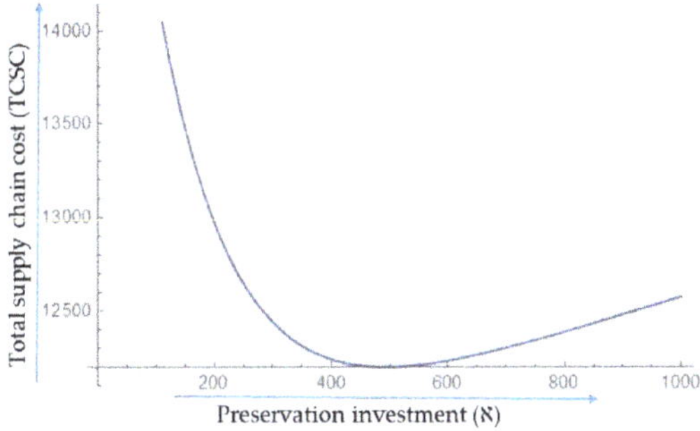

Figure 4. Relationship between total cost and preservation investment.

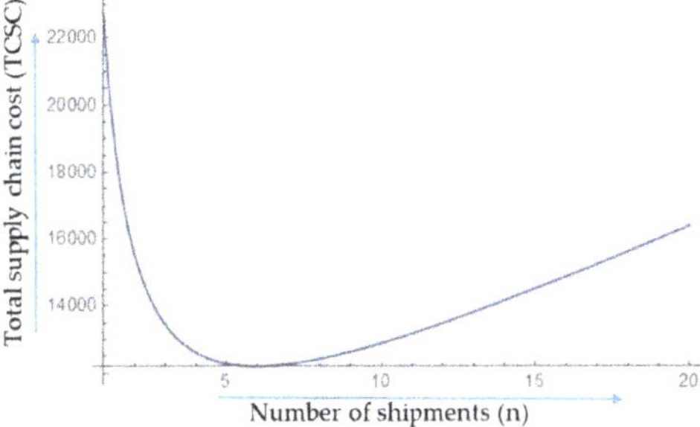

Figure 5. Relationship between total cost and number of shipments.

5.2.2. Inter-Relationship of Decision Variables

In this section, we develop the comparative study of decision variables. Figure 6 shows the impact of investment parameter $\aleph$ on the optimal lot size q; with the increase in investment, the lot size increases as well.

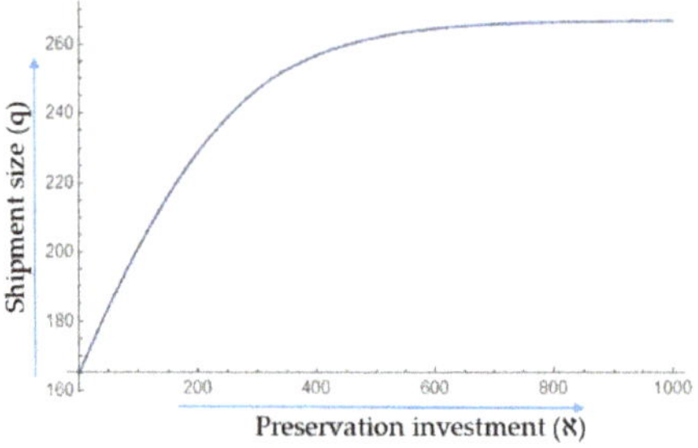

Figure 6. Impact of preservation investment on optimal shipment size.

Figure 7 shows the impact of the number of shipments n on the optimal lot size q; as the number of shipments is increased, the shipment lot size decreases accordingly.

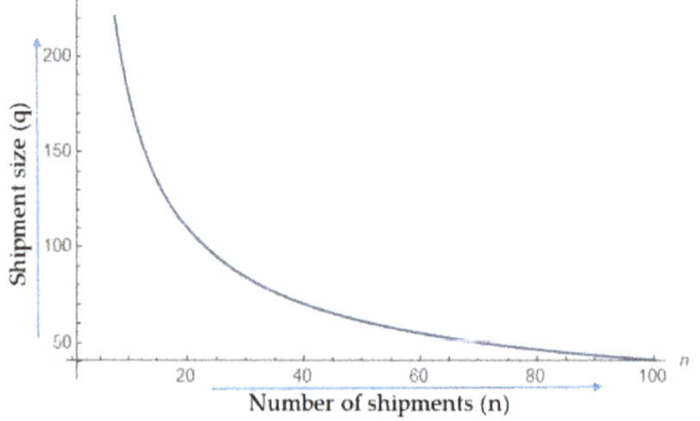

Figure 7. Impact of number of shipments on optimal shipment size.

Figure 8 shows the impact of shipments size q on the optimal preservation investment $\aleph$; as the shipment quantity increases, the preservation investments needs to be increased respectively.

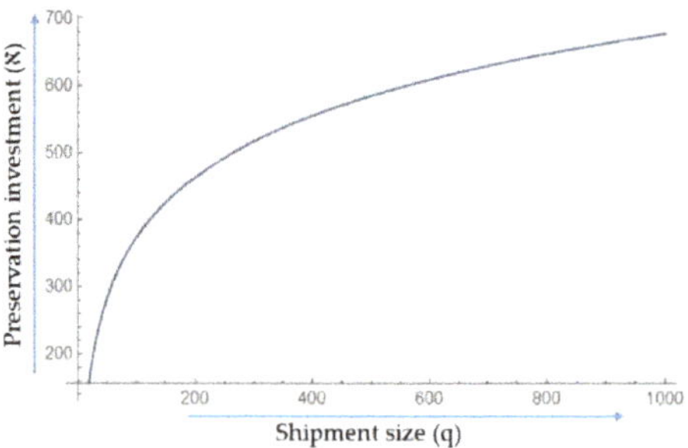

Figure 8. Impact of shipment size on optimal preservation investment.

Figure 9 shows the impact of shipment frequency n on the optimal preservation investment $\aleph$; increasing shipment frequency increases the production lot size, which increases the required preservation investments.

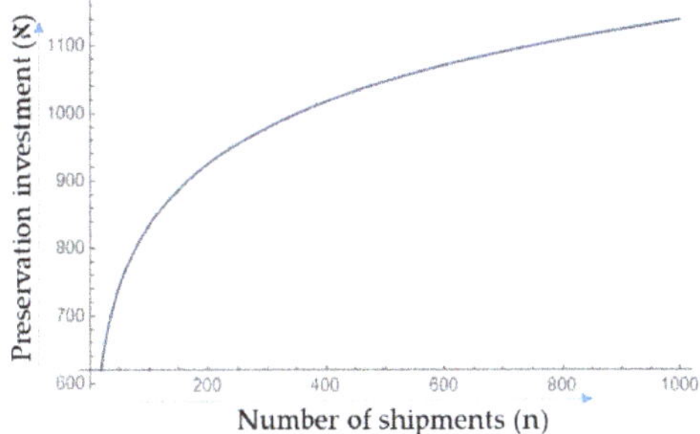

Figure 9. Impact of number of shipments on optimal preservation investment.

Figure 10 shows the impact of shipments quantity q on the optimal number of shipments n. This relation is obvious, as increasing the shipment frequency would decrease the time between two shipments, thus reducing shipment size.

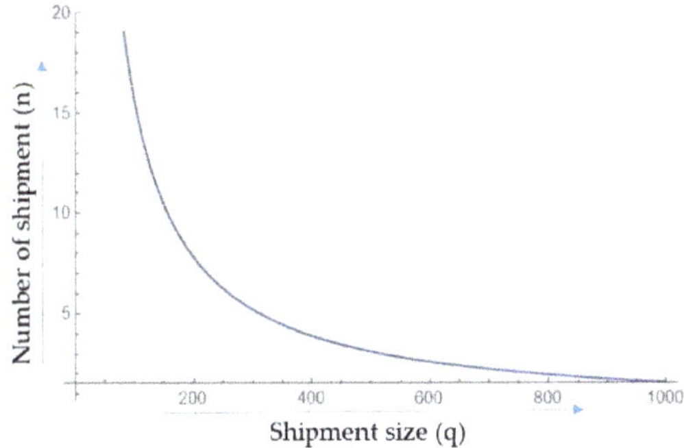

Figure 10. Impact of changing shipment size on optimal number of shipments.

Figure 11 shows the impact of preservation investment $\aleph$ on the optimal number of shipments n; as the preservation investment increases, the shipment frequency increases consequently.

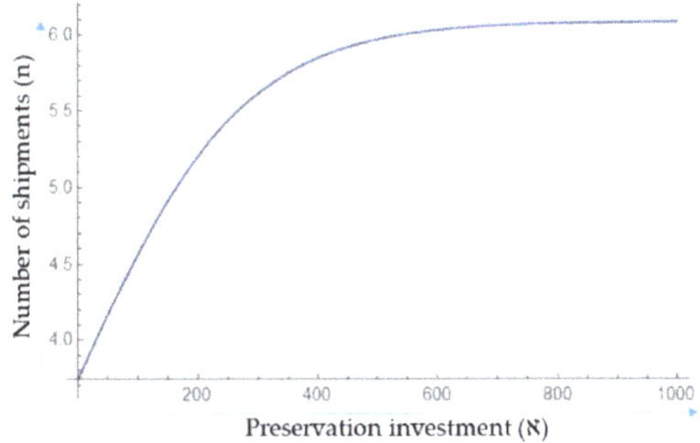

Figure 11. Impact of changing preservation investment on optimal number of shipments.

5.3. Managerial Insights

- Several important managerial insights are drawn from the sensitivity analysis and comparative study to assist supply chain managers in decision making.
- Supply chain managers should consider the investment in preservation technology; as it not only reduces the total cost, but also decreases waste generation in the supply chain. The results of our

study confer a high decrease in the amount of product loss and generated waste with preservation and encouraged investment for deterioration prevention strategies. Without preservation, the firm faces product shortages because of deterioration, in addition, extra waste disposal cost adds up in a system.

- The sensitivity analysis represents the amount of preservation investment decrease with a decrease in the effectiveness of the technology. It also shows that the preservation cost for products depends on their deterioration cost per unit. Consequently, a product with more monetary value and high deterioration cost needs higher preservation investments. However, the total cost of the supply chain increases with a decrease in preservation investment. Therefore, supply chain managers are suggested to consider the deterioration cost of specific products and the most effective preservation technology accordingly.
- The results further showed that preservation leads to an increase in the optimal lot size, which consequently decreases the number of orders per year and ordering cost per year in a supply chain. Managers should consider the ordering costs for products in deciding the investment plans for preservation technology, as higher preservation investment may be required for products that have higher ordering costs in a supply chain.
- Also, the optimal size of shipments increases in a supply chain with an increase in deterioration prevention investment. Hence, supply chain managers should consider preservation investment to increase shipment size instead of shipment frequency, especially when transportation costs are high for the system.
- Another important parameter that should be considered by managers is the retailer holding cost. The higher holding cost of the retailer increases the number of shipments, and hence, increases the transportation cost per year of the supply chain. Furthermore, it also reduces the optimal lot size, which increases the ordering cost of the retailer. Hence, investment in deterioration prevention is crucial for supply chain management with a higher holding cost at the retailer's end.
- Finally, this study suggests that supply chain managers should regulate the environmental and financial impacts of their supply chain networks through deterioration prevention techniques. As with preservation investment, the quantity of decayed/deteriorated products reduces, and the consumption of resources also reduces which makes preservation technology a concluding sustainable approach for perishable products.

6. Conclusions

Product deterioration compromises the ability of any firm to achieve its economic and environmental goals. Economically, firms face extra manufacturing costs to catch up the shortage faced by deterioration. While, environmentally, product deterioration causes extra resource consumption that has no beneficiary output for manufacturing firms, the deteriorated items also cause pollution and landfills backlash. Deteriorating products in a multi-firm network multiply the loss, as deterioration occurs at every stage of the supply chain network. This study extended the existing literature on stochastic deteriorating inventory models in supply chain management with preservation technology investment in production and transportation. A two-echelon supply chain is considered, where the deterioration rate of integrated inventory follows a continuous probability distribution. This paper considered investments in preservation technology by the manufacturer to control the deterioration rate and solid waste generation, and its impacts on supply chain were studied. The retailer considers the transportation of products under a particular condition to increase the effectiveness of the preservation technology. For instance, refrigerated transportation is adopted by the retailer, which means the deteriorating products are transported by the vehicle installed with freezing units. Both parties properly dispose of the waste generated from deteriorating items and endure waste disposal costs.

The tradeoff between preservation investment in supply chain performance is very promising. The results showed that investing in preservation technology reduces the total cost of the system by 13%, because of the controlled deterioration rate and reduced deteriorated products in both the manufacturer's and retailer's inventories. Furthermore, preservation technology reduces waste generation in the proposed supply chain from 235 to 8 deteriorated products per cycle. In the proposed supply chain, these results confer a high decrease in the amount of product loss and generated waste. This also leads to improving the environmental performance of the supply chain. The results further showed that investment in the preservation leads to an increase in the optimal lot size q, hence decreasing the number of orders per year, which reduces the ordering cost per year. This suggests that preservation investment is even more beneficial in supply chains where product ordering costs are high. Also, the optimal size of shipments increases with an increase in deterioration prevention investment. Hence, for a supply chain with high transportation costs, preservation investment is necessary to increase shipment size instead of shipment frequency.

The sensitivity analysis shows that the preservation cost for products depends on their deterioration cost per unit. Deterioration cost per unit symbolizes the commercial value of products and also the cost of replacing a product in the given setup. Consequently, a product with more monetary value and high deterioration cost needs higher preservation investments as well. It was also found from sensitivity analysis, that an increase in buyer holding cost increases the number of shipments, and hence, increases the transportation cost per year. Furthermore, it also reduces the optimal lot size, which increases the ordering cost of the retailer. Hence, investment in deterioration prevention is crucial for supply chain management with a higher holding cost at the retailer's end. The managerial insights help supply chain managers to decide on the required amount of investment and the type of technology for considered deteriorating products. The model can be extended to the multi-item supply chain system with different preservation investment policies. Stochastic or seasonal demand can be another possible extension, as for deteriorating items, demand usually varies with time. Another possible extension is considering shared or joint preservation practices in resilient supply chain management, a comprehensive example is provided by Mari et al. [45] with disruption risks.

Author Contributions: Conceptualization, B.S. and M.U.; Methodology, B.S. and I.A.; Software, I.A. and M.U.; Validation, B.S.; Formal analysis, I.A. and M.U.; Investigation, B.S. and M.U.; Resources, B.S. and I.A.; Data curation, B.S. and M.U.; Writing—original draft preparation, I.A.; Writing—review and editing, B.S. and M.U.; Visualization, I.A. and M.U.; Supervision, B.S.

Funding: This research received no external funding.

Conflicts of Interest: The authors declare no conflict of interest.

Appendix A

The Hessian matrix $H(q, \aleph)$ at given n becomes;

$$H = \begin{pmatrix} \dfrac{\partial^2 TCSC(q, \aleph, n)}{\partial q^2} & \dfrac{\partial^2 TCSC(q, \aleph, n)}{\partial q \partial \aleph} \\ \dfrac{\partial^2 TCSC(q, \aleph, n)}{\partial \aleph \partial q} & \dfrac{\partial^2 TCSC(q, \aleph, n)}{\partial \aleph^2} \end{pmatrix}$$

At optimal solution ($q^* = 261.7$ units, $\aleph^* = \$492.5$), the Hessian matrix can be written as,

$$H = \begin{pmatrix} 0.101 & -0.004 \\ -0.004 & 0.007 \end{pmatrix}$$

We can see that the hessian matrix is positive semidefinite, as all Eigen values of the Hessian matrix are positive. Hence, the obtained results ($q^* = 261.7$ units, $\aleph^* = \$492.5/$Cycle and $n^* = 6/$Cycle) are optimal.

Figure A1 illustrates the graph of $H(q, \aleph)$, we can see with decision parameters q and $\aleph$ that the Hessian matrix $H(q, \aleph)$ is positive for all input points.

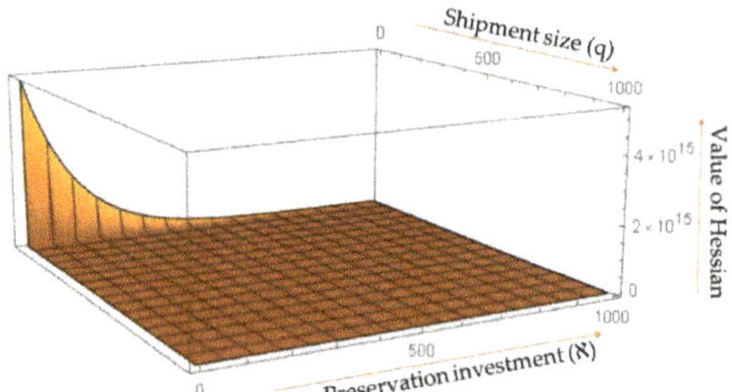

Figure A1. Graph of H (q, ℵ).

The convexity of total cost TCSC(q*, ℵ*, n*) is also illustrated graphically in Figure A2. The graph depicts that TCSC has an optimum value at lot size q* = 260, and then the total cost starts to increase. Also, without any preservation investment when ℵ = 0, the cost is maximum, but with an increase in investment as ℵ > 0, the unit time total cost starts to decrease.

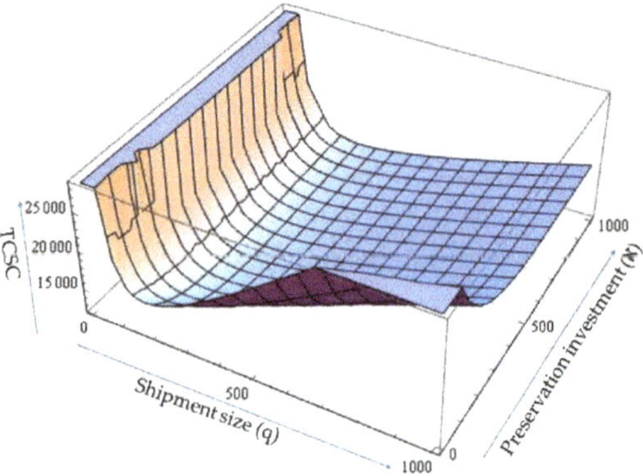

Figure A2. Graph of TCSC (q, ℵ) at n*.

References

1. Arribas, I.; Vegas, I.; García, V.; de la Villa, R.V.; Martínez-Ramírez, S.; Frías, M. The deterioration and environmental impact of binary cements containing thermally activated coal mining waste due to calcium leaching. *J. Clean. Prod.* **2018**, *183*, 887–897. [CrossRef]
2. Ghare, P.M. A model for an exponentially decaying inventory. *J. Ind. Eng.* **1963**, *14*, 238–243.
3. Covert, R.P.; Philip, G.C. An EOQ model for items with Weibull distribution deterioration. *AIIE Trans.* **1973**, *5*, 323–326. [CrossRef]
4. Sarkar, B.; Saren, S.; Cárdenas-Barrón, L.E. An inventory model with trade-credit policy and variable deterioration for fixed lifetime products. *Ann. Oper. Res.* **2015**, *229*, 677–702. [CrossRef]
5. Philip, G.C. A generalized EOQ model for items with Weibull distribution deterioration. *AIIE Trans.* **1974**, *6*, 159–162. [CrossRef]
6. Mukhopadhyay, S.; Mukherjee, R.N.; Chaudhuri, K.S. Joint pricing and ordering policy for a deteriorating inventory. *Comput. Ind. Eng.* **2004**, *47*, 339–349. [CrossRef]
7. Hung, K.C. An inventory model with generalized type demand, deterioration and backorder rates. *Eur. J. Oper. Res.* **2011**, *208*, 239–242. [CrossRef]
8. Sarkar, B.; Sarkar, S. An improved inventory model with partial backlogging, time varying deterioration and stock-dependent demand. *Econ. Model.* **2013**, *30*, 924–932. [CrossRef]
9. Shah, B.J.; Shah, N.H.; Shah, Y.K. EOQ model for time-dependent deterioration rate with a temporary price discount. *Asia-Pac. J. Oper. Res.* **2005**, *22*, 479–485.

10. Tadikamalla, P.R. An EOQ inventory model for items with gamma distributed deterioration. *AIIE Trans.* **1978**, *10*, 100–103. [CrossRef]

11. Guido, R.C. Practical and useful tips on discrete wavelet transforms [sp tips & tricks]. *IEEE Signal Process. Mag.* **2015**, *32*, 162–166.

12. Guariglia, E. Entropy and fractal antennas. *Entropy* **2016**, *18*, 84. [CrossRef]

13. Sarkar, B. An EOQ model with delay in payments and time varying deterioration rate. *Math. Compt. Model.* **2012**, *55*, 367–377. [CrossRef]

14. Ahmad, W.; Sarkar, B.; Ullah, M. Impact of Reparation for Imperfect Quality Items having Shortages in the System under Multi-Trade-Credit-Period. *DJ J. Eng. Appl. Math.* **2018**, *5*, 1–16. [CrossRef]

15. Ning, Y.; Rong, L.; Liu, J. Inventory models for fresh agriculture products with time-varying deterioration rate. *Ind. Eng. Manag. Syst.* **2013**, *12*, 23–29. [CrossRef]

16. Guariglia, E. Harmonic sierpinski gasket and applications. *Entropy* **2018**, *20*, 714. [CrossRef]

17. Sarkar, B.A. Production-inventory model with probabilistic deterioration in two-echelon supply chain management. *App. Math. Model.* **2013**, *37*, 3138–3151. [CrossRef]

18. Iqbal, M.W.; Sarkar, B.A. Model for imperfect production system with probabilistic rate of imperfect production for deteriorating products. *DJ J. Eng. Appl. Math.* **2018**, *4*, 1–2. [CrossRef]

19. Sett, B.K.; Sarkar, B.; Goswami, A. A two-warehouse inventory model with increasing demand and time varying deterioration. *Sci. Iran.* **2012**, *19*, 1969–1977. [CrossRef]

20. Sett, B.K.; Sarkar, S.; Sarkar, B.; Yun, W.Y. Optimal replenishment policy with variable deterioration for fixed-lifetime products. *Sci Iran. E Ind Eng.* **2016**, *23*, 2318.

21. Sarkar, M.; Sarkar, B. An economic manufacturing quantity model with probabilistic deterioration in a production system. *Econ. Model.* **2013**, *31*, 245–252. [CrossRef]

22. Bakker, M.; Riezebos, J.; Teunter, R.H. Review of inventory systems with deterioration since 2001. *Eur. J. Oper. Res.* **2012**, *221*, 275–284. [CrossRef]

23. Sherbrooke, CC. METRIC: A multi-echelon technique for recoverable item control. *Oper. Res.* **1968**, *16*, 122–141. [CrossRef]

24. Tayyab, M.; Sarkar, B.; Ullah, M. Sustainable Lot Size in a Multistage Lean-Green Manufacturing Process under Uncertainty. *Mathematics* **2019**, *7*, 20. [CrossRef]

25. Sarkar, B.; Ullah, M.; Kim, N. Environmental and economic assessment of closed-loop supply chain with remanufacturing and returnable transport items. *Comput. Ind. Eng.* **2017**, *111*, 148–163. [CrossRef]

26. Thyberg, K.; Tonjes, D.J. Drivers of food waste and their implications for sustainable policy development. *Resour. Conserv. Recycl.* **2016**, *106*, 110–123. [CrossRef]

27. Tayyab, M.; Sarkar, B.; Yahya, B. Imperfect Multi-Stage Lean Manufacturing System with Rework under Fuzzy Demand. *Mathematics* **2019**, *7*, 13. [CrossRef]

28. Tibben-Lembke, R.S. Strategic use of the secondary market for retail consumer goods. *Calif. Manag. Rev.* **2004**, *46*, 90–104. [CrossRef]

29. Murr, D.P.; Morris, L.L. Effect of storage temperature on postharvest changes in mushrooms. *J. Am. Soc. Hortic. Sci.* **1975**, *2*. [CrossRef]

30. Zauberman, G.; Ronen, R.; Akerman, M.; Fuchs, Y. Low pH treatment protects litchi fruit color. *Sym. Trop. Fruit Int. Trade* **1989**, *4*, 309–314. [CrossRef]

31. Yang, P.C.; Wee, H.M. A collaborative inventory system with permissible delay in payment for deteriorating items. *Math. Comput. Model.* **2006**, *43*, 209–221. [CrossRef]

32. Ho, J.C.; Solis, A.O.; Chang, Y.L. An evaluation of lot-sizing heuristics for deteriorating inventory in material requirements planning systems. *Comp. Oper. Res.* **2007**, *34*, 2562–2575. [CrossRef]

33. Habib, M.S.; Sarkar, B.; Tayyab, M.; Saleem, M.W.; Hussain, A.; Ullah, M.; Omair, M.; Iqbal, M.W. Large-scale disaster waste management under uncertain environment. *J. Clean. Prod.* **2018**, *19*, 200–222. [CrossRef]

34. Nye, T.J.; Jewkes, E.M.; Dilts, D.M. Optimal investment in setup reduction in manufacturing systems with WIP inventories. *Eur. J. Oper. Res.* **2001**, *135*, 128–141. [CrossRef]

35. Lee, H.H. A cost/benefit model for investments in inventory and preventive maintenance in an imperfect production system. *Comput. Ind. Eng.* **2005**, *48*, 55–68. [CrossRef]

36. Lin, L.C.; Hou, K.L. An inventory system with investment to reduce yield variability and set-up cost. *J. Oper. Res. Soc.* **2005**, *56*, 67–74. [CrossRef]

37. Hsu, P.H.; Wee, H.M.; Teng, H.M. Optimal ordering decision for deteriorating items with expiration date and uncertain lead time. *Comput. Ind. Eng.* **2007**, *52*, 448–458. [CrossRef]
38. Li, J.; Min, K.J.; Otake, T.; Van Voorhis, T. Inventory and investment in setup and quality operations under return on investment maximization. *Eur. J. Oper. Res.* **2008**, *185*, 593–605. [CrossRef]
39. Affisco, J.F.; Paknejad, M.J.; Nasri, F. Quality improvement and setup reduction in the joint economic lot size model. *Eur. J. Oper. Res.* **2002**, *142*, 497–508. [CrossRef]
40. Uçkun, C.; Karaesmen, F.; Savaş, S. Investment in improved inventory accuracy in a decentralized supply chain. *Int. J. Prod. Econ.* **2008**, *113*, 546–566. [CrossRef]
41. Sarkar, B.; Majumder, A.; Sarkar, M.; Kim, N.; Ullah, M. Effects of variable production rate on quality of products in a single-vendor multi-buyer supply chain management. *Int. J. Adv. Manuf. Technol.* **2018**, *99*, 567–581. [CrossRef]
42. Lee, H.H. The investment model in preventive maintenance in multi-level production systems. *Int. J. Prod. Econ.* **2008**, *112*, 816–828. [CrossRef]
43. Iqbal, M.W.; Sarkar, B. Application of preservation technology for lifetime dependent products in an integrated production system. *J. Ind. Manag. Optim.* **2018**, *17*, 563–576.
44. Wong, W.C.; Chee, E.; Li, J.; Wang, X. Recurrent Neural Network-Based Model Predictive Control for Continuous Pharmaceutical Manufacturing. *Mathematics* **2018**, *6*, 242. [CrossRef]
45. Mari, S.I.; Memon, M.S.; Ramzan, M.B.; Qureshi, S.M.; Iqbal, M.W. Interactive Fuzzy Multi Criteria Decision Making Approach for Supplier Selection and Order Allocation in a Resilient Supply Chain. *Mathematics* **2019**, *7*, 137. [CrossRef]

© 2019 by the authors. Licensee MDPI, Basel, Switzerland. This article is an open access article distributed under the terms and conditions of the Creative Commons Attribution (CC BY) license (http://creativecommons.org/licenses/by/4.0/).

 *mathematics*

Article

Low Carbon Supply Chain Coordination for Imperfect Quality Deteriorating Items

Yosef Daryanto [1,2], **Hui Ming Wee** [1,*] and **Gede Agus Widyadana** [3]

[1] Department of Industrial and Systems Engineering, Chung Yuan Christian University, 200 Chung-Pei Rd., 32023 Chung-li, Taoyuan, Taiwan; yosef.daryanto@uajy.ac.id

[2] Department of Industrial Engineering, Universitas Atma Jaya Yogyakarta, Jl. Babarsari 43, 55281 Yogyakarta, Indonesia

[3] Department of Industrial Engineering, Petra Christian University, Surabaya, East Java 60236, Indonesia; gede@peter.petra.ac.id

* Correspondence: weehm@cycu.edu.tw

Received: 19 January 2019; Accepted: 20 February 2019; Published: 5 March 2019

Abstract: Nowadays, many countries have implemented carbon pricing policies. Hence, the industry adapts to this policy while striving for its main goal of maximizing financial benefits. Here, we study a single manufacturer–retailer inventory decision considering carbon emission cost and item deterioration for an imperfect production system. This study examines two models considering two cases of quality inspection. The first is when the buyer performs the quality inspection, and the second is when the quality inspection becomes the vendor's responsibility so that no defective products are passed to the buyer. Carbon emission costs are incorporated under a carbon tax policy, and we consider the carbon footprint from transporting and warehousing the items. The objective is to jointly optimize the delivery quantity and number of deliveries per production cycle that minimize the total cost and reduce the total carbon emissions. This study provides solution procedures to solve the models, as well as two numerical examples.

Keywords: supply chain inventory; imperfect quality; inspection; carbon emission; deteriorating items

---

1. Introduction

Supply chain coordination has a favorable effect on inventory replenishment decisions. Supply chain coordination can be realized through information sharing and joint decision-making. Coordination brings many advantages such as lower inventory-related costs and quality improvement [1]. This study considers supply chain management coordination and examines its effect on both economic and environmental performance. This study proposes supply chain inventory models that consider carbon emission costs and the existence of defective items under different inspection coordination mechanisms. Further, the models also consider the effect of item deterioration. In real life, many inventory items deteriorate over time due to spoilage, physical depletion, or obsolescence.

Due to increasing pressure from legislation, customers, and other organizations, business and industry are striving for more eco-friendly operation. The production, distribution, consumption, and other post-consumption processes of a product are sources of carbon emission. Therefore, the concept of a low-carbon supply chain has gained massive interest among researchers and industry practitioners [2,3]. The objective is to control and reduce CO_2 emissions (the major part of greenhouse gas emission) from the supply chain. Recently, Kazemi et al. [4] considered the effect of carbon emissions on several economic order quantity (EOQ) models. Sarkar et al. [5] considered warehouse emissions in the EOQ model with a rework for the defective items. The model also considered partial backorder and multi-trade-credit-period. Taleizadeh et al. [6] proposed economic production quantity

(EPQ) models that considered carbon emissions. Recently, Sarkar et al. [7,8] and Daryanto and Wee [9] incorporated a carbon tax in a supply chain total cost model. Wahab et al. [10], Jauhari et al. [11], Sarkar et al. [12], Jauhari [13], Gautam and Khanna [14], and Tiwari et al. [15] incorporated both carbon emissions and imperfect quality in a low-carbon supply chain model. The quality inspection is performed by the buyer, and the defective products are sent back to the vendor or sold into the secondary market at a discounted price.

Table 1 illustrates the research gap by comparing this paper with the existing literature. This study focuses on supply chain inventory models for a system that contains imperfect quality items. The decisions of the supply chain dealing with the defective items in the imperfect production processes affect carbon emissions, because defective item processing also adds to the total emissions. Moreover, the loss due to imperfect quality and deterioration also forces the manufacturer to produce more products to satisfy customer demand per period, resulting in the increase in carbon emission from production, holding, and distribution. The objectives of these studies are to simultaneously minimize the total cost and reduce carbon emissions. This paper also contributes to low-carbon supply chain models by considering two cases of quality inspection. In the first case, the buyer performs the quality inspection, and in the second case, the quality inspection becomes the vendor's responsibility. The first model extends the studies of Wahab et al. [10], Jauhari et al. [11], Sarkar et al. [12], Jauhari [13], and Gautam and Khanna [14] to consider the effect of deterioration. In addition to the fixed and variable inspection costs, the model also extends Tiwari et al.'s [15] model by introducing weight and distance-dependent transportation cost and emission variables. The second model extends the first model by introducing an inspection option to prevent defective products from being shipped to the buyer. This model reduces the expected total costs and emission costs of the supply chain.

Table 1. Gap analysis with existing literature.

Authors	Imperfect Quality		Deteriorating Item	Variable Transportation Cost	Carbon Emission
	Vendor's Inspection	Buyer's Inspection			
Huang (2002)		√			
Goyal et al. (2003)		√			
Wee et al. (2006)		√	√		
Wahab et al. (2011)		√			√
Benjaafar et al. (2013)					√
Lee and Kim (2014)		√	√		
Bazan et al. (2014)	√				
Bozorgi et al. (2014)					√
Jauhari et al. (2014)		√			√
Bozorgi (2016)					√
Ghosh et al. (2016)					√
Sarkar et al. (2016b)		√		√	√
Yu and Hsu (2017)		√			
Sarkar et al. (2017)	√			√	
Toptal and Çetinkaya (2017)					√
Bouchery et al. (2017)					√
Dwicahyani et al. (2017)					√
Wangsa (2017)					√
Li et al. (2017)					√
Anvar et al. (2018)					√
Hariga et al. (2018)					√
Ji et al. (2018)					√
Wang and Ye (2018)					√
Gosh et al. (2018)					√
Ma et al. (2018)					√
Darom et al. (2018)					√
Jauhari (2018)		√			√
Gautam and Khanna (2018)		√			√
Wangsa and Wee (2018)				√	
Tiwari et al. (2018)		√	√		√
Kundu and Chakrabarti (2019)					√
This paper	√	√	√	√	√

This study incorporates carbon emissions, item deterioration, and defective percentage to guide the supply chain managers to make the inventory decisions on the delivery size and the number of deliveries per cycle. This introduction section is followed by reviews of previous related studies in

Section 2. Then, Section 3 defines the problem, assumptions, and notations in this study. Section 4 presents two mathematical models. Section 5 provides two numerical examples and the sensitivity analysis to find some insights from the proposed models. In the end, Section 6 summarizes the findings and discusses some opportunities for further research.

2. Literature Review

This study incorporates both economic and emission costs in a two-echelon supply chain production–inventory model assuming that defective products exist in each delivered lot. This section presents the existing literature that supports this study.

2.1. Imperfect Quality Inventory Model

In many industries, production systems are imperfect, producing a certain percentage of defective products. Rosenblatt and Lee [16] and Porteus [17] studied the relationship between the optimal lot size and quality performance. Rosenblatt and Lee [16] studied the optimal production cycle through considering the proportion of defective items, while Porteus [17] related the model to the opportunity for quality improvement and setup cost reduction through investment. Salameh and Jaber [18] incorporated defective items into the EPQ model and considered the screening time and cost. Many other researchers have continued the research on the EOQ and EPQ models with imperfect quality. Those researchers assume that at the end of the screening period, or the end of the cycle, the defective products will be sold at a lower price.

Other researchers bring the effect of imperfect quality items into the integrated vendor–buyer or multi-echelon inventory model. Huang [19] considered imperfect quality and assumed that the vendor provides a product warranty for the defective items. The buyer conducts a 100% inspection, and at the end, the vendor treats the faulty items. Goyal et al. [20] extended the previous model with a single production and multiple deliveries containing defective products. They assumed that the buyer sells the defective items at a discounted price. Figure 1 illustrates the scenario. Wee et al. [21] considered imperfect quality, shortage backorder, and item deterioration in an integrated production–inventory model. Lee and Kim [22] also developed an integrated production–inventory model of imperfect quality deteriorating items.

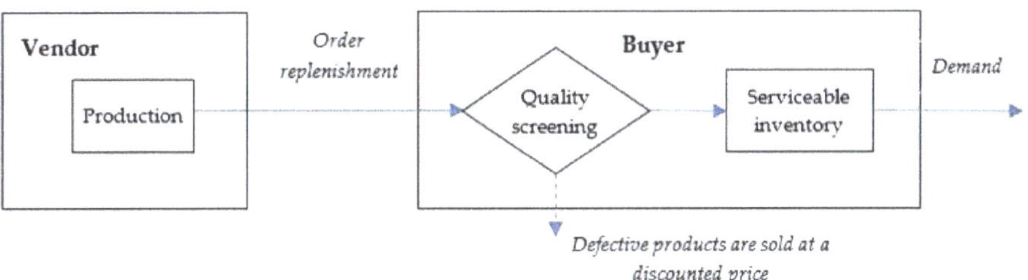

Figure 1. Illustration of an integrated inventory model for imperfect quality items where defective products are sold at a discounted price.

Bazan et al. [23] studied the effect of imperfect quality in different vendor–buyer inventory models. In their study, the vendor performs the inspection and considers one of three possible decisions regarding the defective items: (1) scrap off, (2) salvage at a discounted price, and (3) rework. Figure 2 illustrates the scenario when the vendor performs the inspection. Sarkar et al. [24] studied the integrated inventory model with two-stage inspection by the vendor considering the rework process and variable transportation cost. Yu and Hsu [25] developed a production–inventory model in which the defective items are returned to the vendor immediately for rework.

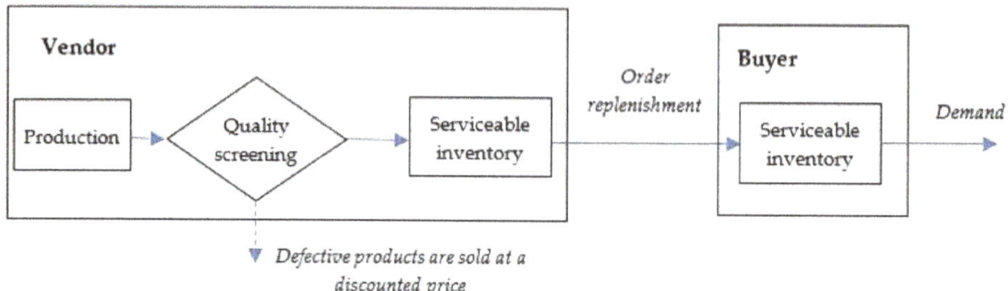

Figure 2. Illustration of an integrated inventory model for imperfect quality items with vendor's inspection.

2.2. Low-Carbon Supply Chain Management

Research on low-carbon supply chain management has increased rapidly in recent years and has been marked by a surge in the amount of literature in this area. Much of this research revealed that supply chain collaboration and the adjustment on operational decisions could reduce carbon emissions without significantly increasing their costs. Wahab et al. [10] studied the optimal shipment size and number of shipments for a two-echelon supply chain with carbon emission cost from transporting the inventory. The emissions are affected by the distance traveled, vehicle fuel efficiency, and the actual shipment weight. Benjaafar et al. [26] modified the traditional supply chain model by associating the carbon footprint from placing an order to the supplier, production setup, production process, and inventory holding. Fahimnia et al. [27] studied the impact of carbon pricing on a closed-loop supply chain through a case study.

Bozorgi et al. [28] considered carbon emissions from transporting and storing cold items that require temperature-controlled trucks and freezers. Bozorgi [29] extended the previous model considering multi-product cold items under limited capacity. Hariga et al. [30] incorporated carbon emissions from transporting and storing the cold items in a three-echelon supply chain. Ghosh et al. [31] considered carbon emissions from production, inventory holding, and transportation in a vendor–buyer supply chain under a single setup and multiple deliveries policy. Toptal and Çetinkaya [32] studied the effect of supply chain coordination and carbon emissions on vendor–buyer order quantity under lot-for-lot delivery. Bouchery et al. [33] examined different supply chain coordination configurations considering carbon emissions under limited vehicle capacity. Dwicahyani et al. [34] incorporated carbon emission costs, energy cost, and waste disposal for a two-echelon supply chain with remanufacturing. Li et al. [35] considered joint carbon tax and cap-and-trade policies for a two-echelon supply chain production–distribution model with transportation outsourcing. Wangsa [36] incorporated the government's penalties and incentive policies to reduce carbon emissions.

Anvar et al. [37] considered emissions from transportation and inventory-holding activities in a one-supplier multi-retailer supply chain. Hariga et al. [38] considered carbon tax and carbon cap policies for a two-echelon supply chain with vendor-managed consignment inventory partnership. The model incorporated emissions from the ordering process, production setup, and holding the inventory. Ji et al. [39] considered a carbon reduction investment from the supplier to get higher customer demand. Wang and Ye [40] compared the effect of considering carbon emissions on just-in-time and economic order quantity decisions for two-echelon supply chain inventory models. Ghosh et al. [41] considered a carbon tax regulation to minimize the total expected cost of a supply chain under stochastic demand and shortage backorder. Ma et al. [42] considered the effect of the carbon tax scheme between suppliers and buyer for production, procurement, and pricing decisions. Darom et al. [43] developed a manufacturer–retailer inventory model considering disruption risks and recovery with safety stock and the effect of carbon emission costs. The model considered carbon emissions from the transportation activities for a better recovery plan. Huang et al. [44] studied inventory and pricing decisions considering carbon emission, production disruption, and controllable deterioration using preservation technology. Recently, Daryanto et al. [45] proposed a low-carbon three-echelon supply chain inventory model considering item deterioration. Kundu and Chakrabarti [46] developed a low-carbon supply chain inventory model taking into account the effect of inflation and the time-value of money. Other researchers incorporated carbon emissions in supplier selection and order allocation [47–49].

3. Problem Definition, Assumption, and Notations

3.1. Problem Definition

This study considers a manufacturer–retailer supply chain that produces one type of item sold solely through one channel. The retailer orders n deliveries of equal lot size (Q) per cycle. The manufacturer implements single-setup multiple-deliveries (SSMD). Hence, it produces nQ units of item per production cycle. This study develops two models considering two cases of quality inspection. (1) In the first, the buyer performs a complete quality inspection process. (2) In the second, the vendor performs the quality inspection so that no defective products are passed to the buyer. The defective items are sold at a discounted price with no additional cost in both scenarios. Both the vendor and the buyer consider the carbon emission costs in their decision to comply with the carbon tax regulation. Model 1 is an extension of Tiwari et al.'s [15] model by introducing weight and distance-dependent transportation costs in addition to the fixed and variable inspection costs. Later, the second model extends the first model by studying another inspection option to reduce the expected total costs and emissions.

3.2. Assumption and Notation

This study explores real-life problems of cost minimization and carbon emission reduction under certain assumptions of a controlled situation. The assumptions are listed below, while the notations are presented in Table 2.

1. The retailer's demand rate and the manufacturer's production rate are known and constant.
2. The manufacturer implements a single-setup multiple-deliveries (SSMD) policy. Based on the retailer's order, the manufacturer produces nQ units of item per production cycle to reduce the setup time and cost. Then, it delivers the item in an equal lot sizes and constant time intervals [50].
3. The replenishment is instantaneous.
4. The items deteriorate in the manufacturer and retailer's inventory. The deterioration rate for both the manufacturer and retailer are equal and constant.
5. The defective percentage, u, has a uniform distribution where $0 \leq \alpha < \beta < 1$.
6. Good products are always available during the quality inspection as $x > D$.
7. The retailer (in Model 1) and the manufacturer (in Model 2) perform a 100% quality inspection to ensure an excellent service.
8. The fixed inspection cost per cycle is constant, whether performed by the buyer or the manufacturer.
9. Carbon emissions come from the fuel and electricity consumption during transporting and holding the inventory.
10. Shortage is not considered.
11. The additional fuel consumption is a linear function of truckloads. Figure 3 illustrates the linear fuel consumption model, which is similar to that of Hariga et al. [30] with an example of the dataset from Volvo Corporation [51].

Table 2. List of notations.

Symbol	Definition
D	demand rate (unit/year);
P	production rate (unit/year);
R	production quantity; $R = PT_1$;
θ	deterioration rate; $(0 \leq \theta < 1)$;
u	the probability of defective products per delivery lot size;
x	quality screening rate (unit/year);
i_c	fixed quality inspection cost (\$/cycle);
u_c	unit inspection cost (\$/unit);
c	retailer's ordering cost (\$/order);
h_d	retailer's holding cost (\$/unit/year);
d_d	retailer's deteriorating cost (\$/unit);
s	manufacturer's setup cost (\$/order);
h_p	manufacturer's holding cost (\$/unit/year);
d_p	manufacturer's deteriorating cost (\$/unit);
t_f	manufacturer's fixed transportation cost per delivery (\$/delivery);
t_v	fuel price for manufacturer's variable transportation cost (\$/liter);
d	distance traveled from vendor to buyer (km);
w	product weight (ton/unit);
c_1	average vehicle fuel consumption when empty (liter/km);
c_2	average additional fuel consumption per ton of load (liter/km/ton);
T_x	carbon emission tax (\$/tonCO$_2$);
F_e	average emissions from fuel combustion (tonCO$_2$/liter);
E_e	average emissions from electricity generation (tonCO$_2$/kWh);
e_1	transportation emission cost (\$/km); $e_1 = c_1 F_e T_x$;
e_2	average additional transportation emission cost per unit product (\$/unit/km); $e_2 = c_2 w F_e T_x$;
c_c	average warehouse energy consumption per unit product (kWh/unit/year);
w_e	warehouse emissions cost per unit product (\$/unit/year); $w_e = e_c E_e T_x$;
T	cycle length;
T_1	production period for the manufacturer in each cycle;
T_2	nonproduction period for the manufacturer in each cycle;
T_i	inspection time per delivery for the retailer;
T_b	inventory cycle length per delivery for the retailer; $T_b = T/n$;
$I_p(t)$	manufacturer's inventory level at time t;
$I_{pd}(t)$	manufacturer's inventory for defective products at time t;
$I_d(t)$	retailer's inventory level at time t;
ETC_d	retailer's expected total cost per year (\$/year);
ETC_p	manufacturer's expected total cost per year (\$/year);
ETC	joint expected total cost per year (\$/year);
ETE_d	retailer's expected total carbon emissions per year (tonCO$_2$/year);
ETE_p	manufacturer's expected total carbon emissions per year (tonCO$_2$/year);
ETE	joint expected total carbon emissions per year (tonCO$_2$/year);

Decision variables

Q	delivery lot size (unit);
n	number of deliveries per order (positive integer).
*	indicates optimal solution

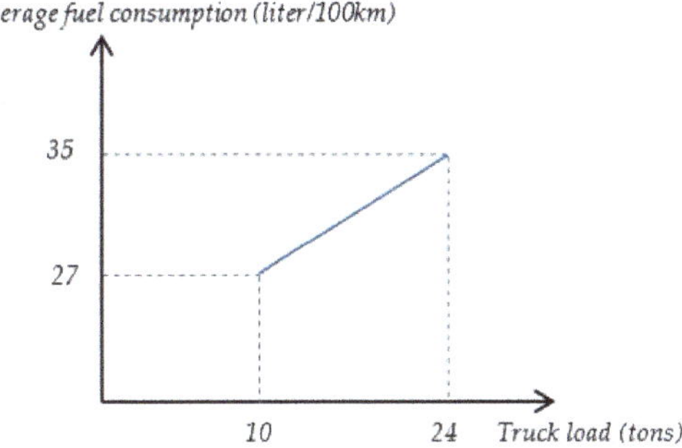

Figure 3. Linear function of average fuel consumption vs. total weight for regional traffic (Data source: Volvo Corporation, 2018).

4. Model Development

This section provides model development for two inspection cases.

4.1. Model Development with Retailer Inspection

This sub-section presents model development when quality inspection becomes the retailer's responsibility. The model is adapted from Tiwari et al. [15]. However, this study considers weight and distance-dependent transportation cost.

The inventory level of the manufacturer and the retailer is illustrated in Figure 4. In one production cycle, the manufacturer produces PT_1 units of the item, and delivers all the produced items to the retailer n times with a constant lot size Q. Right after receiving each lot, the retailer starts the quality inspection that ends at T_i. At T_i, uQ units of defective products will be removed from the inventory. During the period $[0, T/n]$, the retailer's inventory decreases due to demand.

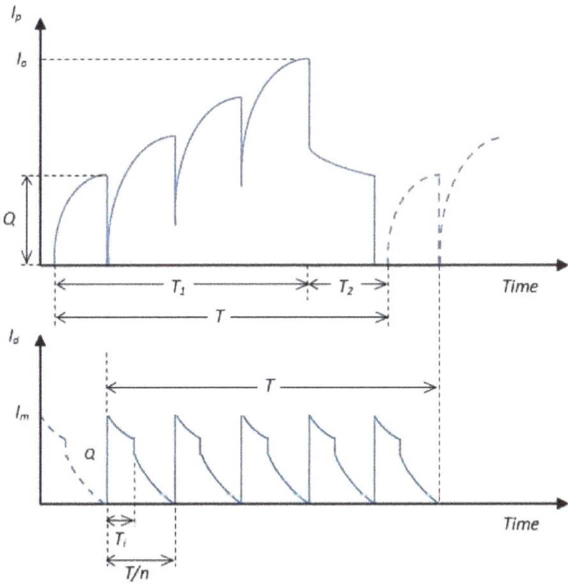

Figure 4. Manufacturer's and retailer's inventory model for constantly deteriorating items with retailer's inspection for $n = 5$.

4.1.1. Retailer Cost and Emission

As c is the retailer's ordering cost, the ordering cost per year is given by c/T (Lee and Kim [22], Yang and Wee [52]). After a lot arrived, the 100% quality inspection starts and then finishes at T_i. As there are fixed inspection costs per delivery, i_c, and unit inspection costs, u_c (Sarkar et al. [12]), the inspection cost per year is given by:

$$i_c \frac{n}{T} + u_c Q \frac{n}{T} = \frac{n}{T}(i_c + u_c Q) \tag{1}$$

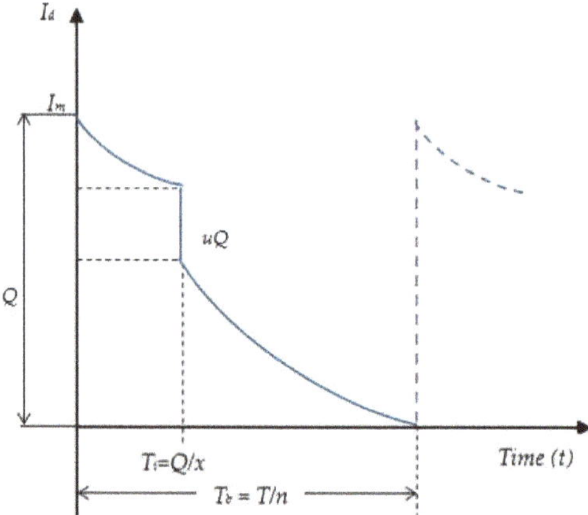

Figure 5. Retailer's inventory model with imperfect quality per delivery cycle.

Figure 5 illustrates the retailer's on-hand inventory per delivery cycle. The inventory level, considering deterioration and imperfect quality, has been studied by researchers such as Jaggi et al. [53] and results in the following equations:

$$I_d(t) = Qe^{-\theta t} + \frac{D}{\theta}\left(e^{-\theta t} - 1\right), \, 0 \le t \le Q/x \tag{2}$$

$$I_d(t) = Qe^{-\theta t} + \frac{D}{\theta}\left(e^{-\theta t} - 1\right) - uQ, \, Q/x \le t \le T/n \tag{3}$$

At $t = T/n$, $I_d(T/n) = 0$, therefore equation (3) becomes:

$$Qe^{-\theta T/n} + \frac{D}{\theta}\left(e^{-\theta T/n} - 1\right) - uQ = 0$$

Solving the equation for Q, one has:

$$Q = -\frac{D\left(e^{-\theta T/n} - 1\right)}{\theta\left(e^{-\theta T/n} - u\right)} = \frac{D\left(e^{-\theta T/n} - 1\right)}{\theta\left(u - e^{-\theta T/n}\right)} \text{ which is } = \frac{D\left(e^{\theta T/n} - 1\right)}{\theta\left(1 - ue^{\theta T/n}\right)} \text{ (Jaggi et al. [53])}$$

Hence:

$$Q = I_d(0) = \frac{D\left(e^{\theta T/n} - 1\right)}{\theta\left(1 - ue^{\theta T/n}\right)} \tag{4}$$

Further, the on-hand inventory per year for the retailer is:

$$\frac{n}{T}\left[\int_0^{Q/x} I_d(t)dt + \int_{Q/x}^{T/n} I_d(t)dt\right]$$

$$\frac{n}{T}\left[\frac{Q}{\theta}\left(1-e^{-\frac{\theta Q}{x}}\right)-\frac{D}{\theta^2}\left(\frac{Q\theta}{x}+e^{-\frac{\theta Q}{x}}-1\right)-\frac{1}{\theta}\left(e^{-\frac{\theta T}{n}}-e^{-\frac{\theta Q}{x}}\right)\left(Q+\frac{D}{\theta}\right)-\left(\frac{T}{n}-\frac{Q}{x}\right)\left(uQ+\frac{D}{\theta}\right)\right] \tag{5}$$

We assumed that storing the items requires electrical energy with a certain amount of carbon footprint. Therefore, substituting Equations (4) with (5) and considering the retailer's holding costs and emissions, the inventory holding and emission cost per year becomes:

$$(h_d+w_e)\frac{n}{T}\left[\frac{D}{\theta^2\left(ue^{\frac{\theta T}{n}}-1\right)^2}\left(\frac{uD}{x}\left(1+e^{\frac{2\theta T}{n}}-2e^{\frac{\theta T}{n}}\right)\right.\right.$$
$$\left.\left.+u\left(1-e^{\frac{2\theta T}{n}}+ue^{\frac{2\theta T}{n}}-ue^{\frac{\theta T}{n}}\right)+\frac{\theta T}{n}\left(u-1+ue^{\frac{\theta T}{n}}-u^2e^{\frac{\theta T}{n}}\right)+e^{\frac{\theta T}{n}}-1\right)\right] \tag{6}$$

Considering the expected probability value of the defective products ($E[u]$), the average warehouse energy consumption per unit product (e_c), and the average emission from electricity generation (Ee), from Equation (6), the retailer's expected carbon footprint (ETE_d) per year of holding the inventory is:

$$ETE_d \quad = (e_cE_e)\frac{n}{T}\left[\frac{D}{\theta^2\left(E[u]e^{\frac{\theta T}{n}}-1\right)^2}\left(\frac{E[u]D}{x}\left(1+e^{\frac{2\theta T}{n}}-2e^{\frac{\theta T}{n}}\right)\right.\right.$$
$$+E[u]\left(1-e^{\frac{2\theta T}{n}}+E[u]e^{\frac{2\theta T}{n}}-E[u]e^{\frac{\theta T}{n}}\right)$$
$$\left.\left.+\frac{\theta T}{n}\left(E[u]-1+E[u]e^{\frac{\theta T}{n}}-E[u]^2e^{\frac{\theta T}{n}}\right)+e^{\frac{\theta T}{n}}-1\right)\right] \tag{7}$$

The retailer's deteriorating cost per year is:

$$\frac{d_dn}{T}\left(Q-uQ-\frac{DT}{n}\right)=d_d\left(\frac{(1-u)n}{T}\frac{D\left(e^{\theta T/n}-1\right)}{\theta\left(1-ue^{\theta T/n}\right)}-D\right) \tag{8}$$

The retailer's total cost is the sum of the ordering, inspection, deteriorating, inventory holding, and emission costs. Therefore, considering the probability of the defective products, the expected total cost per year is:

$$ETC_d \quad = \frac{c}{T}+\frac{n}{T}\left(i_c+u_c\frac{D\left(e^{\frac{\theta T}{n}}-1\right)}{\theta\left(1-E[u]e^{\frac{\theta T}{n}}\right)}\right)+d_d\left(\frac{(1-E[u])n}{T}\frac{D\left(e^{\theta T/n}-1\right)}{\theta\left(1-E[u]e^{\theta T/n}\right)}-D\right)$$
$$+(h_d+w_e)\frac{nD}{T\theta^2\left(E[u]e^{\frac{\theta T}{n}}-1\right)^2}\left(\frac{E[u]D}{x}\left(1+e^{\frac{2\theta T}{n}}-2e^{\frac{\theta T}{n}}\right)\right.$$
$$+E[u]\left(1-e^{\frac{2\theta T}{n}}+E[u]e^{\frac{2\theta T}{n}}-E[u]e^{\frac{\theta T}{n}}\right)$$
$$\left.+\frac{\theta T}{n}\left(E[u]-1+E[u]e^{\frac{\theta T}{n}}-E[u]^2e^{\frac{\theta T}{n}}\right)+e^{\frac{\theta T}{n}}-1\right) \tag{9}$$

4.1.2. Manufacturer Cost and Emission

After the arrival of the retailer's order, the manufacturer starts the production of nQ units of the item at a production rate P. Since s is the setup cost per production cycle, the manufacturer's setup cost per year is s/T.

The first delivery occurs as soon as the quantity is met. The following deliveries occur at T/n intervals. The transportation cost belongs to the manufacturer and consist of fixed and variable costs. Swenseth and Godfrey [54], Nie et al. [55], Rahman et al. [56], and Wangsa and Wee [57] incorporated the variable transportation cost, which is affected by the shipping distance and truckloads. The manufacturer's transportation cost per delivery is given by:

$$t_f+2dc_1t_v+dQwc_2t_v \tag{10}$$

The first element is the fixed transportation setup cost. The second element calculates the transportation cost of an empty truck. As the truck goes from the manufacturer to the retailer and then goes back, the distance is multiplied by two. Then, the transportation cost for the truckload is calculated, which depends on the delivery distance and quantity, product weight, additional fuel consumption per ton per km, and the fuel price. Substituting Equation (4) to (10), the manufacturer's transportation cost per year is given by:

$$\frac{n}{T}\left(t_f + 2dc_1 t_v + d\frac{D\left(e^{\theta T/n} - 1\right)}{\theta\left(1 - ue^{\theta T/n}\right)}wc_2 t_v\right) \tag{11}$$

Wahab et al. [10] identified that the emissions from transportation were affected by the delivery distance, actual shipment weight, fuel consumption per km, and CO_2 emissions per liter of fuel. Therefore, the amount of the manufacturer's carbon emission per year as the result of transportation activity can be derived as follows:

$$\frac{n}{T}(2dc_1 + dQwc_2)F_e = \frac{n}{T}\left(2dc_1 + d\frac{D\left(e^{\theta T/n} - 1\right)}{\theta\left(1 - ue^{\theta T/n}\right)}wc_2\right)F_e \tag{12}$$

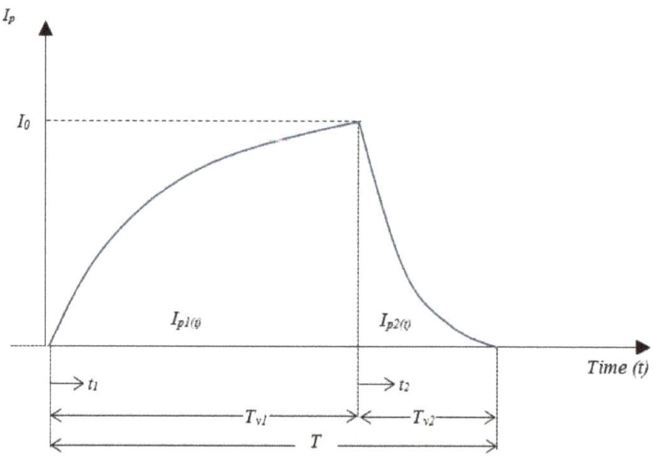

Figure 6. $I_{p1}(t_1)$ and $I_{p2}(t_2)$ vs. time.

As depicted in Figure 6, Lee and Kim [22] studied a similar inventory model for the manufacturer using Yang and Wee's [52] approach. Both production and consumption occur during T_1, while only consumption occurs during T_2. Hsu [58] suggested some revision considering the effect of defective products. Due to defective products and the retailer's quality inspection, the effective demand rate for the manufacturer becomes $D/(1\text{-}u)$. Therefore, the inventory functions are as follows:

$$I_{p1}(t_1) = \frac{P - (D/(1-u))}{\theta}\left(1 - e^{-\theta\,t_1}\right), \ 0 \le t_1 \le T_1 \tag{13}$$

$$I_{p2}(t_2) = \frac{(D/(1-u))}{\theta}\left(e^{\theta\,(T_2-t_2)} - 1\right), 0\ t_2 \le T_2 \tag{14}$$

From the boundary condition $I_{p1}(T_1) = I_{p2}(0)$ and following Misra's [59] approximation:

$$\left(P - \frac{D}{1-u}\right)T_1\left(1 - \frac{1}{2}\theta T_1\right) = \left(\frac{D}{1-u}\right)T_2\left(1 + \frac{1}{2}\theta T_2\right)$$

Hence, one has:

$$T_1 \approx \frac{D}{(1-u)P-D} T_2 \left(1 + \frac{1}{2}\theta T_2\right) \tag{15}$$

$$T \approx \frac{T_2}{(1-u)P-D} \left((1-u)P + \frac{1}{2}D\theta T_2\right) \tag{16}$$

From Yang and Wee [52], the manufacturer's inventory per cycle is:

$$\int_0^{T_1} I_{p1}(t_1)dt_1 + \int_0^{T_2} I_{p2}(t_2)dt_2 - n \int_0^{T/n} I_d(t)dt \tag{17}$$

Hence, the manufacturer's holding cost per year is:

$$\begin{aligned}
\frac{h_p}{T} &\left[\frac{P-(D/(1-u))}{\theta}T_1 + \frac{P-(D/(1-u))}{\theta^2}\left(e^{-\theta\,T_1}-1\right) \right. \\
&\left. - \frac{(D/(1-u))T_2}{\theta} - \frac{(D/(1-u))}{\theta^2}\left(1-e^{\theta\,T_2}\right) \right. \\
&-n\left(\frac{D}{\theta^2\left(ue^{\frac{\theta T}{n}}-1\right)^2}\left(\frac{uD}{x}\left(1+e^{\frac{2\theta T}{n}}-2e^{\frac{\theta T}{n}}\right)+u\left(1-e^{\frac{2\theta T}{n}}+ue^{\frac{2\theta T}{n}}-ue^{\frac{\theta T}{n}}\right)\right.\right. \\
&\left.\left.\left. +\frac{\theta T}{n}\left(u-1+ue^{\frac{\theta T}{n}}-u^2e^{\frac{\theta T}{n}}\right)+e^{\frac{\theta T}{n}}-1\right)\right)\right]
\end{aligned} \tag{18}$$

The amount of a manufacturer's carbon emissions per year as the result of warehousing activity can be derived as follows:

$$\begin{aligned}
\frac{e_c E_c}{T} &\left[\frac{P-(D/(1-u))}{\theta}T_1 + \frac{P-(D/(1-u))}{\theta^2}\left(e^{-\theta\,T_1}-1\right) \right. \\
&\left. - \frac{(D/(1-u))T_2}{\theta} - \frac{(D/(1-u))}{\theta^2}\left(1-e^{\theta\,T_2}\right) \right. \\
&-n\left(\frac{D}{\theta^2\left(ue^{\frac{\theta T}{n}}-1\right)^2}\left(\frac{uD}{x}\left(1+e^{\frac{2\theta T}{n}}-2e^{\frac{\theta T}{n}}\right)+u\left(1-e^{\frac{2\theta T}{n}}+ue^{\frac{2\theta T}{n}}-ue^{\frac{\theta T}{n}}\right)\right.\right. \\
&\left.\left.\left. +\frac{\theta T}{n}\left(u-1+ue^{\frac{\theta T}{n}}-u^2e^{\frac{\theta T}{n}}\right)+e^{\frac{\theta T}{n}}-1\right)\right)\right]
\end{aligned} \tag{19}$$

The manufacturer's carbon emissions per year come from Equations (12) and (19). Therefore, the manufacturer's carbon emission cost is:

$$\begin{aligned}
&\frac{n}{T}\left(2de_1 + d\frac{D\left(e^{\theta T/n}-1\right)}{\theta\left(1-ue^{\theta T/n}\right)}e_2\right) \\
&+\frac{w_e}{T}\left[\frac{P-(D/(1-u))}{\theta}T_1 + \frac{P-(D/(1-u))}{\theta^2}\left(e^{-\theta\,T_1}-1\right) - \frac{(D/(1-u))T_2}{\theta} - \frac{(D/(1-u))}{\theta^2}\left(1-e^{\theta\,T_2}\right) \right. \\
&-n\left(\frac{D}{\theta^2\left(ue^{\frac{\theta T}{n}}-1\right)^2}\left(\frac{uD}{x}\left(1+e^{\frac{2\theta T}{n}}-2e^{\frac{\theta T}{n}}\right)+u\left(1-e^{\frac{2\theta T}{n}}+ue^{\frac{2\theta T}{n}}-ue^{\frac{\theta T}{n}}\right)\right.\right. \\
&\left.\left.\left. +\frac{\theta T}{n}\left(u-1+ue^{\frac{\theta T}{n}}-u^2e^{\frac{\theta T}{n}}\right)+e^{\frac{\theta T}{n}}-1\right)\right)\right]
\end{aligned} \tag{20}$$

The loss due to deterioration in the manufacturer's inventory is the total production during the period T_1, minus the total delivered products to the retailer's inventory. Therefore, the manufacturer's deteriorating cost per year is:

$$\frac{d_p}{T}\left(PT_1 - n\left(\frac{D\left(e^{\theta T/n}-1\right)}{\theta\left(1-ue^{\theta T/n}\right)}\right)\right) \tag{21}$$

From Equations (11), (18), (20), and (21), as well as the setup cost, and considering the probability of the defective products, the manufacturer's expected total cost per year is:

$$
\begin{aligned}
ETC_p \quad &= \frac{S}{T} + \frac{n}{T}\left(t_f + 2dc_1 t_v + d\frac{D\left(e^{\theta T/n}-1\right)}{\theta\left(1-E[u]e^{\theta T/n}\right)}wc_2 t_v \right) \\
&+ \frac{(h_p+w_e)}{T}\left[\frac{P-(D/(1-E[u]))}{\theta}T_1 + \frac{P-(D/(1-E[u]))}{\theta^2}\left(e^{-\theta\, T_1}-1\right) \right. \\
&- \frac{(D/(1-E[u]))T_2}{\theta} - \frac{(D/(1-E[u]))}{\theta^2}\left(1-e^{\theta\, T_2}\right) \\
&-n\left(\frac{D}{\theta^2\left(E[u]e^{\frac{\theta T}{n}}-1\right)^2}\left(e^{\frac{\theta T}{n}}-1+\frac{E[u]D}{x}\left(1+e^{\frac{2\theta T}{n}}-2e^{\frac{\theta T}{n}}\right) \right.\right. \\
&\left.+E[u]\left(1-e^{\frac{2\theta T}{n}}+E[u]e^{\frac{2\theta T}{n}}-E[u]e^{\frac{\theta T}{n}}\right) \right. \\
&\left.\left.+\frac{\theta T}{n}\left(E[u]-1+E[u]e^{\frac{\theta T}{n}}-E[u]^2 e^{\frac{\theta T}{n}}\right)\right)\right] \\
&+\frac{n}{T}\left(2de_1+d\frac{D\left(e^{\theta T/n}-1\right)}{\theta\left(1-E[u]e^{\theta T/n}\right)}e_2\right)+\frac{d_p}{T}\left(PT_1-n\left(\frac{D\left(e^{\theta T/n}-1\right)}{\theta\left(1-ue^{\theta T/n}\right)}\right)\right)
\end{aligned} \tag{22}
$$

4.1.3. The Integrated Manufacturer and Retailer Cost Function

In an integrated decision, the manufacturer and the retailer jointly specify n, which minimizes the expected total cost (ETC). The ETC is the sum of ETC_d and ETC_p in Equations (9) and (22).

Using Taylor's series expansion for a small value of $\theta T/n$, θT_1, and θT_2, we can solve the cost function by assuming e^x as $1 + x + x^2/2 + x^3/6$. Furthermore, the expected total emissions (ETE) per year from the manufacturer and the retailer can be derived from Equations (7), (12), and (19).

4.1.4. Methodology and Solution Search

The objective is to determine the optimal number of deliveries (n^*) that minimize the expected total cost function (ETC). The value of n^* and the respective T, T_1, and T_2 will lead us to the optimal delivery quantity Q^* and production quantity R. The proposed procedure to derive the positive integer decision variable n is adapted from Tiwari et al. [15] and Yang and Wee [52] as follows:

Step 1. Substitute the T_1 and T functions in Equations (15) and (16) into ETC;

Step 2. Input all the known parameters;

Step 3. Set $n = 1$;

Step 4. Derive the partial derivative of ETC with respect to T_2 and set it to zero. Solve the equation to find the value of T_2;

Step 5. Use the known n and T_2 to find the value of T_1 and T using Equations (15) and (16).

Step 6. Derive the corresponding ETC;

Step 7. If $ETC(n) > ETC(n-1)$, then $n^* = n-1$ and go to Step 8; otherwise, set $n = n+1$ and go back to Step 4;

Step 8. Use n^* and the corresponding T^* to find Q^* from Equation (4) and calculate $R = PT_1{}^*$.

4.2. Model Development with Manufacturer Inspection

This sub-section presents the model development when quality inspection becomes the manufacturer's responsibility. The purpose of such a policy is to prevent transporting defective products. The manufacturer performs a quality inspection of all the produced products and keeps the defective products separately until the end of the production period T_1. The defective products will be sold at a discounted price to the secondary market. The inventory level for both the manufacturer (I_p) and the retailer (I_d), including the manufacturer's inventory of the defective products (I_{pd}), is illustrated in Figure 7. In this model, the I_{pd} is accumulated during the T_1 period. Besides, I_d decreases solely due to demand.

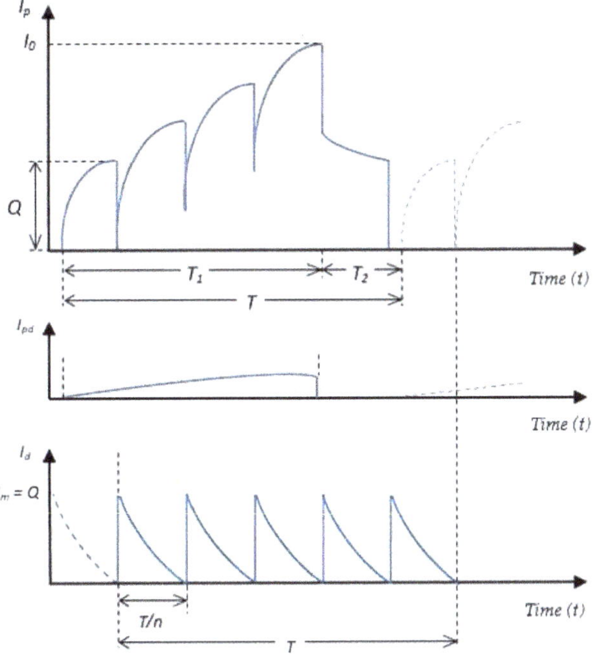

Figure 7. Manufacturer's and retailer's inventory model for constant deteriorating items with manufacturer's inspection for $n = 5$.

4.2.1. Retailer Cost and Emission

The retailer's total cost is the sum of the ordering, inventory holding, deteriorating, and emission costs. The ordering cost per year is given by c/T, which is similar to Model 1 (Section 4.1.1). During the T/n period, I_d decreases due to demand and deterioration. From Yang and Wee [52], the inventory function is as follows:

$$I_d(t) = \frac{D}{\theta}\left(e^{\theta(\frac{T}{n} - t)} - \right), 0\ t \le T/n \tag{23}$$

and

$$Q = I_d(0) = \frac{D}{\theta}\left(e^{\frac{\theta T}{n}} - 1\right) \tag{24}$$

and the holding cost per year is:

$$h_d\frac{n}{T}\left(\int_0^{T/n} I_d(t)dt\right) = h_d\frac{n}{T}\left(\frac{D}{\theta}\left(\frac{1}{\theta}\left(e^{\frac{\theta T}{n}} - 1\right) - \frac{T}{n}\right)\right) \tag{25}$$

Further, the deteriorating cost per year is:

$$d_d\frac{n}{T}\left(Q - D\frac{T}{n}\right) = d_d\frac{n}{T}\left(\frac{D}{\theta}\left(e^{\frac{\theta T}{n}} - 1\right) - \frac{DT}{n}\right) \tag{26}$$

$$ETE_d = e_c E_e\left(\frac{n}{T}\frac{D}{\theta}\left(\frac{1}{\theta}\left(e^{\frac{\theta T}{n}} - 1\right) - \frac{T}{n}\right)\right) \tag{27}$$

Equation (27) shows the retailer's carbon emission (ETE_d) from holding the inventory. Therefore, the retailer's carbon emission cost per year is:

$$w_e\left(\frac{n}{T}\frac{D}{\theta}\left(\frac{1}{\theta}\left(e^{\frac{\theta T}{n}}-1\right)-\frac{T}{n}\right)\right) \tag{28}$$

The retailer's expected total cost per year (ETC_d) becomes:

$$ETC_d = \frac{c}{T} + (h_d + w_e)\frac{n}{T}\left(\frac{D}{\theta}\left(\frac{1}{\theta}\left(e^{\frac{\theta T}{n}}-1\right)-\frac{T}{n}\right)\right) + d_d\frac{n}{T}\left(\frac{D}{\theta}\left(e^{\frac{\theta T}{n}}-1\right)-\frac{DT}{n}\right) \tag{29}$$

4.2.2. Manufacturer Cost and Emission

Due to some percentage of defective products, the production rate of the perfect product is $(1 - u)P$. The manufacturer's setup cost per year is s/T. In this model, the manufacturer will have an additional inspection cost. Since the total number of products being produced per production cycle is PT_1, considering a fixed inspection cost per cycle (i_c) and unit inspection cost (u_c), the manufacturer's inspection cost per year is:

$$\frac{i_c}{T} + \frac{u_c PT_1}{T} \tag{30}$$

The manufacturer's transportation function is similar to Equation (10); therefore, the transportation cost and emissions per year become:

$$\frac{n}{T}\left(t_f + 2dc_1 t_v + d\frac{D}{\theta}\left(e^{\frac{\theta T}{n}}-1\right)wc_2 t_v\right) \tag{31}$$

$$\frac{n}{T}\left(2dc_1 + d\frac{D}{\theta}\left(e^{\frac{\theta T}{n}}-1\right)wc_2\right)F_e \tag{32}$$

From Figure 6, the inventory differential equations are:

$$dI_{p1}(t_1) = ((1-u)P - D)dt_1 - \theta I_{p1}(t_1)dt_1,\ 0 \le t_1 \le T_1$$

$$dI_{p2}(t_2) = -Ddt_2 - \theta I_{p2}(t_2)dt_2,\ 0 \le t_2 \le T_2$$

For the boundary condition for $t_1 = 0$, $I_1(0) = 0$ and for $t_2 = 0$, $I_2(0) = I_o$ and for $t_2 = T_2$, $I_2(T_2) = 0$, the manufacturer's inventory functions for the good products are:

$$I_{p1}(t_1) = \frac{(1-u)P - D}{\theta}\left(1 - e^{-\theta\,t_1}\right),\ 0 \le t_1 \le T_1 \tag{33}$$

$$I_{p2}(t_2) = \frac{D}{\theta}\left(e^{\theta\,(T_2 - t_2)} - 1\right),\ 0\ t_2 \le T_2 \tag{34}$$

From the boundary condition $I_{p1}(T_1) = I_{p2}(0)$, we have the following equation:

$$\frac{((1-u)P - D)}{\theta}\left(1 - e^{-\theta\,T_1}\right) = \frac{D}{\theta}\left(e^{\theta T_2} - 1\right) \tag{35}$$

From Taylor's series expansion and the assumption of $\theta T \ll 1$, following Misra's [59] approximation, one has:

$$((1-u)P - D)T_1\left(1 - \frac{1}{2}\theta T_1\right) = DT_2\left(1 + \frac{1}{2}\theta T_2\right)$$

$$T_1 \approx \frac{D}{(1-u)P - D}T_2\left(1 + \frac{1}{2}\theta T_2\right) \tag{36}$$

$$T \approx \frac{T_2}{(1-u)P - D}\left((1-u)P + \frac{1}{2}D\theta T_2\right) \tag{37}$$

Therefore, the manufacturer's inventory for good products becomes:

$$\int_0^{T_1} \frac{(1-u)P - D}{\theta}\left(1 - e^{-\theta\,t_1}\right)dt_1 + \int_0^{T_2} \frac{D}{\theta}\left(e^{\theta\,(T_2 - t_2)} - 1\right)dt_2 - n\left[\frac{D}{\theta}\left(\frac{1}{\theta}\left(e^{\frac{\theta T}{n}} - 1\right) - \frac{T}{n}\right)\right] \tag{38}$$

Besides, there is an inventory of defective products. From Figure 6, the inventory differential equation for the defective products is:

$$dI_{pd}(t_1) = uPdt_1 - \theta I_{pd}(t_1)dt_1,\ 0 \le t_1 \le T_1$$

For the boundary condition for $t_1 = 0$, $I_1(0) = 0$, the manufacturer's inventory function for the defective products is:

$$I_{pd}(t_1) = \frac{uP}{\theta}\left(1 - e^{-\theta\,t_1}\right),\ 0 \le t_1 \le T_1$$

Therefore, the manufacturer's inventory of the defective products becomes:

$$\int_0^{T_1} \frac{uP}{\theta}\left(1 - e^{-\theta\,t_1}\right)dt_1 \tag{39}$$

Hence, the manufacturer's holding cost per year is:

$$\frac{h_p}{T}\left(\frac{(1-u)P - D}{\theta}T_1 + \frac{(1-u)P - D}{\theta^2}\left(e^{-\theta T_1} - 1\right) - \frac{DT_2}{\theta} - \frac{D}{\theta^2}\left(1 - e^{\theta T_2}\right) \right.$$
$$\left. -n\left(\frac{D}{\theta}\left(\frac{1}{\theta}\left(e^{\frac{\theta T}{n}} - 1\right) - \frac{T}{n}\right)\right) + \frac{uPT_1}{\theta} + \frac{uP}{\theta^2}\left(e^{-\theta T_1} - 1\right)\right) \tag{40}$$

Therefore, based on Equations (32) and (40), the manufacturer's carbon emission cost and the total expected carbon emissions per year can be calculated as follows:

$$\frac{n}{T}\left(2de_1 + d\frac{D}{\theta}\left(e^{\frac{\theta T}{n}} - 1\right)e_2\right)$$
$$+ \frac{w_e}{T}\left(\frac{(1-u)P - D}{\theta}T_1 + \frac{(1-u)P - D}{\theta^2}\left(e^{-\theta T_1} - 1\right) - \frac{DT_2}{\theta} \right.$$
$$- \frac{D}{\theta^2}\left(1 - e^{\theta T_2}\right) - n\left(\frac{D}{\theta}\left(\frac{1}{\theta}\left(e^{\frac{\theta T}{n}} - 1\right) - \frac{T}{n}\right)\right) + \frac{uPT_1}{\theta}$$
$$\left. + \frac{uP}{\theta^2}\left(e^{-\theta T_1} - 1\right)\right) \tag{41}$$

$$\begin{aligned}
ETE_p &= \frac{n}{T}\left(2dc_1 + d\frac{D}{\theta}\left(e^{\frac{\theta T}{n}} - 1\right)wc_2\right)F_e \\
&+ \frac{e_c E_e}{T}\left(\frac{(1-u)P - D}{\theta}T_1 + \frac{(1-u)P - D}{\theta^2}\left(e^{-\theta T_1} - 1\right) - \frac{DT_2}{\theta}\right. \\
&- \frac{D}{\theta^2}\left(1 - e^{\theta T_2}\right) - n\left(\frac{D}{\theta}\left(\frac{1}{\theta}\left(e^{\frac{\theta T}{n}} - 1\right) - \frac{T}{n}\right)\right) + \frac{uPT_1}{\theta} \\
&\left. + \frac{uP}{\theta^2}\left(e^{-\theta T_1} - 1\right)\right)
\end{aligned} \tag{42}$$

The number of deteriorated items in the manufacturer's inventory is the total production during the period T_1, minus the total products delivered to the buyer and the inventory of the defective products. Therefore, the manufacturer's deteriorating cost per year is:

$$\frac{d_p}{T}\left((1-u)PT_1 - n\left(\frac{D}{\theta}\left(e^{\frac{\theta T}{n}} - 1\right)\right) + \left(uPT_1 - \frac{uP}{\theta}\left(1 - e^{-\theta\,T_1}\right)\right)\right) \tag{43}$$

Considering the additional inspection cost and the probability of the defective products, the manufacturer's expected total cost per year is:

$$
\begin{aligned}
ETC_p \;=\; &\tfrac{s}{T} + \tfrac{i_c}{T} + \tfrac{u_c PT_1}{T} + \tfrac{n}{T}\left(t_f + 2dc_1 t_v + d\tfrac{D}{\theta}\left(e^{\frac{\theta T}{n}} - 1\right)wc_2 t_v\right)\\
&+ \tfrac{n}{T}\left(2de_1 + d\tfrac{D}{\theta}\left(e^{\frac{\theta T}{n}} - 1\right)e_2\right)\\
&+ \tfrac{(h_p + w_e)}{T}\left(\tfrac{(1 - E[u])P - D}{\theta}T_1 + \tfrac{(1 - E[u])P - D}{\theta^2}\left(e^{-\theta T_1} - 1\right)\right.\\
&\left. - \tfrac{DT_2}{\theta} - \tfrac{D}{\theta^2}\left(1 - e^{\theta T_2}\right) - n\left(\tfrac{D}{\theta}\left(\tfrac{1}{\theta}\left(e^{\frac{\theta T}{n}} - 1\right) - \tfrac{T}{n}\right)\right) + \tfrac{E[u]PT_1}{\theta}\right.\\
&\left. + \tfrac{E[u]P}{\theta^2}\left(e^{-\theta T_1} - 1\right)\right)\\
&+ \tfrac{d_p}{T}\left((1 - E[u])PT_1 - n\left(\tfrac{D}{\theta}\left(e^{\frac{\theta T}{n}} - 1\right)\right)\right.\\
&\left. + \left(E[u]PT_1 - \tfrac{E[u]P}{\theta}\left(1 - e^{-\theta T_1}\right)\right)\right)
\end{aligned}
\tag{44}
$$

4.2.3. The Integrated Manufacturer and Retailer Cost Function

The *ETC* of the integrated system is the sum of Equations (29) and (44). Using Taylor's series expansion for a small value of $\theta T/n$, θT_1, and θT_2, we can solve the cost function by assuming e^x as $1 + x + x^2/2 + x^3/6$. Furthermore, the *ETE* can be derived from Equation (27) and Equation (42).

4.2.4. Methodology and Solution Search

Similar to Model 1, the objective is to determine the optimal number of deliveries (n^*) that minimize the expected total cost function *ETC*. The proposed procedure to search for the optimum solution is as follows:

Step 1. Substitute the T_1 and T functions in Equations (36) and (37) into *ETC*;

Step 2. Input all the known parameters;

Step 3. Set $n = 1$;

Step 4. Derive the partial derivative of *ETC* with respect to T_2 and set it to zero. Solve the equation to find the value of T_2;

Step 5. Use the known n and T_2 to find the value of T_1 and T using Equations (36) and (37).

Step 6. Derive the corresponding *ETC*;

Step 7. If $ETC(n) > ETC(n\text{-}1)$ then $n^* = n - 1$ and go to step 8, otherwise set $n = n + 1$ and back to Step 4;

Step 8. Use n^* and the corresponding T^* to find Q^* from Equation (24) and calculate $R = PT_1{}^*$.

5. Numerical Example and Management Insights

5.1. Numerical Example 1

The values of the parameters are considered by adopting data from Yang and Wee [52], Hariga et al. [30], and Tiwari et al. [15] as P = 2,000,000 units/year, D = 500,000 units/year, x = 1,725,000 unit/year, i_c = \$500/delivery, u_c = \$0.5/unit, c = \$2,000/order, s = \$100,000/setup, h_d = \$60/unit/year, h_p = \$40/unit/year, d_d = \$600/unit, d_p = \$400/unit, θ = 0.1, d = 100 km, t_f = \$1000/delivery, t_v = \$0.75/liter, w = 0.01 ton/unit, c_1 = 27 L/100 km, c_2 = 0.57 L/100 km/ton truckload, e_c = 1.44 kWh/unit/year, T_x = \$75/tonCO$_2$, F_e = 2.6 × 10^{-3} tonCO$_2$/L (US. EPA [60]), E_e = 0.5 × 10^{-3} tonCO$_2$/kWh (McCarthy [61]), and u is uniformly distributed in which α = 0 and β = 0.04, with $E[u]$ = 0.02.

The minimum value of joint expected total cost can be obtained at n^* = 7 with T_2 = 0.0651856, T_1 = 0.0223966, and T = 0.0875822, as shown in Table 3. The *ETC* is \$2,834,922/year, which is from Equation (4), the optimum Q is 6,387.7 units. The optimum R is 44,793.2 units, and the *ETE* is 30.598 tonCO$_2$/year. If the supply chain solely minimizes the total amount of carbon footprint, the decision is to perform a single-setup single-delivery (SSSD) as n = 1 with *ETE* = 18.460 tonCO$_2$/year (saving

39.7%). However, this situation increases the *ETC* into \$3,366,391 (18.7%). Figure 8 shows the convexity of *ETC* when $n = 7$.

Table 3. Expected total cost for different n in Model 1.

n	$T_2(10^{-5})$	$T_1(10^{-5})$	$T(10^{-5})$	ETC_d	ETC_p	ETC	ETE
1	4960	1703	6663	2,348,991	1,017,400	3,366,391	18.460
2	5641	1937	7578	1,463,323	1,568,493	3,031,816	21.333
3	5961	2048	8009	1,122,012	1,799,193	2,921,205	23.501
4	6161	2117	8278	941,509	1,930,509	2,872,018	25.422
5	6306	2167	8473	830,673	2,017,748	2,848,421	27.216
6	6422	2206	8628	756,370	2,081,526	2,837,896	28.935
7 *	6518	2240	8758	703,611	2,131,311	2,834,922	30.598
8	6603	2269	8872	664,620	2,172,067	2,836,687	32.218
9	6680	2295	8976	634,952	2,206,651	2,841,603	33.805
10	6751	2320	9071	611,889	2,236,823	2,848,712	35.360

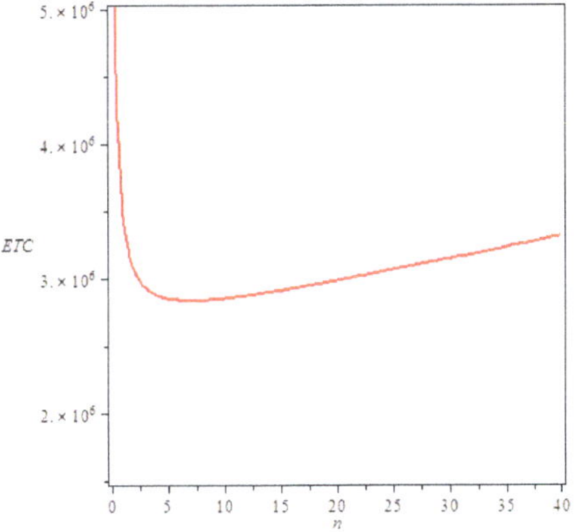

Figure 8. Graphical representation of expected total cost (*ETC*) for a fixed n in Model 1.

When the probability of defective products, unit inspection costs, carbon tax, and variable transport cost are equal to zero ($E[u] = u_c = T_x = t_v = 0$), the results are $n = 7$, $T = 0.08791$, and $ETC = \$2,559,246$ which are similar to the results of Yang and Wee [52].

5.2. Numerical Example 2

We consider the parameters in numerical example one and solve it using the Model 2 results of the following values. The minimum value of joint expected total cost can be obtained at $n^* = 9$ and $T = 0.08869$, as shown in Table 4. The Q, R, and ETC are 4,929.6 units, 45,360.7 units, and \$2,782,396/year, respectively. The *ETE* is 33.52 tonCO$_2$/year.

Table 4. Expected total cost for different n in Model 2.

n	$T_2(10^{-5})$	$T_1(10^{-5})$	$T(10^{-5})$	ETC_d	ETC_p	ETC	ETE
1	4977	1709	6686	2,041,005	1,312,795	3,353,800	18.22
2	5653	1941	7595	1,167,506	1,844,193	3,011,699	21.07
3	5963	2048	8011	827,178	2,067,759	2,894,937	23.21
4	6151	2113	8264	644,684	2,195,273	2,839,957	25.12
5	6282	2158	8441	530,652	2,280,088	2,810,741	26.91
6	6384	2193	8577	425,568	2,342,142	2,794,711	28.63
7	6467	2222	8689	395,714	2,390,607	2,786,322	30.29
8	6538	2246	8784	352,240	2,430,297	2,782,747	31.92
9 *	6601	2268	8869	318,411	2,463,985	2,782,396	33.52
10	6658	2288	8945	290,923	2,493,378	2,784,301	35.11

Table 5 provides the cost comparison between the two models for the result of examples one and two. The number of deliveries per cycle (n) is higher in Model 2 (when the manufacturer performs the quality inspection), while the delivery lot size (Q) is lower. All the retailer's cost components decrease, while all the manufacturer's cost components increase, except for the setup cost. In total, the ETC of Model 2 is 1.85% lower than the ETC in Model 1. However, the retailer's total costs were reduced by 54.7%, while the manufacturer's total costs increased by 15.6%. Considering this situation, cost-saving compensation from the retailer to the manufacturer is an alternative solution so that both parties take advantage of the implementation of the second inspection policy. Based on Goyal [62]:

$$z = \frac{ETC_{d-case1}}{ETC_{case1}}$$

z is the retailer's cost coefficient. Therefore, ETC_d and $ETCp$ after cost-saving compensation are

$$ETC_d{}^a = zETC_{case2}$$

$$ETC_p{}^a = (1-z)ETC_{case2}$$

Table 5. Comparison between Model 1 and Model 2.

Decision Variables and Cost Items	Model 1	Model 2	Saving (%)
Number of deliveries per cycle (n^*)	7	9	
Cycle time (T)	0.08758	0.08869	
Delivery lot size (Q); units	6387.7	4929.6	
Ordering cost ($)	22,835.7	22,550.7	1.25
Inspection cost ($)	295,230.7	0	100
Inventory holding cost ($)	190,027.5	147,863.7	22.2
Deteriorating cost ($)	195,346.3	147,863.7	24.3
Emission cost ($)	171.0	133.1	22.2
Total retailer's cost per year ($)	703,611.2	318,909.1	54.7
Total retailer's cost per year after compensation ($)		*690,574.4*	*1.85*
Setup cost ($)	1,141,784.6	1,127,534.2	1.25
Inspection cost ($)	0	261,366.3	−100
Transportation cost ($)	85,289.5	107,690.0	−26.3
Inventory holding cost ($)	539,976.9	567,658.3	−5.13
Deteriorating cost ($)	362,122.2	397,345.0	−9.73
Emission cost ($)	2,138.0	2390.9	−11.8
Total manufacturer's cost per year ($)	2,131,311.2	2,463,984.8	−15.6
Total manufacturer's cost per year after compensation ($)		*2,091,821.6*	*1.85*
Expected total cost ($)	2,834,922.4	2,782,396.0	1.85
Expected total emission (tonCO$_2$/year)	30.598	33.523	−9.56

Hence, the retailer and manufacturer total cost per year become \$690,574.4 and \$2,091,821.6, respectively. Finally, by using this compensation policy, the cost decreases \$13,036.7 for the retailer and \$39,489.6 for the vendor, or 1.85% for both parties. Table 5 also shows that the *ETE* of Model 2 is 33.52 tonCO$_2$/year, which is 9.55% higher than the *ETE* in Model 1. We can obtain both the cost-saving and emissions-reducing objectives from Table 4, as there is a chance to reduce the *ETE* of Model 2. For $n = 7$, the *ETE* is 30.292, and the *ETC* is \$2,781,779 in which now the *ETE* and the *ETC* are 1.0% and 1.87% lower than those of Model 1, respectively. Thus, the objectives of cost efficiency and carbon footprint level reduction can be obtained simultaneously. The new comparison between the two models is presented in Table 6. It is also observed that the total delivered products to the retailer in Model 2 after adjustment (Model 2 ^{adj}) is less than those in Model 1.

Table 6. Comparison between Model 1 and the adjusted Model 2.

Decision Variables and Cost Items	Model 1	Model 2 ^{adj}	Saving (%)
Number of deliveries per cycle (n^*)	7	7	
Cycle time (T)	0.08758	0.08704	
Delivery lot size (Q); units	6,387.7	6221.2	
Ordering cost (\$)	22,835.7	22,977.4	−0.62
Inspection cost (\$)	295,230.7	0	100
Inventory holding cost (\$)	190,027.5	186,596.1	1.80
Deteriorating cost (\$)	195,346.3	186,596.1	4.48
Emission cost (\$)	171.0	167.9	1.81
Total retailer's cost per year (\$)	703,611.2	396,337.5	43.7
Total retailer's cost per year after compensation (\$)		*690,428.8*	*1.87*
Setup cost (\$)	1,141,784.6	1,148,869.4	−0.62
Inspection cost (\$)	0	261,461.4	−100
Transportation cost (\$)	85,289.5	85,764.5	−0.55
Inventory holding cost (\$)	539,976.9	529,446.7	1.95
Deteriorating cost (\$)	362,122.2	357,812.1	1.20
Emission cost (\$)	2,138.0	2117.7	0.95
Total manufacturer's cost per year (\$)	2,131,311.2	2,385,471.8	−11.9
Total manufacturer's cost per year after compensation (\$)		*2,091,380.5*	*1.87*
Expected total cost (\$)	2,834,922.4	2,781,809.3	1.87
Expected total emission (tonCO$_2$/year)	30.598	30.292	1.00

Sensitivity analysis is performed by increasing or decreasing the value of the parameter by ±25% and ±50% from the original values, as shown in Table 7. The results confirm that the second model is superior to the first model in terms of total cost. The number of deliveries per cycle (n) is sensitive to changes in parameters P, D, s, h_d, h_p, d_d, d_p, and t_f. As the values of parameters P, h_p, d_p, and t_f increase, the smaller the value of n. Contradictory conditions occur for parameters D, s, h_d, and d_d. The expected total cost is highly sensitive to the changes in parameters P, D, θ, s, u_c, h_d, h_p, d_d, d_p, and t_f, and almost insensitive to the changes in other parameters.

It is observed that when the deterioration rate (θ) increases, the expected total cost and the number of deliveries increase, but the delivery quantity decreases. When the probability of defective products (u) increases, the expected total cost increases very slightly, especially when the inspection is performed by the vendor. When the carbon tax (T_x) increases, the number of deliveries remains stable. Otherwise, the delivery quantity and expected total cost increase are very small.

Table 7. Sensitivity analysis of the two models.

Parameter	Value Change	Model 1					Model 2				
		n^*	T	Q^a	ETC	%CTC	n^*	T	Q^b	ETC	%CTC
$P = 2{,}000{,}000$	+50%	7	0.0816	5949.3	3,024,249	6.68	8	0.0819	5120.3	2,966,117.1	6.60
	+25%	7	0.0839	6117.1	2,948,615	4.01	8	0.0842	5263.2	2,892,821.2	3.97
	0	7	0.0876	6387.7	2,834,922	0	9	0.0887	4929.6	2,782,396.0	0
	−25%	8	0.0959	6120.6	2,643,862	−6.74	9	0.0957	5322.0	2,596,679.1	−6.67
	−50%	8	0.1147	7321.0	2,253,495	−20.5	11	0.1153	5296.5	2,217,427.4	−20.3
$D = 500{,}000$	+50%	8	0.0816	7811.6	3,190,858	12.5	10	0.0822	6168.2	3,136,392.1	12.7
	+25%	8	0.0841	6705.8	3,044,190	7.38	9	0.0840	5833.6	2,989,178.5	7.43
	0	7	0.0876	6387.7	2,834,922	0	9	0.0887	4929.6	2,782,396.0	0
	−25%	7	0.0958	5242.7	2,548,150	−10.1	8	0.0962	4510.9	2,499,238.8	−10.2
	−50%	6	0.1100	4683.3	2,151,051	−24.1	8	0.1120	3501.7	2,108,977.7	−24.2
$c = 2000$	+50%	7	0.0880	6415.9	2,846,315	0.40	9	0.0891	4951.6	2,793,646.2	0.40
	+25%	7	0.0878	6401.8	2,840,625	0.20	9	0.0889	4940.6	2,788,027.5	0.20
	0	7	0.0876	6387.7	2,834,922	0	9	0.0887	4929.6	2,782,396.0	0
	−25%	7	0.0874	6373.5	2,829,207	−0.20	9	0.0885	4918.6	2,776,752.0	−0.20
	−50%	7	0.0872	6359.3	2,823,479	−0.40	9	0.0883	4907.5	2,771,095.4	−0.41
$s = 100{,}000$	+50%	9	0.1074	6090.8	3,348,784	18,1	11	0.1081	4917.3	3,292,159.7	18.3
	+25%	8	0.0979	6250.1	3,104,546	9.51	10	0.0988	4945.1	3,049,823.5	9.61
	0	7	0.0876	6387.7	2,834,922	0	9	0.0887	4929.6	2,782,396.0	0
	−25%	6	0.0760	6464.7	2,529,737	−10.8	8	0.0773	4834.4	2,480,006.3	−10.9
	−50%	5	0.0625	6383.9	2,169,316	−23.5	6	0.0631	5257.7	2,122,805.7	−23.7
$i_c = 500$	+50%	7	0.0883	6437.1	2,854,827	0.70	9	0.0888	4935.1	2,785,213.3	0.10
	+25%	7	0.0879	6412.4	2,844,894	0.35	9	0.0887	4932.4	2,783,805.1	0.05
	0	7	0.0876	6387.7	2,834,922	0	9	0.0887	4929.6	2,782,396.0	0
	−25%	7	0.0872	6362.9	2,824,912	−0.35	9	0.0886	4926.8	2,780,986.3	−0.05
	−50%	7	0.0869	6338.0	2,814,883	−0.71	9	0.0886	4924.1	2,779,575.6	−0.10
$u_c = 0.5$	+50%	7	0.0876	6387.5	2,962,557	4.50	9	0.0887	4929.0	2,910,260.4	4.60
	+25%	7	0.0876	6387.6	2,898,740	2.25	9	0.0887	4929.3	2,846,328.1	2.30
	0	7	0.0876	6387.7	2,834,922	0	9	0.0887	4929.6	2,782,396.0	0
	−25%	7	0.0876	6387.8	2,771,105	−2.25	9	0.0887	4929.9	2,718,463.9	−2.30
	−50%	7	0.0876	6387.9	2,707,289	−4.50	9	0.0887	4930.2	2,654,531.7	−4.60
$h_d = 60$	+50%	9	0.0872	4947.9	2,916,253	2,87	11	0.0880	4003.7	2,848,604.1	2.38
	+25%	8	0.0873	5571.5	2,878,457	1.54	10	0.0883	4014.8	2,817,636.9	1.27
	0	7	0.0876	6387.7	2,834,922	0	9	0.0887	4929.6	2,782,396.0	0
	−25%	6	0.0882	7502.8	2,782,702	−1.84	7	0.0885	6327.8	2,739,318.4	−1.55
	−50%	5	0.0893	9119.2	2,716,341	−4.18	5	0.0889	8901.6	2,680,745.8	−3.65
$h_p = 40$	+50%	5	0.0779	7955.9	3,075,129	8.47	5	0.0776	7768.5	3,033,547.3	9.03
	+25%	6	0.0823	7002.8	2,962,650	4.51	7	0.0827	5908.5	2,915,155.6	4.77
	0	7	0.0876	6387.7	2,834,922	0	9	0.0887	4929.6	2,782,396.0	0
	−25%	8	0.0940	5999.5	2,691,834	−5.05	10	0.0951	4758.0	2,634,200.0	−5.33
	−50%	9	0.1020	5786.3	2,531,836	−10.7	13	0.1051	4044.5	2,468,523.7	−11.3
$d_d = 600$	+50%	9	0.0872	49441	2,918,318	2,94	11	0.0880	4003.7	2,848,604.1	2.38
	+25%	8	0.0873	5569.0	2,879,619	1.58	10	0.0883	4416.4	2,817,636.9	1.27
	0	7	0.0876	6387.7	2,834,922	0	9	0.0887	4929.6	2,782,396.0	0
	−25%	6	0.0882	7507.4	2,781,144	−1.90	7	0.0885	6327.8	2,739,318.4	−1.55
	−50%	3	0.0872	14852.7	2,703,928	−4.62	5	0.0889	8901.6	2,680,745.8	−3.65
$d_p = 400$	+50%	7	0.0820	5981.9	3,010,061	6,18	6	0.0806	6721.3	2,957,798.8	6.30
	+25%	7	0.0847	6174.8	2,923,924	3.14	7	0.0839	5999.6	2,875,189.0	3.33
	0	7	0.0876	6387.7	2,834,922	0	9	0.0887	4929.6	2,782,396.0	0
	−25%	8	0.0922	5884.4	2,738,923	−3.39	10	0.0933	4668.1	2,679,457.1	−3.70
	−50%	9	0.0976	5538.2	2,632,492	−7.14	12	0.0997	4155.5	2,566,509.0	−7.76
$\theta = 0.1$	+50%	7	0.0794	5794.4	3,100,001	9.35	9	0.0804	4470.6	3,041,924.7	9.33
	+25%	7	0.0832	6069.2	2,970,737	4.79	9	0.0843	4683.2	2,915,364.1	4.78
	0	7	0.0876	6387.7	2,834,922	0	9	0.0887	4929.6	2,782,396.0	0
	−25%	7	0.0927	6762.9	2,691,447	−5.06	9	0.0939	5219.8	2,641,938.0	−5.05
	−50%	7	0.0989	7213.9	2,538,850	−10.4	8	0.0992	5516.8	2,492,404.9	−10.4
$t_f = 1000$	+50%	6	0.0874	7440.5	2,872,436	1.32	7	0.0883	6308.3	2,826,288.0	1.58
	+25%	7	0.0883	6437.1	2,854,827	0.70	8	0.0886	5542.6	2,805,413.7	0.83
	0	7	0.0876	6387.7	2,834,922	0	9	0.0887	4929.6	2,782,396.0	0
	−25%	8	0.0879	5612.0	2,814,046	−0.74	10	0.0885	4425.0	2,756,197.9	−0.94
	−50%	9	0.0880	4991.1	2,790,972	−1.55	12	0.0884	3685.7	2,725,660.0	−2.04
$t_v = 0.75$	+50%	7	0.0876	6391.8	2,837,605	0.09	9	0.0888	4933.7	2,785,501.1	0.11
	+25%	7	0.0876	6389.7	2,836,264	0.05	9	0.0887	4931.6	2,783,948.8	0.06
	0	7	0.0876	6387.7	2,834,922	0	9	0.0887	4929.6	2,782,396.0	0
	−25%	7	0.0876	6385.7	2,833,581	−0.05	9	0.0886	4927.6	2,780,842.7	−0.06
	−50%	7	0.0875	6383.6	2,832,239	−0.09	9	0.0886	4925.5	2,779,289.1	−0.11

Table 7. *Cont.*

Parameter	Value Change	Model 1					Model 2				
		n^*	T	Q^a	ETC	%CTC	n^*	T	Q^b	ETC	%CTC
$d = 100$	+50%	7	0.0877	6392.8	2,838,309	0.12	9	0.0888	4934.7	2,786,313.0	0.14
	+25%	7	0.0876	6390.3	2,836,616	0.06	9	0.0887	4932.2	2,784,355.0	0.07
	0	7	0.0876	6387.7	2,834,922	0	9	0.0887	4929.6	2,782,396.0	0
	−25%	7	0.0875	6385.1	2,833,229	−0.06	9	0.0886	4927.0	2,780,436.7	−0.07
	−50%	7	0.0875	6382.6	2,831,535	−0.12	9	0.0886	4924.5	2,778,476.4	−0.14
$E[u] = 0.02$	+50%	7	0.0874	6437.8	2,843,908	0.32	9	0.0887	4931.3	2,784,137.7	0.063
	+25%	7	0.0875	6412.6	2,839,414	0.16	9	0.0887	4930.5	2,783,248.6	0.031
	0	7	0.0876	6387.7	2,834,922	0	9	0.0887	4929.6	2,782,396.0	0
	−25%	7	0.0877	6363.1	2,830,435	−0.16	9	0.0887	4928.7	2,781,579.0	−0.029
	−50%	7	0.0878	6338.7	2,825,951	−0.32	9	0.0887	4927.7	2,780,796.5	−0.058
$w = 0.01$	+50%	7	0.0876	6387.7	2,835,956	0.04	9	0.0887	4929.6	2,783,408.9	0.036
	+25%	7	0.0876	6387.7	2,835.439	0.02	9	0.0887	4929.6	2,782,902.6	0.018
	0	7	0.0876	6387.7	2,834,922	0	9	0.0887	4929.6	2,782,396.0	0
	−25%	7	0.0876	6387.7	2,834,405	−0.02	9	0.0887	4929.6	2,781,889.5	−0.018
	−50%	7	0.0876	6.387.7	2,833,889	−0.04	9	0.0887	4929.6	2,781,383.0	−0.036
$c_1, c_2 = 0.27,$ 0.0057	+50%	7	0.0876	6388.7	2,835,627	0.02	9	0.0887	4930.5	2,783,244.8	0.031
	+25%	7	0.0876	6388.2	2,835,275	0.01	9	0.0887	4930.0	2,782,838.6	0.016
	0	7	0.0876	6387.7	2,834,922	0	9	0.0887	4929.6	2,782,396.0	0
	−25%	7	0.0876	6387.2	2,834,570	−0.01	9	0.0887	4928.9	2,782,026.3	−0.013
	−50%	7	0.0876	6386.7	2,834,218	−0.02	9	0.0887	4928.4	2,781,620.2	−0.028
$e_c = 1.44$	+50%	7	0.0876	6386.6	2,835,372	0.016	9	0.0887	4928.7	2,782,845.7	0.016
	+25%	7	0.0876	6387.1	2,835,148	0.008	9	0.0887	4929.2	2,782,620.8	0.008
	0	7	0.0876	6387.7	2,834,922	0	9	0.0887	4929.6	2,782,396.0	0
	−25%	7	0.0876	6388.3	2,834,698	−0.008	9	0.0887	4930.0	2,782,171.2	−0.008
	−50%	7	0.0876	6388.8	2,834,472	−0.016	9	0.0887	4930.5	2,781,946.2	−0.016
$T_x = 75$	+50%	7	0.0876	6387.7	2,836,077	0.04	9	0.0887	4929.8	2,783,658.0	0.045
	+25%	7	0.0876	6387.7	2,835,500	0.02	9	0.0887	4929.7	2,783,027.1	0.023
	0	7	0.0876	6387.7	2,834,922	0	9	0.0887	4929.6	2,782,396.0	0
	−25%	7	0.0876	6387.7	2,834,345	−0.02	9	0.0887	4929.5	2,781,764.9	−0.023
	−50%	7	0.0876	6387.6	2,833,768	−0.04	9	0.0887	4929.4	2,781,134.2	−0.045

6. Conclusions and Future Research

This study considers a two-echelon supply chain consisting of a manufacturer and a retailer where the production activities are resulting in a certain percentage of defective products. The supply chain entities are willing to reduce their environmental impact by coordinating the delivery quantity and number of deliveries per cycle. The effect of carbon emissions, item deterioration, and two choices of inspection are examined. The models are illustrated with two numerical examples, and the results give some insights. This study is an initial exploratory study that attempts to provide a mathematical solution for a controlled situation; it may be applied to handle larger problems of cost minimization and carbon emission reduction in the future.

From the research finding, it is observed that the numbers of delivered products from the manufacturer are less when the inspection is performed by the vendor. As a result, the total cost of the supply chain is less, because the total inventory-holding cost and the total deteriorating cost are decreasing. However, the vendor's total cost becomes higher when it performs the inspection. Therefore, the retailer needs to compensate a certain amount of cost-saving to the manufacturer so that both parties take advantage.

The research finding also revealed that although the total cost is less when the inspection is performed by the vendor, it does not guarantee a reduction in emissions. However, both the cost-saving and emission-reducing objectives can still be obtained simultaneously by reducing the level of cost savings. In this situation, there is a tradeoff between cost savings and reduction in carbon emissions.

Although this study addresses some practical aspects of a supply chain scenario to deal with lower carbon emissions, the scope has a wide opportunity to be extended. Applying the approach in a three-echelon supply chain or more is one opportunity. Future works can consider the possibility of reworking the defective products, the capacity of the vehicle and storage facility, and investment to reduce the carbon emissions. This study assumes a 100% inspection by the manufacturer. Future research may assume a sampling inspection by the manufacturer, as well as incorporate the issue of imperfect quality inspection.

Author Contributions: Conceptualization, Y.D.; Methodology, Y.D.; Software, Y.D.; Validation, G.A.W. and H.M.W.; Formal Analysis, Y.D.; Investigation, Y.D.; Resources, H.M.W.; Writing—Original Draft Preparation, Y.D.; Writing—Review & Editing, H.M.W. and G.A.W.; Visualization, Y.D.; Supervision, H.M.W.

Funding: This research received no external funding.

Acknowledgments: The authors express their gratitude to the editor and the two anonymous reviewers for their comments and valuable suggestions to improve this paper. The first author also thanks to the United Board for Christian Higher Education in Asia for financing his study.

Conflicts of Interest: The authors declare no conflict of interest.

References

1. Glock, C.H. The joint economic lot size problem: A review. *Int. J. Prod. Econ.* **2012**, *135*, 671–686. [CrossRef]
2. Luo, Z.; Gunasekaran, A.; Dubey, R.; Childe, S.J.; Papadopoulos, T. Antecedents of low carbon emissions supply chains. *Int. J. Clim. Chang. Strateg. Manag.* **2017**, *9*, 707–727. [CrossRef]
3. Das, C.; Jharkharia, S. Low carbon supply chain: A state-of-the-art literature review. *J. Manuf. Technol. Manag.* **2018**, *29*, 398–428. [CrossRef]
4. Kazemi, N.; Abdul-Rashid, S.H.; Ghazilla, R.A.R.; Shekarian, E.; Zanoni, S. Economic order quantity models for items with imperfect quality and emission considerations. *Int. J. Syst. Sci. Oper. Logist.* **2018**, *5*, 99–115. [CrossRef]
5. Sarkar, B.; Ahmed, W.; Choi, S.B.; Tayyab, M. Sustainable inventory management for environmental impact through partial backordering and multi-trade-credit period. *Sustainability* **2018**, *10*, 4761. [CrossRef]
6. Taleizadeh, A.A.; Soleymanfar, V.R.; Govindan, K. Sustainable economic production quantity models for inventory system with shortage. *J. Clean. Prod.* **2018**, *174*, 1011–1020. [CrossRef]
7. Sarkar, B.; Ganguly, B.; Sarkar, M.; Pareek, S. Effect of variable transportation and carbon emission in a three-echelon supply chain model. *Transp. Res. Part E Logist. Transp. Rev.* **2016**, *91*, 112–128. [CrossRef]
8. Sarkar, B.; Ahmed, W.; Kim, N. Joint effects of variable carbon emission cost and multi-delay-in-payments under single-setup-multiple-delivery policy in a global sustainable supply chain. *J. Clean. Prod.* **2018**, *185*, 421–445. [CrossRef]
9. Daryanto, Y.; Wee, H.M. Single vendor-buyer integrated inventory model for deteriorating items considering carbon emission. In Proceedings of the 8th International Conference on Industrial Engineering and Operations Management (IEOM), Bandung, Indonesia, 6–8 March 2018; pp. 544–555.
10. Wahab, M.I.M.; Mamun, S.M.H.; Ongkunarak, P. EOQ models for a coordinated two-level international supply chain considering imperfect items and environmental impact. *Int. J. Prod. Econ.* **2011**, *134*, 151–158. [CrossRef]
11. Jauhari, W.A.; Pamuji, A.S.; Rosyidi, C.N. Cooperative inventory model for vendor-buyer system with unequal-sized shipment, defective items and carbon emission cost. *Int. J. Logist. Syst. Manag.* **2014**, *19*, 163–186. [CrossRef]
12. Sarkar, B.; Saren, S.; Sarkar, M.; Seo, Y.W. A Stackelberg game approach in an integrated inventory model with carbon-emission and setup cost reduction. *Sustainability* **2016**, *8*, 1244. [CrossRef]
13. Jauhari, W.A. A collaborative inventory model for vendor-buyer system with stochastic demand, defective items and carbon emission cost. *Int. J. Logist. Syst. Manag.* **2018**, *29*, 241–269. [CrossRef]
14. Gautam, P.; Khanna, A. An imperfect production inventory model with setup cost reduction and carbon emission for an integrated supply chain. *Uncertain Supply Chain Manag.* **2018**, *6*, 271–286. [CrossRef]
15. Tiwari, S.; Daryanto, Y.; Wee, H.M. Sustainable inventory management with deteriorating and imperfect quality items considering carbon emissions. *J. Clean. Prod.* **2018**, *192*, 281–292. [CrossRef]
16. Rosenblatt, M.J.; Lee, H.L. Economic production cycles with imperfect production processes. *IIE Trans.* **1986**, *18*, 48–55. [CrossRef]
17. Porteus, E.L. Optimal lot sizing, process quality improvement and setup cost reduction. *Oper. Res.* **1986**, *34*, 137–144. [CrossRef]

18. Salameh, M.K.; Jaber, M.Y. Economic production quantity model for items with imperfect quality. *Int. J. Prod. Econ.* **2000**, *64*, 59–64. [CrossRef]

19. Huang, C.K. An integrated vendor-buyer cooperative inventory model for items with imperfect quality. *Prod. Plan. Control* **2002**, *13*, 355–361. [CrossRef]

20. Goyal, S.K.; Huang, C.K.; Chen, K.C. A simple integrated production policy of an imperfect item for vendor and buyer. *Prod. Plan. Control* **2003**, *14*, 596–602. [CrossRef]

21. Wee, H.M.; Yu, J.C.P.; Wang, K.J. An integrated production-inventory model for deteriorating items with imperfect quality and shortage backordering considerations. In Proceedings of the International Conference on Computational Science and Its Applications (ICCSA), Glasgow, UK, 8–11 May 2006; Springer: Berlin/Heidelberg, Germany, 2006; pp. 885–897.

22. Lee, S.; Kim, D. An optimal policy for a single vendor single-buyer integrated production-distribution model with both deteriorating and defective items. *Int. J. Prod. Econ.* **2014**, *147*, 161–170. [CrossRef]

23. Bazan, E.; Jaber, M.Y.; Zanoni, S.; Zavanella, L.E. Vendor managed inventory (VMI) with consignment stock (CS) agreement for a two-level supply chain with an imperfect production process with/without restoration interruptions. *Int. J. Prod. Econ.* **2014**, *157*, 289–301. [CrossRef]

24. Sarkar, B.; Shaw, B.K.; Kim, T.; Sarkar, M.; Shin, D. An integrated inventory model with variable transportation cost, two-stage inspection, and defective items. *J. Ind. Manag. Optim.* **2017**, *13*, 1975–1990. [CrossRef]

25. Yu, H.F.; Hsu, W.K. An integrated inventory model with immediate return for defective items under unequal-sized shipments. *J. Ind. Prod. Eng.* **2017**, *34*, 70–77. [CrossRef]

26. Benjaafar, S.; Li, Y.; Daskin, M. Carbon footprint and the management of supply chains: Insights from simple models. *IEEE Trans. Autom. Sci. Eng.* **2013**, *10*, 99–116. [CrossRef]

27. Fahimnia, B.; Sarkis, J.; Dehghanian, F.; Banihashemi, N.; Rahman, S. The impact of carbon pricing on a closed-loop supply chain: An Australian case study. *J. Clean. Prod.* **2013**, *59*, 210–225. [CrossRef]

28. Bozorgi, A.; Pazour, J.; Nazzal, D. A new inventory model for cold items that considers costs and emissions. *Int. J. Prod. Econ.* **2014**, *155*, 114–125. [CrossRef]

29. Bozorgi, A. Multi-product inventory model for cold items with cost and emission consideration. *Int. J. Prod. Econ.* **2016**, *176*, 123–142. [CrossRef]

30. Hariga, M.; As'ad, R.; Shamayleh, A. Integrated economic and environmental models for a multi stage cold supply chain under carbon tax regulation. *J. Clean. Prod.* **2017**, *166*, 1357–1371. [CrossRef]

31. Ghosh, A.; Jha, J.K.; Sarmah, S.P. Optimizing a two-echelon serial supply chain with different carbon policies. *Int. J. Sustain. Eng.* **2016**, *9*, 363–377. [CrossRef]

32. Toptal, A.; Çetinkaya, B. How supply chain coordination affects the environment: A carbon footprint perspective. *Ann. Oper. Res.* **2017**, *250*, 487–519. [CrossRef]

33. Bouchery, Y.; Ghaffari, A.; Jemai, Z.; Tan, T. Impact of coordination on cost and carbon emissions for a two-echelon serial economic order quantity problem. *Eur. J. Oper. Res.* **2017**, *260*, 520–533. [CrossRef]

34. Dwicahyani, A.R.; Jauhari, W.A.; Rosyidi, C.N.; Laksono, P.W. Inventory decisions in a two-echelon system with remanufacturing, carbon emission, and energy effects. *Cogent Eng.* **2017**, *4*, 1–17. [CrossRef]

35. Li, J.; Su, Q.; Ma, L. Production and transportation outsourcing decisions in the supply chain under single and multiple carbon policies. *J. Clean. Prod.* **2017**, *141*, 1109–1122. [CrossRef]

36. Wangsa, I.D. Greenhouse gas penalty and incentive policies for a joint economic lot size model with industrial and transport emissions. *Int. J. Ind. Eng. Comput.* **2017**, *8*, 453–480.

37. Anvar, S.H.; Sadegheih, A.; Zad, M.A.V. Carbon emission management for greening supply chains at the operational level. *Environ. Eng. Manag. J.* **2018**, *17*, 1337–1347.

38. Hariga, M.; Babekian, S.; Bahroun, Z. Operational and environmental decisions for a two-stage supply chain under vendor managed consignment inventory partnership. *Int. J. Prod. Res.* **2018**. [CrossRef]

39. Ji, S.; Zhao, D.; Peng, X. Joint decisions on emission reduction and inventory replenishment with overconfidence and low-carbon preference. *Sustainability* **2018**, *10*, 1119.

40. Wang, S.; Ye, B. A comparison between just-in-time and economic order quantity models with carbon emissions. *J. Clean. Prod.* **2018**, *187*, 662–671. [CrossRef]

41. Ghosh, A.; Sarmah, S.P.; Jha, J.K. Collaborative model for a two-echelon supply chain with uncertain demand under carbon tax policy. *Sādhanā* **2018**, *43*, 144. [CrossRef]

42. Ma, X.; Ho, W.; Ji, P.; Talluri, S. Coordinated pricing analysis with the carbon tax scheme in a supply chain. *Decis. Sci.* **2018**, *49*, 863–900. [CrossRef]

43. Darom, N.A.; Hishamuddin, H.; Ramli, R.; Nopiah, Z.M. An inventory model of supply chain disruption recovery with safety stock and carbon emission consideration. *J. Clean. Prod.* **2018**, *197*, 1011–1021. [CrossRef]

44. Huang, H.; He, Y.; Li, D. Pricing and inventory decisions in the food supply chain with production disruption and controllable deterioration. *J. Clean. Prod.* **2018**, *180*, 280–296. [CrossRef]

45. Daryanto, Y.; Wee, H.M.; Astanti, R.D. Three-echelon supply chain model considering carbon emission and item deterioration. *Transp. Res. Part E Logist. Transp. Rev.* **2019**, *122*, 368–383. [CrossRef]

46. Kundu, S.; Chakrabarti, T. A fuzzy rough integrated multi-stage supply chain inventory model with carbon emissions under inflation and time-value of money. *Int. J. Math. Oper. Res.* **2019**, *14*, 123–145. [CrossRef]

47. Shalke, P.N.; Paydar, M.M.; Hajiaghaei-Keshteli, M. Sustainable supplier selection and order allocation through quantity discounts. *Int. J. Manag. Sci. Eng. Manag.* **2018**, *13*, 20–32.

48. Moheb-Alizadeh, H.; Handfield, R. An integrated chance-constrained stochastic model for efficient and sustainable supplier selection and order allocation. *Int. J. Prod. Res.* **2018**, *56*, 6890–6916. [CrossRef]

49. Moheb-Alizadeh, H.; Handfield, R. Sustainable supplier selection and order allocation: A novel multi-objective programming model with a hybrid solution approach. *Comput. Ind. Eng.* **2019**, *129*, 192–209. [CrossRef]

50. Cao, W.; Hu, Y.; Li, C.; Wang, X. Single setup multiple delivery model of JIT system. *Int. J. Adv. Manuf. Technol.* **2007**, *33*, 1222–1228. [CrossRef]

51. Volvo Truck Corporation. Emissions from Volvo's Truck. Issue 3. 09 March 2018. Available online: http://www.volvotrucks.com/content/dam/volvo/volvo-trucks/markets/global/pdf/our-trucks/Emis_eng_10110_14001.pdf (accessed on 25 September 2018).

52. Yang, P.C.; Wee, H.M. Economic ordering policy of deteriorated item for vendor and buyer: An integrated approach. *Prod. Plan. Control* **2000**, *11*, 474–480. [CrossRef]

53. Jaggi, C.K.; Goel, S.K.; Mittal, M. Economic order quantity model for deteriorating items with imperfect quality and permissible delay in payment. *Int. J. Ind. Eng. Comput.* **2011**, *2*, 237–248. [CrossRef]

54. Swenseth, S.R.; Godfrey, M.R. Incorporating transportation costs into inventory replenishment decisions. *Int. J. Prod. Econo.* **2002**, *77*, 113–130. [CrossRef]

55. Nie, L.; Xu, X.; Zhan, D. Incorporating transportation costs into JIT lot splitting decisions for coordinated supply chains. *J. Adv. Manuf. Syst.* **2006**, *5*, 111–121. [CrossRef]

56. Rahman, M.N.A.; Leuveano, R.A.C.; bin Jafar, F.A.; Saleh, C.; Deros, B.M.; Mahmood, W.M.F.W.; Mahmood, W.H.W. Incorporating logistic costs into a single vendor-buyer JELS model. *Appl. Math. Model.* **2016**, *40*, 10809–10819. [CrossRef]

57. Wangsa, I.D.; Wee, H.M. An integrated vendor-buyer inventory model with transportation cost and stochastic demand. *Int. J. Syst. Sci. Oper. Logist.* **2018**, *5*, 295–309.

58. Hsu, L.F. Erratum to: An optimal policy for a single-vendor single-buyer integrated production-distribution model with both deteriorating and defective items. [Int. J. Prod. Econ. 147 (2014) 161–170]. *Int. J. Prod. Econ.* **2016**, *178*, 187–188. [CrossRef]

59. Misra, R.B. Optimum production lot size model for a system with deteriorating inventory. *Int. J. Prod. Res.* **1975**, *13*, 495–505. [CrossRef]

60. The United States Environmental Protection Agency (US. EPA). Emission Facts: Average Carbon Dioxide Emissions Resulting from Gasoline and Diesel Fuel. February 2005. Available online: https://nepis.epa.gov/ (accessed on 25 September 2018).

61. McCarthy, J.E. EPA Standards for Greenhouse Gas Emissions from Power Plants: Many Questions, Some Answers. CRS Report for Congress, 7–5700. 2013. Available online: http://nationalaglawcenter.org/wp-content/uploads/assets/crs/R43127.pdf (accessed on 25 September 2018).

62. Goyal, S.K. An integrated inventory model for a single supplier-single customer problem. *Int. J. Prod. Res.* **1977**, *15*, 107–111. [CrossRef]

© 2019 by the authors. Licensee MDPI, Basel, Switzerland. This article is an open access article distributed under the terms and conditions of the Creative Commons Attribution (CC BY) license (http://creativecommons.org/licenses/by/4.0/).

Article

Constrained FC 4D MITPs for Damageable Substitutable and Complementary Items in Rough Environments

Sharmistha Halder Jana [1,*], Biswapati Jana [2], Barun Das [3] and Goutam Panigrahi [4] and Manoranjan Maiti [5]

[1] Department of Mathematics, Midnapore College (Autonomous), Midnapore 721101, India
[2] Department of Computer Science, Vidyasagar University, Midnapore 721102, India; biswapatijana@gmail.com
[3] Department of Mathematics, Sidho Kanho Birsha University, Purulia 723104, India; bdasskbu@gmail.com
[4] Department of Mathematics, National Institute of Technology, Durgapur 713209, India; panigrahi_goutam@rediffmail.com
[5] Department of Mathematics, Vidyasagar University, Midnapore 721102, India; mmaiti2005@yahoo.co.in
* Correspondence: sharmistha792010@gmail.com; Tel.: +91-7872227694

Received: 2 January 2019; Accepted: 27 February 2019; Published: 19 March 2019

Abstract: Very often items that are substitutable and complementary to each other are sent from suppliers to retailers for business. In this paper, for these types of items, fixed charge (FC) four-dimensional (4D) multi-item transportation problems (MITPs) are formulated with both space and budget constraints under crisp and rough environments. These items are damageable/breakable. The rates of damageability of the items depend on the quantity transported and the distance of travel i.e., path. A fixed charge is applied to each of the routes (independent of items). There are some depots/warehouses (origins) from which the items are transported to the sales counters (destinations) through different conveyances and routes. In proposed FC 4D-MITP models, per unit selling prices, per unit purchasing prices, per unit transportation expenditures, fixed charges, availabilities at the sources, demands at the destinations, conveyance capacities, total available space and budget are expressed by rough intervals, where the transported items are substitutable and complementary in nature. In this business, the demands for the items at the destinations are directly related to their substitutability and complementary natures and prices. The suggested rough model is converted into a deterministic one using lower and upper approximation intervals following Hamzehee et al. as well as Expected Value Techniques. The converted model is optimized through the Generalized Reduced Gradient (GRG) techniques using LINGO 14 software. Finally, numerical examples are presented to illustrate the preciseness of the proposed model. As particular cases, different models such as 2D, 3D FCMITPs for two substitute items, one item with its complement and two non substitute non complementary items are derived and results are presented.

Keywords: four-dimensional transportation problem; fixed charge; space constraint; budget constraint; substitute and complimentary items; rough interval

1. Introduction

Due to globalization, nowadays, transportation of commodities from sources to the destinations by road is getting more important. The advent of transportation problem (TP) was mainly based on real-life problems. Many numbers of real-life goods carrying problems are easily framed as transportation problems. The credit of first transportation problem goes to Hitchcock [1], which is a particular case of Linear Programming Problem (LPP). This optimizing problem consists of two main constraints i.e., source and destination constraints. However, it is very common knowledge that, in practical situations, these two constraints are not enough to formulate the problem perfectly since there exist other constraints, namely mode of transport, type of products, the distance of path traveled, etc. Due to the presence of these additional real-life constraints, the conventional transportation problems (2D-TPs) are modified to a solid transportation problem (STP-3D-TPs), which was first developed by Schell [2]. After the advent of STP, it has swiftly gained a very important place for research and development in transportation. Researchers like Yang et al. [3], Kocken et al. [4], etc. were on the development of STP, in which conveyances at different sources are considered.

In general circumstances, the length of a route remains unchanged in a normal transportation problem, so it does not make any significant change in the minimization of cost or time. However, reality dictates that there may be a number of available option/paths for the transportation of an item from source to destination. The common knowledge suggests that the cost per unit for transportation or other fixed charges related to routes are of different values. Thus, it can be seen that a three-dimensional transportation problem (3D-TP) is converted to a 4D-TP. Such a 4D-TP was considered by Halder et al. [5].

Nowadays, the business of multi -items is quite popular. Normally, in business, it is thought that if one item brings loss, another item may produce the profit. Moreover, due to the constantly changing business environment, the parameters related to transportation are imprecise. This situation has been recently dealt with in two recent investigations Das et al. [6] and Bera et al. [7]. Again, the transportation of multi-items may be substitutable and/or complementary because more options/choices of items bring more customers. This is true for supply chain models. Recently, Sarkar and Lee [8] and sarkar et al. [9] presented optimum pricing strategies for complementary products using game the orectic approach. Khanna et al. [10] suggested supply chain with customer-based two-level credit policics undcr an impcrfcct quality environment. Recently, Halder et al. [5] also solved an FCSTP for substitute and breakable items in crisp and fuzzy environments.

1.1. Scope of the Paper

Considering the above facts, we formulate the transportation policies of some damageable substitute and/or complementary items from some sources to destinations using some convenient vehicles and appropriate routes where the parameters of the problem are deterministic or imprecise and some realistic resource constraints are imposed.

In this investigation, there are different types of conveyance at different sources and different routes/paths available to travel from a source to a destination. Along some paths, fixed charges such as toll taxes, festival collections, etc. are collected. At each destination, retailers have some limitations on space and budget. The transportation costs, fixed charges, availabilities, demands, resources are imprecise and expressed by rough intervals. As the items are substitutable and complementary, items' demands are influenced by each other's prices. The demand for an item is reduced due to its own and complementary item's prices but increases due to its substitute's price. With these features, the model is formulated as a constrained FC 4DMITP problem and solved using Generalised Reduced Gradient (GRG) method (using LINGO 14.0). As particular cases, transportation policies for several models (i.e., 2D, 3D TPs for different combinations of items) are derived. The model are illustrated numerically. Some sensitivity analyses are also performed. Results of an earlier investigation (Bera et al. [7]) are obtained as a particular case. The novelties of the present investigation are as follows:

- As earlier discussed, TP and STP with various types of constraints are considered by several researchers. However, few researchers have considered 4D-TPs and 4D-MITPs. Moreover, 4D-MITP with rough parameters is an updated contribution.
- The items are complementary and substitutable in nature, that is, demands of the items are appropriately affected by their selling prices.
- The most important issues of this paper are to analyze how the travel distances are related to profit maximization when manufacturing companies transport both complementary and substitute

items that are differentiated by distance including the fixed charge of the path and damageability. The importance of route on profit is illustrated.

- Until now, in transportation, no one has considered the space constraint at the destinations along with the budget constant. The idea of space constraint is introduced here.

- The earlier researchers gave attention to minimization of aggregate transportation expenditure and very few have realized the importance of consideration of total profit instead of total cost/expanses.

- As particular cases, several earlier transportation models are deduced from the present model.

1.2. Structure of the Paper

The structure/organization of the paper is split as:

Section 1: Introduction

Section 2: Notations and Assumptions for the proposed model are given.

Section 3: Model description and formulation

Section 4: Numerical Experiments

Section 5: Particular Cases

Section 6: Sensitivity Analyses

Section 7: A discussion of the models on the basis of numerical results are presented.

Section 8: Practical implication is described.

Section 9: Conclusions drawn.

1.3. Literature Review

Though the transportation problem is quite an age-old one, several researchers are still working in this area as it is important due to national highways within the country and international transportation using cargos and ships. Hirsch and Dantzig [11] first introduced the concept of a fixed charge transportation problem in 1968. In each country, there are different major highways connecting the cities/ports and, for the maintenance, some charges (fixed) are collected from the vehicles plying on these roads. This real-life scenery has been depicted in the fixed charge TPs (FCTP). Kowalski and Lev [12] developed the FCTP as a nonlinear programming problem of practical interest in business and industry. Gen et al. [13] discussed the bi-criteria solid transportation problem with equality constraints solved by a genetic algorithm. Several researchers like Verma et al. [14], Shafaat and Goyal [15], Saad and Abbas [16] and others worked on transportation problems.

Though there are a lot of works on damageable/deteriorating items in inventory (cf. Sarkar and Iqbal [17] and Sarkar et al. [18]), only limited works have been done on damageable/breakable items in transportation. This is because transportation time is not normally considered in TP. However, the damageability may also depend on the route lengths which are different for different routes. Pramanik et al. [19] presented a multi-objective solid transportation problem with reliability for damageable items in a random-fuzzy environment.

Though Sarkar et al. [20] considered a power function of the delivery quantity as a unit transportation cost in inventory, normally constant unit transportation cost is assumed in TP to formulate it as an LPP. Ojha et al. [21] presented a solid transportation problem with nonlinear transportation cost and fuzzy resources, demand and conveyances. However, in the present problem, transportation cost is considered constant. Depending on different aspects, unit transportation costs, resources, demands, available budget and storing capacity at destination, etc. fluctuate due to uncertainty in judgement, lack of evidence, insufficient information, etc. Thus, a transportation model becomes more realistic if these parameters are assumed to be flexible /imprecise in nature i.e., uncertain in non-stochastic sense which may be represented by fuzzy, rough [22], type-2-fuzzy, fuzzy stochastic numbers. There are several investigations on TPs with fuzzy parameters. Verma et al. [14], Bit et al. [23], Jimenez and Verdegay [24], Li and Lai [25], Dey et al. [26] and others solved fuzzy multi-objective TP(s) using a Fuzzy compromise programming approach. Yang and Liu [3] investigated a fixed charge STP with fuzzy costs, fuzzy supplies, fuzzy demands and fuzzy conveyance capacities using an expected value method. There are also many investigations on TP with rough parameters. Recently, Kundu et al. [27] presented an STP considering crisp and rough costs. Tao et al. [28] applied rough multiple objective programming in STP. Das et al. [6] developed a profit-maximizing STP with parameters represented by rough intervals. Recently, Bera et al. [7] solved multi-item 4D TPs under budget constraints using rough interval values for TP parameters. The year-wise investigations of various TPs and STPs with different variations are recorded in the following table.

In spite of all these investigations, there are still some lacunas in making the investigated problem more realistic ones. Some of these gaps such as substitutable and complementary items, both budget and space constraint, damageable units, 4D MITP, fixed charge, profit maximization, etc. have been taken into account in the present investigation.

1.4. Motivation

Although there are so many existing kinds of literature available in the field of transportation problem (cf in Table 1), but still there are some lacunas. Most of the researchers developed two-dimensional or three-dimensional problems that motivated us to form a four-dimensional real transportation problem.

Table 1. Year wise investigations of TP and STP.

References'	Different Kind of TP	Item	Fixed Charge	Space Constraint	Budget Constraint	Different Kind of Environment
Hitchcock et al. [1]	2-dimension	one	×	×	-	crisp
Schell [2]	3-dimension	one	×	×	-	crisp
Haley [29]	3-dimension	one	×	×	-	crisp
Hirsch and Dantzig [11]	2-dimension	one	✓	×	×	crisp
Verma et al. [14]	2-dimension	one	×	×	×	fuzzy
Shafaat and Goyal [15]	2-dimension	one	×	×	×	crisp
Saad and Abbas [16]	2-dimension	one	×	×	×	fuzzy
Jimenez et al. [24]	3-dimension	one	×	×	×	fuzzy
Tao et al. [28]	3-dimension	one	×	×	×	rough
Ojha et al. [30]	2-dimension	multi-item	×	×	✓	fuzzy
Liu et al. [31]	3-dimension	one	×	×	×	type-2 fuzzy
Kundu et al. [27]	3-dimension	multi-item	×	×	×	type-2 fuzzy
Yang et al. [3]	2-dimension	one	✓	×	×	fuzzy
Giri et al. [32]	3-dimension	multi-item	✓	×	×	fuzzy
Kocken et al. [4]	3-dimension	one	×	×	×	fuzzy
Das et al. [6]	3-dimension	one	×	×	×	rough
Present Paper	4-dimensional	multi-item	✓	✓	✓	rough

In modern business, multifarious items are preferable for transportation rather than the single item. In the long term, transportation system fixed charges like (road tax etc.) are very common to real life; this is also a motivation which leads to consideration of fixed charge. Finally, in today's constantly fluctuating life, uncertainty is the only certain factor in our life. In this concept, the different coefficients are not fixed to its deterministic values. It must take an interval. For example, demand of an item is not fixed, but it is represented by its lower and upper limits b_l, b_u, respectively. Again, this limit also may fluctuate, so the necessity of rough interval arise. In this regard, the demand for an item roughly lies on $([\underline{b}_l, \underline{b}_u], [\overline{b}_l, \overline{b}_u])$, respectively.

In reality, most of the problems are not described specifically but described with uncertainty. This may due to the fluctuation of daily life, lack of information, etc. Until now, most of the impreciseness are described by interval values.

In interval analysis, a parameter is represented by lower and upper limits a, b and the possibility of existence of every point within this limits $[a, b]$ are same. However, in reality, it is not so, where the lower 'a' and upper limits 'b' occurred very occasionally. Most of the time, there exists a normal interval $[c, d]$, where $a \leqslant c \leqslant d \leqslant b$. Thus, vagueness here is described by two nested intervals $([a, b], [c, d])$. Such representation is well known as a rough interval.

For example, if we go through "Air Quality Index (AQI)" measurement of the city Ananda Vihar, Delhi in the year 2017, the lower and the upper intervals of this AQI (measured by PM2.5) were $[78, 404]$. The AQI 78 μg/m^3 found in the day of July 2017 (during rainy season) and the AQI is 404 μg/m^3 is found on the day of Dewali (November 2017), whereas the normal range of AQI (PM2.5) is [150, 290]. In this regard, the rough representation of AQI (PM2.5) of Ananda Vihar New Delhi is ([78, 404], [150, 290]).

2. Notation and Assumptions

2.1. Notations

2.1.1. Parameters

For r-th item,

- A_i^r: quantity of homogeneous merchandise available at i-th source.
- D_{0j}^r: market demand at j-th goal.
- D_j^r: actual demand at j-th goal.
- e_{kp}: quantity of the merchandise which can be carried by k-th conveyance along p-th route.
- C_{ijpk}^r: per unit transportation price from i-th origin to j-th goal by k-th vehicle via p-th route.
- S_j^r: selling expenses at the j-th destination.
- P_i^r: purchasing price at the i-th origin.
- f_{ijpk}: fixed transportation cost for shipping units from i-th source/origin to j-th goal/destination by k-th vehicles along p-th route.
- λ_{ijpk}^r: rate of breakability per unit distance from i-th source to j-th goal via p-th route and k-th conveyance.
- $Budj$: total budget at the j-th goal point.
- ds_{ijp}: distance from i-th origin to j-th goal along p-th route.
- SP^r: required space for r-th item.
- SPC_j: available space to the j-th retailer.
- α: power of the route length, related with the frangibility.
- $\beta_j, \theta_j^1, \theta_j^{11}$ and τ: Price sensitivity of products.

2.1.2. Decision Variable

- x_{ijpk}^r: the transported quantity from i-th source/origin to j-th goal/destination by k-th vehicle along p-th route (decision variable).

2.1.3. Indices

- R: total number of items.
- M: total number of origin/sources.
- N: total number of goal/destinations.
- L: total number of route/paths.
- K: total number of vehicles/conveyance.

2.2. Assumptions

(i) Particulars are breakable and carried from sources to goals using a vehicle through a path. Broken/damaged amounts depend on conveyance and path.

(ii) Particulars are substitutable and complementary to each other. In case of a substitute item, the demand is negative and is positive when the items are complementary nature:

$$\left. \begin{array}{l} D_j^1 = D_{0j}^1 - \beta_j S_j^1 + \tau \beta_j S_j^2 - \theta_j^1 \beta_j S_j^3 \\ D_j^2 = D_{0j}^2 + \tau \beta_j S_j^1 - \beta_j S_j^2 - \theta_j^{11} \beta_j S_j^3 \\ D_j^3 = D_{0j}^3 - \theta_j^1 \beta_j S_j^1 - \theta_j^{11} \beta_j S_j^2 - \beta_j S_j^3 \end{array} \right\}.$$

3. Model Description and Formulation

Let us define a four-dimensional transportation problem in this way. A merchant/wholesaler having different M godowns/storing houses (sources) at different locations and different N sales counters/shops (destinations) in different cities. The merchant buys R number of items of which some are substitutes and others complementary to each other and stores at the godowns which are of finite capacities. There are different K conveyances and L routes for transportation of goods from sources to destinations.

The schematic diagram of this 4D-MITP is given in Figure 1. The conveyances are of finite capacities and charge different charges for transportation along different routes. There are some fixed costs/charges (toll taxes, etc.) along each route and items are damageable. This damageability depends on type of conveyance and the distance of traveled by that conveyance. At each destination, the items are known and, for that, some amount of items are stored as per the capacities of the go-downs. The merchant can use any conveyance and route for transportation and the above mentioned transportation parameters are crisp or rough interval numbers. At the destination, the merchant sells the good items at some prices. The merchant has a limitation on his initial expenditure. Now, the problem for the merchant is to decide the amount of quantities to be transported from different sources to different destinations by the appropriate conveyance through the appropriate route so that total profit out of this business is at a maximum subject to the constraints.

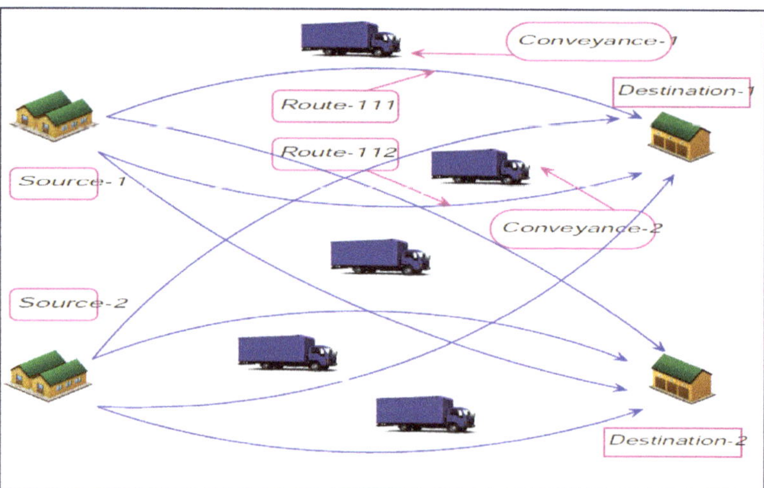

Figure 1. Pictorial representation of a four-dimensional transportation model.

Thus, TP is developed as a profit maximization problem. In this process of business, there are total purchase charge cost, fixed charge cost, transportation cost, revenue in the objective functions and five constraints on source, destination, conveyance capacity, space and budget. Therefore, as per the availability of type of the input parameters with the help of considerable assumptions and notations, we formulate two different models: the crisp model and rough model as given below.

3.1. Model-I: Crisp Model

$$\text{Maximum Profit = Total Revenue - Total Purchase Cost - Total Transportation cost - Total fixed charge}$$

$$\left.\begin{aligned}
&\text{Maximize } Z = \sum_{r=1}^{R}\sum_{i=1}^{M}\sum_{j=1}^{N}\sum_{p=1}^{L}\sum_{k=1}^{K}\left\{S_j^r(1-\lambda_{ijpk}^r ds_{ijp}^\alpha) - P_i^r - C_{ijpk}^r ds_{ijp}\right\}x_{ijpk}^r - \sum_{i=1}^{M}\sum_{j=1}^{N}\sum_{p=1}^{L}\sum_{k=1}^{K} w(x_{ijpk}^r)f_{ijpk}^r\\[4pt]
&\text{subject to } \sum_{j=1}^{N}\sum_{p=1}^{L}\sum_{k=1}^{K} x_{ijpk}^r \le A_i^r, \quad \forall\ i = 1,2,\dots,M, r = 1,2,\dots,R\\[4pt]
&\sum_{i=1}^{M}\sum_{p=1}^{L}\sum_{k=1}^{K}(1-\lambda_{ijpk}^1 ds_{ijp}^\alpha)x_{ijpk}^1 \ge D_{0j}^1 - \beta_j s_j^1 + \tau\beta_j s_j^2 - \theta_j^1\beta_j s_j^3 \quad \forall\ j=1,2,\dots,N, r=1,2,\dots,R\\[4pt]
&\sum_{i=1}^{M}\sum_{p=1}^{L}\sum_{k=1}^{K}(1-\lambda_{ijpk}^2 ds_{ijp}^\alpha)x_{ijpk}^2 \ge D_{0j}^2 + \tau\beta_j S_j^1 - \beta_j S_j^2 - \theta_j^{11}\beta_j S_j^3 \quad \forall\ j=1,2,\dots,N, r=1,2,\dots,R\\[4pt]
&\sum_{i=1}^{M}\sum_{p=1}^{L}\sum_{k=1}^{K}(1-\lambda_{ijpk}^3 ds_{ijp}^\alpha)x_{ijpk}^3 \ge D_{0j}^3 - \theta_j^1\beta_j S_j^1 - \theta_j^{11}\beta_j S_j^2 - \beta_j S_j^3 \quad \forall\ j=1,2,\dots,N, r=1,2,\dots,R\\[4pt]
&\sum_{r=1}^{R}\sum_{i=1}^{M}\sum_{j=1}^{N} x_{ijpk}^r \le e_{kp} \hspace{4.2cm} \forall\ k=1,2,\dots,K, p=1,2,\dots,L\\[4pt]
&\sum_{r=1}^{R}\sum_{i=1}^{M}\sum_{p=1}^{L}\sum_{k=1}^{K}\left[1-\lambda_{ijpk}^r ds_{ijp}^\alpha\right]x_{ijpk}^r * SP^r \le SPC_j^r \quad \forall\ j=1,2,\dots,N,\\[4pt]
&\sum_{r=1}^{R}\sum_{i=1}^{M}\sum_{p=1}^{L}\sum_{k=1}^{K}\left[P_i^r + C_{ijpk}^r ds_{ijp}\right]x_{ijpk}^r + \sum_{i=1}^{M}\sum_{p=1}^{L}\sum_{k=1}^{K} w(x_{ijpk}^r)f_{ijpk} \le Bud_j \quad \forall\ j=1,2,\dots,N,\\[4pt]
&\text{and } x_{ijpk}^r \ge 0\ \forall\ i=1,2,\dots,M, j=1,2,\dots,N, p=1,2,\dots,L, k=1,2,\dots,K, r=1,2,\dots,R
\end{aligned}\right\}. \tag{1}$$

3.2. Model-II: Rough Model

If the parameters, selling expenses, cost of transportation, purchasing prices, fixed charges, availability, demand, capacities, available space and permittable budget are found as rough intervals i.e.,

$$\left.\begin{aligned}
&\breve{S}_j^1 = ([\underline{S}_{jl}^1,\underline{S}_{ju}^1],[\overline{S}_{jl}^1,\overline{S}_{ju}^1]), \breve{S}_j^2 = ([\underline{S}_{jl}^2,\underline{S}_{ju}^2],[\overline{S}_{jl}^2,\overline{S}_{ju}^2]), \breve{S}_j^3 = ([\underline{S}_{jl}^3,\underline{S}_{ju}^3],[\overline{S}_{jl}^3,\overline{S}_{ju}^3]), \breve{P}_i^r = ([\underline{P}_{il}^r,\underline{P}_{iu}^r],[\overline{P}_{il}^r,\overline{P}_{iu}^r]) \breve{C}_{ijpk}^r = ([\underline{C}_{ijpkl},\underline{C}_{ijpku}],[\overline{C}_{ijpkl},\overline{C}_{ijpku}]),\\
&\breve{f}_{ijpku}],[\overline{f}_{ijpkl},\overline{f}_{ijpku}]), \breve{A}_i^r = ([\underline{A}_{il}^r,\underline{A}_{iu}^r],[\overline{A}_{il}^r,\overline{A}_{iu}^r]) \breve{D}_j^1 = ([\underline{D}_{0jl}^1,\underline{D}_{0ju}^1],[\overline{D}_{0jl}^1,\overline{D}_{0ju}^1]) \breve{D}_j^2 = ([\underline{D}_{0jl}^2,\underline{D}_{0ju}^2],[\overline{D}_{0jl}^2,\overline{D}_{0ju}^2]), \breve{D}_j^3 = ([\underline{D}_{0jl}^3,\\
&\underline{D}_{0ju}^3],[\overline{D}_{0jl}^3,\overline{D}_{0ju}^3]) \breve{e}_{kp} = ([\underline{e}_{kpl},\underline{e}_{kpu}],[\overline{e}_{kpl},\overline{e}_{kpu}]), S\breve{P}C_j = ([\underline{SPC}_{jl},\underline{SPC}_{ju}],[\overline{SPC}_{jl},\overline{SPC}_{ju}]) B\breve{u}d_j = ([\underline{Bud}_{jl},\underline{Bud}_{ju}],[\overline{Bud}_{jl},\overline{Bud}_{ju}])\\
&\forall\ i=1,2,\dots,M,\ j=1,2,\dots,N,\ k=1,2,\dots,K,\ r=1,2,\dots,L.
\end{aligned}\right\} \tag{2}$$

Thus, after the introduction of these rough parameters, the above crisp model (1) is translated to the following rough model:

$$\left.\begin{aligned}
&\text{Maximize } Z = \sum_{r=1}^{R}\sum_{i=1}^{M}\sum_{j=1}^{N}\sum_{p=1}^{L}\sum_{k=1}^{K}\left\{\left([\underline{S}_{jl}^r,\underline{S}_{ju}^r],[\overline{S}_{jl}^r,\overline{S}_{ju}^r]\right)(1-\lambda_{ijpk}^r ds_{ijp}^\alpha) - \left([\underline{P}_{il}^r,\underline{P}_{iu}^r],[\overline{P}_{il}^r,\overline{P}_{iu}^r]\right)\right.\\
&\left.- \left([\underline{C}_{ijpkl}^r,\underline{C}_{ijpku}^r],[\overline{C}_{ijpkl}^r,\overline{C}_{ijpku}^r]\right)ds_{ijp}\right\}x_{ijpk}^r - \sum_{i=1}^{M}\sum_{j=1}^{N}\sum_{p=1}^{L}\sum_{k=1}^{K} w(x_{ijpk}^r)\left([\underline{f}_{ijpkl},\underline{f}_{ijpku}],[\overline{f}_{ijpkl},\overline{f}_{ijpku}]\right)\\[4pt]
&\text{subject to } \sum_{j=1}^{N}\sum_{p=1}^{L}\sum_{k=1}^{K} x_{ijpk}^r \le \left([\underline{A}_{il}^r,\underline{A}_{iu}^r],[\overline{A}_{il}^r,\overline{A}_{iu}^r]\right) \quad \forall\ i=1,2,\dots,M, r=1,2,\dots,R\\[4pt]
&\sum_{i=1}^{M}\sum_{p=1}^{L}\sum_{k=1}^{K}(1-\lambda_{ijpk}^\alpha ds_{ijp}^\alpha)x_{ijpk}^1 \ge \left\{\left([\underline{D}_{0jl}^1,\underline{D}_{0ju}^1],[\overline{D}_{0jl}^1,\overline{D}_{0ju}^1]\right) - \beta_j\left([\underline{S}_{jl}^1,\underline{S}_{ju}^1],[\overline{S}_{jl}^1,\overline{S}_{ju}^1]\right)\right.\\
&\left.+\tau\beta_j\left([\underline{S}_{jl}^2,\underline{S}_{ju}^2],[\overline{S}_{jl}^2,\overline{S}_{ju}^2]\right) - \theta_j^1\beta_j\left([\underline{S}_{jl}^3,\underline{S}_{ju}^3],[\overline{S}_{jl}^3,\overline{S}_{ju}^3]\right)\right\}, \quad \forall\ j=1,2,\dots,N, r=1,2,\dots,R\\[4pt]
&\sum_{i=1}^{M}\sum_{p=1}^{L}\sum_{k=1}^{K}(1-\lambda_{ijpk}^2 ds_{ijp}^\alpha)x_{ijpk}^2 \ge \left\{\left([\underline{D}_{0jl}^2,\underline{D}_{0ju}^2],[\overline{D}_{0jl}^2,\overline{D}_{0ju}^2]\right) + \tau\beta_j\left([\underline{S}_{jl}^1,\underline{S}_{ju}^1],[\overline{S}_{jl}^1,\overline{S}_{ju}^1]\right)\right.\\
&\left.-\beta_j\left([\underline{S}_{jl}^2,\underline{S}_{ju}^2],[\overline{S}_{jl}^2,\overline{S}_{ju}^2]\right) - \theta_j^{11}\beta_j\left([\underline{S}_{jl}^3,\underline{S}_{ju}^3],[\overline{S}_{jl}^3,\overline{S}_{ju}^3]\right)\right\}, \quad \forall\ j=1,2,\dots,N, r=1,2,\dots,R\\[4pt]
&\sum_{i=1}^{M}\sum_{p=1}^{L}\sum_{k=1}^{K}(1-\lambda_{ijpk}^3 ds_{ijp}^\alpha)x_{ijpk}^3 \ge \left\{\left([\underline{D}_{0jl}^3,\underline{D}_{0ju}^3],[\overline{D}_{0jl}^3,\overline{D}_{0ju}^3]\right) - \beta_j\theta_j^1\left([\underline{S}_{jl}^1,\underline{S}_{ju}^1],[\overline{S}_{jl}^1,\overline{S}_{ju}^1]\right)\right.\\
&\left.-\beta_j\theta_j^{11}\left([\underline{S}_{jl}^2,\underline{S}_{ju}^2],[\overline{S}_{jl}^2,\overline{S}_{ju}^2]\right) - \beta_j\left([\underline{S}_{jl}^3,\underline{S}_{ju}^3],[\overline{S}_{jl}^3,\overline{S}_{ju}^3]\right)\right\}, \quad \forall\ j=1,2,\dots,N, r=1,2,\dots,R\\[4pt]
&\sum_{r=1}^{R}\sum_{i=1}^{M}\sum_{j=1}^{N} x_{ijpk}^r \le \left([\underline{e}_{kpl},\underline{e}_{kpu}],[\overline{e}_{kpl},\overline{e}_{kpu}]\right) \quad \forall\ k=1,2,\dots,K, p=1,2,\dots,L\\[4pt]
&\sum_{r=1}^{R}\sum_{i=1}^{M}\sum_{p=1}^{L}\sum_{k=1}^{K}\left[1-\lambda_{ijpk}^r ds_{ijp}^\alpha\right]x_{ijpk}^r * SP^r \le \left([\underline{SPC}_{jl}^r,\underline{SPC}_{ju}^r],[\overline{SPC}_{jl}^r,\overline{SPC}_{ju}^r]\right) \quad \forall\ j=1,2,\dots,N,\\[4pt]
&\sum_{r=1}^{R}\sum_{i=1}^{M}\sum_{p=1}^{L}\sum_{k=1}^{K}\left\{\left([\underline{P}_{il}^r,\underline{P}_{iu}^r],[\overline{P}_{il}^r,\overline{P}_{iu}^r]\right) + \left([\underline{C}_{ijpkl}^r,\underline{C}_{ijpku}^r],[\overline{C}_{ijpkl}^r,\overline{C}_{ijpku}^r]\right)ds_{ijp}\right\}x_{ijpk}^r\\
&+\sum_{i=1}^{M}\sum_{p=1}^{L}\sum_{k=1}^{K} w(x_{ijpk}^r)\left([\underline{f}_{ijpkl},\underline{f}_{ijpku}],[\overline{f}_{ijpkl},\overline{f}_{ijpku}]\right) \le \left([\underline{Bud}_{jl},\underline{Bud}_{ju}],[\overline{Bud}_{jl},\overline{Bud}_{ju}]\right)\\
&\forall\ j=1,2,\dots,N\\[4pt]
&\text{and } x_{ijpk}^r \ge 0\ \forall\ i=1,2,\dots,M, j=1,2,\dots,N, p=1,2,\dots,L, k=1,2,\dots,K, r=1,2,\dots,R.
\end{aligned}\right\} \tag{3}$$

According to the discussion of Linear Programming issues with Rough Interval Coefficient (LPRIC) in Appendix A.1, at first, we get two different transportation problems (TPIC-1, TPIC-2) with interval coefficients—and again, as discussed, TPIC-1 and TPIC-2 broken as TP-1, TP-2 and TP-3, TP-4. The above stated consideration can be graphically represented in Figure 2.

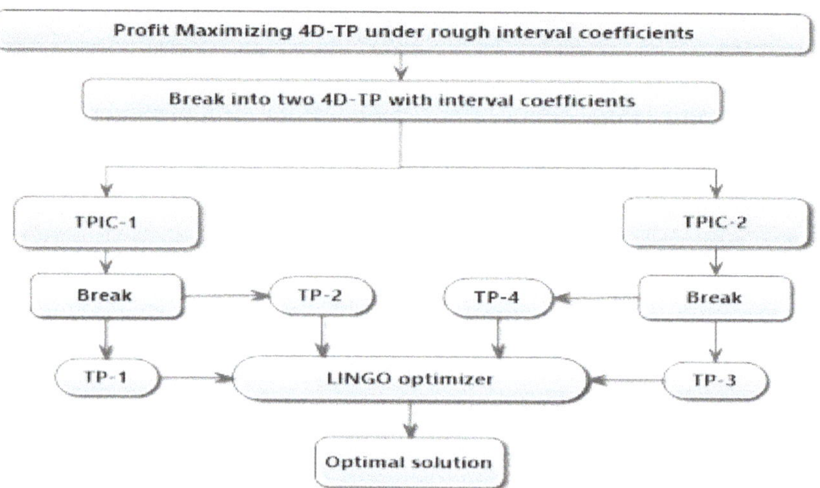

Figure 2. Optimization flowchart of 4D-TP under rough interval coefficients.

3.3. The Mathematical Form of TPIC-1

$$
\begin{aligned}
\text{Maximize } Z^1 = &\sum_{r=1}^{R}\sum_{i=1}^{M}\sum_{j=1}^{N}\sum_{p=1}^{L}\sum_{k=1}^{K} \left\{ [\underline{S}_{jl}^{r}, \underline{S}_{ju}^{r}](1 - \lambda_{ijpk}^{r}ds_{ijp}^{\alpha}) - [\underline{P}_{il}^{r}, \underline{P}_{iu}^{r}] - [\underline{C}_{ijpkl}^{r}, \underline{C}_{ijpku}^{r}]ds_{ijp} \right\}x_{ijpk}^{r} \\
&- \sum_{i=1}^{M}\sum_{j=1}^{N}\sum_{p=1}^{L}\sum_{k=1}^{K} w(x_{ijpk}^{r})[\underline{f}_{ijpkl}, \underline{f}_{ijpku}]
\end{aligned}
$$

$$
\text{s.t } \sum_{j=1}^{N}\sum_{p=1}^{L}\sum_{k=1}^{K} x_{ijpk}^{r} \leq [\underline{A}_{il}^{r}, \underline{A}_{iu}^{r}] \quad \forall\ i = 1,2,...,M, r = 1,2,...,R
$$

$$
\sum_{i=1}^{M}\sum_{p=1}^{L}\sum_{k=1}^{K} (1 - \lambda_{ijpk}^{1}ds_{ijp}^{\alpha})x_{ijpk}^{1} \geq [\underline{D}_{0jl}^{1}, \underline{D}_{0ju}^{1}] - \beta_{j}[\underline{S}_{jl}^{1}, \underline{S}_{ju}^{1}] + \tau\beta_{j}[\underline{S}_{jl}^{2}, \underline{S}_{ju}^{2}] - \theta_{j}^{1}\beta_{j}[\underline{S}_{jl}^{3}, \underline{S}_{ju}^{3}],
$$
$$
\forall\ j = 1,2,...,N, r = 1,2,...,R
$$

$$
\sum_{i=1}^{M}\sum_{p=1}^{L}\sum_{k=1}^{K} (1 - \lambda_{ijpk}^{2}ds_{ijp}^{\alpha})x_{ijpk}^{2} \geq [\underline{D}_{0jl}^{2}, \underline{D}_{0ju}^{2}] + \tau\beta_{j}[\underline{S}_{jl}^{1}, \underline{S}_{ju}^{1}] - \beta_{j}[\underline{S}_{jl}^{2}, \underline{S}_{ju}^{2}] - \theta_{j}^{11}\beta_{j}[\underline{S}_{jl}^{3}, \underline{S}_{ju}^{3}],
$$
$$
\forall\ j = 1,2,...,N, r = 1,2,...,R
$$

$$
\sum_{i=1}^{M}\sum_{p=1}^{L}\sum_{k=1}^{K} (1 - \lambda_{ijpk}^{3}ds_{ijp}^{\alpha})x_{ijpk}^{3} \geq [\underline{D}_{0jl}^{3}, \underline{D}_{0ju}^{3}] - \beta_{j}\theta_{j}^{1}[\underline{S}_{jl}^{1}, \underline{S}_{ju}^{1}] - \beta_{j}\theta_{j}^{11}[\underline{S}_{jl}^{2}, \underline{S}_{ju}^{2}] - \beta_{j}[\underline{S}_{jl}^{3}, \underline{S}_{ju}^{3}],
$$
$$
\forall\ j = 1,2,...,N, r = 1,2,...,R
$$

$$
\sum_{r=1}^{R}\sum_{i=1}^{M}\sum_{j=1}^{N} x_{ijpk}^{r} \leq [\underline{e}_{kpl}, \underline{e}_{kpu}] \quad \forall\ k = 1,2,...,K, p = 1,2,...,L
$$

$$
\sum_{r=1}^{R}\sum_{i=1}^{M}\sum_{p=1}^{L}\sum_{k=1}^{K} \left[1 - \lambda_{ijpk}^{r}ds_{ijp}^{\alpha}\right] x_{ijpk}^{r} * SP^{r} \leq \left([\underline{SPC}_{jl}^{r}, \underline{SPC}_{ju}^{r}] \right) \quad \forall\ j = 1,2,...,N,
$$

$$
\sum_{r=1}^{R}\sum_{i=1}^{M}\sum_{p=1}^{L}\sum_{k=1}^{K} \{[\underline{P}_{il}^{r}, \underline{P}_{iu}^{r}] + [\underline{C}_{ijpkl}^{r}, \underline{C}_{ijpku}^{r}]ds_{ijp}\}x_{ijpk}^{r} + \sum_{i=1}^{M}\sum_{p=1}^{L}\sum_{k=1}^{K} w(x_{ijpk}^{r})[\underline{f}_{ijpkl}, \underline{f}_{ijpku}]
$$
$$
\leq [\underline{Bud}_{jl}, \underline{Bud}_{ju}] \quad \forall\ j = 1,2,...,N
$$

$$
\text{and } x_{ijpk}^{r} \geq 0 \ \forall\ i = 1,2,...,M, j = 1,2,...,N, p = 1,2,...,L, k = 1,2,...,K, r = 1,2,...,R.
$$

$$\tag{4}$$

3.4. The Mathematical Form of TPIC-2

$$
\begin{aligned}
\text{Maximize } Z^2 = {}& \sum_{r=1}^{R}\sum_{i=1}^{M}\sum_{j=1}^{N}\sum_{p=1}^{L}\sum_{k=1}^{K}\{[\overline{S}^r_{jl},\overline{S}^r_{ju}](1-\lambda^r_{ijpk}ds^\alpha_{ijp})-[\overline{P}^r_{il},\overline{P}^r_{iu}]-[\overline{C}^r_{ijpkl},\overline{C}^r_{ijpku}]ds_{ijp}\}x^r_{ijpk} \\[4pt]
& -\sum_{i=1}^{M}\sum_{j=1}^{N}\sum_{p=1}^{L}\sum_{k=1}^{K}w(x^r_{ijpk})[\overline{f}_{ijpkl},\overline{f}_{ijpku}] \\[4pt]
\text{s.t } & \sum_{j=1}^{N}\sum_{p=1}^{L}\sum_{k=1}^{K}x^r_{ijpk}\le[\overline{A}^r_{il},\overline{A}^r_{iu}] \quad \forall\ i=1,2,...,M, r=1,2,...,R \\[4pt]
& \sum_{i=1}^{M}\sum_{p=1}^{L}\sum_{k=1}^{K}(1-\lambda^1_{ijpk}ds^\alpha_{ijp})x^1_{ijpk}\ge[\overline{D}^1_{0jl},\overline{D}^1_{0ju}]-\beta_j[\overline{S}^1_{jl},\overline{S}^1_{ju}]+\tau\beta_j[\overline{S}^2_{jl},\overline{S}^2_{ju}]-\theta^1_j\beta_j[\overline{S}^3_{jl},\overline{S}^3_{ju}], \\[2pt]
& \hspace{6cm} \forall\ j=1,2,...,N, r=1,2,...,R \\[4pt]
& \sum_{i=1}^{M}\sum_{p=1}^{L}\sum_{k=1}^{K}(1-\lambda^2_{ijpk}ds^\alpha_{ijp})x^2_{ijpk}\ge[\overline{D}^2_{0jl},\overline{D}^2_{0ju}]+\tau\beta_j[\overline{S}^1_{jl},\overline{S}^1_{ju}]-\beta_j[\overline{S}^2_{jl},\overline{S}^2_{ju}]-\theta^{11}_j\beta_j([\overline{S}^3_{jl},\overline{S}^3_{ju}], \\[2pt]
& \hspace{6cm} \forall\ j=1,2,...,N, r=1,2,...,R \\[4pt]
& \sum_{i=1}^{M}\sum_{p=1}^{L}\sum_{k=1}^{K}(1-\lambda^3_{ijpk}ds^\alpha_{ijp})x^3_{ijpk}\ge[\overline{D}^3_{0jl},\overline{D}^3_{0ju}]-\beta_j\theta^1_j[\overline{S}^1_{jl},\overline{S}^1_{ju}]-\beta_j\theta^{11}_j[\overline{S}^2_{jl},\overline{S}^2_{ju}]-\beta_j[\overline{S}^3_{jl},\overline{S}^3_{ju}], \\[2pt]
& \hspace{6cm} \forall\ j=1,2,...,N, r=1,2,...,R \\[4pt]
& \sum_{r=1}^{R}\sum_{i=1}^{M}\sum_{j=1}^{N}x^r_{ijpk}\le[\overline{e}_{kpl},\overline{e}_{kpu}] \quad \forall\ k=1,2,...,K, p=1,2,...,L \\[4pt]
& \sum_{r=1}^{R}\sum_{i=1}^{M}\sum_{p=1}^{L}\sum_{k=1}^{K}\left[1-\lambda^r_{ijpk}ds^\alpha_{ijp}\right]x^r_{ijpk}*SP^r\le\left([\overline{SPC}^r_{jl},\overline{SPC}^r_{ju}]\right) \quad \forall\ j=1,2,...,N, \\[4pt]
& \sum_{r=1}^{R}\sum_{i=1}^{M}\sum_{p=1}^{L}\sum_{k=1}^{K}\{[\overline{P}^r_{il},\overline{P}^r_{iu}]+[\overline{C}^r_{ijpkl},\overline{C}^r_{ijpku}]ds_{ijp}\}x^r_{ijpk}+\sum_{i=1}^{M}\sum_{p=1}^{L}\sum_{k=1}^{K}w(x^r_{ijpk})[\overline{f}_{ijpkl},\overline{f}_{ijpku}] \\[2pt]
& \hspace{4cm} \le[\overline{Bud}_{jl},\overline{Bud}_{ju}]\ \forall\ j=1,2,...,N \\[4pt]
\text{and } & x^r_{ijpk}\ge0\ \forall\ i=1,2,...,M, j=1,2,...,N, p=1,2,...,L, k=1,2,...,K, r=1,2,...,R.
\end{aligned} \tag{5}
$$

3.4.1. TP-1

$$
\begin{aligned}
\text{Maximize } \underline{Z}^l = {}& \sum_{r=1}^{R}\sum_{i=1}^{M}\sum_{j=1}^{N}\sum_{p=1}^{L}\sum_{k=1}^{K}\left\{\underline{S}^r_{jl}(1-\lambda^r_{ijpk}ds^\alpha_{ijp})-\underline{P}^r_{iu}-\underline{C}^r_{ijpku}ds_{ijp}\right\}x^r_{ijpk}-\sum_{i=1}^{M}\sum_{j=1}^{N}\sum_{p=1}^{L}\sum_{k=1}^{K}w(x^r_{ijpk})\underline{f}_{ijpku} \\[4pt]
\text{s.t } & \sum_{j=1}^{N}\sum_{p=1}^{L}\sum_{k=1}^{K}x^r_{ijpk}\le\underline{A}^r_{il} \quad \forall\ i=1,2,...,M, r=1,2,...,R \\[4pt]
& \sum_{i=1}^{M}\sum_{p=1}^{L}\sum_{k=1}^{K}(1-\lambda^1_{ijpk}ds^\alpha_{ijp})x^1_{ijpk}\ge\left\{\underline{D}^1_{0ju}-\beta_j\underline{S}^1_{jl}+\tau\beta_j\underline{S}^2_{ju}-\theta^1_j\beta_j\underline{S}^3_{jl}\right\}, \quad \forall\ j=1,2,...,N, r=1,2,...,R \\[4pt]
& \sum_{i=1}^{M}\sum_{p=1}^{L}\sum_{k=1}^{K}(1-\lambda^2_{ijpk}ds^\alpha_{ijp})x^2_{ijpk}\ge\left\{\underline{D}^2_{0ju}+\tau\beta_j\underline{S}^1_{ju}-\beta_j\underline{S}^2_{jl}-\theta^{11}_j\beta_j\underline{S}^3_{jl}\right\}, \quad \forall\ j=1,2,...,N, r=1,2,...,R \\[4pt]
& \sum_{i=1}^{M}\sum_{p=1}^{L}\sum_{k=1}^{K}(1-\lambda^3_{ijpk}ds^\alpha_{ijp})x^3_{ijpk}\ge\left\{\underline{D}^3_{0ju}-\beta_j\theta^1_j\underline{S}^1_{jl}-\beta_j\theta^{11}_j\underline{S}^2_{jl}-\beta_j\underline{S}^3_{jl}\right\}, \quad \forall\ j=1,2,...,N, r=1,2,...,R \\[4pt]
& \sum_{r=1}^{R}\sum_{i=1}^{M}\sum_{j=1}^{N}x^r_{ijpk}\le\underline{e}_{kpl}\ \ k=1,2,...,K, p=1,2,...,L \\[4pt]
& \sum_{r=1}^{R}\sum_{i=1}^{M}\sum_{p=1}^{L}\sum_{k=1}^{K}\left[1-\lambda^r_{ijpk}ds^\alpha_{ijp}\right]x^r_{ijpk}*SP^r\le\underline{SPC}^r_{jl} \quad \forall\ j=1,2,...,N, \\[4pt]
& \sum_{r=1}^{R}\sum_{i=1}^{M}\sum_{p=1}^{L}\sum_{k=1}^{K}\left\{\underline{P}^r_{iu}+\underline{C}^r_{ijpku}ds_{ijp}\right\}x^r_{ijpk}+\sum_{i=1}^{M}\sum_{p=1}^{L}\sum_{k=1}^{K}w(x^r_{ijpk})\underline{f}_{ijpku}\Big)\le\underline{Bud}_{jl}, \quad \forall\ j=1,2,...,N \\[4pt]
\text{and } & x^r_{ijpk}\ge0\ \forall\ i=1,2,...,M, j=1,2,...,N, p=1,2,...,L, k=1,2,...,K, r=1,2,...,R.
\end{aligned} \tag{6}
$$

3.4.2. TP-2

$$\text{Maximize } \underline{Z}^u = \sum_{r=1}^{R}\sum_{i=1}^{M}\sum_{j=1}^{N}\sum_{p=1}^{L}\sum_{k=1}^{K}\left\{\underline{S}_{ju}^r(1-\lambda_{ijpk}^r ds_{ijp}^\alpha) - \underline{P}_{il}^r - \underline{C}_{ijpkl}^r ds_{ijp}\right\}x_{ijpk}^r - \sum_{i=1}^{M}\sum_{j=1}^{N}\sum_{p=1}^{L}\sum_{k=1}^{K}w(x_{ijpk}^r)\underline{f}_{ijpkl}$$

$$\text{s.t } \sum_{j=1}^{N}\sum_{p=1}^{L}\sum_{k=1}^{K}x_{ijpk}^r \leq \underline{A}_{iu}^r \quad \forall\ i=1,2,\dots,M, r=1,2,\dots,R$$

$$\sum_{i=1}^{M}\sum_{p=1}^{L}\sum_{k=1}^{K}(1-\lambda_{ijpk}^1 ds_{ijp}^\alpha)x_{ijpk}^1 \geq \left\{\underline{D}_{0jl}^1 - \beta_j\underline{S}_{ju}^1 + \tau\beta_j\underline{S}_{jl}^2 - \theta_j^1\beta_j\underline{S}_{ju}^3\right\}, \quad \forall\ j=1,2,\dots,N, r=1,2,\dots,R$$

$$\sum_{i=1}^{M}\sum_{p=1}^{L}\sum_{k=1}^{K}(1-\lambda_{ijpk}^2 ds_{ijp}^\alpha)x_{ijpk}^2 \geq \left\{\underline{D}_{0jl}^2 + \tau\beta_j\underline{S}_{jl}^1 - \beta_j\underline{S}_{ju}^2 - \theta_j^{11}\beta_j\underline{S}_{ju}^3\right\}, \quad \forall\ j=1,2,\dots,N, r=1,2,\dots,R$$

$$\sum_{i=1}^{M}\sum_{p=1}^{L}\sum_{k=1}^{K}(1-\lambda_{ijpk}^3 ds_{ijp}^\alpha)x_{ijpk}^3 \geq \left\{\underline{D}_{0jl}^3 - \beta_j\theta_j^1\underline{S}_{ju}^1 - \beta_j\theta_j^{11}\underline{S}_{ju}^2 - \beta_j\underline{S}_{ju}^3\right\}, \quad \forall\ j=1,2,\dots,N, r=1,2,\dots,R$$

$$\sum_{r=1}^{R}\sum_{i=1}^{M}\sum_{j=1}^{N}x_{ijpk}^r \leq \underline{e}_{kpu} \quad \forall\ k=1,2,\dots,K, p=1,2,\dots,L$$

$$\sum_{r=1}^{R}\sum_{i=1}^{M}\sum_{p=1}^{L}\sum_{k=1}^{K}\left[1-\lambda_{ijpk}^r ds_{ijp}^\alpha\right]x_{ijpk}^r * SP^r \leq \underline{SPC}_{ju}^r \quad \forall\ j=1,2,\dots,N,$$

$$\sum_{r=1}^{R}\sum_{i=1}^{M}\sum_{p=1}^{L}\sum_{k=1}^{K}\left\{\underline{P}_{il}^r + \underline{C}_{ijpkl}^r ds_{ijp}\right\}x_{ijpk}^r + \sum_{i=1}^{M}\sum_{p=1}^{L}\sum_{k=1}^{K}w(x_{ijpk}^r)\underline{f}_{ijpkl} \leq \underline{Bud}_{ju} \quad \forall\ j=1,2,\dots,N$$

$$\text{and } x_{ijpk}^r \geq 0\ \forall\ i=1,2,\dots,M, j=1,2,\dots,N, p=1,2,\dots,L, k=1,2,\dots,K, r=1,2,\dots,R. \tag{7}$$

3.4.3. TP-3

$$\text{Maximize } \overline{Z}^l = \sum_{r=1}^{R}\sum_{i=1}^{M}\sum_{j=1}^{N}\sum_{p=1}^{L}\sum_{k=1}^{K}\left\{\overline{S}_{jl}^r(1-\lambda_{ijpk}^r ds_{ijp}^\alpha) - \overline{P}_{iu}^r - \overline{C}_{ijpku}^r ds_{ijp}\right\}x_{ijpk}^r - \sum_{i=1}^{M}\sum_{j=1}^{N}\sum_{p=1}^{L}\sum_{k=1}^{K}w(x_{ijpk}^r)\overline{f}_{ijpku}$$

$$\text{subject to } \sum_{j=1}^{N}\sum_{p=1}^{L}\sum_{k=1}^{K}x_{ijpk}^r \leq \overline{A}_{il}^r \quad \forall\ i=1,2,\dots,M, r=1,2,\dots,R$$

$$\sum_{i=1}^{M}\sum_{p=1}^{L}\sum_{k=1}^{K}(1-\lambda_{ijpk}^1 ds_{ijp}^\alpha)x_{ijpk}^1 \geq \left\{\overline{D}_{0ju}^1 - \beta_j\overline{S}_{jl}^1 + \tau\beta_j\overline{S}_{ju}^2 - \theta_j^1\beta_j\overline{S}_{jl}^3\right\}, \quad j=1,2,\dots,N, r=1,2,\dots,R$$

$$\sum_{i=1}^{M}\sum_{p=1}^{L}\sum_{k=1}^{K}(1-\lambda_{ijpk}^2 ds_{ijp}^\alpha)x_{ijpk}^2 \geq \left\{\overline{D}_{0ju}^2 + \tau\beta_j\overline{S}_{ju}^1 - \beta_j\overline{S}_{jl}^2 - \theta_j^{11}\beta_j\overline{S}_{jl}^3\right\}, \quad \forall\ j=1,2,\dots,N, r=1,2,\dots,R$$

$$\sum_{i=1}^{M}\sum_{p=1}^{L}\sum_{k=1}^{K}(1-\lambda_{ijpk}^3 ds_{ijp}^\alpha)x_{ijpk}^3 \geq \left\{\overline{D}_{0ju}^3 - \beta_j\theta_j^1\overline{S}_{jl}^1 - \beta_j\theta_j^{11}\overline{S}_{jl}^2 - \beta_j\overline{S}_{jl}^3\right\}, \quad \forall\ j=1,2,\dots,N, r=1,2,\dots,R$$

$$\sum_{r=1}^{R}\sum_{i=1}^{M}\sum_{j=1}^{N}x_{ijpk}^r \leq \overline{e}_{kpl} \quad k=1,2,\dots,K, p=1,2,\dots,L$$

$$\sum_{r=1}^{R}\sum_{i=1}^{M}\sum_{p=1}^{L}\sum_{k=1}^{K}\left[1-\lambda_{ijpk}^r ds_{ijp}^\alpha\right]x_{ijpk}^r * SP^r \leq \overline{SPC}_{jl}^r \quad \forall\ j=1,2,\dots,N,$$

$$\sum_{r=1}^{R}\sum_{i=1}^{M}\sum_{p=1}^{L}\sum_{k=1}^{K}\left\{\overline{P}_{iu}^r + \overline{C}_{ijpku}^r ds_{ijp}\right\}x_{ijpk}^r + \sum_{i=1}^{M}\sum_{p=1}^{L}\sum_{k=1}^{K}w(x_{ijpk}^r)\overline{f}_{ijpku} \leq \overline{Bud}_{jl} \quad \forall\ j=1,2,\dots,N$$

$$\text{and } x_{ijpk}^r \geq 0\ \forall\ i=1,2,\dots,M, j=1,2,\dots,N, p=1,2,\dots,L, k=1,2,\dots,K, r=1,2,\dots,R. \tag{8}$$

3.4.4. TP-4

$$\text{Maximize } \overline{Z}^u = \sum_{r=1}^{R}\sum_{i=1}^{M}\sum_{j=1}^{N}\sum_{p=1}^{L}\sum_{k=1}^{K}\left\{\overline{S}_{ju}^r(1-\lambda_{ijpk}^r ds_{ijp}^\alpha) - \overline{P}_{il}^r - \overline{C}_{ijpkl}^r ds_{ijp}\right\}x_{ijpk}^r - \sum_{i=1}^{M}\sum_{j=1}^{N}\sum_{p=1}^{L}\sum_{k=1}^{K}w(x_{ijpk}^r)\overline{f}_{ijpkl}$$

$$\text{subject to } \sum_{j=1}^{N}\sum_{p=1}^{L}\sum_{k=1}^{K}x_{ijpk}^r \leq \overline{A}_{iu}^r \quad \forall\ i=1,2,\dots,M, r=1,2,\dots,R$$

$$\sum_{i=1}^{M}\sum_{p=1}^{L}\sum_{k=1}^{K}(1-\lambda_{ijpk}^1 ds_{ijp}^\alpha)x_{ijpk}^1 \geq \left\{\overline{D}_{0jl}^1 - \beta_j\overline{S}_{ju}^1 + \tau\beta_j\overline{S}_{jl}^2 - \theta_j^1\beta_j\overline{S}_{ju}^3\right\}, \quad \forall\ j=1,2,\dots,N, r=1,2,\dots,R$$

$$\sum_{i=1}^{M}\sum_{p=1}^{L}\sum_{k=1}^{K}(1-\lambda_{ijpk}^2 ds_{ijp}^\alpha)x_{ijpk}^2 \geq \left\{\overline{D}_{0jl}^2 + \tau\beta_j\overline{S}_{jl}^1 - \beta_j\overline{S}_{ju}^2 - \theta_j^{11}\beta_j\overline{S}_{ju}^3\right\}, \quad \forall\ j=1,2,\dots,N, r=1,2,\dots,R$$

$$\sum_{i=1}^{M}\sum_{p=1}^{L}\sum_{k=1}^{K}(1-\lambda_{ijpk}^3 ds_{ijp}^\alpha)x_{ijpk}^3 \geq \left\{\overline{D}_{0jl}^3 - \beta_j\theta_j^1\overline{S}_{ju}^1 - \beta_j\theta_j^{11}\overline{S}_{ju}^2 - \beta_j\overline{S}_{ju}^3\right\}, \quad \forall\ j=1,2,\dots,N, r=1,2,\dots,R$$

$$\sum_{r=1}^{R}\sum_{i=1}^{M}\sum_{j=1}^{N}x_{ijpk}^r \leq \overline{e}_{kpu} \quad k=1,2,\dots,K, p=1,2,\dots,L$$

$$\sum_{r=1}^{R}\sum_{i=1}^{M}\sum_{p=1}^{L}\sum_{k=1}^{K}\left[1-\lambda_{ijpk}^r ds_{ijp}^\alpha\right]x_{ijpk}^r * SP^r \leq \overline{SPC}_{ju}^r \quad \forall\ j=1,2,\dots,N,$$

$$\sum_{r=1}^{R}\sum_{i=1}^{M}\sum_{p=1}^{L}\sum_{k=1}^{K}\left\{\overline{P}_{il}^r + \overline{C}_{ijpkl}^r ds_{ijp}\right\}x_{ijpk}^r + \sum_{i=1}^{M}\sum_{p=1}^{L}\sum_{k=1}^{K}w(x_{ijpk}^r)\overline{f}_{ijpkl} \leq \overline{Bud}_{ju} \quad \forall\ j=1,2,\dots,N$$

$$\text{and } x_{ijpk}^r \geq 0\ \forall\ i=1,2,\dots,M, j=1,2,\dots,N, p=1,2,\dots,L, k=1,2,\dots,K, r=1,2,\dots,R. \tag{9}$$

3.4.5. Approach-2

After use of the Expected Value Method, the Rough transportation problem (Model-II) mathematically reduces to:

$$
\begin{aligned}
\text{Maximize } E\!\left[Z\right] = E\Bigg[&\sum_{r=1}^{R}\sum_{i=1}^{M}\sum_{j=1}^{N}\sum_{p=1}^{L}\sum_{k=1}^{K} \Big\{ \left([\underline{S}^r_{jl},\underline{S}^r_{ju}],[\overline{S}^r_{jl},\overline{S}^r_{ju}]\right)(1 - \lambda^r_{ijpk}ds^\alpha_{ijp}) - \left([\underline{P}^r_{il},\underline{P}^r_{iu}],[\overline{P}^r_{il},\overline{P}^r_{iu}]\right) \\
&- \left([\underline{C}^r_{ijpkl},\underline{C}^r_{ijpku}],[\overline{C}^r_{ijpkl},\overline{C}^r_{ijpku}]\right)ds_{ijp} \Big\}x^r_{ijpk} - \sum_{i=1}^{M}\sum_{j=1}^{N}\sum_{p=1}^{L}\sum_{k=1}^{K} w(x^r_{ijpk})\left([\underline{f}_{ijpkl},\underline{f}_{ijpku}],[\overline{f}_{ijpkl},\overline{f}_{ijpku}]\right) \Bigg] \\[4pt]
\text{subject to } &\sum_{j=1}^{N}\sum_{p=1}^{L}\sum_{k=1}^{K} x^r_{ijpk} \le E\left[\left([\underline{A}^r_{il},\underline{A}^r_{iu}],[\overline{A}^r_{il},\overline{A}^r_{iu}]\right)\right] \quad \forall\ i = 1,2,...,M, r = 1,2,...,R \\
&\sum_{i=1}^{M}\sum_{p=1}^{L}\sum_{k=1}^{K}(1 - \lambda^1_{ijpk}ds^\alpha_{ijp})x^1_{ijpk} \ge E\Big[\Big\{ \left([\underline{D}^1_{0jl},\underline{D}^1_{0ju}],[\overline{D}^1_{0jl},\overline{D}^1_{0ju}]\right) - \beta_j\left([\underline{S}^1_{jl},\underline{S}^1_{ju}],[\overline{S}^1_{jl},\overline{S}^1_{ju}]\right) \\
&\qquad + \tau\beta_j\left([\underline{S}^2_{jl},\underline{S}^2_{ju}],[\overline{S}^2_{jl},\overline{S}^2_{ju}]\right) - \theta^1_j\beta_j\left([\underline{S}^3_{jl},\underline{S}^3_{ju}],[\overline{S}^3_{jl},\overline{S}^3_{ju}]\right) \Big\}\Big], \quad \forall\ j = 1,2,...,N, r = 1,2,...,R \\
&\sum_{i=1}^{M}\sum_{p=1}^{L}\sum_{k=1}^{K}(1 - \lambda^2_{ijpk}ds^\alpha_{ijp})x^2_{ijpk} \ge E\Big[\Big\{ \left([\underline{D}^2_{0jl},\underline{D}^2_{0ju}],[\overline{D}^2_{0jl},\overline{D}^2_{0ju}]\right) + \tau\beta_j\left([\underline{S}^1_{jl},\underline{S}^1_{ju}],[\overline{S}^1_{jl},\overline{S}^1_{ju}]\right) \\
&\qquad - \beta_j\left([\underline{S}^2_{jl},\underline{S}^2_{ju}],[\overline{S}^2_{jl},\overline{S}^2_{ju}]\right) - \theta^{11}_j\beta_j\left([\underline{S}^3_{jl},\underline{S}^3_{ju}],[\overline{S}^3_{jl},\overline{S}^3_{ju}]\right) \Big\}\Big], \quad \forall\ j = 1,2,...,N, r = 1,2,...,R \\
&\sum_{i=1}^{M}\sum_{p=1}^{L}\sum_{k=1}^{K}(1 - \lambda^3_{ijpk}ds^\alpha_{ijp})x^3_{ijpk} \ge E\Big[\Big\{ \left([\underline{D}^3_{0jl},\underline{D}^3_{0ju}],[\overline{D}^3_{0jl},\overline{D}^3_{0ju}]\right) - \beta_j\theta^1_j\left([\underline{S}^1_{jl},\underline{S}^1_{ju}],[\overline{S}^1_{jl},\overline{S}^1_{ju}]\right) \\
&\qquad - \beta_j\theta^{11}_j\left([\underline{S}^2_{jl},\underline{S}^2_{ju}],[\overline{S}^2_{jl},\overline{S}^2_{ju}]\right) - \beta_j\left([\underline{S}^3_{jl},\underline{S}^3_{ju}],[\overline{S}^3_{jl},\overline{S}^3_{ju}]\right) \Big\}\Big], \quad \forall\ j = 1,2,...,N, r = 1,2,...,R \\
&\sum_{r=1}^{R}\sum_{i=1}^{M}\sum_{j=1}^{N} x^r_{ijpk} \le E\left[\left([\underline{e}_{kpl},\underline{e}_{kpu}],[\overline{e}_{kpl},\overline{e}_{kpu}]\right)\right] \quad \forall\ k = 1,2,...,K, p = 1,2,...,L \\
&\sum_{r=1}^{R}\sum_{i=1}^{M}\sum_{p=1}^{L}\sum_{k=1}^{K}\left[1 - \lambda^r_{ijpk}ds^\alpha_{ijp}\right]x^r_{ijpk}*SP^r \le E\left[\left([\underline{SPC}^r_{jl},\underline{SPC}^r_{ju}],[\overline{SPC}^r_{jl},\overline{SPC}^r_{ju}]\right)\right] \quad \forall\ j = 1,2,...,N, \\
&\sum_{r=1}^{R}\sum_{i=1}^{M}\sum_{p=1}^{L}\sum_{k=1}^{K} E\Big[\Big\{ \left([\underline{P}^r_{il},\underline{P}^r_{iu}],[\overline{P}^r_{il},\overline{P}^r_{iu}]\right) + \left([\underline{C}^r_{ijpkl},\underline{C}^r_{ijpku}],[\overline{C}^r_{ijpkl},\overline{C}^r_{ijpku}]\right)ds_{ijp} \Big\}x^r_{ijpk} \\
&\qquad + \sum_{i=1}^{M}\sum_{p=1}^{L}\sum_{k=1}^{K} w(x^r_{ijpk})\left([\underline{f}_{ijpkl},\underline{f}_{ijpku}],[\overline{f}_{ijpkl},\overline{f}_{ijpku}]\right)\Big] \le E\left[\left([\underline{Bud}_{jl},\underline{Bud}_{ju}],[\overline{Bud}_{jl},\overline{Bud}_{ju}]\right)\right] \\
&\qquad \forall\ j = 1,2,...,N \\
&\text{and } x^r_{ijpk} \ge 0\ \forall\ i = 1,2,...,M, j = 1,2,...,N, p = 1,2,...,L, k = 1,2,...,K, r = 1,2,...,R.
\end{aligned}
\tag{10}
$$

The problem (16) can be written as:

$$
\begin{aligned}
\text{Maximize } Z = &\sum_{r=1}^{R}\sum_{i=1}^{M}\sum_{j=1}^{N}\sum_{p=1}^{L}\sum_{k=1}^{K} \Big[\{\eta(\underline{S}^r_{jl} + \underline{S}^r_{ju}) + (1-\eta)(\overline{S}^r_{jl} + \overline{S}^r_{ju})\}/2 * (1 - \lambda^r_{ijpk}ds^\alpha_{ijp}) \\
&- \{\eta(\underline{P}^r_{il} + \underline{P}^r_{iu}) + (1-\eta)(\overline{P}^r_{il} + \overline{P}^r_{iu})\}/2 - \{\eta(\underline{C}^r_{ijpkl} + \underline{C}^r_{ijpku}) + (1-\eta)(\overline{C}^r_{ijpkl} + \overline{C}^r_{ijpku})\}/2 * ds_{ijp}\Big]x^r_{ijpk} \\
&- \sum_{i=1}^{M}\sum_{j=1}^{N}\sum_{p=1}^{L}\sum_{k=1}^{K} w(x^r_{ijpk})\Big[\{\eta(\underline{f}_{ijpkl} + \underline{f}_{ijpku}) + (1-\eta)(\overline{f}_{ijpkl} + \overline{f}_{ijpku})\}/2\Big] \\[4pt]
\text{subject to } &\sum_{j=1}^{N}\sum_{p=1}^{L}\sum_{k=1}^{K} x^r_{ijpk} \le \Big[\{\eta(\underline{A}^r_{il} + \underline{A}^r_{iu}) + (1-\eta)(\overline{A}^r_{il} + \overline{A}^r_{iu})\}/2\Big] \quad \forall\ i = 1,2,...,M, r = 1,2,...,R \\
&\sum_{i=1}^{M}\sum_{p=1}^{L}\sum_{k=1}^{K}(1 - \lambda^1_{ijpk}ds^\alpha_{ijp})x^1_{ijpk} \ge \Big[\{\eta(\underline{D}^1_{0jl} + \underline{D}^1_{0ju}) + (1-\eta)(\overline{D}^1_{0jl} + \overline{D}^1_{0ju})\}/2 - \beta_j\{\eta(\underline{S}^1_{jl} + \underline{S}^1_{ju}) + (1-\eta)(\overline{S}^1_{jl} + \overline{S}^1_{ju})\}/2 \\
&\qquad + \tau\beta_j\{\eta(\underline{S}^2_{jl} + \underline{S}^2_{ju}) + (1-\eta)(\overline{S}^2_{jl} + \overline{S}^2_{ju})\}/2 - \theta^1_j\beta_j\{(\underline{S}^3_{jl} + \underline{S}^3_{ju}) + (1-\eta)(\overline{S}^3_{jl} + \overline{S}^3_{ju})\}/2\Big], \quad \forall\ j = 1,2,...,N, r = 1,2,...,R \\
&\sum_{i=1}^{M}\sum_{p=1}^{L}\sum_{k=1}^{K}(1 - \lambda^2_{ijpk}ds^\alpha_{ijp})x^2_{ijpk} \ge \Big[\{\eta(\underline{D}^2_{0jl} + \underline{D}^2_{0ju}) + (1-\eta)(\overline{D}^2_{0jl} + \overline{D}^2_{0ju})\}/2 + \tau\beta_j\{\eta(\underline{S}^1_{jl} + \underline{S}^1_{ju}) + (1-\eta)(\overline{S}^1_{jl} + \overline{S}^1_{ju})\}/2 \\
&\qquad - \beta_j\{\eta(\underline{S}^2_{jl} + \underline{S}^2_{ju}) + (1-\eta)(\overline{S}^2_{jl} + \overline{S}^2_{ju})/2 - \theta^{11}_j\beta_j\{\eta(\underline{S}^3_{jl} + \underline{S}^3_{ju}) + (1-\eta)(\overline{S}^3_{jl} + \overline{S}^3_{ju})\}/2\Big], \quad \forall\ j = 1,2,...,N, r = 1,2,...,R \\
&\sum_{i=1}^{M}\sum_{p=1}^{L}\sum_{k=1}^{K}(1 - \lambda^3_{ijpk}ds^\alpha_{ijp})x^3_{ijpk} \ge \Big[\{\eta(\underline{D}^3_{0jl} + \underline{D}^3_{0ju}) + (1-\eta)(\overline{D}^3_{0jl} + \overline{D}^3_{0ju})\}/2 - \beta_j\theta^1_j\{\eta(\underline{S}^1_{jl} + \underline{S}^1_{ju}) + (1-\eta)(\overline{S}^1_{jl} + \overline{S}^1_{ju})\}/2 \\
&\qquad - \beta_j\theta^{11}_j\{\eta(\underline{S}^2_{jl} + \underline{S}^2_{ju}) + (1-\eta)(\overline{S}^2_{jl} + \overline{S}^2_{ju})\}/2 - \beta_j\{\eta(\underline{S}^3_{jl} + \underline{S}^3_{ju}) + (1-\eta)(\overline{S}^3_{jl} + \overline{S}^3_{ju})\}/2\Big], \quad j = 1,2,...,N, r = 1,2,...,R \\
&\sum_{r=1}^{R}\sum_{i=1}^{M}\sum_{j=1}^{N} x^r_{ijpk} \le \Big[\eta(\underline{e}_{kpl} + \underline{e}_{kpu}) + (1-\eta)(\overline{e}_{kpl} + \overline{e}_{kpu})]/2)\Big] \quad \forall\ k = 1,2,...,K, p = 1,2,...,L \\
&\sum_{r=1}^{R}\sum_{i=1}^{M}\sum_{p=1}^{L}\sum_{k=1}^{K}\Big[1 - \lambda^r_{ijpk}ds^\alpha_{ijp}\Big]x^r_{ijpk}*SP^r \le \Big[\{\eta(\underline{SPC}^r_{jl} + \underline{SPC}^r_{ju}) + (1-\eta)(\overline{SPC}^r_{jl} + \overline{SPC}^r_{ju})\}/2\Big] \quad \forall\ j = 1,2,...,N, \\
&\sum_{r=1}^{R}\sum_{i=1}^{M}\sum_{p=1}^{L}\sum_{k=1}^{K}\Big[\{\eta(\underline{P}^r_{il} + (1-\eta)\underline{P}^r_{iu}) + (\overline{P}^r_{il} + \overline{P}^r_{iu})\}/2 + \{\eta(\underline{C}^r_{ijpkl} + \underline{C}^r_{ijpku}) + (1-\eta)(\overline{C}^r_{ijpkl} + \overline{C}^r_{ijpku})\}/2 * ds_{ijp}\Big]x^r_{ijpk} + \\
&\sum_{i=1}^{M}\sum_{p=1}^{L}\sum_{k=1}^{K} w(x^r_{ijpk})\Big[\{\eta(\underline{f}_{ijpkl} + \underline{f}_{ijpku}) + (1-\eta)(\overline{f}_{ijpkl} + \overline{f}_{ijpku})\}/2\Big] \le \Big[\{\eta(\underline{Bud}_{jl} + \underline{Bud}_{ju}) + (1-\eta)(\overline{Bud}_{jl} + \overline{Bud}_{ju})\}/2\Big] \\
&\qquad \forall\ j = 1,2,...,N \\
&\text{and } x^r_{ijpk} \ge 0\ \forall\ i = 1,2,...,M, j = 1,2,...,N, p = 1,2,...,L, k = 1,2,...,K, r = 1,2,...,R.
\end{aligned}
\tag{11}
$$

4. Numerical Experiments

Let us consider a real life transportation system where number of sources = 2 (i.e., $M = 2$), number of destination = 2 (i.e., $N = 2$), number of vehicles = 2 (i.e., $K = 2$) and number of routes = 2 ($L = 2$), number of items = 3 (i.e., $R = 3$). The input values of unit transportation costs, fixed charges are of crisp and rough interval forms which are presented in Tables 2 and 3. The amount of breakability in a crisp environment is also presented in Table 2. Here, it is assumed that there are three types of products. The 1st and 2nd ones are the substitute, the 1st and 3rd are complementary, and the 2nd and 3rd are also complimentary.

Table 2. Unit cost of transportation (C_{ijpk}^r), fixed charges (f_{ijpk}), breakability (λ_{ijpk}^r) for proposed Models.

		p		1				2		
		k	1		2		1		2	
		i/j	1	2	1	2	1	2	1	2
						Model-I				
						(Crisp Model)				
C_{ijkp}^1	1		0.5	2.1	1.7	1.12	0.85	0.5	1.38	1.28
	2		1.21	1.28	2.05	0.6	1.65	2.72	1.66	1.78
C_{ijkp}^2	1		1.5	2.25	2.0	1.3	0.08	0.7	1.05	1.25
	2		1.05	2.15	2.2	0.9	1.35	2.25	1.85	0.98
C_{ijkp}^3	1		1.3	1.2	1.91	1.2	0.9	0.94	0.73	0.93
	2		1.22	1.3	2	0.7	1.8	3	1.8	1.8
λ_{ijkp}^1	1		0.01	0.01	0.01	0.01	0.02	0.02	0.02	0.02
	2		0.01	0.01	0.01	0.01	0.02	0.02	0.02	0.02
λ_{ijkp}^2	1		0.01	0.01	0.01	0.01	0.02	0.02	0.02	0.02
	2		0.01	0.01	0.01	0.01	0.02	0.02	0.02	0.02
λ_{ijkp}^3	1		0.01	0.01	0.01	0.01	0.02	0.02	0.02	0.02
	2		0.01	0.01	0.01	0.01	0.02	0.02	0.02	0.02
f_{ijkp}	1		1.9	1.7	1.1	1.3	1.2	1.4	0.3	1.8
	2		1.2	1.85	0.9	1.5	1.5	2.1	2.1	1.6
						Model-II				
						(Rough Model)				
$\check{C}_{ijkp}^1$	1		([1, 2], [0.7, 2.3])	([0.5, 1.5], [0.5, 2])	([1.1, 2.3], [1, 2.5])	([0.8, 1.3], [0.5, 2])	([1.4, 2.3], [1.2, 2.8])	([0.7, 1], [0.5, 1.5])	([1.2, 2], [1.2, 2.3])	([1.2, 2.4], [1, 3])
	2		([1.2, 2.5], [1, 3])	([0.9, 2], [0.5, 3.2])	([1.1, 2], [0.7, 2.5])	([1, 2], [0.9, 2.6])	([1.1, 2.2], [1, 2.7])	([1, 2.5], [0.3, 3])	([0.9, 2.1], [0.7, 2.8])	([1.3, 2.5], [1.1, 3.5])
$\check{C}_{ijkp}^2$	1		([1, 2], [0.7, 2.3])	([0.5, 1.5], [0.5, 2])	([1.1, 2.3], [1, 2.5])	([0.8, 1.3], [0.5, 2])	([1.4, 2.3], [1.2, 2.8])	([0.7, 1], [0.5, 1.5])	([1.2, 2], [1.2, 2.3])	([1.2, 2.4], [1, 3])
	2		([1.2, 2.5], [1, 3])	([0.9, 2], [0.5, 3.2])	([1.1, 2], [0.7, 2.5])	([1, 2], [0.9, 2.6])	([1.1, 2.2], [1, 2.7])	([1, 2.5], [0.3, 3])	([0.9, 2.1], [0.7, 2.8])	([1.3, 2.5], [1.1, 3.5])
$\check{C}_{ijkp}^3$	1		([1.3, 2.7], [1, 3])	([1.7, 2.8], [1.2, 3.1])	([1.5, 2.5], [1.4, 3.1])	([1.6, 2.4], [1.5, 3.5])	([1.3, 3], [1.2, 3.5])	([1.4, 2], [1.1, 2.5])	([1.5, 2.6], [1.3, 3])	([1.8, 2.7], [1.6, 3.4]
	2		([1.5, 2.5], [1, 3.5])	([1.3, 3], [1.1, 3.7])	([1.7, 2.7], [1.5, 3])	([1.4, 3], [1.2, 3.2])	([2, 2.7], [1.8, 3])	([1.3, 2.2], [1, 2.5])	([1.6, 2.1], [1.5, 2.8])	([1.6, 2.5], [1.4, 2.8])
$\check{f}_{ijkp}$	1		([0.5, 1.5], [0.5, 2])	([0.7, 1], [0.6, 1.5])	([0.7, 1.3], [0.4, 2])	([0.8, 1.6], [0.3, 1.7])	([0.7, 1.4], [0.6, 2])	([0.8, 1.8], [0.3, 2.3])	([1, 1.5], [0.8, 1.8])	([0.9, 1.8], [0.7, 2.1])
	2		([0.2, 0.5], [0.1, 1])	([0.3, 0.6], [0.2, 1])	([0.9, 1.7], [0.6, 1.8])	([1, 1.5], [0.5, 1.6])	([0.9, 1.5], [0.4, 1.9])	([0.5, 1.6], [0.2, 2])	([0.4, 0.9], [0.2, 1])	([1, 1.7], [0.5, 2])

The other parametric magnitudes such as unit purchasing price and unit selling price, capacities of conveyances, available space, budget, sources and demands of the Models-I and -II are given in Tables 3 and 4. Let the unit space for 1st, 2nd and 3rd items are two units, three units and two units.

Table 3. Parametric values for the Models.

Models	Source	Demand	Capacities. of Conveyance.
	$(A_1^1, A_1^2, A_1^3$ $A_2^1, A_2^2, A_2^3)$	$(D_{01}^1, D_{01}^2, D_{01}^3$ $D_{02}^1, D_{02}^2, D_{02}^3)$	(e_1^1, e_1^2, e_1^3) (e_2^1, e_2^2, e_2^3)
-I	(90, 80, 85 85, 75, 70)	(75, 73, 70, 65, 63, 60)	(90, 85, 80 85, 70, 75)
-II	{([89.5, 90], [88.6, 91]), ([79.6, 80], [78.5, 81.3]), ([84.7, 85], [83.5, 86]), ([84.4, 85], [83.6, 86]), ([74, 75], [73.6, 76]), ([69.4, 70], [68.7, 71])}	{([74, 75], [73.6, 76]), ([72.6, 73], [72, 74]), ([69.4, 70], [68.7, 71]), ([64.5,65], [64, 66]), ([62.4, 63], [62, 64]), ([59.4, 60], [59, 61])}	{([89, 90], [88, 91]), ([84, 85], [83, 86.3]), ([79, 80], [78, 81.3]), ([84, 85], [83, 86.3]), ([69, 70], [68, 71]), ([74, 75], [73, 76])}

Table 4. Parametric values for the Models.

Models	Purchasing Costs	Unit Selling Price	Budget	
	$(P_1^1, P_1^2, P_1^3$ $P_2^1, P_2^2, P_2^3)$	$(S_1^1, S_1^2, S_1^3$ $S_2^1, S_2^2, S_2^3)$	(Bud_1, Bud_2, Bud_3)	(SPC_1, SPC_2, SPC_3)
-I	(9, 8, 10 6, 7, 8.5)	(52, 30, 24, 46, 28, 20)	(3000, 2500, 2700)	(510, 520, 525)
-III	{([8.3, 9], [8, 10]), ([7.5, 8], [7, 9]), ([9, 10], [8, 10.5]), ([5.5, 6], [5, 7]), ([6.5, 7], [6, 7.8]), ([8, 8.5], [7.5, 9])}	{([51.5, 52], [51, 53]), ([29, 30], [28, 31]), ([23.5, 24], [23, 25]), ([45.5,45], [45.3, 46]), ([27.4, 28], [27, 29]), ([19.5, 20], [19.1, 21])}	{([8250,5000], [7500,8600]), ([7500,8600], [4500,4600]), ([4400,4700]), ([4400,4700])}	{([509.6,510.3], [509, 511]), ([519.4,520], [519, 521]), ([524, 525]), [523,526])}

By considering the general demand function, it is necessary to define β_j, θ_j^1, θ_j^{11} and τ. These parameters are defined as follows: $\beta_j = 0.1$, $\theta_j^1 = 0.052$, $\theta_j^{11} = 0.051$ and $\tau = 0.05$. The following Table 5 represent the distances from different origins to destinations by route-1 and route-2.

Table 5. Distances (ds_{ijp}) of the routes for the Models.

Route	Destination	Origin-1	Origin-2
1	1	35	30
	2	40	25
2	1	30	35
	2	25	30

Now, the problem is to determine the optimal policy of transportation to maximize the profit. Under the objective of maximization of profit subjected to the traditional constraints, available space, and budget constraints for the above given input data, the crisp model is solved using Generalised Reduced Gradient (GRG) method and rough model by (i) Hamzehee et al and (ii) Expected Value Techniques along with GRG method. The results are presented in Table 6 (Crisp Model and Rough Model by Hamzehee Method) and Table 7 (Rough Model by Expected Value Technique). The convergence of the GRG method is also well established (shown in Appendix A.5) [33].

Table 6. Optimum results for Model-I and Model-II (Hamzehee Method) by Lingo.

Models	Crisp Model	4D Rough Model			
		TP-1	TP-2	TP-3	TP-4
Optimal profit	6966.775	5293.95	8058.61	4109.11	9130.81
		Set-1	Set-2	Set-3	Set-4
	$x11111 = 63.07$	$x11221 = 14.04$	$x11122 = 48.7$	$x11111 = 54.4$	$x11111 = 52.44$
	$x11122 = 52.75$	$x12112 = 46$	$x11221 = 65.3$	$x11221 = 32.2$	$x11112 = 25.41$
	$x11222 = 8.75$	$x12211 = 29.96$	$x11222 = 49.6$	$x11222 = 17.9$	$x11222 = 19.9$
	$x21121 = 67.15$	$x21121 = 50.46$	$x12211 = 22.1$	$x12211 = 0.2$	$x12112 = 9.09$
	$x21122 = 47.1$	$x21122 = 21.84$	$x12221 = 42.2$	$x21121 = 20.3$	$x12211 = 62.24$
	$x22211 = 27.74$	$x21211 = 33.04$	$x21111 = 44$	$x21122 = 19.9$	$x21121 = 54.71$
	$x22212 = 39.9$	$x21222 = 45.16$	$x21121 = 9.80$	$x22211 = 31.31$	$x21122 = 65.91$
	$x22213 = 26.9$	$x21223 = 26.07$	$x21123 = 14.80$	$x22213 = 2.23$	$x21123 = 27.09$
	$x11213 = 7.9$	$x21223 = 27.16$	$x22123 = 23.80$	$x22113 = 62.04$	$x22123 = 15.09$
	$x11123 = 15.91$	$x11123 = 15.91$	$x11123 = 15.91$	$x22212 = 18.65$	$x21113 = 28.51$
	others are zero	others are zero	$x21122 = 33.4$	other are zero	$x21211 = 8.29$
			$x22212 = 14.9$		$x22212 = 31.21$
			other are zero		other are zero

Table 7. Optimum results for Model-II (Expected Value Technique) by Lingo.

$\eta = 0.0$	$\eta = 0.2$	$\eta = 0.4$	$\eta = 0.5$	$\eta = 0.6$	$\eta = 0.8$	$\eta = 1.0$
6351.76	6393.95	6412.61	6463.11	6488.81	6504.54	6523.54
$x11111 = 32.75$	$x11111 = 32.84$	$x11111 = 32.86$	$x11111 = 28.85$	$x11111 = 28.14$	$x11111 = 28.09$	$x11111 = 28.92$
$x11222 = 30.33$	$x11222 = 30.33$	$x22212 = 27.46$	$x11222 = 30.13$	$x11222 = 30.02$	$x11222 = 30.33$	$x11222 = 30.33$
$x12211 = 45.65$	$x12211 = 45.576$	$x12211 = 45.52$	$x12211 = 45.5$	$x12211 = 45.48$	$x12211 = 45.44$	$x12211 = 45.40$
$x21121 = 52.20$	$x21121 = 52.16$	$x21121 = 52.13$	$x21111 = 39.2$	$x21113 = 39.09$	$x21113 = 42.76$	$x21113 = 42.76$
$x21122 = 10.8$	$x21122 = 18.75$	$x21122 = 10.712$	$x21121 = 32.3$	$x21121 = 32.24$	$x21121 = 54.05$	$x21121 = 54.21$
$x21223 = 30.74$	$x21223 = 30.44$	$x21223 = 30.04$	$x21123 = 14.9$	$x21123 = 14.71$	$x21123 = 24.01$	$x21123 = 24.32$
$x22111 = 23.9$	$x22111 = 23.13$	$x22111 = 23.80$	$x21222 = 23.14$	$x21222 = 23.37$	$x21222 = 23.41$	$x21222 = 33.51$
$x22213 = 32.9$	$x22213 = 32.06$	$x22213 = 32.16$	$x22213 = 32.16$	$x22213 = 32.22$	$x22113 = 32.8$	$x22113 = 41.05$
and all others variables are zero	and all others variables are zero	$x22212 = 18.9$ and all others variables are zero	$x22212 = 27.46$ and all others variables are zero	$x21211 = 27.49$ $x22212 = 5.61$ and all others are zero	and all others are zero	and all others are zero

To compare the results for existence of different constraints, we present the following Table 8 for $\eta = 0.5$.

Table 8. Optimum results for existence of different constraints (Expected Value Technique) by Lingo.

Results without Space and Budget	Results with Space and without Budget	Results without Space and with Budget
$x11111 = 42.75$	$x11221 = 22.84$	$x11211 = 19.06$
$x11212 = 13.33$	$x11222 = 32.04$	$x21222 = 53.33$
$x12211 = 25.65$	$x12211 = 45.57$	$x22212 = 14.9$
$x21121 = 52.20$	$x21121 = 52.16$	$x21121 = 32.13$
$x21122 = 10.8$	$x21122 = 10.75$	$x21122 = 10.71$
$x21223 = 57.74$	$x21223 = 30.44$	$x21213 = 24.54$
$x22112 = 43.9$	$x22112 = 53.16$	$x22112 = 53.80$
$x22213 = 32.9$	$x22213 = 32.16$	$x22213 = 32.16$
$x12213 = 12.9$	$x12213 = 12.9$	$x21122 = 27.4$
$x11123 = 21.17$	$x11123 = 11.17$	$x11123 = 31.12$
and all others variables are zero	and all others variables are zero	and all others variables are zero
Optimal profit 6579.135	Optimal profit 6529.07	Optimal profit 6513.513

5. Particular Cases

5.1. Three-Dimensional TP Model

If we take the system where there is only one route only, i.e., $L = 1$ in the Model-II (Rough Model), then the converted transportation problem is termed as a solid transportation problem (STP). The corresponding three dimensional TP-1, TP-2, TP-3, TP-4 is obtained by substituting $L = 1$ in the TP-1, TP-2, TP-3, TP-4 of the four-dimensional transportation Model. The results of the 3DTP (i.e., STP) model are given in Table 9.

Table 9. Optimum results of the STP Model ignoring route.

823.76	1263.95	323.61	2623.21
x1111 = 14.22	x1111 = 0.82	x1111 = 16.52	x1111 = 18.04
x1112 = 0.75	x1112 = 4.19	x1112 = 4.43	x1112 = 9.12
x1121 = 0.05	x1121 = 6.26	x1121 = 0.72	x1121 = 0.21
x1122 = 11.225	x1122 = 12.06	x1122 = 0.92	x1122 = 11.42
x1211 = 8.25	x1211 = 0.36	x1211 = 0.26	x1211 = 0.53
x1212 = 0.65	x1212 = 0.16	x1212 = 0.36	x1212 = 0.43
x1222 = 0.25	x1222 = 5.06	x1222 = 0.16	x1221 = 8.12
x2111 = 0.51	x2111 = 0.06	x2111 = 0.91	x1222 = 0.09
x2112 = 0.62	x2112 = 8.03	x2112 = 0.82	x2111 = 5.05
x2121 = 29.74	x2121 = 0.4	x2121 = 4.24	x2112 = 27.3
x2122 = 5.9	x2122 = 0.44	x2122 = 0.80	x2121 = 8.44
x2211 = 0.65	x2211 = 5.26	x2211 = 0.6	x2122 = 5.32
x2213 = 10.25	x2213 = 0.96	x2213 = 0.03	x2213 = 14.76
x2221 = 0.35	x2221 = 4.81	x2221 = 6.6	x2212 = 5.75
x2223 = 5.05	x2223 = 5.76	x2223 = 7.6	x2223 = 9.2
others zero	others zero	others zero	others zero

If we take $K = 1$, then we get another three-dimensional transportation Model. Like above, the corresponding 3D TP-1, TP-2, TP-3, TP-4 are found by substituting $K = 1$ in the TP-1, TP-2, TP-3, TP-4 of the four-dimensional transportation Model. The optimum results of this STP model are given in Table 10.

Table 10. Optimum results of the STP Model ignoring conveyance.

623.76	1103.95	383.61	2056.32
x1111 = 0.24	x1111 = 0.84	x1111 = 0.57	x1111 = 14.17
x1112 = 0.76	x1112 = 0.11	x1112 = 0.33	x111 = 0.42
x1121 = 0.05	x1121 = 0.26	x1121 = 0.62	x1121 = 0.43
x1122 = 0.24	x1122 = 0.26	x1122 = 0.92	x1122 = 0.97
x1211 = 0.23	x1211 = 0.46	x1211 = 0.36	x1211 = 0.64
x1212 = 0.54	x1212 = 0.16	x1212 = 0.46	x1212 = 0.54
x1222 = 0.25	x1222 = 14.06	x1222 = 0.26	x1223 = 0.17
x2111 = 0.51	x2111 = 0.06	x2111 = 0.91	x1333 = 0.65
x2112 = 0.62	x2112 = 42.03	x2112 = 0.82	x1213 = 0.42
x2121 = 32.74	x2121 = 0.4	x2121 = 44.24	x2121 = 41.53
x2122 = 20.9	x2122 = 0.44	x2122 = 0.80	x2122 = 10.42
x2211 = 7.65	x2211 = 10.26	x2211 = 0.6	x2211 = 0.95
x2213 = 0.25	x2213 = 0.96	x2213 = 0.03	x221 = 0.94
x2221 = 0.35	x2221 = 25.11	x2221 − 1.6	x2221 − 0.79
x2223 = 0.05	x2223 = 8.76	x2223 = 12.6	x2222 = 0.57
others zero	others zero	others zero	others zero

5.2. 2D-TP Model

If we take $L = 1$ and $K = 1$ in the Model-II (Rough Model), then we get the 2D-transportation problem. The corresponding 2D TP-1, TP-2, TP-3, TP-4 are found in the following Table 11.

Table 11. Optimum solutions for 2D-TP Model.

4113.06	8093.95	453.61	8243.93
x111 = 63	x111 = 64.82	x111 = 63.32	x111 = 62
x112 = 52	x112 = 19.26	x112 = 2.62	x112 = 17.2
x121 = 50.0	x121 = 16.36	x121 = 0.67	x211 = 8
x122 = 5.0	x122 = 0.70	x122 = 0.32	x223 = 23
x211 = 6	x211 = 23.8	x211 = 2.18	x222 = 26.21
x213 = 5.5	x213 = 5.0	x213 = 2.23	other zero
x221 = 16.5	x221 = 8.0	x221 = 4.15	
x223 = 19.5	x223 = 27.4	x223 = 4.32	
other zero	other zero	other zero	

5.3. 4D-TPs with Different Natures of the Items

Results for the different combinations of substitute and complementary items are shown in Table 12.

Table 12. Results for the different combinations of substitute and complementary items.

Problem	Independent	Substitute	Complementary	Profit for Model-1
1	Item-1, Item-2, Item-3			7388.841
1		Item-1, Item-2	Item-3	6966.775
2		Item-1, Item-2		5497.527
3			Item-1, Item-3	6021.859
4			Item-2, Item-3	5952.067

6. Sensitivity Analyses

Taking $\beta_j = 0.1, 0.2, ..., 0.7$, and keeping θ_j^1, θ_j^{11} and τ as same, the sensitivity analyses for the Model-1 are presented in Table 13.

Table 13. Sensitivity analysis (w.r.t β_j).

Set	τ	$\theta_j^1 = \theta_j^{11}$	β_j	Maximum Profit	Demand
	0.052	0.052	0.1	6966.68	379.542
	0.052	0.052	0.2	6519.775	359.084
	0.052	0.052	0.3	6072.868	338.627
1	0.052	0.052	0.4	5625.962	318.1696
	0.052	0.052	0.5	5179.055	297.712
	0.052	0.052	0.6	4732.149	277.254
	0.052	0.052	0.7	4285.242	256.79

If price of items are sensitives with $\tau = 0.1, 0.2, ..., 0.7$, and $\beta_j = 0.5$, keeping θ_j^1 and θ_j^{11} as unchanged, the sensitivity analyses for the Model-1 are presented in Table 14.

Table 14. Sensitivity Analysis (w.r.t τ).

Set	τ	$\theta_j^1 = \theta_j^{11}$	β_j	Maximum Profit	Demand
	0.1	0.052	0.5	5223.015	298.09
	0.2	0.052	0.5	5314.598	298.89
	0.3	0.052	0.5	5406.181	299.69
1	0.4	0.052	0.5	5497.76	300.49
	0.5	0.052	0.5	5589.34	301.29
	0.6	0.052	0.5	5680.93	302.09
	0.7	0.052	0.5	5772.513	303.89

7. Discussion

Tables 6–10 represent the optimal solutions for given assumed values, which are self explained. From the above observation of TP-1 and TP-2, we can conclude that it yields satisfactory results and that of TP-3 and TP-4 yield almost satisfactory results. From the above observation of Table 7, we can easily infer that, with the increasing value of η , the profit increases. The profit is varied for the substitute and complementary nature of the items observed from Table 11. When items are independent, then profit is maximum that is \$7388.841. When item-1 and item-2 are substituted in nature with each other and item-3 is complementary nature then profit is \$6966.775. However, when the item-1 and item-2 are the substitutes in nature to each other, then profit is minimum that is \$5499.527.

Table 11 shows the optimal profit of model-1 for the nature of the items. This table helps the manager to make the decisions of the type of items he/she should use for his/her business policy. For the current consideration, it is observed that the existence of the complementary item is more profitable than the existence of the substitute item.

From Table 12, it is seen that the demand parameter τ, θ_j^1, θ_j^{11} and β_j have an effect on the amount of demand. As the parameter β_j is the negative indicator of the demand expression, its increase in value causes decreasing demand and optimum profit.

Similarly, Table 13 gives reverse calculation as τ is a positive (+ve) parameter present in the demand expression. The Figures 3 and 4 have also established the same conclusion in Tables 12 and 13. It is to be noticed that the coefficients of selling prices and amounts of gains do not depend on the actual amounts of these factors. Profit depends only on the deviation of the factors.

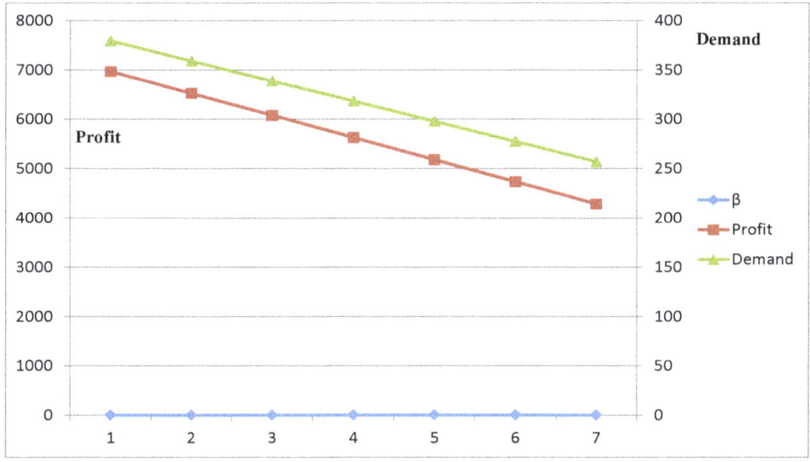

Figure 3. Profit vs. demand with respect to β.

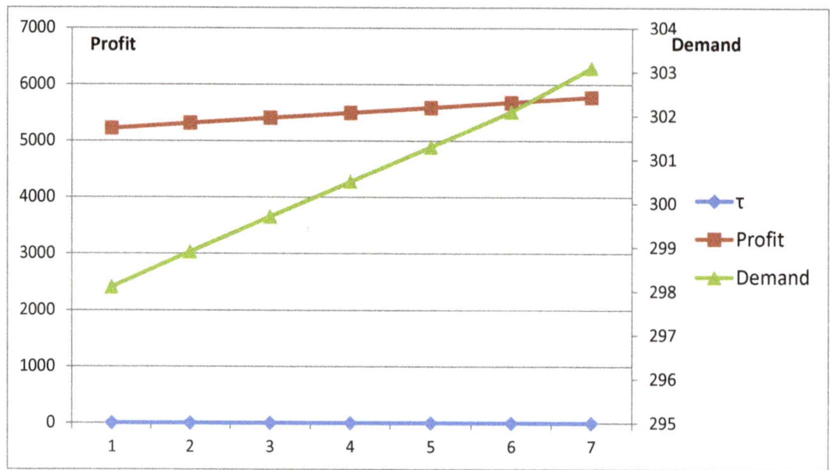

Figure 4. Profit vs. demand with respect to τ.

Thus, with the constant degree of substitution and complementary, profit and demand are constant though the measures of responsiveness of both merchandise are dissimilar for all cases.

Figure 5 draws a comparison between the profit versus price of sensitivities β_j and τ. It is observed that, when the price of sensitivity of product (β_j) is increasing, then profit is decreasing and, when τ is increasing, then profit is increasing. The maximum profit is dependent on both β_j and τ. Thus, it is essential to trade-off between these sensitivities parameters to achieve maximum profit when substitute and complimentary items are transported.

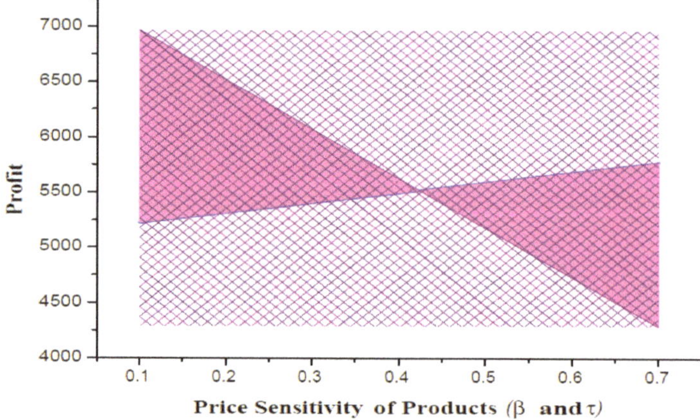

Figure 5. Profit vs. demand with respect to β and τ.

7.1. Discussion for Particular Models

From Table 8, when we consider only the first route of Model-II, the profit range of a corresponding 3D TP is $([823.76, 1263.95], [323.61, 2623.21])$ and, from Table 9, when we take only the first conveyance of Model-II, the profit range of corresponding 3D TP is $([623.76, 1103.95], [383.61, 2056.32])$. Again when we consider only the first route and first conveyance in the Model-II, the profit range of corresponding 2D TP is $([4113.06, 8093.95], [453.61, 8243.93])$.

7.2. Results of Bera et al.'s Model

If we don't consider the damageability, substitutability, complementary and space constraint i.e., if we put $\lambda^r_{ijpk} = 0 = \beta_j = \theta^1_j = \theta^{11}_j = \tau$ and remove the space constraint in Equation (1), it becomes the same type expressions as Bera et al. [7] for three items.

8. A Real-Life Problem

M/S. Sargar studio has two shops (studios) at Midnapore Sadar, Paschim Midnapur and Digha, Purba Medinipur, India. The owner of the studio brings the photographic materials-two types of plastics and one type wooden frame material (by weight) from Kolkata, West Bengal, and Bhubaneswar, Orissa, India. There are two conveyances—bus and lorries—between sources and destinations. Two different routes for transportation are also available. Here, plastic sheets (one high quality and other one little low) are substitutes and the wooden frame is complementary. The relevant data and the results are given in Table 15. The company has sufficient storing spaces at destinations along with the sufficient budget. Here, we take the crisp values for the parameters given by the studio company (Table 16).

Table 15. Unit cost of transportation (C^r_{ijpk}), fixed charges (f_{ijpk}), breakability (λ^r_{ijpk}) for proposed Models.

	p		1				2		
	k	1		2		1		2	
	i/j	1	2	1	2	1	2	1	2
				Model-I					
				(Crisp Model)					
C^1_{ijkp}	1	0.15	1.1	1.2	0.12	0.85	0.5	1.38	1.28
	2	0.21	1.3	2.5	0.6	1.65	1.72	0.66	0.78
C^2_{ijkp}	1	1.5	2.25	2.0	1.3	0.08	0.7	1.05	1.25
	2	0.25	0.25	1.2	1.9	1.35	1.25	1.85	0.98
C^3_{ijkp}	1	1.3	1.2	1.91	1.2	0.9	0.94	0.73	0.93
	2	0.52	1.53	1.2	0.7	1.8	0.53	1.9	0.98
λ^1_{ijkp}	1	0.01	0.01	0.02	0.03	0.01	0.02	0.01	0.02
	2	0.02	0.01	0.03	0.04	0.02	0.01	0.03	0.02
λ^2_{ijkp}	1	0.04	0.03	0.01	0.02	0.05	0.01	0.02	0.03
	2	0.05	0.04	0.02	0.03	0.02	0.01	0.03	0.02
λ^3_{ijkp}	1	0.01	0.01	0.01	0.01	0.02	0.02	0.02	0.02
	2	0.04	0.05	0.02	0.03	0.01	0.02	0.04	0.02
f_{ijkp}	1	1.1	1.4	1.5	1.0	1.2	1.4	0.3	1.8
	2	1.3	1.5	0.9	1.1	1.3	2.1	2.1	1.02

Table 16. Parametric values for the Models.

Models	Source	Demand	Capacities. of Conveyance.	Purchasing Costs	Unit Selling Price
	$(A^1_1, A^2_1, A^3_1$ $A^1_2, A^2_2, A^3_2)$	$(D^1_{01}, D^2_{01}, D^3_{01}$ $D^1_{02}, D^2_{02}, D^3_{02})$	(e^1_1, e^2_1, e^3_1) (e^1_2, e^2_2, e^3_2)	$(P^1_1, P^2_1, P^3_1$ $P^1_2, P^2_2, P^3_2)$	$(S^1_1, S^2_1, S^3_1$ $S^1_2, S^2_2, S^3_2)$
I	(85, 75, 80 83, 72, 68)	(60, 70, 75, 62, 64, 59)	(86, 89, 90 81, 65, 70)	(45, 48, 51 36, 47, 58.5)	(132, 145, 124, 126, 135, 137)

Out put results are without constraints are x11111 = 67.07, x11122 = 51.75, x21121 = 66.15, x21122 = 57.1, x22211 = 16.54, x22212 = 49.9, x22213 = 16.9, x11213 = 7.9, x11123 = 15.91 others are zero. Max z = 6947.52.

9. Conclusions

Renowned scientific discoveries and Research always had a practical application in the real world. Standing in the 21st century where machines, technology and industries govern the civilization, the need for sophisticated and easy techniques in the industry for optimizing products and profit is of immense importance. To further proceed and bring new ideas, we investigate a four-dimensional multifarious item transportation issue with rough interval coefficient, which maximizes the profit considering substitute and complementary products. The concept of rough interval and its properties are discussed briefly. In recent times, the rough interval has played an important role in different techniques handling vagueness for decision-making problems. The models we have discussed comprises of all earlier factors that are all considered as rough intervals. While solving the proposed model, at first, we discuss a solving procedure of a general linear programming problem with rough intervals and then we solve our model according to the concept of the above model. The rough problem is disintegrated into four classical transportation cases. We derive some special cases like a three-dimensional model and two-dimensional model to realize the problem more practically. Few examples are illustrated using LINGO 14.0 software showcasing the smooth working of the technique. We have considered a single profit maximization problem which can be extended to become a multi-objective optimization problem if we simultaneously maximize profit and minimize time. Moreover, the travelling salesman problem (TSP), Inventory routing problem, Travelling purchase problem, etc. are NP hard problems and are represented as LPPs and formulated as both single and multi-objective problems (cf. Khanra et al. [34], Manbera [35], etc.). In these problems, the travelling cost, stay charge, etc. can be taken as rough intervals and Hamzehee et al. [36] can be applied to convert to crisp models, which can be solved by heuristic methods, Genetic-Algorithms, Ant Colony Optimization, etc.

The drawback of the present methodology is that it provides a wide-range of possible profit values. The average of these possible values may give an approximate single value.

Author Contributions: S.H.J., B.D. and M.M. determined this problem and proposed the method for solutions. B.J. and G.P. did computer programming and mathematical calculations. B.J., S.H.J. and B.D. wrote the paper. M.M. and S.H.J. thoroughly checked the paper and gave interpretation to the results.

Funding: This research received no external funding.

Conflicts of Interest: The authors declare no conflict of interest.

Appendix A

In this section, some definitions and properties on Rough Intervals (RI) are given. For more details, see Hamzehee et al. [36]. An RI can be treated as a qualitative amount from a vague concept specified on a variable x in R [37], which is abstracted in the following definition.

Appendix A.1. Rough Intervals and Its Algebra

The algebraic operations applied on Rough Intervals (RIs) are taken from Moore's Interval Arithmetic [14]. Rebolledo [13] discussed the rough interval arithmetic very significantly.

Let $Q = ([\underline{a}_l, \underline{a}_u], [\overline{a}_l, \overline{a}_u])$ and $T = ([\underline{b}_l, \underline{b}_u], [\overline{b}_l, \overline{b}_u])$ represents the rough intervals. Then, the different operations on these are:

Addition: $Q + T = ([\underline{a}_l + \underline{b}_l, \underline{a}_u + \underline{b}_u], [\overline{a}_l + \overline{b}_l, \overline{a}_u + \overline{b}_u])$,

Subtraction: $Q - T = ([\underline{a}_l - \underline{b}_u, \underline{a}_u - \underline{b}_l], [\overline{a}_l - \overline{b}_u, \overline{a}_u - \overline{b}_l])$,

Negation: $-Q = ([-\underline{a}_u, \underline{a}_l], [-\underline{a}_u, \underline{a}_l])$.

Appendix A.2. Expected Value of a Rough Interval (Shu and Edmund [38])

Let an event X be uttered by $\{x \mid \psi(x) \in T\}$, whereas ψ is a mathematical function in universe U in $\mathbb{R}$, $T \subseteq \mathbb{R}$ and X is estimated by $(\underline{X}, \overline{X})$ consorting to the similar relation in $\mathbb{R}$. The corresponding lower expected measure of X is given by

$$\underline{E}[\psi] = \int_0^\infty \underline{Appro}\{\psi \geq y\}dy - \int_{-\infty}^0 \underline{Appro}\{\psi \leq y\}dy \ \ where \ \ \underline{Appro}(X) = \frac{|X \cap \underline{X}|}{\underline{X}}$$

and the corresponding upper expected measure of X is given by

$$\overline{E}[\psi] = \int_0^\infty \overline{Appro}\{\psi \geq y\}dy - \int_{-\infty}^0 \overline{Appro}\{\psi \leq y\}dy \ \ where \ \ \overline{Appro}(X) = \frac{|X|}{\overline{X}}$$

and the corresponding expected value of X is given by

$$E[X] = \int_0^\infty Appro\{\psi \geq y\}dy - \int_{-\infty}^0 Appro\{\psi \leq y\}dy.$$

Proposition (Shu and Edmund [38])

For any predetermined parameter η, chosen by the decision maker's (DM's) preference, $E(X) = \eta\underline{E}(X) + (1 - \eta)\overline{E}(X)$.

Appendix A.3. General Linear Programming Problem in Rough Interval Environments

A Linear Programming Problem (LPP) in which the coefficients of the objective function and the constraints are rough intervals:

$$\left.\begin{aligned}
&\text{Maximize } Z = \sum_{j=1}^N ([\underline{c}_{jl}, \underline{c}_{ju}], [\overline{c}_{jl}, \overline{c}_{ju}])x_j \\
&\text{s.t } \sum_{j=1}^N ([\underline{a}_{ijl}, \underline{a}_{iju}], [\overline{a}_{ijl}, \overline{a}_{iju}])x_j \leq ([\underline{b}_{il}, \underline{b}_{iu}], [\overline{b}_{il}, \overline{b}_{iu}]) \ \forall \ i = 1,2,...m, \\
&\sum_{j=1}^N ([\underline{d}_{kjl}, \underline{d}_{kju}], [\overline{d}_{kjl}, \overline{d}_{kju}])x_j \geq ([\underline{e}_{kl}, \underline{e}_{ku}], [\overline{e}_{kl}, \overline{e}_{ku}]) \ \ \forall \ k = 1,2,...p, \\
&\text{and } x_j \geq 0 \ \ \forall \ j = 1,2,...n,
\end{aligned}\right\} \quad (A1)$$

where $([\underline{c}_{jl}, \underline{c}_{ju}], [\overline{c}_{jl}, \overline{c}_{ju}])$, $([\underline{a}_{ijl}, \underline{a}_{iju}], [\overline{a}_{ijl}, \overline{a}_{iju}])$, $([\underline{b}_{il}, \underline{b}_{iu}], [\overline{b}_{il}, \overline{b}_{iu}])$, $([\underline{d}_{kjl}, \underline{d}_{kju}], [\overline{d}_{kjl}, \overline{d}_{kju}])$ and $([\underline{e}_{kl}, \underline{e}_{ku}], [\overline{e}_{kl}, \overline{e}_{ku}])$ $(i = 1,2,...,m; \ j = 1,2,...,n; \ k = 1,2,...,p)$ are rough intervals. Let $x = (x_1, x_2, ..., x_n)^t$ be the vector form of decision variables.

Now, according to Hamzehee et al. [36], here we state two theorems which help us to find the rough optimal range of the problem (1).

Let us consider two LPIC problems which are as follows.

Appendix A.3.1. LPIC-1

$$\left.\begin{aligned}
&\text{Maximize } Z = \sum_{j=1}^N [\underline{c}_{jl}, \underline{c}_{ju}]x_j \\
&\text{s.t } \sum_{j=1}^N ([\underline{a}_{ijl}, \underline{a}_{iju}])x_j \leq ([\underline{b}_{il}, \underline{b}_{iu}]) \ \ \forall \ i = 1,2,...,m, \\
&\sum_{j=1}^N ([\underline{d}_{kjl}, \underline{d}_{kju}])x_j \geq ([\underline{e}_{kl}, \underline{e}_{ku}]) \ \ \forall \ k = 1,2,...,p, \\
&\text{and } x_j \geq 0 \ \ \forall \ j = 1,2,...,n.
\end{aligned}\right\} \quad (A2)$$

Appendix A.3.2. LPIC-2

$$\left.\begin{aligned}
\text{Maximize } Z &= \sum_{j=1}^{N} \left([\bar{c}_{jl}, \bar{c}_{ju}]\right) x_j \\
\text{s.t } &\sum_{j=1}^{N} \left([\bar{a}_{ijl}, \bar{a}_{iju}]\right) x_j \leq \left([\bar{b}_{il}, \bar{b}_{iu}]\right) \quad \forall \ i = 1, 2, \ldots, m, \\
&\sum_{j=1}^{N} \left([\bar{d}_{kjl}, \bar{d}_{kju}]\right) x_j \geq \left([\bar{e}_{kl}, \bar{e}_{ku}]\right) \quad \forall \ k = 1, 2, \ldots, p, \\
&\text{and } x_j \geq 0 \quad \forall \ j = 1, 2, \ldots, n.
\end{aligned}\right\} \quad (A3)$$

The best optimal solution of the problem (1) according to Hamzehee et al. [36] and Chinneck and Ramadan [39] is i.e., LPIC-1 is found by solving:
LP1:

$$\left.\begin{aligned}
\text{Maximize } Z &= \sum_{j=1}^{N} \underline{c}_{ju} x_j \\
\text{s.t } &\sum_{j=1}^{N} \underline{a}_{ijl} x_j \leq \underline{b}_{iu} \quad \forall \ i = 1, 2, \ldots, m, \\
&\sum_{j=1}^{N} \underline{d}_{kju} x_j \geq \underline{e}_{kl}, \quad \forall \ k = 1, 2, \ldots, p, \\
&\text{and } x_j \geq 0 \quad \forall \ j = 1, 2, \ldots, n.
\end{aligned}\right\} \quad (A4)$$

The best optimal solution of the problem (1) i.e., LPIC-1 is found by solving:
LP2:

$$\left.\begin{aligned}
\text{Maximize } Z &= \sum_{j=1}^{N} \underline{c}_{jl} x_j \\
\text{s.t } &\sum_{j=1}^{N} \underline{a}_{iju}) x_j \leq \underline{b}_{il} \quad \forall \ i = 1, 2, \ldots, m, \\
&\sum_{j=1}^{N} \underline{d}_{kjl} x_j \geq \underline{e}_{ku}) \quad \forall \ k = 1, 2, \ldots, p, \\
&\text{and } x_j \geq 0 \quad \forall \ j = 1, 2, \ldots, n.
\end{aligned}\right\} \quad (A5)$$

The best optimal solution of the problem (1) i.e., LPIC-2 is found by solving:
LP3:

$$\left.\begin{aligned}
\text{Maximize } Z &= \sum_{j=1}^{N} \bar{c}_{ju} x_j \\
\text{s.t } &\sum_{j=1}^{N} \bar{a}_{ijl} x_j \leq \bar{b}_{iu} \quad \forall \ i = 1, 2, \ldots, m, \\
&\sum_{j=1}^{N} \bar{d}_{kju} x_j \geq \bar{e}_{kl} \quad \forall \ k = 1, 2, \ldots, p, \\
&\text{and } x_j \geq 0 \quad \forall \ j = 1, 2, \ldots, n.
\end{aligned}\right\} \quad (A6)$$

The best optimal solution of the problem (1) i.e., LPIC-2 is found by solving:
LP4:

$$\left.\begin{aligned}
&\text{Maximize } Z = \sum_{j=1}^{N} \bar{c}_{jl} x_j \\
&\text{s.t } \sum_{j=1}^{N} \bar{a}_{iju}) x_j \leq \bar{b}_{il} \quad \forall\ i = 1, 2, ..., m, \\
&\sum_{j=1}^{N} \bar{d}_{kjl} x_j \geq \bar{e}_{ku} \quad \forall\ k = 1, 2, ..., p, \\
&\text{and } x_j \geq 0 \quad \forall\ j = 1, 2, ..., n.
\end{aligned}\right\} \tag{A7}$$

Appendix A.3.3. Types of Solutions

There are three possible types of solutions of the problem LPIC-1 and LPIC-2 which are listed below:

- If LP-1 and LP-2 (LP-3 and LP-4) have optimal solutions, then the problem LPIC-1 (LPIC-2) has a finite bounded surely optimal (possibly optimal) range. If the maximizing value of LP-1 and LP-2 (LP-3 and LP-4) are respectively $\underline{z}_l, \underline{z}_u (\bar{z}_l, \bar{z}_u)$, then the surely optimal range (possibly optimal range) of LPIC-1 (LPIC-2) is $[\underline{z}_l, \underline{z}_u]([\bar{z}_l, \bar{z}_u])$
- If LP-2 (LP-4) is unbounded, then LPIC-1 (LPIC-2) is unbounded.
- If LP-1 (LP-3) is infeasible, then LPIC-1 (LPIC-2) is infeasible.

Appendix A.4. Algorithm for Conversion from a Rough LPP to a Crisp LPP

The steps for converting a Rough LPP to a Crisp LPP are

Step 1: Decompose the problem (1) into two linear programming part where interval coefficients are considered i.e., LPIC-1 and LPIC-2 which are given by (2) and (3), respectively.

Step 2: In this step, LPIC-1 again transformed into two linear programming problem where crisp coefficients are considered i.e., LP-1 and LP-2 given by respectively the problems (5) and (6) and similarly from LPIC-2 as LP-3 and LP-4 given by (7) and (8), respectively.

Step 3: In this step, LP-1, LP-2, LP-3 and LP-4 to find $\underline{z}_l$, $\underline{z}_u$, $\bar{z}_l$ and $\bar{z}_u$ respectively. $[\underline{z}_l, \underline{z}_u]$ and $[\bar{z}_l, \bar{z}_u]$ are the solutions of LPIC-1 and LPIC-2 respectively, which are called surely optimal range and possibly optimal range.

Step 4: Here three possible consequences are given below:

- If LPIC-1 and LPIC-2 have an optimal range, then the LPRIC problem (1) has an optimal range $([\underline{z}_l, \underline{z}_u], [(\bar{z}_l, \bar{z}_u)])$ which is a rough interval, compared to the surely optimal range
- If LPIC-1 has boundless range, then the LPRIC problem (1) has boundless range.
- If LPIC-2 is infeasible, then the LPRIC problem (1) has an infeasible solution space.

Appendix A.5. Convergence of the GRG Method

Let us consider the general nonlinear programming problem

$$minimize\ f(x) \tag{A8}$$
$$x \in R^n : g(x) = 0,$$
$$x \geq 0,$$

where $f : \mathbb{R}^n \to \mathbb{R}$ and $g : \mathbb{R}^n \to \mathbb{R}^m$ are continuously differentiable functions. Any nonlinear programming theory is based on the LAGRANGE function of (19), i.e., on

$$L(x, u, \bar{u}) = f(x) - g(x)^T u - x^T \bar{u}, \tag{A9}$$

where $x \in \mathbb{R}^n$, $u \in \mathbb{R}^m$ and $\bar{u} \in \mathbb{R}^n$. Using this LAGRANGE function, the necessary optimality or Kuhn–Tucker conditions are stated in the form

$$\nabla_N L(x, u, \bar{u}) = 0, \tag{A10}$$
$$g(x) = 0,$$
$$x \geq 0,$$
$$\bar{u} \geq 0,$$
$$x^T \bar{u} = 0,$$

which are to be approximated by the GRG algorithm. Note that the first condition can be used to eliminate $\bar{u}$, since we get $\bar{u} = \nabla f(x) - \nabla g(x)^T u$. A further function which will play an important role for proving a global convergence theorem is the augmented LAGRANGE function

$$\Psi_r(x, v, \bar{v}) := f(x) - g(x)^T v + \frac{1}{2} r g(x)^T g(x) - \bar{v}_J^T x_J + \frac{1}{2} r x_J^T x_J - \frac{1}{2r} \bar{v}_K^T \bar{v}_K, \tag{A11}$$

where $r \in \mathbb{R}$, $x \in \mathbb{R}^n$ $v \in \mathbb{R}^m$ and $\bar{v} \in \mathbb{R}^n$.

Assume that the GRG Algorithm produces a sequence of iterates $x^k, v^k, \bar{v}^k, \bar{u}^k, r^k$.

First, we note again that the boundedness of $\{x^k\}$ implies the boundedness of $\{u^k\}$, $\{v^k\}$ and $\{\bar{v}^k\}$.

Again, we denote this set of iteration indices by $\mathbb{K}$. Since only finitely many different index sets R_k and F_k are allowed, there is an infinite subset $\mathbb{K}'$ of $\mathbb{K}$ and index sets R, F with $R_k = R \subset N$, $F_k = F \subset N$, for all $k \in \mathbb{K}'$. For the same reason, we find a constant index set $D \subset N$, i.e., we can also assume that $D_k = D$ and $I_{\bar{k}} = I := N\,D$ for all $k \in \mathbb{K}'$.

Thus, there are accumulation points x^*, u^* and $\bar{u}^*$ of $\{x^k\}, \{u^k\}, and \{\bar{u}^k\}$, respectively, so that

$$\lim_{k \in \mathbb{K}''} x^k = x^*,$$

$$\lim_{k \in \mathbb{K}''} u^k = u^*,$$

$$\lim_{k \in \mathbb{K}''} \bar{u}^k = \bar{u}^*$$

Then, for an infinite subset $\mathbb{K}$

$$\nabla_D L(x^k, u^k, \bar{u}^k) = \nabla_D f(x^k) - \nabla_D g(x^k)^T u^k = 0, \tag{A12}$$
$$\nabla_F L(x^k, u^k, \bar{u}^k) = \nabla_F f(x^k) - \nabla_F g(x^k)^T u^k = 0, \qquad \text{for all } k \in \mathbb{K}''.$$

We conclude that

$$\nabla_N L(x^*, u^*, \bar{u}^*) = 0.$$

In addition, we have $x^* \geq 0$, $x^{*T} \bar{u}^* = 0$, $g(x^*) = 0$ and $\bar{u}^* \geq 0$
x^*, u^* and $\bar{u}^* = 0$ define a Kuhn–Tucker point for problem (19).

References

1. Hitchcok, F.L. The distribution of a product form several sources to numerous localities. *J. Math. Phys.* **1941**, *20*, 224–230. [CrossRef]
2. Schell, E.D. Distribution of a product by several properties. In Proceedings of the 2nd Symposium in Linear Programming, DCS/Comptroller, HQ US Air Force, Washington, DC, USA, 24–25 October 1955; pp. 615–642.
3. Yang, L.; Liu, P.; Li, S.; Gao, Y.; Ralescu, D.A. Reduction methods of type-2 uncertain variables and their applications to solid transportation problem. *Inf. Sci.* **2015**, *291*, 204–237. [CrossRef]

4. Kocken, H.G.; Sivri, M. A simple parametric method to generate all optimal solutions of fuzzy solid transportation problem. *Appl. Math. Model.* **2016**, *40*, 4612–4624. [CrossRef]

5. Halder, S.; Das, B.; Panigrahi, G.; Maiti, M. Some special fixed charge solid transportation problems of substitute and breakable items in crisp and fuzzy environments. *Comput. Ind. Eng.* **2017**, *111*, 272–281. [CrossRef]

6. Das, A.; Bera, U.K.; Maiti, M. A Profit Maximizing Solid Transportation Model Under a Rough Interval Approach. *IEEE Trans. Fuzzy Syst.* **2017**, *25*, 485–498. [CrossRef]

7. Bera, S.; Giri, P.K.; Jana, D.K.; Basu, K.; Maiti, M. Multi-item 4D-TPs under budget constraint using rough interval. *Appl. Soft Comput.* **2018**, *71*, 364–385. [CrossRef]

8. Sarkar, M.; Lee, Y.H. Optimum pricing strategy for complementary products with reservation price in a supply chain model. *J. Ind. Manag. Optim.* **2017**, *13*, 1553–1586. [CrossRef]

9. Sarkar, M.; Hur, S.; Sarkar, B. Effects of Variable Production Rate and Time-Dependent Holding Cost for Complementary Products in Supply Chain Model. *Math. Probl. Eng.* **2017**, *2017*, 2825103. [CrossRef]

10. Khanna, A.; Kishore, A.; Sarkar, B.; Jaggi, C. Supply Chain with Customer-Based Two-Level Credit Policies under an Imperfect Quality Environment. *Mathematics* **2018**, *6*, 299. [CrossRef]

11. Hirsch, W.M.; Dantzig, G.B. The fixed charge problem. *Naval Res. Logist. NRL* **1968**, *15*, 413–424. [CrossRef]

12. Kowalski, K.; Lev, B. On step fixed-charge transportation problem. *Omega* **2008**, *36*, 913–917. [CrossRef]

13. Gen, M.; Li, Y.Z. Spanning tree-based genetic algorithm for bicriteria transportation problem. *Comput. Ind. Eng.* **1998**, *35*, 531–534. [CrossRef]

14. Verma, R.; Biswal, M.P.; Biswas, A. Fuzzy programming technique to solve multi-objective transportation problems with some non-linear membership functions. *Fuzzy Sets Syst.* **1997**, *91*, 37–43. [CrossRef]

15. Shafaat, A.; Goyal, S.K. Resolution of degeneracy in transportation problems. *J. Oper. Res. Soc.* **1988**, *39*, 411–413. [CrossRef]

16. Saad, O.M.; Abbas, S.A. A parametric study on transportation problem under fuzzy environment. *J. Fuzzy Math.* **2003**, *11*, 115–124.

17. Iqbal, M.W.; Sarkar, B. Recycling of lifetime dependent deteriorated products through different supply chains. *RAIRO-Oper. Res.* **2019**, *53*, 129–156. [CrossRef]

18. Sarkar, B.; Mandal, B.; Sarkar, S. Preservation of deteriorating seasonal products with stock-dependent consumption rate and shortages. *J. Ind. Manag. Optim.* **2017**, *13*, 187–206. [CrossRef]

19. Pramanik, S.; Maity, K.; Jana, D.K. A multi-objective solid transportation problem with reliability for damageable items in random fuzzy environment. *Int. J. Oper. Res.* **2018**, *31*, 1–23. [CrossRef]

20. Sarkar, B.; Shaw, B.K.; Kim, T.; Mitali, S.; Shin, D. An integrated inventory model with variable transportation cost, two-stage inspection, and defective items. *J. Ind. Manag. Optim.* **2017**, *13*, 1975–1990. [CrossRef]

21. Ojha, A.; Mondal, S.K.; Maiti, M. A solid transportation problem with partial nonlinear transportation cost. *J. Appl. Comput. Math.* **2014**, *3*, 1–6.

22. Ishfaq, N.; Sayed, S.; Akram, M.; Smarandache, F. Notions of Rough Neutrosophic Digraphs. *Mathematics* **2018**, *6*, 18. [CrossRef]

23. Bit, A.K.; Biswal, M.P.; Alam, S.S. Fuzzy programming approach to multiobjective solid transportation problem. *Fuzzy Sets Syst.* **1993**, *57*, 183–194. [CrossRef]

24. Jiménez, F.; Verdegay, J.L. Solving fuzzy solid transportation problems by an evolutionary algorithm based parametric approach. *Eur. J. Oper. Res.* **1999**, *117*, 485–510. [CrossRef]

25. Li, T.H.S.; Su, Y.T.; Lai, S.W.; Hu, J.J. Walking motion generation, synthesis, and control for biped robot by using PGRL, LPI, and fuzzy logic. *IEEE Trans. Syst. Man Cybern. Part B Cybern.* **2011**, *41*, 736–748. [CrossRef]

26. Dey, A.; Pal, A.; Pal, T. Interval type 2 fuzzy set in fuzzy shortest path problem. *Mathematics* **2016**, *4*, 62. [CrossRef]

27. Kundu, P.; Kar, S.; Maiti, M. Fixed charge transportation problem with type-2 fuzzy variables. *Inf. Sci.* **2014**, *255*, 170–186. [CrossRef]

28. Tao, Z.; Xu, J. A class of rough multiple objective programming and its application to solid transportation problem. *Inf. Sci.* **2012**, *188*, 215–235. [CrossRef]

29. Haley, K.B. New methods in mathematical programming—The solid transportation problem. *Oper. Res.* **1962**, *10*, 448–463. [CrossRef]

30. Ojha, A.; Das, B.; Mondal, S.K.; Maiti, M. A multi-item transportation problem with fuzzy tolerance. *Appl. Soft Comput.* **2013**, *13*, 3703–3712. [CrossRef]

31. Liu, P.; Yang, L.; Wang, L.; Li, S. A solid transportation problem with type-2 fuzzy variables. *Appl. Soft Comput.* **2014**, *24*, 543–558. [CrossRef]
32. Giri, P.K.; Maiti, M.K.; Maiti, M. Fully fuzzy fixed charge multi-item solid transportation problem. *Appl. Soft Comput.* **2015**, *27*, 77–91. [CrossRef]
33. Schittkiowski, K. On the convergence of a generalized reduced gradient algorithm for nonlinear programming problems. Optimization **1986**, *17*, 731–755. [CrossRef]
34. Khanra, A.; Maiti, M.K.; Maiti, M. Profit maximization of TSP through a hybrid algorithm. *Comput. Ind. Eng.* **2015**, *88*, 229–236. [CrossRef]
35. Manerba, D.; Mansini, R.; Riera-Ledesma, J. The traveling purchaser problem and its variants. *Eur. J. Oper. Res.* **2017**, *259*, 1–18. [CrossRef]
36. Hamzehee, A.; Yaghoobi, M.A.; Mashinchi, M. Linear programming with rough interval coefficients. *J. Intell. Fuzzy Syst.* **2014**, *26*, 1179–1189.
37. Rebolledo, M. Rough intervals—Enhancing intervals for qualitative modeling of technical systems. *Artif. Intell.* **2006**, *170*, 667–685. [CrossRef]
38. Xiao, S.; Lai, E.M.K. A rough programming approach to power-balanced instruction scheduling for VLIW digital signal processors. *IEEE Trans. Signal Process.* **2008**, *56*, 1698–1709. [CrossRef]
39. Chinneck, J.W.; Ramadan, K. Linear programming with interval coefficients. *J. Oper. Res. Soc.* **2000**, *51*, 209–220. [CrossRef]

© 2019 by the authors. Licensee MDPI, Basel, Switzerland. This article is an open access article distributed under the terms and conditions of the Creative Commons Attribution (CC BY) license (http://creativecommons.org/licenses/by/4.0/).

Article

A Two-Echelon Supply Chain Management With Setup Time and Cost Reduction, Quality Improvement and Variable Production Rate

Bikash Koli Dey [1,†], **Biswajit Sarkar** [2,*] and **Sarla Pareek** [1,†]

[1] Department of Mathematics and Statistics, Banasthali Vidyapith, Banasthali, Rajasthan 304 022, India; bikashkolidey@banasthali.in or bikashkolidey@gmail.com (B.K.D.); psarla13@gmail.com (S.P.)

[2] Department of Industrial & Management Engineering, Hanyang University, Ansan Gyeonggi-do 15588, Korea

[*] Correspondence: bsarkar@hanyang.ac.kr or bsbiswajitsarkar@gmail.com; Tel.: +82-010-7498-1981; Fax: +82-31-400-5959

[†] These authors contributed equally to this work.

Received: 7 December 2018; Accepted: 11 February 2019; Published: 3 April 2019

Abstract: This model investigates the variable production cost for a production house; under a two-echelon supply chain management where a single vendor and multi-retailers are involved. This production system goes through a long run system and generates an out-of-control state due to different issues and produces defective items. This model considers the reduction of the defective rate and setup cost through investment. A discrete investment for setup cost reduction and a continuous investment is considered to reduce the defective rate and to increase the quality of products. Setup and processing time are dependent on lead time in this model. The model is solved analytically to find the optimal values of the production rate, safety factors, optimum quantity, lead time length, investment for setup cost reduction, and the probability of the production process going out-of-control. An efficient algorithm is constructed to find the optimal solution numerically and sensitivity analysis is given to show the impact of different parameters. A case study and different cases are also given to validate the model.

Keywords: lead-time reduction; production modelling; optimization; inventory control; backorder

---

1. Introduction

In the current competitive business world each and every company would like to make more profit with less investment. The concept of a basic production model was introduced by Taft [1]. To celebrate a century of the economic order quantity model, Cárdenas-Barrón et al. [2] have written about Ford Whitman Harris' model.

A two-echelon supply chain with both buyers and a vendor was developed by Sarkar [3] with several types of deterioration. In this modern business environment, a single vendor fulfils the demand of several customers. Thus, the model of a single-vendor and multiple buyers is a realistic approach these days. In the view of literature, Goyal [4] first optimized the joint cost for a single buyer and single vendor. This research was extended by Banerjee [5]. Again by considering single-setup multi-delivery Goyal [6] extended Banerjee's [5] model. Chakraborty and Bhuiya [7] developed an inventory model with a fuzzy service level constraint. A fuzzy stochastic optimization technique was used for solving their model. In 1996, Ouyang et al. [8] proposed an integrated model in which they considered backorders and variable lead times. The concept of a controllable lead time was introduced by Ouyang et al. [9] with discrete crashing cost. An integrated model with vendor's setup cost reduction was proposed by Sarkar and Majumder [10], where a distribution free approach was incorporated to solve the model. The concept of distribution free was introduced by Gallego and Moon [11]. In recent years, Sarkar et al. [12] proposed a two-echelon supply chain model with an improvement in a product's quality. A selling-price-dependent integrated model with reduced setup cost was proposed by Dey et al. [13]. Recently, Majumder et al. [14] proposed a supply chain model for variable production costs with a variable production rate.

Banerjee and Burton [15] discussed a comparison between coordinated and independent replenishment policies in a single-vendor multi-buyer supply chain model. Banerjee and Banerjee [16] developed a multi-buyer inventory model using an electronic data interchange with an order-up-to inventory control policy. Sarmah et al. [17] considered a single-supplier multi-buyer coordinated supply chain model with a trade credit policy. A variable production cost for inventory model was used by Khouja and Mehrez [18] and Tripathi et al. [19]. Under the time value of money, Chakrabarty et al. [20] developed an inventory model for defective items. Hoque [21] introduced three different single-vendor multi-buyer models by synchronizing the production flow with equal and unequal-sized batch transfers for the first two models and the last model, respectively. Jha and Shankar [22] developed a single-vendor multi-buyer constrained non-linear model under a service level constraint and solved it using the Lagrange multiplier method. Glock and Kim [23] studied the effect of forward integration in a multi-retailer supply chain under retailer competition.

To improve customer service and to reduce stock out loss, it is important to reduce lead time. Liao and Shyu [24] first incorporated a probabilistic inventory model by assuming a lead time as a unique decision variable. Ben-Daya and Rauf [25] considered an inventory model as an extension of Liao and Shyu's [24] model, where lead time was one of the decision variables. Ben-Daya's and Rauf's [25] model dealt with no shortages and continuous lead time. Ouyang et al. [8] extended Ben-Daya's and Rauf's [25] model by assuming a discrete lead time and shortages. Pan and Yang [26] analyzed an integrated inventory model with a controllable lead time. Annadurai and Uthayakumar [27] developed a periodic review inventory model under a controllable lead time and lost sales reduction.

Lo et al. [28] developed an integrated production–inventory model for an imperfect production process and they considered Weibull distribution deterioration under inflation. Poisson distributed lead time was considered by Huang et al. [29]. Recently, Tayyeb and Sarkar [30] discussed a multi-stage cleaner production system, where the defective rate is random. The impact of a random defective rate was calculated by Kang et al. [31] for a production model.

A time-dependent deterioration with partial backlogging was calculated by Mishra [32]. A stochastic lead time demand was considered by Khan et al. [33]. In this model, the effect of a learning and screening error for a production model is considered. An imperfect production and two-stage assembly system in an economic manufacturing quantity model were introduced by Chang et al. [34]. Cárdenas–Barrón et al. [35] provided an improved solution to the replenishment policy in an economic manufacturing quantity model. A multi-delivery policy and rework were also considered in this model. In 2017, Debata and Acharya [36] developed an inventory model under the consideration of a partial backorder. All researchers used different types of deteriorations, but a probabilistic deterioration in a two-echelon supply chain management (SCM) was considered by Sarkar [3], who minimized the cost of whole SCM in this model by using an algebraic solution methodology.

An economic manufacturing quantity (EMQ) model was discussed by Sana and Chaudhuri [37] under an imperfect production process. In reality, backlogging has a huge impact in any production model. Wee et al. [38] proposed an alternative approach to derive an inventory model with a rework process for a single-stage manufacturing system with planned backorders. Sarkar et al. [39] revisited the production model with the rework process in a single-stage manufacturing system with planned backorders. Three different distribution functions were used for the model. A just-in-time production process for an integrated model was developed by Das Roy et al. [40]. Recently, Kim et al. [41] proposed an integrated model with backorders, where they used an improved technique to calculate imperfect items when a process has gone through a long-run process.

It is true that any firm can use a discrete investment to reduce setup time, but this model proposes discrete investment for reducing ordering cost. Two continuous investments are used to reduce setup cost and to reduce the probability of an "in-control" to "out-of-control" state in a long-run process rather than the reduction of setup time. The investment for reducing setup cost is also considered as continuous, which is also quite realistic for an imperfect production model. Many production companies would like to sell more of their products, thus, they aim to produce more reliable products compared to others. Retailers always want more profitable products. Most of today's customers want more quality products, they do not consider the cost. Most customers want quality products, thus, the the quality of products is one of the main targets of most production industries. The quality of product can be improved by some investment discussed by Sarkar and Moon [42]. They also reduced the setup cost for an imperfect production process in this model. Cárdenas-Barrón et al. [43] developed an economic production model with an improved solution procedure. In this model, they also considered rework and multiple shipments. An imperfect production model with stochastic demand was formulated by Pal et al. [44]. A warranty for defective products was also provided, which increased the good-will of the companies. The capacity for holding the product is limited. Regarding this, Sana [45] developed an inventory model under the consideration of stochastic demand. Basically, most researchers considered that the holding cost for any production company is fixed but in reality this is not always true. A nonlinear holding cost for a newsvendor problem was considered by Pal et al. [46]. They considered a distribution-free approach.

Different researchers have developed different types of models under consideration of imperfect production, multi-product production systems with safety stock, and improved quality production processes under setup cost reduction (see for reference Sarkar et al. [12]). However, no one has developed a model for a single vendor-multi-buyer with consideration of a partial backorder, normally distributed lead time, shortages, and a variable production cost along with discrete investment for reduced setup cost for the vendor and an investment for improvement of the quality of the manufacturing process. There is a big research gap in this direction, which is fulfilled by this proposed research.

This research is based on a daily problem; basically in this research model, the lead time and total system cost are reduced. The lead time is dependent on production time and transportation time; let us suppose if one orders through an online delivery system (like pizza), the customer would like to have it as soon as possible. For this type of case, the lead time can be reduced by reducing production time and reducing transportation time. This is the theme along which this work is considered; that the lead time does not follow any distribution. Several researchers have reduced lead time with different considerations, but the consideration of the reduction of production and transportation time, along with a variable production rate for a multiple buyer, single retailer is a novel attempt.

See Table 1 for the contributions of previous authors.

Table 1. Contributions of previous authors.

Author(s)	Buyer	Production Rate	Backorder	Lead Time Crashed	Investment
Ouyang et al. [8]	Single	Constant	Planed	Yes	NA
Sarkar and Majumder [10]	Single	Constant	NA	Yes	NA
Dey et al. [13]	Single	Constant	NA	Yes	Continious
Majumder et al. [14]	Multi	Variable	Partial	Yes	NA
Banerjee and Banerjee [16]	Multi	Constant	NA	NA	NA
Ben-Daya and Rauf [25]	Single	Constant	NA	Yes	NA
Sana and Chaudhuri [37]	Single	Constant	NA	NA	NA
Sarkar et al. [39]	Single	Constant	Planned	NA	NA
This paper	Multi	Variable	Partial	Yes	Continious

"NA" stands for Not Applicable.

2. Problem Definition, Notation, and Assumptions

The problem, which is solved by this model, along with notations and assumptions are briefly described in this section.

2.1. Problem Definition

This model is concerned with a two-echelon supply chain model, where multiple buyers take a single type of products from a single vendor. The rate of production is considered variable along with a variable production cost. A discrete investment is used by the vendor to reduce the setup cost. As the production process runs through a long-run process, after certain time period, it starts to produce defective items. To prevent this, a continuous investment is also added in this model. A Partial backlogging is also consider for buyers, as there are shortages and a lead time crashing cost is used to reduce the lead time of buyers. A distribution-free case is considered, where the lead time is crashed in two ways: reducing production time and by reducing transportation time. It is considered that a single-setup-multi-delivery (SSMD) policy is used by the vendor for shipping the product to different buyers.

2.2. Notation

2.2.1. For Buyers:

The notation of decision variables and parameters for buyers are as follows:

Decision Variables

q_i order quantity for buyer i (units)

k_i safety factor for buyer i

I investment for ordering cost reduction $I = I_{bi}$ ($/order)

Q delivery lot size of vendor such that $Q = \sum_{i=1}^{n} q_i$

A_v setup cost per setup ($/setup)

m number of lots (same for all buyers) delivered to each buyer in one production cycle (positive integer)

θ probability of the production process which may go to *out-of-control* state

Parameters

n number of buyers

d_i average demand per unit time (units)

$A_{0_{bi}}$ initial ordering cost of the buyer per order ($/order)

S_i safety stock for buyer i (units)

A_{bi} ordering cost of the buyer per order ($/order)

h_{bi} holding cost per unit per time ($/unit/unit time)

σ_i standard deviation of the demand

π_i stockout cost per unit of shortage ($/unit)

π_{0i} marginal profit per unit item for buyer i ($/unit)

C_{T_i} transportation cost per lot ($/shipment)

t_{T_i} transportation time (time unit)

t_{s_i} setup and transportation time (time unit)

α_i the fraction of the transportation time t_{T_i} and setup time i.e., $\alpha = \dfrac{t_{T_i}}{t_{s_i}}$

2.2.2. For Vendor

The notation for parameter of the vendor are as follows:

Parameters

P	production rate per unit time (units)
A_{v_0}	initial setup cost for vendor per setup (\$/setup)
h_v	holding cost per unit per unit time \$/unit/unit time)
$C_v(P)$	unit production cost per unit (\$/unit)
β	annual fractional cost of capital investment

Other Notation

X_i	normally distributed lead time demand for buyer i with mean $d_i L_i$ and standard deviation $\sigma_i \sqrt{L_i}$
$E(\cdot)$	mathematical expectation
x^+	maximum value of x and 0

2.3. Assumptions

1. This is a single-vendor, multiple buyer SCM model.
2. The vendor supplies a total of Q quantity to fulfil the demand of each buyer, such that $Q = \sum_{i=1}^{n} q_i$.
3. The mQ quantities are produced by the vendor or manufacturer against the order of q_i quantity for i buyers, and the shipment is in quantity Q over m times. The shipment procedure follows the relation $q_i = d_i \frac{Q}{D}$, i.e., $\frac{q_i}{d_i} = \frac{Q}{D}$.
4. Inventory is continuously reviewed by each buyer. According to this policy, an order is placed whenever the level of inventory decreases to a particular inventory level (reorder point).
5. Ordering cost for each buyer is not always constant. However if the ordering cost is reduced during each order, a continuous investment is not needed. Thus, a discrete investment function is used to reduce the ordering cost for each buyer (see for instance Huang et al., 2011) specifically, $A(I_{bi}) = A_{0_{bi}} e^{-r_i I_{bi}}$, where $i = 0, 1, ..., n$ and $I_{0_{bi}} = 0$.
6. In reality, it is not possible for an industry manager to find out the exact distribution function of lead time demand and to solve the lead time problem. Whenever the previous data are known, then the mean and the standard deviation can be calculated. The buyer's model considers a (Q, S) continuous review inventory model with demand D. The demand D during lead time $L(P, Q)$ follows an unknown distribution having a known mean $DL(P, Q)$ and standard deviation σ_i, where $L(P, Q)$ is the lead time and depends upon setup and transportation time (ts) as well as and processing time $(\frac{q_i}{P})$ i.e., $L(P, Q) = t_{s_i} + \frac{q_i}{P} =$ setup and transportation time and processing time. Lead time of the first shipment is proportional to the lot size produced by the vendor.
7. A partial backorder is considered with a backorder ratio α_i for the retailer i.
8. A customer prefers to never wait to get a product from retailer. Thus, the retailer faces a problem of lost sales, which has a direct effect in market. The lead time has two parts: setup time and transportation time. It is now essential to reduce the lead time to save markets' demand. To reduce this lead time some cost is needed as the lead time crashing cost. The setup and transportation time consists of n mutual components with a normal distribution b_i, the minimum duration a_i, and the crashing cost $C_i = 1, 2, ..., n$, where

$$\sum_{i=1}^{n} b_i \ \leq \ t_{s_i} \leq \sum_{i=1}^{n} a_i = t_{s_{max}}.$$

That indicates the setup time components $1, 2, 3, ..., j$ crashed to their minimum duration i.e.,

$$t_{s,j} = \sum_{i=j+1}^{n} a_i - \sum_{i=1}^{j} b_i$$

for all $j = 1, 2, ..., n$. The crashing cost for the setup and transportation time is

$$C_{R_i(t_{s_i})} = C_j(t_{s,j-1} - t_{s_i}) + \sum_{i=1}^{j+1} C_i(a_i - b_i)$$

9. If lead time is high, then lost sale increases, which causes a huge loss to the industry. Instead of using a single safety factor, it is beneficial to use a double safety factor. The known mean and standard deviation of lead time demand are $DL(P, Q)$ and $DL(t_{T_i})$, respectively and the corresponding standard deviation as $\sigma_i \sqrt{L(P, Q)}$ and $\sigma_i \sqrt{L(t_{T_i})}$, respectively. Thus, the safety stock for the first batch is represented as:

$$S_i = k_1 \sigma_i \sqrt{L(P, Q)} = k_{1_i} \sqrt{t_{s_i} + \frac{Q}{P}}$$

and the safety stock for the second batch to onwards is defined as

$$S_i = k_{2_i} \sigma_i \sqrt{L(t_{T_i})} = k_{2_i} \sigma_i \sqrt{t_{T_i}}$$

which gives relation between safety factors as

$$k_{2_i} = k_{1_i} \sqrt{\frac{t_{s_i} + \frac{Q}{P}}{t_{T_i}}}$$

for batches $2, 3, ..., m$.

10. The lead time crashing cost entirely belongs to the buyer's cost component.

11. Many production models in literature consider a fixed or constant setup cost for vendor, but in reality, it is possible to reduce the setup cost using a continuous investment function (see for reference Sarkar and Moon [42]).

12. In a long-run system, the process changes to an out-of-control state from an in-control state, and as a result, the defective items are produced, which need to be improved via an investment function.

13. The time horizon is infinite.

3. Mathematical Model

In this section the supply chain model is developed and the joint total cost $JTEC$ of the vendor and the buyer is minimized. The vendor produces $Q = q_i$ items for n buyers, the demand of buyer's is $D = d_i$. The vendor uses a single-setup-multi-delivery (SSMD) policy to transport the required items, ordered by buyers and uses m lots to delivery all products. This shipment m must be an integer, thus this problem becomes a mixed-integer programming problem. The main purpose of this model is to optimize the total cost, along with optimized ordered quantity Q, numbers of lots m, different types of investment function to reduce the total cost such as a discrete investment I for reduced ordering cost, two continuous investment A_v, and θ to reduce setup cost and the probability of production process going into an out-of-control state. Finally a modified algorithm is developed to obtain the numerical result. Basically, two players as a vendor and multi-retailer are considered in this model. Two different models for buyers and the vendor are formulated as follows.

3.1. Mathematical Model of Buyers

This is a multi-buyer model where a bunch of buyers n order $Q = q_i$, $(i = 1, 2, ..., n)$ quantity from a single vendor. To reduce the ordering cost, buyers use a discrete investment I_{b_i}. For a more realistic result, a safety stock k_i is used by buyers. The demand of buyers is d_i, which is obviously less than the production rate of the vendor. A distribution free approach is considered in lead time reduction. In this model, the lead time is reduced by two way: one by reducing production time t_{s_i} and the other by reducing transportation time t_{T_i}. The parameter r_i is the reordered point for buyers and $\frac{q_i}{d_i}$ is the expected cycle time for each buyer and $\frac{(m-1)q_i}{d_i}$ is the total cycle length for the buyers. In this model, the buyer's cost component are as follows.

Reduced ordering cost through an investment.

To receive the particular product from the vendor, each buyer should invest some costs to order the product, which known as the ordering cost. It is found that the ordering cost may differ in real life. For example, there are many sim cards providers for mobile in India and the charge to make a phone call is different for different provider. One can use a discrete investment to reduce the ordering cost for a buyer. Thus the ordering cost for buyer i is given by

$$\frac{A_{0_{bi}} e^{-r_i I_{bi} d_i}}{q_i} + \frac{I_{bi} d_i}{q_i}$$

as the expected cycle time for each buyer is $\frac{q_i}{d_i}$.

Holding cost.

Each buyer in the SCM continuously reviews the inventory level. As a result, (q_i) order is placed by buyer i only when the level of inventory reaches to a specified indicator that is the reorder point (r_i) (see Figure 1). Therefore, the approximated average inventory for buyer i over the time cycle is given by

$$\frac{q_i}{2} + r_i - d_i L_i.$$

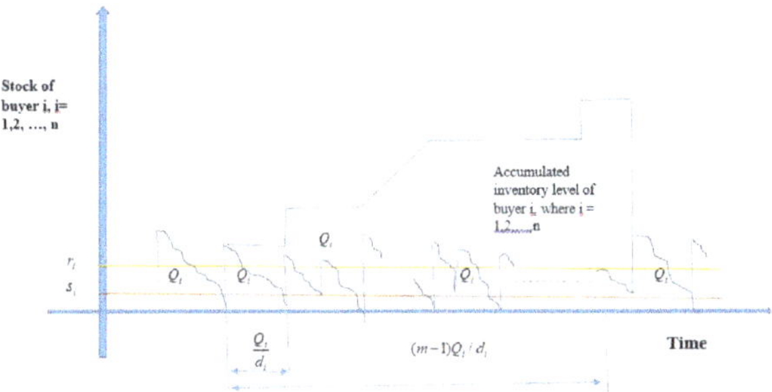

Figure 1. Inventory position for the buyer.

Now, the reorder point r_i can be expressed as $d_i L_i + k_i \sigma_i \sqrt{L_i}$, which results in the average inventory for the i-th buyer being

$$\frac{q_i}{2} + k_i \sigma_i \sqrt{t_{s_i} + \frac{q_i}{P}}$$

Hence, the holding cost for buyer i per unit time is

$$h_{bi}\left[\frac{q_i}{2} + k_i\sigma_i\sqrt{t_{s_i} + \frac{q_i}{P}}\right].$$

Shortage cost.

As the production machine produces defective items in the long-run, shortages must occurs, the backorder quantity for buyers are $E(x_{1_i} - r_{1_i})^+$ and $E(x_{2_i} - r_{2_i})^+.$, then the shortage cost per item per unit time is given by

$$\frac{d_i\pi_i}{q_i}E(x_{1_i} - r_{1_i})^+ + \frac{d_i\pi_i(m-1)}{q_i}E(x_{2_i} - r_{2_i})^+.$$

Transportation cost.

The cost for transportation of buyers is given by

$$\frac{mC_{T_i}d_i}{q_i}.$$

Lead time crashing cost.

Some of the most realistic research these days is to satisfy customers by reducing the lead time when an extra cost is added by the production manager. The lead time can be reduced in two ways, by reducing production and transportation time. According to the assumptions, the lead time crashing cost per unit time can be expressed as

$$\frac{md_iC_{R_i}(t_{s_i})}{q_i}$$

The total expected cost for buyer i is TEC_{bi} = ordering cost + holding cost + shortage cost + transportation cost + lead time crashing cost

Thus, TEC_{bi} leads to the following expression:

$$\begin{aligned}
TEC_{bi}(q_i, k_i, L_i) &= \left[\frac{A_{0_{bi}}e^{-r_i l_{bi}d_i}}{q_i} + \frac{l_{bi}d_i}{q_i} + h_{bi}\left\{\frac{q_i}{2} + k_i\sigma_i\sqrt{t_{s_i} + \frac{q_i}{P}}\right\} + \frac{d_i\pi_i}{q_i}E(x_{1_i} - r_{1_i})^+ \right. \\
&\quad \left. + \frac{d_i\pi_i(m-1)}{q_i}E(x_{2_i} - r_{2_i})^+ + \frac{md_iC_{T_i}}{q_i} + \frac{md_iC_{R_i}(t_{s_i})}{q_i}\right].
\end{aligned} \tag{1}$$

For the distribution-free approach, a lemma was proved by Gallego and Moon [47], in which they proved that "if the distribution G of demand D is unknown, then,

$$E(D - Q)^+ \leq \frac{[\sigma^2 + (Q - \mu)^2]^{\frac{1}{2}} - (Q - \mu)}{2}.$$

The above expression is tight for every Q if there exist a distribution $G^* \in \zeta$, where ζ is the worst possible distribution.

According to Gallego and Moon's [47], lemma where the least favorable distribution $G \in \zeta$, one can obtain

$$E(x_{1_i} - R_{1_i})^+ \leq \frac{\sqrt{\sigma_i^2 L_i(P, q_i) + (R_{1+i} - d_i L_i(P, q_i))^2} - (R_{1_i} - d_i L_i(P, q_i))}{2}$$

$$= \frac{\sigma_i}{2} \sqrt{t_s + \frac{q_i}{P}} [\sqrt{1 + k_{1_i}^2} - k_{1_i}]$$

$$E(x_{2_i} - R_{2_i})^+ \leq \frac{\sqrt{\sigma_i^2 L_i(t_{T_i}) + (R_{2_i} - d_i L_i(t_{T_i}))^2} - (R_{2_i} - d_i L_i(t_{T_i}))}{2}$$

$$= \frac{\sigma_i}{2} \sqrt{t_{T_i}} \left[\sqrt{1 + k_{1_i}^2 \frac{t_{s_i} + \frac{q_i}{P}}{t_{T_i}}} - k_{1_i} \sqrt{\frac{t_{s_i} + \frac{q_i}{P}}{t_{T_i}}} \right].$$

Then, Equation (1) can be rewritten as

$$\begin{aligned}
TEC_{bi}(Q, k_i, P, m, I_{bi}) &= \left[\frac{d_i}{q_i} \left(A_{0_{bi}} e^{-r_i I_{bi}} + I_{bi} + m C_{T_i} \right) + h_{bi} \left\{ \frac{q_i}{2} + k_i \sigma_i \sqrt{t_{s_i} + \frac{q_i}{P}} \right\} \right. \\
&+ \frac{d_i \pi_i \sigma_i}{2 q_i} \sqrt{t_s + \frac{q_i}{P}} [\sqrt{1 + k_{1_i}^2} - k_{1_i}] + \frac{m d_i C_{R_i}(t_{s_i})}{q_i} \\
&\left. + \frac{d_i \pi_i \sigma_i (m-1)}{2 q_i} \sqrt{t_{T_i}} \left[\sqrt{1 + k_{1_i}^2 \frac{t_{s_i} + \frac{q_i}{P}}{t_{T_i}}} - k_{1_i} \sqrt{\frac{t_{s_i} + \frac{q_i}{P}}{t_{T_i}}} \right] \right]
\end{aligned}$$

$$\begin{aligned}
TEC_{bi}(Q, k_i, P, m, I_{bi}) &= \left[\frac{d_i}{q_i} \left(A_{0_{bi}} e^{-r_i I_{bi}} + I_{bi} + m C_{T_i} \right) + h_{bi} \left\{ \frac{q_i}{2} + k_i \sigma_i \sqrt{t_{s_i} + \frac{q_i}{P}} \right\} \right. \\
&+ \frac{d_i \pi_i}{q_i} \left(\frac{\sigma_i}{2} \left[\sqrt{t_s + \frac{q_i}{P}} [\sqrt{1 + k_{1_i}^2} - k_{1_i}] \right] + \frac{m C_{R_i}(t_{s_i})}{\pi_i} \right. \\
&\left. \left. + \frac{\sigma_i (m-1)}{2} \sqrt{t_{T_i}} \left[\sqrt{1 + k_{1_i}^2 \frac{t_{s_i} + \frac{q_i}{P}}{t_{T_i}}} - k_{1_i} \sqrt{\frac{t_{s_i} + \frac{q_i}{P}}{t_{T_i}}} \right] \right) \right].
\end{aligned} \tag{2}$$

3.2. Mathematical Model for the Vendor

To fulfill the buyer's demand, the vendor produces Q quantity at a production rate P and the production cost $C_v(P)$. As it is too difficult to guess how much production is needed, in this model, a variable production rate P with the variable production cost $C_v(P)$ is considered for the vendor. The vendor uses a single-setup-multi-delivery (SSMD) policy to transport the items to each buyer. Thus m shipment is considered for a single-setup-multi-delivery (SSMD) policy. Thus the total cycle time for vendor is $\frac{mQ}{D}$. Two continuous investments are considered by the vendor to reduce the total system cost. An investment A_v is used to reduce the setup cost of the vendor and another investment θ is used to reduce the chance of a system out-of-control state from in-control state. As in long-run system, the production process may move from an in-control to out-of-control state due to the labour problems, machinery problems etc. To reduce this chance, a continuous investment θ is introduced by the vendor. In this model, the following costs component are used for vendor:

Setup cost with an investment.

The setup cost for the vendor per unit time is $\frac{DA_v}{mQ}$. But a continuous investment is introduced to reduce the setup cost. Hence, after introducing continuous investment, the total setup cost for the vendor is given by

$$b \ln \left[\frac{A_{v0}}{A_v} \right] + \frac{A_v D}{mQ}$$

Holding cost.

The average inventory of the vendor is

$$\left[\left\{ mQ \left(\frac{Q}{P} + (m-1)\frac{Q}{D} \right) - \frac{m^2 Q^2}{2P} \right\} - \left\{ \frac{Q^2}{D}(1 + 2 + \ldots + (m-1)) \right\} \right] \frac{D}{mQ}$$

$$= \frac{Q}{2} \left[m \left(1 - \frac{D}{P} \right) - 1 + \frac{2D}{P} \right].$$

(see Figure 2)

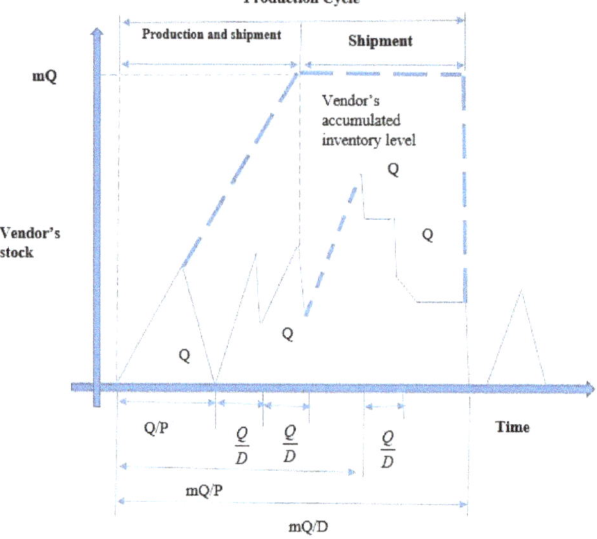

Figure 2. Inventory position for the vendor.

Therefore, the holding cost per unit time for the vendor is

$$h_v \frac{Q}{2} \left[m \left(1 - \frac{D}{P} \right) - 1 + \frac{2D}{P} \right].$$

Investment.

To improve the quality of the product, the vendor uses some investment. Thus, the investment for quality improvement is given by

$$B \ln \left(\frac{\theta_0}{\theta} \right).$$

Total investment.

Thus, the total investment for the reduced setup cost and improved the quality of the product is given by:

$$b \ln \left[\frac{A_{v0}}{A_v} \right] + B \ln \left[\frac{\theta_0}{\theta} \right] \quad = \quad \gamma - B \ln \theta - b \ln A_v$$

where, $\gamma = B \ln \theta_0 + b \ln A_{v0}$ and $0 < \theta \le \theta_0, \ 0 < A_v \le A_{v0}.$

Production/material cost.

Production cost $C_v(P)$ of the vendor assumed to be a function of P. The production cost is of the form is:

$$C_v(P) \quad = \quad \left(\frac{\zeta_1}{P} + \zeta_2 P \right).$$

The unit production cost is $P^* = \sqrt{\frac{\zeta_2}{\zeta_1}}$ Therefore, the total expected cost of to the vendor is expressed as TEC_v = setup cost + holding cost + material cost + investment cost i.e.,

$$TEC_v(m, Q, P, A_v, \theta) \quad = \quad \frac{A_v D}{mQ} + \frac{Q}{2} h_v \left[m \left(1 - \frac{D}{P} \right) - 1 + \frac{2D}{P} \right]$$

$$+ \quad C_v(P)D + \frac{SDmQ\theta}{2} + \beta \left(\gamma - B \ln \theta - b \ln A_v \right). \tag{3}$$

In order to obtain centralized decisions for both the vendor and buyers to minimize the entire supply chain cost, the total cost expression of both ends must be combined. Therefore, the joint total expected cost for both vendor and the buyers ($JTEC$) is obtained as follows

$$JTEC(Q, k_i, m, \theta, P, I, A_v)$$

$$= \sum_{i=1}^{n} \left[\frac{D}{Q} \left(A_{0_{bi}} e^{-r_i I_{bi}} + I_{bi} + \frac{A_v}{m} + mC_{T_i} \right) + h_{bi} \left\{ \frac{Q}{2D} d_i + k_i \sigma_i \sqrt{t_{s_i} + \frac{Q}{P}} \right\} \right.$$

$$+ \quad \frac{D\pi_i}{Q} \left(\frac{\sigma_i}{2} \left[\sqrt{t_s + \frac{q_i}{P}} [\sqrt{1 + k_{1_i}^2} - k_{1_i}] \right] + \frac{mC_{R_i}(t_{s_i})}{\pi_i} + \frac{\sigma_i(m-1)}{2} \sqrt{t_{T_i}} \left[\sqrt{1 + k_{1_i}^2 \frac{t_{s_i} + \frac{q_i}{P}}{t_{T_i}}} \right. \right.$$

$$- \quad \left. \left. k_{1_i} \sqrt{\frac{t_{s_i} + \frac{q_i}{P}}{t_{T_i}}} \right) \right] + \frac{Q}{2} h_v \left[m \left(1 - \frac{D}{P} \right) - 1 + \frac{2D}{P} \right] + DC_v(P)$$

$$+ \quad \frac{SDmQ\theta}{2} + \beta \left(\gamma - B \ln \theta - b \ln A_v \right). \tag{4}$$

4. Solution Methodology

The main aim of this model is to minimize the optimum value of the decision variable such as the total joint cost can be minimized. This is an unconstrained minimization problem along with an integer programming problem. To find the optimum values of decision variable one needs to calculate the first order derivative of the objective function with respect to the decision variables and then equate them to zero. Now, according to the assumptions, m is an integer and therefore, can be treated as a discrete decision variable. One can use the analytic discrete optimization method to find the optimum value of m.

To find the optimum value of the other decision variables, one can use the classical optimization technique, which gives the global optimum value. To do this, after calculating derivatives with respect to k_i, θ, I, A_v, Q and P, one can obtain

$$\frac{\partial JTEC(Q,k,m,\theta,P,I,A_v)}{\partial k_i} = \sum_{i=1}^{n} \left(\sqrt{\frac{q_i}{P}} + t_{s_i} \right) \sigma_i \left[h_{bi} + \frac{d_i \pi_i}{2q_i} \left\{ \left(\frac{k_i}{\sqrt{1+k_i^2}} - 1 \right) \right. \right.$$
$$\left. \left. + (m-1) \left(\frac{k_i \sqrt{(t_{s_i} + \frac{q_i}{P})}}{\sqrt{t_{T_i} + (t_{s_i} + \frac{q_i}{P})k_i^2}} - 1 \right) \right\} \right],$$

$$\frac{\partial JTEC(Q,k,m,\theta,P,I,A_v)}{\partial k_i} = \sum_{i=1}^{n} \left(\sqrt{\frac{q_i}{P}} + t_{s_i} \right) \sigma_i \left[h_{bi} + \frac{d_i \pi_i}{2q_i} \left\{ \left(\frac{k_i}{\sqrt{1+k_i^2}} - 1 \right) \right. \right.$$
$$\left. \left. + (m-1) \left(\frac{k_i \sqrt{(t_{s_i} + \frac{q_i}{P})}}{\sqrt{t_{T_i} + (t_{s_i} + \frac{q_i}{P})k_i^2}} - 1 \right) \right\} \right],$$

$$\frac{\partial JTEC(Q,k,m,\theta,P,I,A_v)}{\partial \theta} = \frac{SDmQ}{2} - \frac{B\beta}{\theta} \qquad ,$$

$$\frac{\partial JTEC(Q,k,m,\theta,P,I,A_v)}{\partial I} = \frac{(1 - A_{v_0} re^{-rI})D}{mQ},$$
$$\frac{\partial JTEC(Q,k,m,\theta,P,I,A_v)}{\partial A_v} = \frac{D}{mQ} - \frac{b\beta}{A_v},$$
$$\frac{\partial JTEC(Q,k,m,\theta,P,I,A_v)}{\partial Q} = -\frac{\tau_1}{Q^2} + \frac{\tau_2}{Q} + \tau_3,$$
$$\frac{\partial JTEC(Q,k,m,\theta,P,I,A_v)}{\partial P} = -\frac{\tau_4}{P^2} + D\zeta_2 \tag{5}$$

(See Appendix A for the values of τ_i, $i = 1, 2, 3, 4$.)

For a fixed positive integer m, the values of Q, k_i, P, I, A_v, and θ can be obtained by equating every individual equation of the system in Equation (5) to zero. Then, one can obtain the optimum result Q^*, k_i^*, P^*, I^*, A_v^* and θ^* as follows:

$$Q^* = \frac{\tau_1}{\tau_3 Q + \tau_2}, \tag{6}$$

$$k_i^* = \frac{m(D\pi_i - 2h_{bi}Q)}{D\pi_i \left(\frac{1}{\sqrt{1+k_i^2}} + \frac{(m-1)\sqrt{\left(\frac{Q}{P} + t_{s_i}\right)}}{\sqrt{t_{T_i} + \left(\frac{Q}{P} + t_{s_i}\right)k_i^2}} \right)}, \tag{7}$$

$$P^* = \sqrt{\frac{\tau_4}{D\zeta_2}}, \tag{8}$$

$$\theta^* = \frac{2B\beta}{SDmQ}, \tag{9}$$

$$I^* = \frac{1}{r} \log(r A_{v_0}),\tag{10}$$

$$A_v^* = \frac{bm\beta Q}{D}.\tag{11}$$

Lemma 1. *For the fixed value of m, the condition is sufficient at the optimum value of the decision variables* Q^*, k_i^*, P^*, I^*, A_v^* *and* θ^*, *i.e., all principal minor of the Hessian matrix is greater than zero for the optimum value of the decision variables* Q^*, k_i^*, P^*, I^*, A_v^* *and* θ^*.

Proof. See Appendix B for the proof of the lemma. □

Solution Algorithm

A closed form solution of this mathematical model is very difficult to obtain. One can use the fixed point iteration technique to create a suitable algorithm in order to solve the model.

Step 1 Set $m = 1$, and input all the values of the parameters.
Step 2 For all buyers $i = 1, 2, ..., n$, assign the values of all parameters and perform the following steps.
Step 3 For every combination of $L_{i,r}, r = 1, 2, ..., N_i, i = 1, 2, ..., n$ perform steps 3a–3e.
Step 3a Set $k_i^{j1} = 0$ for each buyer i.
Step 3b Substitute k_i^{j1}, $(i = 1, 2, ..., n)$ into Equation (6) and evaluate Q^{j1}.
Step 3c Utilize Q^{j1} to determine the value of (k_i^{j2}) for each i from (7).
Step 3d Using the value of (k_i^{j2}), obtain the value of k_i^{j2} from the normal distribution table.
Step 3e Repeat steps 3b to 3d until no changes occur in the values of Q^j and k_i^j and denote these values as Q^{j*} and k_i^{j*}, respectively.
Step 4 Evaluate the value of P^{j*}, I^{j*}, θ^{j*}, and A_v^{j*} from Equations (8), (10), (9), and (11), respectively, using the value of Q^{j*}.
Step 5 Denote the latest updated values of Q^j, k_i^j, P^j, I θ^j, and A_v^j as Q^{j**}, k_i^{j**}, P^{j**} I^{j**}, θ^{j**}, and A_v^{j**} respectively.
Step 6 Obtain $JTEC(Q^{j**}, k_i^{j**}, P^{j**}, I^{j**}, \theta^{j**}, A_v^{j**}, m)$ and
$Min_{j=1,2,...,N_i} JATC(Q^{j**}, k_i^{j**}, P^{j**}, I^{j**}, \theta^{j**}, A_v^{j**}, m)$ for all i.
Step 7 Set $m = m + 1$.
If $JTEC(Q_m^{**}, k_{im}^{**}, P_m^{**}, I_m^{**}, \theta_m^{**}, A_{vm}^{j**}, m) \leq JATC(Q_{m-1}^{**}, k_{m-1}^{**}, I_{m-1}^{**}, \theta_{m-1}^{**}, Av_{m-1}, m-1)$, repeat steps 2–4. Otherwise, go to Step 6.
Step 8 Set $JTEC(Q_m^{**}, k_m^{**}, I_m^{**}, \theta_m^{**}, A_v^{**}, m) = JTEC(Q_{m-1}^{**}, k_{m-1}^{**}, S_{m-1}^{**}, \theta_{m-1}^{**}, A_{vm-1}^{**}, m-1)$.
Then, $(Q^{**}, k^{**}, L^{**}, I^{**}, \theta^{**}, A_v^{**}, m^{**})$ is the optimal solution.

5. Numerical Analysis

In this section, some numerical examples are provided to validate the model. The parametric values of demand, holding cost, initial ordering cost, stockout cost, marginal profit, annual fractional cost for three different buyer's are given in Table 2, and parametric values for the vendor are given in Table 3. The parametric values are taken from Majumder et al. [14]. By using the software Matlab R2015a, one can obtain the optimum results which are shown in Table 4.

From Table 4, one can easily find that the total system cost is minimized when the batch size is 4, which can be obtained by analytic discrete optimization technique, the optimum quantity is 595.65 units, the investment for reducing setup cost per unit is 253.13 (\$/order), the optimum production rate is 704.48, the optimum setup cost is 1152.89 (\$/setup) and the optimum production cost per unit is 2.34 (\$/unit). Using those optimum values, the total system cost was \$2225.18.

Based on the above results this model is more beneficial compared to the Sarkar and Majumder [10], Sarkar and Moon [42], and Kim and Sarkar's [48] model. In Sarkar and Majumder's [10] model, the total system cost was \$6994.4, in Sarkar and Moon's [42] model the total system cost was

$3500.73, whereas in Kim and Sarkar's [48] model this total system cost was $1961.21, with a constant production rate, but in this current model the production rate is variable.

Table 2. Parametric value.

d_i(unit/week)	95, 92, 92	$A0_{bi}$($/order)	332, 315, 314
h_{bi}($/unit/week)	3.4, 2.8, 3.5	π_i($/unit)	30, 25, 20
π_{0i}($/unit)	150, 140, 152	σ_i	5, 7, 9
C_i	10, 30, 70	b_i	0.05, 0.08, 0.04
t_{S_i}	0.03, 0.04, 0.03	a_i	0.1, 0.15, 0.1

Table 3. Parametric value.

ξ_1	ξ_2	S ($/setup)	A_{v_0} ($/unit/setup)	θ_0	r	t_T	C_r ($/shipment)	β	B	b	h_v ($/unit/week)
0.06	0.00333	1	1257	0.0001	0.01	1.9	100	1.5	1300	90	2.5

Table 4. Summary of optimal values.

m	Q (units)	k_1	k_2	k_3	P	θ	I ($/order)	A_v ($/setup)	$C(p)$	TEC ($)
5	595.65	18.70	18.78	18.85	704.48	0.00005	253.13	1152.89	2.34	2225.18

5.1. Special Case I: When No Investment Is Used

When there is no investment, that is $I = 0$, then the total system cost TEC is $179162813414.03. It is found that without investment, the system cost is huge compared to the use of investment. Thus, if one uses investment then the total system cost is remarkably reduced.

5.2. Special Case II: When No Quality Improvement Is Considered

If $\theta = 0$, that is the probability of the production process which may go to an out-of-control state is zero, then the system cost TEC is $5928.30, thus the investment for reducing the probability of the production process, which may go to an *out-of-control* state is also reduces the total system cost.

5.3. Special Case III: When Setup Cost Is Fixed

If the setup cost is fixed, then the total system cost TEC is $404516898.87. Thus, it is clear that use of investment to reduce setup cost and is highly beneficial to any industry.

6. Sensitivity Analysis

One can easily find the effect of a change of parameters to the total cost by the sensitivity shown in Table 5. This table is formulated for a change in parameter -10%, -5%, 5%, and 10%. From Table 5, it is easily concluded that

- A small change in ordering cost for buyers has a great effect in total cost of the SCM system.
- A small change in the initial setup cost also has an impact on total cost. Setup cost is more effective for this model. With very little change in setup cost, there is a huge change in total cost. From the sensitivity table it is clear that setup cost is more sensitive for this model.
- Scaling parameter B is lightly sensitive for the total cost in this model.

Table 5. Sensitivity analysis for different key parameters.

Parameters	Changes(in %)	TEC^N	Parameters	Changes(in %)	TEC^N
A_{b1}	-10%	-18.09	B	-10%	$+24.95$
	-5%	-9.39		-5%	$+12.59$
	$+5\%$	$+10.09$		$+5\%$	-12.81
	$+10\%$	$+20.90$		$+10\%$	-25.83
A_{v_0}	-10%	-41.97			
	-5%	-25.18			
	$+5\%$	$+35.14$			
	$+10\%$	$+81.59$			

The effect of change in total cost are shown graphically in the Figures 3–5.

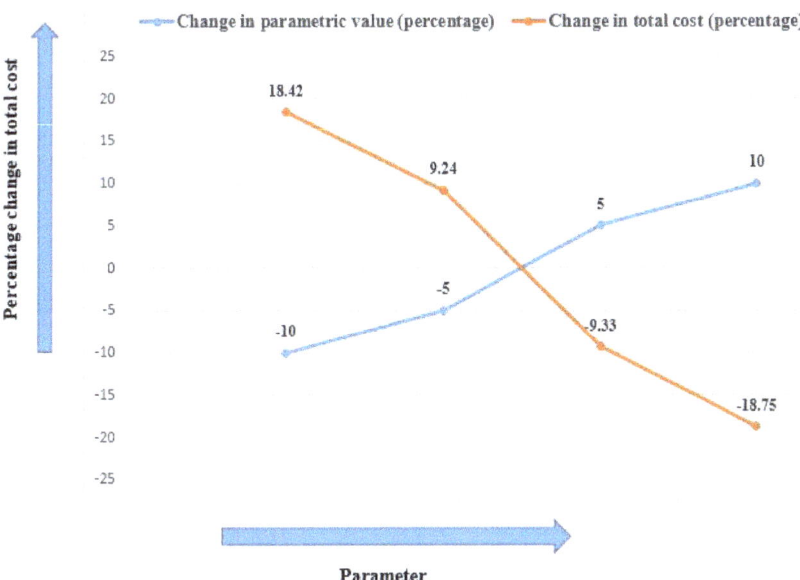

Figure 3. Changes of parameter A_{b_1} versus percentage change in total cost.

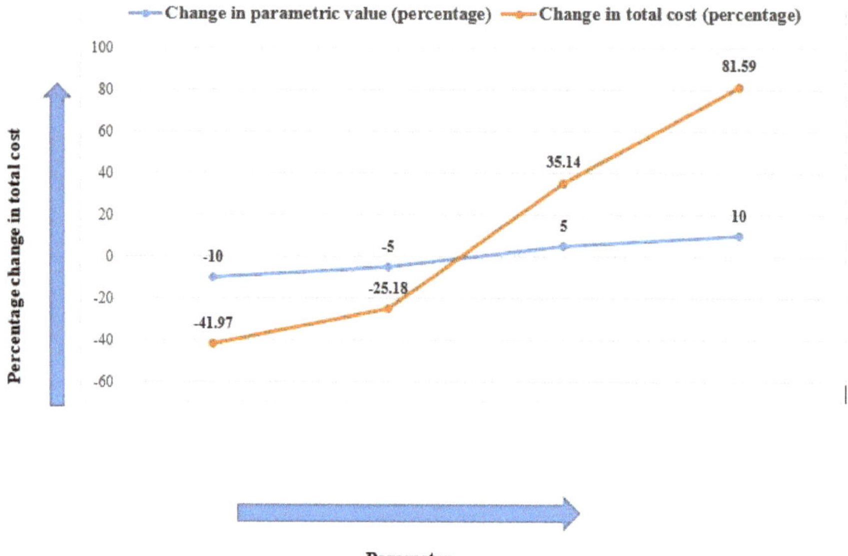

Figure 4. Changes of parameter A_v versus percentage change in total cost.

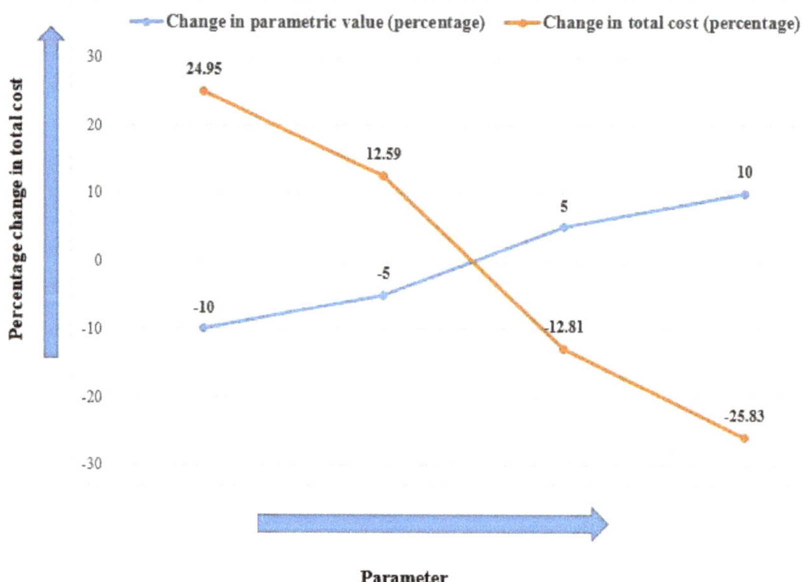

Figure 5. Changes of parameter B versus percentage change in total cost.

Case Study

A real case study was also done to validate this model. The model was tested on real data from a company, located in West Bengal, India. They happily accepted the proposal to allow access to data from their company and the model is validated with the real data. The results were found in the similar direction of the research. The results of the proposed model with the real data were considered after the normalization of each data, as without normalization, those data cannot fit with the proposed model. The basic normalization towards mean was used for the purpose. Using sample mean, sample variance, and histogram, and finally the confirmation test of the distribution function. The input data from the company is given in Tables 6 and 7 and the results are given in Table 8. The company was really satisfied with results. But if they will use the findings of the proposed model. Thus, the proposed strategy effects a major savings of the company. Based on the findings, the company may change their production planning.

Table 6. Parametric value.

d_i(unit/week)	95, 92, 92	A_{bi}($/setup)	332, 320, 313
h_{bi}($/unit/week)	3.4, 2.8, 3.5	π_i($/unit)	30, 25, 20
π_{0i}($/unit)	150, 140, 152	σ_i	5, 7, 9
β	1.5	b_i	0.05, 0.08, 0.04
C_i	10, 30, 70	a_i	0.1, 0.15, 0.1
t_{S_i}	0.03, 0.04, 0.03		

Table 7. Parametric value.

ξ_1	ξ_2	S ($/setup)	A_{v_0} ($/unit/setup)	θ_0	r	t_T	C_r ($/shipment)	B	h_v ($/unit/week)
0.06	0.00333	1	1198	0.0001	0.01	1.9	100	100	2.5

Table 8. Optimum result for case study.

m	Q (units)	k_1	t_s	P	θ	I ($/order)	A_v ($/setup)	$C(p)$	TEC ($)
4	598.91	18.74	1.0	705.27	0.00004	248.32	1159.17	2.35	4800.00

7. Managerial Insights

This model developed a single vendor-multi-buyer SCM model, where the single vendor produces a single item and sends it to multiple buyers using a SSMD policy. The production rate and production cost are variable which is quite realistic. Three types of investments are used to reduce cost and improve the quality of the production system. Based on different variables such as lead time, order quantity, reorder point, production cost, production rate, and number of shipments, investments decision are made. The managerial insights for this model are as follows:

- The production rate are considered as variable, which is more realistic rather than constant. Industries can use variable production for cost savings or earning more profit.
- The production cost is also variable which also a more realistic.
- The company would like to reduce the total cost of their production system. For this, a continuous investment is made to reduce the setup cost of whole production system along with a discrete investment to reduce the ordering cost.
- To control the long-run system, a continuous investment is made such that the production quality can be improved.
- By increasing the lead time crashing cost, a manager can reduce the lead time to upgrade the service label for the customer.

8. Concluding Remarks

This research developed an SCM model, where a single vendor produces a single type of item and sends it to multiple buyers using an SSMD policy. Contradictory to the literature, a variable production rate with variable production cost was used. The production cost depends on the production rate. From the numerical result, it is found that the optimum result is obtained when the number of shipments is four. It is also concluded that some investment in setup cost, reduced the setup cost of the whole system and some investment was done to improve the quality of the production system for the long run. As this model considered defective items, the inspection process (see for reference Sarkar [49]) will be a very interesting finding as a future extension. This model can also be extended by considering the autonomation policy for inspection along with different types of warehousing. This model can be extended in future with unreliability for the vendor. Another very interesting extension of this model would be considering a multi-echelon model with multi-buyer and multi-vendor for multiple product or an assembled product.

Author Contributions: Conceptualization, B.S.; methodology, B.S. and B.K.D.; software, B.K.D.; validation, B.S., B.K.D. and S.P.; formal analysis, B.K.D; investigation, B.S.; resources, B.S. and S.P.; data curation, B.K.D., B.S. and S.P.; writing—original draft preparation, B.K.D.; writing—review and editing, B.S., B.K.D., S.P.; visualization, B.S., and B.K.D.; supervision, S.P., and B.S.

Acknowledgments: This research was supported by Basic Science Research Program through the National Research Foundation of Korea (NRF) funded by the Ministry of Education, Science and Technology (Project Number: 2017R1D1A1B03033846).

Conflicts of Interest: The authors declare no conflict of interest.

Appendix A

$$
\tau_1 = D\left(A_{0_{bi}}e^{-r_i I_{bi}} + \frac{A_v}{m} + mC_{T_i} + I_{bi}\right) + \sum_{i=1}^{n} d_i \pi_i \sigma_i \left[\frac{mC_{R_i}(t_{s_i})}{\pi_i \sigma_i} + \frac{1}{2}\sqrt{\frac{q_i}{P} + t_{s_i}}\left(\sqrt{1 + k_i^2} - k_i\right)\right.
$$
$$
\left. + \; \frac{1}{2}(m-1)\left(\sqrt{t_{T_i} + (\frac{q_i}{P} + t_{s_i})k_i^2} - \sqrt{\frac{q_i}{P} + t_{s_i}}\right)\right]
$$

$$
\tau_2 = \sum_{i=1}^{n} \frac{d_i \pi_i \sigma_i}{4P}\left[\frac{\sqrt{1+k_i^2} - k_i}{\sqrt{\frac{q_i}{P} + t_{s_i}}} + (m-1)k_i\left(\frac{k_i}{\sqrt{t_{T_i} + (\frac{q_i}{P} + t_{s_i})k_i^2}} - \frac{1}{\sqrt{\frac{q_i}{P} + t_{s_i}}}\right)\right]
$$

$$
\tau_3 = \sum_{i=1}^{n} h_{bi}\left(\frac{d_i}{2D} + \frac{k_i \sigma_i}{2P\sqrt{\frac{q_i}{P} + t_{si}}}\right) + \frac{1}{2}mS\theta D + \frac{1}{2}h_v\left(\frac{2D}{P} + m\left(1 - \frac{D}{P}\right) - 1\right)
$$

$$
\tau_4 = \left[D\zeta_1 - \frac{h_v QD(m-2)}{2} + \frac{D\pi_i \sigma_i k_i^2(m-1)}{4\sqrt{t_{T_i} + \left(\frac{Q}{P} + t_{s_i}\right)k_i^2}}\right.
$$
$$
\left. + \; \frac{\sigma_i}{2\sqrt{\left(\frac{Q}{P} + t_{s_i}\right)}}\left(h_{bi}Qk_i - \frac{D\pi_i}{2}(k_i m - \sqrt{1 + k_i^2})\right)\right]
$$

Appendix B

Proof of Lemma 1

This paper computes the Hessian matrix at the optimal values for a given m as follows:

$$|H(JTEC)| = \begin{vmatrix} \frac{\partial^2 JTEC(.)}{\partial I^2} & \frac{\partial^2 JTEC(.)}{\partial I\partial\theta} & \frac{\partial^2 JTEC(.)}{\partial I\partial A_v} & \frac{\partial^2 JTEC(.)}{\partial Q\partial P} & \frac{\partial^2 JTEC(.)}{\partial I\partial k} & \frac{\partial^2 JTEC(.)}{\partial I\partial Q} \\ \frac{\partial^2 JTEC(.)}{\partial\theta\partial I} & \frac{\partial^2 JTEC(.)}{\partial\theta^2} & \frac{\partial^2 JTEC(.)}{\partial\theta\partial A_v} & \frac{\partial^2 JTEC(.)}{\partial\theta\partial P} & \frac{\partial^2 JTEC(.)}{\partial\theta\partial k} & \frac{\partial^2 JTEC(.)}{\partial\theta\partial Q} \\ \frac{\partial^2 JTEC(.)}{\partial A_v\partial I} & \frac{\partial^2 JTEC(.)}{\partial A_v\partial\theta} & \frac{\partial^2 JTEC(.)}{\partial A_v^2} & \frac{\partial^2 JTEC(.)}{\partial A_v\partial P} & \frac{\partial^2 JTEC(.)}{\partial A_v\partial k} & \frac{\partial^2 JTEC(.)}{\partial A_v\partial Q} \\ \frac{\partial^2 JTEC(.)}{\partial P\partial I} & \frac{\partial^2 JTEC(.)}{\partial P\partial\theta} & \frac{\partial^2 JTEC(.)}{\partial P\partial A_v} & \frac{\partial^2 JTEC(.)}{\partial P^2} & \frac{\partial^2 JTEC(.)}{\partial P\partial k} & \frac{\partial^2 JTEC(.)}{\partial P\partial Q} \\ \frac{\partial^2 JTEC(.)}{\partial k\partial I} & \frac{\partial^2 JTEC(.)}{\partial k\partial\theta} & \frac{\partial^2 JTEC(.)}{\partial k\partial A_v} & \frac{\partial^2 JTEC(.)}{\partial k\partial P} & \frac{\partial^2 JTEC(.)}{\partial k^2} & \frac{\partial^2 JTEC(.)}{\partial k\partial Q} \\ \frac{\partial^2 JTEC(.)}{\partial Q\partial I} & \frac{\partial^2 JTEC(.)}{\partial Q\partial\theta} & \frac{\partial^2 JTEC(.)}{\partial Q\partial A_v} & \frac{\partial^2 JTEC(.)}{\partial Q\partial P} & \frac{\partial^2 JTEC(.)}{\partial Q\partial k} & \frac{\partial^2 JTEC(.)}{\partial Q^2} \end{vmatrix}$$

where $JTEC(.) = JTEC(Q, k, \theta, P, I, A_v)$.

The second order partial derivatives at the optimal values are

$$\frac{\partial^2 JTEC(.)}{\partial Q^2} = \frac{2\tau_1}{Q^3} - \frac{\tau_2}{Q^2}$$

$$\frac{\partial^2 JTEC(.)}{\partial P^2} = \frac{2\tau_4}{P^3}$$

$$\frac{\partial^2 JTEC(.)}{\partial I^2} = \frac{A_{0_{bi}} r^2 I e^{-rI}}{Q}$$

$$\frac{\partial^2 JTEC(.)}{\partial\theta^2} = \frac{B\beta}{\theta^2}$$

$$\frac{\partial^2 JTEC(.)}{\partial A_v^2} = \frac{b\beta}{A_v^2}$$

$$\frac{\partial^2 JTEC(.)}{\partial Q\partial I} = \frac{\partial^2 JTEC(.)}{\partial I\partial Q} = -\frac{(1 - A_{0_{bi}} re^{-rI})D}{Q^2}$$

$$\frac{\partial^2 JTEC(.)}{\partial K^2} = \frac{1}{2}D\pi_i\sigma_i\sqrt{\frac{Q}{P} + t_{si}}\left[\frac{1}{(1+k_i^2)^{\frac{3}{2}}} + (m-1)\sqrt{\frac{\frac{Q}{P} + t_{s_i}}{t_{T_i} + \left(\frac{Q}{P} + t_{s_i}\right)k_i^2}}\left(1\right.\right.$$

$$\left.\left. - \frac{\left(\frac{Q}{P} + t_{s_i}\right)k_i^2}{t_{T_i} + \left(\frac{Q}{P} + t_{s_i}\right)k_i^2}\right)\right]$$

$$\frac{\partial^2 JTEC(.)}{\partial Q\partial P} = \frac{\partial^2 JTEC(.)}{\partial P\partial Q} = \frac{1}{2}h_v\left(-\frac{2D}{P^2} + \frac{mD}{P^2}\right) + h_{bi}\left(\frac{Qk_i\sigma_i}{4P^3(\frac{Q}{P} + t_{s_i})^{\frac{3}{2}}} - \frac{k_i\sigma_i}{2P^2\sqrt{\frac{Q}{P} + t_{s_i}}}\right)$$

$$+ \frac{1}{Q}D\pi_i\left(\frac{Q(-k_i + \sqrt{1+k_i^2})\sigma_i}{8P^3(\frac{Q}{P} + t_{s_i})^{\frac{3}{2}}} - \frac{(-k_i + \sqrt{1+k_i^2})\sigma}{4P^2\sqrt{\frac{Q}{P} + t_{s_i}}}\right.$$

$$+ \frac{1}{2}(m-1)\sqrt{tT_i}\left(-\frac{Qk_i}{4P^3(\frac{\frac{Q}{P}+t_{s_i}}{t_{T_i}})^{\frac{3}{2}}t_{T_i}^2} + \frac{k_i}{2P^2\sqrt{\frac{\frac{Q}{P}+t_{s_i}}{t_{T_i}}}t_{T_i}} + \frac{Qk_i^4}{4P^3t_{T_i}^2(1+\frac{(\frac{Q}{P}+t_{s_i})k_i^2}{t_{T_i}})^{\frac{3}{2}}}\right.$$

$$\left.\left. - \frac{k_i^2}{2P^2t_{T_i}\sqrt{1+\frac{(\frac{Q}{P}+t_{s_i})k_i^2}{t_{T_i}}}}\right)\sigma_i\right)$$

$$- \frac{D\pi_i\left(-\frac{Q(-k_i+\sqrt{1+k_i^2})\sigma_i}{4P^2\sqrt{\frac{Q}{P}+t_{s_i}}} + \frac{1}{2}(m-1)\sqrt{tT_i}\left(\frac{Qk_i}{2P^2\sqrt{\frac{\frac{Q}{P}+t_{s_i}}{t_{T_i}}}t_{T_i}} - \frac{Qk_i^2}{2P^2t_{T_i}\sqrt{1+\frac{(\frac{Q}{P}+t_{s_i})k_i^2}{t_{T_i}}}}\right)\sigma_i\right)}{Q^2}$$

$$\frac{\partial^2 JTEC(.)}{\partial Q \partial \theta} = \frac{\partial^2 JTEC(.)}{\partial \theta \partial Q} = \frac{mSD}{2}$$

$$\frac{\partial^2 JTEC(.)}{\partial Q \partial A_v} = \frac{\partial^2 JTEC(.)}{\partial A_v \partial Q} = -\frac{D}{mQ^2}$$

$$\frac{\partial^2 JTEC(.)}{\partial P \partial k} = \frac{\partial^2 JTEC(.)}{\partial k \partial P} = \frac{1}{2P^2}\left[-\frac{h_{bi}Q\sigma_i}{\sqrt{\frac{Q}{P}+t_{si}}} + D\pi_i\sigma_i\left[-\frac{\frac{k_i}{\sqrt{1+k_i^2}-1}}{2\sqrt{\frac{Q}{P}+t_{s_i}}} \right.\right.$$

$$+ (m-1)\left(\frac{1}{2\sqrt{\frac{Q}{P}+t_{si}}} + \frac{(\frac{Q}{P}+t_{si})k_i^2}{2\left(t_{T_i}+(\frac{Q}{P}+t_{si})k_i^2\right)^{\frac{3}{2}}} - \frac{k_i}{\sqrt{t_{T_i}+(\frac{Q}{P}+t_{si})k_i^2}} \right)\Big]\Big]$$

$$\frac{\partial^2 JTEC(.)}{\partial P \partial I} = \frac{\partial^2 JTEC(.)}{\partial I \partial P} = 0$$

$$\frac{\partial^2 JTEC(.)}{\partial P \partial \theta} = \frac{\partial^2 JTEC(.)}{\partial \theta \partial P} = 0$$

$$\frac{\partial^2 JTEC(.)}{\partial P \partial A_v} = \frac{\partial^2 JTEC(.)}{\partial A_v \partial P} = 0$$

$$\frac{\partial^2 JTEC(.)}{\partial k \partial I} = \frac{\partial^2 JTEC(.)}{\partial I \partial k} = 0$$

$$\frac{\partial^2 JTEC(.)}{\partial k \partial \theta} = \frac{\partial^2 JTEC(.)}{\partial \theta \partial k} = 0$$

$$\frac{\partial^2 JTEC(.)}{\partial k \partial A_v} = \frac{\partial^2 JTEC(.)}{\partial A_v \partial k} = 0$$

$$\frac{\partial^2 JTEC(.)}{\partial Q \partial k} = \frac{\partial^2 JTEC(.)}{\partial k \partial Q} = \frac{h_{bi}\sigma_i}{2P\sqrt{\frac{Q}{P}+t}} + \frac{1}{Q}D\pi_i\left(\frac{(\frac{k_i}{\sqrt{1+k_i^2}-1})\sigma_i}{4P\sqrt{\frac{Q}{P}+t_{si}}} \right.$$

$$+ \frac{1}{2}(m-1)\sqrt{t_{T_i}}\left(-\frac{1}{2P\sqrt{\frac{\frac{Q}{P}+t_{si}}{t_{T_i}}}t_{T_i}} - \frac{(\frac{Q}{P}+t_{si})k_i^3}{2Pt_{T_i}^2(1+\frac{(\frac{Q}{P}+t_{si})k_i^2}{t_{T_i}})^{\frac{3}{2}}} + \frac{k_i}{Pt_{T_i}\sqrt{1+\frac{(\frac{Q}{P}+t)k_i^2}{t_{T_i}}}} \right)\sigma_i \right)$$

$$- \frac{D\pi_i\left(\frac{1}{2}\sqrt{\frac{Q}{P}+t_{s_i}}(-1+\frac{k_i}{\sqrt{1+k_i^2}})\sigma_i + \frac{1}{2}(-1+m)\sqrt{t_{T_i}}(-\sqrt{\frac{\frac{Q}{P}+t_{si}}{t_{T_i}}} + \frac{(\frac{Q}{P}+t)\alpha}{t_{T_i}\sqrt{1+\frac{(\frac{Q}{P}+t_{s_i})k_i^2}{t_{T_i}}}})\sigma_i \right)}{Q^2}$$

$$\frac{\partial^2 JTEC(.)}{\partial I \partial \theta} = \frac{\partial^2 JTEC(.)}{\partial \theta \partial I} = 0$$

$$\frac{\partial^2 JTEC(.)}{\partial I \partial A_v} = \frac{\partial^2 JTEC(.)}{\partial A_v \partial I} = 0$$

$$\frac{\partial^2 JTEC(.)}{\partial \theta \partial A_v} = \frac{\partial^2 JTEC(.)}{\partial A_v \partial \theta} = 0$$

Now, the first order principal minor at the optimal vales is given by

$$det(H_{11}) = det\left(\frac{\partial^2 JTEC(.)}{\partial I^2} \right) = \frac{A_{0_{bi}}r^2 I_{bi}e^{-rI_{bi}}}{Q} > 0.$$

The first principal minor is grater than zero since all the parameters and variables are positive:

The second order principal minor of $H(JTEC)$ is

$$
\det(H_{22}) = \det \begin{pmatrix} \frac{\partial^2 JTEC(.)}{\partial I^2} & \frac{\partial^2 JTEC(.)}{\partial I \partial \theta} \\ \frac{\partial^2 JTEC(.)}{\partial I \partial \theta} & \frac{\partial^2 JTEC(.)}{\partial \theta^2} \end{pmatrix}
$$

$$
= \frac{A_{0_{bi}} r^2 I_{bi} e^{-r I_{bi}}}{Q} \frac{B\beta}{\theta^2} - 0 > 0
$$

The second principal minor is also grater then zero as $\frac{\partial^2 JTEC(.)}{\partial I^2} > 0$, $\frac{\partial^2 JTEC(.)}{\partial \theta^2} > 0$, and $\frac{\partial^2 JTEC(.)}{\partial I \partial \theta} = 0$.

The third order principal minor of $H(JTEC)$ is given by

$$
\det(H_{33}) = \det \begin{pmatrix} \frac{\partial^2 JTEC(.)}{\partial I^2} & \frac{\partial^2 JTEC(.)}{\partial I \partial \theta} & \frac{\partial^2 JTEC(.)}{\partial I \partial A_v} \\ \frac{\partial^2 JTEC(.)}{\partial \theta \partial I} & \frac{\partial^2 JTEC(.)}{\partial \theta^2} & \frac{\partial^2 JTEC(.)}{\partial \theta \partial A_v} \\ \frac{\partial^2 JTEC(.)}{\partial I \partial A_v} & \frac{\partial^2 JTEC(.)}{\partial A_v \partial \theta} & \frac{\partial^2 JTEC(.)}{\partial A_v^2} \end{pmatrix}
$$

$$
= \frac{A_{0_{bi}} r^2 I_{bi} e^{-r I_{bi}}}{Q} \frac{B\beta}{\theta^2} \frac{b\beta}{A_v^2} > 0.
$$

Thus, third principal minor is also grater than zero, as all three terms $\frac{\partial^2 JTEC(.)}{\partial I^2}$, $\frac{\partial^2 JTEC(.)}{\partial \theta^2}$, $\frac{\partial^2 JTEC(.)}{\partial A_v^2}$ are positive and others terms are zero.

The forth order principal minor of $H(JTEC)$ is

$$
\det(H_{44}) = \begin{vmatrix} \frac{\partial^2 JTEC(.)}{\partial I^2} & \frac{\partial^2 JTEC(.)}{\partial I \partial \theta} & \frac{\partial^2 JTEC(.)}{\partial I \partial A_v} & \frac{\partial^2 JTEC(.)}{\partial I \partial P} \\ \frac{\partial^2 JTEC(.)}{\partial I \partial \theta} & \frac{\partial^2 JTEC(.)}{\partial \theta^2} & \frac{\partial^2 JTEC(.)}{\partial \theta \partial A_v} & \frac{\partial^2 JTEC(.)}{\partial \theta \partial P} \\ \frac{\partial^2 JTEC(.)}{\partial A_v \partial I} & \frac{\partial^2 JTEC(.)}{\partial \theta \partial A_v} & \frac{\partial^2 JTEC(.)}{\partial A_v^2} & \frac{\partial^2 JTEC(.)}{\partial A_v \partial P} \\ \frac{\partial^2 JTEC(.)}{\partial P \partial I} & \frac{\partial^2 JTEC(.)}{\partial P \partial \theta} & \frac{\partial^2 JTEC(.)}{\partial P \partial A_v} & \frac{\partial^2 JTEC(.)}{\partial P^2} \end{vmatrix}
$$

$$
= \frac{A_{0_{bi}} r^2 I_{bi} e^{-r I_{bi}}}{Q} \frac{B\beta}{\theta^2} \frac{b\beta}{A_v^2} \frac{2\tau_4}{P^3} > 0.
$$

Thus fourth principal minor is grater than zero as all four terms are positive and all others terms are zero.

The fifth order principal minor of $H(JTEC)$ is given by

$$
\det(H_{55}) = \det \begin{pmatrix} \frac{\partial^2 JTEC(.)}{\partial I^2} & \frac{\partial^2 JTEC(.)}{\partial I \partial \theta} & \frac{\partial^2 JTEC(.)}{\partial I \partial A_v} & \frac{\partial^2 JTEC(.)}{\partial I \partial P} & \frac{\partial^2 JTEC(.)}{\partial I \partial k} \\ \frac{\partial^2 JTEC(.)}{\partial I \partial \theta} & \frac{\partial^2 JTEC(.)}{\partial \theta^2} & \frac{\partial^2 JTEC(.)}{\partial \theta \partial A_v} & \frac{\partial^2 JTEC(.)}{\partial \theta \partial P} & \frac{\partial^2 JTEC(.)}{\partial \theta \partial k} \\ \frac{\partial^2 JTEC(.)}{\partial A_v \partial I} & \frac{\partial^2 JTEC(.)}{\partial \theta \partial A_v} & \frac{\partial^2 JTEC(.)}{\partial A_v^2} & \frac{\partial^2 JTEC(.)}{\partial A_v \partial P} & \frac{\partial^2 JTEC(.)}{\partial A_v \partial k} \\ \frac{\partial^2 JTEC(.)}{\partial P \partial I} & \frac{\partial^2 JTEC(.)}{\partial P \partial \theta} & \frac{\partial^2 JTEC(.)}{\partial P \partial A_v} & \frac{\partial^2 JTEC(.)}{\partial P^2} & \frac{\partial^2 JTEC(.)}{\partial P \partial k} \\ \frac{\partial^2 JTEC(.)}{\partial k \partial I} & \frac{\partial^2 JTEC(.)}{\partial k \partial \theta} & \frac{\partial^2 JTEC(.)}{\partial k \partial A_v} & \frac{\partial^2 JTEC(.)}{\partial P \partial k} & \frac{\partial^2 JTEC(.)}{\partial k^2} \end{pmatrix}.
$$

$$
= \frac{A_{0_{bi}} r^2 I_{bi} e^{-r I_{bi}}}{Q} \frac{B\beta}{\theta^2} \frac{b\beta}{A_v^2} \det \begin{pmatrix} \frac{\partial^2 JTEC(.)}{\partial P^2} & \frac{\partial^2 JTEC(.)}{\partial P \partial k} \\ \frac{\partial^2 JTEC(.)}{\partial k \partial P} & \frac{\partial^2 JTEC(.)}{\partial k^2} \end{pmatrix}.
$$

Now, if

$$\frac{2\tau_4}{P^3} > \frac{1}{2P^2}\left[-\frac{h_{bi}Q\sigma_i}{\sqrt{\frac{Q}{P}+t_{si}}} + D\pi_i\sigma_i\left[-\frac{\frac{k_i}{\sqrt{1+k_i^2}}-1}{2\sqrt{\frac{Q}{P}+t_{s_i}}}\right.\right.$$

$$\left.\left. + (m-1)\left(\frac{1}{2\sqrt{\frac{Q}{P}+t_{si}}} + \frac{(\frac{Q}{P}+t_{si})k_i^2}{2\left(t_{T_i}+(\frac{Q}{P}+t_{si})k_i^2\right)^{\frac{3}{2}}} - \frac{k_i}{\sqrt{t_{T_i}+(\frac{Q}{P}+t_{si})k_i^2}}\right)\right]\right]$$

$$\frac{1}{2}D\pi_i\sigma_i\sqrt{\frac{Q}{P}+t_{si}}\left[\frac{1}{\left(1+k_i^2\right)^{\frac{3}{2}}} + (m-1)\sqrt{\frac{\frac{Q}{P}+t_{s_i}}{t_{T_i}+\left(\frac{Q}{P}+t_{s_i}\right)k_i^2}}\left(1 - \frac{\left(\frac{Q}{P}+t_{s_i}\right)k_i^2}{t_{T_i}+\left(\frac{Q}{P}+t_{s_i}\right)k_i^2}\right)\right]$$

$$> \frac{1}{2P^2}\left[-\frac{h_{bi}Q\sigma_i}{\sqrt{\frac{Q}{P}+t_{si}}} + D\pi_i\sigma_i\left[-\frac{\frac{k_i}{\sqrt{1+k_i^2}}-1}{2\sqrt{\frac{Q}{P}+t_{s_i}}}\right.\right.$$

$$\left.\left. + (m-1)\left(\frac{1}{2\sqrt{\frac{Q}{P}+t_{si}}} + \frac{(\frac{Q}{P}+t_{si})k_i^2}{2\left(t_{T_i}+(\frac{Q}{P}+t_{si})k_i^2\right)^{\frac{3}{2}}} - \frac{k_i}{\sqrt{t_{T_i}+(\frac{Q}{P}+t_{si})k_i^2}}\right)\right]\right],$$

then by the formula $xy > z^2$, if $x > z$, and $y > z$, one can get

$$\det\begin{pmatrix} \frac{\partial^2 JTEC(.)}{\partial P^2} & \frac{\partial^2 JTEC(.)}{\partial P\partial k} \\ \frac{\partial^2 JTEC(.)}{\partial k\partial P} & \frac{\partial^2 JTEC(.)}{\partial k^2} \end{pmatrix} = \frac{\partial^2 JTEC(.)}{\partial P^2} \times \frac{\partial^2 JTEC(.)}{\partial k^2} - \left(\frac{\partial^2 JTEC(.)}{\partial k\partial P}\right)^2 > 0.$$

Thus, one can state that fifth order principal minor is grater than zero as this is the products of four positive terms and all others terms are zero. Now, the sixth order minor i.e., the full Hessian is of the form

$$|H_{66}| = \begin{vmatrix} \frac{\partial^2 JTEC(.)}{\partial I^2} & \frac{\partial^2 JTEC(.)}{\partial I\partial\theta} & \frac{\partial^2 JTEC(.)}{\partial I\partial A_v} & \frac{\partial^2 JTEC(.)}{\partial Q\partial P} & \frac{\partial^2 JTEC(.)}{\partial I\partial k} & \frac{\partial^2 JTEC(.)}{\partial I\partial Q} \\ \frac{\partial^2 JTEC(.)}{\partial\theta\partial I} & \frac{\partial^2 JTEC(.)}{\partial\theta^2} & \frac{\partial^2 JTEC(.)}{\partial\theta\partial A_v} & \frac{\partial^2 JTEC(.)}{\partial\theta\partial P} & \frac{\partial^2 JTEC(.)}{\partial\theta\partial k} & \frac{\partial^2 JTEC(.)}{\partial\theta\partial Q} \\ \frac{\partial^2 JTEC(.)}{\partial A_v\partial I} & \frac{\partial^2 JTEC(.)}{\partial A_v\partial\theta} & \frac{\partial^2 JTEC(.)}{\partial A_v^2} & \frac{\partial^2 JTEC(.)}{\partial A_v\partial P} & \frac{\partial^2 JTEC(.)}{\partial A_v\partial k} & \frac{\partial^2 JTEC(.)}{\partial A_v\partial Q} \\ \frac{\partial^2 JTEC(.)}{\partial P\partial I} & \frac{\partial^2 JTEC(.)}{\partial P\partial\theta} & \frac{\partial^2 JTEC(.)}{\partial P\partial A_v} & \frac{\partial^2 JTEC(.)}{\partial P^2} & \frac{\partial^2 JTEC(.)}{\partial P\partial k} & \frac{\partial^2 JTEC(.)}{\partial P\partial Q} \\ \frac{\partial^2 JTEC(.)}{\partial k\partial I} & \frac{\partial^2 JTEC(.)}{\partial k\partial\theta} & \frac{\partial^2 JTEC(.)}{\partial k\partial A_v} & \frac{\partial^2 JTEC(.)}{\partial k\partial P} & \frac{\partial^2 JTEC(.)}{\partial k^2} & \frac{\partial^2 JTEC(.)}{\partial k\partial Q} \\ \frac{\partial^2 JTEC(.)}{\partial Q\partial I} & \frac{\partial^2 JTEC(.)}{\partial Q\partial\theta} & \frac{\partial^2 JTEC(.)}{\partial Q\partial A_v} & \frac{\partial^2 JTEC(.)}{\partial Q\partial P} & \frac{\partial^2 JTEC(.)}{\partial Q\partial k} & \frac{\partial^2 JTEC(.)}{\partial Q^2} \end{vmatrix}$$

$$= \frac{\partial^2 JTEC(.)}{\partial I^2} \times \begin{vmatrix} \frac{\partial^2 JTEC(.)}{\partial\theta^2} & \frac{\partial^2 JTEC(.)}{\partial\theta\partial A_v} & \frac{\partial^2 JTEC(.)}{\partial\theta\partial P} & \frac{\partial^2 JTEC(.)}{\partial\theta\partial k} & \frac{\partial^2 JTEC(.)}{\partial\theta\partial Q} \\ \frac{\partial^2 JTEC(.)}{\partial A_v\partial\theta} & \frac{\partial^2 JTEC(.)}{\partial A_v^2} & \frac{\partial^2 JTEC(.)}{\partial A_v\partial P} & \frac{\partial^2 JTEC(.)}{\partial A_v\partial k} & \frac{\partial^2 JTEC(.)}{\partial A_v\partial Q} \\ \frac{\partial^2 JTEC(.)}{\partial P\partial\theta} & \frac{\partial^2 JTEC(.)}{\partial P\partial A_v} & \frac{\partial^2 JTEC(.)}{\partial P^2} & \frac{\partial^2 JTEC(.)}{\partial P\partial k} & \frac{\partial^2 JTEC(.)}{\partial P\partial Q} \\ \frac{\partial^2 JTEC(.)}{\partial k\partial\theta} & \frac{\partial^2 JTEC(.)}{\partial k\partial A_v} & \frac{\partial^2 JTEC(.)}{\partial k\partial P} & \frac{\partial^2 JTEC(.)}{\partial k^2} & \frac{\partial^2 JTEC(.)}{\partial k\partial Q} \\ \frac{\partial^2 JTEC(.)}{\partial Q\partial\theta} & \frac{\partial^2 JTEC(.)}{\partial Q\partial A_v} & \frac{\partial^2 JTEC(.)}{\partial Q\partial P} & \frac{\partial^2 JTEC(.)}{\partial Q\partial k} & \frac{\partial^2 JTEC(.)}{\partial Q^2} \end{vmatrix}$$

$$- \frac{\partial^2 JTEC(.)}{\partial I\partial Q} \times |H_{55}|.$$

Now, one have to calculate

$$
\frac{\partial^2 JTEC(.)}{\partial I^2} \times \frac{\partial^2 JTEC(.)}{\partial \theta^2} \times
\begin{vmatrix}
\frac{\partial^2 JTEC(.)}{\partial A_v^2} & \frac{\partial^2 JTEC(.)}{\partial A_v \partial P} & \frac{\partial^2 JTEC(.)}{\partial A_v \partial k} & \frac{\partial^2 JTEC(.)}{\partial A_v \partial Q} \\
\frac{\partial^2 JTEC(.)}{\partial P \partial A_v} & \frac{\partial^2 JTEC(.)}{\partial P^2} & \frac{\partial^2 JTEC(.)}{\partial P \partial k} & \frac{\partial^2 JTEC(.)}{\partial P \partial Q} \\
\frac{\partial^2 JTEC(.)}{\partial k \partial A_v} & \frac{\partial^2 JTEC(.)}{\partial k \partial P} & \frac{\partial^2 JTEC(.)}{\partial k^2} & \frac{\partial^2 JTEC(.)}{\partial k \partial Q} \\
\frac{\partial^2 JTEC(.)}{\partial Q \partial A_v} & \frac{\partial^2 JTEC(.)}{\partial Q \partial P} & \frac{\partial^2 JTEC(.)}{\partial Q \partial k} & \frac{\partial^2 JTEC(.)}{\partial Q^2}
\end{vmatrix}
$$

$$
\begin{vmatrix}
\frac{\partial^2 JTEC(.)}{\partial A_v^2} & \frac{\partial^2 JTEC(.)}{\partial A_v \partial P} & \frac{\partial^2 JTEC(.)}{\partial A_v \partial k} & \frac{\partial^2 JTEC(.)}{\partial A_v \partial Q} \\
\frac{\partial^2 JTEC(.)}{\partial P \partial A_v} & \frac{\partial^2 JTEC(.)}{\partial P^2} & \frac{\partial^2 JTEC(.)}{\partial P \partial k} & \frac{\partial^2 JTEC(.)}{\partial P \partial Q} \\
\frac{\partial^2 JTEC(.)}{\partial k \partial A_v} & \frac{\partial^2 JTEC(.)}{\partial k \partial P} & \frac{\partial^2 JTEC(.)}{\partial k^2} & \frac{\partial^2 JTEC(.)}{\partial k \partial Q} \\
\frac{\partial^2 JTEC(.)}{\partial Q \partial A_v} & \frac{\partial^2 JTEC(.)}{\partial Q \partial P} & \frac{\partial^2 JTEC(.)}{\partial Q \partial k} & \frac{\partial^2 JTEC(.)}{\partial Q^2}
\end{vmatrix}
= \frac{\partial^2 JTEC(.)}{\partial A_v^2} \times
\begin{vmatrix}
\frac{\partial^2 JTEC(.)}{\partial P^2} & \frac{\partial^2 JTEC(.)}{\partial P \partial k} & \frac{\partial^2 JTEC(.)}{\partial P \partial Q} \\
\frac{\partial^2 JTEC(.)}{\partial k \partial P} & \frac{\partial^2 JTEC(.)}{\partial k^2} & \frac{\partial^2 JTEC(.)}{\partial k \partial Q} \\
\frac{\partial^2 JTEC(.)}{\partial Q \partial P} & \frac{\partial^2 JTEC(.)}{\partial Q \partial k} & \frac{\partial^2 JTEC(.)}{\partial Q^2}
\end{vmatrix}
$$

$$
+ \frac{D}{mQ^2} \times
\begin{vmatrix}
\frac{\partial^2 JTEC(.)}{\partial P \partial A_v} & \frac{\partial^2 JTEC(.)}{\partial P^2} & \frac{\partial^2 JTEC(.)}{\partial P \partial k} \\
\frac{\partial^2 JTEC(.)}{\partial k \partial A_v} & \frac{\partial^2 JTEC(.)}{\partial k \partial P} & \frac{\partial^2 JTEC(.)}{\partial k^2} \\
\frac{\partial^2 JTEC(.)}{\partial Q \partial A_v} & \frac{\partial^2 JTEC(.)}{\partial Q \partial P} & \frac{\partial^2 JTEC(.)}{\partial Q \partial k}
\end{vmatrix}.
$$

Now, from the previous argument one can find that

$$
\begin{vmatrix}
\frac{\partial^2 JTEC(.)}{\partial P^2} & \frac{\partial^2 JTEC(.)}{\partial P \partial k} & \frac{\partial^2 JTEC(.)}{\partial P \partial Q} \\
\frac{\partial^2 JTEC(.)}{\partial k \partial P} & \frac{\partial^2 JTEC(.)}{\partial k^2} & \frac{\partial^2 JTEC(.)}{\partial k \partial Q} \\
\frac{\partial^2 JTEC(.)}{\partial Q \partial P} & \frac{\partial^2 JTEC(.)}{\partial Q \partial k} & \frac{\partial^2 JTEC(.)}{\partial Q^2}
\end{vmatrix}
> 0 \text{ and, }
\begin{vmatrix}
\frac{\partial^2 JTEC(.)}{\partial P \partial A_v} & \frac{\partial^2 JTEC(.)}{\partial P^2} & \frac{\partial^2 JTEC(.)}{\partial P \partial k} \\
\frac{\partial^2 JTEC(.)}{\partial k \partial A_v} & \frac{\partial^2 JTEC(.)}{\partial k \partial P} & \frac{\partial^2 JTEC(.)}{\partial k^2} \\
\frac{\partial^2 JTEC(.)}{\partial Q \partial A_v} & \frac{\partial^2 JTEC(.)}{\partial Q \partial P} & \frac{\partial^2 JTEC(.)}{\partial Q \partial k}
\end{vmatrix}
> 0.
$$

Hence $|H_{66}| > 0$, since $\frac{\partial^2 JTEC(.)}{\partial I^2} > 0$, $\frac{\partial^2 JTEC(.)}{\partial I \partial Q} < 0$, and $|H_{55}| > 0$. Thus, sixth principal minor $|H_{66}|$ is greater then zero as this is the sum of positive terms.

Hence one can state that all principal minor that is $|H_{11}|$, $|H_{22}|$, $|H_{33}|$, $|H_{44}|$, $|H_{55}|$, $|H_{66}|$ are grater than zero for the optimal values of the decision variables, which is the sufficient condition for the global optimum result of this model.

Hence the Lemma 1 is proved.

References

1. Taft, E.W. The most economical production lot. *Iron Age* **1918**, *101*, 1410–1412.
2. Cárdenas-Barrón, L.E.; Chung, K.J.; Treviño-Garza, G. Celebrating a century of the economic order quantity model in honor of Ford Whitman Harris. *Int. J. Prod. Econ.* **2014**, *155*, 1–7.
3. Sarkar, B. A production-inventory model with probabilistic deterioration in two-echelon supply chain management. *Appl. Math. Model.* **2013**, *37*, 3138–3151. [CrossRef]
4. Goyal, S.K. An integrated inventory model for a single supplier-single customer problem. *Int. J. Prod. Res.* **1976**, *15*, 107–111. [CrossRef]
5. Banerjee, A. A joint economic-lot-size model for purchaser and vendor. *Decis. Sci.* **1986**, *17*, 292–311. [CrossRef]
6. Goyal, S.K. Economic ordering policy for deteriorating items over an infinite time horizon. *Eur. J. Oper. Res.* **1987**, *28*, 298–301. [CrossRef]
7. Chakraborty, D.; Bhuiya, S.K. A Continuous Review Inventory Model with Fuzzy Service Level Constraint and Fuzzy Random Variable Parameters. *Int. J. Appl. Comput. Math.* **2017**, *3*, 3159–3174. [CrossRef]
8. Ouyang, L.Y.; Yeh, N.C.; Wu, K.S. Mixture inventory model with backorders and lost sales for variable lead time. *J. Oper. Res. Soc.* **1996**, *47*, 829–832. [CrossRef]
9. Ouyang, L.Y.; Wu, K.S.; Ho, C.H. Integrated vendor-buyer cooperative models with stochastic demand in controllable lead time. *Int. J. Prod. Econ.* **2004**, *92*, 255–266. [CrossRef]

10. Sarkar B.; Majumder A. Integrated vendor-buyer supply chain model with vendor's setup cost reduction. *Appl. Math. Comput.* **2013**, *224*, 362–371. [CrossRef]

11. Moon, I.; Gallego, G. Distribution free procedures for some inventory models. *J. Oper. Res. Soc.* **1994**, *45*, 651–658. [CrossRef]

12. Sarkar, B.; Majumder, A.; Sarkar, M.; Dey, B.K.; Roy, G. Two-echelon supply chain model with manufacturing quality improvement and setup cost reduction. *J. Ind. Manag. Optim.* **2017**, *13*, 1085–1104. [CrossRef]

13. Dey, B.K.; Sarkar, B.; Sarkar, M.; Pareek, S. An integrated inventory model involving discrete setup cost reduction, variable safety factor, selling-price dependent demand, and investment. *Rairo Oper. Res.* **2019**, *53*, 39–57. [CrossRef]

14. Majumder, A.; Jaggi, C.K.; Sarkar, B. A multi-retailer supply chain model with backorder and variable production cost. *Rairo Oper. Res.* **2018.**, in press. [CrossRef]

15. Banerjee, A.; Burton, J.S. Coordinated versus independent inventory replenishment policies for a vendor and multiple buyers. *Int. J. Prod. Econ.* **1994**, *35*, 215–222. [CrossRef]

16. Banerjee, A.; Banerjee, S. A coordinated order-up-to inventory control policy for a single supplier and multiple buyers using electronic data interchange. *Int. J. Prod. Econ.* **1994**, *35*, 85–91. [CrossRef]

17. Sarmah, S.P.; Acharya, D.; Goyal, S.K. Coordination of a single-manufacturer/multi-buyer supply chain with credit option. *Int. J. Prod. Econ.* **2008**, *111*, 676–685. [CrossRef]

18. Khouja, M.; Mehrez, A. Economic production lot size model with variable production rate and imperfect quality. *J. Oper. Res. Soc.* **1994**, *45*, 1405–1417. [CrossRef]

19. Tripathi, R.P.; Pareek, S.; Kaur, M. Inventory Model with Exponential Time-Dependent Demand Rate, Variable Deterioration, Shortages and Production Cost. *Int. J. Appl. Comput. Math.* **2017**, *3*, 1407–1419. [CrossRef]

20. Chakrabarty, R.; Roy, T.; Chaudhuri, K.S. A Production: Inventory Model for Defective Items with Shortages Incorporating Inflation and Time Value of Money. *Int. J. Appl. Comput. Math.* **2017**, *3*, 195–212. [CrossRef]

21. Hoque M.A. Synchronization in the single-manufacturer multi-buyer integrated inventory supply chain. *Eur. J. Oper. Res.* **2008**, *188*, 811–825. [CrossRef]

22. Jha, J.K.; Shanker, K. Two-echelon supply chain inventory model with controllable lead time and service level constraint. *Comput. Ind. Eng.* **2009**, *57*, 1096–1104. [CrossRef]

23. Glock, C.H.; Kim T. The effect of forward integration on a single-vendor–multi-retailer supply chain under retailer competition. *Int. J. Prod. Econ.* **2015**, *164*, 179–192. [CrossRef]

24. Liao, C.J.; Shyu, C.H. An analytical determination of lead time with normal demand. *Int. J. Oper. Prod. Manag.* **1991**, *11*, 72–78. [CrossRef]

25. Ben-Daya, M.; Raouf A. Inventory models involving lead time as a decision variable. *J. Oper. Res. Soc.* **1994**, *45*, 579–582. [CrossRef]

26. Pan, J.C.H.; Yang, J.S. A study of an integrated inventory with controllable lead time. *Int. J. Prod. Res.* **2002**, *40*, 1263–1273. [CrossRef]

27. Annadurai, K.; Uthayakumar, R. Reducing lost-sales rate in *(T,R,L)* inventory model with controllable lead time. *Appl. Math. Model.* **2010**, *34*, 3465–3477. [CrossRef]

28. Lo, S.T.; Wee, H.M.; Huang, W.C. An integrated production-inventory model with imperfect production processes and Weibull distribution deterioration under inflation. *Int. J. Prod. Econ.* **2007**, *106*, 248–260. [CrossRef]

29. Huang, C.K.; Cheng, T.L.; Kao, T.C.; Goyal, S.K. An integrated inventory model involving manufacturing setup cost reduction in compound poisson process. *Int. J. Prod. Res.* **2011**, *49*, 1219–1228. [CrossRef]

30. Tayyab, M.; Sarkar, B. Optimal batch quantity in a cleaner multi-stage lean production system with random defective rate. *J. Clean. Prod.* **2016**, *139*, 922–934. [CrossRef]

31. Kang, C.W.; Ullah, M.; Sarkar, B.; Hussain, I.; Akhtar, R. Impact of random defective rate on lot size focusing work-in-process inventory in manufacturing system. *Int. J. Prod. Res.* **2017**, *55*, 1748–1766. [CrossRef]

32. Mishra, U. An EOQ Model with Time Dependent Weibull Deterioration, Quadratic Demand and Partial Backlogging. *Int. J. Appl. Comput. Math.* **2017**, *2*, 545–563. [CrossRef]

33. Khan, M.; Hussain, M.; Cárdenas-Barrón, L.E. Learning and screening errors in an EPQ inventory model for supply chains with stochastic lead time demands. *Int. J. Prod. Res.* **2017**, *55*, 4816–4832. [CrossRef]

34. Chang, H.J.; Su, R.H.; Yang, C.T.; Weng, M.W. An economic manufacturing quantity model for a two-stage assembly system with imperfect processes and variable production rate. *Comput. Ind. Eng.* **2012**, *63*, 285–293. [CrossRef]

35. Cárdenas-Barrón, L.E.; Taleizadeh, A.A.; Treviño-Garza, G. An improved solution to replenishment lot size problem with discontinuous issuing policy and rework, and the multi-delivery policy into economic production lot size problem with partial rework. *Expert Syst. Appl.* **2012**, *39*, 13540–13546.

36. Debata, S.; Acharya, M. An Inventory Control for Non-instantaneous Deteriorating Items with Non-zero Lead Time and Partial Backlogging Under Joint Price and Time Dependent Demand. *Int. J. Appl. Comput. Math.* **2017**, *3*, 1318–1393. [CrossRef]

37. Sana, S.S.; Chaudhuri, K.S. An EMQ model in an imperfect production process. *Int. J. Syst. Sci.* **2010**, *41*, 635–646. [CrossRef]

38. Wee, H.M.; Wang, W.T.; Cárdenas-Barrón, L.E. An alternative EPQ model with rework process at a single-stage manufacturing system with planned backorders. *Comput. Ind. Eng.* **2013**, *64*, 748–755. [CrossRef]

39. Sarkar, B.; Cárdenas-Barrón, L.E.; Sarkar, M.; Singgih M.L. An economic production quantity model with random defective rate, rework process and backorders for a single stage production system. *J. Manuf. Syst.* **2014**, *33*, 423–435. [CrossRef]

40. Das Roy, M.; Sana, S.S.; Chaudhuri, K.S. An integrated producer–buyer relationship in the environment of EMQ and JIT production systems. *Int. J. Prod. Res.* **2012**, *50*, 5597–5614.

41. Kim, M.; Kim, J.; Sarkar, B.; Sarkar, M.; Iqbal, M.W. An improved way to calculate imperfect items during long-run production in an integrated inventory model with backorders. *J. Manuf. Syst.* **2018**, *47*, 153–167. [CrossRef]

42. Sarkar, B.; Moon, I. Improved quality, setup cost reduction, and variable backorder costs in an imperfect production process. *Int. J. Prod. Econ.* **2014**, *155*, 204–213. [CrossRef]

43. Cárdenas-Barrón, L.E.; Sarkar, B.; Treviño-Garza, G. An improved solution to the replenishment policy for the EMQ model with rework and multiple shipments. *Appl. Math. Model.* **2013**, *37*, 5549–5554.

44. Pal, B.; Sana, S.S.; Chaudhuri, K. A mathematical model on EPQ for stochastic demand in an imperfect production system. *J. Manuf. Syst.* **2013**, *32*, 260–270. [CrossRef]

45. Sana, S.S. An EOQ model for stochastic demand for limited capacity of own warehouse. *Ann. Oper. Res.* **2015**, *233*, 383–399. [CrossRef]

46. Pal, B.; Sana, S.S.; Chaudhuri, K. A distribution-free newsvendor problem with nonlinear holding cost. *Int. J. Syst. Sci.* **2015**, *46*, 1209–1277. [CrossRef]

47. Gallego, G.; Moon, I. The Distribution free Newsboy problem: Review and extensions. *J. Oper. Res. Soc.* **1993**, *44*, 825–834. [CrossRef]

48. Kim, S.J.; Sarkar, B. Supply Chain Model with Stochastic Lead Time, Trade-Credit Financing, and Transportation Discounts. *Math. Probl. Eng.* **2017**, *2017*, 6465912. [CrossRef]

49. Sarkar, B. Supply chain coordination with variable backorder, inspections, and discount policy for fixed lifetime products. *Math. Probl. Eng.* **2016**, *2016*, 6318737. [CrossRef]

 © 2019 by the authors. Licensee MDPI, Basel, Switzerland. This article is an open access article distributed under the terms and conditions of the Creative Commons Attribution (CC BY) license (http://creativecommons.org/licenses/by/4.0/).

Article

How Does a Radio Frequency Identification Optimize the Profit in an Unreliable Supply Chain Management?

Rekha Guchhait [1,2], Sarla Pareek [1] and Biswajit Sarkar [2,*

[1] Department of Mathematics & Statistics, Banasthali Vidyapith, Rajasthan 304022, India; rg.rekhaguchhait@gmail.com (R.G.); psarla13@gmail.com (S.P.)
[2] Department of Industrial & Management Engineering, Hanyang University, Ansan, Gyeonggi-do 15588, Korea
* Correspondence: bsarkar@hanyang.ac.kr; Tel.: +82-10-7498-1981

Received: 29 March 2019; Accepted: 23 May 2019; Published: 29 May 2019

Abstract: Competition in business is higher in the electronics sector compared to other sectors. In such a situation, the role of a manufacturer is to manage the inventory properly with optimized profit. However, the problem of unreliability within buyers still exists in real world scenarios. The manufacturer adopts the radio frequency identification (RFID) technology to manage the inventory, which can control the unreliability, the inventory pooling effect, and the investment on human labor. For detecting RFID tags, a reasonable number of readers are needed. This study investigates the optimum distance between any two readers when using the optimum number of readers. As a vendor managed inventory (VMI) policy is utilized by the manufacturer, a revenue sharing contract is adopted to prevent the loss of buyers. The aim of this study is to maximize the profits of a two-echelon supply chain management under an advanced technology system. As the life of electronic gadgets is random, it may not follow any specific type of distribution function. The distribution-free approach helps to solve this issue when the mean and the standard deviation are known. The Kuhn-Tucker methodology and classical optimization are used to find the global optimum solution. The numerical analysis demonstrates that the manufacturer can earn more profit in coordination case after utilizing revenue sharing and the optimum distance between readers optimizing cost related to the RFID system. Sensitivity analysis is performed to check the sensibility of the parameters.

Keywords: supply chain management; inventory control; distribution-free approach; revenue sharing; radio frequency identification; information asymmetry

---

1. Introduction

Instead of a traditional business system, supply chain management (SCM) provides different kinds of business policies in terms of inventory management. The vendor managed inventory (VMI) is one of these in which the manufacturer takes full responsibility of the existing inventory at the buyer's position. Dong and Xu [1] found opportunities where buyers received more profit than the manufacturer. The manufacturer's profit may vary according to the business policy, where the short-term and long-term VMI affects the SCM, which were decided by them. They concluded that the short-term VMI can be a competitor for coordination business policy. In any business, the forecasting uncertainty is a major issue and Guo et al. [2] developed a method to reduce the supply chain forecasting uncertainty through information sharing via macro prediction which can reduce the system robustness. However, it is possible that not all information is shared by both parties. Then, unreliability occurs in the business system due to information asymmetry (Mukhopadhyay et al. [3];

Yan and Pei [4]; Xiao and Xu [5]). An information basically flows in the upward direction of SCM. The lack of information of the manufacturer may cause insufficient supply of products which can affect the inventory and production process. The situation is even more complicated when an imperfect production process takes place (Sarkar [6]). The rework of defective products was considered by Cárdenas-Barrón et al. [7] for an imperfect production process. They developed an improved algorithm to find the optimum lot size and replenish the defective production system. Cleaner production can be formed by discarding defective products, which was established by Tayyab and Sarkar [8]. Those defective products were reworked up to good quality through additional investment. This work was extended by multi-stage cleaner production by Kim and Sarkar [9] using budget constraints. There are several researchers who worked on imperfect products, reworking, and deterioration (Guchhait et al. [10], Majumder et al. [11], Tiwari et al. [12]). Finally, Sarkar [13] introduced an exact duration for reworking within a multi-stage multi-cycle production system. However, there is a lack of literature regarding RFID, i.e., RFID was not used to maintain the inventory pooling effect. Reworking was considered by Sarkar et al. [14] in a material requirement planning (MRP) system.

Production quantity mainly depends upon the market demand. In reality, it cannot always be the case that data related with demand are available. If no known distribution function is followed by the demand or no data are available, then instead of taking any arbitrary probability distribution, the distribution-free (DF) approach is used (Gallego and Moon [15], Sarkar et al. [16], Guchhait et al. [17]). This method was invented by Scarf [18]. Due to the complex calculations, it was not understandable to people in the industry at that time. Later, this approach was simplified by Gallego and Moon [15]. This method is used by Sarkar et al. [19] for a consignment stock-based newsvendor model. They allowed a fixed-fee payment technique to prevent loss from any participant. There are multiple manufacturers and retailers available for a single-type of products. Based on advertisements given by the manufacturer, retailers opted to choose their manufacturers. For the random demand, the variable production rate is useful (Sarkar et al. [20]) for modeling uncertain demand. A service level can help avoid shortages (Moon et al. [21]) and backorder (Sarkar [22]) due to the uncertain random demand. Partial trade credit for deteriorating items in the inventory model was discussed by Tiwari et al. [23]. For any industry, it may be that they need to analyze their previous data. Tiwari et al. [24] provided a big data analysis of SCM from 2010 to 2016.

Competitive markets in the business industry becoming more intense everyday. To handle this situation, companies prefer to adopt smart technologies within the SCM. The fast movement of products for the electronic industry is a key feature since competition is very high in the electronics sector. The implementation of technology instead of labor-based production is helpful not only for fast production, but also to profit gain. The use of RFID technology in SCM for managing inventory has been studied by several researchers. A wireless sensing problem for coverage was first studied by Meguerdichian et al. [25]. Zhang and Hou [26] investigated how many readers need to be implemented to provide a complete coverage of a search area. The coverage area sensing radius and transmitting radius were discussed by Hefeeda and Ahmadi [27]. They established that probabilistic sensing coverage can function as deterministic coverage. Dias [28] implemented RFID for a multi-agent system. Sarac et al. [29] surveyed the literature and found several implementation and usages of RFID in different sectors of SCM. They found that inventory loss can be reduced with increased efficiency of the system and real-time information of the inventory. Kim and Glock [30] investigated the effectiveness of an RFID tracking system for container management and found that the return rate of container was increased after using RFID. A four-echelon SCM was studied by Sari [31] to examine the effects of collaboration. They found through simulation that the integrated RFID technology is more beneficial for good collaboration between participants. Besides SCM, warehouse efficiency can be improved using RFID technology (Biswal et al. [32]). In the production sector, RFID improves the efficiency and maintenance, as investigated by Chen et al. [33]. They established that operation time can be increased by up to 89% and that the labor cost is reduced significantly by using RFID. Even, remanufacturing companies can get benefit from RFID via just-in-time (JIT) features or transiting towards a closed-loop SCM (Tsao et al. [34]).

From literature, it is found in most of the studies that RFID is used in SCM to prevent inventory shrinkage as well as minimize the operation time of the system, reduction of lead time, and labor consumption (Ustundag and Tanyas [35]; Jaggi et al. [36]) and improve the efficiency. However, the reason behind this efficiency improvement by RFID is not discussed in the literature. This study introduces for the first time the RFID distance function $f(d)$ based on the sensing and transmitting radii. The distance between two readers can be optimized and thus, the number of RFID readers can be found to increase the efficiency. Based on the transmitting and sensing radii, two types of readers are used by the manufacturer, namely Type 1 and Type 2. To understand the complete search capacity of a Type 1 reader, the area is divided into sub-areas that are under the coverage of Type 2 readers. This combined system may enhances the system accuracy and provides strong coverage of the sensing and transmitting areas. Table 1 gives the contribution of different authors in the literature. This study shows benefits for the buyer in the optimum order quantity, optimizes distance the between two readers, and optimizes the service given by the buyers. The rest of the study is designed as Section 2 gives the details about the mathematical model. Section 3 gives the results of the numerical experiment and Section 4 provides a discussion of results. Section 5 concludes this study. Associated references are attached in the References section.

Table 1. Comparison of author's contribution.

Author(s)	Model Type	Business Policy	Unreliability	RFID
Dong and Xu [1]	stochastic	VMI	NA	NA
Guo et al. [2]	stochastic	macro prediction market	NA	NA
Mukhopadhyay et al. [3]	deterministic	mixed channel	information	NA
Yan and Pei [4]	deterministic	mixed channel	information	NA
Xiao and Xu [5]	deterministic	VMI	NA	NA
Sarkar [6]	stochastic	production model	reliable	NA
Guchhait et al. [10]	deterministic	traditional	NA	NA
Majumder et al. [11]	deterministic	traditional	NA	NA
Gallego and Moon [15]	stochastic (DF)	inventory model	NA	NA
Scarf [18]	stochastic (DF)	inventory model	NA	NA
Sarkar et al. [19]	stochastic (DF)	CP	NA	NA
Moon et al. [21]	stochastic (DF)	inventory model	NA	NA
Tiwari et al. [23]	deterministic	SCM	NA	NA
Meguerdicihian et al. [25]	networking	NA	NA	sensing
Zhang and Hou [26]	networking	NA	NA	sensing
Hefeeda and Ahmadi [27]	networking	NA	NA	coverage
Dias et al. [28]	survey	SCM	NA	survey
Sarac et al. [29]	value chain	survey	NA	survey
Kim and Glock [30]	stochastic	closed-loop	NA	tracking
Shin et al. [37]	stochastic (DF)	inventory	NA	NA
This model	stochastic (DF)	VMI	information	distance and readers

2. Problem Definition, Notation, and Assumptions

This section describes the problem definition for this study. Associated assumptions and notation are given here.

2.1. Problem Definition

A two-echelon supply chain model is considered under the newsvendor framework where participants are in a VMI contract. The inventory of the whole system is controlled by the manufacturer. Controlling the inventory manually by human labor is a time consuming task, as the manufacturer takes full responsibility of the full business of all buyers. To do this, the manufacturer installs smart RFID technology. The number of RFID readers is needed by the manufacturer such that the inventory can be controlled in a proper way within a minimum time duration. The number of readers depends on the sensing distance between two readers. Thus, the distance between readers is optimized for RFID investment. Buyers are not reliable with respect to the manufacturer's business. Buyers provide services to the customers, and therefore an unreliable SCM is formed as a single-manufacturer multi-buyer. The goal of the newsvendor model is to maximize profit for the buyer without incurring any storage or redundancy costs. However, the buyer is unable to decide on the optimum order quantity, where there should not be any understock or overstock costs. For that, the manufacturer takes the full responsibility of the buyers to for profits through the VMI strategy. Even though the manufacturer tries their best to help the buyer, the buyer is unreliable in nature and may provide wrong information regarding the demand to manufacturer. To mitigate this matter, the RFID technology is installed allowing the manufacturer to obtain more profit.

2.2. Notation

The following notation (Table 2) is used in the present study.

Table 2. Notation in this study.

Index	
i	number of buyers i, $i = 1, 2, ..., n$

Decision variables	description
δ_i	service by buyer i
q_i	order quantity of buyer i per cycle (units/cycle), $Q = \sum_{i=1}^{n} q_i$
d	distance between two RFID readers

Parameters	description
p_i	selling price of buyer i per unit under RFID effect ($/unit)
d_i	demand of buyer i per cycle (unit/cycle)
μ_i	mean value of demand d_i, $\mu = \sum_{i=1}^{n} \mu_i$
σ_i	standard deviation
l, b	length and breadth of the search area (m)
S_t	transmission radius of Type 1 reader (m)
S_s	sensing radius of Type 1 reader (m)
ρ	decay parameter for sensing
c_1, c_2	costs of Type 1 and Type 2 reader per unit ($/unit)
λ	maximum threshold value of Type 1 reader
θ	threshold parameter $(0 < \theta < 1)$
w	purchasing/wholesale cost per unit under the RFID effect ($/unit)
π_m	goodwill lost cost of manufacturer per unit with RFID ($/unit)
π_{ri}	buyer i's goodwill lost cost per unit under consideration of RFID ($/unit)
η_i	service investment of buyer i ($)
δ_i	service by buyer i
ζ_i	customer satisfaction cost of buyer i ($/unit)
h_m	holding cost of manufacturer under RFID effect ($/unit/unit time)
h_{ri}	buyer i's holding cost with RFID ($/unit/unit time)

Others	description
$f(d)$	cost of RFID per cycle ($/cycle)
$E(\cdot)$	expected value
ETP	expected total profit of the coordinate case per cycle ($/cycle)
ETP_r	buyer's expected total profit per cycle ($/cycle)
ETP_m	manufacturer's expected total profit per cycle ($/cycle)

2.3. Assumptions

The following assumptions are used for this model.

1. A two-echelon SCM is considered for a single-type of electronic products, where the inventory is managed by a manufacturer through a VMI contract. To ensure the profit of the buyers, a revenue sharing policy for coordination case is used by the manufacturer. The finished products are sent to the n buyers.

2. Buyers are not reliable enough and they are not sharing data to the manufacturer. It forms an information asymmetry in the business system. The manufacturer losses some information about market and installs the RFID system to solve the unreliability issue.

3. As VMI recommends that the supreme controlling authority is the manufacturer and the manufacturer decides to use RFID technology for controlling the unreliability issues. Hence, the manufacturer decides the whole deployment for the design of installing RFID reader, which can be done by the third-party. As the manufacturer cannot reach to the retailer's place in each and every moment, the technology will support to solve the issue of the unreliability. Those support will be taken from the third-party by investing some fixed cost. That fixed cost is inserted within the cost of Type 1 and Type 2 reader. Therefore, the RFID reader deployment cannot be specified within the modelling part of the manufacturer. However, the design of RFID reader can be added for the entering gate or any other place, but it depends on the third-party who is dealing with the whole area for covering the RFID. Therefore, through VMI, it is not the responsibility for the manufacturer to check the design for the installed RFID readers as this is a paid service from the third-party. Two types of reader are used to give a complete coverage of the search area. The total search area is divided into subareas and each subarea is covered by Type 1 readers, based on a disk sensing model. Each subarea is again divided into small search areas that are covered by Type 2 readers, based on an exponential coverage protocol. The frequency range of the readers is measured for usual road transport.

4. It may not be possible that the demand pattern always follows some distribution function. As data are random, it is assumed that the market demand is uncertain and does not follow any particular type of distribution. The known mean is μ_i and the standard deviation is σ_i (Shin et al. [37]).

5. The planning horizon is [0,T] and the lead time is negligible.

3. Mathematical Modelling

A VMI contract policy for the electronic industry is discussed for a single-manufacturer and multi-buyer newsvendor model. The optimum number of RFID readers, which can cover the optimized distance, can provide maximum profit to the supply chain for a long time. As implementation of RFID requires a huge investment, a reasonable demand rate is expected for the manufacturer. However, the market demand (d_i) for buyer i is uncertain, it cannot be predicted. The demand (d_i) for buyer i can be represented by a random variable where the mean is (μ_i) and (σ_i) is the standard deviation which both are known. As d_i does not follow any specific distribution function, this problem can be solved using the DF approach. The surplus and shortage amount can be calculated by the lemma of Gallego and Moon [15]. The required surplus amount is

$$E(q_i - d_i)^+ \leq \frac{1}{2}\left[\sqrt{\sigma_i^2(\mu_i - q_i)^2} + (q_i - \mu_i)\right], \mu_i < q_i \tag{1}$$

and the shortage amount is

$$E(d_i - q_i)^+ = \frac{1}{2}\left[\sqrt{\sigma_i^2(\mu_i - q_i)^2} + (\mu_i - q_i)\right], q_i < \mu_i, \text{ for } F \in \mathbb{F}. \tag{2}$$

3.1. Structure of the Proposed RFID System

The total search area is covered by the RFID tracking system. The cost regarding RFID depends on the number of readers. The concept of VMI is that the manufacturer will manage the whole inventory of the retailer as some unreliable issues are coming from retailer's side. To overcome these issues, the manufacturer introduces RFID technology with the minimum investment for it. Therefore, within the total area of the retailer, how much inventory are these, that should be verified by RFID readers. Therefore, it is not essential to use always powerful RFID readers like as Type 1 or similarly it is not recommended also that always low powerful Type 2 reader should be used. Hence, an optimization is needed to optimize the optimum number of Type 1 and Type 2 reader within the whole area. That is why, this model recommended two types of RFID reader for the sensing and coverage model: the disk sensing model and the exponential coverage model. The entire search area is divided into subareas which are covered by the Type 1 reader. This Type 1 reader has a higher sensing power for coverage, which uses the disk sensing model. Each subarea is divided into subareas those are covered by two Type 2 readers. Type 2 readers have low sensing power and use an exponential coverage protocol system. The connectivity between the sensing radius and transmitting radius is given by the condition $2S_s \leq S_t$ (for instance, see Zhang and Hou [26]).

If l_1 is the length and b_1 is the breadth of each subdivided area, then from the properties of right-angled triangle (Figure 1), it is follows that

$$l_1^2 + b_1^2 = c^2, \text{ i.e., } l_1^2 + b_1^2 = 4S_t^2, \text{ i.e., } S_t = \sqrt{\frac{l_1^2 + b_1^2}{4}}.$$

For each square foot area, $l_1 = b_1$, which implies that

$$S_t = \frac{l_1}{\sqrt{2}}, \text{ i.e., } l_1 = \sqrt{2}S_t.$$

Therefore, if the length and the breadth of the total search area are l and b, respectively, the total number of Type 1 reader is $\left\lceil \dfrac{l}{\sqrt{2}S_t} \right\rceil \left\lceil \dfrac{b}{\sqrt{2}S_t} \right\rceil$.

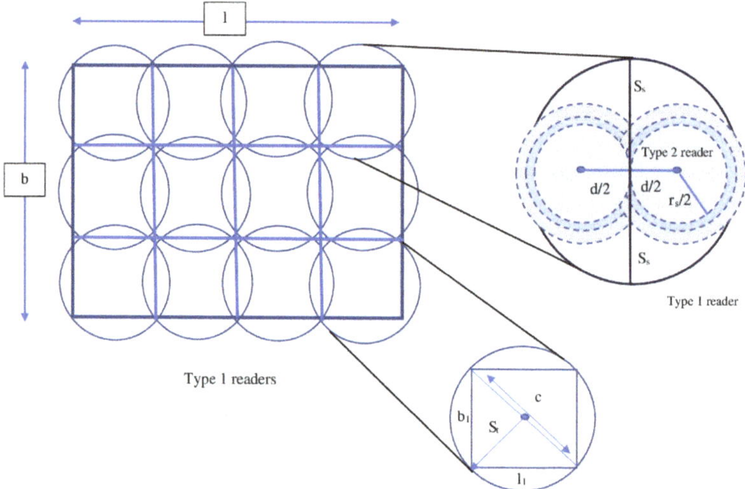

Figure 1. Execution of Type 1 and Type 2 readers for a search area.

Now, each subdivided area of sensing radius S_s is divided into two areas with sensing radius r_s. The maximum distance between two Type 2 readers is d, i.e., $r_s = \dfrac{d}{2}$. Now, from the exponential coverage protocol (Hefeeda and Ahmadi [27]), the maximum distance d between two Type 2 readers is smaller than $\sqrt{3}\left(\lambda - \dfrac{Log\left[1 - \sqrt[3]{1-\theta}\right]}{\rho}\right)$, i.e.,

$$d \leq \sqrt{3}\left(\lambda - \frac{Log\left[1 - \sqrt[3]{1-\theta}\right]}{\rho}\right).$$

The area of the circle for sensing radius S_s is πS_s^2. The area of circle of sensing radius r_s is $\pi r_s^2 = \dfrac{\pi d^2}{4}$. Therefore, the number of Type 2 readers for each subdivided area of Type 1 reader is

$$\left\lceil \frac{\pi S_s^2}{\frac{\pi d^2}{4}} \right\rceil = \left\lceil \frac{4S_s^2}{d^2} \right\rceil.$$ Hence, the total number of Type 2 readers for all Type 1 readers is

$$\left\lceil \frac{4S_s^2}{d^2} \right\rceil \left\lceil \frac{l}{\sqrt{2}S_t} \right\rceil \left\lceil \frac{b}{\sqrt{2}S_t} \right\rceil.$$

3.2. Manufacturer's Model

In reality, it is not always the case that all buyers are reliable enough to share all information to the manufacturer. To prevent the piracy on the inventory inaccuracy, the manufacturer invests in RFID technology even though this may reduce the profit margins. However, there may be long-term benefits compensate the shrinkage of inventory. Still, there may be some ambiguity regarding information due to information asymmetry.

3.2.1. RFID Cost

The total area is covered by $\left\lceil \dfrac{l}{\sqrt{2}S_t} \right\rceil \left\lceil \dfrac{b}{\sqrt{2}S_t} \right\rceil$ Type 1 readers. This area is again subdivided and is covered by Type 2 readers. If c_1 is the cost of each Type 1 reader and c_2 is for each Type 2. A fixed cost is included within c_1 and c_2 which the manufacturer pays as an investment. Then the required RFID cost is given by

$$f(d) = c_1 \left\lceil \frac{l}{\sqrt{2}S_t} \right\rceil \left\lceil \frac{b}{\sqrt{2}S_t} \right\rceil + c_2 \left\lceil \frac{4S_s^2}{d^2} \right\rceil \left\lceil \frac{l}{\sqrt{2}S_t} \right\rceil \left\lceil \frac{b}{\sqrt{2}S_t} \right\rceil$$

subject to the conditions

$$d \leq \sqrt{3}\left(\lambda - \frac{Log\left[1 - \sqrt[3]{1-\theta}\right]}{\rho}\right)$$

$$S_t \geq 2S_s.$$

(3)

Therefore, the RFID cost per cycle is $\dfrac{f(d)\mu}{Q}$, where $D = \sum_i d_i$ and $\mu = E(D)$.

3.2.2. Production Cost and Wholesale Price

If the manufacturer produces a lot size Q per cycle then the production cost of those products is given by cQ. When the manufacturer sells products as a wholesale price w per unit, then the wholesale price is given by wQ.

3.2.3. Holding Cost

The situation of holding products is created when the demand (d_i) is less than the ordered quantity wq_i. If h_{r_i} is the unit holding cost of buyer i, the holding cost is $h_{r_i}E(q_i - d_i)^+$, $d_i \leq q_i$. As the manufacturer pays both the holding cost of the buyers and the manufacturer (h_m), the total holding cost of the manufacturer is given by $\sum_i E(q_i - d_i)^+ \left(h_{r_i} + h_m \right)$, $d_i \leq q_i$.

3.2.4. Goodwill Lost Cost

A goodwill lost cost (π_m) is allowed since the manufacturer takes the responsibility for the products for the whole supply chain, where shortage affects the goodwill of manufacturer. The cost expression for goodwill loss is given by $\sum_i \pi_m E(q_i - d_i)^+$, $q_i < d_i$.

Including the RFID cost, the expected total profit of the manufacturer is given by the following expression

$$
\begin{aligned}
ETP_m(q_i, d) = (w - c)Q - \sum_i \frac{1}{2}(h_{r_i} + h_m)\left[\sqrt{\sigma_i^2 + (\mu_i - q_i)^2} + (q_i - d_i) \right] - \frac{\pi_m}{2}\sum \left[\sqrt{\sigma_i^2 + (\mu_i - q_i)^2} \right. \\
\left. + (d_i - q_i) \right] - \frac{f(d)\mu}{Q}
\end{aligned}
\tag{4}
$$

subject to the conditions

$$
d \leq \sqrt{3}\left(\lambda - \frac{Log\left[1 - \sqrt[3]{1 - \theta}\right]}{\rho} \right)
$$

$$
S_t \geq 2S_s.
$$

3.3. Buyer's Model

Buyers are unreliable resulting in information asymmetry. As this is a dependent business policy and the manufacturer is responsible for both inventory supervision and holding inventory for buyers, all information should be known to the manufacturer. However, today's business systems are very complex and buyers are unreliable at sharing information their own business strategy. Buyer i buys the electronic products from the manufacturer and sells them in the market. To increase market demand, the buyers provide facilities to the customers without telling the manufacturer meaning that an unreliable supply chain system is formulated.

3.3.1. Revenue

p_i is the unit selling price of the electronic products. Now, two types of situation may arise, where the demand (d_i) is more than the ordered quantity (q_i) or vice-versa. Then the selling price can be found as

$$
\begin{cases} p_i d_i & d_i \leq q_i \\ p_i q_i & q_i < d_i. \end{cases}
$$

3.3.2. Purchasing Cost and Goodwill Lost Cost

If w is the unit purchasing cost for the ordered quantity q_i, then the purchasing cost is given by wq_i. When the reverse situation arises i.e., the demand is more than the ordered quantity, backordering occurs, meaning that some goodwill for buyer i is lost. The goodwill lost cost is given by $\pi_{r_i}E(d_i - q_i)^+$, $q_i < d_i$ where π_{r_i} is the unit goodwill lost cost of buyer i.

3.3.3. Service Cost

The buyer provides extra services (δ_i) to attract customers, which requires extra money to invests (η_i). Customer satisfaction is involved in this situation. If the service is appropriate and satisfactory to the customers, the purpose of giving service is fulfilled. On the other hand, if some customers are not happy with the given service or buyer is incapable to give the standard service, customers may not want to buy products from that buyer as customers have multiple choices to buy the same product. This is the opposite situation of the service, i.e., $(1 - \delta_i)$. Thus, it creates some monetary loss to the buyer, which is indicated as customer satisfaction cost. It has the inverse relation with the provided service. Whenever the service increases, the customer satisfaction increases and thus the cost $(1 - \delta_i)^2\zeta_i$, related to the customer satisfaction decreases. If η_i is the service cost and ζ_i is the customer satisfaction cost, the relative cost is given by $\dfrac{\eta_i\delta_i^2}{2} + (1 - \delta_i)^2\zeta_i$. Therefore, the expected total profit of buyer i is

$$ETP_{ri} = \begin{cases} p_i\mu_i - wq_i - \dfrac{\eta_i\delta_i^2}{2} - (1 - \delta_i)^2\zeta_i, & d_i \leq q_i \\[2ex] p_iq_i - wq_i - \pi_{ri}E(d_i - q_i)^+ - \dfrac{\eta_i\delta_i^2}{2} - (1 - \delta_i)^2\zeta_i, & q_i < d_i. \end{cases} \tag{5}$$

The total profit of buyer is given by

$$ETP_r(q_i, \delta_i) = \sum p_i(\mu_i + q_i) - wQ - \frac{1}{2}\sum \pi_{ri}\left(\sqrt{\sigma i^2 + (\mu_i - d_i)^2} + \mu_i - q_i\right) - \sum \frac{\eta_i\delta_i^2}{2} - \sum(1 - \delta_i^2)\zeta_i. \tag{6}$$

Therefore, the expected total profit of SCM is given by

$$ETP(q_i, \delta_i, d) = \sum_i p_i(\mu_i + q_i) - cQ - \frac{1}{2}\sum_i(h_{ri} + h_m)\left[\sqrt{\sigma_i^2 + (\mu_i - q_i)^2} + (q_i - \mu_i)\right] - \frac{1}{2}\sum_i(\pi_{ri} + \pi_m)$$

$$\left[\sqrt{\sigma_i^2 + (\mu_i - q_i)^2} + (\mu_i - q_i)\right] - \sum_i \frac{\eta_i\delta_i^2}{2} - \sum_i(1 - \delta_i)^2\zeta_i - \frac{f(d)\mu}{Q} \tag{7}$$

subject to the conditions

$$d \leq \sqrt{3}\left(\lambda - \frac{Log\left[1 - \sqrt[3]{1 - \theta}\right]}{\rho}\right)$$

$$S_t \geq 2S_s.$$

3.4. Solution Methodology

The solution is found for both the coordination and non-coordination cases. The model is solved by using classical optimization techniques. The necessary conditions give the optimum results for the corresponding decision variables and the sufficient conditions give the stability of the solutions. The constraint function of the manufacturer is modified and transferred into an unconstrained function using the Kuhn-Tucker (KT) method. The modified function is given by

$$LETP_m = (w - c)Q - \sum_i(h_{ri} + h_m)E(q_i - d_i)^+ - \pi_m\sum_i E(d_i - q_i)^+ - \frac{\mu}{Q}\left(c_1\left\lceil\frac{l}{\sqrt{2S_t}}\right\rceil\left\lceil\frac{b}{\sqrt{2S_t}}\right\rceil\right.$$

$$\left. + c_2\left\lceil\frac{4S_s^2}{d^2}\right\rceil\left\lceil\frac{l}{\sqrt{2S_t}}\right\rceil\left\lceil\frac{b}{\sqrt{2S_t}}\right\rceil\right) + \lambda_1\left[\sqrt{3}\left(\lambda - \frac{log(1 - \sqrt[3]{1 - \theta})}{\rho}\right) - d\right] + \lambda_2(2S_s - S_t). \tag{8}$$

Then, the total profit of the entire SCM is

$$LETP = \sum_i p_i(\mu_i + q_i) - cQ - \sum_i (h_{r_i} + h_m)E(q_i - d_i)^+ - \sum_i (\pi_m + \pi_{r_i})E(d_i - q_i)^+ - \frac{\mu}{Q}\left(c_1\left\lceil\frac{l}{\sqrt{2}S_t}\right\rceil\right.$$
$$\left.\left\lceil\frac{b}{\sqrt{2}S_t}\right\rceil + c_2\left\lceil\frac{4S_s^2}{d^2}\right\rceil\left\lceil\frac{l}{\sqrt{2}S_t}\right\rceil\left\lceil\frac{b}{\sqrt{2}S_t}\right\rceil\right) + \lambda_1\left[\sqrt{3}\left(\lambda - \frac{\log(1 - \sqrt[3]{1-\theta})}{\rho}\right) - d\right] \tag{9}$$
$$+ \lambda_2(2S_s - S_t).$$

3.4.1. Non-Coordination Case

The necessary conditions of optimization provide the optimum values of the decision variable for the manufacturer. The value of the decision variable (q_i) is computed by

$$\frac{\partial LETP_m}{\partial q_i} = (w - c) + \frac{\mu_i - q_i}{2\sqrt{\sigma_i^2 + (\mu_i - q_i)^2}}(h_{r_i} + h_m + \pi_m) - \frac{1}{2}(h_{r_i} + h_m - \pi_m) + \frac{f(d)\mu}{Q^2} = 0,$$
$$\text{i.e., } q_i = \mu_i \pm \frac{\sigma_i \Gamma_1}{\sqrt{1 - \Gamma_1^2}}, \tag{10}$$

where

$$\Gamma_1 = \frac{2\left(w - c + \dfrac{f(d)\mu}{Q^2}\right) - h_{r_i} - h_m + \pi_m}{h_{r_i} + h_m + \pi_m}.$$

Therefore, $Q = \sum_i q_i$ (from Equation (10)) gives the optimum order quantity for the manufacturer. The optimum distance is given by the following value of d.

$$\frac{\partial LETP_m}{\partial d} = -\lambda_1 + \frac{8c_2S_s^2}{d^3}\left\lceil\frac{l}{\sqrt{2}S_t}\right\rceil\left\lceil\frac{b}{\sqrt{2}S_t}\right\rceil\frac{\mu}{Q} = 0,$$
$$\lambda_1 = \frac{8\mu c_2 S_s^2 \left\lceil\dfrac{l}{\sqrt{2}S_t}\right\rceil\left\lceil\dfrac{b}{\sqrt{2}S_t}\right\rceil}{Qd^3\left(\lambda - \dfrac{\log(1 - \sqrt[3]{1-\theta})}{\rho}\right)}. \tag{11}$$
$$\frac{\partial LETP_m}{\partial \lambda_1} = \sqrt{3}\left(\lambda - \frac{\log(1 - \sqrt[3]{1-\theta})}{\rho}\right) - d = 0,$$
$$\text{i.e., } d = \sqrt{3}\left(\lambda - \frac{\log(1 - \sqrt[3]{1-\theta})}{\rho}\right).$$

Equation (11) provides the optimum distance between readers. The sufficient conditions prove that the above results represent global solutions.

$$\frac{\partial^2 LETP_m}{\partial q_i^2} = -\frac{1}{2}\frac{(h_{r_i} + h_m + \pi_m)\sigma_i^2}{\left(\sigma_i^2 + (\mu_i - q_i)^2\right)^{\frac{3}{2}}} - \frac{2f(d)\mu}{Q^3} < 0,$$
$$\frac{\partial^2 LETP_m}{\partial d^2} = -\frac{24c_2 S_s^2 \mu}{Qd^4}\left\lceil\frac{l}{\sqrt{2}S_t}\right\rceil\left\lceil\frac{b}{\sqrt{2}S_t}\right\rceil < 0,$$
$$\frac{\partial^2 LETP_m}{\partial q_i d_i} = -\frac{\mu\lambda_1}{Q^2} < 0,$$

and
$$\begin{pmatrix} \dfrac{\partial LETP_m}{\partial q_i^2} & \dfrac{\partial LETP_m}{\partial q_i d_i} \\[2mm] \dfrac{\partial LETP_m}{\partial d_i q_i} & \dfrac{\partial LETP_m}{\partial d_i^2} \end{pmatrix} = \left(\frac{1}{2} \frac{(h_{r_i} + h_m + \pi_m)\sigma_i^2}{\left(\sigma_i^2 + (\mu_i - q_i)^2\right)^{\frac{3}{2}}} + \frac{2f(d)\mu}{Q^3} \right) \left(\frac{24c_2 S_s^2 \mu}{Qd^4} \left[\frac{l}{\sqrt{2S_t}} \right] \left[\frac{b}{\sqrt{2S_t}} \right] \right)$$

$$+ \frac{\mu \lambda_1}{Q^2} > 0.$$

All criterion for the sufficient conditions of a Hessian matrix are satisfied proving the stability of the optimum solution. Therefore, the values of the decision variables are the optimum for the manufacturer.

The optimum values of the decision variables for the buyer are given by the following necessary conditions for optimization.

$$\frac{\partial ETP_r}{\partial q_i} = 0,$$

$$\text{i.e., } q_i = \mu_i \pm \frac{\sigma_i \Gamma_2}{\sqrt{1 - \Gamma_2^2}}, \tag{12}$$

where

$$\Gamma_2 = \frac{2(p_i - w) + \pi_{r_i}}{\pi_{r_i}}.$$

The optimum order quantity for the buyer i is given by Equation (12). Equation (13) gives the optimum service provided by the buyer i to customers.

$$\frac{\partial ETP_r}{\partial \delta_i} = -\eta_i \delta_i + 2\zeta_i(1 - \delta_i) = 0,$$

$$\text{i.e., } \delta_i = \frac{2\zeta_i}{\eta_i + 2\zeta_i}. \tag{13}$$

This sufficient condition proves the global nature of the solution.

$$\frac{\partial ETP_r}{\partial q_i^2} = - \frac{\pi_{r_i} \sigma_i^2}{2\left(\sigma_i^2 + (\mu_i - q_i)^2\right)^{\frac{3}{2}}} < 0,$$

$$\frac{\partial ETP_r}{\partial \delta_i^2} = -\eta_i - 2\zeta_i < 0,$$

$$\frac{\partial^2 ETP_r}{\partial \delta_i q_i} = 0,$$

$$\text{i.e., } \begin{pmatrix} \dfrac{\partial ETP_r}{\partial q_i^2} & \dfrac{\partial ETP_r}{\partial q_i \delta_i} \\[2mm] \dfrac{\partial ETP_r}{\partial \delta_i q_i} & \dfrac{\partial ETP_r}{\partial \delta_i^2} \end{pmatrix} = \frac{\pi_{r_i} \sigma_i^2 (\eta_i + 2\zeta_i)}{2\left(\sigma_i^2 + (\mu_i - q_i)^2\right)^{\frac{3}{2}}} > 0.$$

The Algorithm 1 is developed to find the numerical results from theory. The following steps help to solve the model numerically.

Algorithm 1:	
Step 1	Input all values of all relevant parameters. Set the value of i.
Step 2	Set the initial values of q_i for manufacturer and buyers.
Step 3	Write down the values of q_i from the Equation (10) and d from the Equation (11) for manufacturer.
	For buyers, the values of q_i and δ_i are given by Equation (12) and (13), respectively.
Step 4	Find the value of q_i, δ_i, and d using the values from Step 1 and Step 2.
Step 4.a	If $q_i \geq q_{i+1}$ and $\delta_i \geq \delta_{i+1}$, then terminate the process. The optimum values are obtained as $q_i{}^*$, $\delta_i{}^*$, and d^*.
Step 4.b	Else if $q_i < q_{i+1}$ and $\delta_i < \delta_{i+1}$, go to Step 4.
Step 4.c	Increment of i as $i = i + 1$.
Step 5	Stop.

3.4.2. Coordination Case

The results for the joint profit of the entire SCM are given by the following necessary conditions.

$$\frac{\partial ETP}{\partial q_i} = p_i - c + \frac{f(d)\mu}{Q^2} + \frac{\mu_i - q_i}{2\sqrt{\sigma_1^2 + (\mu_i - q_i)^2}}(h_{r_i} + h_m + \pi_{r_i} + \pi_m) = 0$$

$$-\frac{1}{2}(h_{r_i} + h_m - \pi_{r_i} - \pi_m),\tag{14}$$

$$\text{i.e., } q_i = \mu_i \pm \frac{\sigma_i \Gamma_3}{\sqrt{1 - \Gamma_3^2}},$$

where

$$\Gamma_3 = \frac{2\left(p_i - c + \dfrac{f(d)\mu}{Q^2}\right) - h_{r_i} - h_m + \pi_{r_i} + \pi_m}{h_{r_i} + h_m + \pi_{r_i} + \pi_m},$$

$$\text{and } \frac{\partial ETP}{\partial \delta_i} = -\eta_i \delta_i - 2(1 - \delta_i)\zeta_i(-1) = 0,\tag{15}$$

$$\text{i.e., } \delta_i = \frac{2\zeta_i}{\eta_i + 2\zeta_i}.$$

The optimum order quantity is given by Equation (14) and service is given by Equation (15). Using the necessary conditions, one has

$$\frac{\partial LETP}{\partial \lambda_1} = \sqrt{3}\left(\lambda - \frac{\log(1 - \sqrt[3]{1 - \theta})}{\rho}\right) - d = 0,$$

$$\text{i.e., } d = \sqrt{3}\left(\lambda - \frac{\log(1 - \sqrt[3]{1 - \theta})}{\rho}\right),$$

$$\frac{\partial LETP}{\partial d} = -\lambda_1 + \frac{8c_2 S_s^2}{d^3}\left[\frac{l}{\sqrt{2S_t}}\right]\left[\frac{b}{\sqrt{2S_t}}\right]\frac{\mu}{Q} = 0,\tag{16}$$

$$\text{i.e., } \lambda_1 = \frac{8c_2 S_s^2 \mu\left[\dfrac{l}{\sqrt{2S_t}}\right]\left[\dfrac{b}{\sqrt{2S_t}}\right]}{Qd^3}.$$

Equation (16) gives the optimum distance between two RFID readers. From the sufficient conditions, it can be concluded that since the second order derivatives are negative definite and the values of the Hessian matrix alternate, the required values of the decision variables are global.

$$\frac{\partial^2 ETP}{\partial q_i^2} = -\frac{1}{2}\frac{(h_{r_i} + h_m + \pi_{r_i} + \pi_m)\sigma_i^2}{\left(\sigma_i^2 + (\mu_i - q_i)^2\right)^{\frac{3}{2}}} - \frac{2f(d)\mu}{Q^3} < 0,$$

$$\frac{\partial^2 ETP}{\partial \delta_i^2} = -\eta_i^2 - 2\zeta_i < 0,$$

$$\frac{\partial^2 ETP}{\partial d^2} = -\frac{24c_2 S_s^2}{d^4}\left[\frac{l}{\sqrt{2S_t}}\right]\left[\frac{b}{\sqrt{2S_t}}\right]\frac{\mu}{Q} < 0.$$

Now, the calculation of the principal minors gives

$$H_1 = \begin{pmatrix} \dfrac{\partial^2 ETP}{\partial q_i^2} & \dfrac{\partial^2 ETP}{\partial q_i \partial \delta_i} \\[2mm] \dfrac{\partial^2 ETP}{\partial \delta_i \partial q_i} & \dfrac{\partial^2 ETP}{\partial \delta_i^2} \end{pmatrix} = \left[\frac{\left(h_{r_i} + h_m + \pi_{r_i} + \pi_m\right)\sigma_i^2}{2(\sigma_i^2 + (\mu_i - q_i)^2)^{\frac{3}{2}}} + \frac{2f(d)\mu}{Q^3}\right]\left[\eta_i^2 + 2\zeta_i\right] > 0,$$

$$H_2 = \begin{vmatrix} \dfrac{\partial^2 ETP}{\partial q_i^2} & \dfrac{\partial^2 ETP}{\partial q_i \partial \delta_i} & \dfrac{\partial^2 ETP}{\partial q_i \partial d} \\[2mm] \dfrac{\partial^2 ETP}{\partial \delta_i \partial q_i} & \dfrac{\partial^2 ETP}{\partial \delta_i^2} & \dfrac{\partial^2 ETP}{\partial \delta_i \partial d} \\[2mm] \dfrac{\partial^2 ETP}{\partial d \partial q_i} & \dfrac{\partial^2 ETP}{\partial d \partial \delta_i} & \dfrac{\partial^2 ETP}{\partial d^2} \end{vmatrix} = -\,(\eta_i + 2\zeta_i)\left[\left(\frac{(h_{r_i} + h_m + \pi_{r_i} + \pi_m)\sigma_i^2}{2\left(\sigma_i^2 + (\mu_i - q_i)^2\right)^{\frac{3}{2}}} + \frac{2f(d)\mu}{Q^3}\right)\right.$$

$$\left.\frac{24c_2 S_s^2}{d^4}\left[\frac{l}{\sqrt{2S_t}}\right]\left[\frac{b}{\sqrt{2S_t}}\right] - \left\{\frac{8\mu c_2 S_s^2}{Q^2 d^3}\left[\frac{l}{\sqrt{2S_t}}\right]\left[\frac{b}{\sqrt{2S_t}}\right]\right\}^2\right] < 0.$$

Lemma 1. *The values of the coordinated case are optimum if the Hessian matrix of third order (H_2) has a value less than zero, i.e., $H_2 < 0$. The required criteria is given by*

$$\left(\frac{\dfrac{(h_{r_i} + h_m + \pi_{r_i} + \pi_m)Q^2\sigma_i^2}{3} + \dfrac{2f(d)}{Q}}{2\mu\left(\sigma_i^2 + (\mu_i - q_i)^2\right)^{\frac{3}{2}}}\right)\frac{3}{d} > \left[\frac{l}{\sqrt{2S_t}}\right]\left[\frac{b}{\sqrt{2S_t}}\right].$$

This Algorithm 2 helps to find the numerical results. The following steps are required as follows.

Algorithm 2:

Step 1	Input all parametric values. Set the value of i.
Step 2	Set the initial values of q_i.
Step 3	Write down the values of q_i from the Equation (14), δ_i from Equation (15), and d from the Equation (16).
Step 4	Find the value of q_i, δ_i, and d using the values from Step 1 and Step 2.
Step 4.a	If $q_i \geq q_{i+1}$ and $\delta_i \geq \delta_{i+1}$, then terminate the process. The optimum values are obtained as $q_i{}^*$, $\delta_i{}^*$, and d^*.
Step 4.b	Else $q_i < q_{i+1}$ and $\delta_i < \delta_{i+1}$, go to Step 4.
Step 4.c	i as $i = i + 1$.
Step 5	Stop.

3.5. Revenue Sharing (RS)

Instead of a traditional policy, the manufacturer and multiple buyers are involved in a VMI contract. It is the manufacturer's role to support buyer such that the buyers so that they do not face losses due to the contract. Thus, a revenue sharing policy for coordinated supply chain is incurred by the manufacturer. If α $(0 < \alpha < 1)$ is the sharable revenue by the manufacturer from the total profit, then the sharing mechanism for the coordinated case is αETP. The rest of the profit is accounted for by the manufacturer as he invests more in the business.

4. Numerical Experiment

Numerical experiments are used to validate this study numerically. Supportive data are taken from Sarkar et al. [19] and Xiao and Xu [5]. Some data are taken from an industry visit in West Bengal, India, which justifies the industry using this policy for their business. Two examples are provided here.

Example 1. *Table 3 gives all input values of the related parameters and Table 4 provides the optimum results for Example 1.*

Table 3. Input values of the parameters for Example 1.

Parameters	Values	Parameters	Values	Parameters	Values
n	2	(π_{r_1}, π_{r_2})	\$(10, 11)/unit	π_m	\$20/unit
(p_1, p_2)	\$(33, 34)/unit	(σ_1, σ_2)	(200, 202)	h_m	\$0.33c/unit/year
(μ_1, μ_2)	(200, 210)unit/year	(η_1, η_2)	\$(2, 3)	c	\$19/unit
(h_{r_1}, h_{r_2})	\$(0.21c, 0.23c)/unit/year	(ζ_1, ζ_2)	(0.7, 0.8)	w	\$30/unit
(c_1, c_2)	\$(140, 90)/reader	ρ	0.032	λ	10
l	200 m	b	200 m	S_s	50 m

Table 4. Optimum results from the numerical analysis for Example 1.

Coordination Case					
Variables	**Optimum Values**	**Number of Readers**	**Optimum Values**	**Results**	**Optimum Values**
$(q_1{}^*, q_2{}^*)$	(201.05, 210.82) unit	Type 1	4	d^*	85.56 m
$(s_1{}^*, s_2{}^*)$	(0.41, 0.35)	Type 2	8	ETP	\$19,783.46/cycle
				RFID cost	\$293.48/cycle
Non-Coordination Case					
Manufacturer					
$(q_1{}^*, q_2{}^*)$	(212.86, 251.57) unit	Type 1	4	d^*	85.56 m
RFID cost	\$82.62/cycle	Type 2	8	ETP_m	\$4431.10/cycle
Buyers					
$(q_1{}^*, q_2{}^*)$	(200, 210) unit	(s_1, s_2)	(0.41, 0.35)	ETP_r	\$15,179.07/cycle

Therefore, \$19,783.46 is the total profit of the entire supply chain. After gaining profit from the business, the manufacturer shares the revenue $\alpha = 0.45$ (Xiao and Xu [5]) of the total profit with the buyers, i.e., the manufacturer shares \$8902.56 with the two buyers. Thus, a (\$19,783.46 − \$8902.56) = \$10,880.90 profit is earned by the manufacturer from the VMI contract policy. The required number of Type 1 readers is 4 and the number of Type 2 readers is 8, which cover the total search area.

Example 2. *Table 5 gives all input values of the related parameters and Table 6 provides the optimum results for Example 2.*

Table 5. Input values of the parameters for Example 2.

Parameters	Values	Parameters	Values	Parameters	Values
n	2	(π_{r_1}, π_{r_2})	\$(6, 8)/unit	π_m	\$12/unit
(p_1, p_2)	\$(32, 30) /unit	(σ_1, σ_2)	(200, 202)	h_m	\$0.30c/unit/year
(μ_1, μ_2)	(190, 195) unit/year	(η_1, η_2)	\$(1.8, 1.5)	c	\$18/unit
(h_{r_1}, h_{r_2})	\$(0.18c, 0.19c)/unit/year	(ζ_1, ζ_2)	(0.6, 0.5)	w	\$27/unit
(c_1, c_2)	\$(138, 100)/reader	ρ	0.059	λ	12
l	210 m	b	190 m	S_s	45 m

Table 6. Optimum results from the numerical analysis for Example 2.

Coordination Case

Variables	Optimum Values	Number of Readers	Optimum Values	Results	Optimum Values
$(q_1{}^*, q_2{}^*)$	(190.07, 195.19) unit	Type 1	4	d^*	63.37 m
$(s_1{}^*, s_2{}^*)$	(0.40, 0.40)	Type 2	12	ETP	\$17,122.01/cycle
				RFID cost	\$434.65/cycle

Non-Coordination Case

Manufacturer

Variables	Optimum Values	Number of Readers	Optimum Values	Results	Optimum Values
$(q_1{}^*, q_2{}^*)$	(216.36, 277.90) unit	Type 1	4	d^*	63.37 m
RFID cost	\$82.45/cycle	Type 2	12	ETP_m	\$3439.95/cycle

Buyers

Variables	Optimum Values	Number of Readers	Optimum Values	Results	Optimum Values
$(q_1{}^*, q_2{}^*)$	(190, 195) unit	(s_1, s_2)	(0.40, 0.40)	ETP_r	\$13,464.34/cycle

\$17,122.01 is the total profit of the entire supply chain for Example 2. The manufacturer shares the revenue $\alpha = 0.45$ (Xiao and Xu [5]) of the total profit with the buyers, i.e., the manufacturer shares \$7704.90 with the two buyers for the coordination business policy. Thus, a ($17,122.01 − $7704.90) = \$9417.11 profit is earned by manufacturer from the VMI contract policy. The total search area is covered by 4 number of Type 1 readers and 12 number of Type 2 readers.

Comparative Study of the Coordination and Non-Coordination Cases

From Table 7, it is seen that, manufacturer and buyer's profit in the coordination case are higher than the non-coordination case for both of the examples. The results conclude that the coordination VMI is more beneficial for both business participants. It is seen that the coordination policy is beneficial for both the manufacturer and the total supply chain profit, whereas buyers get more profit in the non-coordination policy than then coordination case. The shared revenue to the buyers in the coordinated case is less than the profit earned from the non-coordination case. As in the non-coordination policy, buyers can move freely according to their surrounding phenomenon, but in the coordination policy, the joint profit for the entire supply chain is more important for a long-term business rather than an individual one. Even though the profit of buyers is less in the coordination case, they do not face any loss from the business. In both cases of coordination and non-coordination policy, the manufacturer needs same number of readers as the area of the manufacturer is fixed for both of the cases.

Table 7. Comparative study between the coordination and non-coordination cases.

	Example 1		Example 2	
Participant(s)	Coordination Case	Non-Coordination Case	Coordination Case	Non-Coordination Case
Manufacturer	$10,880.90	$4431.10	$9417.11	$3439.95
Buyers	$8902.56	$15,179.07	$7704.9	$13,464.34
SCM	$19,783.46	$19,610.17	$17,122.01	$16,904.29

5. Discussion

Service is provided to the customers by buyers. This extra service makes an effect to the customers of satisfaction that they are happy and satisfied after buying products from that buyer. Whenever the service level increases, the satisfaction increases.

The sensitivities of the cost parameters of Example 1 over the total profit are depicted in Table 8. It is found that the manufacturing cost c is the most profit sensitive parameter relative to the others. Positive percentage changes of the parameter are more sensitive than negative changes, i.e., profit loss will be more whenever the cost increases. For the holding cost of the manufacturer (h_m), whenever h_m decreases and increases, the total profit decreases and increases, respectively. Negative percentage changes of h_m result in a smaller q_i, which leads to an increased RFID cost, i.e., decreasing h_m increases the radio frequency cost per cycle. The holding cost of the buyers and the shortage costs of the manufacturer and buyers have the same type of positive and negative changes. The service investment of the buyers has the usual impact on total profit, where increasing the investment causes less profit and vice-versa.

Table 8. Sensitivity analysis of the key parameters of Example 1.

Parameters	Percentage Changes	Changes in Profit (%)	Parameters	Percentage Changes	Changes in Profit (%)
h_{r_1}	-20	0.008	h_{r_2}	-20	0.02
	-10	0.005		-10	0.011
	$+10$	-0.008		$+10$	-0.13
	$+20$	-0.02		$+20$	-0.03
h_m	-20	-2.52	π_m	-20	0.002
	-10	-2.41		-10	0.001
	$+10$	-0.04		$+10$	-0.002
	$+20$	-0.09		$+20$	-0.005
π_{r_1}	-20	0.002	π_{r_2}	-20	0.0008
	-10	0.001		-10	0.0003
	$+10$	-0.001		$+10$	-0.0001
	$+20$	-0.002		$+20$	-0.0003
η_1	-20	0.0002	η_2	-20	0.0002
	-10	0.00007		-10	0.00008
	$+10$	-0.0001		$+10$	-0.0001
	$+20$	-0.0002		$+20$	-0.0002
c	-20	4.75			
	-10	1.14			
	$+10$	-4.29			
	$+20$	-26.88			

6. Conclusions and Future Recommendations

The measurement of the distance between two RFID readers could lead an SCM towards sustainability, which not only helps to prevent inventory shrinkage, but also helps to collect used products via RFID tags and readers. The distance between two readers was optimized, and based on this an industry manager can decide how many readers are needed to cover the whole search area. Results confirmed that RFID could be profitable for a VMI contract. This business policy was shown to be beneficial for the entire supply chain for the coordinated case. Besides that, a non-coordinated business policy provided profit to both the manufacturer and the buyers. This study ensured that the manufacturer need not be worried about the installation of smart technology by themselves. The manufacturer was benefited from a third-party provider and can mitigate the problems of unreliability within the SCM. Implementation of an RFID system was beneficial for the electronics industry by reducing e-waste and reusing products and parts. However, this study did not consider the reuse of tags of used products, which can be an immediate extension for waste reduction. Within this study, it was assumed that the coverage area for Type 1 and Type 2 readers is perfectly circular. In general, it may not be circular always. Using any other geometrical shape or any non-geometrical shape, the number of the readers can be increased or decreased. Those will be further extensions of this model. This study did not consider any obstacles and interference sources within the range of the RFID readers. Therefore, using one or more obstacles or interference can change the number of Type 1 and Type 2 readers as Type 1 readers are more powerful than Type 2 readers. This study can be extended by optimizing the utilization of human labor and a comparative study can be made of human labor over autonomation. Another realistic scenario is imperfect production for which an autonomation policy can help reduce the unclear scarp faster than human labor.

Author Contributions: Conceptualization, methodology, software, validation, writing—original draft preparation, R.G.; formal analysis, data curation, visualization, supervision, S.P. and B.S.; investigation, resources, writing—review and editing, B.S.

Funding: This research was supported by the Basic Science Research Program through the National Research Foundation of Korea (NRF) funded by the Ministry of Education, Science and Technology (Project Number: 2017R1D1A1B03033846).

Conflicts of Interest: The authors declare no conflicts of interest.

Abbreviations

The following abbreviations are used in this manuscript:

SCM	Supply chain management
VMI	Vendor managed inventory
RFID	Radio frequency identification
DF	Distribution-free
RS	Revenue sharing
JIT	Just-in-time

References

1. Dong, Y.; Xu, K. A supply chain model of vendor managed inventory. *Trans. Res. Part E Logist. Trans. Rev.* **2002**, *38*, 75–95. [CrossRef]
2. Guo, Z.; Fang, F.; Whinston, A.B. Supply chain information sharing in a macro prediction market. *Decis. Support Syst.* **2006**, *42*, 1944–1958. [CrossRef]
3. Mukhopadhyay, S.K.; Yao, D.Q.; Yue, X. Information sharing of value-adding retailer in a mixed channel hi-tech supply chain. *J. Bus. Res.* **2008**, *61*, 950–958. [CrossRef]
4. Yan, R.; Pei, Z. Information asymmetry, pricing strategy and firm's performance in the retailer- multi-channel manufacturer supply chain. *J. Bus. Res.* **2011**, *64*, 377–384. [CrossRef]
5. Xiao, T.; Xu, T. Coordinating price and service level decisions for a supply chain with deteriorating item under vendor managed inventory. *Int. J. Prod. Econ.* **2013**, *145*, 743–752. [CrossRef]
6. Sarkar, B. An inventory model with reliability in an imperfect production process. *App. Math. Comput.* **2012**, *218*, 4881–4891. [CrossRef]
7. Cárdenas-Barrón, L.E.; Sarkar, B.; Treviño-Garza, G. Easy and improved algorithms to joint determination of the replenishment lot size and number of shipments for an EPQ model with rework. *Math. Comput. Appl.* **2013**, *18*, 132–138. [CrossRef]

8. Tayyab, M.; Sarkar, B. Optimal batch quantity in a cleaner multi-stage lean production system with random defective rate. *J. Clean. Prod.* **2016**, *139*, 922–934. [CrossRef]

9. Kim, M.S.; Sarkar, B. Multi-stage cleaner production process with quality improvement and lead time dependent ordering cost. *J. Clean. Prod.* **2017**, *144*, 572–590. [CrossRef]

10. Guchhait, R.; Sarkar, M; Sarkar, B.; Pareek, S. Single-vendor multi-buyer game theoretic model under multi-factor dependent demand. *Int. J. Invent. Res.* **2018**, *4*, 303–332. [CrossRef]

11. Majumder, A.; Guchhait, R.; Sarkar, B. Manufacturing quality improvement and setup cost reduction in a vendor-buyer supply chain model. *Eur. J. Ind. Eng.* **2017**, *11*, 588–612. [CrossRef]

12. Tiwari, S.; Cárdenas-Barrón, L.E.; Goh, M.; Shaikh, A.A. Joint pricing and inventory model for deteriorating items with expiration dates and partial backlogging under two-level partial trade credits in supply chain. *Int. J. Prod. Econ.* **2018**, *200*, 16–36. [CrossRef]

13. Sarkar, B. Mathematical and analytical approach for the management of defective items in a multi-stage production system. *J. Clean. Prod.* **2019**, *218*, 896–919. [CrossRef]

14. Sarkar, B.; Guchhait, R.; Sarkar, M.; Cárdenas-Barrón, L.E. How does an industry manage the optimum cash flow within a smart production system with the carbon footprint and carbon emission under logistics framework? *Int. J. Prod. Econ.* **2019**, *213*, 243–257. [CrossRef]

15. Gallego, G.; Moon, I. The distribution free newsboy problem: Review and extensions. *J. Oper. Res. Soc.* **1993**, *44*, 825–834. [CrossRef]

16. Sarkar, B.; Guchhait, R.; Sarkar, M.; Pareek, S.; Kim, N. Impact of safety factors and setup time reduction in a two-echelon supply chain management. *Robot. Comput.-Integr. Manuf.* **2019**, *55*, 250–258. [CrossRef]

17. Guchhait, R.; Pareek, S.; Sarkar, B. *Application of Distribution-Free Approach in Integrated and Dual-Channel Supply Chain under Buyback Contract*; IGI Global: Hershey, PA, USA, 2018; Chapter 21, pp. 303–332.

18. Scarf, H. A min-max solution of an inventory problem. In *Studies in the Mathematical Theory of Inventory and Production*; Arrow, K.J., Karlin, S., Scarf, H.E., Eds.; Standford University Press: Redwood City, CA, USA, 1958, p. 910.

19. Sarkar, B.; Zhang, C.; Majumder, A.; Sarkar, M.; Seo, Y.W. A distribution free newsvendor model with consignment policy and retailer's royalty reduction. *Int. J. Prod. Res.* **2018**, *56*, 5025–5044. [CrossRef]

20. Sarkar, B.; Majumder, A.; Sarkar, M.; Kim, N.; Ullah, M. Effects of variable production rate on quality of products in a single-vendor multi-buyer supply chain management. *Int. J. Adv. Manuf. Technol.* **2018**, *99*, 567–581. [CrossRef]

21. Moon, I.; Shin, E.; Sarkar, B. Min–max distribution free continuous-review model with a service level constraint and variable lead time. *Appl. Math. Comput.* **2014**, *229*, 310–315. [CrossRef]

22. Sarkar, B. Supply chain coordination with variable backorder, inspections, and discount policy for fixed lifetime products. *Math. Probl. Eng.* **2016**, *2016*, 6318737. [CrossRef]

23. Tiwari, S.; Jaggi, C.K.; Gupta, M.; Cárdenas-Barrón, L.E. Optimal pricing and lot-sizing policy for supply chain system with deteriorating items under limited storage capacity. *Int. J. Prod. Econ.* **2018**, *200*, 278–290. [CrossRef]

24. Tiwari, S.; Wee, H.M.; Daryanto, Y. Big data analytics in supply chain management between 2010 and 2016: Insights to industries. *Comput. Ind. Eng.* **2018**, *115*, 319–330. [CrossRef]

25. Meguerdichian, S.; Koushanfar, F.; Potkonjak, M.; Srivastava, M.B. Coverage problems in wireless ad-hoc sensor networks. In Proceedings of the IEEE INFOCOM 2001, Anchorage, AK, USA, 22–26 April 2001; pp. 1380–1387. [CrossRef]

26. Zhang, H.; Hou, J.C. Maintaining sensing coverage and connectivity in large sensor networks. *Ad Hoc Sens. Wirel. Netw.* **2005**, *1*, 89–124. [CrossRef]

27. Hefeeda, M.; Ahmadi, H. A probabilistic coverage protocol for wireless sensor networks. In Proceedings of the 2007 IEEE International Conference on Network Protocols, Beijing, China, 16–19 October 2007; pp. 41–50. [CrossRef]

28. Dias, J.C.Q.; Calado J.M.F.; Luís Osório, L.F.; Morgado, L.F. RFID together with multi-agent systems to control global value chains. *Annu. Rev. Control* **2009**, *33*, 185–195. [CrossRef]

29. Sarac, A.; Absi, N.; Dauzère-Pérès, S. A literature review on the impact of RFID technologies on supply chain management. *Int. J. Prod. Econ.* **2010**, *128*, 77–95. [CrossRef]

30. Kim, T.; Glock, C.H. On the use of RFID in the management of reusable containers in closed-loop supply chains under stochastic container return quantities. *Trans. Res. Part E Logist. Trans. Rev.* **2014**, *64*, 12–27. [CrossRef]

31. Sari. K. Exploring the impacts of radio frequency identification (RFID) technology on supply chain performance. *Eur. J. Oper. Res.* **2010**, *217*, 174–183. [CrossRef]

32. Biswal. A.K.; Jenamani, M.; Kumar, S.K. Warehouse efficiency improvement using RFID in a humanitarian supply chain: Implications for Indian food security system. *Trans. Res. Part E Logist. Trans. Rev.* **2018**, *109*, 205–224. [CrossRef]

33. Chen. J.C.; Cheng, C.H.; Huang, P.B. Supply chain management with lean production and RFID application: A case study. *Exp. Syst. Appl.* **2013**, *40*, 3389–3397. [CrossRef]

34. Tsao. Y.C.; Linh, V.T.; Lu, J.C. Closed-loop supply chain network designs considering RFID adoption. *Comput. Ind. Eng.* **2017**, *113*, 716–726. [CrossRef]

35. Ustundag. A.; Tanyas, M. The impacts of radio frequency identification (RFID) technology on supply chain costs. *Trans. Res. Part E Logist. Trans. Rev.* **2009**, *45*, 716–726. [CrossRef]

36. Jaggi. A.S.; Sawhney, R.S.; Balestrassi, P.P.; Simonton, J.; Upreti, G. An experimental approach for developing radio frequency identification (RFID) ready packaging. *J. Clean. Prod.* **2014**, *85*, 371–381. [CrossRef]

37. Shin, D.; Guchhait, R.; Sarkar, B.; Mittal, M. Controllable lead time, service level constraint, and transportation discounts in a continuous review inventory model. *RAIRO Oper. Res.* **2016**, *50*, 921–934. [CrossRef]

© 2019 by the authors. Licensee MDPI, Basel, Switzerland. This article is an open access article distributed under the terms and conditions of the Creative Commons Attribution (CC BY) license (http://creativecommons.org/licenses/by/4.0/).

Article

Joint Inventory and Pricing Policy for an Online to Offline Closed-Loop Supply Chain Model with Random Defective Rate and Returnable Transport Items

Biswajit Sarkar [1], Mehran Ullah [1] and Seok-Beom Choi [2,*

[1]   Department of Industrial & Management Engineering, Hanyang University, Ansan Gyeonggi-do 15588, Korea; bsarkar@hanyang.ac.kr (B.S.); mehrandirvi@gmail.com (M.U.)
[2]   Department of International Business, Cheju Halla University, Jeju-do 63092, Korea
*   Correspondence: sbchoi@cau.ac.kr; Tel.: +82-10-3854-2765

Received: 14 April 2019; Accepted: 20 May 2019; Published: 1 June 2019

Abstract: Environmental deterioration is one of the current hot topics of the business world. To cope with the negative environmental impacts of corporate activities, researchers introduced the concept of closed-loop supply chain (CLSC) management and remanufacturing. This paper studies joint inventory and pricing decisions in a multi-echelon CLSC model that considers online to offline (O2O) business strategy. An imperfect production process is examined with a random defective rate that follows a probability distribution. The results show that the O2O channel increases the profit of the system. For the defective rate, three different distributions are considered and three examples are solved. The results of the three examples conclude that the highest profit is generated when the defective rate follows a uniform distribution. Furthermore, based on the salvage value of defective items, two cases were studied. Results and sensitivity analysis show that the increase in defective rate does not reduce total profit in every situation, as perceived by the existing literature. Sensitivity analysis and numerical examples are given to show robustness of the model and draw important managerial insights.

Keywords: CLSC management; O2O channel; random defective rate; hybrid manufacturing-remanufacturing strategy

---

1. Introduction

The widespread usage of the internet and competitive business environment have resulted in the development of a dual channel sales strategy called the O2O strategy [1]. Companies are integrating online and offline sales channels to compete in the market by increasing service level, and to attract more demand. The literature termed integrating offline and online channels as a multi-channel context [2]. Traditional enterprises can deliver diverse types of products and services by incorporating online sales channels for their businesses. The number of companies taking advantage of this opportunity is increasing. This enables potential customers to browse the catalogs, price information, availability of the products, and even order the products before visiting the physical stores. Therefore, online sales channels can improve the sales of physical (offline) stores. A case study conducted by Chang et al. [3] provided factual proof that integrating offline and online sales channels increases the sales order. However, O2O sales channel requires the customer acquisition cost to convince the potential consumer or customer to purchase a specific product; literature considers this cost a primary business metric for O2O channeling. It determines the worth of the end customer to the business and, through it, the return on investment from the customers can be obtained. This enables businesses to analyze their investment decisions on a single customer to improve profitability [4]. Organizations make investments for online marketing and offline marketing of their products. Then, to compute the customer acquisition cost, each customer, who buys the product or service, is asked about the channel of information through which he came to know about that product. In this way, organizations collect a handful of data for a specific period and calculate customer acquisition cost for both types of channels.

Empirical evidence shows that the O2O channel plays a vital role for value creation in the business strategy [4]. However, O2O brings many management complexities and costs into supply chain management. These costs include online and offline marketing investments, subsidies, commissions, administration costs, and bonuses. Successful investment in O2O channeling is directly associated with future corporate profits. It increases customer retention and expands market shares by capturing new customers [5]. To incorporate an e-commerce strategy, this study considers a supply chain management model that captures extra market demand through the O2O channel and considers the required investment and maintenance cost for the O2O channel. In this context, this paper studies joint inventory and pricing decisions in a multi-echelon closed-loop supply chain management with remanufacturing and returnable transport items (RTI). Even though many papers had explored CLSC management, they either did not focus on O2O or they did not consider all other important features such as the consideration of transportation packaging management. In fact, from our literature review, we find that none of the existing studies examine the packaging management system and investment decisions in a CLSC system with O2O channel. Notice that the most closely related paper is Sarkar et al. [6], however, they did not consider dual-channel structure. Furthermore, they also neglect defective production and assume that all the produced products are perfect items, which is an unrealistic assumption. Moreover, they only studied inventory policies and considered constant demand, while this paper studies joint inventory and pricing policies with price and investment dependent demand.

2. Literature Review

This section gives an overview of the existing literature relevant to this paper. This paper contributes to research in the field of CLSC management, O2O channel in CLSC management, and RTI management and design, therefore, we focus on the related literature in these three areas.

Because of the increasing appreciation of environmental sustainability, CLSC management becomes a very important topic. More specifically, the remanufacturing process has grown into a big business. According to Bulmus et al. [7], \$43 billion valued products were sold from remanufacturing used products in the U.S. in 2011. Many giant well-established enterprises, including Xerox, Kodak, and HP, have extensively incorporated their remanufacturing processes into their regular production lines and operations [8]. Research in this field dates back to the 1960s, with the first work reported from Schrady [9]. This paper studied an inventory model with single manufacturing batch followed by many repair batches, this policy is called (R, I) policy. Later on, Richter [10] and Richter [11] extended Schrady's model and introduced a hybrid manufacturing-remanufacturing system. According to Richter assumptions, market demand is fulfilled by manufacturing new products from raw material, and by repairing used products. Continuing his work, Richter [12] found that a complete recovery or complete disposal is the optimal choice. Many others extended this work with similar results including Richter and Dobos [13], Dobos and Richter [14], and Dobos and Richter [15]. Maiti and Giri [16] considered the used product quality in CLSC management and examined analytical models in five different scenarios including the centralized, vertical Nash, and three decentralized cases. Recently, Moshtagh and Taleizadeh [17] considered return rate with quality and used three distinct distributions to model the return rate. Tian and Zhang [18] considered disassembly scheduling and pricing of returned goods with price based yield. For a more comprehensive review of CLSC management, it is suggested that readers study Govindan and Soleimani [19] and Diallo et al. [20].

Today, with the progress of the Internet, online shopping is gaining popularity; the consumer buys products both through online and offline channels. The online purchasing channel has various advantages such as energy and time savings. Consumers can also compare the product prices among different websites and choose the best product. However, some customers like to observe, feel and examine finished products before buying, which could not be accomplished without the traditional offline channel. Therefore, organizations operate both the channels to get the associated advantages. Many business giants like Lenovo, Sony, IBM, Dell, Nike, and HP use this dual-channel policy to improve their sales [21–23]. In the literature, many articles have examined the impact of introducing the online channel on performances of companies and supply chain management. For example, Tsay and Agrawal [24] examined the channel coordination policies to reduce channel disputes between supply chain members. Yao and Liu [25] used Stackelberg and Bertrand games to investigate channel conflict and competition among suppliers and retailers. Huang and Swaminathan [26] andYan [27] also used game theory to investigate optimal pricing decisions in O2O supply chain management. Dan et al. [28] studied dual-channel supply chain management with retailer manipulation of the retail services; they found that the optimal pricing strategies can be affected by retail service. Huang et al. [22] considered pricing policies for an O2O supply chain management including demand disruptions. Li et al. [29] reviewed the difference between single-channel and dual-channel supply chain management. The O2O business model is characterized by high transportation demand because of last mile supply systems; furthermore, the returns from CLSC model also increase the transportation demand of the system; therefore, energy efficient transportation becomes a primary part of O2O supply chain. Bányai et al. [30] and Bányai [31] studied energy efficiency in first mile and last mile logistics systems. Despite all these excellent research works, papers that deal with product remanufacturing, reuse, and transportation packaging with the imperfect production system in the O2O supply chains are sparse in the literature. This study bridges this gap by studying a dual-channel CLSC model with remanufacturing, and returnable transport items considering random imperfection in the production system.

Solid waste generation is one of the main environmental threats posed by supply chain management [32,33]. RTIs are used by organizations to reduce solid waste of supply chain management [34,35]. In addition, disposable packaging also increases the cost of the product. Livingstone and Sparks [36] stated that packaging is responsible for 10–40% of product prices. Moreover, RTI also increases the protection and security of the finished products and improves working conditions and manual handling [37]. Historically, these RTIs were studied independently from inventory management, for a detailed review, readers may study [38]. Nevertheless, contemporary research recognized packaging as a vital component of supply chain inventory management. In this regard, [39] studied RTI and inventory policies jointly. They investigated a two-level CLSC inventory management and used RTIs to transport finished goods among supply chain members. Glock and Kim [40] studied RTI and inventory management in a single-vendor multi-retailer CLSC structure. They also considered RTI design and considered RTI size and the required number of RTIs as decision variables. Sarkar et al. [6] extended Glock and Kim [40] with RTI design and management policy in hybrid manufacturing-remanufacturing CLSC management. Others, for example, Kelle and Silver [41] proposed several methods to forecast the expected demand and returns of RTI; and Goh and Varaprasad [42] developed a system for evaluating the container return distribution by considering RTIs shrinkage.

Many authors, for example,Lee [43], Gupta and Chakraborty [44], and Tayi and Ballou [45] studied imperfect items in inventory models. Among these, the last two did not consider shortages due to reworking. Lee et al. [46] and Glock and Jaber [47] presented imperfect production models without considering rework of defective products. Salameh and Jaber [48] studied an inventory model for imperfect quality items where these items are withdrawn from stock, resulting in lower holding cost per unit time. Jaber et al. [49] consider policies for handling imperfect quality products. Ouyang et al. [50] and Cárdenas-Barrón [51] developed inventory models and considered back ordering, while Eroglu and Ozdemir [52] considered shortages. Ben-Daya [53] likewise extended the inventory model with planned maintenance schedules. Sarkar et al. [54] considered quality of products in multi-echelon supply chain model. Sarkar [55] provided different approaches for defective product management. An improved way to calculate imperfect items is given by Kim et al. [56], and Sett et al. [57] determined optimal buffer inventory in imperfect production system. Sarkar et al. [58] studied warranty, optimal run time with inspection errors in the imperfect production system, and Kang et al. [59] discussed optimal ordering policies for the imperfect production system. Recently, Khanna et al. [60] considered two-level credit policies with imperfect quality, Tayyab et al. [61] considered process uncertainty, and Tayyab et al. [62] studied the multistage imperfect production system. The majority of the existing literature considered rework or scrap option for defective items, however, this paper considers salvage option for defective items.

3. Problem Definition

This paper considers a three-layer dual-channel hybrid CLSC model with RTI design and management. The manufacturer fulfills market demand through multiple retailers. To increase market share, each retailer opens an online selling channel, for which the retailer pays information cost and transportation cost. Furthermore, the demand rate jointly depends on finished product price and investment in the online channel. The manufacturer produced two types of products, the manufactured products from raw material, and the remanufactured products from used products collected from consumers. An imperfect production system is considered where the perfect items are sold into the primary market, and the defective items are marketed into a secondary market with a lower price. Finished products are transported in reusable secondary packages called RTIs that are owned by a third party logistics (3PL) provider. The 3PL also collects used products from the end customer and delivers them to the manufacturer for remanufacturing. The whole scenario is depicted in Figure 1.

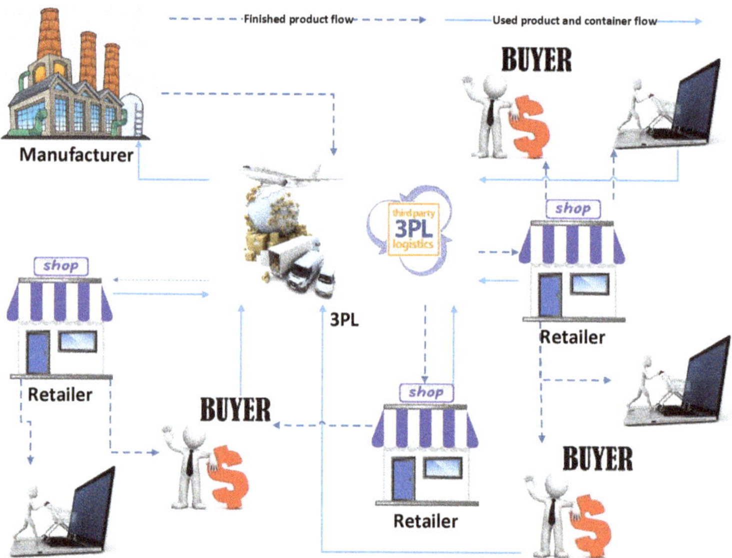

Figure 1. Logistics diagram of the proposed closed-loop supply chain (CLSC).

It is assumed that the total initial deterministic demand (offline market share) for the ith retailer is a_i, which follows demand price sensitivity, b, in a linear relationship. Other possible relationships, in which the demand function can be modeled, are quadratic and exponential. However, linear demand curves are widely used in the literature because of their simplicity and more importantly their sufficiency to catch important managerial implications. Thus, the total price-dependent deterministic demand at the ith retailer is $a_i - bp$. Now we consider the second part of demand, the online channel demand; which for the ith retailer can be given as kIN_i. Where k is the parameter and IN_i is the potential demand. Thus the net realized demand at retailer i is $d_i = a_i - bp + k\text{IN}_i$. It is obvious that the maximum price in the market would be $p = a/b$.

4. Mathematical Model

This paper considered a centralized supply chain system. To model the proposed centralized system, this section first calculates total revenue of the supply chain followed by the cost of individual player and, finally, the net profit is calculated by subtracting the costs of all players from total revenue.

4.1. Total Revenue of the Supply Chain

The proposed system generates revenue by selling the finished products of perfect quality in the dual-channel primary market and imperfect items in the secondary market. Total revenue can be expressed mathematically as,

$$REV = \sum_{i=1}^{n} pd_i + cwr \sum_{i=1}^{n} d_i \tag{1}$$

The first term in (1) shows revenue generated by n retailers, while the second term shows revenue generated from selling imperfect items.

4.2. Retailer's Model

Total cost of the ith retailer consists ordering cost $\frac{A_i}{T}$, holding cost $\frac{d_i h_i}{2}$, which can be calculated from Figure 2. Investment or information cost of O2O channel $\frac{\text{IN}_i}{T}$, and transportation cost for the proposed O2O channel $\frac{TR_i}{T}$ are also part of retailer cost function. For n retailers the total cost can be stated as,

$$TC_1 = \sum_{i=1}^{n} \frac{A_i + \text{IN}_i + TR_i}{T} + \sum_{i=1}^{n} \frac{1}{2} T d_i h_i. \tag{2}$$

4.3. Manufacturer's Model

The manufacturer produces finished products based on the market demand in a single setup per cycle, for which the setup cost per unit time is $\frac{S_m}{T}$. In each cycle, the manufacturer orders used products, for which the ordering cost is $\frac{A_m}{T}$. The manufacturer inventory diagram is shown in Figure 2, the holding cost for the finished product is formulated in Appendix A.1 and can be written as follows,

$$h_f \sum_{i=1}^{n-1} d_{i+1} l_i + \frac{d^2 T}{2(1-r)P_m}.$$

The manufacturer inspects each product for which total inspection cost is $ck\sum_{i=1}^{n} d_i$, and $h_u d\tau \left(1 - \frac{\sum_{i=1}^{n} d_i}{2(1-r)P_m}\right)$ is the holding cost of used products. During production, $r\%$ of the total items produced are defective, to avoid shortages the manufacturer outsources the defective quantity with total cost $ow * r * \sum_{i=1}^{n} d_i$. τd is the remanufactured quantity and $d(1 - \tau)$ is the quantity of new products. Therefore, the cost of remanufacturing used products is $bd\tau (C_r)$, and the cost of manufacturing is $C_m d(1 - \tau)$. The finished products quality is $0 < q \leq 1$, to attain this quality, the manufacturer total quality cost is $d(1 - \tau)q^2 C_q$ for manufacturing and $d\tau C_q \left(q^2 - q_r^2\right)$ for remanufacturing. The manufacturer incurs goodwill lost cost for producing low quality products which is $dg(1 - q)$. Finally, the total emission cost from production activities can be expressed as $e_m \sum_{i=1}^{n} C_{ec} d_i$. Hence, the total cost per unit time for the manufacturer (TC_2) can be expressed as follows:

$$TC_2 = \frac{A_m + S_m}{T} + h_f\left(\sum_{i=1}^{n-1} l_i \sum_{k=i+1}^{n} d_k + \frac{d^2 T}{2(1-r)P_m}\right) + (ck + ow * r)\sum_{i=1}^{n} d_i + h_u d\tau\left(1 - \frac{\sum_{i=1}^{n} d_i}{2(1-r)P_m}\right)$$
$$+ \quad d(1-\tau)C_m + d\tau C_r + d\tau C_q\left(q^2 - q_r^2\right) + dq^2(1-\tau)C_q + dg(1-q) + e_m \sum_{i=1}^{n} C_{ec}d_i. \tag{3}$$

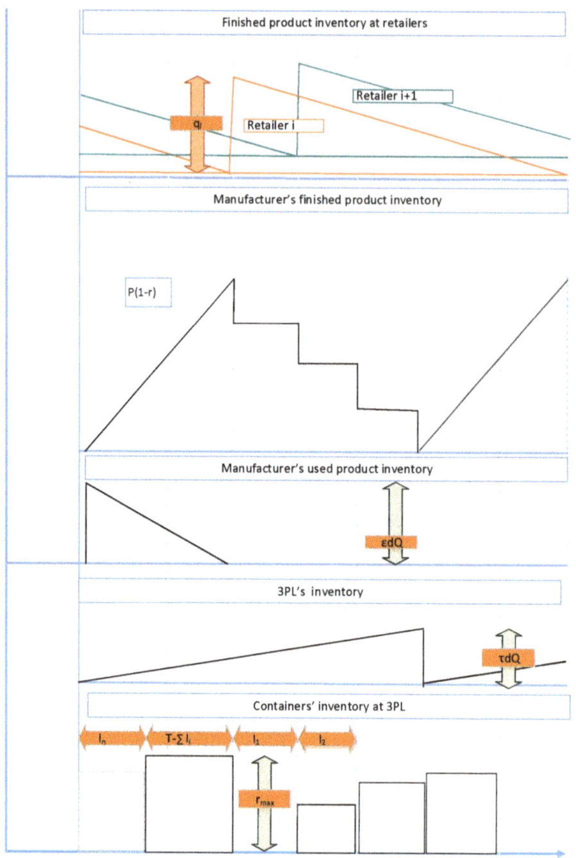

Figure 2. Logistics diagram of the proposed CLSC.

4.4. 3PL's Model

3PL's total cost comprises of collection setup cost per cycle $\frac{S_3}{T}$. The collection cost paid to customers is $\tau A_t \sum_{i=1}^{n} d_i$. The average holding cost of used product at 3PL is $h_u \frac{1}{2} d\tau T$. 3PL also provides transportation services, total transportation cost comprised of three parts: (1) finished products transportation from manufacturer to retailer i, $\left(\frac{d_i l_{im}}{\lambda}\right)$, (2) empty RTI transportation from retailer i to 3PL $\frac{d_i l_{ij}}{\lambda}$, and (3) used products transportation from 3PL to manufacturer $\frac{dl_{jm}}{\lambda}$. For n retailers total transportation cost is the sum of forward and reverse transportation costs represented by $C_t\left(\sum_{i=1}^{n} \frac{d_i l_{ij}}{\lambda} + \sum_{i=1}^{n} \frac{d_i l_{im}}{\lambda} + \frac{dl_{jm}}{\lambda}\right)$. Holding cost of RTIs is given by $h_R\left(\sum_{i=1}^{n} \frac{l_i(d_{\max} - d_i)}{\lambda} + \frac{d_1\left(T - \sum_{i=1}^{n} l_i\right)}{\lambda}\right)$, see Appendix A.2. RTI management cost is $C_\lambda d_{\max}\lambda^{s-1}$. For setting up collection centers and arranging the inspection system for determining the quality of the EOL/EOU products, 3PL makes an initial investment which can be expressed mathematically as $\frac{\gamma\tau^2}{T}$. Finally, 3PL's cost also consists of emissions cost from transportation, which is determined based on the traveled distance, and emissions from fuel required per unit distance for a specific truck is considered to model transportation emissions. Therefore, total emissions due to transportation between retailer i and manufacturer m can be modeled as $\frac{d_i l_{im}}{t_c}(e_r g_m)$, similarly emissions from transportation between retailer i and 3PL j can be written by $\left.\frac{d_i l_{ij}}{t_c}\right) e_r g_m$, and emissions from 3PL j to manufacturer m is given by $\frac{d\tau l_{jm}}{t_c}$. Now for n retailers, a manufacturer, and a 3PL, total emission cost from transportation can be expressed mathematically as $(C_{ec} e_r g_m)(\sum_{i=1}^{n} \frac{d_i l_{im}}{t_c} + \sum_{i=1}^{n} \frac{d_i l_{ij}}{t_c} + \frac{d\tau l_{jm}}{t_c})$. Total cost per unit of time for the 3PL TC_3 comprises of the above parts and can be expressed as,

$$
\begin{aligned}
TC_3 \;=\;& \frac{S_3}{T} + \tau A_t \sum_{i=1}^{n} d_i + \frac{d\tau Th_u}{2} + C_t\left(\sum_{i=1}^{n} \frac{d_i l_{ij}}{\lambda} + \sum_{i=1}^{n} \frac{d_i l_{im}}{\lambda} + \frac{d l_{jm}}{\lambda} \right) + C_\lambda d_{\max} \lambda^{s-1} \\
+\;& h_R\left(\sum_{i=1}^{n} \frac{l_i (d_{\max} - d_i)}{\lambda} + \frac{d_1 (T - \sum_{i=1}^{n} l_i)}{\lambda} \right) + c_d(\tau - \tau)d + \frac{\gamma \tau^2}{T} + C_{ec} e_r g_m \qquad (4) \\
& \sum_{i=1}^{n} \frac{\tau d_i l_{ij}}{t_c} + C_{ec} e_r g_m \sum_{i=1}^{n} \frac{d_i l_{im}}{t_c} + \frac{d\tau C_{ec} e_r g_m l_{jm}}{t_c}
\end{aligned}
$$

4.5. Net Profit of the Supply Chain

Net profit of the supply chain can be calculated by subtracting cost of retailer (2), cost of manufacturer (3), and cost of 3PL (4) from total generated revenue (1), such that

$$
\Pi = Rev - TC_1 - TC_2 - TC_3
$$

Using the transformation from demand function, total profit can be expressed as follows,

$$
\begin{aligned}
\Pi \;=\;& cwr \sum_{i=1}^{n} (a_i - bp + kIN_i) + \sum_{i=1}^{n} p\,(a_i - bp + kIN_i) - \left(\sum_{i=1}^{n} \frac{Th_i}{2}(a_i - bp + kIN_i) + \sum_{i=1}^{n} \right. \\
& \frac{A_i + IN_i + TR_i}{T} \bigg) - \left(h_f \left(\sum_{i=1}^{n-1} l_i (a_{i+1} - bp + kIN_i) + \frac{T(\sum_{i=1}^{n}(a_i - bp + kIN_i))^2}{2(1-r)P_m} \right) + \frac{A_m + S_m}{T} \right. \\
+\;& h_u \sum_{i=1}^{n} \tau\,(a_i - bp + kIN_i)\left(1 - \frac{\sum_{i=1}^{n}(a_i - bp + kIN_i)}{2(1-r)P_m} \right) + ((ck + ow * r)) \sum_{i=1}^{n} a_i - bp + kIN_i + (C_m * 1 - \tau \\
& \sum_{i=1}^{n} (a_i - bp + kIN_i)) + \tau C_q \sum_{i=1}^{n} (q^2 - q_r^2)(a_i - bp + kIN_i) + (\tau \sum_{i=1}^{n} C_r (a_i - bp + kIN_i)) + (C_q \\
& q^2(1 - \tau) \sum_{i=1}^{n} (a_i - bp + kIN_i)) + e_m \sum_{i=1}^{n} C_{ec}(a_i - bp + kIN_i) + \sum_{i=1}^{n} g(1 - q)(a_i - bp + kIN_i) \bigg) \qquad (5) \\
-\;& \left(\tau A_t \sum_{i=1}^{n} (a_i - bp + kIN_i) + \frac{T}{2}(\tau h_u \sum_{i=1}^{n} (a_i - bp + kIN_i)) + \frac{S_3}{T} + C_t \left(\sum_{i=1}^{n} \frac{l_{im}(a_i - bp + kIN_i)}{\lambda} \right. \right. \\
+\;& \sum_{i=1}^{n} \frac{\tau l_{ij}(a_i - bp + kIN_i)}{\lambda} + \frac{l_{jm} \sum_{i=1}^{n} \tau(a_i - bp + kIN_i)}{\lambda} \bigg) + (C_\lambda \lambda^{s-1}(a_{\max} - bp + kIN_{a_{max}})) \\
+\;& \frac{h_R((T - \sum_{i=1}^{n} l_i)(a_{\max} - bp + kIN_{a_{max}}))}{\lambda} + \frac{\gamma \tau^2}{T} + C_{ec} e_r g_m \sum_{i=1}^{n} \frac{\tau l_{ij}(a_i - bp + kIN_i)}{t_c} + C_{ec} e_r g_m \\
& \sum_{i=1}^{n} \frac{l_{im}(a_i - bp + kIN_i)}{t_c} + \frac{\sum_{i=1}^{n} \tau C_{ec} e_r g_m l_{jm}(a_i - bp + kIN_i)}{t_c} \bigg)
\end{aligned}
$$

Assuming T, λ, τ, and p as any real nonzero numbers, then, $\exists$ a unique T^*, λ^*, τ^* and p^* that maximizes the total profit and satisfies the following first-order conditions (FOC):

$$
\frac{\partial (\Pi)}{\partial p} = 0, \qquad (6)
$$

$$
\frac{\partial (\Pi)}{\partial \tau} = 0, \qquad (7)
$$

$$
\frac{\partial (\Pi)}{\partial T} = 0, \qquad (8)
$$

$$
\frac{\partial (\Pi)}{\partial \lambda} = 0. \qquad (9)
$$

Furthermore, (Π) is concave, if and only if at T^*, λ^*, τ^* and p^* the hessian of (Π) is negative semidefinite.

Proof. See Appendix B for the proof of concavity of objective function. $\square$

Until now, the model considers random defective rates of production, but no probability distribution is considered. Now, one can test the model under different distributions: uniform, triangular, and beta. Thus, if r follows a uniform distribution, then the probability density function $f(r)$ for the uniform distribution with parameter x and y is,

$$f(r) = \begin{cases} \frac{1}{x-y} & x \leq r \leq y \\ \\ 0 & otherwise \end{cases}$$

and its expected value is $E[r] = \frac{y+x}{2}$.

Similarly, if r follows a triangular distribution, then the probability density function for the triangular distribution with parameter x, z, and y is

$$f(r) = \begin{cases} 0 & r < x \\ \frac{2(r-x)}{(y-x)(z-x)} & x \leq r < z \\ \frac{2}{(y-x)} & r = z \\ \frac{2(y-r)}{(y-x)(y-z)} & z < r \leq y \\ 0 & y < r \end{cases}$$

and its expected value is $E[r] = \frac{x+y+z}{3}$.

Finally, if r follows a beta distribution, then the probability density function for the beta distribution with parameter λ and β is

$$f(r) = \left\{ \frac{r^{\lambda-1}(1-r)^{\beta-1}}{B(\lambda,\beta)} \quad \lambda > 0, \beta > 0, 0 \leq r < 1 \right.$$

and its expected value is $E[r] = \frac{\lambda}{\lambda+\beta}$.

4.6. Solution Procedure

Our aim is to derive the optimal values of the decision variables and maximize the total net profit under all the three distributions. The solution procedure is as follows:

Step 1.0 Select the distribution function.

Step 1.1 Plug the expected value of r in (5).

Step 2.0 Set $\tau = 0$, $\lambda = 1$, and $T = 1$ find p^* from FOC given in (6).

Step 2.1 Using $p = p^*$, $\lambda = 1$, and $T = 1$ find τ^* from FOC given in (7).
Step 2.2 Using $p = p^*$, $\lambda = 1$, and $\tau = \tau^*$ find T^* from FOC given in (8).
Step 2.3 Using $p = p^*$, $T = T^*$, and $\tau = \tau^*$ find λ^* from FOC given in (9).
Step 2.4 Using $\tau = \tau^*$, $\lambda = \lambda^*$, and $T = T^*$ find p^* from FOC given in (6).
Step 2.5 Using $p = p^*$, $\lambda = \lambda^*$, and $T = T^*$ find τ^* from FOC given in (7).
Step 2.6 Using $p = p^*$, $\lambda = \lambda^*$, and $\tau = \tau^*$ find T^* from FOC given in (8).
Step 2.7 Using $p = p^*$, $T = T^*$, and $\tau = \tau^*$ find λ^* from FOC given in (9).
Step 2.8 Repeat Steps 2.4 - 2.7 until the values of τ^*, λ^*, T^*, and p^* stop changing.
Step 2.9 Update the optimal values τ^*, λ^*, T^*, and p^*.

Step 3.0 Set $\lambda = \lfloor \lambda^* \rfloor$, calculate total profit from (5).

Step 4.0 Set $\lambda = \lceil \lambda^* \rceil$, calculate total profit from (5).
Step 5.0 If $\Pi(\tau^*, \lfloor \lambda^* \rfloor, T^*, p^*) < \Pi(\tau^*, \lceil \lambda^* \rceil, T^*, p^*)$, update $\lambda^* = \lfloor \lambda^* \rfloor$, else update $\lambda^* = \lceil \lambda^* \rceil$.

    Step 5.1 Repeat Steps 2.4–2.6 until the values of τ^*, T^*, and p^* stop changing.

Step 6.0 Find total profit from (5).
Step 7.0 Repeat the process for all the three probability distributions.

5. Numerical Experiment

The developed model is tested with numerical experiment and sensitivity analysis of the input parameters. The numerical experiment considers a three-layer supply chain network consisting of four retailers, a manufacturer, and a 3PL. The values of input parameters are obtained from [6] and are given in Table 1.

Table 1. General Input parameter values for numerical examples.

A_m	20	S_3	10	n	4	γ	2000	A_t	20	C_λ	0.5	h_R	5	s	2		
C_m	110	C_r	84	C_q	5	q_r	0.3	g_m	0.25	e_r	0.01	C_{ec}	18	g	10		
C_t	0.004	l_{jm}	50	t_c	80	e_m	0.002066	P_m	10,000	h_f	5.2	h_u	2.5	S_m	60		
l_1	0.009	l_2	0.008	l_3	0.007	l_4	0.008	l_m	373	l_{2m}	226	l_{3m}	216	l_{4m}	371		
l_j	373	l_{2j}	226	l_{3j}	216	l_{4j}	371	h_1	8.	h_2	7.4	h_3	8.2	h_4	8.1		
A_1	63	A_2	51	A_3	39	A_4	63	IN_1	500	IN_2	489	IN_3	623	IN_4	585		
TR_1	300	TR_2	283	TR_3	250	TR_4	310	a_1	1200	a_2	720	a_3	520	a_4	600		
b	0.9	q	0.9	ck	5	cw	200	k	0.4	x	0.15	y	0.4	z	0.2		
λ	0.15	β	0.4	ow	115												

The optimal results for all the three distribution are given in Table 2. We obtain the optimal cycle time as $T = 0.64$ years (for uniform distribution), $T = 0.66$ years (for triangular distribution) and $T = 0.67$ years (for beta distribution). The optimal container capacity is $\lambda = 3$ units (for uniform distribution), $\lambda = 4$ units (for triangular distribution) and $\lambda = 5$ units (for beta distribution). For optimal remanufacturing rate, the results are $\tau = 65$ percent (for uniform distribution), $\tau = 69$ percent (for triangular distribution) and $\tau = 73$ percent (for beta distribution). The optimal prices are $p = \$581.21$ (for uniform distribution), $p = \$582.6$ (for triangular distribution) and $p = \$581.4$ (for beta distribution). From the total profit, it is clear that the highest profit is yielded by uniform distribution $\$828,278$ followed by the beta distribution $\$827,970$. The least profit is obtained from triangular distribution, which is $\$824,728$.

Table 2. Optimal results of three examples.

	Example 1	Example 2	Example 3
Distribution of Defective Rate	Uniform	Triangular	beta
$\Pi(\$)$	$828,278$	$824,728$	$827,970$
$T^*(days)$	0.634	0.66	0.67
$\lambda^*(units)$	3	4	5
$\tau^*(\%)$	0.65	0.69	0.73
$P^*(\$)$	581.21	582.64	581.43

5.1. Sensitivity Analysis

A sensitivity analysis is performed for all the key parameters and the results are compiled in Tables 3–8. Table 3 provides sensitivity analysis of key parameters related to manufacture and 3PL. From the results, the following insights are obtained:

- The most influential parameter for all the three examples is manufacturing cost. Decreasing the manufacturing cost increases the profit of the system. However, the percentage change in profit to both positive and negative changes in manufacturing cost is asymmetric. Decreasing manufacturing cost by 50 percent increases the total profit by around 11%, on the other hand, an equal increase reduces the profit by less than one percent. Thus, it can be concluded that the profit is more sensitive to negative changes in manufacturing cost compared to the positive changes. Hence, in hybrid systems, supply chain managers must focus on technologies to reduce manufacturing cost to improve overall profitability.

- Almost similar results are shown by remanufacturing cost. Decreasing remanufacturing cost by 50% increases profit by more than 8% for all examples, while increasing remanufacturing cost by 50% reduces profit by less than one percent.
- In the case of investment for the collection of used product, the models showed opposite asymmetry. In this case, the profit is more sensitive towards positive changes in the investment cost. A 50% increase in investment cost reduces total profit by 0.37% and a 50% decrease only increases the profit by 0.16%. Hence it can be concluded that the sensitivity of profit towards positive change is almost double of the negative change.
- An interesting result is obtained from the sensitivity of production rate. The results are different for different examples. For beta distribution, the percent change is almost symmetric in both directions. For uniform distribution, the results are less asymmetric, however, for triangular distribution, the results show a higher degree of asymmetry compared to the other two models. This shows that for supply chains, in which the defective rate follows a triangular distribution, increasing the production rate will not increase the profit. However, in the case of uniform and beta, distribution production rate can be used as a tool to increase the profit of the supply chain.
- Among the three examples, the model with triangular distribution is more sensitive towards manufacturer's setup cost compared to uniform and beta distributions. This means that for supply chain systems, where the defective rate follows a triangular distribution, managers must focus on reduction of the manufacturer's setup cost.

Table 3. Sensitivity analysis for parameters related to manufacturer and 3PL.

Parameter	% Change in Value	% Change in Π (Uniform)	% Change in Π % (Triangular)	% Change in Π (beta)
C_m	-50%	11.19	11.28	11.19
	-25%	5.45	5.472	5.43
	$+25\%$	-0.086	-0.106	-0.08
	$+50\%$	-0.123	-0.138	-0.14
C_r	-50%	8.603	8.625	8.582
	-25%	4.254	4.259	4.256
	$+25\%$	-0.159	-0.17	-0.151
	$+50\%$	-0.163	-0.17	-0.163
γ	-50%	0.169	0.138	0.13
	-25%	0.065	0.0591	0.07
	$+25\%$	-0.033	-0.043	-0.03
	$+50\%$	-0.374	-0.13	-0.37
P_m	-50%	-0.037	-0.035	-0.031
	-25%	0.001	-0.012	0.001
	$+25\%$	0.022	0.001	0.022
	$+50\%$	0.028	0.013	0.028
S_m	-50%	0.019	0.004	0.019
	-25%	0.017	0.0026	0.002
	$+25\%$	0.001	-0.002	0.001
	$+50\%$	0.009	-0.106	-0.008

Sensitivity of all other parameters, related to manufacturer and 3PL, is given in Table A1, Appendix C.

This paper assumed that the demand depends upon price and investment in O2O channel that attracts more customers. Table 4 provides a sensitivity analysis of the demand parameters b and k. The following insights are obtained from the results:

- The profit of the system increases with a decrease in the demand parameter b, and decreases with a decrease in k. However, the percent change is different for both the parameters.
- Considering the demand parameter b, a 50% decrease produces a 114.8% increase for both uniform and beta distributions. The increase for the triangular distribution is, however, slightly higher, 115.3%. Similarly, on the positive side, triangular distribution is also slightly more sensitive compared to the beta distribution and uniform distribution.
- In the case of demand parameter k, the triangular distribution shows a 21.03% decrease to 50% decrease in k, which is slightly higher than beta distribution, 20.92%, and uniform distribution, 20.90%. On the positive side, however, the results are comparable with slight differences. For a 50% increase in k, the total profit increased by 23.3% for the uniform distribution, 23.48% for the triangular distribution, and 23.38% for the beta distribution.

Table 4. Sensitivity analysis for demand parameters.

Parameter	% Change in Value	% Change in II (Uniform)	% Change in II % (Triangular)	% Change in II (beta)
b	−50%	114.8	115.3	114.8
	−25%	38.2	38.59	38.26
	+25%	−22.7	−23.46	−22.87
	+50%	−38.09	−38.46	−38.19
k	−50%	−20.90	−21.03	−20.92
	−25%	−10.7	−10.81	−10.76
	+25%	11.38	11.42	11.40
	+50%	23.3	23.48	23.38

This paper assumes random defective rates. It is a common assumption in the traditional modeling approaches, that higher defective rates generate low profits. However, this is not the case for this paper. Traditional modeling approaches either considered rework or scrap option for defective products. However, our modeling approach considered that defective products are salvaged. Now two situations arise: (1) salvage value of the defective product is more than the outsourcing cost, (2) salvage value of defective products is less than the outsourcing cost. In the first case, the generated revenue is more than cost and hence decreasing the defective rate decreases the total profit. While in the second case, the generated revenue from defective products is less than the total production cost, hence, the total profit increases with the decrease in defective rate.

5.1.1. Case 2

To investigate this phenomenon in detail, we consider a second numerical example. All parametric values are taken from example 1, except that salvage value cw is considered less than the outsourcing cost of the products. In this particular case, we considered $cw = 100 < ow = 115$. The results of case 2 are compiled in Table 5. The results show that the profit of the supply chain is reduced; however, this is obvious as the revenue from defective product salvaging is reduced by 50%. Among the three different distributions, the highest profit is generated by triangular distribution compared to the previous case, in which triangular distribution generated the lowest profit.

Table 5. Optimal results of three examples, Case 2.

	Example 1	Example 2	Example 3
Distribution of Defective Rate	Uniform	Triangular	beta
$\Pi(\$)$	781, 267	781, 911	781, 326
$T^*(days)$	0.681203	0.681946	0.681272
$\lambda^*(units)$	5	5	5
$\tau^*(\%)$	0.712269	0.710318	0.712085
$P^*(\$)$	594.779	594.59	594.762

In order to clearly understand the relationship of salvage value, defective percentage, and profit of the system, a sensitivity analysis for defective rate parameters is performed for both the examples. The results of sensitivity are summarized in Tables 6 and 7. The results very clearly show that the sensitivity of profit is reversed for the second case. Although the negative and positive changes in the defective rate result in asymmetric changes in the total profit, the trend remains the same. The following important insights can be drawn from the sensitivity analysis:

- For the uniform distribution, increasing the parameter x increases the profit of the system under Case 1 and decreases it under Case 2. Increasing the distribution parameter x increases the expected value of the defective rate, this means, increasing the defective rate increases the profit of the system in Case 1 and decreases the profit in Case 2. The results are also similar for triangular distribution, in which the expected value increases with increasing the parameter x. However, x is more effective in uniform distribution compared to triangular distribution, this is because the net effect in the expected value is more in uniform distribution compared to triangular distribution.
- Increasing the parameter y also increases the expected value of the defective rate. Therefore, increasing the parameter y increases the profit of the system under Case 1 and decreases profit under Case 2, for both uniform and triangular distributions. The impact of y is more than double compared to the impact of x for uniform distribution; for triangular distribution, the impact of x, on profit, is about 50% the effect of y.
- The parameter z only applies to the triangular distribution, therefore, it has no effect on uniform distribution. For triangular distribution, increasing z increases the profit of Case 1 and decreases the profit of Case 2.
- For beta distribution, increasing λ increases the expected value of the defective rate, therefore, it increases the profit in Case 1 and decreases it in Case 2. On the other hand, increasing β decreases the expected value of the defective rate and, therefore, when β is increased, the profit of Case 1 is reduced and the profit of Case 2 is increased.

From the above results it is clear that, for Case 1, total profit increases with an increase in the defective rate, while for Case 2, the total profit decreases with an increase in the defective rate. The cost of outsourcing can also be considered as the cost of shortage and lost sales. Whatever be the situation or assumptions, if the cost is more than salvage value, supply chain managers need not worry about the defective rate. However, when the cost of shortage, outsourcing, or lost sales is more than the salvage value of imperfect items, supply chain managers should focus on reduction of defective rate.

Table 6. Sensitivity analysis for parameter related to defective rate (uniform and triangular distributions).

Parameter	% Changes in Value	Case 1, $cw > ow$		Case 2, $cw < ow$	
		% Change in Π (Uniform)	% Change in Π (Triangular)	% Change in Π (Uniform)	% Change in Π (Triangular)
x	−50%	−0.65	−0.44	0.13	0.09
	−25%	−0.32	−0.22	0.001	0.05
	+25%	0.33	0.21	−0.06	−0.04
	+50%	0.67	0.44	−0.14	−0.08
y	−50%	−1.76	−0.60	0.34	0.23
	−25%	−0.88	−0.29	0.16	0.12
	+25%	0.88	0.59	−0.16	−0.10
	+50%	1.77	1.15	−0.35	−0.21
z	−50%	0.00	−1.35	0.00	0.12
	−25%	0.00	−0.67	0.00	0.06
	+25%	0.00	0.29	0.00	−0.05
	+50%	0.00	0.59	0.00	−0.10

Table 7. Sensitivity analysis for parameter related to defective rate (beta distribution).

Parameter	% Changes in Value	Case 1, $cw > ow$	Case 2, $cw < ow$
		% Change in Π (beta)	% Change in Π (beta)
λ	−50%	−2.04	0.39
	−25%	−0.95	0.18
	+25%	0.78	−0.15
	+50%	1.40	−0.29
β	−50%	2.78	−0.54
	−25%	1.04	−0.18
	+25%	−0.74	0.15
	+50%	−1.30	0.25

Finally, we considered the impacts of investment cost in the O2O channel on total profit. Sensitivity analysis is performed for information cost and transportation cost. The results are compiled in Table 8. The results clearly show that rise in investment cost increases total profit. The effect is slightly asymmetric; the profit is more sensitive towards positive changes compared to negative changes.

Table 8. Sensitivity analysis of O2O channel parameters.

Parameter	% Change in Value	% Change in Π (Uniform)	% Change in Π % (Triangular)	% Change in Π (beta)
IN_i	−50%	−20.76	−20.83	−20.7
	−25%	−10.67	−10.72	−10.6
	+25%	11.293	11.34	11.29
	+50%	23.142	23.27	23.15
TR_i	−50%	0.080	0.091	0.080
	−25%	0.064	−0.016	0.064
	+25%	−0.043	−0.058	−0.04
	+50%	−0.090	−0.096	−0.09

6. Conclusions

This paper studies joint pricing and inventory policies for a multi-echelon CLSC model with an O2O sales channel. In order to reduce solid waste generation in supply chain, RTI is used to transport finished products from the manufacturer to retailers. A 3PL collects used products from the consumer and delivers them to the manufacturer for remanufacturing. Furthermore, carbon emissions from transportation and production activities are also considered to improve the environmental performance of supply chain management. Imperfect production is investigated with a random defective rate, whereas the defective rate follows three different probability distributions. Traditionally, supply chain inventory management considered rework or scrap option for the defective product. However, this model considered that defective products are salvaged, which generated extra revenue. Based on this assumption, two cases were developed, and the results showed that unlike traditional modeling approaches, total profit increases with an increase in defective rate when the salvage value of defective products is more than the cost of outsourcing. When the cost of outsourcing is higher compared to the salvage value, our model converged to traditional modeling approaches, in which profit decreases with an increase in defective rate. The results further showed that in CLSC models the most important parameters are manufacturing cost and remanufacturing cost, their decrease produces more positive impacts compared to the negative impacts caused by their increments. Moreover, the sensitivity analysis provided that investment in the O2O channel leads to a substantial increase in total profit. This research can be extended in many directions, for example, as we considered deterministic demand, considerations of stochastic demand would be a more realistic example. Furthermore, the modeling approach of defective items needs to be explored further in the presences of shortage cots, lost sales and back ordering. Another important extension of this model is to consider deteriorated products, such as the study done by Ullah et al. [63].

Author Contributions: Conceptualization, B.S. and M.U.; Methodology, B.S. and M.U.; Software, M.U.; Validation, B.S.; Formal analysis, B.S. and M.U.; Investigation, B.S. and M.U.; Resources, B.S. and M.U.; Data curation, B.S. and M.U.; Writing—original draft preparation, M.U.; Writing—review and editing, M.U.; Visualization, B.S. and M.U.; Supervision, B.S. and S.-B.C.

Funding: This research received no external funding.

Conflicts of Interest: The authors declare no conflict of interest.

Abbreviations

The following notation are used to develop mathematical model.

Notation

Decision variables

T	replenishment cycle time (year)
τ	remanufacturing rate (percentage of demand)
p	price of the finished product
λ	capacity of a single container (units)

Parameters

d_i	net demand realized by the ith retailer (units/unit time)
IN_i	demand from online channel at i^{th} retailer (units/unit time)
b	demand parameter
k	demand parameter
TR_i	Fixed cost of online channel per unit time (\$/unit time)
h_i	holding cost of produced products at i^{th} retailer (\$/unit/unit time)
l_i	lead time of the ith retailer
A_i	ordering cost per order of ith retailer (\$/order)
n	number of retailers
P_m	production rate of the manufacturer (units/unit time)
S_m	manufacturer setup cost per setup (\$/setup)
S_r	3PL setup cost per setup for collecting used products(\$/setup)
A_m	manufacturer ordering cost per order being placed to 3PL (\$/order)
h_f	holding cost of produced products for the manufacturer (\$/unit/unit time)
h_u	holding cost of collected products (\$/unit/unit time)
C_m	manufacturing cost (\$/product)
C_r	remanufacturing cost (\$/product)

C_q	quality improving cost (\$/product)
g	goodwill lost cost for the manufacturer (\$/product)
A_t	average incentives paid to consumer for collected products (\$/unit)
C_t	transportation cost per container per unit distance (\$/container/unit distance)
l_{im}	distance between the manufacturer and retailer i (kilometers)
l_{ij}	distance between retailer i and 3PL (kilometers)
l_{jm}	distance between 3PL and the manufacturer (kilometers)
C_λ	cost of managing RTIs, including depreciation and repair per unit capacity of RTI (\$/RTI capacity)
s	scaling factor for how the container capacity affects the cost of managing a container
h_R	holding cost of containers at 3PL (\$/unit/unit time)
γ	effective investment by 3PL to collects EOL/EOU products in dollars (\$)
e_m	carbon emissions per product from manufacturing (kg/product)
e_r	carbon emissions per gallon of fuel used by truck (kg/gallon)
g_m	fuel required per mile (gallon/miles)
C_{ec}	carbon tax per unit of emitted carbon (\$/unit of emitted carbon)
C_q	quality upgradation cost (\$/product)
t_c	truck capacity (number of products)
cw	salvage value of defective products (\$/unit)
ow	cost of out sourcing one product (\$/unit), $ow > C_m$
ck	inspection cost per product (\$/unit)
r	defective rate (random variable)
TC_1	cost of multiple retailers (\$/cycle)
TC_2	cost of manufacturer (\$/cycle)
TC_3	cost of 3PL (\$/cycle)

The following abbreviations are used in this manuscript.

O2O	Online to offline
CLSC	Closed-loop supply chain
RTI	Returnable transport item
3PL	Third party logistics

Appendix A

Appendix A.1

From Figure 2, manufacturer inventory can be written as,

$$\frac{Q^2}{2(1-r)P_m} + l_1\left(Q - q_1\right) + l_2\left(Q - q_1 - q_2\right)\ldots\ldots\ldots + l_{n-1}\left(Q - q_1 - q_2\ldots\ldots - q_{n-1}\right)$$

$$Q = (q_1 + q_2 + q_3 + q_4\ldots)\cdot(+q_{n-1}) + q_n$$

$$\frac{Q^2}{2(1-r)P_m} + T\sum_{i=1}^{n-1} l_i d_i$$

Therefore, the time-weighted average inventory is

$$\frac{d^2 T}{2(1-r)P_m} + \sum_{i=1}^{n-1} l_i d_{i+1}$$

Appendix A.2

The average inventory of RTI at 3PL can be calculated with the help of Figure 2.

$$= r_{max}\left(T - \sum_{i=1}^{n} l_i\right) + \sum_{i=1}^{n} l_i\left(r_{max} - r_i\right)$$

$$= \frac{d_{max}T}{\lambda}\left(T - \sum_{i=1}^{n} l_i\right) + \sum_{i=1}^{n} l_i\left(\frac{d_{max}T}{\lambda} - \frac{T}{\lambda}d_i\right).$$

Therefore, the time weighted average inventory of RTI at 3PL

$$= \frac{d_{max}}{\lambda}\left(T - \sum_{i=1}^{n} l_i\right) + \sum_{i=1}^{n} l_i\left(\frac{d_{max}}{\lambda} - \frac{1}{\lambda}d_i\right).$$

Appendix B

If,

$$a_{1,1} = \frac{\partial}{\partial T}\frac{\partial \Pi}{\partial T}, \; a_{1,2} = \frac{\partial}{\partial \tau}\frac{\partial \Pi}{\partial T}, \; a_{1,3} = \frac{\partial}{\partial \lambda}\frac{\partial \Pi}{\partial T}, \; a_{1,4} = \frac{\partial}{\partial p}\frac{\partial \Pi}{\partial T}, \; a_{2,1} = \frac{\partial}{\partial T}\frac{\partial \Pi}{\partial \tau}, \; a_{2,2} = \frac{\partial}{\partial \tau}\frac{\partial \Pi}{\partial \tau}, \; a_{2,3} = \frac{\partial}{\partial \lambda}\frac{\partial \Pi}{\partial \tau}, \; a_{2,4} = \frac{\partial}{\partial p}\frac{\partial \Pi}{\partial \tau}, \; a_{3,1} = \frac{\partial}{\partial T}\frac{\partial \Pi}{\partial \lambda}, \; a_{3,2} = \frac{\partial}{\partial \tau}\frac{\partial \Pi}{\partial \lambda}, \; a_{3,3} = \frac{\partial}{\partial \lambda}\frac{\partial \Pi}{\partial \lambda}, \; a_{3,4} = \frac{\partial}{\partial p}\frac{\partial \Pi}{\partial \lambda}, \; a_{4,1} = \frac{\partial}{\partial T}\frac{\partial \Pi}{\partial p}, \; a_{4,2} = \frac{\partial}{\partial \tau}\frac{\partial \Pi}{\partial p}, \; a_{4,3} = \frac{\partial}{\partial \lambda}\frac{\partial \Pi}{\partial p}, \; a_{4,4} = \frac{\partial}{\partial p}\frac{\partial \Pi}{\partial p},$$

Then the Hessian matrix of (5) can be expressed as

$$H = \begin{pmatrix} a_{1,1} & a_{1,2} & a_{1,3} & a_{1,4} \\ a_{2,1} & a_{2,2} & a_{2,3} & a_{2,4} \\ a_{3,1} & a_{3,2} & a_{3,3} & a_{3,4} \\ a_{4,1} & a_{4,2} & a_{4,3} & a_{4,4} \end{pmatrix}$$

Appendix B.1. When Defective Rate Follows Uniform Distribution

The Hessian matrix for example 1, when defective rate follows uniform distribution, at $T = 0.634$; $\lambda = 3$; $\tau = 0.65$; $p = 581.21$ is,

$$\begin{pmatrix} -33524.3 & 4001.85 & 487.243 & 23.4258 \\ 4001.85 & -6199.8 & 300.356 & -9.77975 \\ 487.243 & 300.356 & -503.327 & -0.693627 \\ 23.4258 & -9.77975 & -0.693627 & -7.20307 \end{pmatrix}$$

The first four principle minors are -33524.3, $+1.91829 \times 10^8$, -9.08852×10^{10}, and $+6.5227 \times 10^{11}$. Hence total profit is strictly concave at $T = 0.634$; $\lambda = 3$; $\tau = 0.65$; $p = 581.21$.

Appendix B.2. When Defective Rate Follows Triangular Distribution

The Hessian matrix for example 2, when defective rate follows triangular distribution, at $T = 0.66$; $\lambda = 4$; $\tau = 0.69$; $p = 582.64$ is,

$$\begin{pmatrix} -31987.9 & 4059.47 & 273.633 & 23.041 \\ 4059.47 & -6060.61 & 168.46 & -10.0465 \\ 273.633 & 168.46 & -216.887 & -0.208652 \\ 23.041 & -10.0465 & -0.208652 & -7.20295 \end{pmatrix}$$

The first four principle minors are -31987.9, $+1.77387 \times 10^8$, -3.6737×10^{10}, and $+2.63656 \times 10^{11}$. Hence total profit is strictly concave at $T = 0.66$; $\lambda = 4$; $\tau = 0.69$; $p = 582.64$.

Appendix B.3. When Defective Rate Follows Beta Distribution

The Hessian matrix for example 3, when defective rate follows beta distribution, at $T = 0.67$; $\lambda = 5$; $\tau = 0.73$; $p = 581.43$ is,

$$\begin{pmatrix} -31332.3 & 4222.72 & 175.343 & 23.1492 \\ 4222.72 & -5970.15 & 108.056 & -10.3247 \\ 175.343 & 108.056 & -113.68 & 0.0186797 \\ 23.1492 & -10.3247 & 0.0186797 & -7.20296 \end{pmatrix}$$

The first four principle minors are -31332.3, $+1.69227 \times 10^8$, -1.85284×10^{10}, and $+1.32964 \times 10^{11}$. Hence total profit is strictly concave at $T = 0.67$; $\lambda = 5$; $\tau = 0.73$; $p = 581.43$.

Appendix C.

Table A1. Sensitivity analysis for parameters related to manufacturer and 3PL.

Parameter	% Change in Value	% Change in Π (Uniform)	% Change in Π % (Triangular)	% Change in Π (beta)
h_f	-50%	0.059	0.010	0.059
	-25%	0.037	0.0211	0.003
	$+25\%$	-0.006	-0.027	-0.007
	$+50\%$	-0.029	-0.049	-79.6
h_u	-50%	0.302	0.290	0.302
	-25%	0.143	0.133	0.147
	$+25\%$	-0.09	-0.091	-0.091
	$+50\%$	-0.15	-0.196	-0.152
A_m	-50%	0.001	0.0017	0.001
	-25%	0.008	0.0008	0.0002
	$+25\%$	-0.008	-0.0008	0.013
	$+50\%$	0.012	-0.001	0.002
C_t	-50%	0.039	0.064	0.039
	-25%	0.044	0.030	0.028
	$+25\%$	-0.02	-0.034	-0.020
	$+50\%$	-0.04	-0.057	-0.044
A_t	-50%	2.011	1.999	2.01
	-25%	0.999	0.989	1.00
	$+25\%$	-0.1596	-0.17	-0.20
	$+50\%$	-0.186	-0.17	-0.16
C_λ	-50%	0.131	0.101	0.115
	-25%	0.06	0.04	0.064
	$+25\%$	-0.03	-0.05	-0.030
	$+50\%$	-0.086	-0.09	-0.075
C_q	-50%	0.381	0.345	0.381
	-25%	0.203	0.179	0.192
	$+25\%$	-0.18	-0.182	-0.19
	$+50\%$	-0.37	-0.40	-0.375
q_r	-50%	-0.03	-0.045	-0.033
	-25%	-0.02	-0.021	-0.028
	$+25\%$	0.044	0.030	0.0446
	$+50\%$	0.10	0.085	0.100
ck	-50%	0.522	0.510	0.507
	-25%	0.253	0.2548	0.253
	$+25\%$	-0.256	-0.254	-0.25
	$+50\%$	-0.506	-0.508	-0.50
h_R	-50%	0.0078	0.04	0.007
	-25%	0.0256	-0.024	0.02
	$+25\%$	-0.049	-0.033	-0.02
	$+50\%$	-0.098	-0.089	-0.02

References

1. Cassab, H.; MacLachlan, D.L. Interaction fluency: A customer performance measure of multichannel service. *Int. J. Prod. Perform. Manag.* **2006**, *55*, 555–568. [CrossRef]
2. Yang, S.; Lu, Y.; Chau, P.Y. Why do consumers adopt online channel? An empirical investigation of two channel extension mechanisms. *Decis. Support Syst.* **2013**, *54*, 858–869. [CrossRef]
3. Chang, Y.W.; Hsu, P.Y.; Yang, Q.M. Integration of online and offline channels: A view of O2O commerce. *Internet Res.* **2018**, *28*, 926–945. [CrossRef]
4. Sarkar, B.; Tayyab, M.; Choi, S.B. Product Channeling in an O2O Supply Chain Management as Power Transmission in Electric Power Distribution Systems. *Mathematics* **2019**, *7*, 4. [CrossRef]
5. Livne, G.; Simpson, A.; Talmor, E. Do customer acquisition cost, retention and usage matter to firm performance and valuation? *J. Bus. Financ. Account.* **2011**, *38*, 334–363. [CrossRef]
6. Sarkar, B.; Ullah, M.; Kim, N. Environmental and economic assessment of closed-loop supply chain with remanufacturing and returnable transport items. *Comput. Ind. Eng.* **2017**, *111*, 148–163. [CrossRef]
7. Bulmus, S.C.; Zhu, S.X.; Teunter, R. Competition for cores in remanufacturing. *Eur. J. Oper. Res.* **2014**, *233*, 105–113. [CrossRef]
8. Taleizadeh, A.A.; Sane-Zerang, E.; Choi, T.M. The effect of marketing effort on dual-channel closed-loop supply chain systems. *IEEE Trans. Syst. Man Cybern. Syst.* **2018**, *48*, 265–276. [CrossRef]
9. Schrady, D.A. A deterministic inventory model for reparable items. *Nav. Res. Logist.* **1967**, *14*, 391–398. [CrossRef]
10. Richter, K. The EOQ repair and waste disposal model with variable setup numbers. *Eur. J. Oper. Res.* **1996**, *95*, 313–324. [CrossRef]
11. Richter, K. The extended EOQ repair and waste disposal model. *Int. J. Prod. Econ.* **1996b**, *45*, 443–447. [CrossRef]
12. Richter, K. Pure and mixed strategies for the EOQ repair and waste disposal problem. *Oper. Res. Spekt.* **1997**, *19*, 123–129. [CrossRef]
13. Richter, K.; Dobos, I. Analysis of the EOQ repair and waste disposal problem with integer setup numbers. *Int. J. Prod. Econ.* **1999**, *59*, 463–467. [CrossRef]
14. Dobos, I.; Richter, K. An extended production/recycling model with stationary demand and return rates. *Int. J. Prod. Econ.* **2004**, *90*, 311–323. [CrossRef]
15. Dobos, I.; Richter, K. A production/recycling model with quality consideration. *Int. J. Prod. Econ.* **2006**, *104*, 571–579. [CrossRef]
16. Maiti, T.; Giri, B. A closed loop supply chain under retail price and product quality dependent demand. *J. Manuf. Syst.* **2015**, *37*, 624–637. [CrossRef]
17. Moshtagh, M.S.; Taleizadeh, A.A. Stochastic integrated manufacturing and remanufacturing model with shortage, rework and quality based return rate in a closed loop supply chain. *J. Clean. Prod.* **2017**, *141*, 1548–1573. [CrossRef]
18. Tian, X.; Zhang, Z.H. Capacitated disassembly scheduling and pricing of returned products with price-dependent yield. *Omega* **2019**, *84*, 160–174. [CrossRef]
19. Govindan, K.; Soleimani, H. A review of reverse logistics and closed-loop supply chains: A Journal of Cleaner Production focus. *J. Clean. Prod.* **2017**, *142*, 371–384. [CrossRef]
20. Diallo, C.; Venkatadri, U.; Khatab, A.; Bhakthavatchalam, S. State of the art review of quality, reliability and maintenance issues in closed-loop supply chains with remanufacturing. *Int. J. Prod. Res.* **2017**, *55*, 1277–1296. [CrossRef]
21. Chiang, W.Y.K.; Chhajed, D.; Hess, J.D. Direct marketing, indirect profits: A strategic analysis of dual-channel supply-chain design. *Manag. Sci.* **2003**, *49*, 1–20. [CrossRef]
22. Huang, S.; Yang, C.; Zhang, X. Pricing and production decisions in dual-channel supply chains with demand disruptions. *Comput. Ind. Eng.* **2012**, *62*, 70–83. [CrossRef]
23. Ahmad, W.; Sarkar, B.; Ullah, M. Impact of Reparation for Imperfect Quality Items having Shortages in the System under Multi-Trade-Credit-Period. *DJ J. Eng. Appl. Math.* **2018**, *5*, 1–16. [CrossRef]
24. Tsay, A.A.; Agrawal, N. Channel conflict and coordination in the e-commerce age. *Prod. Oper. Manag.* **2004**, *13*, 93–110. [CrossRef]

25. Yao, D.Q.; Liu, J.J. Competitive pricing of mixed retail and e-tail distribution channels. *Omega* **2005**, *33*, 235–247. [CrossRef]
26. Huang, W.; Swaminathan, J.M. Introduction of a second channel: Implications for pricing and profits. *Eur. J. Oper. Res.* **2009**, *194*, 258–279. [CrossRef]
27. Yan, R. Pricing strategy for companies with mixed online and traditional retailing distribution markets. *J. Prod. Brand Manag.* **2008**, *17*, 48–56. [CrossRef]
28. Dan, B.; Xu, G.; Liu, C. Pricing policies in a dual-channel supply chain with retail services. *Int. J. Prod. Econ.* **2012**, *139*, 312–320. [CrossRef]
29. Li, B.; Zhu, M.; Jiang, Y.; Li, Z. Pricing policies of a competitive dual-channel green supply chain. *J. Clean. Prod.* **2016**, *112*, 2029–2042. [CrossRef]
30. Bányai, T.; Illés, B.; Bányai, Á. Smart scheduling: An integrated first mile and last mile supply approach. *Complexity* **2018**, *2018*, 5180156 [CrossRef]
31. Bányai, T. Real-time decision making in first mile and last mile logistics: How smart scheduling affects energy efficiency of hyperconnected supply chain solutions. *Energies* **2018**, *11*, 1833. [CrossRef]
32. Habib, M.S.; Sarkar, B.; Tayyab, M.; Saleem, M.W.; Hussain, A.; Ullah, M.; Omair, M.; Iqbal, M.W. Large-scale disaster waste management under uncertain environment. *J. Clean. Prod.* **2019**, *212*, 200–222. [CrossRef]
33. Ullah, M.; Sarkar, B. Smart and sustainable supply chain management: A proposal to use rfid to improve electronic waste management. In Proceedings of the International Conference on Computers and Industrial Engineering, Auckland, New Zealand, 2–5 December 2018; Volume 2018.
34. Johansson, O.; Hellström, D. The effect of asset visibility on managing returnable transport items. *Int. J. Phys. Distrib. Logist. Manag.* **2007**, *37*, 799–815. [CrossRef]
35. Witt, C. Transport packaging: Neat and clean or down and dirty. *Mater. Hand. Eng.* **1999**, *54*, 71–73.
36. Livingstone, S.; Sparks, L. The new German packaging laws: Effects on firms exporting to Germany. *Int. J. Phys. Distrib. Logist. Manag.* **1994**, *24*, 15–25. [CrossRef]
37. Kroon, L.; Vrijens, G. Returnable containers: An example of reverse logistics. *Int. J. Phys. Distrib. Logist. Manag.* **1995**, *25*, 56–68. [CrossRef]
38. Glock, C.H. Decision support models for managing returnable transport items in supply chains: A systematic literature review. *Int. J. Prod. Econ.* **2017**, *183*, 561–569. [CrossRef]
39. Kim, T.; Glock, C.H.; Kwon, Y. A closed-loop supply chain for deteriorating products under stochastic container return times. *Omega* **2014**, *43*, 30–40. [CrossRef]
40. Glock, C.H.; Kim, T. Container management in a single-vendor-multiple-buyer supply chain. *Logist. Res.* **2014**, *7*, 112. [CrossRef]
41. Kelle, P.; Silver, E.A. Forecasting the returns of reusable containers. *J. Oper. Manag.* **1989**, *8*, 17–35. [CrossRef]
42. Goh, T.; Varaprasad, N. A statistical methodology for the analysis of the life-cycle of reusable containers. *IIE Trans.* **1986**, *18*, 42–47. [CrossRef]
43. Lee, H.L. Lot sizing to reduce capacity utilization in a production process with defective items, process corrections, and rework. *Manag. Sci.* **1992**, *38*, 1314–1328. [CrossRef]
44. Gupta, T.; Chakraborty, S. Looping in a multistage production system. *Int. J. Prod. Res.* **1984**, *22*, 299–311. [CrossRef]
45. Tayi, G.K.; Ballou, D.P. An integrated production-inventory model with reprocessing and inspection. *Int. J. Prod. Res.* **1988**, *26*, 1299–1315. [CrossRef]
46. Lee, H.H.; Chandra, M.J.; Deleveaux, V. Optimal batch size and investment in multistage production systems with scrap. *Prod. Plan. Cont.* **1997**, *8*, 586–596. [CrossRef]
47. Glock, C.H.; Jaber, M.Y. Learning effects and the phenomenon of moving bottlenecks in a two-stage production system. *Appl. Math. Model.* **2013**, *37*, 8617–8628. [CrossRef]
48. Salameh, M.; Jaber, M. Economic production quantity model for items with imperfect quality. *Int. J. Prod. Econ.* **2000**, *64*, 59–64. [CrossRef]
49. Jaber, M.Y.; Zanoni, S.; Zavanella, L.E. Economic order quantity models for imperfect items with buy and repair options. *Int. J. Prod. Econ.* **2014**, *155*, 126–131. [CrossRef]

50. Ouyang, L.Y.; Chen, C.K.; Chang, H.C. Quality improvement, setup cost and lead-time reductions in lot size reorder point models with an imperfect production process. *Comput. Oper. Res.* **2002**, *29*, 1701–1717. [CrossRef]

51. Cárdenas-Barrón, L.E. Economic production quantity with rework process at a single-stage manufacturing system with planned backorders. *Comput. Ind. Eng.* **2009**, *57*, 1105–1113. [CrossRef]

52. Eroglu, A.; Ozdemir, G. An economic order quantity model with defective items and shortages. *Int. J. Prod. Econ.* **2007**, *106*, 544–549. [CrossRef]

53. Ben-Daya, M. The economic production lot-sizing problem with imperfect production processes and imperfect maintenance. *Int. J. Prod. Econ.* **2002**, *76*, 257–264. [CrossRef]

54. Sarkar, B.; Majumder, A.; Sarkar, M.; Kim, N.; Ullah, M. Effects of variable production rate on quality of products in a single-vendor multi-buyer supply chain management. *Int. J. Adv. Manuf. Technol.* **2018**, *99*, 567–581. [CrossRef]

55. Sarkar, B. Mathematical and analytical approach for the management of defective items in a multi-stage production system. *J. Clean. Prod.* **2019**, *218*, 896–919. [CrossRef]

56. Kim, M.S.; Kim, J.S.; Sarkar, B.; Sarkar, M.; Iqbal, M.W. An improved way to calculate imperfect items during long-run production in an integrated inventory model with backorders. *J. Manuf. Syst.* **2018**, *47*, 153–167. [CrossRef]

57. Sett, B.K.; Sarkar, S.; Sarkar, B. Optimal buffer inventory and inspection errors in an imperfect production system with preventive maintenance. *Int. J. Adv. Manuf. Technol.* **2017**, *90*, 545–560. [CrossRef]

58. Sarkar, B.; Sett, B.K.; Sarkar, S. Optimal production run time and inspection errors in an imperfect production system with warranty. *J. Ind. Manag. Optim.* **2018**, *14*, 267–282. [CrossRef]

59. Kang, C.W.; Ullah, M.; Sarkar, B. Optimum ordering policy for an imperfect single-stage manufacturing system with safety stock and planned backorder. *Int. J. Adv. Manuf. Technol.* **2018**, *95*, 109–120. [CrossRef]

60. Khanna, A.; Kishore, A.; Sarkar, B.; Jaggi, C. Supply Chain with Customer-Based Two-Level Credit Policies under an Imperfect Quality Environment. *Mathematics* **2018**, *6*, 299. [CrossRef]

61. Tayyab, M.; Sarkar, B.; Ullah, M. Sustainable Lot Size in a Multistage Lean-Green Manufacturing Process under Uncertainty. *Mathematics* **2019**, *7*, 20. [CrossRef]

62. Tayyab, M.; Sarkar, B.; Yahya, B. Imperfect Multi-Stage Lean Manufacturing System with Rework under Fuzzy Demand. *Mathematics* **2019**, *7*, 13. [CrossRef]

63. Ullah, M.; Sarkar, B.; Asghar, I. Effects of Preservation Technology Investment on Waste Generation in a Two-Echelon Supply Chain Model. *Mathematics* **2019**, *7*, 189. [CrossRef]

© 2019 by the authors. Licensee MDPI, Basel, Switzerland. This article is an open access article distributed under the terms and conditions of the Creative Commons Attribution (CC BY) license (http://creativecommons.org/licenses/by/4.0/).

Article

Sustainable Supplier Selection Process in Edible Oil Production by a Hybrid Fuzzy Analytical Hierarchy Process and Green Data Envelopment Analysis for the SMEs Food Processing Industry

Chia-Nan Wang [1,2,*], Van Thanh Nguyen [1,3,*], Hoang Tuyet Nhi Thai [3], Ngoc Nguyen Tran [3] and Thi Lan Anh Tran [3]

1   Department of Industrial Engineering and Management, National Kaohsiung University of Science and Technology, Kaohsiung 80778, Taiwan
2   Department of Industrial Engineering and Management, Fortune Institute of Technology, Kaohsiung 81160, Taiwan
3   Department of Industrial Systems Engineering, CanTho University of Technology, Can Tho 900000, Vietnam; thtnhi.htcn0114@student.ctuet.edu.vn (H.T.N.T.); tnnguyen.htcn0115@student.ctuet.edu.vn (N.N.T.); ttlanh.htcn0115@student.ctuet.edu.vn (T.L.A.T.)
*   Correspondence: cn.wang@nkust.edu.tw (C.-N.W.); jenny9121989@gmail.com (V.T.N.)

Received: 12 November 2018; Accepted: 3 December 2018; Published: 4 December 2018

Abstract: Today, business organizations are facing increasing pressure from a variety of sources to operate using sustainable processes. Thus, most companies need to focus on their supply chains to enhance sustainability to meet customer demands and comply with environmental legislation. To achieve these goals, companies must focus on criteria that include CO_2 (carbon footprint) and toxic emissions, energy use and efficiency, wastage generations, and worker health and safety. As in other industries, the food processing industry requires large inputs of resources, which results in several negative environmental effects; thus, decision-makers have to evaluate qualitative and quantitative factors. This work identifies the best supplier for edible oil production in the small and medium enterprise (SME) food processing industry in Vietnam. This study also processes a hybrid multicriteria decision-making (MCDM) model using a fuzzy analytical hierarchy process (FAHP) and green data envelopment analysis (GDEA) model to identify the weight of all criteria of a supplier's selection process based on opinions from company procurement experts. Subsequently, GDEA is applied to rank all potential supplier lists. The primary objective of this work is to present a novel approach which integrates FAHP and DEA for supplier selection and also consider the green issue in edible oil production in uncertain environments. The aim of this research is also to provide a useful guideline for supplier selection based on qualitative and quantitative factors to improve the efficiency of supplier selection in the food industry and other industries. The results reveal that Decision-Making Unit 1 (DMU 1), DMU 3, DMU 7, and DMU 9 are identified as extremely efficient for five DEA models, which are the optimal suppliers for edible oil production. The contributions of this research include a proposed MCDM model using a hybrid FAHP and GDEA model for supplier selection in the SME food processing industry under a fuzzy environment conditions in Vietnam. This research also is part of an evolution of a new hybrid model that is flexible and practical for decision-makers. In addition, the research also provides a useful guideline in supplier selection in the food processing industry and a guideline for supplier selection in other industries.

Keywords: supplier selection process; multicriteria decision making (MCDM); edible oil; fuzzy analytical hierarchy process (FAHP); green data envelopment analysis (GDEA)

---

1. Introduction

The global edible oil market is expected to witness significant growth due to the increasing prevalence of unrefined, unprocessed, healthy, and organic oil. In the years to come, vegetable oil that is low in cholesterol, fat, and calories is likely to receive great attention due to increased health awareness. In addition, major improvements in retail networks, increased crop yields, oil production, and developing economies are among the key factors supporting the growth of the global edible oil market. Furthermore, the growing popularity of canola oil, trans fat-free soybean oil, and emerging olive oil will fuel the global edible oil market [1].

However, under pressure from the global oil market, the importance of edible supply chain management has to develop at the industrial and scientific levels. Challenges faced in the food supply chain are not only a concern with minimizing costs and delivering on time but also achieving sustainable levels of production. According to the Lowell Center for Sustainable Manufacturing, "Sustainability can be viewed as having three parts: environmental, economic and social (including political) with no pollution; social reward and creativity for everyone working; economic viability; conserve energy and natural resources; and safe and healthy for employees, communities and consumers" [2], and global warming, due to the excessive use of fossil fuels, has driven researchers to focus on sustainable energy sources for the future. For clean production systems, biofuel is expanding the domain of renewable and sustainable energy supplies. An efficient and sustainable supply chain plays a pivotal role in ensuring this supply [3]. Thus, to achieve these goals, companies have to pay attention to a lot of criteria that include CO_2 (carbon footprint) and toxic emissions, energy use and efficiency, wastage generations, and worker health and safety. The sustainable manufacturing concept is shown in Figure 1.

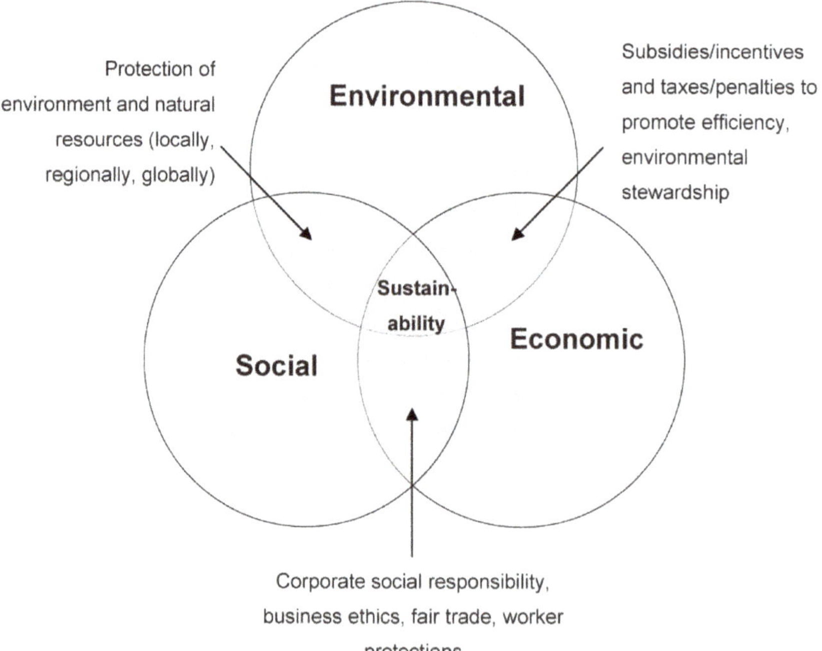

Figure 1. Sustainable manufacturing concept [4].

In order to achieve sustainability in production operations, sustainability needs to be incorporated in all stages of an organization's supply chain. Supplier selection plays an important role in supply chain management, contributing to the success of production-business organizations. Selecting a green supplier and appropriate management, which helps organizations to reduce input costs, improves the quality of goods and services provided to customers and improves market competitiveness.

Many research studies have applied the multicriteria decision-making (MCDM) model to various fields of science and engineering, and this trend has been increasing for many years. One of the fields that the MCDM model has been applied to is for supplier selection, yet few studies, to the best of our knowledge, consider this problem under uncertainly environmental conditions in the food processing industry. For the selection process of sustainable edible oil suppliers, many environmental and economic criteria need to be considered in the assessment process. Particularly in the food processing industry, companies have to deal with higher uncertainties both upstream and downstream of the supply chain. Consequently, the sustainable supplier selection process can be considered as an MCDM. However, most criteria of sustainable supplier selection are evaluated by decision-makers. Thus, to solve this problem, the author proposes a hybrid model using a fuzzy analytical hierarchy process (FAHP) and green data envelopment analysis (GDEA), which is an effective tool for quantifying ambiguous and incomplete information. Initially, the FAHP model identifies the weight of all criteria of the supplier's selection process based on opinions of company procurement experts. Subsequently, GDEA is applied for ranking all potential suppliers' lists.

The remainder of the paper introduces background materials to assist the authors in developing the MCDM model. Then, a hybrid model using the FAHP and GDEA approaches is presented to select the best supplier for edible oil production in the food industry. The results and contributions are discussed at the end of this article.

2. Literature Review

Supplier evaluation and selection problems have attracted serious research attention in the last decade. The role of the supplier selection function of supply chain management in these newer supply chain practices has only been partly explored in the literature. A number of conceptual papers have been published in the last decades that have solved selection problems with mathematical models; for example, Lin et al. [5] applied six main criteria for selecting suppliers.

Jia et al. [6] developed the framework based on the number of factors in the supplier selection process. Mendoza-Fong et al. [7] proposed some criteria for selecting and evaluating suppliers through traditional standards, such as cost, quality, delivery time, and time (JIT), and seeking continuous improvement in processes and products to face competition.

Pearson and Ellram [8] proposed supplier selection and evaluation criteria in small and large electronic firms. Wang et al. [9] applied the multicriteria group decision-making (MCGDM) model for supplier selection in a rice supply chain. In this research, all potential suppliers were to be selected based on financial, delivery and services, qualitative, and environmental management system factors. Zaimes et al. [10] discussed key research opportunities and challenges in the design of supply chains. Deng et at. [11] developed multiple attribute decision making (MADM) with some 2-tuple linguistic Pythagorean fuzzy Hamy mean operators. Wang et al. [12] developed an MADM model with interval-valued 2-tuple linguistic Pythagorean fuzzy information.

Habib and Sarkar [13] proposed an integrated location–allocation model for temporary disaster debris management under an uncertain environment. Stanković [14] used the FAHP model to select criteria and for the assessment of the impact of traffic accessibility on the development of suburbs. Hadi-Vencheh [15] proposed a hybrid model using FAHP–DEA for multiple criteria ABC inventory classification. Ulutas et al. [16] developed an integrated model including FAHP, Fuzzy Technique for Order of Preference by Similarity to Ideal Solution (TOPSIS), and AHP, axiomatic design (AD), and DEA for evaluating and selecting optimal suppliers. Gan et al. [17] used the triangular fuzzy number (TFN), AHP, and DEA approaches for analyzed economic feasibility. Rouyendegh et al. [18] combined a hybrid DEA–Analytic Network Process (ANP) for selecting the process within Iran Amirkabir University.

Based on the literature review and experts' opinion, there are some factors must be considered in edible oil supplier selection process, such as financial, delivery and devices, qualitative, and environmental management systems, and there are many researchers who have applied the MCDM model to various fields of science and engineering, a trend that has been increasing for many years. One of the fields that the MCDM model has been applied to is for supplier selection, yet very few studies consider this problem under uncertainly environmental conditions in the food processing industry. This is the reason the author has proposed a MCDM model in this research.

3. Methodology

In order to build an effective supplier selection model, the implementation process is carried out in three stages, as shown in Figure 2.

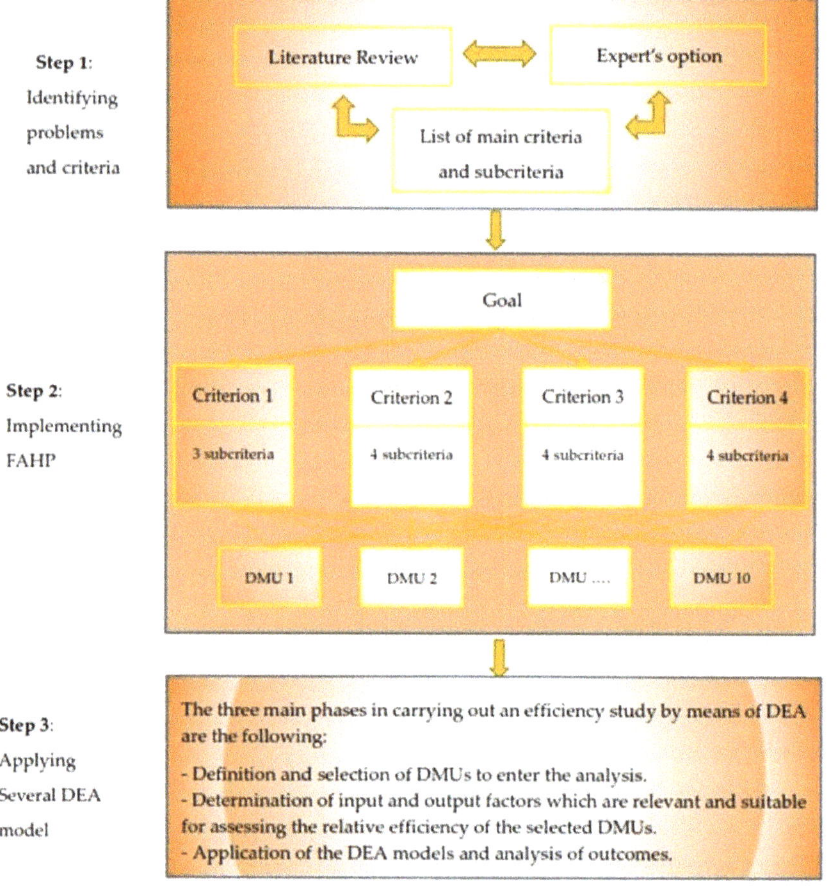

Figure 2. Research process. DMU: Decision-making unit. DEA: Data envelopment analysis.

Stage 1: Identify Problems

Identify problems to analyze and evaluate the current status of a company's selection process; the first step is to understand the procurement process and the supplier selection criteria. Collect sufficient data as criteria for selecting suppliers from experts, articles, and scientific research works related to the problem being studied (e.g., potential suppliers; data on criteria that suppliers meet).

Stage 2: Fuzzy Hybrid Analytical Hierarchy Process (FAHP)

An advantage of AHP is its stability and flexibility regarding changes within and additions to the hierarchy. Moreover, the method is able to rank criteria according to buyer needs, which also leads to more precise decisions concerning supplier selection. After receiving the overall information of each supplier in the first stage, to overcome the disadvantages of the AHP method, we used the FAHP model to determine the weight of all potential edible oil suppliers.

Stage 3: Green Data Envelopment Analysis (GDEA)

The increase in popularity of bibliographies and the large number of investigators who deal with the DEA method indicate the advantages of using the DEA method (opportunities in the estimation of return, evolution and contrast evaluation, capability of referring/marking nonprofit ability for every entry and exit in every unit, which is ideal for examining a large number of units, the requirement of low calculating power, input and outputs (units do not have to be equal)), as the leader of nonparametric methods and the FAHP may not provide a correct solution [19,20]. Thus, the author applied the GDEA model in this step. All potential suppliers were ranked by several DEA models, and optimal suppliers were determined efficient in all proposed DEA models [9].

3.1. Fuzzy Analytic Hierarchy Process (FAHP)

3.1.1. Fuzzy Sets and Fuzzy Number

In 1965, Zadeh developed a new solution to solve existing problems in an uncertain environment. It was called a Fuzzy set, which is a function that shows the dependence degree of one fuzzy number on a set number. The value of the membership function is between [0; 1] [21,22]. The triangular fuzzy number (TFN) can be defined as (o, g, p); o, g, and p $(o \leq g \leq p)$ are parameters, indicating the smallest, the most promising, and the largest value in TFN, respectively. The TFNs are shown in Figure 3.

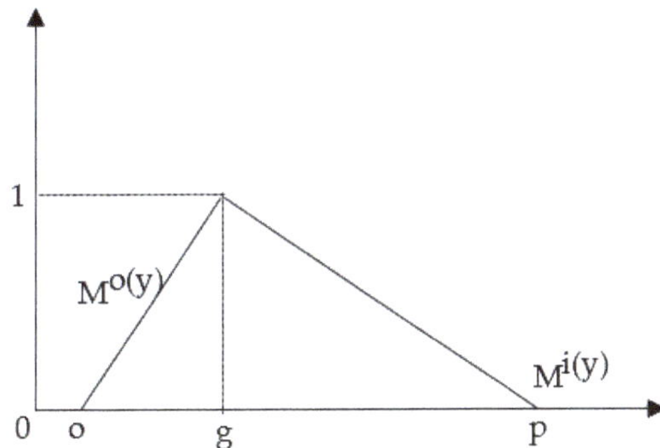

Figure 3. Traingular fuzzy number.

TFN also can be defined as follows:

$$\mu\left(\frac{x}{\widetilde{M}}\right) = \begin{cases} 0, & x < l, \\ \frac{x-o}{g-o} & o \leq x \leq g, \\ \frac{p-x}{p-g} & g \leq x \leq p, \\ 0, & x > p, \end{cases} \tag{1}$$

The representatives of each level of membership given a fuzzy number are as follows:

$$\widetilde{M} = (M^{o(y)}, M^{i(y)}) = [o + (g - o)y, p + (g - p)y], y \in [0, 1], \tag{2}$$

where $o(y)$, $i(y)$ indicates both the left side and the right side of a fuzzy number as:

$$
\begin{aligned}
(o_1, g_1, p_1) + (l_2, g_2, p_2) &= (o_1 + o_2, g_1 + g_2, p_1 + p_2) \\
(o_1, g_1, p_1) - (l_2, g_2, p_2) &= (o_1 - o_2, g_1 - g_2, p_1 - p_2) \\
(o_1, g_1, p_1) \times (l_2, m_2, u_2) &= (o_1 \times o_2, g_1 \times g_2, p_1 \times p_2) \\
\frac{(o_1, g_1, p_1)}{(l_2, g_2, p_2)} &= (o_1/p_2, g_1/g_2, p_1/p_2).
\end{aligned}
\tag{3}
$$

A pairwise comparisons matrix is used to determine the priorities on each level of the hierarchy that are quantified using a $1 \div 9$ scale, which is used by the FAHP method.

3.1.2. Fuzzy AHP

In this paper, AHP using fuzzy logic was applied to define the weight of each potential edible oil supplier. We had eight steps in the FAHP process:

Step 1: Calculation of Triangular Fuzzy Number

All criteria were considered by a pairwise comparison matrix. In place of a numeral value, the FAHP is a range of values that are combined to evaluate the weight of criteria [23]. The fuzzy prioritization method uses this scale in Parkash's study [24]. The fuzzy conversion scale is presented in Table 1.

Table 1. Triangular fuzzy scale (TFS).

Importance Intensity	TFS	Importance Intensity	TFS
1	$(1, 1, 1)$	1/1	$(1, 1, 1)$
2	$(1, 2, 3)$	1/2	$(1/3, 1/2, 1/1)$
3	$(2, 3, 4)$	1/3	$(1/4, 1/3, 1/2)$
4	$(3, 4, 5)$	1/4	$(1/5, 1/4, 1/3)$
5	$(4, 5, 6)$	1/5	$(1/6, 1/5, 1/4)$
6	$(5, 6, 7)$	1/6	$(1/7, 1/6, 1/5)$
7	$(6, 7, 8)$	1/7	$(1/8, 1/7, 1/6)$
8	$(7, 8, 9)$	1/8	$(1/9, 1/8, 1/7)$
9	$(9, 9, 9)$	1/9	$(1/9, 1/9, 1/9)$

Step 2: Calculation of $\widetilde{P}_d$

$$
\widetilde{P}_d = (o_d, g_d, p_d)
\tag{4}
$$

$$
o_d = (o_{d1} \otimes o_{d2} \otimes \ldots \otimes o_{da})^{\frac{1}{a}}, \; d = 1, 2, \ldots a
\tag{5}
$$

$$
g_d = (g_{d1} \otimes g_{d2} \otimes \ldots \otimes g_{da})^{\frac{1}{a}}, \; d = 1, 2, \ldots a
\tag{6}
$$

$$
p_d = (p_{d1} \otimes p_{d2} \otimes \ldots \otimes p_{da})^{\frac{1}{a}}, \; a = 1, 2, \ldots a
\tag{7}
$$

Step 3: Calculation of $\widetilde{P}_Y$

$$
\widetilde{P}_Y = \left(\sum_{d=1}^{a} o_d, \sum_{d=1}^{a} g_d, \sum_{d=1}^{a} p_d \right)
\tag{8}
$$

Step 4: Calculation of $\widetilde{R}$

$$
\widetilde{R} = \frac{\widetilde{P}_d}{\widetilde{P}_Y} = \frac{(o_d, g_d, p_d)}{\sum_{d=1}^{a} l_d, \sum_{d=1}^{a} g_d, \sum_{d=1}^{a} p_a} = \left[\frac{o_d}{\sum_{d=1}^{a} p_d}, \frac{g_d}{\sum_{d=1}^{a} p_d}, \frac{p_d}{\sum_{d=1}^{a} o_d} \right]
\tag{9}
$$

Step 5: Calculation of $td_{\beta o}$

The criteria depend on β cut values, which are defined for the calculated β. The fuzzy priorities shall apply for lower and upper bounds for each β value:

$$Td_{\beta o} = \left(Tdo_{\beta o}, Tdp_{\beta o}\right); d = 1, 2, \ldots a; o = 1, 2, \ldots O \tag{10}$$

Step 6: Calculation of T_{do}, T_{dp}

$$T_{do} = \frac{\sum_{d=1}^{a} \beta(T_{do})_o}{\sum_{o=1}^{O} \beta_o}; d = 1, 2, \ldots a; o = 1, 2, \ldots O \tag{11}$$

$$T_{dp} = \frac{\sum_{d=1}^{a} \beta\left(t_{dp}\right)_o}{\sum_{o=1}^{O} \beta_o}; d = 1, 2, \ldots a; o = 1, 2, \ldots O \tag{12}$$

Step 7: Calculation of T_{wd}

The optimism index (γ) to order to defuzzy was used by combining the upper and the lower bounds values.

$$T_{wd} = \gamma \cdot W_{dp} + (1 - \gamma) \cdot W_{do}; \gamma \in [0, 1] da = 1, 2, \ldots a \tag{13}$$

Step 8: Calculation of T_{dz}

The defuzzification values priorities are normalized by:

$$T_{dz} = \frac{T_{wd}}{\sum_{d=1}^{a} T_{wd}}; d = 1, 2, \ldots a \tag{14}$$

3.2. Data Envelopment Analysis Model

3.2.1. Charnes–Cooper–Rhodes Model (CCR Model)

The Charnes–Cooper–Rhodes model (CCR) is a basic DEA model [25]. The definition of the CCR model is as follows:

$$\max_{e.d} F = \frac{d^F y_0}{w^F x_0},$$

subject to :

$$d^F y_e - w^F x_e \leq 0, \ n = 1, 2, \ldots, l \tag{15}$$
$$d \geq 0$$
$$w \geq 0$$

In addition, the fractional program as a linear program (LP) is as follows [26] if:

$$\max_{d.w} \zeta = d^F y_0$$

subject to :

$$w^F x_0 - 1 = 0 \tag{16}$$
$$d^F y_n - w^F x_n \leq 0, \ n = 1, 2, \ldots, l$$
$$w \geq 0$$
$$d \geq 0$$

The linear program (Equation (1)) is equal to the fractional program (Equation (2)) [27].

The Farrell model of the linear program (Equation (1)) with variable ζ and a nonnegative vector $\alpha = \alpha_1, \alpha_2, \alpha_3, \ldots, \alpha_f$ as [26]:

$$\max \sum_{h=1}^{g} s_i^- + \sum_{i=1}^{k} s_r^+,$$

subject to :

$$\sum_{e=1}^{n} x_{hn}\alpha_h + s_h^- = \zeta x_{h0}, \ h = 1, 2, \ldots, j$$
$$\sum_{e=1}^{n} y_{in}\alpha_n - s_i^+ = y_{i0}, \ i = 1, 2, \ldots, k$$
$$\alpha_n \geq 0, \ e = 1, 2, \ldots, l$$
$$s_h^- \geq 0, b = 1, 2, \ldots, j$$
$$s_i^+ \geq 0, r = 1, 2, \ldots, k$$

$$(17)$$

Avoid the inefficiency border point by invoking a linear program as follows [26]:

$$\max \sum_{h=1}^{g} s_b^- + \sum_{i=1}^{m} s_r^+,$$

subject to :

$$\sum_{n=1}^{l} x_{hn}\alpha_e + s_h^- = \zeta x_{h0}, \ h = 1, 2, \ldots, j$$
$$\sum_{n=1}^{l} y_{in}\alpha_n - s_i^+ = y_{i0}, \ i = 1, 2, \ldots, k$$
$$\alpha_n \geq 0, \ n = 1, 2, \ldots, l$$
$$s_h^- \geq 0, h = 1, 2, \ldots, j$$
$$s_i^+ \geq 0, i = 1, 2, \ldots, k$$

$$(18)$$

In this case, however, note that the choices the s_h^- and s_i^+ do not affect the optimal ζ^*. DMU$_0$ achieves 100% efficiency if and only if both (15) $\zeta = 1$ and (2) $s_h^{-*} = s_i^+ = 0$. Only if both (15) $\zeta^* = 1$ and (16) $s_h^{-*} \neq 0$ and $s_i^+ \neq 0$ for a or i in optimal options, so the performance of DMU$_0$ is inefficient. Therefore, the preceding development amounts to solving the problem as follows [26]:

$$\min\theta - \mu \left(\sum_{h=1}^{g} s_b^- + \sum_{i=1}^{m} s_r^+ \right),$$

subject to :

$$\sum_{h=1}^{l} x_{hn}\alpha_n + s_h^- = \zeta x_{h0}, \ h = 1, 2, \ldots, j$$
$$\sum_{n=1}^{n} y_{in}\alpha_n - s_i^+ = y_{i0}, \ i = 1, 2, \ldots, k$$
$$\alpha_n \geq 0, \ n = 1, 2, \ldots, l$$
$$s_h^- \geq 0, h = 1, 2, \ldots, j$$
$$s_i^+ \geq 0, i = 1, 2, \ldots, j$$

$$(19)$$

In this case, the s_h^- and s_i^+ variables are used to convert the inequalities into tantamount equations. Reducing min ζ at the first phase will resolve the resolution, and then fixing $\zeta = \zeta^*$, where the slacks variables achieve a maximum value, but do not affect the previously determined value of $\xi = \xi^*$. The objective will be converted from max to min, to obtain [26]:

$$\max_{d.w} \xi = \frac{d^F x_0}{w^F y_n},$$

subject to :

$$d^F x_0 \leq w^F y_n, \ n = 1, 2, \ldots, l \tag{20}$$
$$w \geq \varepsilon > 0$$
$$d \geq \varepsilon > 0$$

If the non-Archimedean element is defined and the $\varepsilon > 0$, the input models are as follows [26]:

$$\max_{d.w} \xi = w^F x_0$$

subject to :

$$d^F y_0 = 1 \tag{21}$$
$$w^F x_o - d^F y_n \geq 0, \ n = 1, 2, \ldots, l$$
$$w \geq \varepsilon > 0$$
$$d \geq \varepsilon > 0$$

And:

$$\max \phi - \varepsilon \left(\sum_{h=1}^{g} s_i^- + \sum_{i=1}^{m} s_r^+ \right),$$

subject to :

$$\sum_{n=1}^{l} x_{hn} \alpha_n + s_h^- = x_{h0}, \ h = 1, 2, \ldots, j \tag{22}$$
$$\sum_{n=1}^{l} y_{in} \alpha_n - s_i^+ = \emptyset \, y_{r0}, \ i = 1, 2, \ldots, k$$
$$\alpha_n \geq 0, \ n = 1, 2, \ldots, l$$
$$s_h^- \geq 0, h = 1, 2, \ldots, j$$
$$s_i^+ \geq 0, i = 1, 2, \ldots, k$$

The CCR input-oriented (CCR–I) has the dual multiplier model expressed as [26]:

$$\max z = \sum_{i=1}^{k} \partial_i y_{i0}$$

subject to :

$$\sum_{i=1}^{k} \partial_i y_{in} - \sum_{i=1}^{k} a_i y_{in} \leq 0 \tag{23}$$
$$\sum_{h=1}^{j} a_h x_{h0} = 1$$
$$c_i, a_h \geq \varepsilon > 0$$

The CCR output-oriented (CCR–O) has the dual multiplier model expressed as [26]:

$$\min k = \sum_{h=1}^{j} a_h x_{h0},$$

subject to :

$$\sum_{h=1}^{j} a_h x_{hn} - \sum_{i=1}^{q} \partial_i y_{in} \leq 0 \tag{24}$$
$$\sum_{i=1}^{k} \partial_i y_{i0} = 1$$
$$w_i, d_h \geq \varepsilon > 0$$

3.2.2. Banker Charnes Cooper Model (BCC Model)

Banker et al. introduced an input-oriented BCC model (BCC–I), which is able to assess the efficiency of DMU$_0$ by solving the following linear program:

$$\zeta_H = min\zeta,$$

subject to :

$$\sum_{n=1}^{l} x_{hn}\alpha_n + s_h^- = \zeta x_{h0}, \ h = 1, 2, \ldots, j$$
$$\sum_{n=1}^{l} y_{in}\alpha_n - s_i^+ = y_{i0}, \ i = 1, 2, \ldots, k \qquad (25)$$
$$\sum_{c=1}^{l} \alpha_c = 1$$
$$\alpha_c \geq 0, c = 1, 2, \ldots, l$$

Avoid the inefficiency border point by invoking the linear program as follows [26]:

$$\max \sum_{h=1}^{g} s_b^- + \sum_{i=1}^{m} s_r^+,$$

subject to :

$$\sum_{n=1}^{l} x_{hn}\alpha_n + s_h^- = \zeta x_{h0}, \ h = 1, 2, \ldots, j$$
$$\sum_{n=1}^{l} y_{in}\alpha_n - s_i^+ = y_{i0}, \ i = 1, 2, \ldots, k \qquad (26)$$
$$\sum_{c=1}^{l} \alpha_c = 1$$
$$\alpha_c \geq 0, \ c = 1, 2, \ldots, l$$
$$s_h^- \geq 0, h = 1, 2, \ldots, j$$
$$s_i^+ \geq 0, i = 1, 2, \ldots, k$$

Therefore, this is the first multiplier form to the solve problem as follows [26]:

$$min\zeta - \varepsilon \left(\sum_{h=1}^{g} s_b^- + \sum_{i=1}^{m} s_r^+ \right),$$

subject to :

$$\sum_{n=1}^{l} x_{hn}\alpha_n + s_h^- = \zeta x_{a0}, \ h = 1, 2, \ldots, j$$
$$\sum_{n=1}^{l} y_{in}\alpha_n - s_i^+ = y_{i0}, \ i = 1, 2, \ldots, k \qquad (27)$$
$$\sum_{k=1}^{n} \alpha_k = 1$$
$$\alpha_c \geq 0, \ c = 1, 2, \ldots, l$$
$$d_h^- \geq 0, h = 1, 2, \ldots, j$$
$$d_i^+ \geq 0, i = 1, 2, \ldots, k$$

The second multiplier form given by the linear program is expressed as [26]:

$$\max_{d.w,d_0} \zeta_H = d^F y_0 - d_0,$$

subject to :

$$w^F x_0 = 1 \qquad (28)$$
$$d^F y_n - w^F x_n - d_0 \leq 0, \ n = 1, 2, \ldots, l$$
$$w \geq 0$$
$$d \geq 0$$

The cases f and u are vectors, and the scalar v_0 may be positive or disclaim or zero. Therefore, the dual program [26] has the equivalent BCC fractional program:

$$\max_{d,w} \zeta = \frac{d^F y_0 - d_0}{w^F x_0},$$

subject to :

$$\frac{d^F y_e - d_0}{w^F x_e} \leq 1, \; n = 1, 2, \ldots, l \tag{29}$$
$$w \geq 0$$
$$d \geq 0$$

BCC is effectively the DMU_0 if an optimal solution $(\zeta_B^*, s^{-*}, s^{+*})$ is required in this two-phase process for satisfying $\zeta_B^* = 1$, and has no slack s $s^{-*} = s^{+*} = 0$. Alternatively, the BCC is non-efficient. The BCC effective illustration [26] is the improved activity $(\zeta^* x - s^{-*}, y + s^{+*})$. In addition, a DMU has a minimum input value for any input item, or a maximum output value for any output item.

3.2.3. Slacks-Based Measure Model (SBM Model):

The stacks-based measure model SBM input effective Input-Oriented SBM (SBM–I–C) is Input-oriented SBM under a constant-returns-to-scale assumption [26]:

$$\rho_I^* = \min_{\alpha, s^-, s^+} 1 - \frac{1}{g} \sum_{h=1}^{g} \frac{s_h^-}{x_{he}},$$

subject to :

$$x_{hw} = \sum_{n=1}^{g} x_{hw} \alpha_h + s_h^-, \; h = 1, 2, \ldots j \tag{30}$$
$$y_{wi} = \sum_{n=1}^{g} y_{wi} \alpha_n - s_i^+, \; i = 1, 2, \ldots k$$
$$\alpha_n \geq 0, c(\forall j), \; s_h^- \geq 0 \; (\forall n), s_i^+ \geq 0 (\forall n)$$

The output-oriented SBM effective ρ_O^* of $DMU_z = (x_z, y_z)$ is outlined by SBM–O–C [27]:

$$\frac{1}{\rho_O^*} = \max_{\alpha, s^-, s^+} 1 + \frac{1}{s} \sum_{i=1}^{k} \frac{s_r^+}{y_{iz}},$$

subject to :

$$x_{rh} = \sum_{n=1}^{l} x_{hn} \alpha_n + s_n^- \; (h = 1, \ldots j) \tag{31}$$
$$y_{hr} = \sum_{n=1}^{l} y_{hn} \alpha_n + s_h^+ \; (h = 1, \ldots j)$$
$$\alpha_n \geq 0 (\forall n), \; s_n^- \geq 0 (\forall h), s_n^+ \geq 0 \; (\forall i)$$

4. Case Study

Edible oil suppliers are important in business operations for the small and medium enterprise (SME) food processing industry. The supplier ensures that raw materials are of sufficient quantity, quality, and stability and accuracy to meet the requirements of production and business with a low cost and delivery time. Therefore, selecting good suppliers and managing them are prerequisites for organizing the production of quality products as desired, according to schedules, with reasonable prices and competitiveness in the market and also to obtain supplier support to continue to achieve a higher goal. Thus, with MCDM used in edible oil supplier selection, the decision-maker must consider both qualitative and quantitative factors.

Thus, the main aim of this work is the proposed suppliers' selection processes, using FAHP and DEA with green factors for edible oil suppliers based on financial, delivery, environmental management system, and qualitative factors.

The proposed model was used to rank potential soybean suppliers of a well-known food processing industry in Vietnam. After preliminary evaluation, 10 potential suppliers (decision-making units (DMU)) were selected by interviewing experts and heads of purchasing departments based on product capacity, time of delivery, supplier's location, and unit price. A supplier's list and their symbol in the proposed model are shown in Table 2.

Table 2. The symbol of ten edible Oil suppliers.

No	Name	Symbol
1	Wilmar argo Vietnam Company Limited	DMU1
2	Truong Thinh Joint Stock Company	DMU2
3	Long Gia Company Limited	DMU3
4	Binh Minh Joint Stock Incorporated Company	DMU4
5	Kim Hai Rice Private Business	DMU5
6	Van Nam Export-Production Joint Stock Company	DMU6
7	Thi Hien Joint Stock Company	DMU7
8	Binh Dien Export-Production Joint Stock Company	DMU8
9	Nuy Uyn Joint Stock Company	DMU9
10	Sa Dec Joint Stock Company	DMU10

The list of main criteria and subcriteria for selecting edible oil suppliers from experts, articles, and scientific research works are shown in Table 3.

Table 3. List of main criteria and subcriteria for selecting the best suppliers.

Main Criteria	Subcriteria
C1: Financial	C11: Capital and finance status C12: Prices C13: Transportation cost to the geographical location
C2: Delivery and services	C21: Delivery C22: Customer service C23: Communication system C24: Production capacity
C3: Qualitative	C31: Quality of Products C32: Operational Control C33: Expert labor, technical capabilities, and facilities C34: Business experience and position among competitors
C4: Environmental management system	C41: Environmental emissions (carbon footprint) C42: Environmental planning C43: Environmentally friendly material C44: Environmentally friendly technology

The weight of potential suppliers, defined by the AHP model using fuzzy logic, is shown in Table 4.

Table 4. The weight of each supplier.

No	DMU	Weight
1	DMU1	0.252
2	DMU2	0.087
3	DMU3	0.093
4	DMU4	0.053
5	DMU5	0.089
6	DMU6	0.103
7	DMU7	0.081
8	DMU8	0.085
9	DMU9	0.070
10	DMU10	0.085

Based on literature reviews and experts, five factors were considered in the DEA model, including unit price, carbon footprint, delivery time, quality of edible oil, and qualitative benefits factor. In summary, a graphic of the green DEA model for the analysis of potential suppliers along with two inputs and three outputs is shown in Figure 4.

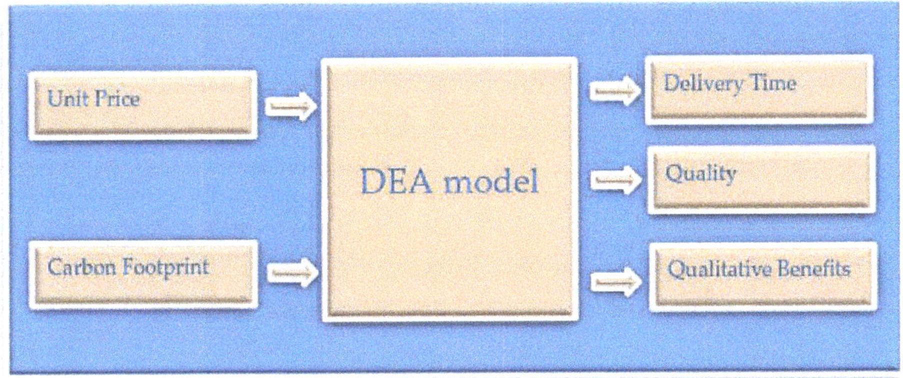

Figure 4. A Graphic of the green data envelopment analysis (GDEA) model.

The data used in the DEA model are shown in Table 5. The unit price and carbon footprint provided by suppliers are inputs. The delivery, and quality of edible oil are outputs of the DEA model; the quality of edible oil was ranked by the team of experts (on a Likert scale 1–9) [28]. Additionally, the results of the FANP model for the ranking of various suppliers on qualitative factors were utilized in the output qualitative benefits of the DEA model [29].

Table 5. Data used in the DEA model.

DMU	(I) Unit Price	(I) Carbon Footprint	(O) Delivery	(O) Quality	(O) Qualitative Benefits
DMU1	0.989	0.953	7	8	0.252
DMU2	0.997	0.978	7	8	0.087
DMU3	0.982	1.310	9	7	0.093
DMU4	1.000	0.978	5	8	0.053
DMU5	0.991	0.998	7	8	0.089
DMU6	1.006	1.010	5	8	0.103
DMU7	0.910	0.999	8	8	0.081
DMU8	1.003	1.013	6	8	0.085
DMU9	0.969	0.665	8	6	0.070
DMU10	0.787	1.001	4	5	0.085

The variables of inputs and outputs for the correlation coefficient matrix should comply with the Isotonicity premise. The results of the Pearson correlation test are in Table 6; all correlation coefficients are positive, so they meet the basic requirements of the DEA model.

Table 6. Pearson correlation coefficient.

	Unit Price	Carbon Footprint	Delivery	Quality	Qualitative Benefits
Unit price	1	0.020	0.332	0.759	0.127
Carbon footprint	0.020	1	0.117	0.203	0.021
Delivery	0.332	0.117	1	0.170	0.105
Quality	0.759	0.203	0.170	1	0.207
Qualitative benefits	0.127	0.021	0.105	0.207	1

5. Results and Discussion

In order to promote the selection of suppliers in the food industry, the selection of suppliers in the edible oil sector is important in achieving supply chain goals. However, the edible oil selection process tends to be incomplete, inaccurate, and vague.

The MCDM model has been applied in various fields of science and engineering; moreover, this trend has been increasing for many years. One of the fields that the MCDM model has been applied to is for supplier selection in edible oil production, yet very few studies consider this problem with green issues and under fuzzy environmental conditions. Thus, we have developed a supplier selection process for the SME food processing industry. In this research, to define the weight of each potential supplier, the AHP model with the combination of fuzzy logic was presented in the first stage of this study; further, the FAHP model focuses on the suppliers' ratings with four main and 15 subcriteria derived from research-relevant documents and experts. As per the literature review, FAHP was proven to be an appropriate method for evaluating and making multicriteria decisions. Then, several DEA models were proposed for ranking edible oil suppliers. The results showed that DMU 1, DMU 3, DMU 7, and DMU 9 were identified as extremely efficient for five DEA models, as shown in Table 7 [30], which have a condition response to the enterprises' supply requirement.

Table 7. Results of the DEA model.

DMU	CCR–I	CCR–O	BCC–I	SBM–I–C	SBM–O–C
DMU 1	1	1	1	1	1
DMU 2	0.981	0.981	0.981	0.975	0.608
DMU 3	1	1	1	1	1
DMU 4	0.980	0.980	0.980	0.973	0.415
DMU 5	0.971	0.970	0.971	0.961	0.621
DMU 6	0.958	0.958	0.958	0.949	0.610
DMU 7	1	1	1	1	1
DMU 8	0.957	0.957	0.957	0.947	0.577
DMU 9	1	1	1	1	1
DMU 10	0.743	0.743	1	0.679	0.598

6. Conclusions

Supplier selection plays an important role in supply chain management and contributes to the success of production-business organizations. Selecting a green supplier and appropriate management, which helps organizations reduce input costs, improves the quality of goods and services provided to customers, and improves competitiveness in the market.

Many studies have applied the MCDM model to various fields of science and engineering, and this trend has been increasing for many years. One of the fields that the MCDM model has been applied to is for supplier selection in supply chain management, such as in Diouf and Kwak [31], Kim and Changhee [32], Wang and Tsai [33], etc. Nevertheless, few studies consider this problem under uncertain environmental conditions, and the edible oil supplier selection process can be considered as an MCDM. Thus, the author proposed a hybrid model using FAHP and GDEA, which is an effective tool for quantifying ambiguous and incomplete information. Initially, the FAHP model identified the weight of all criteria of the supplier's selection process based on the opinion of company procurement experts. Subsequently, GDEA was applied to rank all potential suppliers. The results revealed that DMU 1, DMU 3, DMU 7, and DMU 9 were identified as extremely efficient for five DEA models. Further, the study results suggest that the proposed model is feasible. This study gives businesses more choices in decision-making.

The contribution of this research is proposing new and feasible approaches for supplier evaluation and selection in the food processing industry under a fuzzy environment. This is a useful model in the academic and practical fronts. In addition, this research can be broadened, thus creating a premise for applying supplier selection in other industries and, in particular, extending the model in evaluating and selecting green suppliers.

For future research, it is suggested that applications be increased through the development of new criteria and approached, such as Preference Ranking Organizational Method for Enrichment Evaluation (PROMETHEE) and FANP. A lexicographic optimization algorithm will also be considered for other fields within the food processing industry.

Author Contributions: In this research, C.-N.W. built the research ideas and reviewed the manuscript. V.T.N. designed the frameworks, analyzed the data, and wrote the manuscript. H.T.N.T. and N.N.T. collected the data and wrote the manuscript. T.L.A.T. wrote and formatted the manuscript.

Funding: This research is not received funding.

Acknowledgments: The authors appreciate the support from the National Kaohsiung University of Science and Technology, and Ministry of Sciences and Technology in Taiwan.

Conflicts of Interest: The authors declare no conflicts of interest.

References

1. Transparencymarketresearch. Available online: https://www.transparencymarketresearch.com/edible-oil-market.html (accessed on 12 June 2018).
2. Lowell Center for Sustainable Production. Available online: https://www.uml.edu/research/lowell-center/ (accessed on 12 June 2018).
3. Kuswandari, R. *Assessment of Different Methods for Measuring the Sustainability of Forest Management Retno Kuswandari*; International Institute for Geo-Information Science and Earth Observation: Enschede, The Netherlands, 2004.
4. Prakash, T. *Land Suitability Analysis for Agricultural Crops: A Fuzzy Multi Criteria Decision Making Approach*; International Institute for Geo-Information Science and Earth Observation: Enschede, The Netherlands, 2003.
5. Ahmed, W.; Sarkar, B. Impact of carbon emissions in a sustainable supply chain management for a second generation biofuel. *J. Clean. Prod.* **2018**, *186*, 807–820. [CrossRef]
6. Arkay, E.; Ertek, G.; Buyukozkan, G. Analyzing the solutions of DEA through information visualization and data mining techniques: SmartDEA framework. *Expert Syst. Appl.* **2012**, *39*, 7763–7775. [CrossRef]
7. Diouf, M.; Kwak, C. Fuzzy AHP, DEA, and Managerial Analysis for Supplier Selection and Development; From the Perspective of Open Innovation. *Sustainability* **2018**, *10*, 3779. [CrossRef]
8. Kim, I.; Kim, C. Supply Chain Efficiency Measurement to Maintain Sustainable Performance in the Automobile Industry. *Sustainability* **2018**, *10*, 2852. [CrossRef]
9. Al-Quran, A.; Alkhazaleh, S. Solar Panel Supplier Selection for the PhotovoltaicSystem Design by Using Fuzzy Multi-CriteriaDecision Making (MCDM) Approaches. *Energies* **2018**, *11*, 1989. [CrossRef]
10. Rosen, M.A.; Kishawy, H.A. Sustainable Manufacturing and Design: Concepts, Practices and Needs. *Sustainability* **2012**, *4*, 154–174. [CrossRef]
11. Lin, K.P.; Hung, K.-C.; Lin, Y.-T.; Hsieh, Y.-H. Green Suppliers Performance Evaluation in Belt and Road Using Fuzzy Weighted Average with Social Media Information. *Sustainability* **2017**, *10*, 5. [CrossRef]
12. Jia, P.; Govindan, K.; Choi, T.-M.; Rajendran, S. Supplier Selection Problems in Fashion Business Operations with Sustainability Considerations. *Sustainability* **2015**, *7*, 1603–1619. [CrossRef]
13. Mendoza-Fong, J.R.; García-Alcaraz, J.L.; Díaz-Reza, J.R.; Sáenz Diez Muro, J.C.; Fernández, J.B. The Role of Green and Traditional Supplier Attributes on Business Performance. *Sustainability* **2017**, *9*, 1520. [CrossRef]
14. Pearson, J.N.; Ellram, L.M. Supplier selection and evaluation in small versus large electronics firms. *J. Small Bus. Manag.* **1995**, *33*, 53–65.
15. Wang, C.; Nguyen, V.T.; Duong, D.H.; Do, H.T. A Hybrid Fuzzy Analytic Network Process (FANP) and Data Envelopment Analysis (DEA) Approach for Supplier Evaluation and Selection in the Rice Supply Chain. *Symmetry* **2018**, *10*, 221. [CrossRef]
16. Zaimes, G.G.; Vora, N.; Chopra, S.S.; Landis, A.E.; Khanna, V. Design of Sustainable Biofuel Processes and Supply Chains: Challenges and Opportunities. *Processes* **2015**, *3*, 634–663. [CrossRef]
17. Xiahou, X.; Tang, Y.; Yuan, J.; Chang, T.; Liu, P.; Li, Q. Evaluating Social Performance of Construction Projects: An Empirical Study. *Sustainability* **2018**, *10*, 2329. [CrossRef]

18. Stanković, M.; Gladović, P.; Popović, V.; Lukovac, V. Selection Criteria and Assessment of the Impact of Traffic Accessibility on the Development of Suburbs. *Sustainability* **2018**, *10*, 1977. [CrossRef]
19. Hadi-Vencheh, A.; Mohamadghasemi, A. A fuzzy ahp-dea approach for multiple criteria abc inventory classification. *Expert Syst. Appl.* **2011**, *38*, 3346–3352. [CrossRef]
20. Ulutas, A.; Kiridena, S.; Gibson, P.; Shukla, N. A novel integrated model to measure supplier performance considering qualitative and quantitative criteria used in the supplier selection process. *Int. J. Logist. SCM Syst.* **2012**, *6*, 57–70.
21. Gan, L.; Xu, D.; Hu, L.; Wang, L. Economic feasibility analysis for renewable energy project using an integrated tfn–ahp–dea approach on the basis of consumer utility. *Energies* **2017**, *10*, 2089. [CrossRef]
22. Rouyendegh, B.D.; Erol, S. The dea—Fuzzy anp department ranking model applied in iran amirkabir university. *Acta Polytech. Hung.* **2010**, *7*, 103–114.
23. Ziemba, P.; Wątróbski, J.; Jankowski, J.; Piwowarski, M. Research on the Properties of the AHP in the Environment of Inaccurate Expert Evaluations. In *Selected Issues in Experimental Economics*; Springer: Cham, Switzerland, 2016; pp. 227–243. [CrossRef]
24. Shu, M.S.; Cheng, C.H.; Chang, J.R. Using intuitionistic fuzzy set for fault-tree analysis on printed circuit board assembly. *Microelectron. Reliab.* **2006**, *46*, 2139–2148. [CrossRef]
25. Kahraman, Ç.; Ruan, D.; Ethem, T. Capital budgeting techniques using discounted fuzzy versus probabilistic cash. *Inf. Sci.* **2002**, *42*, 57–76. [CrossRef]
26. Cooper, C.W.; Rhodes, E. Measuring the efficiency of decision making units. *Eur. J. Oper. Res.* **1978**, *2*, 429–444.
27. Farrell, M.J. The Measurement of Productive Efficiency. *J. R. Stat. Soc.* **1957**, *120*, 253–281. [CrossRef]
28. Wen, M. Uncertain Data Envelopment Analysis. In *Uncertainty and Operations Research*; Springer: Berlin/Heidelberg, Germany, 2015.
29. Kumar, A.; Jain, V.; Kumar, S. A comprehensive environment friendly approach for supplier selection. *Omega* **2014**, *42*, 109–123. [CrossRef]
30. Sarkis, J. A methodological framework for evaluating environmentally conscious manufacturing programs. *Comput. Ind. Eng.* **1999**, *36*, 793–810. [CrossRef]
31. Tone, K. A slacks-based measure of efficiency in data envelopment analysis. *Eur. J. Oper. Res.* **2001**, *130*, 498–509. [CrossRef]
32. Deng, X.; Wang, J.; Wei, G.; Lu, M. Models for Multiple Attribute Decision Making with Some 2-Tuple Linguistic Pythagorean Fuzzy Hamy Mean Operators. *Sustainability* **2018**, *6*, 236. [CrossRef]
33. Wang, J.; Wei, G.; Gao, H. Approaches to Multiple Attribute Decision Making with Interval-Valued 2-Tuple Linguistic Pythagorean Fuzzy Information. *Mathematics* **2018**, *6*, 201. [CrossRef]

© 2018 by the authors. Licensee MDPI, Basel, Switzerland. This article is an open access article distributed under the terms and conditions of the Creative Commons Attribution (CC BY) license (http://creativecommons.org/licenses/by/4.0/).

Article

A Model and an Algorithm for a Large-Scale Sustainable Supplier Selection and Order Allocation Problem

Jong Soo Kim [1,*], Eunhee Jeon [1], Jiseong Noh [1] and Jun Hyeong Park [2]

[1] Department of Industrial and Management Engineering, Hanyang University, Erica Campus, Ansan 15588, Korea; jackiejeh@naver.com (E.J.); slaylina@naver.com (J.N.)
[2] KPMG Samjong Accounting Corp., Gangnam Finance Center, 152 Teheran-ro, Gangnam-gu, Seoul 06236, Korea; common123@nate.com
* Correspondence: pure@hanyang.ac.kr

Received: 14 November 2018; Accepted: 11 December 2018; Published: 13 December 2018

Abstract: We consider a buyer's decision problem of sustainable supplier selection and order allocation (SSS & OA) among multiple heterogeneous suppliers who sell multiple types of items. The buyer periodically orders items from chosen suppliers to refill inventory to preset levels. Each supplier is differentiated from others by the types of items supplied, selling price, and order-related costs, such as transportation cost. Each supplier also has a preset requirement for minimum order quantity or minimum purchase amount. In the beginning of each period, the buyer constructs an SSS & OA plan considering various information from both parties. The buyer's planning problem is formulated as a mathematical model, and an efficient algorithm to solve larger instances of the problem is developed. The algorithm is designed to take advantage of the branch-and-bound method, and the special structure of the model. We perform computer experiments to test the accuracy of the proposed algorithm. The test result confirmed that the algorithm can find a near-optimal solution with only 0.82 percent deviation on average. We also observed that the use of the algorithm can increase solvable problem size by about 2.4 times.

Keywords: optimization; integer linear programming; sustainable; supplier selection; order allocation

1. Introduction

Supplier evaluation and selection are important decisions in the management of a supply network [1,2]. After determining suppliers to fill orders, the subsequent decision to allocate orders to chosen suppliers follows. Recent awareness in sustainable supply chain management frequently integrates these decisions with sustainability factors. The concept of sustainability plays an essential role in many organization and industries with respect to environmental protection and social responsibility [3]. As a consequence, sustainable supplier selection and order allocation (SSS & OA) emerges as a hot issue in the area of production and logistics. Huge number of papers have been published for this important decision problem. For example, Kuo et al. [4] developed a supplier selection system through fuzzy AHP and DEA. Their method was successfully applied to an auto lighting system company in Taiwan. The SSS & OA can be included in green supply chain management to improve the performance of a supply chain. Roehrich et al. [5] did such a study for a globalized German-based aircraft interior manufacturer and six key suppliers. There are a few commercial systems having supplier selection and evaluation functions. eSourcing Capability Models developed by ITSqc and CMMI-ACQ, made by SEI, are useful systems in the business area for acquiring products and services [6,7].

This paper studies an SSS & OA problem for a buyer who performs regular replenishment activities with heterogeneous suppliers who sell a few types of items. The system analyzed here is a two-stage supply chain system, which consists of a single buyer controlling inventories using a periodic order-up-to inventory control policy, and multiple heterogeneous suppliers who can supply items in response to orders from the buyer. The buyer sells items to end customers and replenishes items regularly based on the inventory status and future demand forecasts. In response to an order from the buyer, the suppliers transport the ordered amount after a constant lead time.

The problem analyzed in this paper is a buyer's decision problem of selecting suppliers and, at the same time, order allocation for selected suppliers. Based on such replenishment decisions, the buyer considers various system variables and several contract terms, including minimum order quantity (MOQ) and minimum purchase amount (MPA) requirements. The MOQ and MPA specify that suppliers accept only those orders that exceed a predetermined minimum order quantity and minimum order value [8–10]. Additional factors the buyer considers in the decision process include working capital requirement and sustainability factors.

Even though several optimization model variants have been introduced for systems similar to the one analyzed in this paper, a detailed model representing all the important characteristics of the SSS & OA process has not yet been analyzed. To handle larger instances of real decision processes requiring big data and excessive computational capacity, an efficient new solution methodology is also desired to make full use of a developed model. Considering this research need, the current paper introduces a mathematical model and solution methodology, which are constructed by relaxation and ideas from the branch-and-bound method.

2. Literature Review

A large number of studies dealing with the supplier selection problem have been published. A recent survey paper reviewed 370 works in this area [11]. As stated in their review, the subjects of supplier selection problems are very wide, ranging from criteria analysis for supplier selection to multiple criteria inventory control problems. Among numerous topics studied in this area, our review of previous research is narrowly focused on the supplier selection and order allocation problem of a single buyer dealing with multiple items, as well as multiple suppliers requiring MOQ and MPA constraints, working capital requirement constraint, and sustainability features. Thus, the basic forms of research related to this paper can be classified into two sub-areas. The first sub-area is about supplier selection and order allocation, while the second area is the sustainable supplier selection and order allocation. Previous research on the two sub-areas are presented followed by a discussion on the research gaps and contribution of this paper.

2.1. Supplier Selection and Order Allocation

To solve the supplier selection and order allocation problem, Ghorbani et al. [12] proposed a two-phased model. At first, suppliers are evaluated according to both quantitative and qualitative criteria resulting from SWOT analysis. Shannon entropy is used to calculate criteria weights. Then, the results are used as an input for an integer linear programming model to allocate orders to suppliers. Nazari-Shirkouhi et al. [13] provided an integrated linear programming model that aimed to minimize total ordering costs and defective items. Jadidi et al. [14], [15] modeled the supplier selection as a multi-objective optimization model where minimization of price, rejects, and lead-time were considered as three objectives. Sodenkamp et al. [16] proposed a novel meta-approach for collaborative multi-objective supplier selection and order allocation (SSOA)decisions by combining multi-criteria decision analysis and linear programming. The proposed model accounted for suppliers' performance synergy effects within a hierarchical decision-making process. Shabanpour et al. [17] proposed efficiency improvement plans for supplier selection, including goal programming and data envelopment analysis applications to rank sustainable suppliers.

In addition to usual constraints included in the previous research on SSOA, our model includes two other kinds of features practiced in the real world. The first constraint is MOQ/MPA-related practices, and the second is limitation caused by working capital management. Research concerning an SSOA considering the MOQ/MPA requirements was initiated by Robb and Silver [18]. Afterward, several researchers, including Kiesmüller et al. [9], Zhao and Katehakis [19], Zhou et al. [20], and Meena and Sarmah [21] have studied several variants of the SSOA problems with associated requirements. All of these studies could be categorized a basic model, because all studied a single-item problem. More realistic multi-item problems were first analyzed by Zhou [22], and Aktin and Gergin [23]. Recently, Park et al. [10] considered an order allocation problem with the MOQ/MPA requirements and proposed a rolling-horizon implementation strategy for solving a formulated optimization model more efficiently. Their model, however, did not contain a sustainability feature or working capital requirements.

Supply chain models typically only consider the physical transformation activities and disregard the financial implications of those activities. Recently, however, the literature on supply chain management (SCM) became aware of the real-world situation that financing and operational problems are closely connected and, thus, optimizing the two problems jointly could improve the entire performance of a supply chain [24,25]. However, only a few related papers were found on an SSOA with a working capital requirement (WCR). Chao et al. [26] developed recursive equations for a replenishment (order size determination) problem with a cash flow constraint. The problem was for a single item without considering supplier's perspectives, and thus could be categorized as the primitive type of research compared with our current problem. Bendavid et al. [27] studied a buyer's replenishment problem with a single type of item using a more sophisticated flow balance equation for the working capital constraint. Bian et al. [24] presented a new generic working capital requirement model for a single-item lot sizing problem. They presented a mixed integer programming model, including a flow balance equation, for operating working capital requirement (OWCR). To the best of our knowledge, there is no prior work addressing the SSOA problem that also directly considered WCR or OWCR.

2.2. Sustainable Supplier Selection and Order Allocation

The traditional supplier selection and order allocation problem has now been changed to an SSS & OA, where sustainability triple bottom line (3BL) attributes (environmental, economic, and social) are integrated into the selection and allocation processes [28]. The environmental factors can also be evaluated in terms of political, economic, social, technological, and environmental aspects, as can be seen in the well-known method named PESTEL [29]. The literature on sustainable supplier selection is quite rich. A few prior studies include [30–48]. These studies used various kinds of methods, including the AHP, DEMATEL, ANP, TOPSIS, multi-objective GA, DEA, and VIKOR for evaluating and selecting desirable sustainable suppliers. All the above referenced research deals with the question of which sustainable supplier to select. Research dealing with order allocation together with sustainable supplier selection is in its early stages. Only five papers on SSS & OA have been noted during the literature review. Kannan et al. [49] introduced a fuzzy TOPSIS method for supplier selection and a bi-objective model for order allocation. Govindan et al. [50] analyzed a five-echelon supply chain for assigning suppliers for a single product. Aktin and Gergin [23] introduced a mixed integer programming model using 3BL index scores. Problems analyzed in these three papers can be categorized as basic SSS & OA because they considered a single product and single period case with a deterministic demand. Recently, more sophisticated models have been offered by Gören [51] and Ghadimi et al. [1]. The former solved a problem with multiple products and suppliers, and formulated a bi-objective optimization model for a single period decision. The latter analyzed a similar system, but formulated it as a multi-period bi-objective model. However, both of these studies assumed a deterministic demand and did not consider other realistic features, such as transportation lead time or MOQ requirement.

2.3. Research Gap and the Contribution of This Paper

As can be found in the discussion of previous research and also in Table 1, our study is the first attempt to analyze the most realistic and complicated SSS & OA problem representing various important features of a real system, including transportation features (transportation lead times and capacity of the suppliers) and buyer monetary limitations (multi-period working capital flow balances and limitation, time value of money). Given the various aspects we are considering for this analysis, the optimization model introduced in this paper is the most sophisticated of any existing models representing SSS & OA activities. One of the challenges we experienced during the development of such a large-scale model is that none of the existing methods can solve our model to a desired accuracy within a practical time limit. For example, a problem with 20 items and 12 time periods cannot be solved within 24 h time limit. When we consider that real-world problems can include more than 100 items, it is necessary to fill this research gap. In response to this research challenge, a new algorithm specifically aimed to solve such a big model is developed. During a computational experiment, the algorithm is capable of solving such a model within a reasonable computational time with desired accuracy.

Table 1. Comparison of the contributions of different authors.

Author(s)	Problem Type: Supplier Selection	Problem Type: Order Allocation	Model Type: Single Supplier	Model Type: Multiple Supplier	Demand Process: Determi-nistic	Demand Process: Stochastic — Stationary	Demand Process: Stochastic — Non-Stationa-ry	Number of Items: Single Item	Number of Items: Multi-Item	Number of Periods: Single Period	Number of Periods: Multi-Period	New Solution Method	Constraints: MOQ/MPA	Constraints: WCM	Constraints: Sustainability	Transportation Lead Time
Zhao and Katehakis [19]		√	√			√					√					
Zhou et al. [20]		√	√			√		√			√					
Chao et al. [26]		√	√			√		√			√	√		√		
Zhou [22]		√	√			√			√		√		√			
Kiesmüller et al. [9]		√	√			√		√			√		√			√
Kannan et al. [49]	√	√		√	√			√		√					√	
Meena and Sarmah [21]	√	√		√	√			√		√			√			
Govindan et al. [38]	√	√		√		√		√		√		√			√	
Ayhan and Kilic [52]	√	√		√	√				√	√					√	
Trapp and Sarkis [43]	√			√					√	√		√				
Aktin and Gergin [23]	√	√		√	√				√	√					√	
Bendavid et al. [27]		√	√			√		√			√			√		
Gören [51]	√	√		√	√				√		√				√	
Ghadimi et al. [1]	√	√		√	√				√	√					√	
Bian et al. [24]		√	√		√			√			√	√		√		
Park et al. [10]	√	√		√		√	√		√		√		√			√
This model	√	√		√		√	√		√		√	√	√	√	√	√

3. System Description and Assumptions

The system analyzed in this paper involves two or more heterogeneous suppliers and a single buyer. The suppliers are distinguished from each other by the type and selling prices of the items they carry, delivery lead times, and minimum order quantity requirements. The buyer carries multiple types of items which are sold to end customers. The items are replenished to minimize related inventory costs based on a periodic order-up-to inventory control policy. Previous research on inventory control frequently assumed that the end customer demand can be described by a known probability distribution. However, since the future demand for a product can be influenced by unforeseeable events, complete information on future demand distribution may not be available [53]. Considering this kind of real-world situation, this paper assumes that the demand of the end customers may not belong to a theoretical probability distribution. Other assumptions are as follows:

- There is a planned allocation schedule of money for each period during a planning horizon.
- Money remaining at the end of a period is inflated by interest rate and carried forward to the next period.
- Payment for purchase and transportation costs are made as an order is placed.
- Nonzero transportation lead time exists between an order placement and the arrival of the ordered amount.
- Major and minor ordering costs occur when an order is placed.
- The major ordering cost occurs as a fixed amount when an order is placed.
- The minor ordering cost occurs in proportion to an order size.
- A supplier has limited production capacity and thus has an order size limit per order.
- A supplier has a limited number of transportation vehicles.
- Any amount of an item can be purchased at a price higher than supplier's regular price from a spot market.
- 3BL factor scores of each potential supplier are prepared for input to an SSS & OA decision.

Considering the characteristics of each supplier, the buyer must make an SSS & OA decision at the beginning of each period. The objective that the buyer is trying to achieve is to minimize the net present value of the related costs occurring throughout the planning horizon. Required notations are as follows.

Indices:

i	item number, $i = 1, 2, \cdots, I$,
j	3BL index, $j = \text{env}, \text{eco}, \text{soc}$,
k	supplier number, $k = 1, 2, \cdots, K$,
t	period, $t = 1, 2, \cdots, T$, where T denotes the end period of the planning horizon.

Parameters:

$K(i)$	set of suppliers who sell item i, $\forall i$,
$I(k)$	set of items sold by supplier k, $\forall k$,
d_{it}	demand forecast of item i during future period t, $\forall i, t$,
ς_{it}	standard deviation of error of $d_{i,t}$, $\forall i, t$,
v_i	volume of item i, $\forall i$,
wc_t	warehouse capacity of the buyer during period t, $\forall t$,
h_i	holding cost of item i, $\forall i$,
b_i	shortage cost of item i, $\forall i$,
p_{ikt}	unit purchase price for item i paid by the buyer to supplier k during period t, $\forall i, k \in K(i), t$,
p_{it}^s	unit spot market price during period t for item i, $\forall i, t$,
$owcl_t$	operating working capital limit in period t, $\forall t$,
$capt_t$	capital originally allocated to period t, $\forall t$,
ic_t	inventory control related cost (holding plus shortage costs) in period t, $\forall t$,

rc_t	replenishment related cost in period t, $\forall t$,
r	per-period discount (interest) rate.
moq_{ik}	per-period minimum order quantity specified by supplier k for item i, $\forall i, k$,
mpa_k	per-period minimum purchase amount set by supplier k, $\forall k$,
mpl_{ik}	per-period maximum purchase limit for item i specified by supplier k, $\forall i$, $k \in K(i)$,
ma_k	major ordering cost for supplier k, $\forall k$,
mi_{ik}	minor ordering cost for item i for supplier k, $\forall i, k \in K(i)$,
sc_{jk}	jth 3BL factor score of supplier k, $\forall j$, k,
$target_{jt}$	jth 3BL factor target score of period t, $\forall j, t$,
l_k	supplier k's lead time, $\forall k$,
f_{kt}	freight fair per vehicle of supplier k during period t, $\forall t, k$,
vc_k	volume capacity per vehicle of supplier k, $\forall k$,
nv_{kt}	number of vehicles available for transportation of supplier k in period t, $\forall k$,
$\widetilde{x}_{ikt}$	purchase already made at the start of past period t and in delivery of item i from supplier k, $\forall i$, $k \in K(i)$, $t = -1, -2, \cdots, 1 - l_k$,
IP_{i0}	inventory position of item i at the start of planning, $\forall i$,
M	very large number.

Decision variables:

IP_{it}	inventory position of item i at the end of period t, $\forall i$, t,
IP_{it}^{+}	positive part of $IP_{i,t}$, $\forall i$, t,
IP_{it}^{-}	negative part of $IP_{i,t}$, $\forall i$, t,
x_{ik1}	purchase amount of item i from supplier k during the present period (period 1), $\forall i$, $k \in K(i)$,
x_{ikt}	planned purchase amount of item i from supplier k during future period t, $\forall i$, $k \in K(i)$, $t = 2, \cdots, T$,
x_{i1}^{s}	purchase quantity of item i from the spot market for the present period, $\forall i$,
x_{it}^{s}	planned purchase quantity of item i from the spot market for period t, $\forall i$, $t = 2, \cdots, T$,
RL_{it}	replenishment level of item i after the arrival of orders scheduled to arrive at the start of period t, $\forall i, t$,
o_{jt}	positive deviation from target $_{jt}$ in period t,
α_{ikt}^{moq}	binary integer for controlling the minimum order quantity requirement, $\forall i, k, t$,
α_{kt}^{mpa}	binary integer for controlling the minimum purchase amount requirement, $\forall k$, t,
β_{kt}^{MA}	binary integer for controlling major ordering cost, $\forall k$, t,
β_{ikt}^{MI}	binary integer for controlling minor ordering cost, $\forall i, k, t$,
θ_i	safety factor of item i, $\forall i$.

4. Model Formulation

4.1. Relevant Costs

Cost factors included in the total cost of our model are inventory-related costs (holding and shortage costs) and replenishment-related costs (major and minor ordering costs, transportation, and purchase costs). Inventory-related costs are the sum of inventory holding and shortage costs incurred during the planning horizon, and are expressed as in Equation (1).

$$ic_t = \sum_{i=1}^{I} \left(\frac{1}{2} h_i (RL_{it} + IP_{it}^{+}) + b_i IP_{it}^{-} \right), \quad \forall t. \tag{1}$$

Transportation cost of period t is

$$\sum_{k=1}^{K} f_{kt} NV_{kt}.$$

Purchase cost is the sum of the payment to suppliers and spot market.

$$\sum_{i=1}^{I} \sum_{k \in K(i)} p_{ikt} x_{ikt} + \sum_{i=1}^{I} sp_{it} x_{it}^{s}.$$

Major and minor ordering costs are as follows:

$$\sum_{k=1}^{K} ma_{k} \beta_{kt}^{MA} + \sum_{i=1}^{I} \sum_{k \in K(i)} mi_{ik} \beta_{ikt}^{MI}.$$

Replenishment-related cost is the sum of the cost factors in Equation (2).

$$rc_t = \sum_{k=1}^{K} \left(ma_k \beta_{kt}^{MA} + f_{kt} nv_{kt} \right) + \sum_{i=1}^{I} \sum_{k \in K(i)} \left(p_{ikt} x_{ikt} + mi_{ik} \beta_{ikt}^{MI} \right) \\ + \sum_{i=1}^{I} sp_{it} x_{it}^{s}, \quad \forall t. \tag{2}$$

The total cost function of the model (*TC*) is the present value of inventory control cost plus replenishment-related cost incurred during the planning horizon. When we use a discounting factor r to account for the time value of money, the cost function can be written as

$$TC = \sum_{t=1}^{T} \frac{1}{(1+r)^t} (ic_t + rc_t).$$

4.2. Operating Working Capital Requirement

In practice, many firms are financially constrained; therefore, their ability to manage their inventories is directly affected by many factors, including their operating working capitals. To represent this financial constraint, the following equations are included.

$$ic_t + rc_t \leq owcl_t, \quad \forall t, \tag{3}$$

$$owcl_t = capt_t + (1 + \gamma)(owcl_{t-1} - rc_{t-1} - ic_{t-1}), \quad \forall t. \tag{4}$$

Equation (3) specifies that the cost occurring during period t is limited by an operating working capital limit (OWCL) in that period. The equation was based on the cash-to-cash methodology found in Theodore Farris and Hutchison [54], and Hofmann and Kotzab [55]. Consequently, we assumed that the OWCR for replenishing a unit of product depends on the money invested in the related operations, for example, purchasing, setup, transportation, inventory holding, and shortage costs. Also, as in Bian et al. [24], it is assumed that the profit portion of the sales revenue is not accounted for in the OWCR. Profit can be allocated to other higher priority objectives of the firm (e.g., debt reduction, dividend payments, or internal and external investment). Thus, the profit portion of a firm's activities was not represented in our model (e.g., Equations (3) and (4)). Equation (4) models monetary flow during two adjacent periods and ensures that the OWCL in period t equals the sum of the operating working capital (OCM) allocated to period t and the money left in the previous period inflated by interest and forwarded to the current period.

4.3. 3BL Target Constraints

$$\sum_{i=1}^{I} \sum_{k=1}^{K} sc_{jk} \beta_{ikt}^{MI} - o_{jt} = target_{jt}, \quad \forall j, t. \tag{5}$$

As stated in Aktin and Gergin [23], corporate sustainability is concerned with the integration of environmental, economical, and social dimensions, called the triple-bottom-line (3BL), into the company processes. In response to this need, SSS & OA decisions try to combine the 3BL sustainability factors into supplier selection and order allocation activities. A practical way to find good sustainable procurement strategies is to measure sustainability scores for all potential suppliers. Then, the completed 3BL factor scores of each supplier are input to a mathematical model formulated for supplier selection and order allocation. Equation (5) performs this kind of function. It states that all selected suppliers' combined 3BL score should at least equal to a preset target score for environmental, economical, and social dimensions.

4.4. Mathematical Programming Model

In this section, we define a mixed integer programming model to solve the SSS & OA problem. The proposed MIP model can be defined as follows:

MIP1: Min TC

s.t.

$$IP_{it-1} + \sum_{k\in K(i)|l_k=0} x_{ikt} + \sum_{\substack{k \in K(i)|l_k \geq 1 \\ t - l_k \leq 0}} \tilde{x}_{ik,t-l_k} + \sum_{\substack{k \in K(i)|l_k \geq 1 \\ t - l_k > 0}} x_{ik,t-l_k} + x_{it}^s = RL_{it},$$
$$\forall i,\, t, \tag{6}$$

$$RL_{it} - d_{it} = IP_{it},\ \forall i,\, t, \tag{7}$$

$$IP_{it} = IP_{it}^+ - IP_{it}^-,\quad \forall i,\, t, \tag{8}$$

$$\sum_{k=1}^{K}\left(ma_k\beta_{kt}^{MA} + f_{kt}nv_{kt}\right) + \sum_{i=1}^{I}\sum_{k\in K(i)}\left(p_{ikt}x_{ikt} + mi_{ik}\beta_{ikt}^{MI}\right) + \sum_{i=1}^{I}p_{it}^s x_{it}^s$$
$$= rc_t,\ \forall t, \tag{9}$$

$$\sum_{i=1}^{I}\left(\frac{1}{2}h_i\left(RL_{it} + IP_{it}^+\right) + b_i IP_{it}^-\right) = ic_t,\ \forall t, \tag{10}$$

$$ic_t + rc_t \leq owcl_t,\ \forall t, \tag{11}$$

$$owcl_t = capt_t + (1+\gamma)(owcl_{t-1} - rc_{t-1} - ic_{t-1}),\ \forall t, \tag{12}$$

$$IP_{it} \geq \theta_i\xi_{it},\ \forall i,\, t, \tag{13}$$

$$x_{ikt} \leq mpl_{ik},\ \forall i,\, k \in K(i),\, t, \tag{14}$$

$$x_{ikt} \leq M\alpha_{ikt}^{moq},\ \forall i,\, k \in K(i),\, t, \tag{15}$$

$$x_{ikt} \geq moq_{ik} - M(1 - \alpha_{ikt}^{moq}),\ \forall i, k \in K(i), t, \tag{16}$$

$$\sum_{i\in I(k)} p_{ikt}x_{ikt} \leq M\alpha_{kt}^{mpa},\ \forall k, t, \tag{17}$$

$$\sum_{i\in I(k)} p_{ikt}x_{ikt} \geq mpa_k - M(1 - \alpha_{kt}^{mpa}),\ \forall k, t, \tag{18}$$

$$\sum_{i=1}^{I}\sum_{k=1}^{K} sc_{jk}\beta_{ikt}^{MI} - o_{jt} = target_{jt},\ \forall j, t, \tag{19}$$

$$\sum_{i\in I(k)} x_{ikt} \leq M\beta_{kt}^{MA},\ \forall k, t, \tag{20}$$

$$x_{ikt} \leq M\beta_{ikt}^{MI},\ \forall i, k \in K(i), t, \tag{21}$$

$$\sum_{i=1}^{I} v_i RL_{it} \leq wc_t,\ \forall t, \tag{22}$$

$$\sum_{i\in K(i)} v_i x_{ikt} \leq vc_k nv_{kt},\ \forall k, t, \tag{23}$$

$$x_{ikt} \geq 0, \quad \alpha_{ikt}^{moq}, \ \beta_{ikt}^{MI} \ 0 \text{ or } 1, \ \forall i, k \in K(i), t, \tag{24}$$

$$\alpha_{kt}^{mpa}, \ \beta_{kt}^{MA}, \ 0 \text{ or } 1, \quad nv_{kt}, \text{ nonnegative integer}, \ \forall k, \ t, \tag{25}$$

$$x_{it}^{s}, \ RL_{it}, \ IP_{it}^{+}, \ IP_{it}^{-} \geq 0, \quad IP_{it}, \text{ unrestricted}, \ \forall i, \ t, \tag{26}$$

$$o_{jt} \geq 0, \ \forall j, \ t, \tag{27}$$

$$\theta_i, \ \text{unrestricted}, \ \forall i, \tag{28}$$

$$M, \quad \text{large number}. \tag{29}$$

The objective function in Equation (1) is to minimize the present value of the expected total cost, which is the sum of the inventory and replenishment-related costs. Equation (6) enforces that the replenishment level of item i is the sum of the initial inventory, spot market purchases, and orders scheduled to arrive from each supplier during the period. Equation (7) regulates that the net inventory of item i at the end of period t is equal to the inventory position at the start of the period, minus the depletion due to the demand of item i during period t. Equation (8) sets that, at the end of period t, the net inventory of item i, $IP_{i,t}$, is equal to the on-hand inventory level of item i at the end of period t, $IP_{i,t}^{+}$, minus the shortage level of item i at the end of period t, $IP_{i,t}^{-}$.

Equations (9)–(12) enforce the operating working capital limit. Equations (13) and (14) describe the customer service level and maximum purchase limit set by a supplier, respectively. Equations (15) and (16) are for the minimum order quantity requirement. Term moq_{ik} in the constraint is the minimum order size of supplier k, and α_{ikt}^{moq} is a binary variable used to enforce the relationship as planned. The variable M is a very large number used to activate the minimum order constraint only when an order is placed. As a consequence, if the buyer purchases item i from supplier k, the term α_{ikt}^{moq} will become 1 in Equation (15), thereby validating Equation (16) and enforcing the minimum order size requirement. The next constraints, described by Equations (17) and (18), concern the minimum purchase amount requirement. If the buyer purchases an item from supplier k, Equation (17) makes the term α_{kt}^{mpa} equal to 1. Equation (18), in this case, forces the purchase amount to be at least the minimum purchase amount (mpa_k). Equation (19) concerns the 3BL target constraint. Equations (20) and (21) control the major and minor ordering occurrences. The following Equations (22) and (23) address the buyer's warehouse capacity and supplier's transportation capacity, respectively.

MIP1 has $5IKT + 3IT + 4KT + JT + 5T$ constraints and $3IKT + 5IT + 3KT + JT + I$ variables. If a contract problem has a weekly planning grid with a one year planning period (52 weeks) and 10 suppliers with 20 items, 58,136 constraints and 37,086 variables are present. It is possible to solve the size of MIP1 using commercially available software tools (GAMS, LINGO etc.). However, if the size of MIP1 grows considerably large for real-world applications, a prohibitive computational burden will result. In other words, expanding the size of the system requires an excessive computational resource. Considering this difficulty, a faster and reasonably accurate algorithm is needed for real life problems. The next section discusses such an algorithm.

5. Solution Method

5.1. Conceptual View of the Proposed Algorithm

The algorithm introduced in this section is referred to as the branch-and-freeze (BF) algorithm. The logical idea behind the BF algorithm is to solve relaxed problems (sub-problems) of the original problem in a manner similar to the branch-and-bound method. We observed that the MOQ and MPA constraints in Equations (15)–(18) are computationally burdensome because of the binary variables involved and the large number of constraints, amounting to the total $2IKT + KT$. Based on this observation, the sub-problems are created by removing the MOQ and MPA constraints of the original problem, MIP1. When the sub-problem is solved, one of three cases can occur, as illustrated in Figure 1.

The first of the three cases is that a solution to a sub-problem also satisfies the MOQ and MPA constraints of all suppliers, which we call complete feasibility (CF) case. In Figure 1 below, the CF case is represented by the left-most branch. In this case, the solution is also optimal to MIP1. Since, the original problem is solved optimally, the algorithm stops. The second case occurs when the MOQ and MPA constraints are satisfied partially, which is the situation where the solution to a sub-problem satisfies the two constraints of all suppliers up to a certain intermediate period, but not to the end of the planning horizon. This case is named partial feasibility (PF). When PF occurs, the algorithm stores the current output up to the satisfied period, which is called freezing. For the remaining periods that are not frozen, a new condensed problem is generated by adding the MOQ and MPA constraints of the supplier(s) whose constraints were violated in the previous run. In Figure 1 below, this process is denoted by the circle with FC (freezing and condensing). When the condensed problem is solved afterwards, it results in one of the three cases already explained above.

The final case, which is in the right most side of Figure 1, is the complete infeasibility (CI) case, where the sub-problem's output can satisfy none of the suppliers' MOQ and MPA constraints, even at the starting period. If this happens, a new sub-problem is created by adding all of the removed constraints. This process is denoted by a circle with an R (restoring) inside. When a restored problem is solved, one of the same three cases can occur. A node is fathomed when the stop condition is met after CF, or no additional constraint is available for addition after PF or CI. The best feasible solution to the original problem is the best feasible solution found until all end nodes are fathomed. If there is no feasible solution found up to that point, the original problem is infeasible.

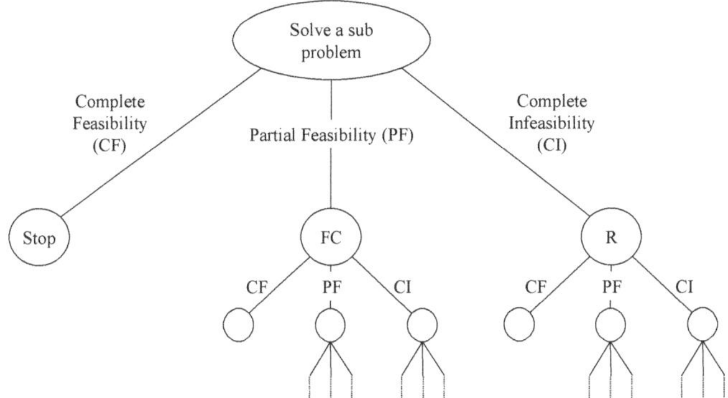

Figure 1. Conceptual view of the branch-and-freeze (BF) algorithm.

5.2. Branch-and-Freeze (BF) Algorithm

The BF algorithm can be described formally as follows:

Step 1: (Initialize)

Let the current period be period 1.
Set the current inventory level, $z_{i,0} = 0$ for $i = 1, 2, \cdots, I$.
Forecast demand for all future periods, $d_{i,t}$ for $t = 1, 2, \cdots, T$.

Step 2: (Generate sub-problem for the first run)

Construct the sub-problem by removing the MOQ and MPA constraints (Equations (15)–(18) from MIP1).

Step 3: (Run sub-problem)

Run the current sub-problem.

Step 4: (Check status and branch)

Step 4.1 Check the output of Step 3. If status is PF or CF, go to Step 4.3.
Step 4.2 (Complete feasibility case)
Algorithm found a feasible solution. Stop.
Step 4.3 (Partial feasibility or complete infeasibility case)
If there is no constraint to add, the given problem is infeasible. Stop.
Go to Step 5 if status is PF. Otherwise, go to Step 6.

Step 5: (Partial feasibility case)

Step 5.1 (Freeze the output)
Freeze the output for the feasible periods.
Step 5.2 (Re-initialize)
Let the starting period be the first infeasible period.
Reset the current inventory level to the net inventory level of the last feasible period.
Reset the forecast of demand from the starting to the end periods of the planning horizon.
Step 5.3 (Prepare a sub-problem)
Prepare a new sub-problem by adding the MOQ or MPA constraints of the supplier(s), which caused infeasibility during the previous run. Go to Step 3.

Step 6: (Complete Infeasibility case)

Prepare a new sub-problem by adding the MOQ or MPA constraints of the supplier(s), which caused infeasibility during the previous run. Go to Step 3.

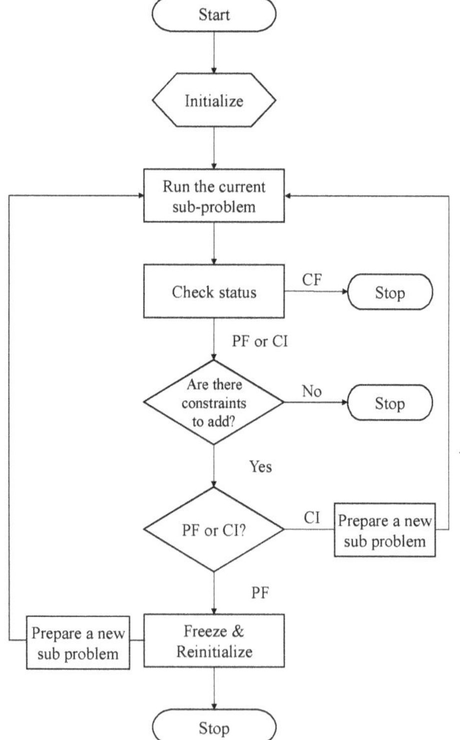

Figure 2. Flow chart of the BF algorithm.

Step 1 is for the initialization required for the first planning run. The constraint relaxation that is required to solve MIP1 without the MOQ and MPA constraints is done in Step 2. After initialization and relaxation, a relaxed version of MIP1 (sub-problem) is solved in Step 3. In Step 4, the output of the previous run is evaluated for status. Based on the status, the algorithm stops or proceeds to Steps 5 or 6 when there is (are) a constraint(s) to add. Figure 2 shows the flow of the algorithm. Figure 3 illustrates an implementation of the algorithm.

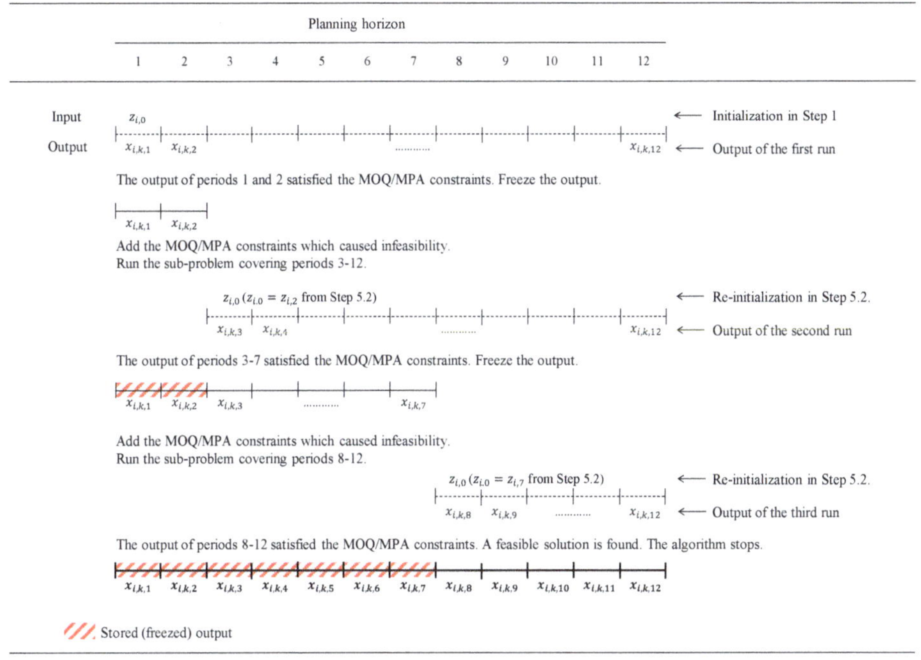

Figure 3. Implementation diagram of the BF algorithm.

6. Numerical Experiments

In this chapter, numerical experiments are carried out with two objectives in mind. The first objective of the numerical experiment is to test the accuracy of the BF algorithm by comparing it with a commercially available software tool (GAMS/XPRESS solver). The second experiment explores the maximum size of MIP1 that can be solved by the BF algorithm and by commercial software tools (GAMS/XPRESS and GAMS/COINGLPK solvers). The results of these two numerical experiments will determine the effectiveness of the BF algorithm. The GAMS used in the numerical experiment is a very popular modeling language containing many powerful solvers. Thus, it is a suitable competitor for verifying the accuracy and identifying the maximum solvable problem size of the BF algorithm. The experiments were performed on a PC with Microsoft 7 OS, 3.4GHz Intel i5 CPU, and 16 GB RAM.

6.1. Accuracy Test of the BF Algorithm

The purpose of this experiment is to identify the accuracy of the BF algorithm by comparing the results obtained using our algorithm with those of the GAMS/XPRESS solver. The comparative experiment is performed with the assumption that the item's demand is generated from a stationary demand process. Most previous studies on inventory management performed experiments by assuming that demand follows a stationary demand process, such as a Poisson or normal distribution [56]. For our problem, many relevant studies, including Robb and Silver [18],

Chen et al. [57], and Zhou [22], also assumed a normal distribution. This experiment is also carried out similarly by assuming a normal distribution assumption. The number of item types is set to five. The actual demand data for each item were generated from five different normal distributions ($N(400, 20^2)$, $N(600, 30^2)$, $N(700, 40^2)$, $N(800, 40^2)$, $N(900, 50^2)$). Each item's demand forecast is prepared using the forecasting module of SPSS.

To obtain the average total cost, the experiment was repeated 10 times for each setting. The average cost obtained in this manner is plugged into the following percent deviation measure to identify the accuracy of the BF algorithm.

$$\text{Percent deviation} = \frac{\text{BF algorithm' cost} - \text{GAMS' cost}}{\text{GAMS' cost}} \times 100.$$

The experimental design is as follows:

- There are 10 suppliers in the system.
- Transportation lead time is zero.
- Each supplier can deliver all five types of items.
- The unit period length is four weeks.
- The planning horizon length is sized to 48 weeks, which amounts to 1 year.

Other input parameters were prepared as shown in Tables 2–6.

Table 2. Input parameters for the comparative experiment.

Warehouse Capacity of Buyer (wc_t)	Very Large Number (M)	Discount Factor (r)	Initial Inventory Level (IP_{i0})
5000.00	10^7	0.01	0

Table 3. Input parameters for each item.

Item	Holding Cost (h_i)	Shortage Cost (b_i)	Volume (v_i)	Spot Market Price (p_{it}^s)
All items	$N(2, 0.1^2)$	$N(15, 1^2)$	2.00	$N(27, 1^2)$

N denotes a normal distribution.

Table 4. Input parameters for each supplier.

Supplier	Minimum Purchase Amount (mpa_k)	Major Ordering Cost (ma_k)
All suppliers	$20 \times 1.0 \times \hat{d}$	$N(250, 10^2)$

$\hat{d}$ is the forecast average for the planning horizon.

Table 5. Input parameters for each item of each supplier.

Supplier	Minimum Order Quantity (mpa_{ik})	Minor Ordering Cost (mi_{ik})
	Item 1 to 5	Item 1 to 5
All suppliers	$N(1, 0.2^2) \times \hat{d}_i$	$N(1.5, 0.2^2)$

$\hat{d}_j$ denotes the average of forecasts for item j at the planning horizon.

Table 6. Input parameters for each period.

Supplier	Purchase Price (p_{ikt}) for All Period	Maximum Purchase Limit (mpl_{ik}) for All Period
	Item 1 to 5	Item 1 to 5
All suppliers	$N(20, 1^2)$	$N(3, 0.5^2) \times \hat{d}_i$

The results of the first experiment are summarized in Table 7. Using the BF algorithm instead of commercial solvers (GAMS/XPRESS solver), the average total discounted cost increased by 0.82%. The reason for this was the inventory and shortage appearing at the end of the planning horizon. However, the proposed BF algorithm can offer a result very close to the optimum solution. Thus, it seems that the BF algorithm is able to find a near-optimal solution, even though it is a heuristic algorithm mainly developed to solve larger instances of the problem which cannot be solved by any other existing tools.

Table 7. Summary of the accuracy test results.

Method	Average Annual Discounted Cost	Average Percent Deviation (%)	Standard Deviation of Percent Deviation	Average CPU Time	Average Number of Sub-Problems Solved
GAMS	$805,041.43	-	-	1.467 s	-
BF algorithm	$811,630.50	0.82%	0.34%	2.920 s	5.5

6.2. Experiment to Estimate the Maximum Solvable Problem Size of the BF Algorithm

Various software tools developed to solve optimization models have a maximum size limit on the problem which can be solved within a reasonable computational time. Considering this limitation, we attempted to estimate the maximum problem size that can be solved by the BF algorithm. More specifically, the maximum size of problems that can be solved by commercial software tools and the BF algorithm was estimated for comparison. In the experiment, GAMS/XPRESS and GAMS/COINGLPK solvers were selected for comparison. The planning horizon of the problem was fixed to 12 time periods, and the number of suppliers was set to 10 times the number of items. The maximum computational time limit was set to 24 h. We increased the number of items until each method could not find a solution within the time limit. Other input parameters were set as shown in Section 5.1 (Tables 2–6), and demand data were generated using a normal distribution.

Table 8. Summary of the results for the maximum problem size test.

GAMS Solver	Size of the Problem That Can Be Solved (Number of Items, Number of Constraints, and Number of Variables)	
	GAMS Solver	**BF Algorithm**
COINGLPK	(12, 92,688, 56,928)	(17, 182,268, 111,233)
XPRESS	(69, 2,892,300, 1,743,045)	(108, 7,054,224, 4,244,544)

The results of the experiment are summarized in Table 8, and show that the BF algorithm could considerably increase the solvable problem size. The BF algorithm using the COINGLPK to solve sub-problems can double the solvable problem size compared with a naive use of the COINGLPK. Moreover, for the XPRESS case, the size increased approximately 2.4 times in terms of the number of constraints. Thus, it is expected that buyers will be aided in effective decision-making upon using this method for solving real-world complex problems.

7. Managerial Implications

7.1. Academic Implications

In this paper, we studied a sustainable supplier selection and order allocation problem. This is the first attempt to develop a model for the most realistic and complicated SSS & OA problem representing various important features of a real system. A new algorithm specifically designed to solve such a large-scale model is developed. The algorithm performed as expected by increasing solvable problem size considerably. In this way, we have done some initiating academic research in SSS & OA that will help researchers study related follow-up problems.

7.2. Managerial Implications

This study provides valuable insights for firms that regularly make a supplier selection and order allocation decisions. The model and solution method of this paper helps managers to make the SSS & OA decision more systematically. They can prepare a cost-minimizing plan quickly and easily after they complete a computerized planning system. The model is flexible and customizable, and can be modified based on the actual needs of a firm. Output of the developed system provides an efficient SSS & OA plan and, also, some useful additional information, which can be used for many what-if analyses. For example, the dual price of Equation (19) is an incremental cost for raising the 3BL target value by one unit. A firm trying to achieve more stringent sustainability performance can use the estimated cost to make an investment decision for improving a production or logistics system for better sustainability. Thus, good implementation of the model and algorithm of this research will result in better decisions on reducing costs, increasing profitability, and improving customer service sustainability. The final result will be enhanced competitiveness and improved financial status.

8. Conclusions

This paper presents models representing an SSS & OA problem for a buyer replenishing from two or more heterogeneous suppliers with MOQ and MPA constraints, operating working capital limits, and a 3BL sustainability target requirement. A mixed-integer programming model can find a cost-minimizing SSS & OA plan of choosing order-fulfilling suppliers and allocate the order amount for each select supplier. Since the size of a completed model for a real-life application is too big to implement it naively, a fast heuristic algorithm, called a BF algorithm, was also developed for such a large-scale implementation.

The logical idea behind the BF algorithm is two-fold, relaxation and branching. Observing that the MOQ and MPA constraints of the model are very computationally burdensome because of the binary variables included and large number of constraints involved, the algorithm creates sub-problems by relaxing (removing) the MOQ and MPA constraints of the original problem. When the sub-problem is created, a procedure similar to the branch-and-bound method is employed to solve the sub-problems efficiently. Several types of experiments were conducted using the GAMS solvers and IBM SPSS statistics package to verify the validity of the proposed model and to test the accuracy of the developed algorithm. The test result confirmed that the algorithm can find a near-optimal solution with only 0.82 percent deviation on average.

Another test was done to find how much larger a model can be solved when using the proposed algorithm compared with a direct one-time use of popular commercial solvers. The test result showed that the use of the BF algorithm can increase solvable problem size by as much as 2.4 times. It was verified that a model with 7 million constraints and 9 million variables can be handled by our algorithm. All in all, the test results can be summarized as the BF algorithm is an effective tool for handling complex real-life applications. Buyers faced with a large-scale system will, thus, be able to handle such large-scale decision problems without much difficulties.

There are some related research topics that require exploration. Further research may incorporate the supplier's perspective into the current problem to extend to a supplier–buyer problem. Also, the single objective of the current model can be extended to allow bi- or multi-objective functions to consider quantitative targets and qualitative preferences at the same time. Then, another kind of solution methodology should be developed to solve such a multi-objective optimization model of realistic size. Finally, the current model describes SSS & OA activities in a two-stage supply chain composed of a single buyer and several suppliers. These simple stages can be extended to a more complex case, such as a three-stage model, including another layer of suppliers or manufacturers.

Author Contributions: Writing and methodology, J.K. and J.P.; Data preparation and experiments, E.J. and J.N.

Funding: This research received no external funding.

Acknowledgments: The authors would like to thank the editors and referees for their valuable comments to enhance the clarity of this paper.

Conflicts of Interest: The authors declare that there is no conflict of interest regarding the publication of this paper.

Data Availability: All data generated or analyzed during this study are included in internet site (http://www.mecors.hanyang.ac.kr/bbs.htm). Requests for material should be made to the corresponding author.

References

1. Ghadimi, P.; Toosi, F.G.; Heavey, C. A multi-agent systems approach for sustainable supplier selection and order allocation in a partnership supply chain. *Eur. J. Oper. Res.* **2018**, *269*, 286–301. [CrossRef]
2. Fazlollahtabar, H. An integration between fuzzy promethee and fuzzy linear program for supplier selection problem: Case study. *J. Appl. Math. Model. Comput.* **2016**, *1*. Available online: http://www.lawarencepress.com/ojs/index.php/JAMMC/article/download/30/543 (accessed on 10 December 2018).
3. Kannan, D. Role of multiple stakeholders and the critical success factor theory for the sustainable supplier selection process. *Int. J. Prod. Econ.* **2018**, *195*, 391–418. [CrossRef]
4. Kuo, R.; Lee, L.; Hu, T.-L. Developing a supplier selection system through integrating fuzzy ahp and fuzzy dea: A case study on an auto lighting system company in taiwan. *Prod. Plan. Control* **2010**, *21*, 468–484. [CrossRef]
5. Roehrich, J.K.; Hoejmose, S.U.; Overland, V. Driving green supply chain management performance through supplier selection and value internalisation: A self-determination theory perspective. *Int. J. Oper. Prod. Manag.* **2017**, *37*, 489–509. [CrossRef]
6. CMMI Product Team. *Cmmi for Acquisition*; Version 1.3.; Software Engineering Institute: Pittsburgh, PA, USA, 2010.
7. ITSqc. Available online: http://www.itsqc.org (accessed on 10 December 2018).
8. Musalem, E.P.; Dekker, R. Controlling inventories in a supply chain: A case study. *Int. J. Prod. Econ.* **2005**, *93*, 179–188. [CrossRef]
9. Kiesmüller, G.P.; De Kok, A.; Dabia, S. Single item inventory control under periodic review and a minimum order quantity. *Int. J. Prod. Econ.* **2011**, *133*, 280–285. [CrossRef]
10. Park, J.H.; Kim, J.S.; Shin, K.Y. Inventory control model for a supply chain system with multiple types of items and minimum order size requirements. *Int. Trans. Oper. Res.* **2018**, *25*, 1927–1946. [CrossRef]
11. Yao, M.; Minner, S. Review of multi-supplier inventory models in supply chain management: An update. 2017. Available online: https://papers.ssrn.com/sol3/papers.cfm?abstract_id=2995134 (accessed on 10 December 2018).
12. Ghorbani, M.; Bahrami, M.; Arabzad, S.M. An integrated model for supplier selection and order allocation; using shannon entropy, swot and linear programming. *Procedia Soc. Behav. Sci.* **2012**, *41*, 521–527. [CrossRef]
13. Nazari-Shirkouhi, S.; Shakouri, H.; Javadi, B.; Keramati, A. Supplier selection and order allocation problem using a two-phase fuzzy multi-objective linear programming. *Appl. Math. Model.* **2013**, *37*, 9308–9323. [CrossRef]
14. Jadidi, O.; Cavalieri, S.; Zolfaghari, S. An improved multi-choice goal programming approach for supplier selection problems. *Appl. Math. Model.* **2015**, *39*, 4213–4222. [CrossRef]
15. Jadidi, O.; Zolfaghari, S.; Cavalieri, S. A new normalized goal programming model for multi-objective problems: A case of supplier selection and order allocation. *Int. J. Prod. Econ.* **2014**, *148*, 158–165. [CrossRef]
16. Sodenkamp, M.A.; Tavana, M.; Di Caprio, D. Modeling synergies in multi-criteria supplier selection and order allocation: An application to commodity trading. *Eur. J. Oper. Res.* **2016**, *254*, 859–874. [CrossRef]
17. Shabanpour, H.; Yousefi, S.; Saen, R.F. Future planning for benchmarking and ranking sustainable suppliers using goal programming and robust double frontiers dea. *Transp. Res. Part D Transp. Environ.* **2017**, *50*, 129–143. [CrossRef]
18. Robb, D.J.; Silver, E.A. Inventory management with periodic ordering and minimum order quantities. *J. Oper. Res. Soc.* **1998**, *49*, 1085–1094. [CrossRef]
19. Zhao, Y.; Katehakis, M.N. On the structure of optimal ordering policies for stochastic inventory systems with minimum order quantity. *Probab. Eng. Inf. Sci.* **2006**, *20*, 257–270. [CrossRef]
20. Zhou, B.; Zhao, Y.; Katehakis, M.N. Effective control policies for stochastic inventory systems with a minimum order quantity and linear costs. *Int. J. Prod. Econ.* **2007**, *106*, 523–531. [CrossRef]

21. Meena, P.; Sarmah, S. Multiple sourcing under supplier failure risk and quantity discount: A genetic algorithm approach. *Transp. Res. Part: Logist. Transp. Rev.* **2013**, *50*, 84–97. [CrossRef]

22. Zhou, B. Inventory management of multi-item systems with order size constraint. *Int. J. Syst. Sci.* **2010**, *41*, 1209–1219. [CrossRef]

23. Aktin, T.; Gergin, Z. Mathematical modelling of sustainable procurement strategies: Three case studies. *J. f Clean. Prod.* **2016**, *113*, 767–780. [CrossRef]

24. Bian, Y.; Lemoine, D.; Yeung, T.G.; Bostel, N.; Hovelaque, V.; Viviani, J.-L.; Gayraud, F. A dynamic lot-sizing-based profit maximization discounted cash flow model considering working capital requirement financing cost with infinite production capacity. *Int. J. Prod. Econ.* **2018**, *196*, 319–332. [CrossRef]

25. Chen, T.-L.; Lin, J.T.; Wu, C.-H. Coordinated capacity planning in two-stage thin-film-transistor liquid-crystal-display (tft-lcd) production networks. *Omega* **2014**, *42*, 141–156. [CrossRef]

26. Chao, X.; Chen, J.; Wang, S. Dynamic inventory management with cash flow constraints. *Nav. Res. Logist.* **2008**, *55*, 758–768. [CrossRef]

27. Bendavid, I.; Herer, Y.T.; Yücesan, E. Inventory management under working capital constraints. *J. Simul.* **2017**, *11*, 62–74. [CrossRef]

28. Azadnia, A.H.; Saman, M.Z.M.; Wong, K.Y. Sustainable supplier selection and order lot-sizing: An integrated multi-objective decision-making process. *Int. J. Prod. Res.* **2015**, *53*, 383–408. [CrossRef]

29. Oxford. What Is a Pestel Analysis? Available online: https://blog.oxfordcollegeofmarketing.com/2016/06/30/pestel-analysis/ (accessed on 10 December 2018).

30. Handfield, R.; Walton, S.V.; Sroufe, R.; Melnyk, S.A. Applying environmental criteria to supplier assessment: A study in the application of the analytical hierarchy process. *Eur. J. Oper. Res.* **2002**, *141*, 70–87. [CrossRef]

31. Lu, L.Y.; Wu, C.; Kuo, T.-C. Environmental principles applicable to green supplier evaluation by using multi-objective decision analysis. *Int. J. Prod. Res.* **2007**, *45*, 4317–4331. [CrossRef]

32. Lee, A.H.; Kang, H.-Y.; Hsu, C.-F.; Hung, H.-C. A green supplier selection model for high-tech industry. *Expert Syst. Appl.* **2009**, *36*, 7917–7927. [CrossRef]

33. Hsu, C.-W.; Hu, A.H. Applying hazardous substance management to supplier selection using analytic network process. *J. Clean. Prod.* **2009**, *17*, 255–264. [CrossRef]

34. Kannan, G.; Pokharel, S.; Kumar, P.S. A hybrid approach using ism and fuzzy topsis for the selection of reverse logistics provider. *Resour. Conserv. Recycl.* **2009**, *54*, 28–36. [CrossRef]

35. Yeh, W.-C.; Chuang, M.-C. Using multi-objective genetic algorithm for partner selection in green supply chain problems. *Expert Syst. Appl.* **2011**, *38*, 4244–4253. [CrossRef]

36. Büyüközkan, G.; Çifçi, G. A novel hybrid mcdm approach based on fuzzy dematel, fuzzy anp and fuzzy topsis to evaluate green suppliers. *Expert Syst. Appl.* **2012**, *39*, 3000–3011. [CrossRef]

37. Shaw, K.; Shankar, R.; Yadav, S.S.; Thakur, L.S. Supplier selection using fuzzy ahp and fuzzy multi-objective linear programming for developing low carbon supply chain. *Expert Syst. Appl.* **2012**, *39*, 8182–8192. [CrossRef]

38. Govindan, K.; Khodaverdi, R.; Jafarian, A. A fuzzy multi criteria approach for measuring sustainability performance of a supplier based on triple bottom line approach. *J. Clean. Prod.* **2013**, *47*, 345–354. [CrossRef]

39. Shen, L.; Olfat, L.; Govindan, K.; Khodaverdi, R.; Diabat, A. A fuzzy multi criteria approach for evaluating green supplier's performance in green supply chain with linguistic preferences. *Resour. Conserv. Recycl.* **2013**, *74*, 170–179. [CrossRef]

40. Dobos, I.; Vörösmarty, G. Green supplier selection and evaluation using dea-type composite indicators. *Int. J. Prod. Econ.* **2014**, *157*, 273–278. [CrossRef]

41. Kannan, D.; Govindan, K.; Rajendran, S. Fuzzy axiomatic design approach based green supplier selection: A case study from singapore. *J. Clean. Prod.* **2015**, *96*, 194–208. [CrossRef]

42. Awasthi, A.; Kannan, G. Green supplier development program selection using ngt and vikor under fuzzy environment. *Comput. Ind. Eng.* **2016**, *91*, 100–108. [CrossRef]

43. Trapp, A.C.; Sarkis, J. Identifying robust portfolios of suppliers: A sustainability selection and development perspective. *J. Clean. Prod.* **2016**, *112*, 2088–2100. [CrossRef]

44. Qin, J.; Liu, X.; Pedrycz, W. An extended todim multi-criteria group decision making method for green supplier selection in interval type-2 fuzzy environment. *Eur. J. Oper. Res.* **2017**, *258*, 626–638. [CrossRef]

45. Gupta, H.; Barua, M.K. Supplier selection among smes on the basis of their green innovation ability using bwm and fuzzy topsis. *J. Clean. Prod.* **2017**, *152*, 242–258. [CrossRef]

46. Luthra, S.; Govindan, K.; Kannan, D.; Mangla, S.K.; Garg, C.P. An integrated framework for sustainable supplier selection and evaluation in supply chains. *J. Clean. Prod.* **2017**, *140*, 1686–1698. [CrossRef]

47. Yu, F.; Yang, Y.; Chang, D. Carbon footprint based green supplier selection under dynamic environment. *J. Clean. Prod.* **2018**, *170*, 880–889. [CrossRef]

48. Banaeian, N.; Mobli, H.; Fahimnia, B.; Nielsen, I.E.; Omid, M. Green supplier selection using fuzzy group decision making methods: A case study from the agri-food industry. *Comput. Oper. Res.* **2018**, *89*, 337–347. [CrossRef]

49. Kannan, D.; Khodaverdi, R.; Olfat, L.; Jafarian, A.; Diabat, A. Integrated fuzzy multi criteria decision making method and multi-objective programming approach for supplier selection and order allocation in a green supply chain. *J. Clean. Prod.* **2013**, *47*, 355–367. [CrossRef]

50. Govindan, K.; Jafarian, A.; Nourbakhsh, V. Bi-objective integrating sustainable order allocation and sustainable supply chain network strategic design with stochastic demand using a novel robust hybrid multi-objective metaheuristic. *Comput. Oper. Res.* **2015**, *62*, 112–130. [CrossRef]

51. Gören, H.G. A decision framework for sustainable supplier selection and order allocation with lost sales. *J. Clean. Prod.* **2018**, *183*, 1156–1169. [CrossRef]

52. Ayhan, M.B.; Kilic, H.S. A two stage approach for supplier selection problem in multi-item/multi-supplier environment with quantity discounts. *Comput. Ind. Eng.* **2015**, *85*, 1–12. [CrossRef]

53. Yang, Y.H.; Kim, J.S. An adaptive joint replenishment policy for items with non-stationary demands. *Oper. Res.* **2018**, 1–20. [CrossRef]

54. Theodore Farris, M.; Hutchison, P.D. Cash-to-cash: The new supply chain management metric. *Int. J. Phys. Distrib. Logist. Manag.* **2002**, *32*, 288–298. [CrossRef]

55. Hofmann, E.; Kotzab, H. A supply chain-oriented approach of working capital management. *J. Bus. Logist.* **2010**, *31*, 305–330. [CrossRef]

56. Agrawal, N.; Smith, S.A. Estimating negative binomial demand for retail inventory management with unobservable lost sales. *Nav. Res. Logist.* **1996**, *43*, 839–861. [CrossRef]

57. Chen, J.; Zhao, X.; Zhou, Y. A periodic-review inventory system with a capacitated backup supplier for mitigating supply disruptions. *Eur. J. Oper. Res.* **2012**, *219*, 312–323. [CrossRef]

© 2018 by the authors. Licensee MDPI, Basel, Switzerland. This article is an open access article distributed under the terms and conditions of the Creative Commons Attribution (CC BY) license (http://creativecommons.org/licenses/by/4.0/).

Article

Application of Optimization to Select Contractors to Develop Strategies and Policies for the Development of Transport Infrastructure

Chia-Nan Wang [1,2], Tien-Muoi Le [1,*] and Han-Khanh Nguyen [3]

[1] Department of Industrial Engineering and Management, National Kaohsiung University of Sciences and Technology, Chien Kung Campus, 415 Chien Kung Road, Kaohsiung 807, Taiwan; cn.wang@fotech.edu.tw
[2] Department of Industrial Engineering and Management, Fortune Institute of Technology, Kaohsiung 83160, Taiwan
[3] Faculty of Economics, Thu Dau Mot University, Number 6, Tran Van On Street, Phu Hoa Ward, Thu Dau Mot City 820000, Binh Duong Province, VietNam; khanhnh@tdmu.edu.vn
* Correspondence: letienmuoinkust2018@gmail.com; Tel.: +886-968-466-956

Received: 9 December 2018; Accepted: 12 January 2019; Published: 18 January 2019

Abstract: Many factors influence the efficiency and quality of transport works. In particular, consultants and construction contractors of these works play important roles, and critical factors directly affect the quality of traffic works. If the quality of consultancy and construction is good, the project will reduce the total investment; if the contractor is good, the completion time of the new project is guaranteed, thus reducing construction costs. The longer the construction time is, the higher the cost of the project. In this study, the authors used optimal algorithms to evaluate past, present, and future contractors' technical, technological, and performance effectiveness. Research results show that bidders are divided into three groups: highly effective bidders, stable contractors, and inefficient groups. Research results for this subject will help the government, regulatory agencies, and investors select good contractors as the basis for developing strategies and policies for the development of transport infrastructure.

Keywords: transport infrastructure; develop strategies; optimal algorithms; business performance

1. Introduction

Contractor Selection Procedure: There are currently many models for selecting contractors for transport infrastructure projects, in which the best bidder selection model by Jyh-Bin Yang and Wei-Chih Wang is utilized (Figure 1).

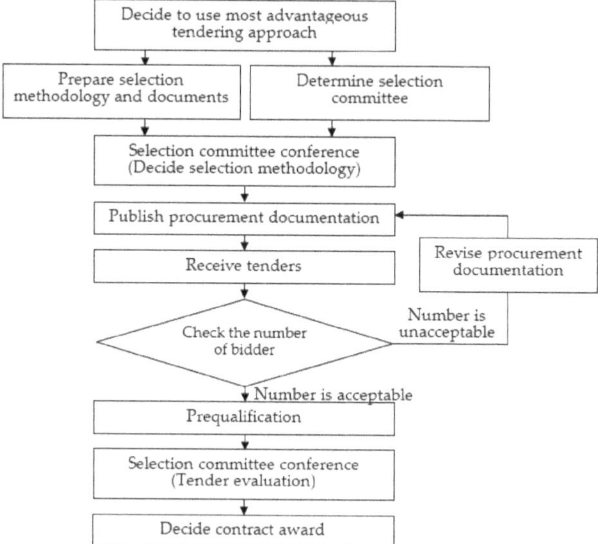

Figure 1. Contractor selection procedure [1].

Choosing a contractor for construction of traffic works is an important issue. Therefore, the contractor must have sufficient professional capacity, human resources, and financial capacity. Depending on the scale, the nature and sources of capital for the construction of traffic works are considered when selecting contractors. In particular, the selection of contractors for transport works is divided into three main forms as follows [2]:

Open bidding: Used to select contractors for construction works in the form of unlimited number of participating contractors.

Restricted bidding: To be used for the selection of contractors for construction of transport works; there are only a number of contractors that meet all the conditions in regard to the capability of construction activities and the capability of practicing the construction of works. Traffic is invited to participate in bidding.

Appointment of contractor: Investment deciders or investors of transport works are entitled to designate contractors directly qualified for construction activities and are capable of practicing construction of traffic works at a reasonable cost.

Each traffic works in different areas require different qualifications, construction techniques, and experience. In order to ensure that construction works are designed to bring economic efficiency for the investor in particular and the community in general, the selection of contractors is made in accordance with the complexity of each project. In this study, the authors use optimal algorithms to forecast and evaluate the economic, technical, and technological efficiency of contractors in the field of construction of traffic works during the period from 2014–2021. This research provides government, managers, and investors with a solid foundation in the development of strategies and policies for the development of transport infrastructure. The results of this research are important to consider when selecting contractors for construction of traffic works in the future.

2. Literature Review

2.1. Related Research

Nowadays, optimal mathematical models have become popular tools for researchers. In particular, there have been many high-impact research projects serving the macroeconomic management decisions of governments and decisions of investors in strategies and policies for developing products for businesses. In particular, research uses forecasting methods and modern data analysis techniques to find the best solutions for businesses that are of special interest to researchers worldwide. In particular, in 2006, Zhou, Ang, and Poh [3] used the Grey model to forecast electricity demand in Singapore. In 2012, Ze Zhao, Jianzhou Wang, Jing Zhao, and Zhongyue Su [4] used the Grey model to forecast annual net income per capita of rural households in China. In 2010, Guang-Ming Shi, Jun Bi, and Jin-Nan Wang [5] evaluated China's industrial energy efficiency based on DEA models. In 2007, Matthias Staat [6] used DEA models to evaluate the performance of hospitals in Germany. There are also many researchers who combine predictive models and performance assessments.

These studies have yielded good results and high applicability. Among them, Chia-Nan Wang, Han-Khanh Nguyen, and Ruei-Yuan Liao [7] combined the Grey model and DEA model to find a strategic partner in the supply chain of the textile and garment industry in Vietnam. In 2011, Jia-Jane Shuai and Wei-Wen Wu [8] used the DEA model and Grey model to evaluate marketing effectiveness through the website of hotels in Taiwan. In those studies, DEA models were used to analyze and evaluate the business situation of enterprises and economic sectors of countries, together with the forecasting models to forecast the development trend of these enterprises, from which to make appropriate decisions. The results show that these studies have yielded good results and achievements, making an important contribution to the economic growth of businesses and countries.

In the construction industry, the application of optimal mathematical models to find solutions to minimize costs and maximize profits for businesses has also been made by many researchers. In this study, the authors used the GM (1, 1) model to forecast the business situation of contractors of traffic works in Vietnam. Optimal mathematical models are used to simultaneously evaluate the three indicators of technical efficiency, technology efficiency, and business performance of past, present, and future contractors, which give the government, regulators, investors, and business leaders the basis for developing strategies and policymaking for the development of transport infrastructure. This is a new and important application in the selection of qualified contractors for the construction of highly efficient and economically efficient transport works.

2.2. Overview of Transport Infrastructure in Vietnam

Roads: Vietnam's transport infrastructure has changed dramatically in recent years. Many large and modern works have been put into operation. Especially, the road infrastructure has been built and improved. The overview of road transport infrastructure in Vietnam is shown in Table 1.

Table 1. Road infrastructure.

Order	Type of Roads	Total Route	Total Length (km)
1	Highways	13	745
2	Route	146	23,816
3	Provincial road	998	27,176
4	District roads	8680	57,294
5	Commune roads	61,402	173,294
6	Urban road	23,495	27,910
7	Other rural roads	168,888	256,377
8	Specialized road	2476	8528

Source: [9].

Railway: The Vietnam national railway infrastructure is as follows [9]:

- Total length of railway: 3161 km.
- Terminal area: 2,029,837 m^2.
- Leased area: 1,316,175 m^2.

Seaway: Maritime transport infrastructure is as follows: There are 44 seaports in the country (14 seaports of type I and IA, 17 seaports of type II, 13 offshore ports of type III). There are 254 harbors with 59.4 km long wharfs and total design capacity of about 500 million tons per year [9].

Inland waterways: At present, Vietnam has 45 inland waterways with a total length of about 7075 km [9].

- Signal systems on the route: 12,539 signaling towers, 18,458 signaling beams, 3070 signaling beams, and 9153 signaling lights.
- Bridges over the route: 251/532 bridges and river crossings located on national inland waterway lanes with less than technical specifications as approved.
- Inland waterway system: By the end of August 2017, Vietnam had 277 ports, of which 220 ports are on national inland waterways and 57 on local inland waterways.

By air: Currently, Vietnam has 21 airports in operation, including eight international airports and 13 domestic airports [9].

3. Data and Methodology

3.1. Data

3.1.1. Contractors Collection

The data collection of contractors to carry out this study plays an important and practical role in the development of strategies and policies for the development of transport infrastructure. Therefore, the authors pay special attention to the conditions of the use of optimal mathematical models. The selected bidders must have complete, continuous data, along with extensive experience in the design and construction of traffic works. These contractors must meet the human, financial, and property requirements. Through the Website of the General Statistics Office, the authors compiled 17 contractors to satisfy the conditions of this study (Table 2).

Table 2. List of contractors.

I_n	Companies
I_1	Construction joint stock company No. 1
I_2	HUD3 investment and construction joint stock company
I_3	Construction joint stock company No. 9
I_4	Coteccons construction joint stock company
I_5	Hoa Binh construction group joint stock company
I_6	Binhduong trade and development joint stock company
I_7	Development investment construction Hoi An joint stock company
I_8	Licogi 14 joint stock company
I_9	Tien Giang investment and construction joint stock company
I_{10}	Construction joint stock company No. 5
I_{11}	Chuong Duong joint stock company
I_{12}	Constrexim No 8 investment and construction joint stock company
I_{13}	Vietnam Construction joint stock company No. 2
I_{14}	Binh Duong construction and civil engineering joint stock company
I_{15}	HUD1 investment & construction joint stock company
I_{16}	Vinaconex 6 joint stock company
I_{17}	Sci E&C joint stock company

Source: [10].

3.1.2. Elements Selection

Considering the characteristics in the business of construction contractors of transport infrastructure with long production cycles, the place of production is frequently changed according to the works, and the production value of the design activities and construction is large. However, contractors still follow basic economic models. Therefore, in this study, the authors use four inputs and two outputs to analyze the economic, technical, and technological performance of the contractor. These inputs include

Total assets (F_1) are the value of all assets of an enterprise, including tangible assets such as buildings, machinery, equipment, materials, products, and goods, along with other intangible assets such as computer software, patent, commercial advantage, copyright, etc.

Equity (F_2) is the source of capital that makes up the assets of an enterprise that is contributed by the owner and investor or formed from the results of its business.

Cost of sales (F_3) is the total cost of producing a finished product. For a contractor to build consultancy and construction of transportation works, the cost of goods sold is the total cost needed to complete the works or services that the contractor receives for consultancy, design, and construction (purchase prices from suppliers, shipping, insurance, etc.).

Enterprise cost management (F_4) reflects the total administrative expense allocated to the finished product and goods sold in the enterprise's reporting year.

The outputs include

After-tax profit (F_5): The total value of a business's annual turnover, determined by the net of operating profit and other profits subtracting the cost corporate income tax.

Net sales (F_6): Represents the total sales of products, goods, and services in the reporting year of the enterprise.

The inputs and outputs used by these authors fully reflect the assets, costs, and profits of the contractors. After compiling data in local currency (VND), the authors convert to international currency (USD). All data are summarized in Tables 3–6.

Table 3. Data in 2014.

I_n	F_1	F_2	F_3	F_4	F_5	F_6
I_1	27,619.71	10,642.98	18,457.90	937.68	515.42	20,056.04
I_2	28,709.94	7639.48	13,874.53	829.40	555.53	15,705.79
I_3	69,303.22	8152.75	25,489.89	1355.45	492.23	29,433.26
I_4	213,612.52	111,010.37	310,891.73	9433.09	15,701.84	335,310.80
I_5	254,920.91	43,753.02	136,519.46	12,303.89	3021.07	154,538.01
I_6	275,092.39	51,467.31	61,864.00	2470.02	4951.20	76,694.83
I_7	8965.29	2328.10	11,151.77	415.67	132.48	11,794.35
I_8	23,157.62	2239.67	7996.86	179.96	830.19	9456.81
I_9	17,820.80	6940.10	19,197.12	1141.27	747.30	23,063.05
I_{10}	96,207.60	14,653.46	57,300.18	956.57	1465.09	61,143.84
I_{11}	35,509.26	11,122.29	9447.72	526.14	348.99	11,005.28
I_{12}	4032.71	1130.60	5697.05	100.94	57.37	5866.17
I_{13}	70,769.76	11,952.00	23,135.13	2340.04	729.60	26,622.03
I_{14}	41,732.93	15,413.33	25,252.74	273.30	941.02	26,854.88
I_{15}	32,607.10	7802.01	30,375.07	986.96	334.76	31,927.61
I_{16}	23,968.93	4944.83	24,042.54	874.82	342.88	25,566.41
I_{17}	12,052.33	2929.92	8139.35	582.72	357.69	9618.24

Source: [10].

Table 4. Data in 2015.

I_n	F_1	F_2	F_3	F_4	F_5	F_6
I_1	25,427.87	10,544.98	15,046.55	546.92	524.69	16,143.51
I_2	27,240.89	7804.60	16,308.62	935.70	690.64	18,736.61
I_3	58,661.12	8387.84	26,583.99	1288.07	486.56	33,167.85
I_4	343,282.14	142,430.07	551,576.25	15,936.92	32,188.75	600,414.22
I_5	320,271.10	47,193.66	210,066.59	5260.52	3638.61	223,054.97
I_6	323,038.50	52,527.06	50,531.61	2336.35	4789.37	68,593.61
I_7	9782.04	2316.94	9064.47	454.06	131.64	9783.93

Table 4. *Cont.*

I_n	F_1	F_2	F_3	F_4	F_5	F_6
I_8	21,538.00	2982.72	26,301.02	372.01	896.48	28,106.62
I_9	26,452.39	8868.83	24,122.09	1319.39	2460.31	30,724.63
I_{10}	99,017.47	13,671.11	59,662.10	788.29	1571.26	62,866.42
I_{11}	35,416.75	11,524.74	12,789.79	594.80	753.06	15,497.41
I_{12}	3941.09	1127.26	4227.08	115.57	43.88	4402.08
I_{13}	68,716.47	12,154.76	26,534.48	1686.70	649.75	29,570.57
I_{14}	78,950.45	15,484.49	20,196.87	307.35	1107.01	22,297.69
I_{15}	27,798.57	7888.80	26,135.77	892.70	368.32	27,642.07
I_{16}	23,237.08	4712.94	22,131.74	746.43	102.21	23,132.23
I_{17}	15,926.25	5473.47	10,984.45	731.45	617.42	12,838.24

Source: [10].

Table 5. Data in 2016.

I_n	F_1	F_2	F_3	F_4	F_5	F_6
I_1	35,109.27	10,485.68	22,603.28	1134.68	564.14	24,390.61
I_2	26,796.59	7970.64	22,050.17	1076.09	703.64	24,479.21
I_3	60,403.74	8407.83	34,716.18	1268.74	609.55	37,280.20
I_4	515,723.84	273,815.34	833,852.12	13,056.99	62,468.41	912,891.79
I_5	502,938.31	80,390.64	418,775.39	16,586.71	24,950.27	472,911.18
I_6	319,702.35	53,657.30	45,777.73	2306.88	5634.81	63,677.86
I_7	10,255.38	2320.89	9588.46	493.99	147.59	10,267.46
I_8	19,557.49	3960.06	7722.72	374.11	1183.62	9963.19
I_9	30,463.49	12,107.59	28,276.70	1286.45	3799.47	36,441.09
I_{10}	87,299.68	14,039.25	61,030.95	945.63	1841.62	64,615.23
I_{11}	32,274.41	12,500.76	11,130.81	568.00	710.45	12,491.93
I_{12}	4615.74	1132.18	5321.93	125.01	42.96	5497.76
I_{13}	111,536.69	12,839.03	39,513.77	2294.98	1322.29	45,818.27
I_{14}	55,524.62	15,639.86	36,510.14	341.96	678.34	37,868.28
I_{15}	28,725.27	7715.34	15,890.36	485.60	142.01	16,929.51
I_{16}	30,738.16	4943.11	30,534.96	847.72	332.38	31,949.23
I_{17}	24,888.24	5590.88	13,841.11	1338.85	177.02	15,710.31

Source: [10].

Table 6. Data in 2017.

I_n	F_1	F_2	F_3	F_4	F_5	F_6
I_1	35,716.50	10,548.01	24,608.44	1811.97	666.61	27,375.57
I_2	32,613.60	8584.32	14,505.26	1242.48	842.67	17,305.95
I_3	74,012.57	8369.17	43,573.89	1199.83	553.81	46,708.38
I_4	697,419.42	320,950.06	1,104,166.34	17,331.82	72,594.78	1,192,729.49
I_5	614,888.03	108,362.54	630,370.59	20,780.77	37,798.35	704,445.13
I_6	345,556.56	54,166.49	40,539.08	2031.60	5903.72	59,339.64
I_7	10,254.63	2289.53	7539.38	476.64	105.86	8201.51
I_8	13,150.60	6368.54	9517.91	658.88	2767.79	14,748.52
I_9	37,817.72	14,129.26	30,026.08	2179.23	4033.59	39,965.81
I_{10}	88,450.18	15,173.54	81,247.45	2393.50	2634.74	86,402.59
I_{11}	41,387.15	12,037.26	8806.98	599.72	1272.61	10,050.29
I_{12}	4187.99	1135.78	5589.87	135.07	43.22	5772.25
I_{13}	99,261.08	13,428.69	81,743.76	3821.78	1348.60	92,106.15
I_{14}	50,846.69	15,659.84	24,190.22	410.00	1286.71	25,434.63
I_{15}	42,474.18	7645.81	22,175.67	749.19	230.35	23,825.16
I_{16}	35,662.03	4940.52	25,106.20	795.23	329.79	26,310.29
I_{17}	32,817.24	6018.81	35,321.30	1315.83	673.69	38,384.97

Source: [10].

3.2. Methodology

3.2.1. Grey Forecasting Model

Deng [11] first introduced the grey system theory in 1982. Since its introduction, grey theories have been used extensively in statistics. Models in this theory require only a limited amount of data in the past to predict and estimate future values [11]. So far, grey system theory has been applied in almost all fields of economics, finance, science and industry, and transportation.

In models, the grey system theory is referred to as GM (1, 1); GM (1, n); GM (2, 1); GM (2, n). The general form is GM (n, m), where n is the order of the differential equations, and m is the number of variables. There are many models, but the most commonly used model is GM (1, 1), as this model is an accurate prediction model.

In this study, the steps in the GM (1, 1) model were carried out in the following steps [11]:

From the original data range:

$$X^{(0)} = \left(x_{(1)}^{(0)}, x_{(2)}^{(0)}, x_{(3)}^{(0)}, \ldots, x_{(n)}^{(0)} \right); n \geq 4. \tag{1}$$

Use the accumulating generation operation (AGO) method to compute $X^{(1)}$ values [11]:

$$X^{(1)} = \left(x_{(1)}^{(1)}, x_{(2)}^{(1)}, x_{(3)}^{(1)}, \ldots, x_{(n)}^{(1)} \right); n \geq 4. \tag{2}$$

In which,

$$\begin{cases} x_{(1)}^{(0)} = x_{(1)}^{(1)} \\ x_{(k)}^{(1)} = \sum_{i=1}^{k} x_{(i)}^{(0)} \end{cases}, (k = 1, 2, 3, \ldots, n) \tag{3}$$

Calculate the mean values $Z^{(1)}$:

$$Z^{(1)} = \left(z_{(2)}^{(1)}, z_{(3)}^{(1)}, z_{(4)}^{(1)}, \ldots, z_{(n)}^{(1)} \right); n \geq 4. \tag{4}$$

In which,

$$Z_{(n)}^{(1)} = \frac{x_{(k)}^{(1)} + x_{(k-1)}^{(1)}}{2}, k= 1, 2, 3, 4 \ldots; n = 2, 3, 4 \tag{5}$$

From the values of $X^{(0)}$, $Z^{(1)}$ the authors obtain the following system of equations:

$$\begin{cases} x_{(2)}^{(0)} + a \times z_{(2)}^{(1)} = b \\ x_{(3)}^{(0)} + a \times z_{(3)}^{(1)} = b \\ x_{(4)}^{(0)} + a \times z_{(4)}^{(1)} = b \end{cases} \tag{6}$$

From the above equations, transformed into the matrix form as follows:

$$B = \begin{bmatrix} -z_{(2)}^{(1)} \\ \ldots \quad \ldots \\ -z_{(n)}^{(1)} \end{bmatrix}; Y_N = \begin{bmatrix} x_{(2)}^{(0)} \\ \ldots \ldots \\ x_{(n)}^{(0)} \end{bmatrix} \tag{7}$$

$$\hat{a} = \begin{bmatrix} a \\ b \end{bmatrix} = \frac{1}{(B^T B)} B^T Y_N \tag{8}$$

Use the values a and b to find the differential equation:

$$\frac{dx_{(k)}^{(1)}}{dt} + ax^{(0)} = b \tag{9}$$

Equation prediction:

$$\hat{x}_{(k+1)}^{(1)} = (\hat{x}_{(1)}^{(0)} - \frac{b}{a}) \times e^{-(ak)} + \frac{b}{a} \tag{10}$$

Use the accumulating generation operation method to calculate predictive values:

$$\hat{x}_{(k+1)}^{(0)} = \hat{x}_{(k+1)}^{(1)} - \hat{x}_{k}^{(1)} \tag{11}$$

The GM (1, 1) model is used in this study to predict the business performance of contractors consulting, designing, and executing traffic infrastructure works for the period 2018–2021.

3.2.2. Mean Absolute Percentage Error

To test the predictive accuracy, the authors used the mean absolute error (ε) calculated according to the formula and convention below [12].

$$\varepsilon = \frac{1}{n}\sum_{i=1}^{n}\frac{|\lambda_i - \hat{\lambda}_i|}{\lambda_i} \tag{12}$$

$$\begin{cases} (\varepsilon) \leq 10\% : \text{Excellent}, 10\% < (\varepsilon) \leq 20\% : \text{Good} \\ 20\% < (\varepsilon) \leq 50\% : \text{Qualified}, (\varepsilon) > 50\% : \text{Unqualified} \end{cases}$$

3.2.3. Malmquist Model

Many researchers have used the Malmquist productivity index (EMPI) to evaluate the effectiveness of many areas of the world [13–18]. In this study, the authors evaluated the technical, technology, and business efficiency of the consultants, design, and construction contractors of traffic infrastructure between the two periods $P(x_k^t, y_k^t)$ and $Q(x_k^{t+1}, y_k^{t+1})$ (as shown in Figure 2). In it, the catch-up index (ECA) is used to evaluate the technical efficiency of the contractor. The frontier-shift index (EFR) is used to evaluate the technology efficiency of the contractors. The Malmquist productivity index (EMPI) is used to evaluate the performance of contractors.

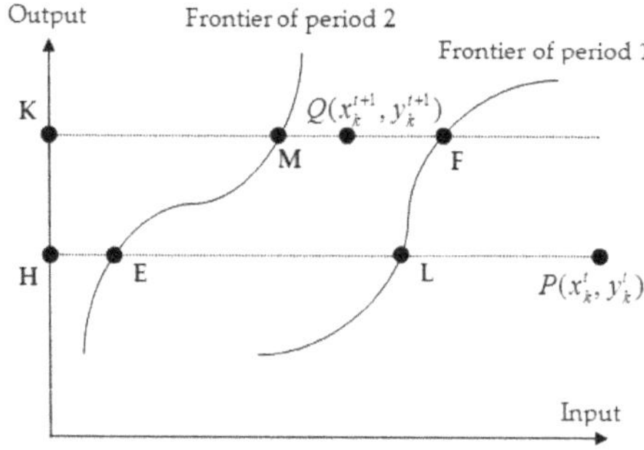

Figure 2. Catch-up terms [18].

The ECA, EFR, and EMPI indexes are used to evaluate the technical efficiency, technology efficiency, and business performance of investors at time $(t + 1)$ is the point $Q(x_k^{t+1}, y_k^{t+1})$, score for where t is the point $P(x_k^t, y_k^t)$ is calculated by the following formula.

$$\text{ECA} = \frac{\text{KM}}{\text{KQ}} \times \frac{\text{HP}}{\text{HL}} = \left[\frac{D_k^{t+1}(x_k^{t+1}, y_k^{t+1})}{D_k^t(x_k^t, y_k^t)} \right] \tag{13}$$

$$\text{EFR} = \sqrt{\frac{\text{HL}}{\text{HE}} \times \frac{\text{KF}}{\text{KM}}} = \left[\frac{D_k^t(x_k^{t+1}, y_k^{t+1})}{D_k^{t+1}(x_k^{t+1}, y_k^{t+1})} \times \frac{D_k^t(x_k^t, y_k^t)}{D_k^{t+1}(x_k^t, y_k^t)} \right]^{\frac{1}{2}} \tag{14}$$

$$\begin{aligned}
\text{EMPI} &= \text{ECA} \times \text{EFR} \\
&= \left[\frac{D_k^{t+1}(x_k^{t+1}, y_k^{t+1})}{D_k^t(x_k^t, y_k^t)} \right] \times \left[\frac{D_k^t(x_k^{t+1}, y_k^{t+1})}{D_k^{t+1}(x_k^{t+1}, y_k^{t+1})} \times \frac{D_k^t(x_k^t, y_k^t)}{D_k^{t+1}(x_k^t, y_k^t)} \right]^{\frac{1}{2}}
\end{aligned} \tag{15}$$

In particular:

The efficiency of investors at point (x_k^t, y_k^t) period t is determined by the following formula [19]:

$$D_k^t(x_k^t, y_k^t) = \text{Max}\,\varphi_k^t \tag{16}$$

$$\sum_{j=1}^{n} \lambda_j x_{ij}^t \leq x_{ik}^t,\ j = 1,\ 2,\ 3, \ldots,\ m \tag{17}$$

$$\sum_{j=1}^{n} \lambda_j y_{rj}^t \geq \varphi_k^t y_{rk}^t,\ r = 1,\ 2,\ 3, \ldots,\ s \tag{18}$$

$$\lambda_j \geq 0 \tag{19}$$

The efficiency of investors at point (x_k^t, y_k^t) period $t+1$ is determined by the following formula [19]:

$$D_k^{t+1}(x_k^t, y_k^t) = \text{Max}\,\varphi_k^{t+1} \tag{20}$$

$$\sum_{j=1}^{n} \lambda_j x_{ij}^{t+1} \leq x_{ik}^t,\ j = 1,\ 2,\ 3, \ldots,\ m \tag{21}$$

$$\sum_{j=1}^{n} \lambda_j y_{rj}^{t+1} \geq \varphi_k^{t+1} y_{rk}^t,\ r = 1,\ 2,\ 3, \ldots,\ s \tag{22}$$

$$\lambda_j \geq 0 \tag{23}$$

The efficiency of investors at point (x_k^{t+1}, y_k^{t+1}) period t is determined by the following formula [19]:

$$D_k^t(x_k^{t+1}, y_k^{t+1}) = \text{Max}\,\varphi_k^t \tag{24}$$

$$\sum_{j=1}^{n} \lambda_j x_{ij}^{t+1} \leq x_{ik}^t,\ j = 1,\ 2,\ 3, \ldots,\ m \tag{25}$$

$$\sum_{j=1}^{n} \lambda_j y_{rj}^t \geq \varphi_k^t y_{rk}^{t+1},\ r = 1,\ 2,\ 3, \ldots,\ s \tag{26}$$

$$\lambda_j \geq 0 \tag{27}$$

The efficiency of investors at point (x_k^{t+1}, y_k^{t+1}) period $t+1$ is determined by the following formula [19]:

$$D_k^{t+1}(x_k^{t+1}, y_k^{t+1}) = \text{Max}\,\varphi_k^{t+1} \tag{28}$$

$$\sum_{j=1}^{n} \lambda_j x_{ij}^{t+1} \leq x_{ik}^{t+1},\ j = 1,\ 2,\ 3, \ldots,\ m \tag{29}$$

$$\sum_{j=1}^{n} \lambda_j y_{rj}^{t+1} \geq \varphi_k^{t+1} y_{rk}^{t+1}, \; r = 1, 2, 3, \ldots, s \tag{30}$$

$$\lambda_j \geq 0 \tag{31}$$

If the ECA, EFR, EMPI indexes are less than 1 (<1), this reflects that the technical, technological, and financial investment of the consultants, and design and construction contractors in the traffic layer during the assessment period were not effective. If these indicators are greater than 1 (>1), it is clear that the bids achieved the above three indicators during this period. These indicators are equal to 1 (=1) reflecting the performance in the comparison period corresponding to the previous period [20].

3.2.4. Correlation Coefficients

The correlation coefficient (k), first introduced in 1895, is a statistical tool used to test correlations between variables. The formula for calculating the correlation coefficient is as follows [21]:

$$k = \frac{\sum\limits_{i=1}^{n} (\alpha_i - \bar{\alpha})(\delta_i - \bar{\delta})}{\left[\sum\limits_{i=1}^{n} (\alpha_i - \bar{\alpha})^2 \sum\limits_{i=1}^{n} (\delta_i - \bar{\delta})^2\right]^{\frac{1}{2}}} \tag{32}$$

The value of k depends on the range $(-1; 1)$. The value of negative k denotes the inverse relationship between the two factors (if the value of this factor increases, then the other factor decreases and vice versa). The value of positive k denotes the positive correlation between the two factors (if the value of this factor increases, the value of the other factor increases; if this value decreases, the value of the other factor will decrease accordingly). The value of $k = 0$ denotes independent factors [22].

4. Results

4.1. Forecasting

Designing and executing traffic infrastructure works for the period 2018–2021, the authors use data from factor F_1 of I_{12} (Table 7) to explain the computational steps in the GM (1, 1) model used in this study (as follows).

Table 7. Data of I_{12}.

Year	F_1	F_2	F_3	F_4	F_5	F_6
2014	4032.71	1130.60	5697.05	100.94	57.37	5866.17
2015	3941.09	1127.26	4227.08	115.57	43.88	4402.08
2016	4615.74	1132.18	5321.93	125.01	42.96	5497.76
2017	4187.99	1135.78	5589.87	135.07	43.22	5772.25

From the original data range: $X^{(0)} = (4032.71; 3941.09; 4615.74; 4187.99)$
Use the accumulating generation operation method to compute $X^{(1)}$ values:

$$X^{(1)} = (4032.71; 7973.80; 12589.54; 16777.53)$$

Calculate the mean values $Z^{(1)}$: $Z^{(1)} = (6003.26; 10281.67; 14683.53)$
From the values of $X^{(0)}$, $Z^{(1)}$ the authors obtain the following system of equations:

$$\begin{cases} 3941.0853 + a \times 6003.26 = b \\ 4615.7374 + a \times 10281.67 = b \\ 4187.9907 + a \times 14683.53 = b \end{cases}$$

From the above equations, transformed into the matrix form as follows:

$$B = \begin{bmatrix} -6003.26 \\ -10281.67 \\ -14683.53 \end{bmatrix}; \ Y_N = \begin{bmatrix} 3941.0853 \\ 4615.7374 \\ 4187.9907 \end{bmatrix}$$

$$\begin{bmatrix} a \\ b \end{bmatrix} = \begin{bmatrix} -0.0278 \\ 3960.8791 \end{bmatrix}$$

Use the values a and b to find the differential equation:

$$\frac{dx_{(k)}^{(1)}}{dt} - 0.0278x^{(0)} = 3960.8791$$

Equation prediction:

$$\hat{x}_{(k+1)}^{(1)} = \left(4032.71 + \frac{3960.8791}{0.0278}\right) \times e^{0.0278k} - \frac{3960.8791}{0.0278}$$
$$= 146510.3755 \times e^{0.0278k} - 142477.6655$$

In turn, the values for $k = 0, 1, 2, ..., 7$ are given, $\hat{x}_{(k+1)}^{(1)}$ by the following values:

$$\hat{x}_{(k+1)}^{(1)} = \{4032.71; 8163.09; 12410.08; 16776.97; 21267.14; 25884.08; 30631.36; 35512.66\} \tag{33}$$

Use the accumulating generation operation method to calculate the predicted values by the formula: $\hat{x}_{(k+1)}^{(0)} = \hat{x}_{(k+1)}^{(1)} - \hat{x}_{k}^{(1)}$ obtains the predicted values for the years in Table 8.

Table 8. Forecast data F_1 of I_{12}.

Year	2014	2015	2016	2017	2018	2019	2020	2021
Forecast	4032.71	4130.38	4246.99	4366.89	4490.17	4616.94	4747.28	4881.30

As calculated above, the authors obtained predictive values for all elements of bidders for the period 2018–2021 which are summarized in Tables 9–12 below:

Table 9. Forecast data in 2018.

I_n	F_1	F_2	F_3	F_4	F_5	F_6
I_1	43,148.97	10,529.26	31,583.34	3002.34	743.78	35,590.60
I_2	34,876.68	8937.01	16,084.48	1427.70	914.95	18,918.35
I_3	81,722.91	8369.65	55,330.65	1166.97	616.75	54,842.26
I_4	970,992.39	464,410.49	1,528,230.15	17,007.63	105,075.32	1,645,842.91
I_5	837,908.16	158,089.65	1,010,373.13	35,425.99	80,927.61	1,146,236.67
I_6	352,912.66	55,106.43	36,489.40	1941.59	6626.52	55,128.51
I_7	10,574.94	2281.91	7373.88	497.44	106.19	8000.74
I_8	11,396.45	9067.29	2843.32	861.83	4424.96	6251.14
I_9	44,823.14	17,802.49	33,790.76	2746.82	5195.65	45,802.04
I_{10}	81,267.23	15,874.31	92,890.06	3753.24	3361.32	99,281.31
I_{11}	43,105.18	12,531.78	7512.81	592.52	1623.51	8090.65
I_{12}	4490.17	1140.29	6514.66	145.95	42.70	6697.85
I_{13}	124,451.91	14,130.21	132,385.34	5576.55	1904.47	149,909.49
I_{14}	37,470.96	15,770.61	30,520.06	469.61	1254.61	31,322.63
I_{15}	51,484.78	7509.46	17,234.57	555.00	116.78	18,751.01
I_{16}	44,165.08	5095.48	28,770.26	844.86	526.54	30,174.41
I_{17}	45,949.20	6265.79	55,965.60	1772.24	573.39	60,413.53

Table 10. Forecast data in 2019.

I_n	F_1	F_2	F_3	F_4	F_5	F_6
I_1	50,268.76	10,530.78	39,335.96	5135.68	841.40	45,135.29
I_2	38,405.69	9380.65	15,372.27	1645.09	1016.03	18,323.39
I_3	92,406.66	8360.35	70,525.62	1126.83	653.54	65,455.69
I_4	1,362,850.22	652,348.51	2,123,825.36	17,859.22	147,332.03	2,274,415.58
I_5	1,128,340.52	231,141.55	1,650,867.09	57,935.96	161,765.74	1,892,158.68
I_6	365,367.83	55,956.38	32,713.77	1815.37	7327.12	51,272.07
I_7	10,823.35	2268.44	6785.52	509.16	96.74	7383.38
I_8	9129.50	13,497.17	1389.64	1206.79	8428.18	3906.00
I_9	53,806.69	22,183.54	37,561.50	3691.75	6451.81	52,051.24
I_{10}	76,607.59	16,738.56	109,851.23	7293.26	4420.41	117,992.54
I_{11}	47,016.95	12,796.42	6273.92	595.05	2228.47	6521.79
I_{12}	4616.94	1144.59	7428.14	157.77	42.37	7609.37
I_{13}	144,387.22	14,849.66	241,883.95	8566.41	2532.30	272,267.60
I_{14}	29,524.41	15,859.39	32,487.51	544.20	1392.92	32,828.68
I_{15}	65,218.21	7392.40	15,507.31	492.67	82.55	17,042.17
I_{16}	54,172.45	5215.07	30,323.31	870.30	770.08	31,838.47
I_{17}	64,451.77	6575.94	111,382.72	2242.08	621.99	115,378.84

Table 11. Forecast data in 2020.

I_n	F_1	F_2	F_3	F_4	F_5	F_6
I_1	58,563.35	10,532.30	48,991.59	8784.90	951.83	57,239.69
I_2	42,291.79	9846.31	14,691.59	1895.59	1128.26	17,747.14
I_3	104,487.13	8351.07	89,893.44	1088.07	692.53	78,123.11
I_4	1,912,847.87	916,341.45	2,951,541.13	18,753.45	206,582.56	3,143,049.80
I_5	1,519,441.39	337,950.12	2,697,381.85	94,748.97	323,352.64	3,123,494.98
I_6	378,262.58	56,819.44	29,328.81	1697.36	8101.78	47,685.40
I_7	11,077.60	2255.05	6244.11	521.17	88.13	6813.66
I_8	7313.48	20,091.29	679.17	1689.81	16,053.07	2440.64
I_9	64,590.75	27,642.71	41,753.01	4961.74	8011.67	59,153.09
I_{10}	72,215.11	17,649.85	129,909.41	14,172.20	5813.20	140,230.21
I_{11}	51,283.70	13,066.65	5239.32	597.59	3058.84	5257.15
I_{12}	4747.28	1148.90	8469.72	170.56	42.05	8644.94
I_{13}	167,515.87	15,605.75	441,951.10	13,159.30	3367.09	494,496.04
I_{14}	23,263.11	15,948.68	34,581.79	630.64	1546.49	34,407.14
I_{15}	82,615.00	7277.16	13,953.15	437.35	58.35	15,489.06
I_{16}	66,447.39	5337.48	31,960.20	896.51	1126.29	33,594.31
I_{17}	90,404.86	6901.43	221673.89	2836.48	674.71	220,352.59

Table 12. Forecast data in 2021.

I_n	F_1	F_2	F_3	F_4	F_5	F_6
I_1	68,226.59	10,533.82	61,017.34	15,027.11	1076.76	72,590.24
I_2	46,571.11	10,335.09	14,041.06	2184.23	1252.90	17,189.01
I_3	118,146.88	8341.79	114,580.07	1050.65	733.84	93,242.00
I_4	2,684,804.91	1,287,167.27	4,101,841.54	19,692.46	289,661.07	4,343,428.75
I_5	2,046,104.07	494,114.03	4,407,301.41	154,953.26	646,347.78	5,156,132.51
I_6	391,612.42	57,695.82	26,294.10	1587.02	8958.35	44349.63
I_7	11,337.82	2241.74	5745.90	533.45	80.28	6287.90
I_8	5858.70	29,907.01	331.94	2366.17	30,576.11	1525.03
I_9	77,536.16	34,445.35	46,412.26	6668.62	9948.66	67,223.91
I_{10}	68,074.49	18,610.76	153,630.09	27,539.30	7644.83	166,658.94
I_{11}	55,937.66	13,342.59	4375.34	600.14	4198.64	4237.74
I_{12}	4881.30	1153.24	9657.34	184.37	41.73	9821.43
I_{13}	194,349.37	16,400.33	807,497.87	20,214.66	4477.08	898,110.26
I_{14}	18,329.65	16,038.46	36,811.08	730.80	1716.98	36,061.50
I_{15}	104,652.34	7163.71	12,554.75	388.24	41.24	14,077.50
I_{16}	81,503.71	5462.76	33,685.45	923.51	1647.25	35,446.97
I_{17}	126,808.59	7243.04	441,175.36	3588.46	731.90	420,833.36

To test the accuracy of the predicted values, the authors used MAPE to ensure that the results of this study were highly reliable. The results in Table 13 are as follows:

Table 13. Errors results.

I_n	MAPE (%)	I_n	MAPE (%)
I_1	3.82	I_{10}	5.12
I_2	5.01	I_{11}	3.73
I_3	1.92	I_{12}	1.78
I_4	5.38	I_{13}	6.85
I_5	15.13	I_{14}	9.90
I_6	0.99	I_{15}	13.16
I_7	3.35	I_{16}	8.08
I_8	13.55	I_{17}	16.47
I_9	3.80	Average $_{\text{all } In}$ (%)	6.94

As shown in Table 13, average $_{\text{all } In}$ = 6.94% indicates that the forecasts for contractors' business performance for the period 2018–2021 are high. In particular, the predicted data for 4/17 of contractors ranged from 10 to 20%, while the predicted data for 13 of 17 contractors had errors of less than 10%. This confirms that the GM (1, 1) model used to predict the contractor's business performance in this study is consistent with high reliability.

4.2. Correlation Coefficient

The correlation coefficients are shown in Table 14. Based on the convention mentioned in Section 3.2.4, it was found that the factors used in this study were positive (the increase in inputs would lead to the increase in output). This is in line with economic law. All coefficients are greater than 0.6, indicating that these factors have a strong correlation.

Table 14. Correlation coefficient.

	Time Period 2014						Time Period 2015					
	F_1	F_2	F_3	F_4	F_5	F_6	F_1	F_2	F_3	F_4	F_5	F_6
F_1	1.00	0.81	0.71	0.79	0.67	0.73	1.00	0.85	0.75	0.75	0.68	0.76
F_2	0.81	1.00	0.96	0.77	0.97	0.96	0.85	1.00	0.96	0.97	0.96	0.96
F_3	0.71	0.96	1.00	0.83	0.96	1.00	0.75	0.96	1.00	0.99	0.96	1.00
F_4	0.79	0.77	0.83	1.00	0.67	0.84	0.75	0.97	0.99	1.00	0.97	0.99
F_5	0.67	0.97	0.96	0.67	1.00	0.96	0.68	0.96	0.96	0.97	1.00	0.96
F_6	0.73	0.96	1.00	0.84	0.96	1.00	0.76	0.96	1.00	0.99	0.96	1.00
	Time Period 2016						Time Period 2017					
F_1	1.00	0.83	0.87	0.92	0.85	0.87	1.00	0.89	0.92	0.93	0.92	0.92
F_2	0.83	1.00	0.97	0.77	0.99	0.97	0.89	1.00	0.97	0.81	0.98	0.96
F_3	0.87	0.97	1.00	0.88	0.99	1.00	0.92	0.97	1.00	0.92	0.99	1.00
F_4	0.92	0.77	0.88	1.00	0.84	0.89	0.93	0.81	0.92	1.00	0.90	0.93
F_5	0.85	0.99	0.99	0.84	1.00	0.99	0.92	0.98	0.99	0.90	1.00	0.99
F_6	0.87	0.97	1.00	0.89	0.99	1.00	0.92	0.96	1.00	0.93	0.99	1.00

This result shows that the data used in this study are consistent with the conditions of use of the optimal mathematical models used in this study. The conclusions of the study have sufficient grounds for assessing the technical, technological, and economic efficiency of contractors in the field of construction of transport infrastructure works.

4.3. Catch-Up Index

The catch-up index reflects the technical efficiency of the consultants, design, and construction contractors for transport infrastructure in Vietnam for the period 2014–2021, as shown in Table 15.

Table 15. The catch-up index.

(ECA)	'14–'15	'15–'16	'16–'17	'17–'18	'18–'19	'19–'20	'20–'21	Average
I_1	0.9956	0.9211	1.0104	1.0768	1.0419	1.0380	1.0180	1.0146
I_2	0.9718	0.9855	0.9172	1.0726	1.0179	1.0063	0.9963	0.9954
I_3	1.0000	0.9404	1.0634	1.0000	0.9599	0.8961	0.9620	0.9745
I_4	1.0000	1.0000	1.0000	1.0000	1.0000	1.0000	1.0000	1.0000
I_5	1.0000	1.0000	1.0000	1.0000	1.0000	1.0000	1.0000	1.0000
I_6	1.0000	1.0000	1.0000	1.0000	1.0000	1.0000	1.0000	1.0000
I_7	0.9546	1.0145	0.9616	1.0450	1.0276	1.0000	1.0000	1.0005
I_8	1.0000	1.0000	1.0000	1.0000	1.0000	1.0000	1.0000	1.0000
I_9	1.0000	1.0000	1.0000	1.0000	1.0000	1.0000	1.0000	1.0000
I_{10}	1.0000	1.0000	0.9769	0.9571	0.9936	1.0014	1.0088	0.9911
I_{11}	0.9935	0.8856	0.8911	0.9988	1.0654	1.2272	1.0000	1.0088
I_{12}	1.0000	1.0000	1.0000	1.0000	1.0000	1.0000	1.0000	1.0000
I_{13}	0.9167	1.0966	1.0490	1.0000	1.0000	1.0000	1.0000	1.0089
I_{14}	1.0000	1.0000	1.0000	0.9437	0.9841	0.9882	0.9828	0.9855
I_{15}	0.9256	0.9967	0.9960	1.0410	1.0017	1.0041	1.0132	0.9969
I_{16}	0.9121	1.0963	0.9424	1.0088	0.9947	0.9936	0.9950	0.9919
I_{17}	0.9588	0.9491	1.0989	1.0000	1.0000	1.0000	1.0000	1.0010
Average	0.9782	0.9933	0.9945	1.0085	1.0051	1.0091	0.9986	0.9982
Max	1.0000	1.0966	1.0989	1.0768	1.0654	1.2272	1.0180	1.0146
Min	0.9121	0.8856	0.8911	0.9437	0.9599	0.8961	0.9620	0.9745
SD	0.0323	0.0523	0.0499	0.0342	0.0232	0.0626	0.0120	0.0092

The results in Table 15 and Figure 3 show that the consultants, design, and construction of transport infrastructure have not achieved technical efficiency in the period 2013–2017. Specifically, the catch-up index in this phase is less than 1 (average ECA 2014–2015 = 0.9782; average ECA 2015–2016 = 0.9933; average ECA 2016–2017 = 0.9945). However, according to the results above, in the period 2017–2020, the contractor has made changes and achieved technical efficiency. In the period 2017–2020, the catch-up index is greater than 1 (average ECA 2017–2018 = 1.0085; average ECA 2018–2019 = 1.0051; average ECA 2019–2020 = 1.0091). In particular, there are strong changes and high efficiency of three contractors (I_1, I_{11}, I_{15}). These bidders have not really achieved technical efficiency in the period 2014–2017, but, from the period 2017–2021, there are solutions to create efficiency (catch-up index is greater than 1). In addition, nine out of 17 contractors have maintained stable technical performance in the period 2017–2021, including I_4, I_5, I_6, I_7, I_8, I_9, I_{12}, I_{13}, I_{17}. However, many contractors have not yet achieved technical efficiency (I_1, I_2, I_{10}, I_{14}, I_{16}). Managers rely on this result to set up timely measures, in line with the actual situation, to improve the technical efficiency of their businesses in the future.

In construction, machinery is always a decisive factor in productivity, quality, and efficiency. Therefore, the frequent research and development of new high-tech techniques play a decisive role. On the other hand, the machinery and equipment in construction are typically of great economic value; further, the maintenance of machinery and equipment, ensuring the machinery is always operating properly and takes advantage of the capacity of machinery equipment, plays a huge role in improving the technical efficiency of contractors.

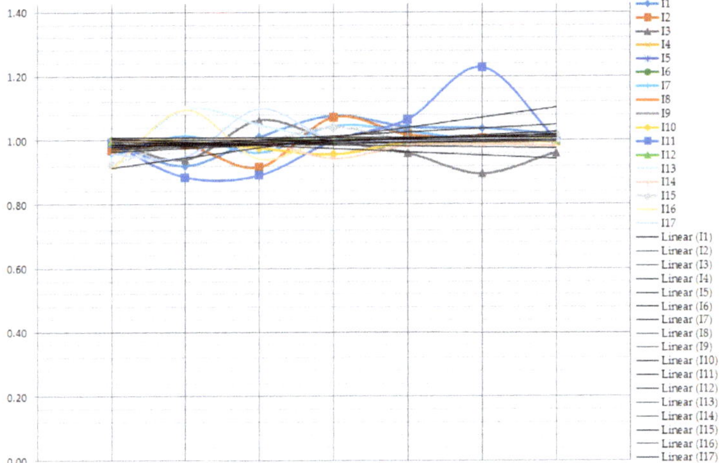

Figure 3. Catch-up index.

4.4. Frontier Index

The CFR index reflects the technological efficiency of construction contractors for transport infrastructure development. The results in Table 16 show that the technology efficiency scores of past contractors (2014–2017) and future predictions (2018–2021) are good. Specifically, the average technology efficiency scores for bidders for the period 2014–2021 are average 2014–2015: 1.0744; average 2015–2016: 1.0764; average 2016–2017: 1.0656; average 2017–2018:1.0872; average 2018–2019: 1.0928; average 2019–2020: 1.0936; average 2020–2021: 1.0955.

Table 16. Frontier index.

(EFR)	'14–'15	'15–'16	'16–'17	'17–'18	'18–'19	'19–'20	'20–'21	Average
I_1	1.0368	1.0291	1.0220	0.9833	1.0017	0.9990	1.0124	1.0120
I_2	1.0407	1.0087	1.0753	0.9297	0.9813	1.0006	1.0128	1.0070
I_3	1.0844	0.9756	1.0096	1.1078	1.1737	1.2170	1.1320	1.1000
I_4	1.4438	1.4294	1.0633	1.2018	1.1780	1.1831	1.1799	1.2399
I_5	1.1949	1.1485	1.1439	1.3345	1.3750	1.4227	1.4220	1.2917
I_6	1.0948	1.1329	1.0830	1.0965	1.0616	1.0301	1.0293	1.0755
I_7	1.0183	0.9836	1.0256	0.9536	0.9802	1.0020	1.0005	0.9948
I_8	0.7654	1.1597	1.7164	0.9066	1.0000	1.0000	1.0000	1.0783
I_9	1.3344	1.2253	1.0100	1.0820	1.0204	1.0138	1.0119	1.0997
I_{10}	1.1063	0.9491	0.9558	1.0619	1.0943	1.0855	1.0147	1.0382
I_{11}	1.0329	1.0480	1.1554	1.1902	0.9499	0.8692	1.0352	1.0401
I_{12}	1.0000	1.1096	1.0000	1.0743	1.0621	1.0599	1.0562	1.0517
I_{13}	1.0637	1.0108	1.0723	1.2639	1.3339	1.3365	1.3396	1.2030
I_{14}	0.8733	1.1119	0.7263	1.0872	1.0017	0.9991	1.0193	0.9741
I_{15}	1.0672	1.0071	1.0069	0.9640	1.0043	1.0025	0.9947	1.0067
I_{16}	1.0757	0.9580	0.9814	1.0202	1.0112	1.0106	1.0087	1.0094
I_{17}	1.0321	1.0111	1.0685	1.2259	1.3474	1.3590	1.3545	1.1998
Average	1.0744	1.0764	1.0656	1.0872	1.0928	1.0936	1.0955	1.0836
Max	1.4438	1.4294	1.7164	1.3345	1.3750	1.4227	1.4220	1.2917
Min	0.7654	0.9491	0.7263	0.9066	0.9499	0.8692	0.9947	0.9741
SD	0.1523	0.1210	0.1920	0.1231	0.1389	0.1545	0.1415	0.0946

Based on the results of the technology efficiency score of the contractors, as shown in Table 16 and Figure 4, the authors created three groups:

1. Highly efficient bidders in the future, including I_3, I_4, I_5, I_6, I_9, I_{10}, I_{12}, I_{13}, I_{16}, I_{17}. These bidders are expected to have a technology efficiency score greater than 1 (>1) over the period 2018–2021.
2. Contractors that maintain stable technology in the future, including I_8. This contractor is expected to be technologically stable (=1) over the period 2018–2021.
3. Contractors not effective in terms of technology in the future, including I_1, I_2, I_7, I_{11}, I_{14}, I_{15}. These bidders are predicted to be technologically efficient with high volatility during the period 2018–2021.

In general, construction contractors of transport works in Vietnam have approached and mastered modern technologies, improving the capacity of construction of transport works to meet the requirements of construction works. Along with that is the use of high-quality materials and application of new technology to bring high economic efficiency and longevity for the works.

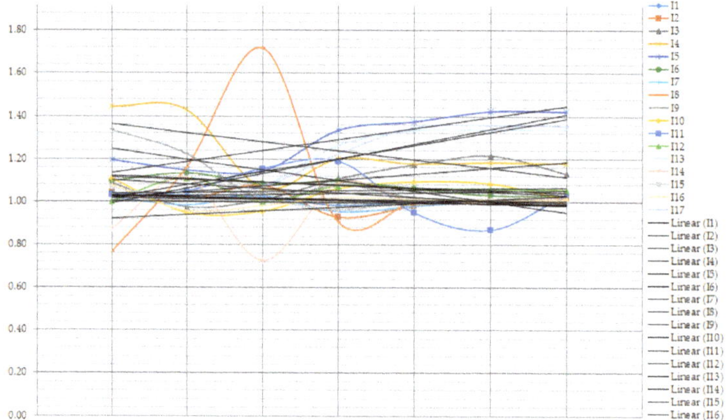

Figure 4. Frontier index.

4.5. Malmquist Index

As shown in Table 17 and Figure 5, EMPI average 2014–2018 = 1.0807, which reflects the business performance of the contractor of traffic works in general and is quite satisfactory. Specifically, according to the forecast results, in the period 2018–2021, 13 out of 17 contractors have achieved high business results, including I_1, I_3, I_4, I_5, I_6, I_9, I_{10}, I_{11}, I_{12}, I_{13}, I_{15}, I_{16}, I_{17}. In addition, 2/17 contractors are expected to maintain stable business performance during this period, including I_7, I_8. However, 2/17 of the contractors whose business results are forecasted to be ineffective include I_2, I_{14}.

Table 17. Malmquist index.

(EMPI)	'14–'15	'15–'16	'16–'17	'17–'18	'18–'19	'19–'20	'20–'21	Average
I_1	1.0322	0.9480	1.0326	1.0588	1.0436	1.0370	1.0307	1.0261
I_2	1.0114	0.9941	0.9863	0.9972	0.9989	1.0070	1.0090	1.0005
I_3	1.0844	0.9174	1.0736	1.1078	1.1266	1.0905	1.0890	1.0699
I_4	1.4438	1.4294	1.0633	1.2018	1.1780	1.1831	1.1799	1.2399
I_5	1.1949	1.1485	1.1439	1.3345	1.3750	1.4227	1.4220	1.2917
I_6	1.0948	1.1329	1.0830	1.0965	1.0616	1.0301	1.0293	1.0755
I_7	0.9721	0.9979	0.9862	0.9965	1.0072	1.0020	1.0005	0.9946
I_8	0.7654	1.1597	1.7164	0.9066	1.0000	1.0000	1.0000	1.0783
I_9	1.3344	1.2253	1.0100	1.0820	1.0204	1.0138	1.0119	1.0997
I_{10}	1.1063	0.9491	0.9338	1.0163	1.0872	1.0870	1.0236	1.0290
I_{11}	1.0262	0.9281	1.0296	1.1888	1.0120	1.0667	1.0352	1.0409
I_{12}	1.0000	1.1096	1.0000	1.0743	1.0621	1.0599	1.0562	1.0517
I_{13}	0.9751	1.1085	1.1248	1.2639	1.3339	1.3365	1.3396	1.2118
I_{14}	0.8733	1.1119	0.7263	1.0260	0.9858	0.9872	1.0018	0.9589
I_{15}	0.9878	1.0038	1.0029	1.0035	1.0060	1.0066	1.0078	1.0026
I_{16}	0.9812	1.0503	0.9250	1.0292	1.0058	1.0041	1.0037	0.9999
I_{17}	0.9895	0.9597	1.1741	1.2259	1.3474	1.3590	1.3545	1.2014
Average	1.0513	1.0691	1.0595	1.0947	1.0972	1.0996	1.0938	1.0807
Max	1.4438	1.4294	1.7164	1.3345	1.3750	1.4227	1.4220	1.2917
Min	0.7654	0.9174	0.7263	0.9066	0.9858	0.9872	1.0000	0.9589
SD	0.1590	0.1316	0.1974	0.1128	0.1318	0.1397	0.1407	0.0973

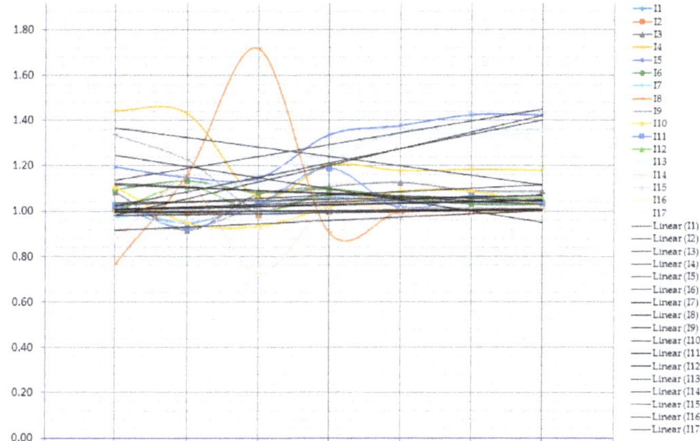

Figure 5. Malmquist index.

In order to maintain and improve business efficiency, contractors themselves must actively create, overcome difficulties, promote advantages, and exploit and make use of favorable conditions and factors of the environment and geographic location, thus combining multiple measures to maximize the use of resources and businesses to achieve optimal efficiency. The economic efficiency of production and business activities in consultancy, design, and construction of traffic works is an integrated category in many fields. In order to improve the economic efficiency of production and business activities, contractors must use the combination of measures from raising the management capacity and managing production and business activities of enterprises in the office to work. In addition to enhancing and improving all the activities within the enterprise, enterprises are always adapting to the changes of the market, i.e., adaptation to each project in different localities. Bidders must regularly maintain and ensure the balance of the relationship between parties from the construction site, so the new office can enhance the sense of responsibility of each person, i.e., enhance the initiative in production to bring high economic efficiency.

4.6. Contractor Selection

The results of forecasting and evaluating the technical, technological, and business efficiency of the consultants, design, and construction contractors of the transport infrastructure are shown in Table 18.

Table 18. Classify contractor.

	Good Efficiency	Efficiency	Inefficiency
ECA	I_1, I_{11}, I_{15}	$I_4, I_5, I_6, I_7, I_8, I_9, I_{12}, I_{13}, I_{17}$	$I_2, I_3, I_{10}, I_{14}, I_{16}$
EFR	$I_3, I_4, I_5, I_6, I_9, I_{10}, I_{12}, I_{13}, I_{16}, I_{17}$	I_8	$I_1, I_2, I_7, I_{11}, I_{14}, I_{15}$
EMPI	$I_1, I_3, I_4, I_5, I_6, I_9, I_{10}, I_{11}, I_{12}, I_{13}, I_{15}, I_{16}, I_{17}$	I_7, I_8	I_2, I_{14}

Based on these results, the government, regulatory authorities in strategy formulation, policy development, selection of consultants, and design and construction contractors have good capacity to implement projects. It is of decisive importance in ensuring the progress and quality of traffic works. The result of this business performance assessment is also a good basis for all self-revising subjects, to see where their businesses are in the overall picture of the construction investment sector to provide timely solutions to improve capacity and more effective implementation of assigned tasks.

5. Conclusions

At present, with the rapid development of the economy, demand for transportation and transportation of goods and people remains large. As a result, road, rail, waterway, and air infrastructure works are being increasingly built. Therefore, the role of contractor consulting, design, and construction of transport infrastructure has become more important and maintains a special position. The main contractor is the decisive factor affecting the quality and progress of construction of the transportation infrastructure. In this study, the authors used a modern, highly accurate forecasting method to forecast contractors' business, design, and construction contractor performance. Also, in this study, the authors used optimized mathematical models to evaluate past, present, and future contractors' technical, technological, and performance effectiveness. The result of this study is a solid basis for the government, regulatory agencies, and investors to use for strategic planning and policy development of transport infrastructure with high efficiency. The best way to do this is through the selection of contractors who have the human, financial, technical, and technological capabilities to meet the requirements of managers and investors.

In addition to the results achieved, this research still has certain limitations: It does not combine with the qualitative factors, weather factors, and government policies. In addition, many of the optimal mathematical models have not been considered in this study. The authors will continue to address these issues in subsequent studies.

Author Contributions: C.-N.W. developed and checked theoretical calculations. T.-M.L. wrote and formatted the manuscript. H.-K.N. contributed to analyzing the data. All authors have read and accepted this research.

Funding: This research received no external funding.

Conflicts of Interest: The authors declare no conflict of interest.

References

1. Yang, J.B.; Wang, W.C. Contractor selection by the most advantageous tendering approach in Taiwan. *J. Chin. Inst. Eng.* **2003**, *26*, 381–387. [CrossRef]
2. Holt, G.D. Which contractor selection methodology? *Int. J. Proj. Manag.* **1998**, *16*, 153–164. [CrossRef]

3. Zhou, P.A.B.W.; Ang, B.W.; Poh, K.L. A trigonometric grey prediction approach to forecasting electricity demand. *Energy* **2006**, *31*, 2839–2847. [CrossRef]

4. Zhao, Z.; Wang, J.; Zhao, J.; Su, Z. Using a Grey model optimized by Differential Evolution algorithm to forecast the per capita annual net income of rural households in China. *Omega* **2012**, *40*, 525–532. [CrossRef]

5. Shi, G.M.; Bi, J.; Wang, J.N. Chinese regional industrial energy efficiency evaluation based on a DEA model of fixing non-energy inputs. *Energy Policy* **2010**, *38*, 6172–6179. [CrossRef]

6. Pilyavsky, A.; Staat, M. Efficiency and productivity change in Ukrainian health care. *J. Prod. Anal.* **2008**, *29*, 143–154. [CrossRef]

7. Wang, C.N.; Nguyen, H.K.; Liao, R.Y. Partner Selection in Supply Chain of Vietnam's Textile and Apparel Industry: The Application of a Hybrid DEA and GM (1,1) Approach. *Math. Probl. Eng.* **2017**. [CrossRef]

8. Shuai, J.J.; Wu, W.W. Evaluating the influence of E-marketing on hotel performance by DEA and grey entropy. *Expert Syst. Appl.* **2011**, *38*, 8763–8769. [CrossRef]

9. Minh-Hue, N. Logistics Vietnam report 2017. Available online: http://nhaxuatbancongthuong.com (accessed on 21 May 2018).

10. Statistics. General Statistics Office of Vietnam. Available online: https://www.gso.gov.vn (accessed on 25 May 2018).

11. Deng, J. Control problems of grey systems. *Syst. Control Lett.* **1982**, *1*, 288–294.

12. Wang, C.N.; Nguyen, H.K. Enhancing Urban Development Quality Based on the Results of Appraising Efficient Performance of Investors—A Case Study in Vietnam. *Sustainability* **2017**, *9*, 1397. [CrossRef]

13. Boles, J.-S.; Donthu, N.; Ritu, L. Salesperson evaluation using relative performance efficiency: The application of data envelopment analysis. *J. Pers. Sell. Sales Manag.* **1995**, *15*, 31–49.

14. Charnes, A.; Cooper, W.-W.; Rhodes, E. Evaluating Program and Managerial Efficiency: An Application of Data Envelopment Analysis to Program Follow Through. *Manag. Sci.* **1981**, *27*, 668–697. [CrossRef]

15. Banker, R.-D.; Charnes, A.; Cooper, W.-W. Some models for estimating technical and scale inefficiencies in data envelopment analysis. *Manag. Sci.* **1984**, *30*, 1078–1092. [CrossRef]

16. Mao, W.; Koo, W.W. Productivity growth, technological progress, and efficiency change in chinese agriculture after rural economic reforms: A DEA approach. *China Econ. Rev.* **1997**, *8*, 157–174. [CrossRef]

17. Macpherson, A.-J.; Principe, P.-P.; Mehaffey, M. Using Malmquist Indices to evaluate environmental impacts of alternative land development scenarios. *Ecological Indicators* **2013**, *34*, 296–303. [CrossRef]

18. Paradi, J.C.; Asmild, M.; Aggarwall, V.; Schaffnit, C. *Performance Evaluation in an Oligopoly Environment: Combining DEA Window Analysis with the Malmquist Index Approach—A Study of the Canadian Banking Industry*; Centre for Management of Technology and Entrepreneurship University of Toronto, University of Toronto: Toronto, ON, Canada, 2011; pp. 31–47.

19. Odeck, J. Identifying traffic safety best practice: An application of DEA and Malmquist indices. *Omega* **2006**, *34*, 28–40. [CrossRef]

20. Wang, C.N.; Nguyen, N.T.; Tran, T.T. Integrated DEA Models and Grey System Theory to Evaluate Past-to-Future Performance: A Case of Indian Electricity Industry. *Sci. World J.* **2015**. [CrossRef] [PubMed]

21. Wang, J. Pearson Correlation Coefficient. In *Encyclopedia of Systems Biology*; Springer: New York, NY, USA, 2013; pp. 97–141.

22. Nahler, G. Pearson correlation coefficient. In *Dictionary of Pharmaceutical Medicine*; Springer: Vienna, Austria, 2017.

 © 2019 by the authors. Licensee MDPI, Basel, Switzerland. This article is an open access article distributed under the terms and conditions of the Creative Commons Attribution (CC BY) license (http://creativecommons.org/licenses/by/4.0/).

Article

A Decision Support System for Dynamic Job-Shop Scheduling Using Real-Time Data with Simulation

Ahmet Kursad Turker *, Adnan Aktepe, Ali Firat Inal, Olcay Ozge Ersoz, Gulesin Sena Das and Burak Birgoren

Department of Industrial Engineering, Kirikkale University, 71451 Campus, Turkey; aaktepe@kku.edu.tr (A.A.); afinal@kku.edu.tr (A.F.I.); ooersoz@hotmail.com (O.O.E.); senadas@kku.edu.tr (G.S.D.); birgoren@kku.edu.tr (B.B.)
* Correspondence: kturker@kku.edu.tr

Received: 26 February 2019; Accepted: 14 March 2019; Published: 19 March 2019

Abstract: The wide usage of information technologies in production has led to the Fourth Industrial Revolution, which has enabled real data collection from production tools that are capable of communicating with each other through the Internet of Things (IoT). Real time data improves production control especially in dynamic production environments. This study proposes a decision support system (DSS) designed to increase the performance of dispatching rules in dynamic scheduling using real time data, hence an increase in the overall performance of the job-shop. The DSS can work with all dispatching rules. To analyze its effects, it is run with popular dispatching rules selected from the literature on a simulation model created in Arena®. When the number of jobs waiting in the queue of any workstation in the job-shop falls to a critical value, the DSS can change the order of schedules in its preceding workstations to feed the workstation as soon as possible. For this purpose, it first determines the jobs in the preceding workstations to be sent to the current workstation, then finds the job with the highest priority number according to the active dispatching rule, and lastly puts this job in the first position in its queue. The DSS is tested under low, normal, and high demand rate scenarios with respect to six performance criteria. It is observed that the DSS improves the system performance by increasing workstation utilization and decreasing both the number of tardy jobs and the amount of waiting time regardless of the employed dispatching rule.

Keywords: industry 4.0; dynamic job-shop scheduling; simulation; decision support systems; internet of things

---

1. Introduction

Production is the transformation of natural resources to value-added products or services to meet consumer needs. In order to manage the production well, it is necessary to meet the consumer demands in terms of price, time, quantity, and quality, while decreasing the inventory levels and increasing the stock cycle speed/service level. At this point, questions such as which product, how many, which features, where, and by whom, should be answered to minimize production costs or maximize profits of an organization.

In a job-shop production environment, production is mostly carried out according to the customer order, the due date of which is usually set by the customer. Since the variety of products is high and the order size is low, production flows (routes) usual change from product to product among universal machines in the workshop. Therefore, the coordination of production resources in a job-shop production environment is often difficult. This challenge necessitates the use of advanced production planning and control systems in job-shop environments.

Scheduling is the planning of the activities in the job-shop at the operational level (day, hour, minute, etc.). When scheduling production, various performance criteria should be considered, such as

(i) timely delivery (minimization of delays), (ii) reduced time spent in the system (minimization of waitin), and (iii) maximization of machine utilization rates [1].

In a job-shop, scheduling of jobs is usually performed with static dispatching rules. This scheduling approach is ineffective because it does not consider dynamic factors like new arriving orders and probabilistic or stochastic real-life problems such as job postponement or machine failures [2]. Moreover, research shows that classical scheduling fails to meet the needs of production environments in practice [3,4]. Thus, advanced scheduling tools are needed to model this dynamic production environment.

One of the tools to deal with such a dynamic production environment is simulation. Simulation allows modeling and analysis of real-life processes and systems in a computer environment in shorter times with lower costs [5–7].

This study proposes a decision support system (DSS) designed to increase the performance of dispatching rules in dynamic scheduling using real-time data, hence to increase the overall performance of the job-shop. To analyze its effects, it is run with popular dispatching rules selected from the literature on a simulation model created in Arena®. When the number of jobs waiting in the queue of any workstation in the job-shop falls to a critical value, the DSS can change the order of schedules in its preceding workstations to feed the workstation as soon as possible. For this purpose, it first determines the jobs in the preceding workstations to be sent to the current workstation, then finds the job with the highest priority number according to the active dispatching rule, and lastly puts this job at the first position in its queue.

The data needed for the DSS includes real-time machine and product status including operating conditions, workstation queue status, workload, etc. The data is collected from production tools that are capable of communicating with each other via the Internet of Things (IoT). The IoT is the network of devices such as vehicles and home appliances that contain electronics, software, sensors, actuators, and connectivity that allows these things to connect, interact, and exchange data. It involves technologies such as Wi-Fi, Bluetooth, and Radio Frequency Identification (RFID). The functioning of a representative job-shop equipped with the IoT technologies is shown in Figure 1. In manufacturing shop floors, use of the IoT has turned machines into smart manufacturing objects that can communicate with each other, enabling access to vast amounts of real data [8]. In manufacturing, IoT could generate so much business value that it is believed to lead to the Fourth Industrial Revolution, which is also referred to as Industry 4.0.

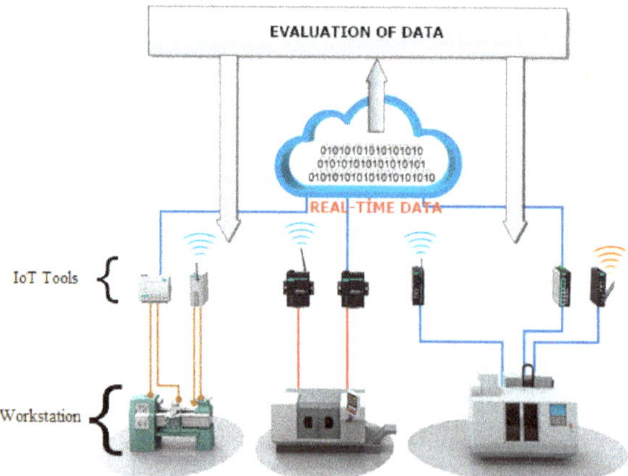

Figure 1. Functioning of the system.

This system can be integrated into an Enterprise Resource Planning (ERP) information system to make scheduling more effective. We believe that improving the efficiency of production management activities with such approaches could lead to an increase in the demand for Industry 4.0 practices.

Before introducing the functioning of the proposed system in Section 4, a literature review and discussion on dispatching rules are presented in Sections 2 and 3, respectively. Later in Section 4, both the designed system and the developed simulation model are discussed. Finally, conclusions and future work are supplied in Section 5.

2. Related Work

In this section, we presented approaches and methods used to determine dispatching rules in different production environments.

Marinho et al. [9] developed a decision support system for a dynamic production scheduling system. They stated that these systems are suitable for small and medium enterprises. In this study, the manufacturing orders are scheduled dynamically. The scheduling was carried out with an earliest/latest finish, minimum/maximum slack, smallest free intervals, and minimum delay heuristics. Deadlines of manufacturing orders and resources occupation were considered in the study. In addition, an interface was developed in the study for helping managers make faster decisions.

Aydin and Oztemel [10] developed an approach for the solution of the dynamic job-shop type scheduling problem by using the agent and the simulated environment. The intelligent agent determines the most appropriate rule in the real time production environment, whereas scheduling is carried out using the rule chosen by the simulation technique. The intelligent agent used in the model is trained with a learning algorithm developed in the study. The results obtained with the developed smart agent are better than shortest processing time (SPT), cost over time (COVERT), and critical ratio (CR) rules.

Li et al. [11], studied on a real time production improvement through bottleneck control. They stated that a dynamic bottleneck control system is developed in order to efficiently use the finite manufacturing resources. Their objective was to achieve a continuous production improvement. The method they developed is applied in an automotive assembly line. As a result, they reduced the downtime of the bottleneck machine.

Heilala et al. [12] developed a simulation-based decision support system as an operative simulation model that enables to handle unforeseen events such as downtime and changes in operations. In their study, they stated that a simulation-based decision support system could be used to help achieve more efficient manufacturing. They discussed that data integration, automated simulation, and the visualization of results in this field.

Mahdavi and Shirazi [13] presented a review of intelligent decision support systems in production planning of flexible manufacturing systems. In their study, they developed real time control of the shop floor. They presented performance criteria for effective and efficient control of the production. They determined the sequence of the jobs in the system using the developed rules.

Sharma and Jain [14] examined the dynamic job-shop type scheduling problem in the stochastic environments by adding the time-dependent preparation times constraint. In the study, makespan, average flow time, maximum flow time, average tardiness, maximum tardiness, number of tardy jobs, total setup time, and average setup time were calculated by the developed algorithm. The JMEDD rule developed in the study gave the best value among all rules. In another study, Sharma and Jain [15] developed four new rules for the same problem that was considered by them. These rules are: (1) TDDSSPT: Shortest (time to due date + setup time + processing time); (2) JTDDSSPT: Same setup time and shortest (time to due date + setup time + processing time); (3) JSLACK: Same setup time and shortest slack time; and (4) JSLACKW: Same setup time and the shortest slack time per unit job.

Different from other works, Zhong et al. [16] used data that is obtained from a production environment using the Radio Frequency Identification (RFID) system. This system collected data and analyzed these data with data mining to determine standard processing times and dispatching rules.

Next, decision trees were used to find that the use of concurrent data improves the determination of dispatching rules.

Kulkarni and Venkatesvaran [17] developed the simulation-based optimization algorithm (SbO) for job-shop type scheduling problems. They developed a hybrid algorithm to solve the problem. The developed algorithm has been tested in deterministic and stochastic environments. Obtained results were better than classical mixed integer programming model.

Zhong et al. [8] carried out a Big Data Analytics for RFID logistics data by defining different behaviors of smart manufacturing objects. They developed physical internet-enabled intelligent shop floor. The task weight was considered in the logistics decision-making. According to results of the application, the highest residence time occurs in a buffer with the value of 40.57% of the total delivery time.

Phanden and Jain [18] developed a genetic algorithm approach that is based on a simulation model. In their study, the model selects a job that becomes a candidate to change the available process plans. The objective of the model is to minimize mean tardiness. Three case studies were conducted in their study. They found that changing the current plan according to algorithm results helped to reduce mean tardiness.

Kuck et al. [19] proposed a data-driven simulation-based optimization algorithm for the control of dynamic production systems. In the present study, it was emphasized that flexibility in production is very important. They have developed an approach for the re-scheduling of production according to the new situations considering factors that may cause confusion, such as the simultaneous arrival of a large amount of orders.

Ersoz et al. [20] tried to reduce the difference between practice and scheduling theory. They adapted the real-time information generated by the process control and control systems, to their planning activities. In the offered system, the dynamic structure of the production environment is immediately perceived and the schedule is updated according to the new conditions. The traceability of the parts increased in the factory. In addition, unnecessary waiting or downtimes have been minimized.

Zhang et al. [21] studied real time job-shop scheduling. In their study, they offered two algorithms: the simulation-based value iteration and simulation based Q learning, which were developed to solve the scheduling problem from the perspective of a Markov decision process (MDP). They also used an intelligent system to estimate value function. The MDP rule is compared with SPT, longest processing time (LPT), first-in-first-out (FIFO), and CR. It is observed that MDP performed better than others.

Bierwirth and Kuhpfahl [22] proposed a new approach by synthesizing the GRASP algorithm with local search methods that minimized the total weighted tardiness in job-shop type scheduling. The model based on critical tree building produced better results in terms of total processing time criterion compared to the conventional GRASP algorithm.

Xiong et al. [23] proposed a simulation-based model for determining dispatching rules in a dynamic scheduling problem where job release times and extended technical priority constraints are included. The proposed algorithm reduced the total tardiness and the number of tardy jobs.

Zhang et al. [24] reviewed the literature on job-shop scheduling problems and discussed new perspectives under Industry 4.0. They reviewed more than 120 papers. They stated that under Industry 4.0, the scheduling problems are dealt with new methods and approaches. According to their findings, scheduling research needs to shift its focus to smart distributed scheduling modeling and optimization. According their evaluation, this can be achieved with two approaches: (1) combining traditional methods and proposing a new method and (2) proposing new algorithms for smart distributed scheduling.

Rossit et al. [25] described the concept of intelligent production that emerged with Industry 4.0 in their work. They have dealt with the issue of smart scheduling, which they believe to have an important place in today's production understanding. They have developed the concept of tolerant scheduling in a dynamic environment in order to prevent the need for re-scheduling in production. Similarly, Tao et al. [26] stated that one of the important studies conducted in the literature in the scope of Industry 4.0 is dynamic scheduling. They examined the recent innovations in production systems and the smart manufacturing approaches, as well as models that came up with Industry 4.0. In the study, the analysis of the data life cycle in production and the use of large data in production are explained. Conceptual models on the use of large data in production is developed and the use of large data for different sectors is explained with examples.

Jiang et al. [27] studied an energy-efficient job-shop scheduling problem. Their aim was to minimize the sum of the energy consumption cost and the completion-time cost. However, the handled problem was considered NP-Hard. Thus, they developed an improved whale optimization algorithm for solving this problem. They used dispatching rules, nonlinear convergence factors, and mutation operation for the improvement of whale optimization algorithm. To show the effectiveness of the algorithm, they performed simulations. According to results of simulations carried out, the algorithm provided advantages in terms of efficiency.

Ortiz et al. [28] analyzed a flexible job-shop problem and proposed a new model for the solution. They formulated a real-world production-scheduling problem and also provided an efficient tool to solve it. They developed a new algorithm that minimizes average tardiness and found better solutions than the existing dispatching rules.

Ding and Jiang [29] discussed the effect of the IoT technology in a manufacturing environment. They state that, with IoT, production data increased but that these data are sometimes discrete, uncorrelated, and hard-to-use. Therefore, they developed a method to use invaluable data. They provided an RFID-based production data analysis method for production control in IoT-enabled smart job-shops. In addition, a big data approach was developed to excavate hidden information and knowledge from the historical production data.

Leusin et al. [30] developed a multi agent system in a cyber-physical system to solve the dynamic job-shop scheduling problem. The proposed solution had self-configuring features in the production line. This was achieved with the use of agents and IoT. Real time data were used for efficient decision making in the job-shop. The model was tested with a real case study. Under different scenarios, they gained results that are more efficient than standard dispatching rules. In addition, the advantages of using dynamic data and IoT in industrial applications are discussed.

Zhang et al. [31] emphasized the importance of learn concepts in operational management, especially the importance of the lean approach in Industry 4.0. In this study, process control theory was used for lean methods. Thus, they proposed Lean-Oriented Optimum-State Control Theory (L-OSCT) in the study. L-OSCT provides dynamic process control in industrial networking systems. The application was carried out in a large-size paint making company to show the effectiveness of the approach.

After examining the relevant literature, it was observed that several models were developed to determine dispatching rules in dynamic job-shop type scheduling. As a contribution, we developed a decision support system for dynamic environments that could work with different dispatching rules. Our aim was to increase the efficiency of the production management and job-shop.

3. Dispatching Rules in Scheduling

Scheduling aims to assign jobs to workstations according to a dispatching rule was done to determine order of processing. These rules are procedures designed to provide good solutions to complex problems in a real-time production environment [4]. Many researchers have proposed various dispatching rules to optimize some performance criteria in a production environment. Since the list of orders to process is updated continuously, the actual problem is dynamic and complex; however, many classical rules do not take this dynamic nature into account. By taking advantage of the modern information technology provided in Industry 4.0, dispatching rules that could handle this dynamic and complex structure could be developed. However, it should be noted that a dispatching rule could not improve all performance criteria at the same time.

When scheduling the jobs according to a specific rule (as described below), a priority value is calculated for each job in the queue. Later, the job with the smallest or the largest value is selected and assigned to the workstation.

Dispatching rules can be classified in various ways. The rules which are assigned according to the conditions in the workshop can be grouped under three major classes as given in Figure 2.

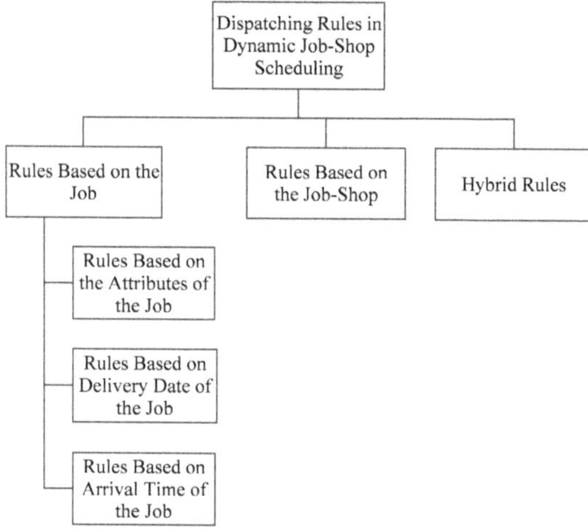

Figure 2. Dispatching rules in job-shop scheduling.

Rules Based on the Job: According to the dispatching rules based on the job, the assignment is done by ignoring the interdependencies between the workstations. Priorities for jobs are determined with respect to some of the attributes and values of the job, such as arrival times, processing times, due dates, etc. Then, the job having the smallest priority value is selected first. Rules such as SPT, EDD, SLACK, and PR can be classified as rules based on the Job.

Rules Based on the Job-Shop: In order to increase the efficiency in smart factories of the future, the components and subsystems must be integrated with each other. For this integration, machines should work intelligently by communicating with other machines. In such a system, process monitoring can be carried out in a comprehensive and effective manner using simultaneous data from the machines and other units. The machines will be able to plan their own production resources. Thus, lean manufacturing and just-in-time manufacturing could be realized.

Until recently, real-time data to show the status of the job-shop were not available. Therefore, dispatching rules that can update the priority values of the jobs dynamically and autonomously were not very common. Work load in next queue (WinQ) rule is a good example for this new class of rules.

Hybrid Rules: Hybrid rules are formed by combining two or more dispatching rules. In this case, a priority value is calculated based on the priority values of these rules. This value is used to specify the job to be assigned.

Dispatching Rules Considered in the Study

The DSS moves in whenever the number of jobs waiting in the queue of any workstation in the job-shop falls to the critical value of one. The DSS is designed to increase the performance of dispatching rules in dynamic scheduling using real time data. For this purpose, five dispatching rules with good job-shop performances, which were demonstrated in the literature, are selected in order to compare the performances of the dispatching rules with and without the DSS.

These rules are explained in this section. Notations are given in Table 1.

Table 1. Notations for Dispatching Rules.

$\pi_{i,j}$	Priority value in the jth operation of job	n_i	Total number of operations of job
d_i	Due date of job	A_i	Arrival time of job
$P_{i,j}$	Processing time of the jth operation of job	t	The time when a dispatching decision is needed

i: Job Index; j: Operation Index; k: Current Operation Step.

Smallest Processing Time (SPT): In this rule, the priority value is the processing time at the workstation. Between the jobs waiting in the queue, the job with the smallest processing time in the queue is selected as the job to be processed first.

$$\pi_{i,k} = P_{i,k}, \tag{1}$$

Earliest Due Date (EDD): In this rule, the priority value of jobs is the due date. Between the jobs waiting in the queue, the job with the earliest due date in the queue is selected as the job to be processed first.

$$\pi_{i,k} = d_i, \tag{2}$$

Shortest Slack Time (SLACK): In this rule, the priority value is obtained by subtracting the remaining total processing time from the remaining time to due date. Between the jobs waiting in the queue, the job with the smallest priority value in the queue is selected as the job to be processed first.

$$\pi_{i,k} = d_i - \left[t + \sum_{j=k}^{n_i} P_{i,j} \right], \tag{3}$$

Priority Ratio (PR): In this rule, the priority value is obtained by dividing the remaining time until the due date by the remaining total processing time. Between the jobs waiting in the queue, the job with the smallest priority value in the queue is selected as the job to be processed first.

$$\pi_{i,k} = (d_i - t) / \sum_{j=k}^{n_i} P_{i,j}, \tag{4}$$

Work load in the next Queue (WinQ): Apart from these static dispatching rules, a dispatching rule that can be adapted to dynamic environments is presented below. The rule WinQ was proposed by Holthaus and Rajendran [32]. This rule determines the priority value of a job by considering the conditions of the job-shop. In this rule, priority value is obtained by looking at total processing times of all jobs in the next workstation, depending on a job's route. Among the jobs waiting in the current queue, the job with the smallest priority value is selected as the next job to be processed. They reported that use of this rule minimizes the average flow time. In cases when the utilization of a job-shop is high, the due dates are determined in A narrow range. This rule also minimizes the tardiness of the jobs and the ratio of the tardy jobs.

4. Decision Support System

A DSS is a computerized information system used to support decision-making in an organization. The DSS proposed in this study can be considered as an automated DSS [33], which automatically collects data from the job-shop, analyze the data, and intervenes in the processing order of the jobs if necessary. In other words, it continuously monitors operations, seeks opportunities to increase the job-shop performance, and formulates a better processing order and implements it.

When dispatching rules are used for scheduling, the rule to be applied is determined before the system starts running and no modification are allowed at any time. In this study, we propose a new dynamic approach where priority of the jobs can be updated in real time depending on the information on the current status of the workstations. The goal is to improve the performance of the system by decreasing idle time in the workstations.

To collect real-time data from the workstations, the status of the machines (idle, busy), queue lengths, and the property of the jobs in the queue (remaining processing time, processing time, due date etc.) are sent to the management module of the system. When the information is received by the DSS, the system makes necessary modifications in the queue order if it is needed. As soon as the number of jobs waiting in a queue at any workstation falls to a critical level (in this study this critical level is chosen as one), the DSS examines all the preceding workstations to detect the jobs that will be sent to this workstation. Then, it determines the job with the highest priority number according to the active dispatching rule. Finally, the DSS places this job to the first order of the its queue so that it will be first one to be processed when the station becomes idle. Thus, idle times will be avoided in the workstations as much as possible.

To explain how this approach works, an example is presented in Figure 3.

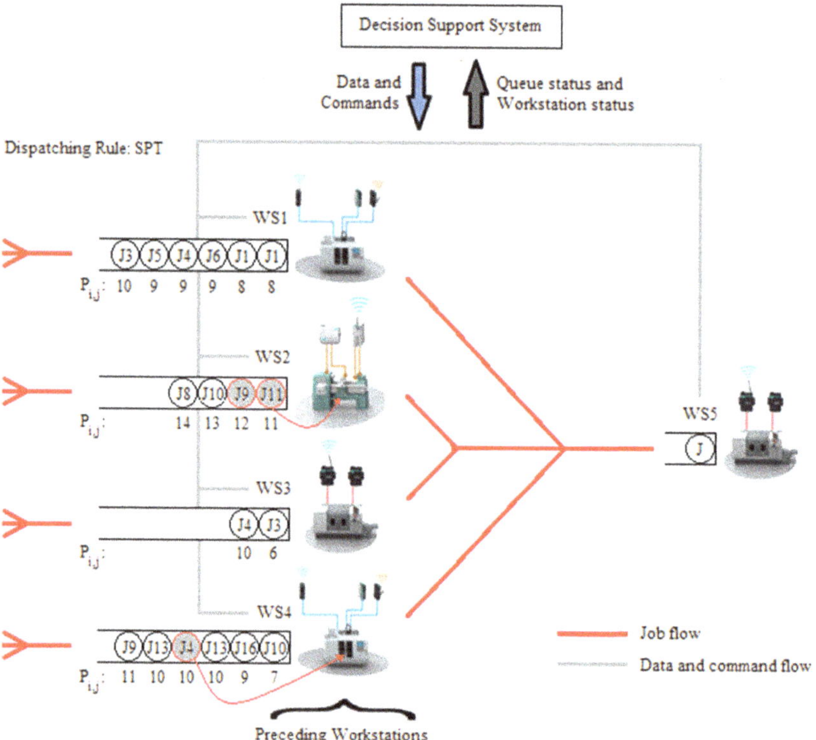

Figure 3. An example of how a Decision Support System (DSS) works.

In this example, suppose that at a specific time point, Figure 3 illustrates the current queue order in the system. Further, suppose that the DSS works under the SPT rule to schedule the jobs in the system and the number of jobs in the queue of WS5 drops to one (a critical level). When the route information is examined, it is determined that WS1, WS2, WS3, and WS4 are the preceding workstations that could send the jobs either to WS5 or to other workstations, depending on the route of the jobs. Jobs that could be processed in the preceding workstations of WS5 are given below:

WS1 = {J1, *J2*, J3, J4, J5, J6}
WS2 = {J7, J8, *J9*, J10, *J11*, *J12*, *J13*, *J14*}
WS3 = {J3, J4, *J5*, J6, J7, J10, J13, J14, J15, *J16*, J17, *J18*, J19, *J20*}
WS4 = {J1, J2, J3, *J4*, *J5*, *J6*, *J7*, *J8*, J9, J10, J13, J14, J15, J16, J17, J18, J19, J20}

The jobs that are underlined in the lists are the ones that will be sent to WS5 after processed in their current workstations. The other jobs will be sent to other workstations according to their routes. The routes and precedence orders can be seen in Table 2 in the Section 5.

The DSS first determines which of these jobs could be processed next in WS5 by checking WS1, WS2, WS3, and WS4. The job numbers colored in gray in each queue in Figure 3 are the jobs that could be processed in WS5, namely J4, J9, and J11. Among them, J4 is selected since it has the smallest processing time (according to SPT). This job is in the WS4 queue, and hence will be processed in the WS4 as soon as it becomes idle.

Table 2. Routes and Unit Processing Times of Parts.

	Operation 1	Operation 2	Operation 3	Operation 4	Operation 5	Operation 6	Operation 7
Part 1	WS1(8)	WS2(12)	WS5(12)	WS4(7)	WS7(13)		
Part 2	WS1(8)	WS5(19)	WS4(12)	WS8(15)	WS6(10)		
Part 3	WS1(10)	WS4(8)	WS3(6)	WS6(8)	WS9(9)		
Part 4	WS1(9)	WS6(10)	WS3(10)	WS4(10)	WS5(14)	WS9(8)	WS10(10)
Part 5	WS1(9)	WS4(10)	WS3(8)	WS5(15)	WS10(11)	WS9(5)	
Part 6	WS1(9)	WS2(10)	WS3(8)	WS4(11)	WS5(10)	WS6(9)	WS10(14)
Part 7	WS2(13)	WS3(11)	WS4(10)	WS5(16)	WS8(18)	WS9(9)	
Part 8	WS2(14)	WS4(14)	WS5(13)	WS7(14)	WS8(18)	WS9(10)	
Part 9	WS2(12)	WS5(9)	WS4(11)	WS7(16)	WS10(14)		
Part 10	WS2(13)	WS3(9)	WS4(7)	WS8(14)	WS7(14)	WS6(10)	
Part 11	WS2(11)	WS5(10)	WS6(10)				
Part 12	WS2(10)	WS5(11)	WS4(10)	WS8(17)	WS9(12)		
Part 13	WS2(12)	WS5(12)	WS4(10)	WS3(8)	WS6(8)	WS9(7)	
Part 14	WS2(13)	WS5(10)	WS4(11)	WS3(8)	WS6(9)	WS7(17)	WS10(12)
Part 15	WS3(10)	WS6(9)	WS4(11)	WS8(17)	WS7(15)		
Part 16	WS3(9)	WS5(10)	WS4(9)	WS7(18)			
Part 17	WS3(8)	WS4(7)	WS8(15)				
Part 18	WS3(9)	WS5(10)	WS4(10)	WS6(8)			
Part 19	WS3(8)	WS4(9)	WS7(15)	WS6(11)			
Part 20	WS3(8)	WS5(11)	WS4(9)	WS8(15)	WS6(12)		

5. The Simulation Model

A simulation model for a job-shop production environment, developed in Arena, is used to assess the effects of the proposed DSS. The virtual job-shop environment starts the production with the arrival of a demand for a job. It is assumed that there is a continuous dynamic job demand arrival. Manufacturing takes place at 10 workstations and real-time data can be obtained from them. Each workstation has a different numbers of machines. The representative layout of this job-shop environment is shown in Figure 1.

The purpose of this simulation study is to analyze the effects of introducing the DSS into a job-shop environment that is normally working with some dispatching rule. In this simulation, some real-life phenomena, such machine failures will not be considered since they are not considered to have a significant influence on the performance of the DSS. The simulation model was formed under the following assumptions:

- Each order has only a single type of job.
- A job cannot be divided and requires different operations in different workstations. For this reason, two operations of the same job cannot be processed at the same time.
- Previous operations of a job must be completed to start a new operation.
- Jobs cannot be cancelled. Each job should be processed until it is completed.
- Machine failures have been ignored.
- Quality control operations are ignored. Thus, waste parts do not occur.
- The time required to move parts between workstations is also ignored.
- Queues in front of workstations are allowed.
- Jobs can wait in a queue for the machine to become idle. On the other hand, machines in workstations can remain idle.

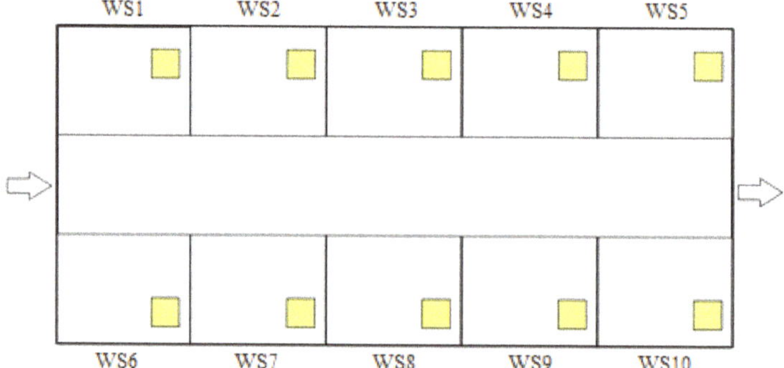

Figure 4. Representative Job-shop Layout Plan.

The routes and the unit processing times of the 20 parts to be produced is obtained from a company that produces spare parts. Data about parts are given in Table 2. It is assumed that the incoming job order can be for any product (each of the 20 parts are equally likely) and the size of the incoming order can be any value between 10 and 30.

When jobs arrive the system, they are directed to the first workstation on their route. If the targeted workstation is idle, the process starts, otherwise it is kept in the queue. If there are jobs waiting in the queue, a priority value for each job is calculated based on the dispatching rule. Then, the job with the highest or lowest priority is selected. Jobs that are processed in a workstation are directed to their next workstation in their route. The processing of a job is completed when all the workstations on the route are visited.

5.1. Determining the Best Working Conditions of the System

The simulation model was set up to represent a job-shop with realistic due dates and a realistic number of machines in workstations, so that job flows are made as smooth as possible without serious bottlenecks.

In order for this setup, the simulation model of the designed system was tested according to the first-in-first-out (FIFO) dispatching rule under exponential job arrival times with various means. The results were compared in terms of the capacity utilization rates of the workstations and the length of the queue formed in front of each workstation (whether the queue lengths increase continuously or cause bottlenecks). It was observed that a mean value of 65 produced satisfactory results. The model was tested with smaller and larger mean values as will be discussed in Section 5.2.

Since some performances criteria are directly related to due dates, due dates were determined as realistically as possible. In real life, due dates are usually set by customers and jobs can have a wide or narrow due-date interval depending on how urgent the production is. This must include a coincidence factor that considers the nature of the orders received (normal or urgent) in determining the due date. In our study, a new equation (Equation (5)) was used to determine the due date.

To obtain due dates (d_i), the proposed model was again run according to the FIFO dispatching rule and the amount of time spent in the system by each job was found. Next, the processing time of each job was subtracted from the amount of time the job spends in the system to find the waiting time of the job in the system. Then the waiting time was divided to the processing time to obtain a ratio. This ratio shows the percentage of processing time to waiting time. An analysis of these ratios showed that their distribution can be approximated by an exponential distribution except for the fact that it does not start from zero. It is known that exponential distribution can produce values between zero and infinity. Suppose that we obtained zero waiting time from this distribution. This means that the due date of the job would be equal to its arrival time plus its processing time. However, in the job-shop, this due date is not realistic since the job with this due date would probably be a tardy job. To avoid this, we add a constant term ($1 + h$) to the formula presented in Equation (5) to produce more realistic due dates. We defined a tightness factor h to transform the value obtained from the exponential distribution.

Accordingly, the newly proposed due date generation rule (Equation (5)) is as follows;

$$d_i = A_i + \left(\sum_{i=1}^{20} \sum_{j=1}^{10} P_{i,j} \right) \cdot (1 + h + Expo(\mu)), \tag{5}$$

where d_i represents the i-th due date and h represents the tightness factor. We determined that h should be 1.5 after trial and error. Below this value, we observed many tardy jobs. Finally, the μ value is set to 1.5.

Before testing our scenarios, the simulation model was run by considering different numbers of machines at each workstation to determine the right number of machines at each workstation. Some runs according to different machine combinations (X, Y, and Z) at each workstation were shown in Table 3. Results showed that best machine combination is the Z, with one machine in WS1, WS9, and WS10, three machines in WS4 and WS5, and two machines in the rest of the stations, which were best in terms of average workstation utilization, average waiting time, and average number of parts waiting.

5.2. Scenarios

The simulation model was run under three different scenarios:

- Scenario 1: Exponential distribution with a mean of 62 is used to represent arrivals of a fast order rate which overloads the job-shop.
- Scenario 2: Exponential distribution with a mean of 65 is used to represent arrivals of a moderate order rate for a balanced structure.
- Scenario 3: Exponential distribution with a mean of 68 is used to represent arrivals of a slower order rate which causes a more relaxed job-shop.

These three scenarios were run by considering SPT, EDD, SLACK, PR, and WinQ dispatching rules in a job-shop supported by the DSS. Two situations in terms of control are considered: with DSS and without DSS (Normal).

In each scenario, 5000 orders are received randomly according to the arrival speed and a simulation run is completed when all the orders were produced. The fact that the same orders were created at the same time for each scenario enabled one-to-one comparisons. Thus, more realistic evaluations were achieved.

Table 3. Results of some runs according to different machine combinations.

	X: 1-1-2-2-2-2-2-1-1-1			Y: 1-2-2-3-2-2-2-2-1-1			Z: 1-2-2-3-3-2-2-2-1-1		
	Average Number Waiting	Average Waiting Time	Average Workstation Utilization	Average Number Waiting	Average Waiting Time	Average Workstation Utilization	Average Number Waiting	Average Waiting Time	Average Workstation Utilization
WS1	0.81	344	0.42	1.13	344	0.58	1.56	344	0.80
WS2	506.64	132,869	0.92	2.34	438	0.65	3.20	438	0.88
WS3	0.55	98	0.48	3.19	407	0.67	6.26	582	0.92
WS4	118.81	15,946	0.73	0.94	90	0.68	3.54	248	0.94
WS5	15.46	2638	0.71	520.83	63443	0.99	3.52	313	0.91
WS6	0.21	43	0.46	0.63	94	0.64	2.78	301	0.88
WS7	0.13	42	0.44	0.40	92	0.62	1.41	237	0.85
WS8	210.25	68,121	0.99	0.65	149	0.70	6.41	1081	0.96
WS9	0.16	58	0.47	0.57	148	0.66	3.03	572	0.90
WS10	0.35	172	0.49	0.58	204	0.69	6.01	1548	0.94

5.3. Results of Scenarios

To reduce the effects of randomness on results, the simulation model was run under 50 replications for each scenario and the averages of performance criteria were obtained.

The scenarios were compared based on the following performance criteria:

- Number of tardy jobs: The timely delivery of orders in the job-shop is of great importance both for customer satisfaction and penalty costs.
- Average waiting time in queue (Avg. waiting time): This criterion attempted to determine whether the waiting periods in the system decreased or not.
- Utilization of the workstations (Avg. utilization): This criterion attempted to determine whether the scenarios increase the efficiency of the system.
- Work in Process (Wip): This criterion attempted to determine whether the scenarios increase the work load within the system.
- Average tardiness (Avg. tardiness): Considering only the number of tardy jobs may lead to incorrect results. It will be better to examine delays together with the average deviations from the due date.
- Average earliness (Avg. earliness): In addition to the late completion of the jobs, it is not desirable to complete jobs early due to inventory holding costs. To avoid these costs jobs should be completed as close as possible to their corresponding due dates.

The simulation results with respect to these performance criteria, are given in the Appendix A. The graphical comparisons of the scenarios based on the criteria are shown in Figures 5–10.

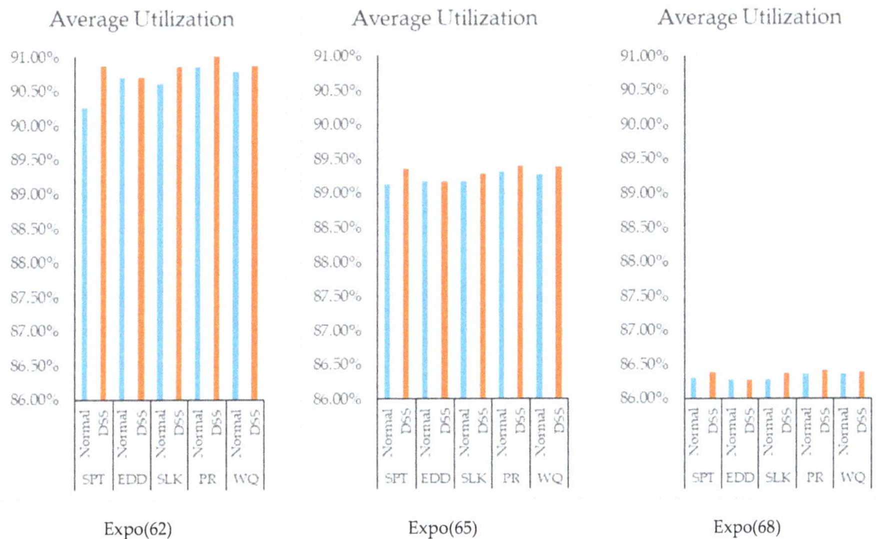

Figure 5. Average utilization.

Since the DSS basically controls the situation of each workstation and updates the sequence of the jobs according to the dispatching rule, an increase in average utilization of each scenario is observed in Figure 5.

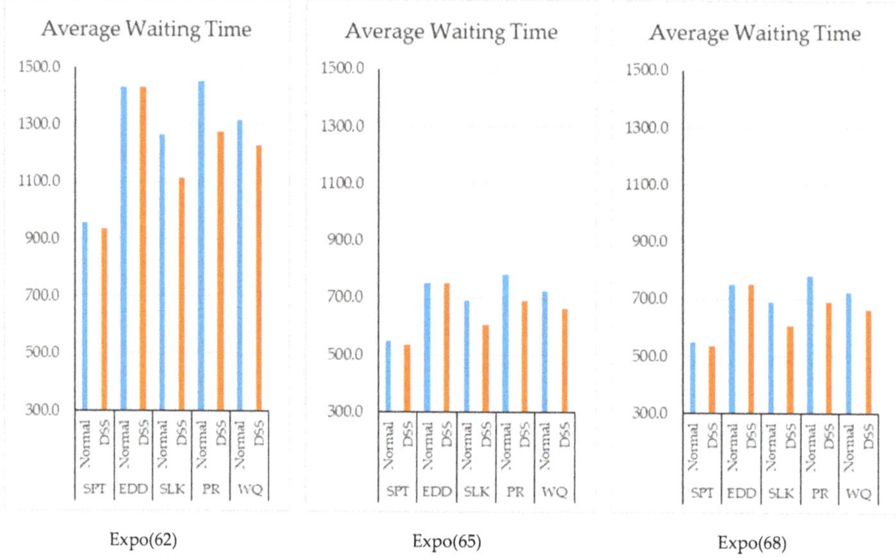

Expo(62)

Expo(65)

Expo(68)

Figure 6. Average waiting time.

The proposed approach also leads to a decrease on average waiting time of the jobs in the system; this is in line with the increase in utilization rates. On the other hand, the amount of decrease observed in each scenario was different. The sharpest decrease in Figure 6 was observed in Scenario 1.

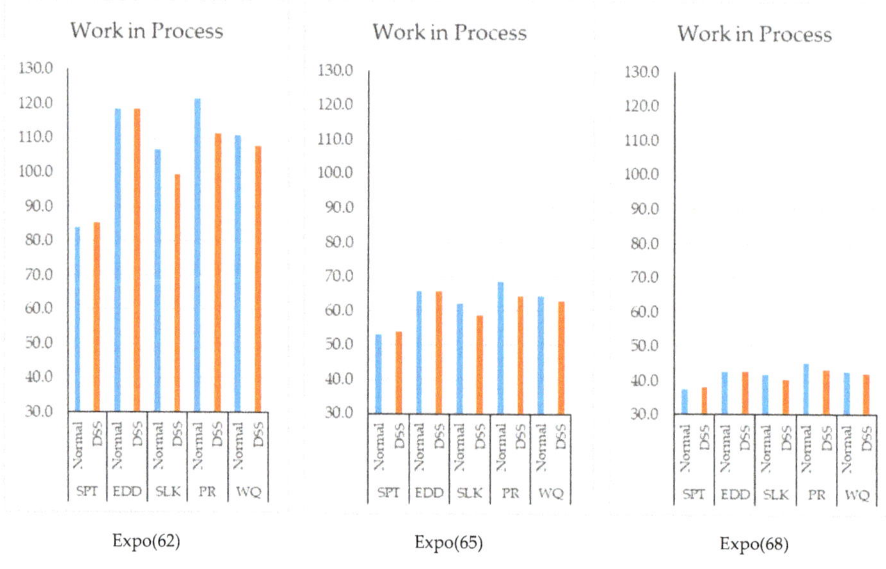

Expo(62)

Expo(65)

Expo(68)

Figure 7. Work in Process.

Because of the decrease in the average waiting times, the number of jobs in the system decreased except for the cases in which the SPT rule is used.

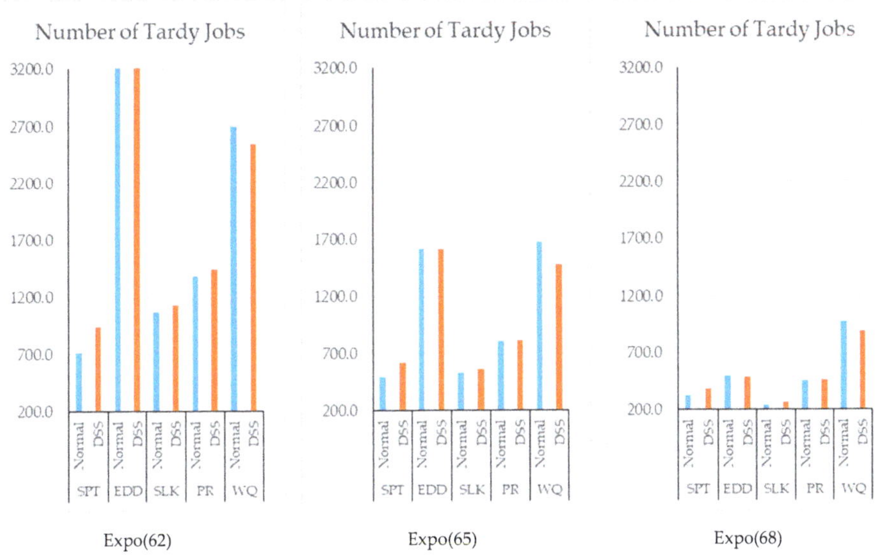

Figure 8. Number of Tardy Jobs.

After the results presented in Figure 8 were evaluated, we did not observe a significant difference in terms of tardy jobs. This is probably because of the randomness in the due dates and the fact that the EDD, the SLACK, and the PR dispatching rules consider the due dates of jobs. On the other hand, a decrease is observed when the WinQ rule is used since this rule is based on the amount of jobs in the system.

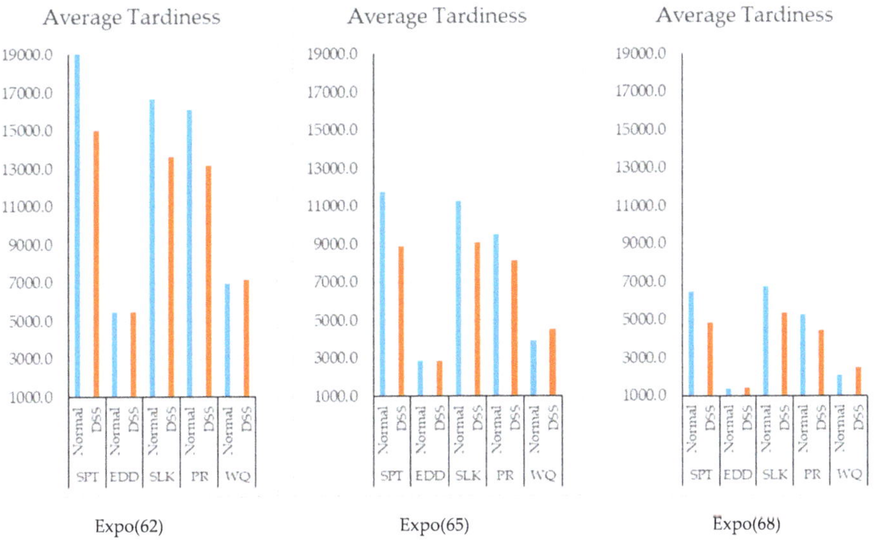

Figure 9. Average Tardiness.

A decrease in the number of tardy jobs was not observed when dispatching rules based on due date (see Figure 9). However, for the cases in which WinQ is considered, a decrease in the average waiting times does not occurred because of the increase in the average utilization of workstations.

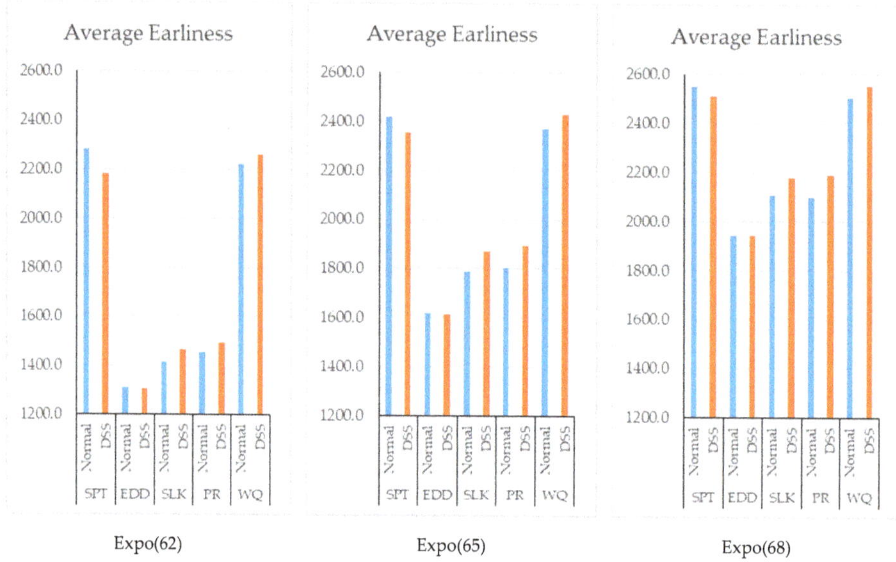

Figure 10. Average Earliness.

After the results in Figure 10 were examined, an increase in average earliness was observed, except for the case in which SPT rule was used. This was an expected outcome since an increase in utilization and a decrease in waiting times is already observed.

6. Results and Discussion

The DSS proposed in this study aims at increasing the performance of dispatching rules in dynamic scheduling using real time data, hence increasing the overall performance of the job-shop. The DSS can work with all dispatching rules. This study shows that a DSS that fed with real time data from a job-shop can help improve several performance criteria. This can be achieved in job-shops equipped with Industry 4.0 hardware and software. The proposed approach improves the performance of a system in terms of number of tardy jobs, average waiting time in queue, average utilization of the workstations, work in process, average tardiness, and earliness as it is discussed in the previous section. As expected, machine utilization increased and thus waiting times in the system and the amount of work-in-process decreased using the DSS.

If the big data collected in real data could be analyzed, inner dynamics between the components of the system could be solved. Understanding the production systems better could help the researchers to develop advanced DSS and the decision makers to make better decisions. Controlling the production environment using real time data, could also help researchers to model the system without making any assumptions. This approach could close the gap between theory and practice and result in realistic solutions. Moreover, it would be possible to validate actual processing times of jobs considering past data about jobs.

Data stored in dynamic databases could also be processed with data mining techniques in order to determine potential problems in the production; to obtain rules that could improve the efficiency of production; to generate rules to effectively control the production; to develop automation based on work intelligence; and to improve quality of products and even to design a production management system that could improve itself.

Since the analysis of this big data would help predict the behavior of the system under different conditions, any subsystem could easily integrate into this system. The integration of artificial intelligence or machine learning based subsystems could constitute smart factories of the future. In this respect, this study demonstrates, at a theoretical level, the benefits to be gained by such subsystem integrations, though technological integration in a real job-shop may take substantial time and effort.

7. Conclusions

In this study, we considered a dynamic job-shop where job lists are updated dynamically because of continuous arrival of new orders. In future studies, various cases including machine breakdowns, cancellation of orders, and quality control processes should be considered. Handling these more complex cases requires collection of more data with an enriched nature. This will enable the decision makers to control the production better by considering real time management of stochastic disturbances to the system. However, to analyze and evaluate this big data better, advanced decision support systems supported by efficient strategies are needed.

Author Contributions: This paper carried out by all authors equally. Conceptualization, O.O.E. and B.B.; Data curation, A.F.I.; Formal analysis, A.F.I.; Investigation, A.A.; Methodology, A.K.T.; Project administration, G.S.D.; Software, A.K.T.; Supervision, A.A.; Validation, A.F.I., G.S.D., and B.B.; Visualization, O.O.E.

Funding: This research received no external funding.

Conflicts of Interest: The authors declare no conflict of interest.

Appendix A

		SPT	EDD	SLACK	PR	WinQ	
Expo(62)	Number of tardy jobs	714.6	3333.8	1070.9	1386.3	2693.1	**Normal**
	Avg. waiting Time	954.3	1430.8	1266.7	1452.2	1317.0	
	Avg. utilization	90.267%	90.701%	90.619%	90.856%	90.786%	
	Work in process	83.9	118.3	106.7	121.6	110.8	
	Avg. tardiness	20570.1	5483.9	16681.9	16124.5	6956.0	
	Avg. earliness time	2281.4	1307.7	1416.2	1453.0	2223.2	
Expo(65)	Number of tardy jobs	492.7	1614.9	525.3	802.2	1669.2	
	Avg. waiting Time	548.9	750.9	690.8	782.4	723.4	
	Avg. utilization	89.127%	89.171%	89.167%	89.309%	89.270%	
	Work in process	53.0	65.6	62.1	68.5	64.3	
	Avg. tardiness	11762.6	2830.6	11249.4	9468.0	3910.0	
	Avg. earliness time	2416.6	1617.9	1785.9	1803.3	2372.1	
Expo(68)	Number of tardy jobs	322.4	492.4	243.8	450.0	972.0	
	Avg. waiting Time	334.9	420.9	404.4	458.1	419.3	
	Avg. utilization	86.291%	86.271%	86.270%	86.353%	86.353%	
	Work in process	37.6	42.7	41.8	45.3	42.8	
	Avg. tardiness	6445.9	1361.2	6730.3	5250.2	2065.9	
	Avg. earliness time	2548.1	1942.7	2109.5	2101.5	2504.0	
Expo(62)	Number of tardy jobs	937.6	3333.8	1132.6	1444.2	2542.5	
	Avg. waiting Time	936.1	1430.1	1115.3	1277.1	1230.1	
	Avg. utilization	90.871%	90.703%	90.849%	91.003%	90.864%	
	Work in process	85.3	118.3	99.2	111.5	107.8	
	Avg. tardiness	14996.8	5481.3	13636.3	13190.4	7138.9	
	Avg. earliness time	2183.4	1305.8	1465.3	1491.7	2261.2	

Expo(65)	Number of tardy jobs	620.0	1612.3	556.1	811.5	1481.2	**DSS**
	Avg. waiting Time	534.5	750.6	605.3	691.0	663.2	
	Avg. utilization	89.356%	89.170%	89.284%	89.402%	89.376%	
	Work in process	54.0	65.6	58.7	64.5	62.9	
	Avg. tardiness	8863.9	2833.7	9087.0	8112.7	4505.2	
	Avg. earliness time	2354.0	1615.3	1868.9	1893.8	2429.7	
Expo(68)	Number of tardy jobs	385.3	489.0	268.1	459.0	887.7	
	Avg. waiting Time	317.6	420.9	357.1	400.4	381.3	
	Avg. utilization	86.383%	86.271%	86.359%	86.407%	86.387%	
	Work in process	38.0	42.7	40.5	43.2	42.1	
	Avg. tardiness	4857.4	1398.7	5368.9	4437.0	2493.2	
	Avg. earliness time	2509.7	1943.3	2176.9	2192.2	2554.3	

References

1. Ersöz, O.Ö.; Türker, A.K. Simultaneous production planning & control with current workstation loading. *Manas J. Soc. Stud.* **2016**, *5*, 5.
2. Elhüseyni, M. *Hipotetik Bir Tekstil Atölyesinin Dinamik Çizelgelenmesinde Yollama Kurallarının Benzetim Tekniğiyle Analizi*; İstanbul Teknik Üniversitesi, Fen Bilimleri Enstitüsü: Istanbul, Turkey, 2012.
3. Azadeh, A.; Negahban, A.; Moghaddam, M. A hybrid computer simulation-artificial neural network algorithm for optimisation of dispatching rule selection in stochastic job shop scheduling problems. *Int. J. Prod. Res.* **2012**, *50*, 551–566. [CrossRef]
4. Larsen, R.; Marco, P. A framework for dynamic rescheduling problems. *Int. J. Prod. Res.* **2018**, *57*, 1–18. [CrossRef]
5. Banks, J.; Carson, J.S.; Nelson, B.L.; Nicol, D.M. *Discrete-Event System Simulation*, 3rd ed.; Printice Hall: Upper Saddle River, NJ, USA, 2001; ISBN 978-0136062127.
6. Law, A.M.; Kelton, W.D. *Simulation Modeling and Analysis*, 2nd ed.; McGraw-Hill International: New York, NY, USA, 1991; ISBN 978-0073401324.
7. Koruca, H.İ.; Özdemir, G.; Aydemir, E.; Çayırlı, M. Bir simülasyon Yazılımı için Esnek İş Akış Planı Editörü Geliştirilmesi; İşlemlerin Gantt Şemasında Çizelgelenmesi. *J. Fac. Eng. Archit. Gazi Univ.* **2010**, *25*, 77–81.
8. Zhong, R.Y.; Chen, C.; Huang, G.Q. Big Data Analytics for Physical Internet-based intelligent manufacturing shop floors. *Int. J. Prod. Res.* **2015**, *55*, 2610–2621. [CrossRef]
9. Marinho, R.; Bragança, A.; Ramos, C. Decision Support System for Dynamic Production Scheduling. In Proceedings of the 1999 IEEE International Symposium on Assembly and Task Planning (ISATP'99) (Cat. No. 99TH8470), Porto, Portugal, 24–24 July 1999; pp. 424–429.
10. Aydın, M.E.; Öztemel, E. Dynamic job-shop scheduling using reinforcement learning agents. *Robot. Auton. Syst.* **2000**, *33*, 169–178. [CrossRef]
11. Li, L.; Chang, Q.; Ni, J.; Biller, S. Real time production improvement through bottleneck control. *Int. J. Prod. Res.* **2009**, *47*, 6145–6158. [CrossRef]
12. Heilala, J.; Montonen, J.; Jarvinen, P.; Kivikunnas, S.; Maantila, M.; Sillanpaa, J.; Jokinen, T. Developing Simulation-Based Decision Support Systems for Customer-driven Manufacturing Operation Planning. In Proceedings of the 2010 Winter Simulation Conference, Baltimore, MD, USA, 5–8 December 2010; pp. 3363–3375.
13. Madhavi, I.; Shirazi, B. A Review of Simulation-based Intelligent Decision Support System Architecture for the Adaptive Control of Flexible Manufacturing Systems. *J. Artif. Intell.* **2010**, *3*, 201–219. [CrossRef]
14. Sharma, P.; Jain, A. Analysis of dispatching rules in a stochastic dynamic job shop manufacturing system with sequence-dependent setup times. *Front. Mech. Eng.* **2014**, *9*, 380–389. [CrossRef]
15. Sharma, P.; Jain, A. New setup-oriented dispatching rules for a stochastic dynamic job shop manufacturing system with sequence-dependent setup times. *Concurr. Eng. Res. Appl.* **2016**, *24*, 58–68. [CrossRef]

16. Zhong, R.Y.; Huang, G.Q.; Dai, Q.Y.; Zhang, T. Mining SOTs and dispatching rules from RFID-enabled real-time shopfloor production data. *J. Intell. Manuf.* **2014**, *25*, 825–843. [CrossRef]

17. Kulkarni, K.; Venkateswaran, J. Hybrid approach using simulation-based optimisation for job shop scheduling problems. *J. Simul.* **2015**, *9*, 312–324. [CrossRef]

18. Phanden, R.K.; Jain, A. Assessing the impact of changing available multiple process plans of a job type on mean tardiness in job shop scheduling. *Int. J. Adv. Manuf. Technol.* **2015**, *80*, 1521–1545. [CrossRef]

19. Kück, M.; Ehm, J.; Freitag, M.; Frazzon, E.M.; Ricardo, P. A Data-Driven Simulation-Based Optimisation Approach for Adaptive Scheduling and Control of Dynamic Manufacturing Systems. *Adv. Mater. Res.* **2016**, *1140*, 449–456. [CrossRef]

20. Ersöz, S.; Türker, A.K.; Aktepe, A. *Üretim Süreçlerinin Optimizasyonunda RFID Teknolojisi ve Uzman Sistem Temelli Tümleşik Yapının ERP Sistemine Entegrasyonu ve FNSS Savunma Sistemleri A.Ş.'de Uygulanması*; San-Tez Project Report; Ankara, Turkey, 2016; Available online: https://adnanaktepe.com/projeler/ (accessed on 14 March 2019).

21. Zhang, T.; Xie, S.; Rose, O. Real-Time Job Shop Scheduling Based on Simulation and Markov Decision Processes. In Proceedings of the 2017 Winter Simulation Conference, Las Vegas, NV, USA, 3–6 December 2017; pp. 3899–3907.

22. Bierwirth, C.; Kuhpfahl, J. Extended GRASP for the job shop scheduling problem with total weighted tardiness objective. *Eur. J. Oper. Res.* **2017**, *261*, 835–848. [CrossRef]

23. Xiong, H.; Fan, H.; Jiang, G.; Li, G. A simulation-based study of dispatching rules in a dynamic job shop scheduling problem with batch release and extended technical precedence constraints. *Eur. J. Oper. Res.* **2017**, *257*, 13–24. [CrossRef]

24. Zhang, J.; Ding, G.; Zou, Y.; Qin, S.; Fu, J. Review of job shop scheduling research and its new perspectives under Industry 4.0. *J. Intell. Manuf.* **2017**. [CrossRef]

25. Rossit, D.A.; Tohme, F.; Frutos, M. Industry 4.0: Smart Scheduling. *Int. J. Prod. Res.* **2018**, *56*. [CrossRef]

26. Tao, F.; Qi, Q.; Liu, A.; Kusiak, A. Data-driven smart manufacturing. *J. Manuf. Syst.* **2018**, *48*, 157–168. [CrossRef]

27. Jiang, T.; Zhang, C.; Zhu, H.; Gu, J.; Deng, G. Energy-Efficient Scheduling for a Job Shop Using an Improved Whale Optimization Algorithm. *Mathematics* **2018**, *6*, 220. [CrossRef]

28. Ortiz, M.A.; Betancourt, L.E.; Negrete, K.P.; Felice, F.D.; Petrillo, A. Dispatching algorithm for production programming of flexible job-shop systems in the smart factory industry. *Ann. Oper. Res.* **2018**, *264*, 409–433. [CrossRef]

29. Ding, K.; Jiang, P. RFID-based production data analysis in an IoT-enabled smart job-shop. *IEEE/CAA J. Autom. Sin.* **2018**, *5*, 128–138. [CrossRef]

30. Leusin, M.E.; Frazzon, E.M.; Maldonado, M.U.; Kück, M.; Freitag, M. Solving the Job-Shop Scheduling Problem in the Industry 4.0 Era. *Technologies* **2018**, *6*, 107. [CrossRef]

31. Zhang, K.; Qu, T.; Zhou, D.; Thürer, M.; Liu, Y.; Nie, D.; Li, C.; Huang, G.Q. IoT-enabled dynamic lean control mechanism for typical production systems. *J. Ambient Intell. Hum. Comput.* **2019**, *10*, 1009–1023. [CrossRef]

32. Holthaus, O.; Rajendran, C. Efficient dispatching rules for scheduling in a job shop. *Int. J. Prod. Econ.* **1997**, *48*, 87–105. [CrossRef]

33. Abdullah, M. A Review of Automated Decision Support System. *J. Fund. Appl. Sci.* **2018**, *10*, 252–257.

 © 2019 by the authors. Licensee MDPI, Basel, Switzerland. This article is an open access article distributed under the terms and conditions of the Creative Commons Attribution (CC BY) license (http://creativecommons.org/licenses/by/4.0/).

Article

Change Point Detection for Airborne Particulate Matter ($PM_{2.5}$, PM_{10}) by Using the Bayesian Approach

Muhammad Rizwan Khan [1] and **Biswajit Sarkar** [2,*]

[1] Department of Industrial Engineering, Hanyang University, 222 Wangsimni-Ro, Seoul 133-791, Korea; mrizwankhan162@gmail.com

[2] Department of Industrial & Management Engineering, Hanyang University, Ansan, Gyeonggi-do 15588, Korea

* Correspondence: bsbiswajitsarkar@gmail.com; Tel.: +82-10-7498-1981; Fax: +82-31-400-5959

Received: 28 February 2019; Accepted: 8 May 2019; Published: 24 May 2019

Abstract: Airborne particulate matter (PM) is a key air pollutant that affects human health adversely. Exposure to high concentrations of such particles may cause premature death, heart disease, respiratory problems, or reduced lung function. Previous work on particulate matter ($PM_{2.5}$ and PM_{10}) was limited to specific areas. Therefore, more studies are required to investigate airborne particulate matter patterns due to their complex and varying properties, and their associated (PM_{10} and $PM_{2.5}$) concentrations and compositions to assess the numerical productivity of pollution control programs for air quality. Consequently, to control particulate matter pollution and to make effective plans for counter measurement, it is important to measure the efficiency and efficacy of policies applied by the Ministry of Environment. The primary purpose of this research is to construct a simulation model for the identification of a change point in particulate matter ($PM_{2.5}$ and PM_{10}) concentration, and if it occurs in different areas of the world. The methodology is based on the Bayesian approach for the analysis of different data structures and a likelihood ratio test is used to a detect change point at unknown time (k). Real time data of particulate matter concentrations at different locations has been used for numerical verification. The model parameters before change point (θ) and parameters after change point (λ) have been critically analyzed so that the proficiency and success of environmental policies for particulate matter ($PM_{2.5}$ and PM_{10}) concentrations can be evaluated. The main reason for using different areas is their considerably different features, i.e., environment, population densities, and transportation vehicle densities. Consequently, this study also provides insights about how well this suggested model could perform in different areas.

Keywords: airborne particulate matter; Bayesian approach; change point detection; likelihood ratio test; time series analysis; air quality

---

1. Introduction

Airborne particulate matter is one of the most dangerous air pollutants and harmful to human health. For the last two decades, information about the negative impacts of PM_{10} (particles less than 10 μm in diameter) and $PM_{2.5}$ (particles less than 2.5 micrometers in diameter) has increased enormously. Exposure to high concentrations of such particles may cause premature death, heart disease, respiratory problems, or reduced lung function through different mechanisms, which include pulmonary and systemic inflammation, accelerated atherosclerosis, and altered cardiac autonomic function (Heroux et al. [1] and Pope et al. [2]). Therefore, to control particulate matter (PM) pollution, and to make effective plans for counter measurements, it is important to measure the efficiency and effectiveness of policies applied by the Ministry of Environment. Every region has developed different kinds of extensive bodies of legislation, which establish air quality standards for key air pollutants to improve the air quality and to satisfy these standards. The European Environment Agency, the United States Environmental Protection Agency, and the Ministry of Environment in South Korea have each set their own air quality standards for all air pollutants. Thus, It is essential to follow established air quality monitoring systems to measure the PM concentrations on an hourly as well as daily basis, because some areas deviate from the established PM standards. This may cause adverse environmental effects and serious health problems.

Until now, a number of statistical methods have been established to model the hazards of PM from air quality standards. A Bayesian multiple change point model was proposed to measure the quantitative efficiency of pollution control programs for air quality, which estimate the hazards of different air pollutants. In the model, it was assumed as a nonhomogeneous Poisson process with multiple change points. The change points were identified, and a rate function was estimated by using a reversible jump MCMC algorithm (Gyarmati-Szabo et al. [3]). In another study, the changes in health effects due to simultaneous exposure to physical and chemical properties of airborne particulate matter were gauged through Bayesian approach and inferences were drawn via the Markov Chain Monte Carlo method (Pirani et al. [4]). A Bayesian approach was introduced to estimate the distributed lag functions in time series models, which can be used to determine the short-term health effects of particulate air pollution on mortality (Welty et al. [5]). Hybrid models were proposed to forecast the PM concentrations for four major cities of China; Beijing, Shanghai, Guangzhou, and Lanzhou (Qin et al. [6]).

A change point detection method for detecting changes in the mean of the one dimensional Gaussian process was proposed on the basis of a generalized likelihood ratio test (GLRT). The important characteristic of this method is that it includes data dependence and covariance of the Gaussian process. However, in case of unidentified covariance, the plug-in GLRT method was suggested which remains asymptotically near optimal (Keshavarz et al. [7]). A new method for acute change point detection was proposed for fractional Brownian motion with a time dependent diffusion coefficient. The likelihood ratio method has been used for change point detection in Brownian motion. A statistical test was also suggested to identify the significance of a calculated critical point (Kucharczyk et al. [8]). The change point detection technique in machine monitoring was suggested, which was based on two stages. In the first stage, irregularities are measured in time series data through the automatic regression (AR) model, and then the martingale statistical test is applied to detect the change point in unsupervised time series data (Lu et al. [9]). An integrated inventory model was developed to determine the optimal lot size and production uptime while considering stochastic machine breakdown and multiple shipments for a single-buyer and single-vendor (Taleizadeh et al. [10]).

A statistical change point algorithm based on nonparametric deviation estimation between time series samples from two retrospective segments was proposed in which the direct density ratio estimation method was applied for deviation measurement through relative Pearson divergence (Liu et al. [11]). A novel statistical methodology for online change point detection was suggested in which data for an uncertain system was composed through an autoregressive model. On the basis of nonparametric estimation of unidentified elements, an innovative CUSUM-like scheme was recommended for change detection. This estimation method could also be updated online (Hilgert et al. [12]). A new methodology, the Karhunen-Loeve expansions of the limit Gaussian processes, was suggested for change point test in the level of a series. Firstly, change point detection in the mean was explained, which later extended to linear and nonlinear regression (Górecki et al. [13]). A Cramer-von Mises type test was presented to test the sudden changes in random fields which was dependent on Hilbert space theory (Bucchia and Wendler [14]). The continuous-review inventory model was developed for Controllable lead time for comparing two models; one with normally distributed lead time demand and the second assumes that there is no specific distribution for lead time demand (Shin et al. [15]).

A new technique was developed to identify the structural changes in linear quantile regression models. When a structural change in the relationship between covariates and response at a specific point exists, it may not be at the centre of response distribution, but at the tail. The traditional mean regression method might not be applicable for change point detection of such structural changes at tails. Subsequently, the proposed technique could be appropriate for it (Zhou et al. [16]). For detection of simultaneous changes in mean and variance, a new methodology called the fuzzy classification maximum likelihood change point (FCML-CP) algorithm was suggested. Multiple change points in the mean and variance of a process can be estimated by this method. This technique is much better than the normal statistical mixture likelihood method because it saves a lot of time (Lu and Chang [17]). A model for Partial Trade-Credit Policy of Retailer was developed in which deterioration of products was assumed as exponentially distributed (Sarkar and Saren [18]).

The Bayesian change point algorithm for sequential data series was introduced which has some uncertain limitations regarding location and number of change points. This algorithm was precisely based on posterior distribution to deduce if a change point has occurred or not. It can also update itself linearly as new data points are observed. Posterior distribution monitoring is the finest way to identify the presence of a new change point in observed data points. Simulation studies illustrate that this algorithm is good for rapid detection of existing change points, and it is also known for a low rate of false detection (Ruggieri and Antonellis [19]). Due to the probabilistic concept of Bayesian change point detection (BCPD), this methodology can overcome threats in identifying the location and number of change points.

The performance of two different methods for change point detection of multivariate data with both single and multiple changes was compared. The results illustrated adequate performance for both Expectation Maximization (EM) and Bayesian methods. However, EM exhibits better performance in case of minor changes and unsuitable priors while the Bayesian method has less computational work to do (Keshavarz and Huang [20]). The Bayesian multiple change point model was suggested for the identification of Distributed Denial of Service (DDoS) flooding attacks in VoIP systems in which Session Initiation Protocol (SIP) is used as signalling mechanism (Kurt et al. [21]). One of the well-known change detection techniques is post classification with multi temporal remote sensing images. An innovative post classification technique with iterative slow feature analysis (ISFA) and Bayesian soft fusion was suggested to acquire accurate and reliable change detection maps. Three steps were suggested in this technique, first was to get the class probability of images through independent classification. After that, a continuous change probability map of multi temporal images was obtained by ISFA algorithm. Lastly, posterior probabilities for the class combinations of coupled pixels were determined through the Bayesian approach to assimilate the class probability with the change probability, which is called Bayesian soft fusion. This technique could be widely applicable in land cover monitoring and change detection at a large scale (Wu et al. [22]).

The Bayesian change point technique was designed to analyse biomarkers time series data in women for the diagnosis of ovarian cancer. The identification of such kind of change points could be used to diagnose the disease earlier (Mariño et al. [23]). The Generalized Extreme Value (GEV) fused lasso penalty function was applied to identify the change point of annual maximum precipitation (AMP) in South Korea. Numerical analysis and applied data analysis were conducted in order to compare performance from the GEV fused lasso and Bayesian change point analysis, which shows that when water resource structures are hydrologically designed the GEV fused lasso method should be used to identify the change points (Jeon et al. [24]). The Bayesian method was recommended to identify the change point occurrence in extreme precipitation data, and the model follows a generalized Pareto distribution. This Bayesian change point detection was inspected for four different situations, one with no change model, second with a shape change model, third with a scale change model, and fourth with both a scale and shape change model. It was determined that unexpected and sustained change points need to be considered in extreme precipitation while making hydraulic design (Chen et al. [25]).

Bayesian change point methodology was presented to identify changes in the temporal event rate for a non-homogeneous Poisson process. This methodology was used to determine if a change in the event rate has occurred or not, the time for change, and the event rate before or after the change. The methodology has been explained through an example of earthquake occurrence in Oklahoma. This spatiotemporal change point methodology can also be used for identifying changes in climate patterns and assessing the spread of diseases. It permits participants to make real time decisions about the influence of changes in event rates (Gupta and Baker [26]). A new Bayesian methodology was recommended to analyze multiple time series with the objective of identifying abnormal regions. A general model was developed and it was shown that Bayesian inference allows independent sampling from the posterior distribution. Copy number variations (CNVs) are identified by using data from multiple individuals. The Bayesian method was evaluated on both real and simulated CNV data to provide evidence that this method is more precise as compare to other suggested methods for analyzing such data (Bardwell and Fearnhead [27]).

All the above mentioned methods are either too complex and complicated for application on random hazards of PM or not applicable to randomness of PM hazards. Therefore, still more studies are required to investigate the PM hazards, due to its complex and varying properties and associated (PM_{10} and $PM_{2.5}$) concentrations and compositions, to investigate the numerical productivity of pollution control programs for air quality. The primary purpose of this research is to develop models for change point detection of particulate matter ($PM_{2.5}$ and PM_{10}) concentrations if it occurs in different areas. The pollutant concentrations before and after a change point has to be critically analyzed so that the proficiency and success of environmental policies for particulate matter ($PM_{2.5}$ and PM_{10}) concentrations can be evaluated. The Bayesian approach is used to analyze random hazards of PM concentrations with a change point at an unknown time (k).

To demonstrate the proposed approach, real time data of random hazards of PM concentrations at different sites has been used. The PM concentrations change point (k), parameters before change point (θ), and parameters after change point (λ) have been comprehensively analyzed by using the Bayesian technique. Thus, simulation models have been constructed for different data structures. The main reason for using different areas is their considerably different features i.e., environment, population densities, and transportation vehicle densities. Consequently, this study also provides insight about how well this suggested model could perform in different areas. The paper is structured as follows: Section 2 refers to problem definitions, explaining assumptions along with notation, and Section 3 shows the formulation of mathematical models. Sections 4 and 5 depict numerical examples and results, respectively, to validate the practical applications of the proposed models. Section 6 discusses the depicted results of previous section, it also explains the managerial insights of results. Finally, Section 7 presents conclusions of this study. Table 1 depicts the comparative study of different authors who have contributed in the direction of research, while the last row of the table portrays the contribution of this research paper. On the other hand, Tables 2 and 3 compares the difference in previous workings and this work.

Table 1. Author's contribution in the direction of research.

Authors	Airborne Particulate Matters ($PM_{2.5}$, PM_{10})	Change-Point Detection	BCPD (Bayesian Change Point Detection)	Time Series Analysis
Lim et al. (2012) [28]	Ambient particulate matter (PM)	-	-	-
Rao et al. (2018) [29]	Particulate matter air pollution	-	-	-
Wei and Meng (2018) [30]	Airborne fine particulate matter ($PM_{2.5}$)	-	-	-
Wang et al. (2017) [31]	Inhaled particulate matter	-	-	-
Heroux et al. (2015) [1]	Ambient air pollutants	-	-	-
Wellenius et al. (2012) [32]	Ambient Air Pollution	-	-	-
Pirani et al. (2015) [4]	Airborne particles	-	-	Bayesian inference within Time series
Li et al. (2013) [33]	Airborne particulate matter	-	-	Time series analysis
Kim et al. (2018a) [34]	Particulate matter	-	-	-
Kim et al. (2017) [35]	Air pollution	-	-	Time series model
Qin et al. (2017) [36]	Air pollution	-	-	Time series analysis
Lee et al. (2015) [37]	Fine and coarse particles	-	-	Time series analysis
Cabrieto et al. (2018) [38]	-	kernel change point detection, correlation changes	-	-
Keshavarz et al. (2018) [7]	-	Generalized likelihood ratio test, One-dimensional Gaussian process	-	-
Kucharczyk et al. (2018) [8]	-	likelihood ratio test, fractional Brownian motion	-	-
Lu et al. (2017) [9]	-	Change-point detection, machine monitoring, Anomaly measure (AR model), Martingale test	-	Time series data
Hilgert et al. (2016) [12]	-	On-line change detection, autoregressive dynamic models, CUSUM-like scheme	-	-
Górecki et al. (2017) [13]	-	Change point detection, heteroscedastic, Karhunen-Loeve expansions	-	Time series data
Bucchia and Wendler (2017) [14]	-	Change point detection, bootstrap, Hilbert space valued random fields, (Cramer–von Mises type test)	-	Time series data
Zhou et al. (2015) [16]	-	Sequential change point detection, linear quantile regression models	-	-
Lu and Chang (2016) [17]	-	Detecting change points, mean/variance shifts, FCML-CP algorithm	-	-

Table 1. *Cont.*

Authors	Airborne Particulate Matters ($PM_{2.5}$, PM_{10})	Change-Point Detection	BCPD (Bayesian Change Point Detection)	Time Series Analysis
Liu et al. (2013) [11]	-	Change point detection, relative density ratio estimation	-	Time series analysis
Ruggieri and Antonellis (2016) [19]	-	-	Bayesian sequential change point detection	-
Keshavarz and Huang (2014) [20]	-	-	Bayesian and Expectation Maximization methods, multivariate change point detection	-
Kurt et al. (2018) [21]	-	-	Bayesian change point model, SIP-based DDoS attacks detection	-
Wu et al. (2017) [22]	-	-	Post classification change detection, iterative slow feature analysis, Bayesian soft fusion	-
Gupta and Baker (2017) [26]	-	-	Spatial event rates, change point, Bayesian statistics, induced seismicity	-
Marino et al. (2017) [23]	-	-	Change point, multiple biomarkers, ovarian cancer	-
Jeon et al. (2016) [24]	-	-	Abrupt change point detection, annual maximum precipitation, fused lasso	-
Bardwell and Fearnhead (2017) [27]	-	-	Bayesian Detection, Abnormal Segments	Time series analysis
Chen et al. (2017) [25]	-	-	Bayesian change point analysis, extreme daily precipitation	-
This study	Airbone Particulate Matter	Change point detection	Bayesian approach and likelihood ratio test for change point detection	Time series data

Table 2. Change point detection.

Authors	Change-Point Detection
Cabrieto et al. (2018) [38]	Detection of correlation changes by applying kernel change point detection on the running correlations
Keshavarz et al. (2018) [7]	Detecting a change in the mean of one-dimensional Gaussian process data in the fixed domain regime based on the generalized likelihood ratio test (GLRT)
Kucharczyk et al. (2018) [8]	Variance change point detection for fractional Brownian motion based on the likelihood ratio test
Lu et al. (2017) [9]	Graph-based structural change detection for rotating machinery monitoring by martingale-test method
Hilgert et al. (2016) [12]	On-line change detection for uncertain autoregressive dynamic models through nonparametric estimation (CUSUM-like scheme)
Górecki et al. (2017) [13]	Change point detection in heteroscedastic time series through Karhunen-Loeve expansions
Bucchia and Wendler (2017) [14]	Change point detection and bootstrap for Hilbert space valued random fields (Cramer-von Mises type test)
Zhou et al. (2015) [16]	A method for sequential detection of structural changes in linear quantile regression models.
Lu and Chang (2016) [17]	Detecting change-points for shifts in mean and variance using fuzzy classification maximum likelihood change-point (FCML-CP) algorithm
Liu et al. (2013) [11]	Change point detection in time-series data by relative density-ratio estimation
This study	Change point detection for airborne particulate matter ($PM_{2.5}$, PM_{10}) through Bayesian approach and likelihood ratio test

Table 3. BCPD (Bayesian change point detection).

Authors	BCPD (Bayesian Change Point Detection)
Ruggieri and Antonellis (2016) [19]	A sequential Bayesian change point algorithm was proposed that provides uncertainty bounds on both the number and location of change points
Keshavarz and Huang (2014) [20]	Bayesian and Expectation Maximization (EM) methods for change point detection problem of multivariate data with both single and multiple changes
Kurt et al. (2018) [21]	A Bayesian change point model for detecting SIP-based DDoS attacks
Wu et al. (2017) [22]	A post-classification change detection method based on iterative slow feature analysis and Bayesian soft fusion
Gupta and Baker (2017) [26]	Estimating spatially varying event rates with a change point using Bayesian statistics: Application to induced seismicity
Marino et al. (2017) [23]	Change-point of multiple biomarkers in women with ovarian cancer
Jeon et al. (2016) [24]	Abrupt change point detection of annual maximum precipitation through fused lasso penalty function by using the Generalized Extreme Value (GEV) distribution
Bardwell and Fearnhead (2017) [27]	Bayesian Detection of Abnormal Segments in Multiple Time Series
Chen et al. (2017) [25]	Bayesian change point analysis for extreme daily precipitation
This study	Change point detection for airborne particulate matter ($PM_{2.5}$, PM_{10}) through Bayesian approach and likelihood ratio test

2. Problem Definition, Notation and Assumptions

2.1. Problem Definition

The major objective of this research is to develop a more precise, well defined and user friendly method for application on random hazards of PM to detect the change point of subjected air pollutant hazards at any unknown time (k) if it occurs at any area across the globe. The existing methods are either too complex and complicated for the application on random hazards of PM due to its complex and varying properties or not applicable to randomness of PM hazards. Therefore, still more studies are required to develop a such kind of methodology, which is easily understandable and appropriate to model the hazards of the PM concentrations from air quality standards that can also detect change points in these hazards. Secondly, this method could be applicable for any kind of time series and data distributions. Analysis of these changes need to be done, whether these change points are favorable or not for the environment. For this, a comparison of subjected pollutant hazards before and after a change point has to be done for the evaluation of pollution control programs adopted by environmental protection agencies. If hazards occurrences increase after the change point, then environmental policies have a negative impact which marks the failure of pollution control program, but if the hazards occurrences reduce after the change point, then it demonstrates the effectiveness of the pollution control program. Thirdly, an alteration in occurrences must be measured to define the new pollution control policies for further improvements in the current level of subjected air pollutant hazards.

For anticipated goals, the Bayesian approach will be used to determine posterior probabilities of pollutant occurrences and the likelihood ratio test will be used for identifying the change point in that Bayesian model. This suggested model would be numerically validated by using real-time data of particulate matters' concentrations in different areas of Seoul, South Korea, observed from January 2004 to December 2013. The change point (k) for for particulate matter ($PM_{2.5}$ and PM_{10}) hazards, the rate before the change point (θ), and the rate after the change point (λ) would be comprehensively analyzed. The central idea for using different regions is their considerably different features i.e., environment, population densities, and transportation vehicle densities. Hence, this study can also be a vision for the implementation of recommended model in different areas.Air quality standards for particular matter $PM_{2.5}$ and PM_{10} are given in Table 4. Results have been determined by following these standards.

Table 4. Region-wise air quality standards for particulate matters ($PM_{2.5}$ and PM_{10}).

Air Quality Standards for $PM_{2.5}$ ($\mu g/m^3$)		Air Quality Standards for PM_{10} ($\mu g/m^3$)	
European Standard ($PM_{2.5}$)	25	European Standard (PM_{10}) 24 h	50
American Standard ($PM_{2.5}$)	35	American Standard (PM_{10}) 24 h	150
Korean Standard ($PM_{2.5}$) 24 h	50	Korean Standard (PM_{10}) 24 h	100

Tables 5 and 6 illustrate some details regarding data collected for Guro, Nowon, Songpa, and Yongsan which exhibit standard-wise and location-wise percentage of polluted days. In case of $PM_{2.5}$, more than 44%, 21% and 8% days are polluted as per European, American and Korean standards respectively, which is alarming. Similarly, in case of PM_{10}, the polluted days concentrations as per European and Korean standards is more than 40% and 6% respectively and it could not be acceptable. Hence, there is a need to control hazards of PM.

Table 5. Particulate matter ($PM_{2.5}$) location-wise data.

Particle Pollution	Criteria ($\mu g/m^3$)	Total Number of Days	Effective Readings	Dangerous Concentration	%Age against Total Days	%Age against Effective Readings
\multicolumn{7}{c}{$PM_{2.5}$ Guro's data (Seoul, South Korea)}						
European Standard	25	2498	2486	1125	45.04%	45.25%
American Standard	35	2498	2486	532	21.30%	21.40%
Korean Standard	50	2498	2486	207	8.29%	8.33%
\multicolumn{7}{c}{$PM_{2.5}$ Nowon's data (Seoul, South Korea)}						
European Standard	25	3228	3031	1371	42.47%	45.23%
American Standard	35	3228	3031	718	22.24%	23.69%
Korean Standard	50	3228	3031	255	7.90%	8.41%
\multicolumn{7}{c}{$PM_{2.5}$ Songpa's data (Seoul, South Korea)}						
European Standard	25	3653	3388	1547	42.35%	45.66%
American Standard	35	3653	3388	795	21.76%	23.47%
Korean Standard	50	3653	3388	281	7.69%	8.29%
\multicolumn{7}{c}{$PM_{2.5}$ Yongsan's data (Seoul, South Korea)}						
European Standard	25	3653	3456	1537	42.08%	44.47%
American Standard	35	3653	3456	746	20.42%	21.59%
Korean Standard	50	3653	3456	280	7.66%	8.10%

Table 6. Particulate matter (PM_{10}) location-wise data.

Particle Pollution	Criteria ($\mu g/m^3$)	Total Number of Days	Effective Readings	Dangerous Concentration	%Age against Total Days	%Age against Effective Readings
			PM_{10} Guro's data (Seoul, South Korea)			
European Standard	50	3653	3540	1580	43.25%	44.63%
American Standard	150	3653	3540	60	1.64%	1.69%
Korean Standard	100	3653	3540	279	7.64%	7.88%
			PM_{10} Nowon's data (Seoul, South Korea)			
European Standard	50	3653	3531	1444	39.53%	40.89%
American Standard	150	3653	3531	41	1.12%	1.16%
Korean Standard	100	3653	3531	239	6.54%	6.77%
			PM_{10} Songpa's data (Seoul, South Korea)			
European Standard	50	3653	3467	1490	40.79%	42.98%
American Standard	150	3653	3467	50	1.37%	1.44%
Korean Standard	100	3653	3467	240	6.57%	6.92%
			PM_{10} Yongsan's data (Seoul, South Korea)			
European Standard	50	3653	3508	1566	42.87%	44.64%
American Standard	150	3653	3508	60	1.64%	1.71%
Korean Standard	100	3653	3508	282	7.72%	8.04%

2.2. Notation

The list of notation to represent the random variables and parameters is as follows:

Indices

i replication or sequence, $i = 1, 2, \ldots$
j position in the chain, $j = 1, 2, \ldots n$

Random variables

Y random process
y variable (Y) at any given point
y_i variable (Y) at point i where $i \in 0, 1, 2 \ldots$

Parameters

k change point in the random process
θ parameter before change point k associated with probability distribution function of random variable Y
λ parameter after change point k associated with probability distribution function of random variable Y

Variables

$Pr(\theta)$ prior distribution for parameter θ
$Pr(\theta|y_i)$ posterior distribution for parameter θ
$Pr(\lambda|y_i)$ posterior distribution for parameter λ
$Pr(y_i|\theta)$ likelihood or sampling model
V mean of the chain or replications (Average of daily pollutant concentrations)
V_{ij} jth observation from the ith replication

V_i	mean of *i*th replication
V	mean of *m* replications
B	between sequence variance represents the variance of replications with the mean of *m* replications
S_i^2	variance for all replications
W	within sequence variance is the mean variance for *m* replications
$Var(V)$	overall estimate of the variance of V in the target distribution
$\sqrt{R}$	estimated potential scale reduction for convergence

2.3. Assumptions

The following assumptions were used for the proposed model:

1. Y represents the number of times an event occurs in time t and Y is always positive real numbers $y \in 1, 2...$ that can be any random value.
2. $Y(0) = 0$ means that no event occurred at time $t = 0$.
3. Time series random data observed on equal interval of lengths.
4. The particulate matter daily concentrations or occurrence of events follow specific random probability distribution function.
5. The particulate matter daily concentrations in any interval of length (t) is a random variable and number of times event occurs is also positive random variable with parameter $(rate = \theta)$.

3. Mathematical Model

3.1. Formulation of Mathematical Model

The probability distribution function of a random variable Y at any given point y in the sample space is given as follows:

$$f(y; \theta) = Pr(Y = y | \theta) \ for \ y \in 1, 2...$$

There could be a single parameter or multiple parameters depending upon the probability distribution function of random variable Y.

The change point for random process Y is being detected by the likelihood ratio test and that is a statistical test used for comparing the goodness of fit for two statistical models; one is null model and other is alternative model. The test is based on the likelihood ratio, which states how many times more likely the data are under one model than the other. This likelihood ratio compared to a critical value used to decide whether to reject the null model.

$$f(\text{Change point} | Y, \text{Eexpectation before change point}, \text{Expectation after change point})$$

$$= \frac{L(Y; \text{Change point}, \text{Expectation before change point}, \text{Expectation after change point})}{\sum_{j=1}^{n} L(Y; j, \text{Change point}, \text{Expectation before change point}, \text{Expectation after change point})} \quad (1)$$

and parameters' comparison before and after the change-point is also being done.

Let the change point in the random process be denoted by k and θ be the random variable parameter before change point k while λ be the random variable parameter after change point k. It can be represented as:

$$y_i \sim pdf(\theta) \ for \ i = 1, 2, ..., k$$

$$y_i \sim pdf(\lambda) \ for \ i = k+1, k+2, ..., n$$

Hence,

$$f(y; \theta) = Pr(Y = y_i | \theta) \text{ for } i = 1, 2, ..., k$$

$$f(y; \lambda) = Pr(Y = y_i | \lambda) \text{ for } i = k+1, k+2, ..., n$$

The joint pdf (probability density function) is the product of marginal pdf. If random variable $Y = y_i$ with parameter θ is modelled, then joint pdf of our sample data will be as below:

$$Pr(Y = y_i | \theta) = \prod_{i=1}^{n} Pr(y_i | \theta) \text{ for } i \in 0, 1, 2, ..., n$$

A class of prior densities is conjugate for the likelihood/sampling model $Pr(y_i | \theta)$ if the posterior distribution is also in the same class. Therefore, prior distribution $Pr(\theta)$ and posterior distribution $Pr(\theta | y_i)$ will follow the same conjugate prior distribution to the likelihood/sampling model $Pr(y_i | \theta)$. However, the likelihood $Pr(y_i | \theta)$ follows the random distribution based on data. Therefore, the prior distribution $Pr(\theta)$ of parameters and posterior distribution $Pr(\theta | y_i)$ of the same parameters must be same and conjugate for Bayesian analysis. Bayes theorem for parameter's θ and λ is as follows:

$$Pr(\theta | y_i) \propto Pr(\theta) Pr(y_i | \theta)$$

$$Pr(\lambda | y_i) \propto Pr(\lambda) Pr(y_i | \lambda)$$

By applying Bayes theorem, the posterior distribution of model parameters θ and λ can be determined

$$Pr(\theta | y_i) = \frac{Pr(y_i | \theta) Pr(\theta)}{Pr(y_i) \text{ for } i \in 1, 2, ..., k}$$

$$Pr(\lambda | y_i) = \frac{Pr(y_i | \lambda) Pr(\lambda)}{Pr(y_i) \text{ for } i \in k+1, k+2, ..., n}$$

As,

$$L(\theta | Y) = f_\theta(Y) = f(Y | \theta)$$

Now, apply likelihood ratio test statistic for change point detection

$$f(k | Y, \theta, \lambda) = \frac{L(Y; k, \text{Expectation before change point, Expectation after change point})}{\sum_{j=1}^{n} L(Y; j, \text{Expectation before change point, Expectation after change point})}$$

And likelihood will be determined as given by:

$$L(Y; k, \theta, \lambda) = \left[exp\left(k((\text{Expectation after } k \text{ point} - \text{Expectation before } k \text{ point}) \right) \left(\frac{\text{Expectation before } k}{\text{Expectation after } k} \right)^{\sum_{i=1}^{k} y_i} \right]$$

The change point k is uniform over y_i. Please note that θ, λ and k are all independent of each other.

3.1.1. Convergence of the Parameters

A single simulation run of a somewhat arbitrary length cannot represent the actual characteristics of the resulting model. Therefore, to estimate the steady-state parameters, the Gelman-Rubin Convergence diagnostic has to be applied in which target parameters are estimated by running multiple sequences of the chain. m replications of the simulation ($m \geq 10$) are made, each of length $n = 1000$. If the target distribution is unimodal then Cowles and Carlin recommends that we must run at least ten chains, as this approach monitors the scalar numbers of interest in the analysis. Therefore, the mean rate of pollutant concentrations is a parameter of interest that is denoted by V.

Scalar summary V = Mean of the chain (Average of daily pollutant concentrations)

Let V_{ij} be the jth observation from the ith replication

$$V_{ij}, \; i = 1, 2, ..., m \; j = 1, 2, ..., n$$

Mean of ith replication

$$V_i = \frac{1}{n} \sum_{j=1}^{n} V_{ij}$$

Mean of m replications

$$V = \frac{1}{m} \sum_{i=1}^{m} V_i$$

The between sequence variance represents the variance of replications with the mean of m replications calculated as follows:

$$B = \frac{n}{m-1} \sum_{i=1}^{m} (V_i - V)^2$$

Variance for all replications is calculated to determine the within sequence variance

$$S_i^2 = \frac{1}{n-1} \sum_{j=1}^{n} (V_{ij} - V)^2$$

The within sequence variance is the mean variance for k replications determined as given below:

$$W = \frac{1}{m} \sum_{i=1}^{m} S_i^2$$

Finally, the within sequence variance and between sequence variance are combined to get an overall estimate of the variance of V in the target distribution

$$Var(V) = \frac{n-1}{n} W + \frac{1}{n} B$$

Convergence is diagnosed by calculating

$$\sqrt{R} = \sqrt{\frac{Var(V)}{W}}$$

This factor $\sqrt{R}$ (estimated potential scale reduction) is the ratio between the upper and lower bound on the space range of V which is used to estimate the factor by which $Var(V)$ could be reduced through more iterations. Further iterations of the chain must be run if the potential scale reduction is high. Run the replications until R is less than 1.1 or 1.2 for all scalar summaries.

3.1.2. Flowchart

The flowchart (Figure 1) for change point k detction, for any random process Y, is given as follows:

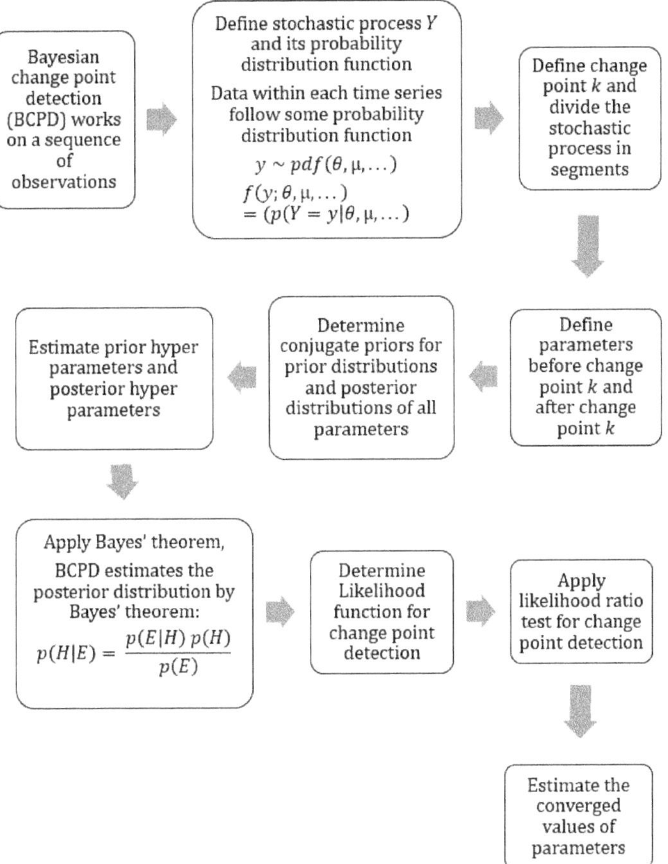

Figure 1. Flowchart for change point (k) detection.

3.2. Comparison Method for Change Point Detection

A change point analysis has been done by using a combination of CUSUM (cumulative sum control chart) and bootstrapping for comparative analysis.

3.2.1. The CUSUM (Cumulative Sum Control Chart) Technique

The CUSUM (cumulative sum control chart) is a sequential analysis technique typically used for monitoring change detection. CUSUM charts are constructed by calculating and plotting a cumulative sum based on the data. The cumulative sums are calculated as follows:

1. First calculate the average.

$$\bar{X} = \left(\frac{X_1 + X_2 + X_3 + ...,}{n} \right)$$

2. Start the cumulative sum at zero by setting $S_0 = 0$
3. Calculate the other cumulative sums by adding the difference between current value and the average to the previous sum, i.e.,

$$S_i = S_{i-1} + \left(X_i - \bar{X} \right)$$

Plot the series and the cumulative sum is not the cumulative sum of the values. Instead it is the cumulative sum of differences between the values and the average. Because the average is subtracted from each value, the cumulative sum also ends at zero.

Interpreting a CUSUM chart requires some practice. Suppose that during a period of time the values tend to be above the overall average. Most of the values added to the cumulative sum will be positive and the sum will steadily increase. A segment of the CUSUM chart with an upward slope indicates a period where the values tend to be above the overall average. Likewise a segment with a downward slope indicates a period of time where the values tend to be below the overall average. A sudden change in direction of the CUSUM indicates a sudden shift or change in the average. Periods where the CUSUM chart follows a relatively straight path indicate a period where the average did not change.

3.2.2. Bootstrap Analysis

A confidence level can be determined for the apparent change by performing a bootstrap analysis. Before performing the bootstrap analysis, an estimator of the magnitude of the change is required. One choice, which works well regardless of the distribution and despite multiple changes, is S_{diff} defined as:

$$S_{diff} = S_{max} - S_{min}$$

$$S_{max} = \max_{i=0,1,2,\ldots,} S_i$$

$$S_{min} = \min_{i=0,1,2,\ldots,} S_i$$

Once the estimator of the magnitude of the change has been selected, the bootstrap analysis can be performed. A single bootstrap is performed by:

1. Generate a bootstrap sample of n units, denoted $X^0_1, X^0_2, X^0_3, \ldots X^0_n$ by randomly reordering the original n values. This is called sampling without replacement.
2. Based on the bootstrap sample, calculate the bootstrap CUSUM, denoted $S^0_0, S^0_1, S^0_2, \ldots S^0_n$.
3. Calculate the maximum, minimum and difference of the bootstrap CUSUM, denoted S^0_{max}, S^0_{min} and S^0_{diff}.
4. Determine whether the bootstrap difference S^0_{diff} is less than the original difference S_{diff}.

The idea behind bootstrapping is that the bootstrap samples represent random reordering of the data that mimic the behavior of the CUSUM if no change has occurred. By performing a large number of bootstrap samples, it can be estimated that how much S_{diff} would vary if no change took place. It would be compared with the S_{diff} value calculated from the data in its original order to determine if this value is consistent with what has been expected if no change occurred. If bootstrap CUSUM charts tend to stay closer to zero than the CUSUM of the data in its original order, this leads one to suspect that a change must have occurred. A bootstrap analysis consists of performing a large number of bootstraps and counting the number of bootstraps for which S^0_{diff} is less than S_{diff}. Let N be the number of bootstrap samples performed and let X be the number of bootstraps for which $S^0_{diff} < S_{diff}$. Then the confidence level that a change occurred as a percentage is calculated as follows:

$$\text{Confidence Level} = 100 \, \frac{X}{N} \, \text{percentage}$$

This is strong evidence that a change did in fact occur. Ideally, rather than bootstrapping, one would like to determine the distribution of S^0_{diff} based on all possible reordering of the data. However, this is generally not feasible. A better estimate can be obtained by increasing the number of bootstrap samples. Bootstrapping results in a distribution free approach with only a single assumption, that of an independent error structure. Both control charting and change-point analysis are based on the mean-shift model. Let $X_1, X-2, X_3, ...$ represent the data in time order. The mean-shift model can be written as

$$X_i = \mu_i + \epsilon_i$$

where μ_i is the average at time i. Generally $\mu_i = \mu_{i-1}$ except for a small number of values of i called the change-points. ϵ_i is the random error associated with the ith value. It is assumed that the ϵ_i are independent with means of zero. Once a change has been detected, an estimate of when the change occurred can be made. One such estimator is the CUSUM estimator. Let m be such that:

$$| S_m | = \max_{i=0,1,2,...,} | S_i |$$

S_m is the point furthest from zero in the CUSUM chart. The point m estimates last point before the change occurred. The point $m+1$ estimates the first point after the change. Once a change has been detected, the data can be broken into two segments, one each side of the change-point, 1 to m and $m+1$ to 24, estimating the average of each segment, and then analyzing the two estimated averages.

4. Numerical Example

The formulated mathematical model has been used for the numerical verification and the validity of the model has also been checked. That is why real-time data of particulate matter hazards for four different sites of Seoul, South Korea has been utilized for this investigation.

4.1. Particulate Matter (PM$_{2.5}$) and (PM$_{10}$) Change Points for Four Different Sites

Two dissimilar cases need to be considered

4.1.1. Case 1—When There Is No Hazard

In this case, there is no hazard and concentrations of particulate matter does not exceed the threshold value of the standards. Therefore, there will be no polluted day and random variable Y would always be $y = 0$. Hence, due to zero hazard in the concentrations of particulate matter, this model has not been applied.

4.1.2. Case 2—When There Are Hazards

In this case, several polluted days for particulate matter ($PM_{2.5}$ and PM_{10}) concentrations are considered as a Poisson process. A counting process is a Poisson counting process with the rate $\theta > 0$. Here, we report the results obtained by applying the method described in Section 3 to the particulate matter ($PM_{2.5}$ and PM_{10}) concentrations for four different sites (Guro, Nowon, Songpa, and Yongsan) in Seoul, South Korea. We used the daily data observed from January 2004 to December 2013 to compute the change point of both pollutants.

$$f(y;\theta) = (Pr(Y = y|\theta)) = Poisson(y,\theta) = \frac{e^{-\theta}\theta^y}{y!} \; for \; y \in 1,2,....,n$$

Poisson distribution is the number of events occurring in a given time period. So in this case, occurrence of the number of polluted days in a month is taken as Poisson distribution. The rate of polluted days for both $PM_{2.5}$ and PM_{10} are given in Tables 7 and 8 respectively.

Table 7. $PM_{2.5}$ Poisson Process.

		$PM_{2.5}$ **Poisson Process (Rate)** θ		
Area	**Distribution**	**European Standards**	**American Standards**	**Korean Standards**
Guro	Poisson	13.720	6.488	2.524
Nowon	Poisson	12.934	6.774	2.406
Songpa	Poisson	12.892	6.625	2.342
Yongsan	Poisson	12.808	6.217	2.333

Table 8. PM_{10} Poisson Process.

		PM_{10} **Poisson Process (Rate)** θ		
Area	**Distribution**	**European Standards**	**American Standards**	**Korean Standards**
Guro	Poisson	13.167	0.500	2.325
Nowon	Poisson	12.033	0.342	1.992
Songpa	Poisson	12.417	0.417	2.000
Yongsan	Poisson	13.050	0.500	2.350

The change point for this Poisson process has to be detected to know whether a change has occurred, the most likely month in which change has occurred, and if the rate of polluted days has increased or decreased after the change point. It has been assumed that the number of polluted days for (particulate matter) $PM_{2.5}$ and PM_{10} concentrations follows a Poisson distribution with a mean rate θ until the month k. After the month k, the polluted days are distributed according to the Poisson distribution with a mean rate λ. It can be represented as:

$$y_i \sim Poisson(\theta) \; for \; i = 1, 2, \dots, k$$

$$y_i \sim Poisson(\lambda) \; for \; i = k+1, k+2, \dots, n$$

Hence,

$$f(y; \theta) = Pr(Y = y_i | \theta) \; for \; i = 1, 2, \dots, k$$

$$f(y; \lambda) = Pr(Y = y_i | \lambda) \; for \; i = k+1, k+2, \dots, n$$

If we model $Y = y_i$ as Poisson with mean rate θ then joint pdf of our sample data will be as below:

$$Pr(Y = y_i | \theta) = \prod_{i=1}^{n} pr(y_i | \theta) = \prod_{i=1}^{n} \frac{e^{-\theta} \theta^{y_i}}{y!} = c(y_1, y_2, \dots y_n) e^{-n\theta} \theta^{\sum y_i} \; i \in 0, 1, 2, \dots, n$$

This means that whatever our conjugate class of densities is, it will have to include terms like $e^{-C_2 \theta} \theta^{C_1}$ for constants C_1 and C_2. The simplest class of such densities, which include these terms and corresponding probability distributions, are known as family of Gamma distributions. Therefore, prior distribution $Pr(\theta)$ and posterior distribution $Pr(Y = \theta | y_1, y_2, \dots y_n)$ will follow a Gamma distribution, but likelihood or sampling model $Pr(y_1, y_2, \dots y_n | \theta)$ follow a Poisson distribution.

Therefore, the prior distributions of θ and λ, uncertain positive quantities θ and λ has $Gamma(a_1, b_1)$ and $Gamma(a_2, b_2)$ distributions respectively, where a_1 is shape parameter and b_1 is rate parameter for θ while a_2 is shape parameter and b_2 is rate parameter for λ

$$Pr(\theta) = Gamma(\theta, a_1, b_1) = \frac{b_1^{a_1} e^{-b_1 \theta} \theta^{a_1 - 1}}{\Gamma(a_1)}$$

$$Pr(\lambda) = Gamma(\lambda, a_2, b_2) = \frac{b_2^{a_2} e^{-b_2 \lambda} \lambda^{a_2 - 1}}{\Gamma(a_2)}$$

Gamma distribution is also conjugate prior of the rate (inverse scale) parameter of the Gamma distribution itself. That is why the rate parameter b_1 and b_2 will also follow a Gamma distribution with different shape and rate parameters as given below:

$$b_1 \sim Gamma(c_1, d_1) \text{ where } c_1 = \text{shape parameter } d_1 = \text{rate parameter}$$

$$b_2 \sim Gamma(c_2, d_2) \text{ where } c_2 = \text{shape parameter } d_2 = \text{rate parameter}$$

By applying Bayes theorem, posterior distributions for rate parameters θ, λ, b_1 and b_2 will be determined in the following way. Likelihood and prior distributions of θ

$$Pr(y_1, y_2, y_3, \ldots, y_n | \theta) \sim Poisson(\theta)$$

$$Pr(\theta) = Gamma(\theta, a_1, b_1)$$

$$Pr(\theta | y_1, y_2, y_3, \ldots, y_n) = \frac{Pr(y_1, y_2, y_3, \ldots, y_n | \theta) Pr(\theta)}{Pr(y_1, y_2, y_3, \ldots, y_n)} = (e^{-b_1\theta}\theta^{a_1-1}) \times (e^{-n\theta}\theta^{\Sigma y_i}) \times c(y_1, y_2, y_3, \ldots, y_n)$$

$$= (e^{-(b_1+n)\theta}\theta^{a_1+\Sigma y_i-1}) \times c(y_1, y_2, y_3, \ldots, y_n, a_1, b_1)$$

$$(\theta | y_1, y_2, y_3, \ldots, y_n) \sim Gamma\left(a_1 + \sum_{i=1}^{n} y_i, b_1 + n\right)$$

This is evidently a Gamma distribution. Hence, the conjugacy of Gamma family for the Poisson sampling model or likelihood is confirmed. Hence, it is concluded from the above that if:

$$\theta \sim Gamma(a_1, b_1)$$

$$Pr(y_1, y_2, y_3, \ldots, y_n | \theta) \sim Poisson(\theta)$$

Then:

$$(\theta | y_1, y_2, y_3, \ldots, y_n) \sim Gamma\left(a_1 + \sum_{i=1}^{n} y_i, b_1 + n\right)$$

Similarly, the posterior distributions of all parameters θ, λ, b_1 and b_2 can be determined as given below:

$$(\theta | y, \lambda, b_1, b_2, k) \sim Gamma\left(a_1 + \sum_{i=1}^{k} y_i, k + b_1\right)$$

$$(\lambda | y, \theta, b_1, b_2, k) \sim Gamma\left(a_2 + \sum_{i=k+1}^{n} y_i, k + b_2\right)$$

$$(b_1 | y, \theta, \lambda, b_2, k) \sim Gamma(a_1 + c_1, \theta + d_1)$$

$$(b_2 | y, \theta, \lambda, b_1, k) \sim Gamma(a_2 + c_2, \lambda + d_2)$$

As Gamma is a two-parameter family of continuous probability distribution. As a result, the function:

$$L(\theta | Y) = f_\theta(Y) = f(Y | \theta)$$

The likelihood ratio test statistic is:

$$f(k | Y, \theta, \lambda) = \frac{L(Y; k, \theta, \lambda)}{\sum_{j=1}^{n} L(Y; j, \theta, \lambda)}$$

The likelihood is determined as given by:

$$L(Y; k, \theta, \lambda) = exp(k(\lambda - \theta)) \, (\theta/\lambda)^{\sum_{i=1}^{k} y_i}$$

For Bayesian approach, MATLAB has been used for change point detection of particulate matter ($PM_{2.5}$ and PM_{10}) data during the study period (2004–2013) for four different sites (Guro, Nowon, Songpa and Yongsan) in Seoul, South Korea. 10 replications of each simulation are made with 1100 observations in each replication. First 100 observations are discarded as a burn-in period. Replication Mean V_i of remaining 1000 observations has been taken for each replication as shown in Tables 9 and 10. Then mean (V) of replication mean has been taken to get the converged values of parameters.

Moreover, the CUSUM charts of polluted days as per European, American and Korean standards are shown in Figures 2–9 for four different sites Guro, Nowon, Songpa and Yongsan in Seoul, South Korea.

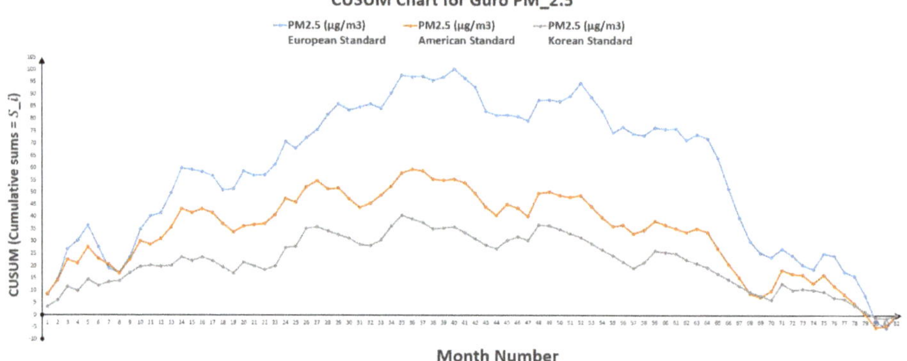

Figure 2. CUSUM chart for Guro $PM_{2.5}$.

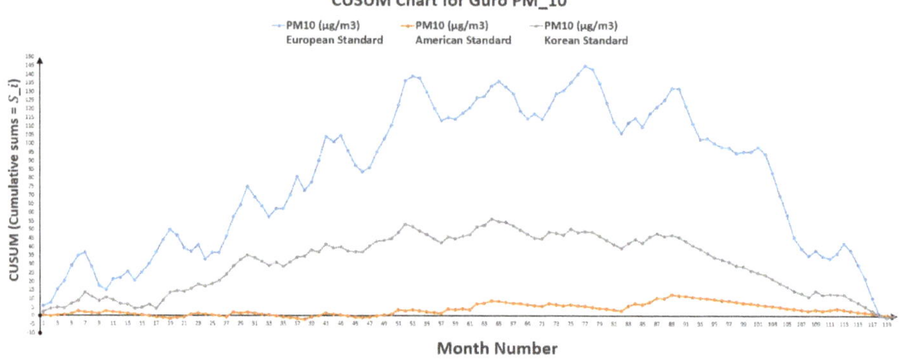

Figure 3. CUSUM chart for Guro PM_{10}.

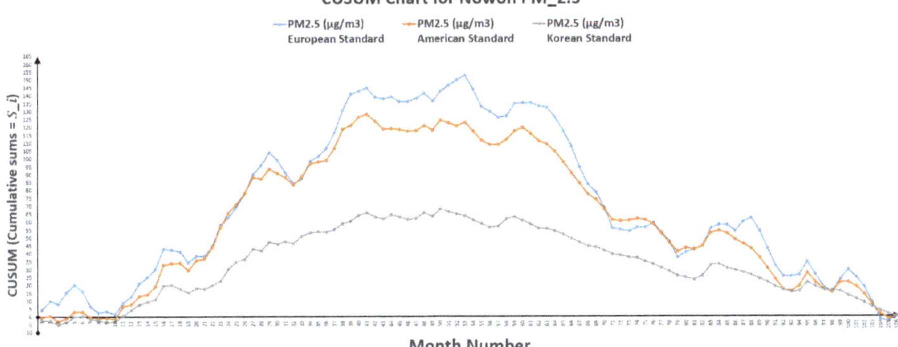

Figure 4. CUSUM chart for Nowon $PM_{2.5}$.

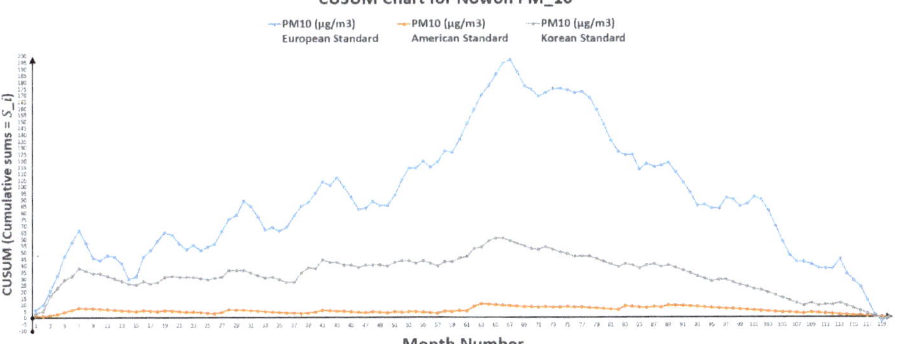

Figure 5. CUSUM chart for Nowon PM_{10}.

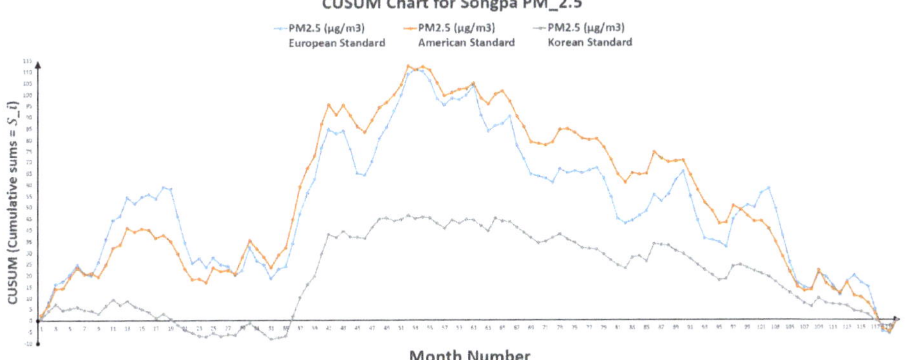

Figure 6. CUSUM chart for Songpa $PM_{2.5}$.

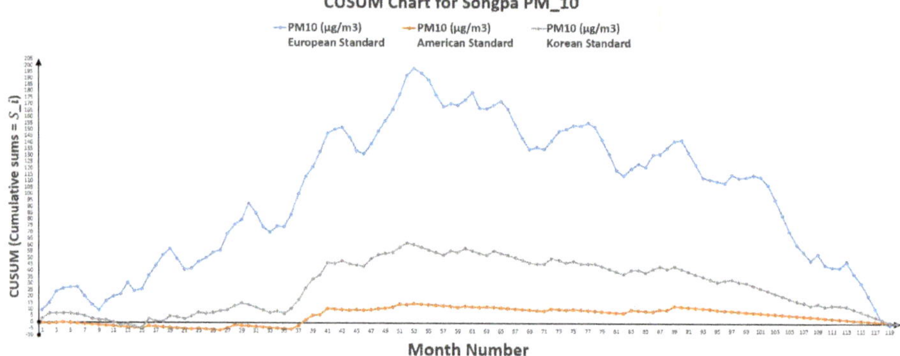

Figure 7. CUSUM chart for Songpa PM_{10}.

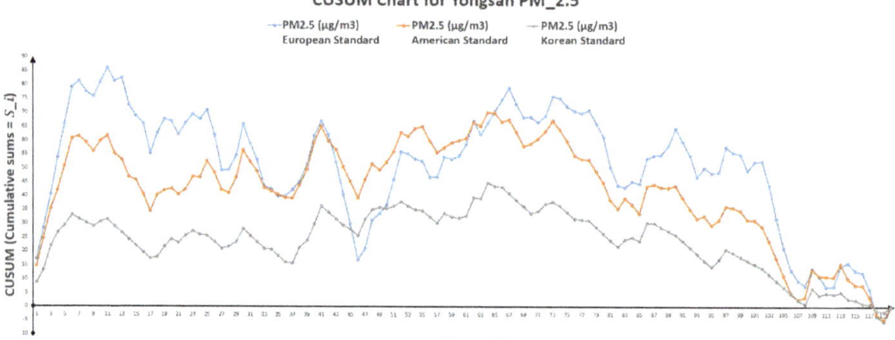

Figure 8. CUSUM chart for Yongsan $PM_{2.5}$.

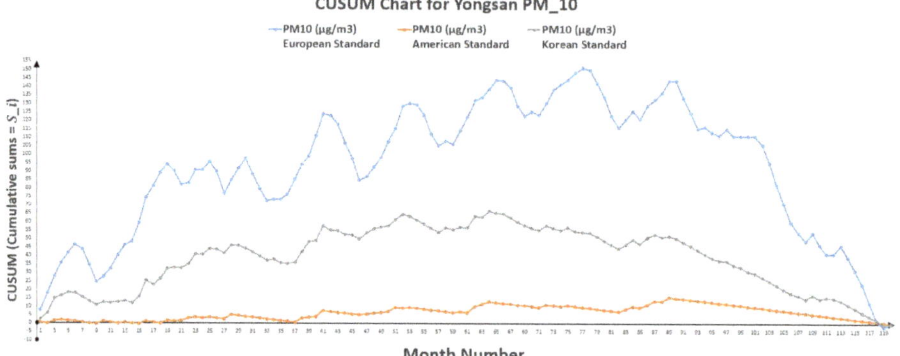

Figure 9. CUSUM chart for Yongsan PM_{10}.

Table 9. $PM_{2.5}$ Converged values of parameters (Bayesian approach).

	Guro			Nowon			Songpa			Yongsan		
Replication Mean	(θ) Replication Mean	(λ) Replication Mean	K Replication Mean	(θ) Replication Mean	(λ) Replication Mean	K Replication Mean	(θ) Replication Mean	(λ) Replication Mean	K Replication Mean	(θ) Replication Mean	(λ) Replication Mean	K Replication Mean
European standards												
V_1	15.96	11.17	43.29	15.88	10.30	49.98	14.93	11.24	53.69	24.80	12.13	6.92
V_2	16.03	11.25	42.53	15.89	10.34	49.83	14.90	11.28	53.81	24.46	12.15	8.07
V_3	16.00	11.18	43.12	15.90	10.32	49.76	14.91	11.25	53.92	24.81	12.07	7.49
V_4	16.01	11.17	43.19	15.90	10.30	49.96	14.91	11.24	54.05	24.77	12.14	7.75
V_5	15.98	11.12	43.52	15.88	10.26	50.13	14.93	11.23	53.59	24.56	12.09	7.48
V_6	15.96	11.14	43.61	15.87	10.28	50.12	14.89	11.24	53.98	24.07	12.08	9.48
V_7	15.95	11.17	43.41	15.84	10.30	50.12	14.88	11.26	53.91	24.61	12.03	8.20
V_8	16.04	11.21	42.76	15.94	10.34	49.65	14.93	11.26	53.73	24.69	12.13	7.87
V_9	16.01	11.19	42.75	15.88	10.30	49.78	14.90	11.24	53.71	24.50	12.10	7.83
V_{10}	15.96	11.15	43.59	15.85	10.30	50.23	14.88	11.25	54.07	24.23	12.12	8.30
(V) Mean of 10 replications	15.99	11.17	43.18	15.88	10.30	49.95	14.90	11.25	53.85	24.55	12.10	7.94
American standards												
V_1	8.35	5.32	33.46	9.82	4.77	42.11	8.73	4.95	53.32	16.12	5.69	6.17
V_2	8.24	5.33	33.67	9.84	4.80	41.76	8.72	4.96	53.31	16.04	5.71	6.17
V_3	8.22	5.30	34.31	9.82	4.78	42.12	8.72	4.95	53.35	16.00	5.69	6.21
V_4	8.29	5.30	34.10	9.84	4.79	41.75	8.72	4.95	53.32	16.08	5.68	6.18
V_5	8.31	5.30	33.49	9.86	4.77	41.67	8.72	4.93	53.32	16.05	5.67	6.17
V_6	8.27	5.31	33.70	9.80	4.77	42.11	8.71	4.94	53.39	15.96	5.68	6.20
V_7	8.24	5.30	34.01	9.88	4.79	41.66	8.70	4.95	53.33	15.92	5.69	6.19
V_8	8.25	5.32	33.70	9.86	4.80	41.61	8.74	4.96	53.25	16.16	5.69	6.16
V_9	8.47	5.33	32.46	9.83	4.78	41.66	8.72	4.94	53.21	15.99	5.68	6.16
V_{10}	8.38	5.33	33.28	9.82	4.79	41.89	8.71	4.95	53.42	15.96	5.69	6.21
(V) Mean of 10 replications	8.30	5.31	33.62	9.84	4.78	41.83	8.72	4.95	53.32	16.03	5.69	6.18

Table 9. *Cont.*

Replication Mean	Guro			Nowon			Songpa			Yongsan		
	(θ) Replication Mean	(λ) Replication Mean	K Replication Mean	(θ) Replication Mean	(λ) Replication Mean	K Replication Mean	(θ) Replication Mean	(λ) Replication Mean	K Replication Mean	(θ) Replication Mean	(λ) Replication Mean	K Replication Mean
						Korean standards						
V_1	3.58	1.63	38.14	3.75	1.21	50.29	3.13	1.64	56.57	7.95	2.05	5.80
V_2	3.57	1.65	37.90	3.74	1.22	50.25	3.15	1.65	56.26	7.95	2.06	5.76
V_3	3.58	1.64	37.97	3.75	1.21	50.23	3.15	1.64	56.34	7.92	2.05	5.80
V_4	3.57	1.62	38.40	3.74	1.20	50.43	3.15	1.63	56.54	8.01	2.06	5.77
V_5	3.56	1.61	38.48	3.74	1.20	50.32	3.14	1.63	56.50	7.94	2.05	5.78
V_6	3.55	1.62	38.63	3.73	1.20	50.52	3.13	1.63	56.87	7.97	2.04	5.77
V_7	3.54	1.62	38.60	3.75	1.20	50.30	3.13	1.64	56.53	7.87	2.05	5.80
V_8	3.59	1.64	37.93	3.75	1.21	50.27	3.16	1.64	56.08	7.98	2.06	5.78
V_9	3.57	1.63	37.84	3.73	1.20	50.43	3.15	1.64	55.86	7.90	2.05	5.79
V_{10}	3.55	1.62	38.69	3.73	1.21	50.36	3.13	1.63	56.86	7.90	2.05	5.79
(V) Mean of 10 replications	3.56	1.63	38.26	3.74	1.21	50.34	3.14	1.64	56.44	7.94	2.05	5.78

Table 10. PM_{10} Converged values of parameters (Bayesian approach).

	Guro			Nowon			Songpa			Yongsan		
							PM_{10} Converged Values of Parameters (Bayesian Approach)					
Replication Mean	(θ) Replication Mean	(λ) Replication Mean	K Replication Mean	(θ) Replication Mean	(λ) Replication Mean	K Replication Mean	(θ) Replication Mean	(λ) Replication Mean	K Replication Mean	(θ) Replication Mean	(λ) Replication Mean	K Replication Mean
European standards												
V_1	14.71	9.20	85.53	14.95	8.36	67.04	16.11	9.49	53.29	14.57	8.41	90.09
V_2	14.75	9.27	85.10	14.93	8.39	67.10	16.08	9.52	53.30	14.57	8.50	89.53
V_3	14.74	9.20	85.49	14.93	8.36	67.16	16.11	9.45	53.29	14.58	8.47	89.63
V_4	14.75	9.19	85.47	14.93	8.35	67.17	16.08	9.47	53.52	14.57	8.39	90.16
V_5	14.76	9.12	85.65	14.92	8.33	67.13	16.06	9.44	53.75	14.56	8.35	89.85
V_6	14.72	9.18	85.59	14.91	8.35	67.20	16.06	9.47	53.43	14.54	8.42	89.98
V_7	14.71	9.18	85.75	14.90	8.36	67.13	16.14	9.48	53.36	14.52	8.41	90.12
V_8	14.77	9.25	85.07	14.96	8.37	67.05	16.08	9.39	53.93	14.61	8.51	89.23
V_9	14.73	9.18	85.29	14.92	8.35	67.03	16.07	9.47	53.26	14.60	8.48	88.67
V_{10}	14.72	9.19	85.67	14.91	8.37	67.16	16.03	9.46	53.91	14.53	8.39	90.18
(V) Mean of 10 replications	14.73	9.20	85.46	14.93	8.36	67.12	16.08	9.46	53.50	14.56	8.43	89.74
American standards												
V_1	0.63	0.12	90.65	0.69	0.13	66.56	0.56	0.07	87.63	0.66	0.04	91.61
V_2	0.63	0.12	90.76	0.63	0.11	72.52	0.55	0.07	89.08	0.66	0.04	91.33
V_3	0.63	0.12	90.67	0.52	0.08	81.79	0.56	0.07	87.86	0.66	0.04	91.46
V_4	0.63	0.12	90.34	0.53	0.09	81.32	0.57	0.07	86.80	0.66	0.04	91.41
V_5	0.63	0.12	90.23	0.76	0.14	60.70	0.56	0.06	88.35	0.67	0.04	91.03
V_6	0.63	0.12	90.48	0.52	0.08	81.91	0.56	0.07	87.98	0.66	0.04	91.50
V_7	0.63	0.12	90.53	0.45	0.06	89.85	0.56	0.06	88.72	0.66	0.05	91.18
V_8	0.63	0.12	90.63	0.82	0.15	56.42	0.56	0.09	88.73	0.66	0.04	91.11
V_9	0.63	0.13	89.33	0.64	0.11	71.37	0.56	0.07	88.19	0.66	0.04	91.69
V_{10}	0.64	0.12	89.79	0.59	0.11	75.61	0.56	0.07	88.26	0.66	0.04	91.29
(V) Mean of 10 replications	0.63	0.12	90.34	0.62	0.11	73.80	0.56	0.07	88.16	0.66	0.04	91.36

Table 10. *Cont.*

	PM_{10} Converged Values of Parameters (Bayesian Approach)											
	Guro			Nowon			Songpa			Yongsan		
Replication Mean	(θ) Replication Mean	(λ) Replication Mean	K Replication Mean	(θ) Replication Mean	(λ) Replication Mean	K Replication Mean	(θ) Replication Mean	(λ) Replication Mean	K Replication Mean	(θ) Replication Mean	(λ) Replication Mean	K Replication Mean
Korean standards												
V_1	2.89	0.91	85.25	2.92	0.89	65.68	3.16	1.09	52.78	3.32	1.11	67.44
V_2	2.90	0.92	85.17	2.91	0.89	65.67	3.17	1.10	52.65	3.33	1.13	66.93
V_3	2.89	0.90	85.96	2.92	0.89	65.69	3.17	1.09	52.79	3.32	1.12	67.52
V_4	2.88	0.88	86.67	2.91	0.88	65.69	3.17	1.09	52.83	3.30	1.08	68.72
V_5	2.90	0.91	84.97	2.91	0.88	65.67	3.16	1.08	52.79	3.30	1.09	68.37
V_6	2.88	0.89	86.13	2.91	0.88	65.72	3.16	1.09	52.75	3.27	1.06	69.89
V_7	2.87	0.87	87.12	2.90	0.89	65.70	3.16	1.09	52.64	3.28	1.08	68.93
V_8	2.89	0.89	86.14	2.92	0.88	65.67	3.18	1.09	52.59	3.33	1.12	67.09
V_9	2.88	0.89	86.27	2.90	0.88	65.63	3.16	1.09	52.56	3.31	1.10	67.66
V_{10}	2.88	0.90	86.17	2.90	0.89	65.74	3.15	1.09	53.15	3.30	1.10	68.08
(V) Mean of 10 replications	2.89	0.90	85.98	2.91	0.89	65.68	3.16	1.09	52.75	3.31	1.10	68.06

However, the bootstraps analysis of European standards has been shown in Figures 10–17 as given below.

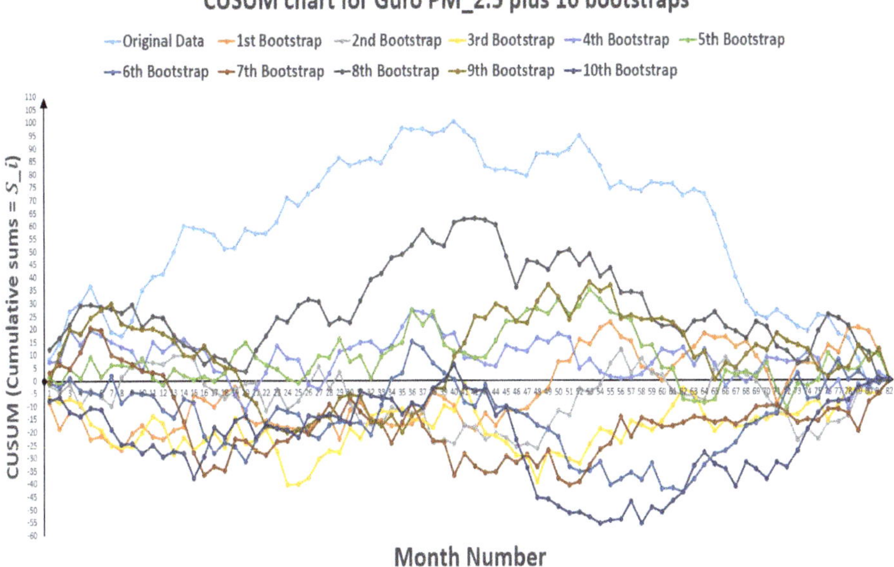

Figure 10. CUSUM chart for Guro $PM_{2.5}$ plus 10 bootstraps (European Standards).

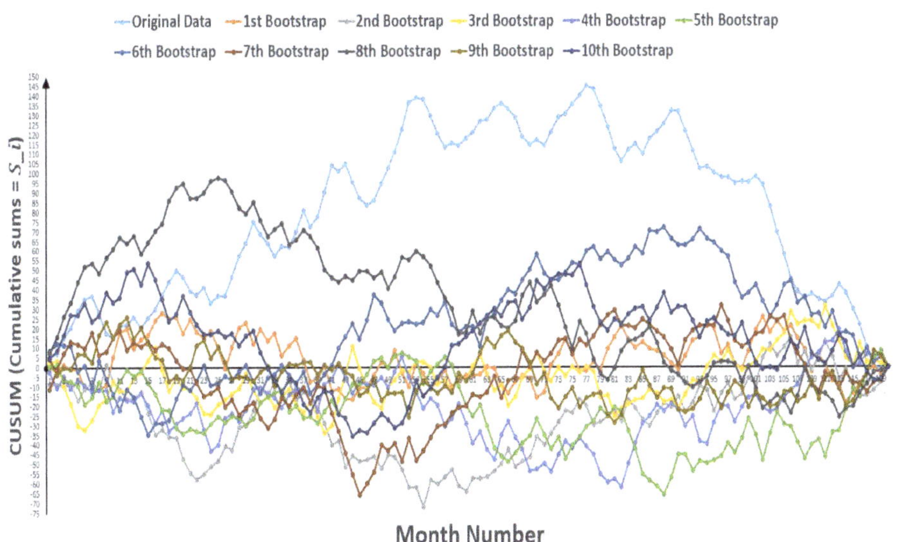

Figure 11. CUSUM chart for Guro PM_{10} plus 10 bootstraps (European Standards).

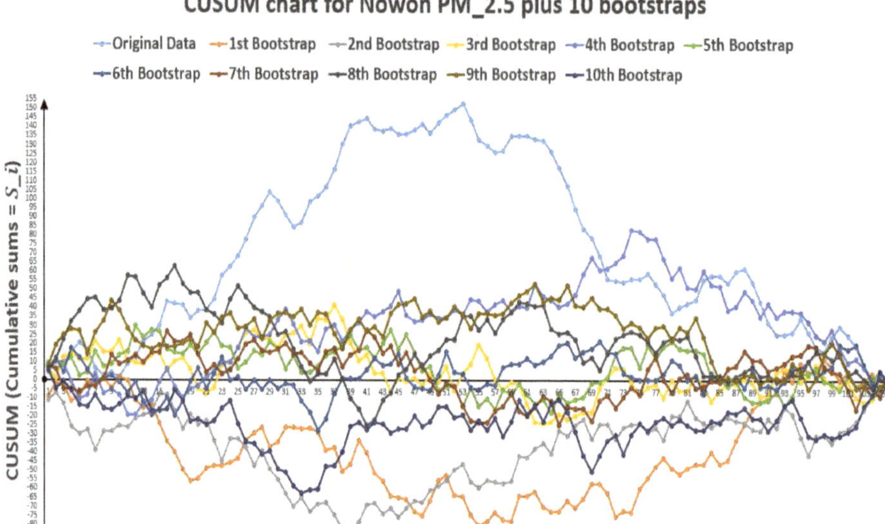

Figure 12. CUSUM chart for Nowon $PM_{2.5}$ plus 10 bootstraps (European Standards).

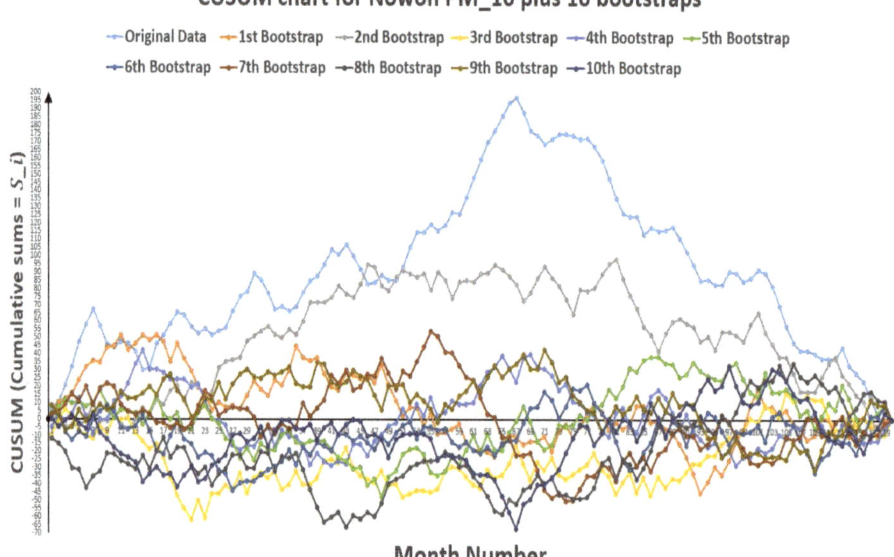

Figure 13. CUSUM chart for Nowon PM_{10} plus 10 bootstraps (European Standards).

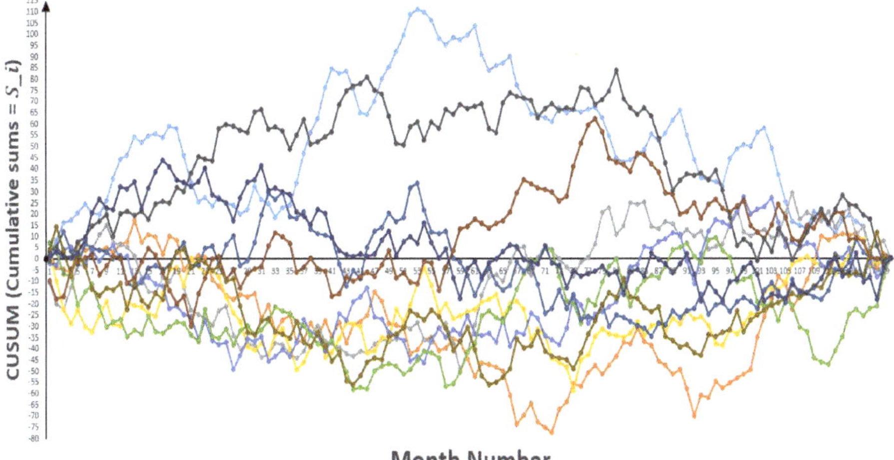

Figure 14. CUSUM chart for Songpa $PM_{2.5}$ plus 10 bootstraps (European Standards).

Figure 15. CUSUM chart for Songpa PM_{10} plus 10 bootstraps (European Standards).

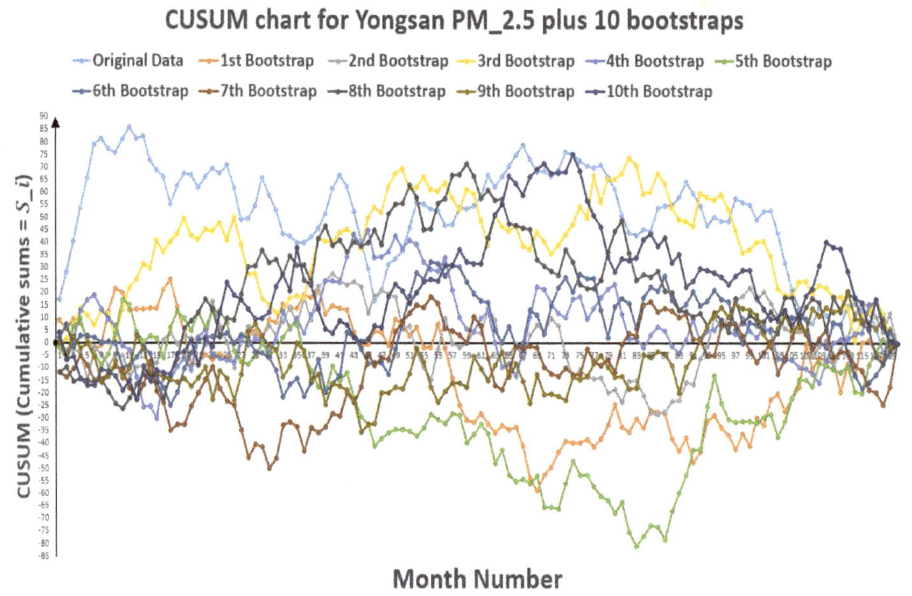

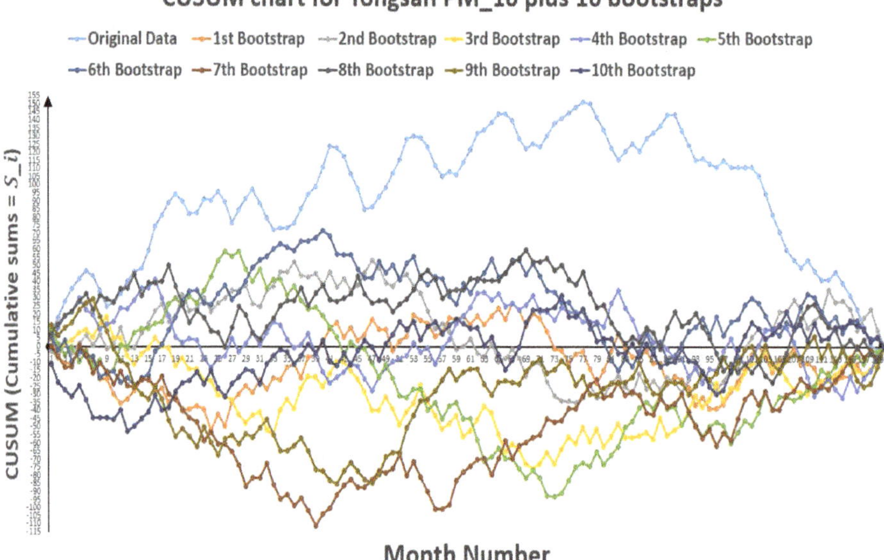

Figure 16. CUSUM chart for Yongsan $PM_{2.5}$ plus 10 bootstraps (European Standards).

Figure 17. CUSUM chart for Yongsan PM_{10} plus 10 bootstraps (European Standards).

The change point k is discrete uniform over $(1, 2, 3...120)$ as there are 120 months in 10 years. Please note that θ, λ and k are all independent of each other.

5. Numerical Results

Two dissimilar approaches have been used to attain the results. First one is Bayesian approach, which is based on probability distributions. It can be applicable to any kind of data distribution. For this, firstly data distributions are defined and then proposed method is applied to acquire the results. This approach is better to apply for random data structures and time series. While the second method is based on CUSUM charts, this technique is directly applicable on the raw data, which is good for deterministic data structures. Summarized forms of particulate matter ($PM_{2.5}$ and PM_{10}) change point (k), the rate before change point (θ) and the rate after change point (λ) during the study period (2004–2013) for four different sites (Guro, Nowon, Songpa and Yongsan) in Seoul, South Korea are given in Tables 11–14. The results have been computed by following the European, American, and Korean Standards as discussed in Table 4.

5.1. $PM_{2.5}$ Change Point (k) through Bayesian Approach

In Table 11, the results obtained through Bayesian approach have been described, where (k) is the predicted change point varies for different areas and different air quality standards. The results indicate the reduction of polluted days after change point (k) for $PM_{2.5}$. While (θ) represents the per month rate of polluted days before change point, (k) and (λ) be the rate of per month polluted days after change point (k).

5.2. $PM_{2.5}$ Last Point before Change (k) and First Point after Change (k + 1) through CUSUM Approach

Table 12 represents the results obtained for $PM_{2.5}$ through CUSUM approach, where (k) is the last point before change and ($k + 1$) be the first point after change point. So, the change point leis somewhere between (k) and ($k + 1$). This method also shows the reduction of polluted days after change point as (θ) represents the per month rate of polluted days before change point and (λ) be the rate of per month polluted days after change point.

5.3. PM_{10} Change Point (k) through Bayesian Approach

Table 13 explains the results obtained for PM_{10} through Bayesian approach. Hence, the expected change point is (k) that differs for different areas and various air quality standards. These results show the reduction of polluted days after change point (k) for PM_{10}. While (θ) is the per month rate of polluted days before change point (k) and (λ) represents the rate of per month polluted days after change point (k).

5.4. PM_{10} Last Point before Change (k) and First Point after Change (k + 1) through CUSUM Approach

The results obtained for PM_{10} through CUSUM approach have been described in Table 14, where the last point before change is (k) and the first point after change point is ($k + 1$). Therefore, the change point leis anywhere between (k) and ($k + 1$). This method also depicts the reduction of polluted days after change point. (θ) represents the per month rate of polluted days before change point and (λ) be the rate of per month polluted days after change point.

Table 11. $PM_{2.5}$ Change Point (k) for European, American and Korean Standards through Bayesian approach.

Area Seoul, South Korea	(θ) Rate before Change Point	(λ) Rate after Change Point	K Change Point	$\sqrt{R}\theta$ Convergence Parameter for (θ)	$Var(V)\theta$ Variance for (θ)	$\sqrt{R}\lambda$ Convergence Parameter for (λ)	$Var(V)\lambda$ Variance for (λ)	$\sqrt{R}K$ Convergence Parameter for K	$Var(V)K$ Variance for K
					Bayesian Approach for Change Point Detection				
				European standards					
Guro	15.99	11.17	43.18	1.000408	0.583	1.000719	0.541	1.000747	59.219
Nowon	15.88	10.30	49.95	0.999943	0.783	1.000094	0.472	1.000036	34.238
Songpa	14.90	11.25	53.85	1.000199	0.302	1.000007	0.180	1.000870	9.719
Yongsan	24.55	12.10	7.94	1.000898	21.961	1.000943	0.477	1.002031	90.053
				American standards					
Guro	8.30	5.31	33.62	1.002540	1.064	1.000105	0.176	1.001256	75.773
Nowon	9.84	4.78	41.83	1.000595	0.296	1.000200	0.089	1.001159	13.007
Songpa	8.72	4.95	53.32	0.999857	0.172	0.999934	0.076	1.000255	2.395
Yongsan	16.03	5.69	6.18	0.999993	5.726	0.999956	0.101	1.000028	0.356
				Korean standards					
Guro	3.56	1.63	38.26	1.000475	0.132	1.000439	0.052	1.000865	41.605
Nowon	3.74	1.21	50.34	0.999959	0.083	1.000377	0.025	0.999978	9.024
Songpa	3.14	1.64	56.44	1.000200	0.068	1.000144	0.031	1.000588	45.887
Yongsan	7.94	2.05	5.78	1.000136	1.498	1.000237	0.018	0.999780	0.339

Table 12. $PM_{2.5}$ Last point before change (k) and First point after change ($k+1$) for European, American and Korean Standards through CUSUM approach.

Area Seoul, South Korea	(θ) Average Rate before Change	(λ) Average Rate after Change	K Last Point before Change	$k+1$ First Point after Change	$\|S_m\|$ Most Extreme Point	S_{max} Highest Point in CUSUM	S_{min} Lowest Point in CUSUM	S_{diff} Magnitude of Change	Confidence Level %
				European standards					
Guro	16.23	11.33	40	41	100.22	100.220	-5.280	105.500	100%
Nowon	15.81	10.06	53	54	152.50	152.500	-3.066	155.566	100%
Songpa	14.98	11.24	53	54	110.74	110.742	-6.108	116.850	100%
Yongsan	20.64	12.02	11	12	86.11	86.108	-5.192	91.300	70%
				American standards					
Guro	8.14	5.20	36	37	59.44	59.439	-5.024	64.463	100%
Nowon	9.90	4.80	41	42	128.28	128.283	-3.321	131.604	100%
Songpa	8.79	4.97	52	53	112.50	112.500	-5.375	117.875	100%
Yongsan	7.31	4.96	64	65	70.13	70.133	-4.783	74.917	100%
				Korean standards					
Guro	3.69	1.66	35	36	40.65	40.646	-1.476	42.122	100%
Nowon	3.76	1.20	50	51	67.72	67.717	-5.217	72.934	100%
Songpa	3.23	1.66	52	53	46.23	46.233	-8.275	54.508	100%
Yongsan	3.03	1.54	64	65	44.67	44.667	-1.667	46.333	100%

Table 13. PM_{10} Change Point (k) for European, American and Korean Standards.

Area Seoul, South Korea	(θ) Rate before Change Point	(λ) Rate after Change Point	K Change Point	$\sqrt{R}\theta$ Convergence Parameter for (θ)	$Var(V)\theta$ Variance for (θ)	$\sqrt{R}\lambda$ Convergence Parameter for (λ)	$Var(V)\lambda$ Variance for (λ)	$\sqrt{R}K$ Convergence Parameter for K	$Var(V)K$ Variance for K
				Bayesian Approach for Change Point Detection					
				European standards					
Guro	14.73	9.20	85.46	1.000072	0.425	1.000427	0.925	0.999931	64.845
Nowon	14.93	8.36	67.12	0.999879	0.464	0.999925	0.327	1.000040	3.184
Songpa	16.08	9.46	53.50	1.000300	0.643	1.001282	0.310	1.003363	9.133
Yongsan	14.56	8.43	89.74	1.000123	0.597	1.000437	1.492	1.001094	75.750
				American standards					
Guro	0.63	0.12	90.34	0.999840	0.009	1.000320	0.005	1.002821	31.264
Nowon	0.62	0.11	73.80	1.041187	0.173	1.036355	0.011	1.053574	1089.508
Songpa	0.56	0.07	88.16	1.000135	0.016	1.000253	0.037	1.000710	174.232
Yongsan	0.66	0.04	91.36	0.999851	0.016	1.000296	0.004	1.000219	32.243
				Korean standards					
Guro	2.89	0.90	85.98	1.000820	0.046	1.001576	0.053	1.003429	59.252
Nowon	2.91	0.89	65.68	0.999881	0.045	1.000137	0.016	1.000338	0.567
Songpa	3.16	1.09	52.75	0.999935	0.066	1.000075	0.017	1.001445	7.194
Yongsan	3.31	1.10	68.06	1.002209	0.077	1.003709	0.050	1.004787	81.344

Table 14. PM_{10} Last point before change (k) and First point after change ($k+1$) for European, American and Korean Standards through CUSUM approach.

Area Seoul, South Korea	(θ) Average Rate before Change	(λ) Average Rate after Change	K Last point before Change	$k+1$ First Point after Change	$\mid S_m \mid$ Most Extreme Point	S_{max} Highest Point in CUSUM	S_{min} Lowest Point in CUSUM	S_{diff} Magnitude of Change	Confidence Level %
CUSUM Approach for Change Point Detection									
European standards									
Guro	15.05	9.79	77	78	145.17	145.167	−0.833	146.000	100%
Nowon	14.97	8.32	67	68	196.77	196.767	−2.967	199.733	100%
Songpa	16.15	9.46	53	54	197.92	197.917	−1.583	199.500	100%
Yongsan	15.01	9.53	77	78	151.15	151.150	−1.950	153.100	100%
American standards									
Guro	0.64	0.10	89	90	12.50	12.500	−2.000	14.500	100%
Nowon	0.51	0.16	63	64	10.48	10.475	0.000	10.475	80%
Songpa	0.70	0.19	53	54	14.92	14.917	−5.833	20.750	100%
Yongsan	0.67	0.00	89	90	15.50	15.500	−0.500	16.000	10%
Korean standards									
Guro	3.20	1.32	64	65	56.20	56.200	−0.675	56.875	100%
Nowon	2.91	0.87	66	67	60.55	60.550	−1.008	61.558	100%
Songpa	3.19	1.09	52	53	62.00	62.000	−4.000	66.000	100%
Yongsan	3.39	1.16	64	65	66.60	66.600	−0.650	67.250	100%

6. Disussion

As the results of two different approaches have been described in the previous section.

1. Bayesian approach is based on probability distributions, which can be applicable on any kind of data distribution. In this case, firstly data distributions are defined and then proposed method is applied to acquire the results. This approach is better to apply for random data structures and time series.
2. CUSUM Approach is directly applied on the raw data, which is good for deterministic data structures.

6.1. Guro (Seoul, South Korea)

Guro is located in the southwestern part of Seoul, and has an important position as a transport link which includes railroads and land routes. The largest digital industrial complex in Korea is also positioned in Guro, centering on research and development activities as well as advanced information and knowledge industries. That is why, the policies of the Ministry of Environment in South Korea have influenced the concentrations of particulate matters ($PM_{2.5}$ and PM_{10}) in Guro and rate of polluted days has reduced in any of the cases.

6.1.1. Bayesian Approach

Bayesian method is better to apply for random time series data. If we look in case of Guro, Table 11 indicates that for $PM_{2.5}$ change-point (k) of polluted days were 43.18, 33.62, and 38.26 according to European, American, and Korean standards respectively. Therefore, a change occurred in the rate of polluted days, but it varied according to standards. At the minimum, the rate of polluted days ($\theta = 15.99$) was reduced 30.14% and the maximum reduction ($\theta = 3.56$) to ($\lambda = 1.63$) was 54.21% in the case of Korean standards. Similarly, Table 13 refers to the reduction of polluted days (θ) to (λ) for PM_{10} after change point (k) which were 85.46, 90.34, and 85.98 according to European, American, and Korean standards, respectively. Moreover, the decrease in the rate of polluted days ($\theta = 14.73$) to ($\lambda = 9.20$) was at least 37.54% for European standards, but it was 80.95% in the case of American standards. Figures 18 and 19 graphically represent the replications of monthly polluted days before and after the change point which are discussed in Tables 9 and 10.

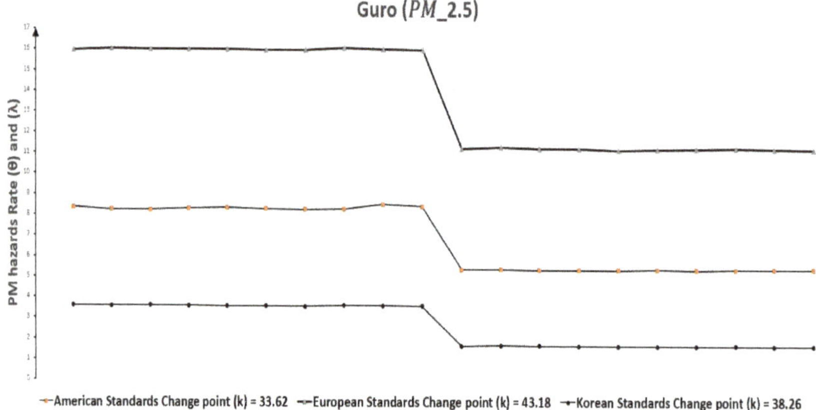

Figure 18. Guro (Seoul, South Korea) monthly polluted days before and after the change point due to $PM_{2.5}$.

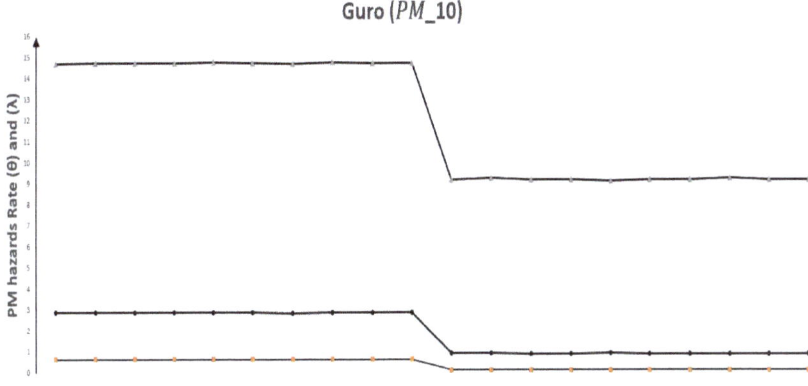

Figure 19. Guro (Seoul, South Korea) monthly polluted days before and after the change point due to PM_{10}.

6.1.2. CUSUM Approach

CUSUM Approach also indicates a reduction in hazards rate from (θ) to (λ) after change. As for Guro, Table 12 also represents the change of $PM_{2.5}$ polluted days through CUSUM approach, which shows that change point occurred between point 40 (k) and 41 ($k + 1$) for European standards, between point 36 (k) and 37 ($k + 1$) for American standards and it lies in-between point 35 (k) and 36 ($k + 1$) for Korean standards. While Table 14 indicates the change of PM_{10} polluted days through CUSUM approach with an indication of change point lies between point 77 (k) and 78 ($k + 1$) for European standards, point 89 (k) and 90 ($k + 1$) for American standards and point 64 (k) and 65 ($k + 1$) for Korean standards.

6.2. Nowon (Seoul, South Korea)

Nowon is located in the northeastern part of the city, and has the highest population density in Seoul with 619,509 persons living in 35.44 km^2, which is surrounded by mountains and forests on the northeast. The policies of the Ministry of Environment in Nowon have improved the rate of polluted days for $PM_{2.5}$ and PM_{10} hazards from θ to λ. Improvement in the reduction of polluted days varies case to case.

6.2.1. Bayesian Approach

Correspondingly, in case of Nowon, Table 11 depicts that change point (k) of polluted days for $PM_{2.5}$ were 49.95, 41.83, and 50.34 according to European, American, and Korean standards respectively. Particularly for this case, the change point was the same according to European and Korean standards, but varied for American standards. The rate of polluted days (θ = 15.88) for European standards showed a minimum decrease of 35.14% after the change point and approached (λ = 10.30), but the maximum decrease was for Korean standards which was 67.65% with (θ = 3.74) and (λ = 1.21). In the same manner, when we study Table 13, it elaborates that for PM_{10}, again there was a reduction in the rate of polluted days after change point (k) of 67.12, 73.80, and 65.68 to European, American, and Korean standards, respectively. That is comparable in cases of European and Korean standards, but a bit different for American standards. In addition, the reduction in the rate of polluted days for the European standard was at least 44.01% from (θ = 14.93) to (λ = 8.36), while the maximum reduction was (θ = 0.62) to (λ = 0.11) 82.25% for American standards. Figures 20 and 21 graphically represent the replications of monthly polluted days before and after the change point which are given in Tables 9 and 10.

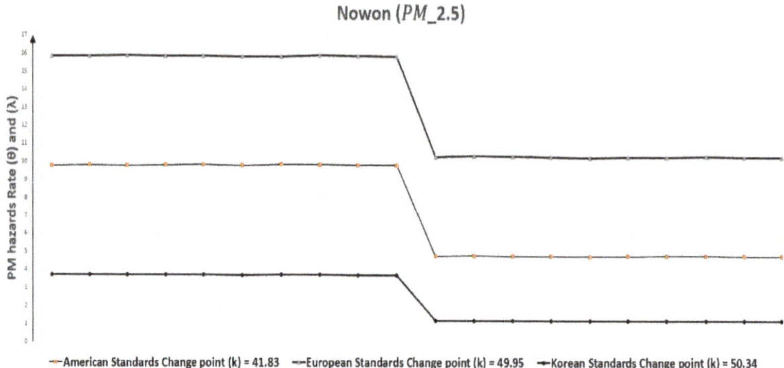

Figure 20. Nowon (Seoul, South Korea) monthly polluted days before and after the change point due to $PM_{2.5}$.

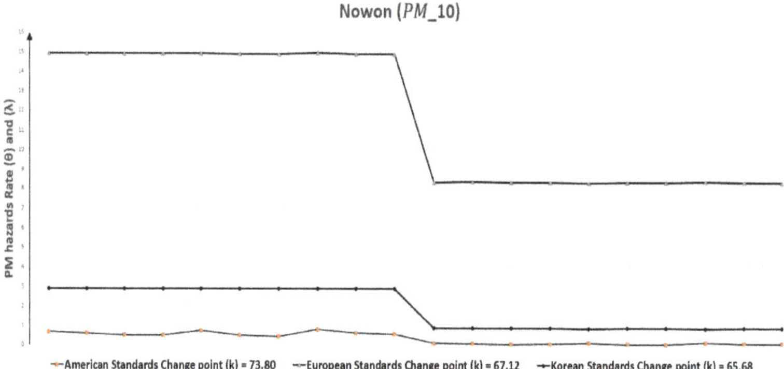

Figure 21. Nowon (Seoul, South Korea) monthly polluted days before and after the change point due to PM_{10}.

6.2.2. CUSUM Approach

Moreover, CUSUM Approach also validates the reduction of PM hazards. In case of Nowon, Table 12 represents the change of $PM_{2.5}$ polluted days through CUSUM approach, which shows that change point occurred between point 53 (k) and 54 ($k + 1$) for European standards, between point 41 (k) and 42 ($k + 1$) for American standards and it lies in-between point 50 (k) and 51 ($k + 1$) for Korean standards. While Table 14 indicates the change of PM_{10} polluted days through CUSUM approach with an indication of change point lies between point 67 (k) and 68 ($k + 1$) for European standards, point 63 (k) and 64 ($k + 1$) for American standards and point 66 (k) and 67 ($k + 1$) for Korean standards.

6.3. Songpa (Seoul, South Korea)

Songpa is located at the southeastern part of Seoul, and has largest population, with 647,000 residents. As per Ministry of Environment policies in Songpa, there is a smaller reduction for the rate of polluted days (θ) to (λ) as compared to Guro and Nowon, but still there is a significant reduction in PM hazards.

6.3.1. Bayesian Approach

Now for Songpa, we can check from Table 11 that change point (k) of polluted days for $PM_{2.5}$ were 53.85, 53.32, and 56.44 for European, American, and Korean standards respectively, which were all similar. The reduction in rate of polluted days was at least 24.50% for European Standards ($\theta = 14.90$)

to (λ = 11.25) while it was highest for Korean standards at 47.77% with (θ = 3.14) and (λ = 1.64). Correspondingly, we can also inspect the improvement in the rate of polluted days from (θ) to (λ) for PM_{10} after change point (k) in Table 13. Change point (k) for the rate of polluted days due to PM_{10} concentration were 53.50, 88.16, and 52.75 according to European, American and Korean standards respectively, which was the same for European and Korean standards. The slightest improvement 41.17% has been in the case of European standards and (θ = 16.08) is converted to (λ = 9.46). On the other hand, if we look at American standards, the rate of polluted days (θ = 0.56) was already low which further decreased 87.5% to (λ = 0.07). Hence, this area is almost a meeting of the PM_{10} concentration requirements for American standards but not for other standards. Figures 22 and 23 graphically represent the replications of monthly polluted days before and after the change point which are given in Tables 9 and 10.

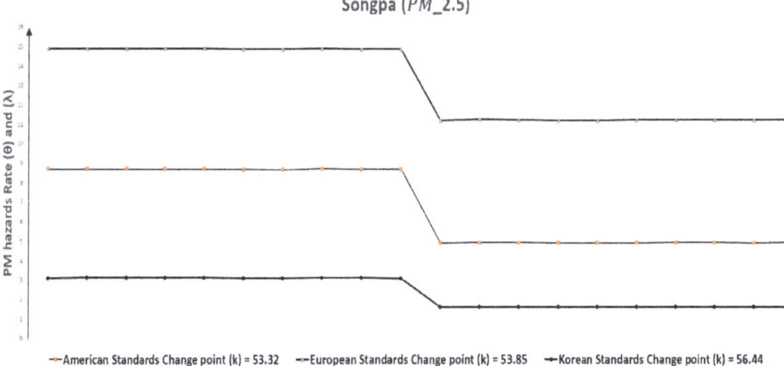

Figure 22. Songpa (Seoul, South Korea) monthly polluted days before and after the change point due to $PM_{2.5}$.

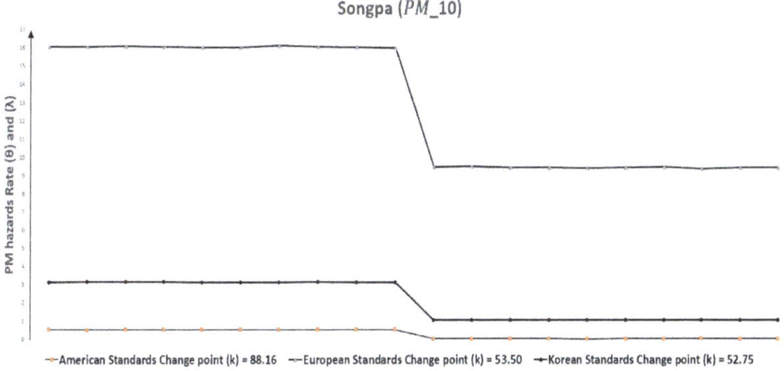

Figure 23. Songpa (Seoul, South Korea) monthly polluted days before and after the change point due to PM_{10}.

6.3.2. CUSUM Approach

As per CUSUM Approach, there is a decrease in PM hazards. Table 12 also represents the change of $PM_{2.5}$ polluted days through CUSUM approach, which shows that change point occurred between point 53 (k) and 54 ($k+1$) for European standards, between point 52 (k) and 53 ($k+1$) for American standards and it lies in-between point 52 (k) and 53 ($k+1$) for Korean standards. While Table 14 indicates the change of PM_{10} polluted days through CUSUM approach with an indication of change point lies between point 53 (k) and 54 ($k+1$) for European standards, point 53 (k) and 54 ($k+1$) for American standards and point 52 (k) and 53 ($k+1$) for Korean standards.

6.4. Yongsan (Seoul, South Korea)

Yongsan is a place in the center of Seoul in which almost 250,000 people reside. Prominent locations in Yongsan includes Yongsan station, electronic market and Itaewon commercial area with heavy traffic and transportation. Consequently, the policies of the Ministry of Environment in Yongsan has affected the particulate matter ($PM_{2.5}$ and PM_{10}) concentrations more than all the previous three locations (Guro, Nowon and Songpa). There is a remarkable decrease in rate of polluted days from (θ) to (λ).

6.4.1. Bayesian Approach

Similarly, in the case of Yongsan, Tables 11 and 13 tell us that the rate of polluted days (θ) for particulate matters ($PM_{2.5}$ and PM_{10}) was the highest in Seoul. The change occurred for $PM_{2.5}$ with the change point (k) 7.94, 6.18, and 5.78 with respect to European, American and Korean standards respectively, which was comparable for all the three standards. There was minimally a 50.71% fall in the rate of polluted days (θ = 24.55) to (λ = 12.10) for European standards, but the reduction in rate of polluted days was a maximum of 74.18% in the case of Korean standards. On the same note, Table 13 indicates that the change in the rate of polluted days has also occurred for PM_{10} concentrations. The change point (k) for it were 89.74, 91.36, and 68.06 for European, American and Korean standards, respectively. Furthermore, at least 42.10% rate of polluted days (θ = 14.56) was reduced to (λ = 8.43) for European standards but its maximum decrease was 93.93% for American standards (θ = 0.66) to (λ = 0.04), although, it is already approaching the requirements of this standard. Figures 24 and 25 graphically represent the replications of monthly polluted days before and after the change point which are given in Tables 9 and 10.

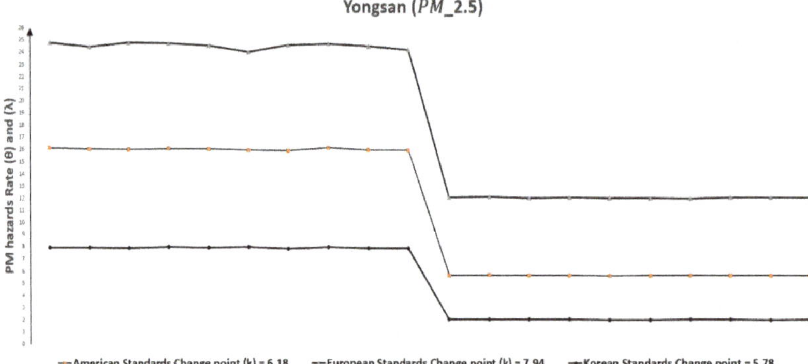

Figure 24. Yongsan (Seoul, South Korea) monthly polluted days before and after the change point due to $PM_{2.5}$.

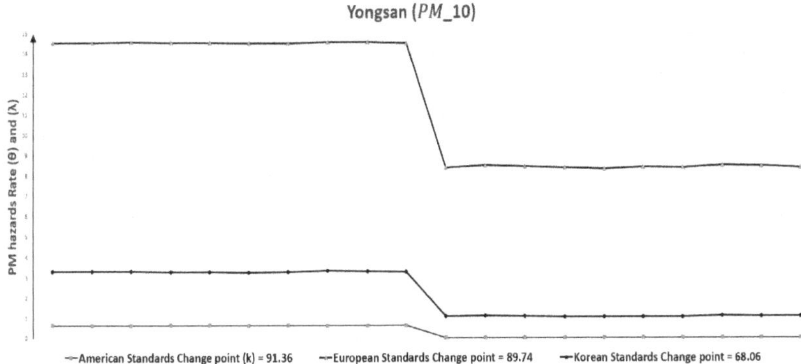

Figure 25. Yongsan (Seoul, South Korea) monthly polluted days before and after the change point due to PM_{10}.

6.4.2. CUSUM Approach

CUSUM Approach is directly applied on the raw data, which should be better for deterministic data structures. It also shows a reduction in PM hazards. In case of Yongsan, Table 12 also represents the change of $PM_{2.5}$ polluted days through CUSUM approach, which shows that change point occurred between point 11 (k) and 12 ($k+1$) for European standards, between point 64 (k) and 65 ($k+1$) for American standards and it lies in-between point 64 (k) and 65 ($k+1$) for Korean standards. While Table 14 indicates the change of PM_{10} polluted days through CUSUM approach with an indication of change point lies between point 77 (k) and 78 ($k+1$) for European standards, point 89 (k) and 90 ($k+1$) for American standards and point 64 (k) and 65 ($k+1$) for Korean standards.

6.5. Strengths

1. This approach is very precise, well defined, user friendly and easily understandable for applications on probability distributions, time series and random data.
2. The above mentioned model is an appropriate approach for detection of change points in random data structures.
3. Good technique for evaluation of process control programs by comparing the parameters before and after change point.

6.6. Limitations

1. Detection of only single change point is given in this model.
2. Further extension is required by making a model for a multiple number of change points for locating changed segments.

6.7. Managerial Insights

1. This model presents a suitable technique to analyze the air quality and pollutant hazards in the air.
2. By detecting change points in particulate matter ($PM_{2.5}$ and PM_{10}) concentrations and analyzing the occurrences of polluted days before and after a change point, environmental protection agencies can understand the role of their legislation efforts, and whether these change points are favorable or not for the environment.
3. A comparison of particulate matter hazards before and after a change point evaluates a pollution control program adopted by environmental protection agencies to make a decision. If these policies need further revision or not for the reduction of death rates and burden of diseases due airborne particulate matter concentrations in the air.

4.   This study of pollutant hazards also defines the current levels of subjected air pollutant in the air which is helpful to make new pollution control policies for further improvements.
5.   This research also brings an intuition to define new goals if previously defined goals have been achieved and also provides a vision if the environmental standards need to be revised, or not to overcome environmental challenges.

7. Conclusions

The main focus of this research work was to elucidate an appropriate change point detection model for occurrences of pollutant hazards due to higher concentrations of particulate matter ($PM_{2.5}$ and PM_{10}) in different locations. The rate of pollutant hazards before and after a change point was also estimated comprehensively to investigate the effectiveness of policies applied by the Ministry of Environment. To verify the model, four major locations (Guro, Nowon, Songpa, and Yongsan) in Seoul, South Korea were selected as study areas due to their different characteristics, such as climate zones, environment, populations and population densities. Three different environmental standards (European, American and Korean) were chosen as threshold values. Then, the model was applied to real time data sets in all cases and conclusions were drawn. The rate before and after the change point of particulate matter concentrations indicated a reduction in polluted days over a 10-year period. The overall results of our study confirm the effective role of legislation efforts used consistently to improve the air quality through the years but pollutant hazards still exist. Hence, further improvements are required to meet set standards to nullify hazards. This study can be further extended by making a multi-parameter change point model for a multiple number of change points considering the fact that different data structures follow different probability distributions.

Author Contributions: Conceptualization, Muhammad Rizwan Khan (M.R.K.) and Biswajit Sarkar (B.S.); methodology, M.R.K.; software, M.R.K.; validation, M.R.K. and B.S.; formal analysis, M.R.K.; investigation, M.R.K.; resources, M.R.K. and B.S.; data curation, M.R.K. and B.S.; writing—original draft preparation, M.R.K.; writing—review and editing, B.S.; visualization, M.R.K. and B.S.; supervision, B.S.; project administration, M.R.K. and B.S.; funding acquisition, B.S.

Funding: This research received no external funding.

Conflicts of Interest: The authors declare no conflict of interest.

References

1.   Héroux, M.E.; Anderson, H.R.; Atkinson, R.; Brunekreef, B.; Cohen, A.; Forastiere, F.; Hurley, F.; Katsouyanni, K.; Krewski, D.; Krzyzanowski, M.; et al. Quantifying the health impacts of ambient air pollutants: recommendations of a WHO/Europe project. *Int. J. Public Health* **2015**, *60*, 619–627. [CrossRef] [PubMed]
2.   Pope, C.A.; Burnett, R.T.; Thurston, G.D.; Thun, M.J.; Calle, E.E.; Krewski, D.; Godleski, J.J. Cardiovascular mortality and long-term exposure to particulate air pollution: Epidemiological evidence of general pathophysiological pathways of disease. *Circulation* **2004**, *109*, 71–77. [CrossRef] [PubMed]
3.   Gyarmati-Szabó, J.; Bogachev, L.V.; Chen, H. Modelling threshold exceedances of air pollution concentrations via non-homogeneous Poisson process with multiple change-points. *Atmos. Environ.* **2011**, *45*, 5493–5503. [CrossRef]
4.   Pirani, M.; Best, N.; Blangiardo, M.; Liverani, S.; Atkinson, R.W.; Fuller, G.W. Analysing the health effects of simultaneous exposure to physical and chemical properties of airborne particles. *Environ. Int.* **2015**, *79*, 56–64. [CrossRef] [PubMed]
5.   Welty, L.J.; Peng, R.; Zeger, S.; Dominici, F. Bayesian distributed lag models: Estimating effects of particulate matter air pollution on daily mortality. *Biometrics* **2009**, *65*, 282–291. [CrossRef]
6.   Qin, S.; Liu, F.; Wang, J.; Sun, B. Analysis and forecasting of the particulate matter (PM) concentration levels over four major cities of China using hybrid models. *Atmos. Environ.* **2014**, *98*, 665–675. [CrossRef]
7.   Keshavarz, H.; Scott, C.; Nguyen, X. Optimal change point detection in Gaussian processes. *J. Stat. Plan. Inference* **2018**, *193*, 151–178. [CrossRef]

8. Kucharczyk, D.; Wyłomańska, A.; Sikora, G. Variance change point detection for fractional Brownian motion based on the likelihood ratio test. *Phys. Stat. Mech. Its Appl.* **2018**, *490*, 439–450. [CrossRef]

9. Lu, G.; Zhou, Y.; Lu, C.; Li, X. A novel framework of change-point detection for machine monitoring. *Mech. Syst. Signal Process.* **2017**, *83*, 533–548. [CrossRef]

10. Taleizadeh, A.A.; Samimi, H.; Sarkar, B.; Mohammadi, B. Stochastic machine breakdown and discrete delivery in an imperfect inventory-production system. *J. Ind. Manag. Optim.* **2017**, *13*, 1511–1535. [CrossRef]

11. Liu, S.; Yamada, M.; Collier, N.; Sugiyama, M. Change-point detection in time-series data by relative density-ratio estimation. *Neural Netw.* **2013**, *43*, 72–83. [CrossRef]

12. Hilgert, N.; Verdier, G.; Vila, J.P. Change detection for uncertain autoregressive dynamic models through nonparametric estimation. *Stat. Methodol.* **2016**, *33*, 96–113. [CrossRef]

13. Górecki, T.; Horváth, L.; Kokoszka, P. Change point detection in heteroscedastic time series. *Econom. Stat.* **2018**, *7*, 63–88. [CrossRef]

14. Bucchia, B.; Wendler, M. Change-point detection and bootstrap for Hilbert space valued random fields. *J. Multivar. Anal.* **2017**, *155*, 344–368. [CrossRef]

15. Shin, D.; Guchhait, R.; Sarkar, B.; Mittal, M. Controllable lead time, service level constraint, and transportation discounts in a continuous review inventory model. *RAIRO-Oper. Res.* **2016**, *50*, 921–934. [CrossRef]

16. Zhou, M.; Wang, H.J.; Tang, Y. Sequential change point detection in linear quantile regression models. *Stat. Probab. Lett.* **2015**, *100*, 98–103. [CrossRef]

17. Lu, K.P.; Chang, S.T. Detecting change-points for shifts in mean and variance using fuzzy classification maximum likelihood change-point algorithms. *J. Comput. Appl. Math.* **2016**, *308*, 447–463. [CrossRef]

18. Sarkar, B.; Saren, S. Partial trade-credit policy of retailer with exponentially deteriorating items. *Int. J. Appl. Comput. Math.* **2015**, *1*, 343–368. [CrossRef]

19. Ruggieri, E.; Antonellis, M. An exact approach to Bayesian sequential change point detection. *Comput. Stat. Data Anal.* **2016**, *97*, 71–86. [CrossRef]

20. Keshavarz, M.; Huang, B. Bayesian and Expectation Maximization methods for multivariate change point detection. *Comput. Chem. Eng.* **2014**, *60*, 339–353. [CrossRef]

21. Kurt, B.; Yıldız, Ç.; Ceritli, T.Y.; Sankur, B.; Cemgil, A.T. A Bayesian change point model for detecting SIP-based DDoS attacks. *Digit. Signal Process.* **2018**, *77*, 48–62. [CrossRef]

22. Wu, C.; Du, B.; Cui, X.; Zhang, L. A post-classification change detection method based on iterative slow feature analysis and Bayesian soft fusion. *Remote. Sens. Environ.* **2017**, *199*, 241–255. [CrossRef]

23. Mariño, I.P.; Blyuss, O.; Ryan, A.; Gentry-Maharaj, A.; Timms, J.F.; Dawnay, A.; Kalsi, J.; Jacobs, I.; Menon, U.; Zaikin, A. Change-point of multiple biomarkers in women with ovarian cancer. *Biomed. Signal Process. Control.* **2017**, *33*, 169–177. [CrossRef]

24. Jeon, J.J.; Sung, J.H.; Chung, E.S. Abrupt change point detection of annual maximum precipitation using fused lasso. *J. Hydrol.* **2016**, *538*, 831–841. [CrossRef]

25. Chen, S.; Li, Y.; Kim, J.; Kim, S.W. Bayesian change point analysis for extreme daily precipitation. *Int. J. Climatol.* **2017**, *37*, 3123–3137. [CrossRef]

26. Gupta, A.; Baker, J.W. Estimating spatially varying event rates with a change point using Bayesian statistics: Application to induced seismicity. *Struct. Saf.* **2017**, *65*, 1–11. [CrossRef]

27. Bardwell, L.; Fearnhead, P. Bayesian detection of abnormal segments in multiple time series. *Bayesian Anal.* **2017**, *12*, 193–218. [CrossRef]

28. Lim, S.S.; Vos, T.; Flaxman, A.D.; Danaei, G.; Shibuya, K.; Adair-Rohani, H.; AlMazroa, M.A.; Amann, M.; Anderson, H.R.; Andrews, K.G.; et al. A comparative risk assessment of burden of disease and injury attributable to 67 risk factors and risk factor clusters in 21 regions, 1990–2010: A systematic analysis for the Global Burden of Disease Study 2010. *Lancet* **2012**, *380*, 2224–2260. [CrossRef]

29. Rao, X.; Zhong, J.; Brook, R.D.; Rajagopalan, S. Effect of particulate matter air pollution on cardiovascular oxidative stress pathways. *Antioxid. Redox Signal.* **2018**, *28*, 797–818. [CrossRef]

30. Wei, T.; Meng, T. Biological Effects of Airborne Fine Particulate Matter (PM2.5) Exposure on Pulmonary Immune System. *Environ. Toxicol. Pharmacol.* **2018**, *60*, 195–201. [CrossRef]

31. Wang, Y.; Xiong, L.; Tang, M. Toxicity of inhaled particulate matter on the central nervous system: neuroinflammation, neuropsychological effects and neurodegenerative disease. *J. Appl. Toxicol.* **2017**, *37*, 644–667. [CrossRef]

32. Wellenius, G.A.; Burger, M.R.; Coull, B.A.; Schwartz, J.; Suh, H.H.; Koutrakis, P.; Schlaug, G.; Gold, D.R.; Mittleman, M.A. Ambient air pollution and the risk of acute ischemic stroke. *Arch. Intern. Med.* **2012**, *172*, 229–234. [CrossRef]
33. Li, P.; Xin, J.; Wang, Y.; Wang, S.; Shang, K.; Liu, Z.; Li, G.; Pan, X.; Wei, L.; Wang, M. Time-series analysis of mortality effects from airborne particulate matter size fractions in Beijing. *Atmos. Environ.* **2013**, *81*, 253–262. [CrossRef]
34. Kim, S.E.; Bell, M.L.; Hashizume, M.; Honda, Y.; Kan, H.; Kim, H. Associations between mortality and prolonged exposure to elevated particulate matter concentrations in East Asia. *Environ. Int.* **2018**, *110*, 88–94. [CrossRef]
35. Kim, S.E.; Honda, Y.; Hashizume, M.; Kan, H.; Lim, Y.H.; Lee, H.; Kim, C.T.; Yi, S.M.; Kim, H. Seasonal analysis of the short-term effects of air pollution on daily mortality in Northeast Asia. *Sci. Total Environ.* **2017**, *576*, 850–857. [CrossRef]
36. Qin, R.X.; Xiao, C.; Zhu, Y.; Li, J.; Yang, J.; Gu, S.; Xia, J.; Su, B.; Liu, Q.; Woodward, A. The interactive effects between high temperature and air pollution on mortality: A time-series analysis in Hefei, China. *Sci. Total Environ.* **2017**, *575*, 1530–1537. [CrossRef]
37. Lee, H.; Honda, Y.; Hashizume, M.; Guo, Y.L.; Wu, C.F.; Kan, H.; Jung, K.; Lim, Y.H.; Yi, S.; Kim, H. Short-term exposure to fine and coarse particles and mortality: A multicity time-series study in East Asia. *Environ. Pollut.* **2015**, *207*, 43–51. [CrossRef]
38. Cabrieto, J.; Tuerlinckx, F.; Kuppens, P.; Wilhelm, F.H.; Liedlgruber, M.; Ceulemans, E. Capturing correlation changes by applying kernel change point detection on the running correlations. *Inf. Sci.* **2018**, *447*, 117–139. [CrossRef]

 © 2019 by the authors. Licensee MDPI, Basel, Switzerland. This article is an open access article distributed under the terms and conditions of the Creative Commons Attribution (CC BY) license (http://creativecommons.org/licenses/by/4.0/).

Article

Stochastic-Petri Net Modeling and Optimization for Outdoor Patients in Building Sustainable Healthcare System Considering Staff Absenteeism

Chang Wook Kang [1], Muhammad Imran [2], Muhammad Omair [3], Waqas Ahmed [2], Misbah Ullah [4] and Biswajit Sarkar [1,*]

[1] Department of Industrial & Management Engineering, Hanyang University, Ansan, Gyeonggido 155 88, Korea; cwkang57@hanyang.ac.kr
[2] Department of Management & HR', NUST Business School, National University of Sciences & Technology, Islamabad 44000, Pakistan; imran.ime13@gmail.com (M.I.); engrwaqas284@gmail.com (W.A.)
[3] Department of Industrial Engineering, Jalozai Campus, University of Engineering & Technology, Peshawar 25000, Pakistan; muhamad.omair87@gmail.com
[4] Department of Industrial Engineering, University of Engineering & Technology, Peshawar 25000, Pakistan; misbah@uetpeshawar.edu.pk
* Correspondence: bsbiswajitsarkar@gmail.com; Tel.: +82-31-400-5259; Fax: +82-31-436-8146

Received: 13 April 2019; Accepted: 29 May 2019; Published: 1 June 2019

Abstract: Sustainable healthcare systems are gaining more importance in the era of globalization. The efficient planning with sustainable resources in healthcare systems is necessary for the patient's satisfaction. The proposed research considers performance improvement along with future sustainability. The main objective of this study is to minimize the queue of patients and required resources in a healthcare unit with the consideration of staff absenteeism. It is a resource-planning model with staff absenteeism and operational utilization. Petri nets have been integrated with a mixed integer nonlinear programming model (MINLP) to form a new approach that is used as a solution method to the problem. The Petri net is the combination of graphical, mathematical technique, and simulation for visualizing and optimization of a system having both continuous and discrete characteristics. In this research study, two cases of resource planning have been presented. The first case considers the planning without absenteeism and the second incorporates planning with the absenteeism factor. The comparison of both cases showed that planning with the absenteeism factor improved the performance of healthcare systems in terms of the reduced queue of patients and improved operational sustainability.

Keywords: stochastic-petri net modeling; sustainable healthcare system; staff absenteeism; patient's queue; mixed integer nonlinear programming

1. Introduction

Sustainability was first introduced by Brundtland in 1987 [1] and it deals with economic, environmental, and social dimensions [2–6]. Efficient healthcare systems ensure social and economic sustainability. The management of healthcare systems involves the management of operations within the facility. Healthcare organizations should encourage social sustainability to set values for corporate social responsibility. To attain this, they should hunt for a balance between patient comfort, care, and economic need to guarantee sustainable development [7]. The patient's satisfaction and comfort are considered as a primary social indicator in the healthcare industry. The rapid increase in diseases has diverged the direction of researchers to sustainable healthcare systems. Promising innovations and sustainable operations are necessary for the healthcare system to achieve the patient's satisfaction level [8]. A healthcare system may be in a hospital or clinic or some temporary medical camp.

Patients visit hospitals for their treatment, but the limited resources and absence of doctors, dispensers, and nurses at healthcare systems creates issues. The severity of disease also varies from patient to patient. Therefore, sometimes, longer wait times in queue results in death. In order to avoid such problems, there should be some planning of resources by building sustainable healthcare systems for a patient's satisfaction. Since this comfort and satisfaction is a basic social aspect and pillar of sustainability. The planning of resources of the healthcare system is extensively reported in the literature [9–13].

Zeinali, et al. [14] addressed the resource planning model in the emergency department. The resources included were nurses, receptionists, and beds. However, they did not consider the physician, which are a core requirement of any healthcare system. In addition, they used a simulation model and performed a what-if analysis for resource management. However, the proposed model addresses the management of resources with the consideration of their absenteeism probability. Yousefi, et al. [15] focused the resource planning in the emergency department using a genetic algorithm. However, the outdoor patient department is not well focused. Feng, et al. [16] studied medical resource allocation in a healthcare system. They used a simulation integrated multi-objective algorithm for the allocation of resources such as beds, nurses, and other medical staff. The processes in healthcare centers are both discrete and continuous. However, simulation models only discrete events and system dynamic models focus on the continuous system. The proposed research integrates the Petri net with mixed integer nonlinear programming for modeling both discrete and continuous states in the outdoor patient department, which has not been well addressed in previous literature.

Most of the researchers have presented some static mathematical models for the planning of healthcare systems and optimized linear programming or heuristics [17–19]. However, these models might not be suitable for dynamic systems due to randomness in patient arrival, the nature of the disease, and treatment time of patients. Hence, healthcare systems should be efficient, sustainable, and flexible to incorporate the staff absenteeism factor. The dynamics of the healthcare system is modeled with either system dynamic (SD) approaches or discrete event simulation (DES). The objects in DES are different entities and discrete, while, in DS, are merged in a population to form a continuous quantity with a continuous state change [20].

A healthcare system has both characteristics DES and SD such as a number of patients in a system are discrete. However, the state change in a resource is a system dynamic. For example, a physician takes different treatment times for different patients, which is a continuous variable. SD methods are good for dynamic complexity, which makes it useful for strategic decision making and DES focuses on the detailed complexity of operational decision-making [21]. The Petri nets introduced by Adam Carl Petri (1962) combines the qualities of DES and SD. In Petri net modeling, the events, event calendar, entities, workflow, and constraints (decisional, operational) are presented in a DES pattern with a strong mathematical background [22]. The discrete event simulation (DES) has been widely used for the performance evaluation of healthcare systems such as utilization of resources, waiting time for patients, patient flow, and patient safety [23–27].

There are two types of resources in the healthcare system. One is an infrastructure of healthcare systems such as a number of beds, statures, and surgery equipment and other are staff members, doctor, nurses, and clerks. This research considers the human resource management, even though some researchers have considered the number of staff personnel required for patient's satisfaction [14,15,28]. Yalçındağ, et al. [29] addressed the planning of surgeons in the healthcare department but did not consider the absenteeism factor. Similarly, En-nahli, et al. [30] optimized the resources in the emergency department without staff absenteeism. However, this research focuses the resource management on the outdoor patient department for patient queue reduction and satisfaction. It was found that the required staff might not be available or on duty to meet the patient service. The absenteeism of staff may result in a long queue for the patient.

The concept of resource planning with absenteeism of staff has not received proper attention. Thus, the viability of such a healthcare system is necessary for sustainability and growth. The mixed integer nonlinear programming-based Petri net have been introduced to model this situation because of its discrete and continuous nature. Petri net itself combine the properties of simulation and mathematical models and it is possible to do what-if analysis. It is a time-consuming process and does not guarantee an optimal solution. However, in this research integration of Petri net with mixed integer nonlinear programming, reduces the computational time and iterations and gives optimal results. The major advantage of using mixed integer nonlinear programming Petri net model is reduced computational time, improved visualization of the flow of entities (patients), and its capability to find the best solution. The proposed mathematical solution and model approaches are useful for the sustainability of healthcare systems. The incorporation of Stochastic-Petri net approach with MINLP for real-time optimization in the scheduling model will enable the attainment of sustainable operational efficiency. The objective of the proposed model is to reduce the cycle time for the patient's satisfaction under the factor of absenteeism. In order to minimize cycle time and improve customer satisfaction, patient queues and resources should be minimized. The rest of the paper is organized as follows. The detailed literature review is in the second section. The third section includes the development of the model. The solution methodology of the research is represented in detail form as in the fourth section. A numerical example of a healthcare center is presented in the fifth section as an application of the proposed model.

2. Literature Review

The two major categories of the industries can be divided into manufacturing and service industry. Researchers are working to enhance the production, quality, and cycle time of the manufacturing firm to support the manufacturers, customers and government [31–33]. However, healthcare systems are essential for the population to provide services. Therefore, management and control is inevitable. Many operational and managerial strategies have been introduced to plan and control the healthcare systems. The planning of the resources especially human resources is critical because of the direct interaction with the patients. Healthcare models developed can be classified into four types including static models, dynamic models, simulation models, and Petri net models.

2.1. Static System Models

Bachouch, Guinet and Hajri-Gabouj [17] addressed the static strategic enterprise-resource-planning model for the healthcare systems and developed a multi-objective goal-programming model with finance, labor, revenue, capacity, and admission resources. Hans, Van Houdenhoven and Hulshof [11] proposed the scheduling model for healthcare systems for minimizing idle time of nurses in a week. Ghazalbash, et al. [34] developed a mathematical model for the scheduling of the operating rooms of teaching hospitals with the objectives of minimizing operation time and operating room idle time. Nasir and Dang [35] solved a flexible home health care routing and scheduling problem with the joint patient and nursing staff selection. Tan, et al. [36] presented a two-stage multi-objective mathematical model for the planning and scheduling of operating rooms considering the patient flow. This is clear from the literature that the static models have mostly focused the operating rooms, nurses, and operating equipment while neglecting the major functional department of healthcare systems such as an outdoor patient ward, where many patients visit on a daily basis. There are various reasons for modeling this department and the most prominent is the daily arrival of patients in healthcare systems. The static mathematical model does not provide the clarity of the dynamic behavior of the systems.

2.2. System Dynamic Models

Since the behavior of healthcare systems is dynamic and this factor should be taken into account while modeling the healthcare systems. Lane, Monefeldt and Rosenhead [13] provided the dynamic model for the accident and emergency department and planned the resources such as beds and the hospital process by minimizing the delay time of the patients. Carayon, et al. [37] improved the quality and patient safety by introducing the system engineering initiative for the patient safety (SEIPS) model, which analyzes the behavior of workers in human factor contexts. Atun [38] discussed the dynamic complexity of healthcare systems by considering the dynamic interactions between innovation and institutions. Fraher, et al. [39] developed the projection model for the forecasting of surgeons for healthcare systems for 2009 to 2028 with consideration of specialty, sex, and age to tackle the dynamic change in surgery complexity. There is plenty of literature on the system dynamic behavior of the healthcare system. The shorter review of dynamic modeling in healthcare is available in References [10,40]. The system dynamic models are helpful when a change in the state or occurrence of the event is continuous. However, it is not necessary that events are continuous in nature. The nature of events might be discrete, too. This limitation of dynamic models is not to consider the discrete change.

2.3. Discrete Event Simulation Models

A simulation is a tool for performance evaluation, planning, and scheduling of the healthcare systems because of its ability to model the discrete events. Rohleder, Lewkonia, Bischak, Duffy and Hendijani [27] reported the DES model for the performance improvement of the orthopedic outpatient clinic. The objective of their research was to minimize the waiting time of patients. Robinson, Radnor, Burgess and Worthington [26] introduced the concept of SimLean in healthcare, which is the process of performance improvement. It is a helpful tool for the planning and scheduling of healthcare systems.

Abo-Hamad and Arisha [41] presented an interactive program with simulation-based decision support systems for the optimization of the healthcare systems for optimum resource utilization. Zeltyn, et al. [42] focused the workforce staffing problems in the emergency department and developed the simulation model to evaluate different alternatives of the emergency department for customer satisfaction. The use of simulation models for the scheduling or planning of the healthcare systems is reported in the literature and some details can be found in the models of References [43–46].

2.4. Petri-Net Models

The Petri nets are a combination of dynamic modeling and simulation incorporating visualization with the help of graphs. These models deal with both discrete and continuous quantities. The planning and scheduling of the healthcare system involve both discrete and continuous variables. The number of patients, the number of physicians, and a number of nurses have a discrete nature. The examination time, medication time, and waiting time are the example of continuous variables. However, some mathematical models have been employed for scheduling they did not consider the dynamics of both discrete and continuous variables [47]. The processes within a healthcare center are random or stochastic. Therefore, in this research, stochastic Petri net modeling is chosen for realistic results.

There are very few articles on healthcare planning and scheduling using Petri nets. Mahulea, et al. [48] modeled the healthcare system by using Petri nets and evaluated the performance measure, but they did not optimize this process for customer satisfaction. Bertolini, et al. [49] analyzed the blood transfusion with component-based timed-arc Petri nets and modeled the healthcare workflow. Hicheur, et al. [50] modeled the flexible healthcare systems by using recursive and workflow Petri nets and evaluated the performance of the healthcare system by analyzing the medical properties of healthcare processes. In all of the above cases, Petri net have been used for flow modeling and process visualization and do not give us any information when the maximum number of patients will be served. However, proposed research not only models the healthcare system using a Petri net but also optimizes the process. In this regard, Petri nets have been integrated with a mixed integer nonlinear programming model that has been utilized to maximize customer satisfaction by reducing the cycle time.

Wang [51] modeled the emergency response in the emergency department by using Petri nets. The proposed methodology considered the stochastic Petri net modeling and optimization for outdoor patients in healthcare systems and the aim is to satisfy each patient with minimum waiting time while considering the staff (doctors, nurses, or cleric) unavailability. The concept of planning and scheduling the resources with consideration of staff absenteeism has not been paid proper attention.

3. Petri Nets Modeling of the Healthcare System

The Petri nets models are the graphical, simulation, and mathematical representation of the system [52]. A typical Petri net model is a bipartite directed graph, which consists of four elements called places, transitions, arcs, and token. Places and transitions are connected with each other with the help of the directed arc [53]. Figure 1 shows a basic Petri net graph and definition of each element in the Petri net is given below.

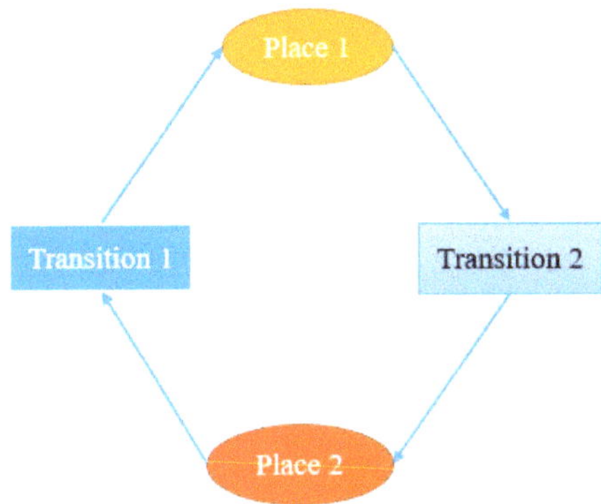

Figure 1. A Petri net graph.

a. Place

A place in the Petri net is a position with two states. A token is a process or event that occurs in place, so a place has a token or not. In a healthcare example, a physician is in a place with two states, either he/she sits idle or examines the patient. Places in Petri nets define discrete variables.

b. Transition

Transition shows the activity or process and its duration within the model. A transition represents continuous variables within the process. For example, patient examination time of the physician is a transition.

c. Arcs

Arcs are used to direct and connect the transitions with places. The sequence of a visit from registration to the medication is an example of art that shows the process plan.

d. Tokens

Tokens are the entities moving along arcs within transitions and places in a Petri net graph. The number of patients within the healthcare center are the example of tokens. The event in Petri net can be defined by introducing a transition between input and output places. The dynamic behavior analysis of Petri net involves availability or unavailability of tokens in the input or output places. If the token is available, then the condition associated with the place is true. Otherwise, it is false.

Definition 1. *A Petri net can be defined with five-tuple* $M = (\rho, \tau, I, O, \mu,)$, *where*

(1) $\rho = \{\sigma_1, \sigma_2, \ldots \ldots \upsilon_m\}$ *is the finite set of places*
(2) $\tau = \{t_1, t_2, \ldots \ldots t_n\}$ *is the finite set of transitions* $\rho \cup \tau \neq \varnothing$, *and* $\rho \cap \tau = \varnothing$
(3) $I : \rho \times \tau \rightarrow M$ *is a function, which acts as an input for defining a directed arc between places and transitions and "M" is set to positive integers.*
(4) $O : \tau \times \rho \rightarrow M$ *is a function, which acts as an output for defining a directed arc between places.*

(5) $\mu : \rho \to M$ *is called an initial marking.*

The assignment of the token to the places in Petri net is called marking. The position of token keeps on changing, according to the input and output directing arcs in Petri net modeling [54].

3.1. Petri Net Graphical Representation

The Petri net models can be explained well with the help of the graph, which is a bipartite directed multigraph. There are two types of nodes in Petri net graphs. A place node, which is usually represented by a circle and transition is represented by a bar or box. To understand the graphical illustration of the Petri net considers a healthcare center. There are two types of outdoor customers or patients. The patients arrive in the healthcare system, register themselves for treatment at the registration desk, and move to the next station where physicians examine them. The doctor completes the examination of the patient. He/she is sent to a dispenser for medicines. Table 1 shows the process plan for patient treatment in a healthcare center.

Table 1. Outdoor patient treatment process plan.

	Registration	Examination	Medication
$i = 1$	t_{11}	t_{12}	t_{13}
$i = 2$	t_{22}	t_{22}	t_{23}

The bipartite directed multigraph for the Petri net model can be developed using the process plan of patient treatment, as shown in Figure 2, which is built using the concept of Figure 1 and process plan in Table 1. It is clear from Figure 2 that there are two types of overlapping circuits called the patient circuit and sequencing circuits. The distribution and number of tokens in net control of its execution. The number of tokens in the patient circuit is equal to the number of patients waiting for service. The sequencing circuit is used for the representation of servers. Figure 2 is the graphical representation of the flow of entities such as patients in a healthcare system. Rectangles in a blue color show the transitions that are used for assigning times to processes. However, places in red circles are used to represent resources. Green-colored and orange-colored tokens represent two different types of patients.

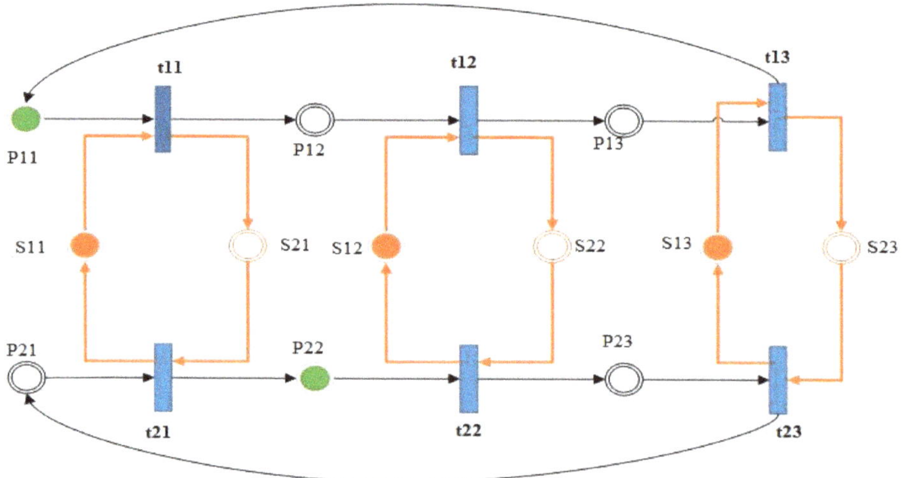

Figure 2. Petri net model of the healthcare system with a process plan.

3.1.1. Model Assumptions

To develop a mathematical model, the following assumptions must be taken into account.

1. The service times are stochastic.
2. One server can serve only one patient at a time.
3. Transitions and places cannot be together so, during any loop, this rule should not be violated.
4. Server interruption by another patient is not allowed when a server is in service of another patient.
5. The patient keeps arriving at the healthcare system randomly.

3.1.2. Mathematical Model

In the sequence loop, the number of tokens is invariants. The total transition time $\tau(\zeta)$ is equal to the sum of transition firing times in a sequence circuit. t_{ij} is the time taken by the server j to serve patient i.

$$\tau(\zeta) = \sum_{i=1}^{I} \sum_{j=1}^{J} t_{ij} \tag{1}$$

The total sum of tokens in a sequential loop or server loop $M(\zeta)$ is in Equation (2). In this case, S_{ij} shows a place of server j for serving patient i.

$$M(\zeta) = \sum_{j=1}^{J} S_{ij} \, \forall_i \tag{2}$$

Aziz, et al. [55] computed cycle time of each sequential circuit $C(\zeta)$ using Equation (3).

$$C(\zeta) = \frac{\tau(\zeta)}{M(\zeta)} \tag{3}$$

However, this cycle time does not consider the absenteeism of the server from the service stations. If there is some probability of absenteeism, then Equation (3) becomes PA, which represents the probability of absenteeism.

$$C(\zeta) = \frac{\tau(\zeta)}{PA \times M(\zeta)} \tag{4}$$

$\tau(\zeta_s)$ is the total service time in one service cycle by only one server to provide the service to a single patient. If the number of the server at each service station is known, then the service rate SR can be calculated using Equation (5).

$$SR = \frac{1}{C(\zeta)} \tag{5}$$

$$SR \times \tau(\zeta_s) = \frac{\tau(\zeta_s)}{C(\zeta_s)} \text{ and } M(\zeta_s) \geq \frac{\tau(\zeta_s)}{C(\zeta_s)} \tag{6}$$

After a finite transition, the system starts to operate at a steady state. As the system achieves the steady state, the service cycle time becomes equal to the maximum cycle time of the servers. The service cycle time reduces with the introduction of more tokens. The number of tokens in the server circuit is normally kept fixed per the capacity of healthcare centers. However, the number of patient tokens depends upon the arrival of patients and planning horizon. Equation (7) shows the required cycle time of the patient circuit $C(R\zeta_p)$,

$$C(R\zeta_p) = \frac{Planning\ horizon}{Number\ of\ patients} = \frac{PH}{NP}, \tag{7}$$

while

$$M(\zeta_s) \geq \frac{\tau(\zeta_s)}{C(R\zeta_p)} \tag{8}$$

Bottleneck server cycle time $C(B\zeta_s)$ and $C(R\zeta_p)$ have two possible cases as follows:

1. $C(B\zeta_s) \geq C(R\zeta_p)$
2. $C(B\zeta_s) < C(R\zeta_p)$

The performance of the system will be influenced because of the maximum number of bottleneck server cycle times $C(B\zeta_s)$ and required cycle times for the patient circuit $C(R\zeta_p)$ [56].

$$C(\zeta_{sMAX}) = Max \left\{ C(B\zeta_s), C(R\zeta_p) \right\} \tag{9}$$

By considering the server bottleneck circuit as well as cycle time as the maximum cycle time, then the Petri net model is formulated as a mixed integer nonlinear programming problem [57].

$$Minimize \sum_{i=1}^{I} \sum_{j=1}^{j} P_{ij} \tag{10}$$

Subject to

$$C(\zeta) \leq C(\zeta_{sMAX}) \; with \; C(\zeta) > C(R\zeta_p) \tag{11}$$

Equation (12) does not consider the absenteeism while Equation (13) considers it.

$$\tau(\zeta) = C(\zeta_{sMAX}) \times M(\zeta) \tag{12}$$

$$\tau(\zeta) = C(\zeta_{sMAX}) \times PA \times M(\zeta) \tag{13}$$

$$C(\zeta_p) \leq C(R\zeta_p) \; \forall \; \zeta_p \quad with \; C(\zeta_p) > C(R\zeta_p) \tag{14}$$

$$\forall \; P_{ij} , \; S_{ij} \quad Integer \; \geq 0$$

The objective function of this model is to minimize the service cycle time. Equation (10) is the objective function. The constraint in Equation (11) ensures that the cycle must be within critical limits defined by the bottleneck cycle time or required cycle time for the patient loop. Equation (12) considers the effect of critical or maximum cycle time on total transition time without considering the absenteeism of the server. However, Equation (13) takes this factor into account. Equation (14) shows patient service achievement.

4. Solution Methodology

The method explained in Section 3 has been explained with the help of a numerical example of a hypothetical healthcare center.

4.1. Numerical Example

The numerical example of the proposed model is demonstrated as follows.

4.1.1. Problem Statement

Consider a healthcare center that provides services to the patients. Patients visiting the healthcare center have to process three steps including registration, medical examination, and medication. There are two servers at the registration boot, three physicians for the treatment of patients, and two servers in the dispensary. There are three possibilities due to the absenteeism of servers.

1. The server at the registration booth may be absent and this situation results in the long queue of patients at the registration booth.

2. The absenteeism of physicians may result in the long queue of patients and the station becomes a bottleneck.
3. If any of the servers from the last station is absent, then it also creates a longer wait time at the last station, which is also the reason for the bottleneck.

Table 2 shows the probability of absenteeism at registration, medical examination, and dispensary. The probability of absenteeism is computed using past data. The data has been collected through a brief interview and brainstorming session with healthcare managers. He suggested the following probability of absenteeism for each resource. Following one-week probability can be projected for one year.

Table 2. Absenteeism probability of servers at each step in the healthcare center.

	Absenteeism Probability		
Days	**Registration**	**Medical Examination**	**Medication**
Monday	~0.03	~0.07	~0.04
Tuesday	~0.05	~0.07	~0.05
Wednesday	~0.05	~0.08	~0.05
Thursday	~0.05	~0.10	~0.05
Friday	~0.12	~0.14	~0.10
	Average = 0.060	Average = 0.092	Average = 0.058

There are two types of patients. Type 1 patients have severe diseases and type 2 patients have less severe diseases. The processing time at all stations is stochastic, which can be modeled with the help of some distribution. Processing time of each server at each station is measured and data is analyzed to find the best fit distribution. In distribution analysis, data was analyzed on a set of distributions such as uniform, lognormal, binomial, and normal distribution. In this data analysis, uniform distribution was selected because it had the highest value of Log-likelihood and minimum value of Akaike information criterion. Table 3 shows the processing time of each process at each step in the healthcare center. Here, U represents un-firm distribution, in the case of medical examination U (3,7), which means that a physician takes three to seven minutes to examine a patient. Data analysis also showed that the number of patients visiting the hospital follows the Poisson distribution. The number of patients visiting in the next month has been generated randomly using a convolution theorem. Table 4 shows the number of patients visiting each month.

Table 3. Outdoor patient treatment process plan.

	Registration	Examination	Medication
$i = 1$	U (1,5)	U (3,7)	U (1,2)
$i = 2$	U (1,5)	U (4,6)	U (1,2)

Table 4. Number of patients visiting the healthcare center each month.

January	February	March	April	May	June	July	August	September	October	November	December
2979	2726	1798	2630	2809	1736	2208	1629	1750	1643	2334	2841

The hospital is interested in the planning of the outdoor patient department (OPD) for minimizing the number of patients waiting in the queue at all servers by determining the number of servers considering their absenteeism.

4.1.2. Problem Solution

The station in the system becomes a bottleneck when there is physician absenteeism at the medical examination center. Considering the server capacity at each station, the service cycle time can be calculated using the equation below.

$$Service\ Cycle\ time = \frac{total\ proces\sin g\ time}{Number\ of\ servers} \tag{15}$$

Service rate is computed using equation as follows.

$$Service\ rate = \frac{1}{Service\ cycle\ time} \tag{16}$$

In this case, the processing time is uniformly distributed. Therefore, the cycle time of each station will also be uniformly distributed. The planning horizon of the hospital is one month and there are 26 working days and, there are 12 operating hours. Therefore, the available time is 18,720 min. Considering the number of patients visiting the hospital from Table 4 and available time in each month, the required cycle time for the patient circuit can be computed using Equation (7). Table 5 shows the required cycle time of a patient circuit.

Table 5. Required patient circuit cycle time for each month.

Month#	January	February	March	April	May	June	July	August	September	October	November	December
$C(R\zeta_p)$	6.3	6.9	10.4	7.1	6.7	10.8	8.5	11.5	10.7	11.4	8.0	6.6

This actual cycle time is not according to the required cycle time. Therefore, the system should be redesigned in such a way that there should be a minimum queue of patients at the station. Mixed integer nonlinear programming model is solved for each required cycle time without considering the absenteeism.

4.1.3. Case for a Numerical Example

Two cases for the numerical example have been presented. The first case is monthly planning without absenteeism and the second case considers the absenteeism.

Case 1: Human Resource Planning without Absenteeism

In this case, the Petri net have been used to optimize and compute the number of servers and queues using a mathematical model consisting of Equations (10)–(12) and Equation (14). The processing time has been generated using Table 3. The required cycle time of each month shown in Table 6 is used for cycle time constraint in the Petri net model.

Table 6. Required resources and Queue length with absenteeism.

Month	$C(R\zeta_p)$	Patients in Queue			Number of Servers Required		
		Station 1	Station 2	Station 3	Station 1	Station 2	Station 3
January	6.3	2	2	2	1	2	2
February	6.9	1	2	1	1	1	2
March	10.4	1	1	1	0	1	1
April	6.71	1	2	1	1	2	2
May	6.70	1	2	1	1	2	2
June	10.8	1	1	1	0	2	1
July	8.50	1	1	1	1	2	1
August	11.5	1	1	1	1	1	0
September	10.7	1	1	1	1	1	0
October	11.4	1	1	1	1	1	1
November	8.0	1	2	1	1	1	1
December	6.60	2	2	2	1	2	2
Average		1.2	1.5	1.2	0.8	1.1	1.3

Case 2: Human Resource Planning with Absenteeism

In this case, the factor of absenteeism is considered for the planning of human resources. Equations (10), (11), (13) and (14) have been used in the model. The processing times are the same in either case.

5. Results and Discussion

Considering Figure 1 and the mathematical model in Section 3.1.2, an excel spreadsheet program is generated and solved using a personal computer with 4 GB RAM and 2.67 GHz processor.

5.1. Results of Case 1

It can be seen from Table 1 that there are two types of patients and three types of servers and service processes consist of three processes so the total number of places is 12 and the total number of transitions is six because there are three processes and two types of patients. There are three types of loops in the Petri net model, which are the patient loop, a sequential loop, and a mixed loop. There are three sequential loops because the number of processes of service is three. The number of patient loops is equal to the type of patients who visit the hospital. In each elementary circuit, the availability and unavailability of each token (server, patient) are represented by binary (0, 1) digits. It is clear from Table 6 that the required cycle time for the month of January is 6.3 min per patient. The problem is solved by using MINLP for the required cycle time of January.

Number of patients waiting at registration $= P11 + P21 = 0 + 1 = 1$

Number of patients waiting at the medical examination center $= P12 + P22 = 0 + 1 = 1$

Number of patients waiting at the dispensary $= P13 + P23 = 2 + 0 = 2$

Similarly, the required number of the server at each station can be calculated using the following formula.

Number of servers at the registration booth $= S11 + S21 = 0 + 2 = 2$

Number of servers at the medical examination center $= S12 + S22 = 1 + 1 = 2$

Required servers at the dispensary $= S13 + S23 = 1 + 0 = 1$

Similarly, the MINLP model is run for each cycle time in Table 5 for each month and the required number of servers and number of patients waiting in a queue obtained from Petri net-MINLP at each station are given in Table 7.

Table 7. Required resources and Queue length without absenteeism.

Month	$C(R\zeta_p)$	Patients in Queue			Number of Servers Required		
		Station 1	Station 2	Station 3	Station 1	Station 2	Station 3
January	6.3	1	2	2	3	2	1
February	6.9	1	2	2	2	1	2
March	10.4	1	1	1	1	0	1
April	7.1	1	2	2	1	1	2
May	6.70	1	2	2	2	2	1
June	10.8	1	1	1	1	1	1
July	8.50	1	1	1	1	1	1
August	11.5	1	1	1	1	1	1
September	10.7	1	1	1	1	1	1
October	11.4	1	1	1	1	1	1
November	8.0	1	2	1	1	1	1
December	6.60	1	2	2	2	2	1
Average		1	1.5	1.4	1.4	1.2	1.2

It is clear from Table 8 that the average queue length at station 1, station 2, and station 3 are 1, 1.5, and 1.4. However, the average number of servers at each station is 1.4, 1.2, and 1.2, respectively.

5.2. Results of Case 2

The aim of this research to design a healthcare center with a factor of resource absenteeism. The same numerical example is used for the planning of the healthcare center with the factor of absents. The objective is to determine the optimal number of resources at each station with a minimum queue length. To do so, the Petri net-mixed integer nonlinear programming (MINLP) model included Equation (13) instead of Equation (12). The Petri net-MINLP is run for each cycle for each month and the required number of resources and queue generated at the station is recorded. Table 6 shows that the average queue length of each station has been reduced while the required number of average resources has increased.

The results of resource planning with absenteeism are more robust and ensure patient satisfaction with reduced queue length. The Petri net-MINLP model used the process plan from Table 1 and employed Figure 2 for determining sequential, patient, and mixed loops. Since there are three stations, it contains three sequential loops. A number of patient loops are two because two types of patients visit the healthcare center. The mixed loops depend on the sequential and patient loops. These loops can be computed using the following Equation (17).

$$Mixed\ loops = Patient\ loops \times sequential\ loops \tag{17}$$

Table 8 shows markings of the circuit with cycle time for the Petri net model for the healthcare system. The binary digits show the availability and unavailability of the token in places, hiring, and firing from the transitions. Two types of cycle time have been used for this Petri net model. Circuit cycle time is known from the process plan and Petri net diagram. However, for the required cycle, the Petri net model is hybridized with MINLP to get the optimal number of servers and queue length.

Table 8. Marking of the circuit with cycle time for the Petri net model for the healthcare system in the month of January without absenteeism.

Transition Time	$U(1,1.5)$	$U(3,7)$	$U(1,2)$	$U(1,1.5)$	$U(4,6)$	$U(1,2)$	$U(1,1.5)$	$U(1,1.5)$	$U(3,7)$	$U(4,6)$	$U(1,2)$	$U(1,2)$				
Transition Name	t11	t12	t13	t21	t22	t23	t11	t21	t12	t22	t13	t23				
Place Name	P11	P12	P13	P21	P22	P23	S11	S21	S12	S22	S13	S23				
No.	1	2	3	4	5	6	7	8	9	10	11	12	$\tau(\zeta)$	L.T	$M(\zeta)lp$	$C(\zeta)lp$
1	0	0	0	0	0	0	1	1	0	0	0	0	8	S.C	1.27	6.30
2	0	0	0	0	0	0	0	0	1	1	0	0	12	S.C	1.90	6.30
3	0	0	0	0	0	0	0	0	0	0	1	1	11	S.C	1.75	6.30
4	1	1	1	0	0	0	0	0	0	0	0	0	15	P.C	2.38	6.30
5	0	0	0	1	1	1	0	0	0	0	0	0	16	P.C	2.54	6.30
6	1	0	0	0	1	1	0	1	0	0	1	0	27	MX	4.29	6.30
7	0	1	1	1	0	0	1	0	0	0	0	1	23	MX	3.65	6.30
8	1	1	0	0	0	1	0	0	0	1	1	0	24	MX	3.81	6.30
9	0	1	0	1	0	1	1	0	0	1	0	0	21	MX	3.33	6.30
10	1	0	1	0	1	0	0	1	1	0	0	0	30	MX	4.76	6.30
11	0	0	1	1	1	0	0	0	1	0	0	1	30	MX	4.76	6.30
Marking	1	1	1	1	1	0	0	1	1	1	1	1	9.8			

L.T: Loop type. S.C: Sequencing circuit. P.C: Patient circuit. MX: Mixed circuit of sequencing and patient. $M(\zeta)lp$: Sum of tokens in circuits in the mixed integer nonlinear programming model (MINLP). $C(\zeta)lp$: Cycle time of the circuit for MI.

5.3. Comparison of Results of Case 1 and Case 2 with the Existing System

In this section, a detailed analysis of servers and queue is presented and results are compared with the existing system. Figure 3 shows a healthcare center and its process plan.

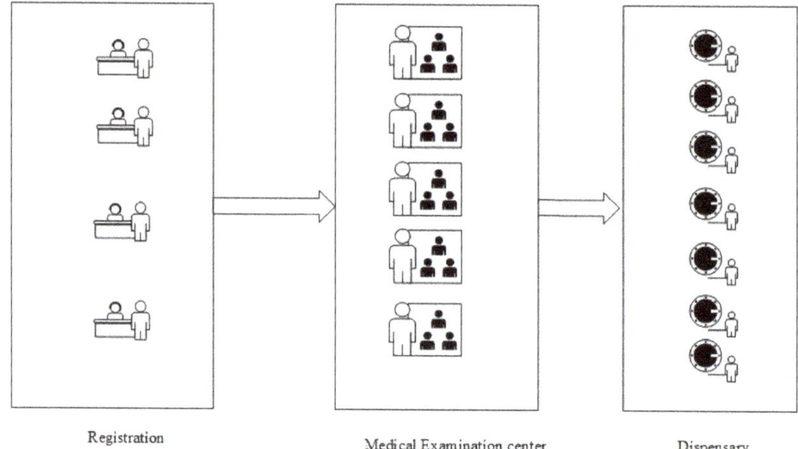

Figure 3. Process plan of a healthcare center.

5.3.1. Analysis of Servers

The healthcare center operated the overutilization of servers. The existing healthcare center contained two servers at station 1, three servers at station 2, and two servers at station 3. Figure 4 shows the number of servers required in both cases with and without the absenteeism case. It is clear from Figure 4 that the existing number of servers at the registration booth are in access. However, the required number of servers in both cases is reduced for the effective utilization of resources. Figure 5 shows the analysis of a number of servers at the medical examination center. It is shown that the required number of servers, in either case, is less than the existing servers.

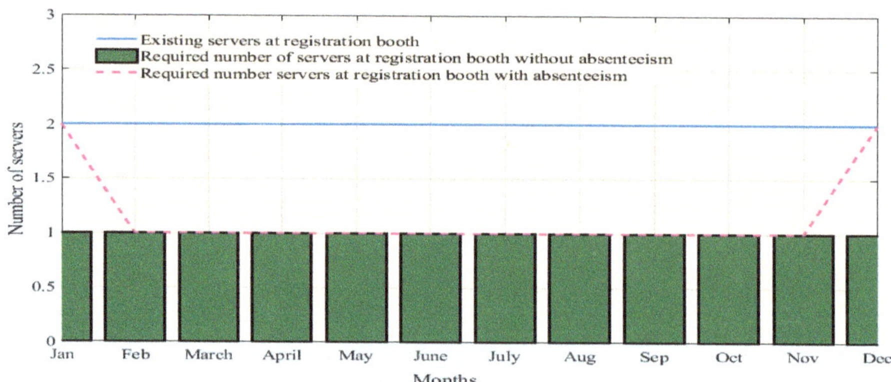

Figure 4. Number of servers at the registration booth during each month.

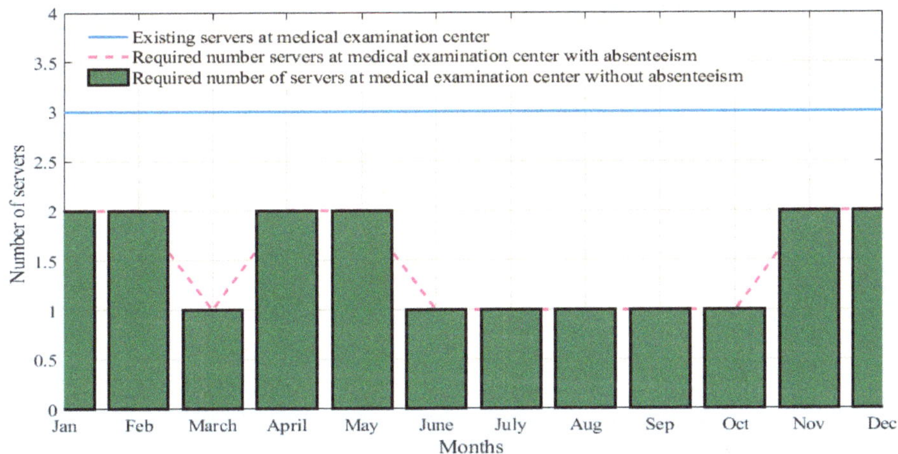

Figure 5. Number of servers at a medical examination center for each month.

Figure 6 shows the number of servers at the dispensary. It can be seen that a number of servers with an absenteeism case are higher from April to June.

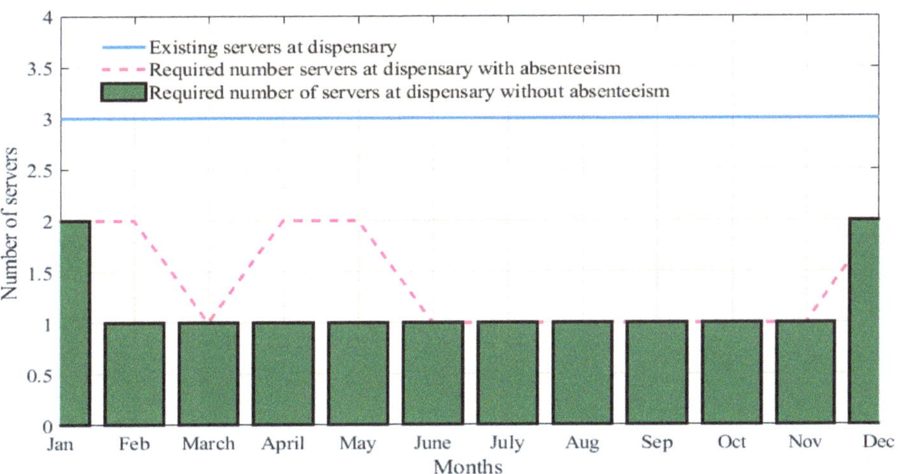

Figure 6. Number of servers at a dispensary for each month.

5.3.2. Analysis of Queue

Considering the case of server absenteeism reduced the queues of patients at all booths in a healthcare center. Figure 7 shows the comparison of queues for both cases of absenteeism and without absenteeism.

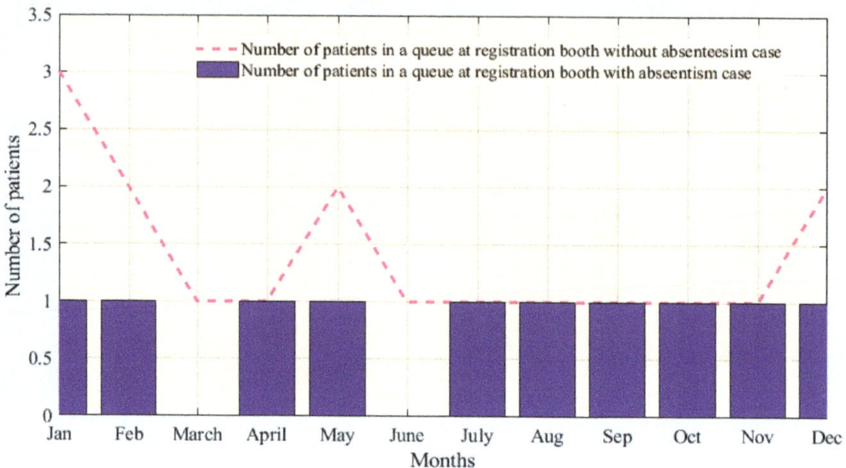

Figure 7. Number of patients waiting for service at the registration booth.

Figure 8 shows the number of patients waiting for medical treatment at the medical examination center. Planning resources as the factor of absenteeism reduced the queue length at the examination center.

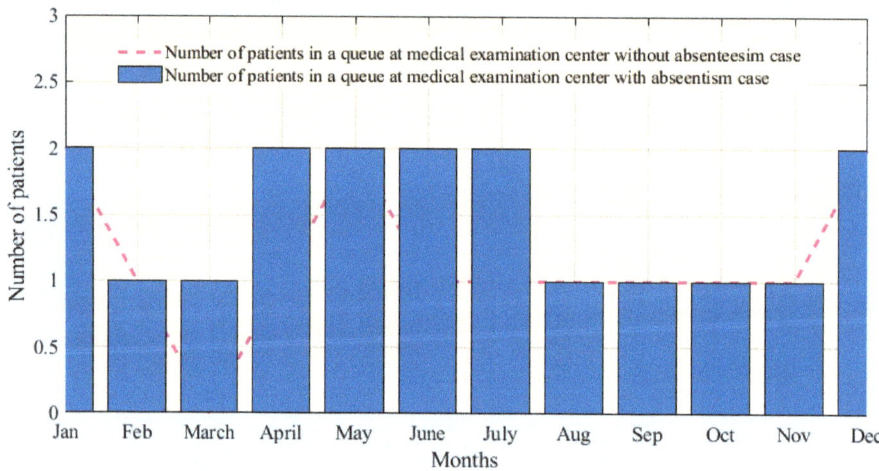

Figure 8. Number of patients waiting for service at the medical examination center.

It is prominent from Figures 7–9 that the length of queue reduced by determining the number of servers at each station. Therefore, this model provides the best approach for the planning of resources in a healthcare system. The consideration of server absenteeism in the planning of resources reduces queue length substantially, which increases resource utilization and patient satisfaction.

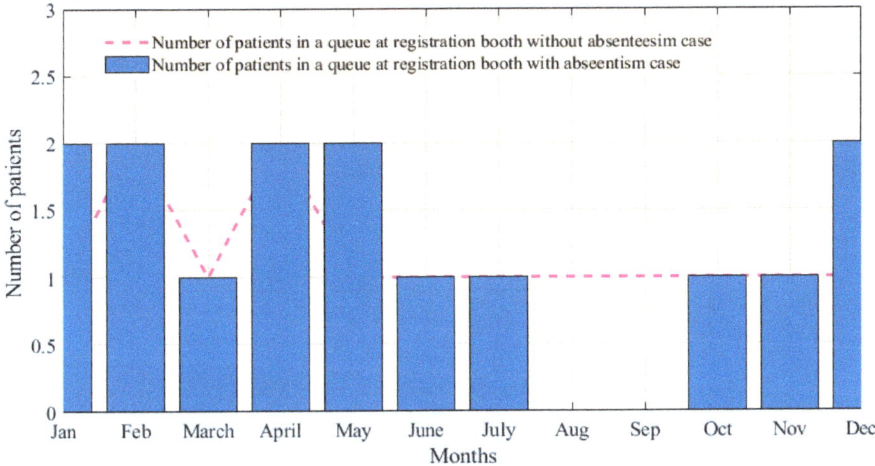

Figure 9. Number of patients waiting for service at the dispensary.

5.4. Petrinet Based Mixed Integer Nonlinear Programming (Petrinet-MINLP)

Most of the real-life problems have a nonlinear nature because of their complexity and involvement of many dependent decision variables. An optimization problem is said to be nonlinear if it contains a nonlinear equation either in the objective function or in a constraint. There are different software packages used for nonlinear optimization such as MATLAB, CPL EX, GAMS, and LINGO. However, in all these packages, an interior point algorithm is used for the optimization in mixed integer nonlinear programming. Mixed integer nonlinear programming is an exact method and guarantees the global optima in the optimization problem (Tawarmalani, Sahinidis, & Sahinidis, 2002). Traditionally, MINLP was used for mathematical optimization.

Petri net is a combination of simulation, mathematical, and graphical methods that are used to represent the real systems in both dynamic and static ways. Traditionally, Petri nets have been used for what-if analysis and could not optimize the complete process in a dynamic environment (Jensen, 2013). The optimization of the Petri net has not been well focused. In this research, we made an attempt to integrate the Petri net with mixed integer nonlinear programming for the optimization of real systems in a dynamic way. In this research, the Petri net-based mixed integer nonlinear programming is used to solve a healthcare resource planning problem. However, this proposed method can be applied to all business problems for their dynamic analysis and optimization.

5.5. Performance Evaluation of Petri Net-Based Mixed-Integer Nonlinear Programming

In this research study, a nonlinear mixed integer nonlinear programming model has been integrated with a Petri net for the dynamic behavior of patients flow in a healthcare center. In order to evaluate the performance of a proposed method, Petri nets have been integrated with genetic algorithm, a pattern search method, and case, which considers that the absenteeism has been solved using a Petri net integrated genetic algorithm and a patient integrated pattern search algorithm. Details of each method is given below.

5.5.1. Petri Net Based Genetic Algorithm

Genetic algorithm is a heuristic optimization method based on the theory of evaluation by Darwin. According to the theory of Darwin, only the fittest individuals survive in the next generation. In a genetic algorithm, an initial solution for an initial population is randomly generated. A randomly generated solution is called chromosome or individuals. The chromosome is composed of genes.

In optimization, a decision variable is treated as a gene and a chromosome is treated as an initial feasible solution. A set of feasible solutions is called a population.

The structure of a chromosome is given below in Figure 10. Crossover and mutation are the two genetic operators. The crossover occurs between two individuals or chromosomes, which is done by exchanging the genes. In this research, we used a two-point cross over. A mutation is because of change in genes in an individual chromosome. Mutation introduces the diversity while chromosomes transfer parental characteristics to the children or generations.

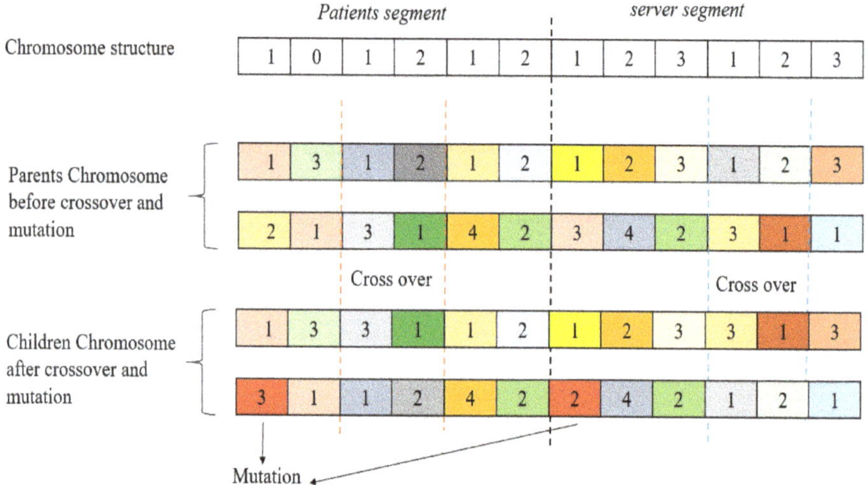

Figure 10. Chromosome structure and genetic operators.

The fittest chromosome in a population is called elite chromosome. Genetic algorithm is a heuristic method and it may or may not give the global optima [58]. In addition, it follows the random search method. Therefore, its computation time is higher than the mixed integer nonlinear programming. In solving the numerical example, the Petri net is integrated with a genetic algorithm [59]. In a Petri net-based genetic algorithm, the population consisted of 200 chromosomes. The mutation function was constraint-dependent. Elitism was $0.25 \times population\ size$ with 80% crossover probability, which was used for the diversity in next generations, and a stochastic uniform selection method was adopted for the selection of individuals or chromosomes for the next or upcoming generation. Termination criteria for the genetic algorithm were the repetition of the same elite in 20 successive generations.

5.5.2. Petri Net-Based Pattern Search Method

The pattern search is a derivative-free search method and its independent of a gradient. In this method, the initial solution is provided and it follows the basic procedure of other optimization methods. This method was introduced in 1961 and emerged in two domains such as the exploratory search and pattern move. The exploratory search follows the direction along the gradient and the pattern move keeps improving its search. The pattern search method involves some parameters that affect its performance for a different problem data set. Kang, Li, and Li (2013) explained the pattern move with the help of Figure 11.

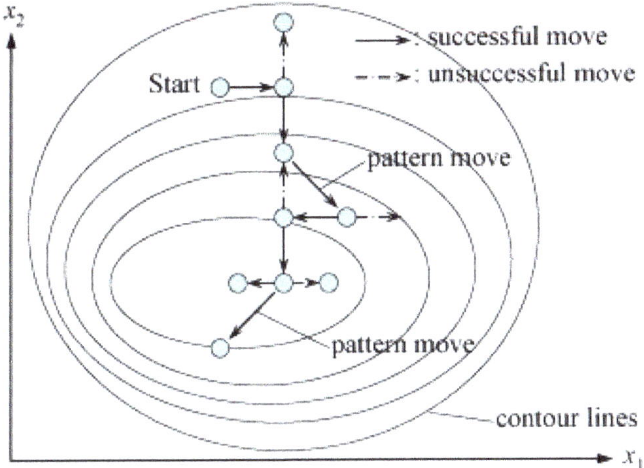

Figure 11. Pattern search illustration.

It is clear that the pattern search method initiates its move to find an appropriate direction for one variable at a time along with the coordinates of the individual from the base point. Once the exploratory search is over, the pattern move starts accelerating along the predetermined direction by the exploratory search. The expansion and contraction during the move are the most influencing parameters in the pattern search. In order to optimize the numerical example, the Petri net was integrated with a pattern search method [60]. In the patient-pattern search method, the same termination criteria were used as in the Petri net-genetic algorithm, expansion, and contraction factors, which were restricted to 2 and 0.5.

5.5.3. Comparison of Methods

The optimization problems can be classified into three categories based on the search method, which include the heuristic method, the direct search method, and the exact method. In this research study, the Petri net has been integrated with each kind of method. In order to evaluate the relative performance of these methods, the same numerical problem presented in Section 4.1.1 is solved with all methods. All methods have been coded in MATLAB (2017) Ra on a personal computer with 8 GB RAM and a 2.0 GHz Processor. The objective function in Equation (10) has been optimized subject to constraints in Equations (11)–(14). Table 9 shows the results of all the algorithms. In order to simplify the results, an average number of servers, and the average number of queues of all stations is considered.

The problem is solved for all required cycle times using both methods and an average number of servers and queues are given in Table 9 for all months of the relevant years. The computational power of methods has been measured in terms of a number of iterations and computation time. The relative performance of each method is computed using a gap analysis. Equation (18) has been used for the computation of the percentage gap.

$$\%Gap = \frac{Achieved\ value - Best\ value}{Achieved\ value} \times 100 \tag{18}$$

Table 9. Results of different algorithms for a case II with an absenteeism factor.

	Petri Net-Genetic Algorithm				Petri Net-Pattern Search				Petri Net-Mixed Integer Nonlinear Programming			
	Average Queue	Average Servers	Iterations	CPU Time (s)	Average Queue	Average Servers	Iterations	CPU Time (s)	Average Queue	Average Servers	Iterations	CPU Time (s)
January	3	4	59	7.12	4	3	4	5.34	2	2	254	2.33
February	3	4	55	6.35	4	4	5	4.99	1	1	258	2.78
March	4	4	416	10.12	5	4	5	4.54	1	1	268	3.1
April	3	3	92	7.87	4	3	5	4.46	1	2	253	2.35
May	3	3	75	7.23	4	4	5	4.25	1	2	265	2.78
Jun	3	3	69	7.15	4	4	5	4.21	1	1	259	3.32
July	3	3	74	7.31	3	3	5	4.29	1	1	270	3.42
August	3	4	89	7.67	3	3	5	4.23	1	1	257	2.45
September	3	3	52	6.45	4	3	5	4.19	1	1	267	2.43
October	4	3	68	7.78	4	2	5	4.89	1	1	268	2.35
November	4	4	67	7.53	5	3	5	4.56	1	1	255	2.34
December	3	4	57	7.23	4	3	5	4.45	2	2	254	2.31
Average	3	4	98	7.48	4	3	5	4.53	1	1	261	2.66

In the case of minimization/maximization problems, the best will be a minimum/maximum value of the specific performance measure. It can be seen from Table 9 that a number of iterations are greater in a Petri net mixed integer nonlinear program.

MINLP is an exact method and evaluates one solution in one iteration. Therefore, a number of iterations are evaluated in a short period of time. However, in the case of the Genetic algorithm and pattern search, it takes more time to evaluate one iteration. In one generation, multiple solutions are evaluated. Average queue length, the average number of servers at the station, the number of iterations, and computational time has been considered as performance measures of the proposed method. Table 10 shows the average performance measures computed from three different methods.

Table 10. Average performance measurement of algorithms.

	Monthly Average Queue	Monthly Average Servers	Average Iterations	Average CPU Time
Petri net-Genetic algorithm	3	4	98	7.5
Petri net-Pattern search	4	3	5	4.5
Petri net-Mixed-integer nonlinear programming	1	1	261	2.7

Using Equation (18) and Table 10, the relative percentage gap has been computed, which is given in Table 11. In order to understand the percentage gap, consider the monthly average queue in Table 10. The best value of the monthly queue is one patient. Therefore, using Equation (18), the percentage gap of the Petri net-genetic algorithm can be computed as follows. Similarly, performance measures of rest of the algorithms are given in Table 11.

$$\% \; Gap = \frac{3-1}{3} \times 100 = 66.67\% \tag{19}$$

Table 11. Relative percentage gap in different methods.

	Monthly Average Queue	Monthly Average Servers	Average Iterations	Average CPU Time	Sum
Petri net-Genetic algorithm	66.67	75.00	94.90	64.00	300.56
Petri net-Pattern Search	75.00	66.67	0.00	40.00	181.67
Petri net-MINLP	0.00	0.00	98.08	0.00	98.08

It is clear that the total percentage gap is minimum in case of Petri net-based mixed integer nonlinear programming as compared to the other methods. The Petri net-genetic algorithm and Petri net-paterren search are the heuristics and follow a random search pattern. However, in mixed integer non-linear programming, the branch and bound algorithm is used in the exact solution method.

5.6. Advantages and Disadvantages of Petri Net Integrated Mixed Integer Non-Linear Programming

The Petri net combines the characteristics of graphics, simulation, and mathematical modeling for the imitation of real systems. Petri nets are good tools for analyzing the dynamic behavior of systems with both continuous and discrete state changes. Integration of optimization with Petri nets provides the following benefits.

1. It enables real-time optimization of systems for more practical results. In traditional literature, a mathematical model is optimized and results are then used in a simulation model for the validation. However, in this research, simulation and optimization takes place at the same time. Therefore, the word real-time optimization is more suitable for this type of optimization.

2. It visualizes the patterns and their flow within the program and highlights the state changes in a system at a different process.
3. It combines the computational power of simulation and mathematics for more robust results.
4. Integration of Petri net with exact optimization reduces computational time and improves the performance.

In addition to the advantages, Petri nets have some disadvantages as well. Complex systems involving more than three processes and entities are hard to model. In Petri nets, model construction is a cumbersome process and needs computational and graphical efforts.

5.7. Managerial Insights

This research is useful for operation managers in healthcare for efficient planning of resources such as physicians and other medical staff. The consideration of the factor of absenteeism makes it more robust for long-term decision-making in determining the optimal number of resources. This planning model can be useful in determining the more accurate number of resources even in case of seasonal diseases or some disease caused by viral viruses or bacteria.

6. Conclusions

This paper presented a Petri net based mixed integer nonlinear programming model for the outdoor patient department (OPD) of the sustainable healthcare center. The integration of the Stochastic-Petri net approach with MINLP for real-time optimization in the planning model enabled the realization of sustainable operational utilization with a long-term competitive advantage. The objective of the research was to minimize the queue length of patients and the required number of servers for a sustainable future. The reason for the queue is the limited resources including unavailability or absenteeism of servers at their stations. The introduction of absenteeism in sustainable planning made this research a novel. A numerical example of a sustainable healthcare center is presented as a case study. The example provided resource planning for the period of one year. The forecast of a number of patients visiting the healthcare system is used to determine the required cycle time for each month. The Petri net mixed integer nonlinear programming is used to solve the model for determining an optimal number of servers at each station of OPD for two cases. The first case considered the determination of resources without absenteeism and the second case considered the determination of resources with absenteeism. It was observed that long queues (two to three patients) are generated if we plan the resource without the absenteeism factor and queue is substantially reduced to one patient only when absenteeism factors are considered. The reduced queue has various benefits associated with it such as maximum resource utilization and improved efficiency of the system with the highest degree of patient satisfaction. In order to evaluate the performance of a Petri net-based mixed integer nonlinear programming model, the same numerical example is solved with other methods such as the Petri net-genetic algorithm and the Petri net-pattern search method. The minimum gap of 98.08% showed that the performance of Petri net-MINLP is better in terms of computation time and function value. This research is useful for the operation manager for resource planning in a healthcare center. This is a planning model for the period of one year and estimates the required number of servers at each station. However, it does not provide any information for controlling the mechanism of servers if the server suddenly leaves the system without serving the patients on a daily basis. The sudden absenteeism control of servers is included in future research.

Author Contributions: Conceptualization, B.S.; methodology and writing original draft preparation, M.I.; writing review and editing, M.O.; formal analysis, W.A.; data curation, M.U.; supervision, C.W.K.

Funding: This research received no external funding.

Conflicts of Interest: The authors declare no conflict of interest.

Abbreviations

The notations and abbreviations used in this model are given below.

Indices:

i	patient type $i = 1, 2 \ldots , I$
j	server type $j = 1, 2 \ldots , J$

Legends:

$\rightarrow$	directed arc shows the flow of token between transition and place
$\circ$	token (patient)

Parameters:

t_{ij}	attendance time of patient "i" at server "j".
ρ_{ij}	a place for waiting of patient "i" to be served by server "j".
S_{ij}	place of server "j" for serving patient "i".
ζ_p	overlapping circuit for each patient
ζ_s	overlapping sequencing circuits
$\tau(\zeta)$	total transition time
$C(\zeta)$	the cycle time of each sequential loop
PA_j	the probability of absenteeism of server "j"
SR	service rate
PH	planning horizon
NP	the expected number of patients
$C(\zeta_{sMAX})$	the maximum cycle time of the bottleneck server or required cycle time for a patient

Decision Variables:

NQ_{ij}	Number of patients " " waiting for service at server "j"
NS_{ij}	Number of servers "j" serving the patient "i"

References

1. Brundtland, G.H.; Khalid, M.; Agnelli, S.; Al-Athel, S. *Our Common Future*; Oxford University Press: Oxford, UK, 1987.
2. Omair, M.; Sarkar, B.; Cárdenas-Barrón, L.E. Minimum Quantity Lubrication and Carbon Footprint: A Step towards Sustainability. *Sustainability* **2017**, *9*, 714. [CrossRef]
3. Habib, M.S.; Sarkar, B.; Tayyab, M.; Saleem, M.W.; Hussain, A.; Ullah, M.; Omair, M.; Iqbal, M.W. Large-scale disaster waste management under uncertain environment. *J. Clean. Prod.* **2019**, *212*, 200–222. [CrossRef]
4. Omair, M.; Noor, S.; Hussain, I.; Maqsood, S.; Khattak, S.; Akhtar, R.; Haq, I.U. Sustainable development tool for Khyber Pakhtunkhwa's dimension stone industry. *Technol. J.* **2015**, *20*, 160–165.
5. Sarkar, B.; Omair, M.; Choi, S.-B. A Multi-Objective Optimization of Energy, Economic, and Carbon Emission in a Production Model under Sustainable Supply Chain Management. *Appl. Sci.* **2018**, *8*, 1744. [CrossRef]
6. Tayyab, M.; Sarkar, B.; Ullah, M. Sustainable Lot Size in a Multistage Lean-Green Manufacturing Process under Uncertainty. *Mathematics* **2018**, *7*, 20. [CrossRef]
7. Carnero, M.C. Assessment of environmental sustainability in health care organizations. *Sustainability* **2015**, *7*, 8270–8291. [CrossRef]
8. Liou, J.J.; Lu, M.-T.; Hu, S.-K.; Cheng, C.-H.; Chuang, Y.-C. A Hybrid MCDM Model for Improving the Electronic Health Record to Better Serve Client Needs. *Sustainability* **2017**, *9*, 1819. [CrossRef]
9. Carpman, J.R.; Grant, M.A. *Design That Cares: Planning Health Facilities for Patients and Visitors*; John Wiley & Sons: Hoboken, NJ, USA, 2016; Volume 142.
10. Dangerfield, B. System Dynamics Applications to European Healthcare Issues. In *Operational Research for Emergency Planning in Healthcare*; Springer: Berlin, Germany, 2016; Volume 2, pp. 296–315.
11. Hans, E.W.; Van Houdenhoven, M.; Hulshof, P.J. A Framework for Healthcare Planning and Control. In *Handbook of Healthcare System Scheduling*; Springer: Berlin, Germany, 2012; pp. 303–320.
12. Hulshof, P.J.; Kortbeek, N.; Boucherie, R.J.; Hans, E.W.; Bakker, P.J. Taxonomic classification of planning decisions in health care: A structured review of the state of the art in OR/MS. *Health Syst.* **2012**, *1*, 129–175. [CrossRef]

13. Lane, D.C.; Monefeldt, C.; Rosenhead, J. Looking in the wrong place for healthcare improvements: A system dynamics study of an accident and emergency department. In *Operational Research for Emergency Planning in Healthcare*; Springer: Berlin, Germany, 2016; Volume 2, pp. 92–121.

14. Zeinali, F.; Mahootchi, M.; Sepehri, M.M. Resource planning in the emergency departments: A simulation-based metamodeling approach. *Simul. Model. Pract. Theory* **2015**, *53*, 123–138. [CrossRef]

15. Yousefi, M.; Yousefi, M.; Ferreira, R.P.M.; Kim, J.H.; Fogliatto, F.S. Chaotic genetic algorithm and Adaboost ensemble metamodeling approach for optimum resource planning in emergency departments. *Artif. Intell. Med.* **2018**, *84*, 23–33. [CrossRef]

16. Feng, Y.-Y.; Wu, I.-C.; Chen, T.-L. Stochastic resource allocation in emergency departments with a multi-objective simulation optimization algorithm. *Health Care Manag. Sci.* **2017**, *20*, 55–75. [CrossRef] [PubMed]

17. Bachouch, R.B.; Guinet, A.; Hajri-Gabouj, S. An integer linear model for hospital bed planning. *Int. J. Prod. Econ.* **2012**, *140*, 833–843. [CrossRef]

18. Barado, J.; Guergué, J.M.; Esparza, L.; Azcárate, C.; Mallor, F.; Ochoa, S. A mathematical model for simulating daily bed occupancy in an intensive care unit. *Crit. Care Med.* **2012**, *40*, 1098–1104. [CrossRef] [PubMed]

19. Ma, G.; Demeulemeester, E. A multilevel integrative approach to hospital case mix and capacity planning. *Comput. Oper. Res.* **2013**, *40*, 2198–2207. [CrossRef]

20. Brailsford, S.; Hilton, N. A Comparison of Discrete Event Simulation and System Dynamics for Modelling Health Care Systems. In *Planning for the Future: Health Service Quality and Emergency Accessibility*; Riley, J., Ed.; University of Southampton: Southampton, UK, 2001.

21. Morecroft, J.; Robinson, S. Explaining Puzzling Dynamics: Comparing the Use of System Dynamics and Discrete-Event Simulation. In Proceedings of the 23rd International Conference of the System Dynamics Society, Boston, MA, USA, 17–21 July 2005; pp. 17–21.

22. Duggan, J. A Comparison of Petri Net and System Dynamics Approaches for Modelling Dynamic Feedback Systems. In Proceedings of the 24th International Conference of the Systems Dynamics Society, Nijmegen, The Netherlands, 23–27 July 2006.

23. Eppich, W.; Howard, V.; Vozenilek, J.; Curran, I. Simulation-based team training in healthcare. *Simul. Healthc.* **2011**, *6*, 14–19. [CrossRef] [PubMed]

24. Hall, R. Patient flow. *AMC* **2014**, *10*, 12.

25. Meiller, Y.; Bureau, S.; Zhou, W.; Piramuthu, S. Adaptive knowledge-based system for health care applications with RFID-generated information. *Decis. Support. Syst.* **2011**, *51*, 198–207. [CrossRef]

26. Robinson, S.; Radnor, Z.J.; Burgess, N.; Worthington, C. SimLean: Utilising simulation in the implementation of lean in healthcare. *Eur. J. Oper. Res.* **2012**, *219*, 188–197. [CrossRef]

27. Rohleder, T.R.; Lewkonia, P.; Bischak, D.P.; Duffy, P.; Hendijani, R. Using simulation modeling to improve patient flow at an outpatient orthopedic clinic. *Health Care Manag. Sci.* **2011**, *14*, 135–145. [CrossRef]

28. Du, G.; Liang, X.; Sun, C.J.S. Scheduling optimization of home health care service considering patients' priorities and time windows. *Sustainability* **2017**, *9*, 253. [CrossRef]

29. Yalçındağ, S.; Cappanera, P.; Scutellà, M.G.; Şahin, E.; Matta, A.J.C.; Research, O. Pattern-based decompositions for human resource planning in home health care services. *Comput. Oper. Res.* **2016**, *73*, 12–26. [CrossRef]

30. En-nahli, L.; Allaoui, H.; Nouaouri, I.J.I.-P. A multi-objective modelling to human resource assignment and routing problem for home health care services. *Sci. Direct* **2015**, *48*, 698–703. [CrossRef]

31. Ganguly, B.; Pareek, S.; Sarkar, B.; Sarkar, M.; Omair, M. Influence of controllable lead time, premium price, and unequal shipments under environmental effects in a supply chain management. *RAIRO-Oper. Res.* **2018**. [CrossRef]

32. Wook Kang, C.; Ullah, M.; Sarkar, M.; Omair, M.; Sarkar, B. A Single-Stage Manufacturing Model with Imperfect Items, Inspections, Rework, and Planned Backorders. *Mathematics* **2019**, *7*, 446. [CrossRef]

33. Omair, M. Effect of Workplace Stress in Short Term and Long Term Controllable Production System. Ph.D. Thesis, Graduate School of Hanyang University, Seoul, Korea, 2019. Available online: http://dcollection. hanyang.ac.kr/common/orgView/000000107873.

34. Ghazalbash, S.; Sepehri, M.M.; Shadpour, P.; Atighehchian, A. Operating room scheduling in teaching hospitals. *Adv. Oper. Res.* **2012**, *2012*, 16. [CrossRef]

35. Nasir, J.A.; Dang, C. Solving a More Flexible Home Health Care Scheduling and Routing Problem with Joint Patient and Nursing Staff Selection. *Sustainability* **2018**, *10*, 148. [CrossRef]

36. Tan, Y.; El Mekkawy, T.; Peng, Q.; Oppenheimer, L. Mathematical programming for the scheduling of elective patients in the operating room department. *Proc. Can. Eng. Educ. Assoc.* **2011**. [CrossRef]

37. Carayon, P.; Wetterneck, T.B.; Rivera-Rodriguez, A.J.; Hundt, A.S.; Hoonakker, P.; Holden, R.; Gurses, A.P. Human factors systems approach to healthcare quality and patient safety. *Appl. Ergon.* **2014**, *45*, 14–25. [CrossRef]

38. Atun, R. Health systems, systems thinking and innovation. *Health Policy Plan.* **2012**, *27*, 4–8. [CrossRef]

39. Fraher, E.P.; Knapton, A.; Sheldon, G.F.; Meyer, A.; Ricketts, T.C. Projecting surgeon supply using a dynamic model. *Ann. Surg.* **2013**, *257*, 867–872. [CrossRef]

40. Lee, C.W.; Kwak, N. Strategic enterprise resource planning in a health-care system using a multicriteria decision-making model. *J. Med. Syst.* **2011**, *35*, 265–275. [CrossRef] [PubMed]

41. Abo-Hamad, W.; Arisha, A. Simulation-based framework to improve patient experience in an emergency department. *Eur. J. Oper. Res.* **2013**, *224*, 154–166. [CrossRef]

42. Zeltyn, S.; Marmor, Y.N.; Mandelbaum, A.; Carmeli, B.; Greenshpan, O.; Mesika, Y.; Wasserkrug, S.; Vortman, P.; Shtub, A.; Lauterman, T. Simulation-based models of emergency departments: Operational, tactical, and strategic staffing. *ACM Trans. Modeling Comput. Simul. (TOMACS)* **2011**, *21*, 24. [CrossRef]

43. Hamrock, E.; Parks, J.; Scheulen, J.; Bradbury, F.J. Discrete event simulation for healthcare organizations: A tool for decision making. *J. Healthc. Manag.* **2013**, *58*, 110. [CrossRef] [PubMed]

44. Jahangirian, M.; Naseer, A.; Stergioulas, L.; Young, T.; Eldabi, T.; Brailsford, S.; Patel, B.; Harper, P. Simulation in health-care: Lessons from other sectors. *Oper. Res.* **2012**, *12*, 45–55. [CrossRef]

45. Zhang, Y.; Puterman, M.L.; Nelson, M.; Atkins, D. A simulation optimization approach to long-term care capacity planning. *Oper. Res.* **2012**, *60*, 249–261. [CrossRef]

46. Ozcan, Y.A.; Tànfani, E.; Testi, A. A Simulation-Based Modeling Framework to Deal with Clinical Pathways. In Proceedings of the Winter Simulation Conference, Phoenix, AZ, USA, 11–14 December 2001; pp. 1190–1201.

47. Granja, C.; Almada-Lobo, B.; Janela, F.; Seabra, J.; Mendes, A. An optimization based on simulation approach to the patient admission scheduling problem using a linear programing algorithm. *J. Biomed. Inform.* **2014**, *52*, 427–437. [CrossRef] [PubMed]

48. Mahulea, C.; Mahulea, L.; García-Soriano, J.-M.; Colom, J.-M. Petri Nets with Resources for Modeling Primary Healthcare Systems. In Proceedings of the 18th International Conference System Theory, Control and Computing (ICSTCC), Sinaia, Romania, 17–19 October 2014; pp. 639–644.

49. Bertolini, C.; Liu, Z.; Srba, J. Verification of Timed Healthcare Workflows Using Component Timed-Arc Petri Nets. In Proceedings of the International Symposium on Foundations of Health Informatics Engineering and Systems, Paris, France, 27–28 August 2012; pp. 19–36.

50. Hicheur, A.; Dhieb, A.B.; Barkaoui, K. Modelling and Analysis of Flexible Healthcare Processes Based on Algebraic and Recursive Petri Nets. In Proceedings of the International Symposium on Foundations of Health Informatics Engineering and Systems, Paris, France, 27–28 August 2012; pp. 1–18.

51. Wang, J. Emergency healthcare workflow modeling and timeliness analysis. *IEEE Trans. Syst. Man Cybern. Part. A Syst. Hum.* **2012**, *42*, 1323–1331. [CrossRef]

52. Zeng, Q.; Lu, F.; Liu, C.; Duan, H.; Zhou, C. Modeling and verification for cross-department collaborative business processes using extended Petri nets. *IEEE Trans. Syst. Man Cybern. Syst.* **2015**, *45*, 349–362. [CrossRef]

53. Mahulea, C.; Mahulea, L.; Soriano, J.M.G.; Colom, J.M. Modular Petri net modeling of healthcare systems. *Flex. Serv. Manuf. J.* **2018**, *30*, 329–357. [CrossRef]

54. Wang, J. Petri Nets for Dynamic Event-Driven System Modeling. *Handb. Dyn. Syst. Modeling* **2007**, *1*.

55. Aziz, M.; Bohez, E.L.; Pisuchpen, R.; Parnichkun, M. Petri Net model of repetitive push manufacturing with Polca to minimise value-added WIP. *Int. J. Prod. Res.* **2013**, *51*, 4464–4483. [CrossRef]

56. Mukhlash, I.; Rumana, W.N.; Adzkiya, D.; Sarno, R. Business Process Improvement of Production Systems Using Coloured Petri Nets. *Bull. Electr. Eng. Inform.* **2018**, *7*, 102–112.

57. Premchaiswadi, W.; Porouhan, P. Process modeling and bottleneck mining in online peer-review systems. *SpringerPlus* **2015**, *4*, 441. [CrossRef] [PubMed]

58. Ren, Y.; Zhang, C.; Zhao, F.; Xiao, H.; Tian, G. An asynchronous parallel disassembly planning based on genetic algorithm. *Eur. J. Oper. Res.* **2018**, *269*, 647–660. [CrossRef]

59. Si, Y.-W.; Chan, V.-I.; Dumas, M.; Zhang, D. A Petri Nets based Generic Genetic Algorithm framework for resource optimization in business processes. *Simul. Model. Pract. Theory* **2018**, *86*, 72–101. [CrossRef]
60. Tax, N.; Sidorova, N.; van der Aalst, W.M.; Haakma, R. LocalProcessModelDiscovery: Bringing Petri Nets to the Pattern Mining World. In Proceedings of the 39th International Conference on Applications and Theory of Petri Nets and Concurrency, Bratislava, Slovakia, 24–29 June 2018; pp. 374–384.

 © 2019 by the authors. Licensee MDPI, Basel, Switzerland. This article is an open access article distributed under the terms and conditions of the Creative Commons Attribution (CC BY) license (http://creativecommons.org/licenses/by/4.0/).

Article

Monitoring the Performance of Petrochemical Organizations in Saudi Arabia Using Data Envelopment Analysis

Hisham Alidrisi [1], **Mehmet Emin Aydin** [2,*], **Abdullah Omer Bafail** [1], **Reda Abdulal** [1] and **Shoukath Ali Karuvatt** [1]

[1] Department of Industrial Engineering, King Abdulaziz University, Jeddah, 21589, Saudi Arabia; hmalidrisi@kau.edu.sa (H.A.); abafail@kau.edu.sa (A.O.B.); rabdelaal@kau.edu.sa (R.A.); drshoukath@hotmail.com (S.A.K.)

[2] Department of Computer Science, University of West of England, Bristol, BS16 1QY, UK

* Correspondence: mehmet.aydin@uwe.ac.uk

Received: 3 May 2019; Accepted: 4 June 2019; Published: 6 June 2019

Abstract: The petrochemical industry plays a crucial role in the economy of the Kingdom of Saudi Arabia. Therefore, the effectiveness and efficiency of this industry is of high importance. Data envelopment analysis (DEA) is found to be more acceptable in measuring the effectiveness of various industries when used in conjunction with non-parametric methods such as multiple regression, analytical hierarchy process (AHP), multidimensional scaling (MDS), and other multiple criteria decision making (MCDM) approaches. In this study, ten petrochemical companies in the Kingdom of Saudi Arabia are evaluated using Banker, Charnes and Cooper (BCC)/Charnes, Cooper, and Rhodes (CCR) models of DEA to compute the technical and super-efficiencies for ranking according to their relative performances. Data were collected from the Saudi Stock Exchange on key financial performance measures, five of which were chosen as inputs and five as outputs. Five DEA models were developed using different input–output combinations. The efficiency plots obtained from DEA were compared with the Euclidean distance scatter plot obtained from MDS. The dimensionality of MDS plots was derived from the DEA output. It was found that the two-dimensional positioning of the companies was congruent in both plots, thus validating the DEA results.

Keywords: data envelopment analysis; benchmarking; petrochemical industries; technical and super-efficiencies; multidimensional scaling; efficiency and scatter plots

1. Introduction

The dynamic nature of the Saudi economy stems from the fluctuating oil prices and the balancing force of the huge foreign exchange reserves, with $734,500 million [1]. Despite the efforts of the Saudi Industrial Property Authority (MODON) to boost industrial production in the country, the oil sector remains the pivot sector in driving the Saudi economy ahead of others. Hence, it is essential that the country adopt bold steps to boost industries related to the oil sector and its spin-off sectors. Although the developments in the oil market are slow-paced currently, there are opportunities to come up with new plans. During an imbalance of supply and demand, any country with low-cost production will have an edge over other countries. In this regard, the petroleum sector of Saudi Arabia is likely to be the most benefited. The petroleum companies need to adopt a steady approach and focus on enhancing their productivity skills [2].

In view of the recent developments reflected and the literature regarding the petroleum sector of the Kingdom of Saudi Arabia (KSA), it is paramount to monitor the performance of the companies within the sector and investigate ways to improve their performances. This is due to the role of the sector being played within the economy of the KSA, which is one of the foremost contributors to the economic growth of the nation. The motivation for this research was founded on these facts. Furthermore, as the KSA is all set for a big leap towards industrialization and globalization, the role of the petroleum sector companies in upholding the already established ranking of the Saudi share market is vital. One effective way of achieving this is the continuous monitoring and controlling of the financial performance of companies.

Realizing the importance of the role of energy efficiency and its measurement in the petroleum and petrochemical industries, many researchers have studied their efficiency problems with the aid of various approaches. However, compared to other sectors, it appears that the number of research attempts is much lower. In a comprehensive literature survey on DEA applications in various sectors [3], the authors comment that only about 15 were found to be related to petroleum companies, out of 4015 research publications surveyed. Probably, this was due to the inaccessibility or unavailability of data on petrochemical industries.

Oliveira et al. [4] applied DEA to study the vulnerability of oil market variations in Venezuela, Brazil, Ecuador, Argentina, Colombia, and Peru, based on 2005 data from the British Petroleum (BP) Statistical Review of World Energy. The efficiency is computed based on usage and dependency of the resources using inputs and outputs (I/Os) variables such as production, consumption, and proved reserves of oil and natural gas. Ranjbar et al. [5] applied window analysis, a dynamic method in DEA, to measure the relative performance of petrochemical industries listed active during the 2003–2010 period at the Tehran Stock Exchange. They used the financial ratios such as current assets, fixed assets, and cost of goods sold as input and return on assets, return on investment, sale, and profit-to-sales ratio as output variables.

Alsahlawi [6], studied the energy efficiency status of six GCC countries using DEA to set the appropriate policies without adverse effects on their economic development strategies. They claim that it is the first attempt in this direction. They used economic-thermodynamic indicators like energy GDP ratio to compare the level of relative efficiency of each GCC country. Another notable research attempt in this area is by Shaverdia et al. [7] on the Iranian petrochemical industries. Jandaghi and Ramshini [8] applied fuzzy AHP (FAHP) and DEA to rank and separate efficient from non-efficient petrochemical companies in Iran. Recently, a few prominent studies have been conducted on the efficiency of the Chinese power industry, as indicated in [9–11].

The main aim of the research reported in this paper is to analyze the financial performance of petrochemical industries in Saudi Arabia to set reference points that would help companies to benchmark their performance. The research includes studying the performances of functioning petrochemical companies via the Saudi share market, where the relevant financial data are collected, including the key economic performance indicators (KPIs) related to petrochemical sector companies for the last five years. The data are studied using variants of data envelopment analysis (DEA) to determine the relative performance efficiencies of the companies, and multidimensional scaling (MDS) is applied to find the relative positions of these companies and enhance the confidence in the results from DEA. This leads to measuring the relative performance of the petroleum companies for the last five years using windows analysis to help readjustment by each company.

The use of DEA in this work is due to its solid and strong mathematical basis, being one of strongest quantitative technique in performance measurement of companies. The use of qualitative approaches such as analytical hierarchy processes are usually due to the lack of relevant numerical data and the complexities emerging from existing data. It is also crucial to select the best-fitting DEA model in order to get high accuracy and soundness in the measurements and the concluded knowledge. Multidimensional scaling (MDS) is another useful distance-based approach that helps draw a baseline of the performances for validation and verification purposes. It also helps visualize the outputs of DEA, making it more comprehensive.

The rest of this paper is organized as follows. Section 2 introduces data envelopment analysis and the models used in this study, alongside an introduction of MDS, while Section 3 explains data analysis using DEA and MDS to reveal the performances. Section 4 provides relevant discussions and recommendations as a result of the study, while Section 5 includes the conclusions.

2. Data Envelopment Analysis (DEA)

The basic technique behind data envelopment analysis was originally conceived by Farrell [12] in his pioneering work on the measurement of the productive efficiency of industries. A basic feature of his method was the distinction between price and technical efficiency. Price measures a firm's success in choosing an optimal set of inputs, while technical efficiency is its success in producing maximum outputs for a given set of inputs, where the performance (the output–input ratio) of a perfectly efficient firm is expected to be 100%. If the input per unit of output is large, the efficiency is indefinitely small. An increase in the input per unit of output of one factor will indicate a lower technical efficiency.

The performance evaluation technique used in [12] comes under the genre of fractional linear programming methods. A DEA model identifies the most efficient decision-making unit (DMU), assigning a score of unity to it, and attributes a measure of inefficiency relative to it for all others. The less efficient organizations are assigned scores between 0 and 1. Thus, DEA does not measure optimal efficiency. Instead, it differentiates the least efficient organization from among the set of all companies. DEA has gained more acceptability in recent years for the evaluation and measurement of the relative efficiency of any type of system with an input and output, including organizations, educational institutions, industrial organizations, etc., provided quality data is available. The theoretical understanding of DEA requires a working knowledge of economics and mathematical programming, and the results are objective, unlike traditional partial productivity measures. Later DEA, as it is known now, was introduced by Charnes et al. [13] based on Farrell's pioneering work [14]. They generalized the single-output-to-single-input ratio definition of efficiency to multiple inputs and outputs. The original DEA model called Charnes, Cooper, and Rhodes (CCR model) suggested that the efficiency of a DMU can be obtained as the maximum of a ratio of weighted outputs to weighted inputs, subject to the condition that the same ratio for all DMUs must be less than or equal to 1. The DEA model must be run n times, once for each unit, to get the relative efficiency of all DMUs. As indicated in [15], the efficiency scores provided by DEA correspond to the economic concept of technical efficiency (TE) instead of the conventional partial efficiency (PE) of the output-to-input ratio.

The framework of DEA is adapted from multi-input, multi-output production functions and applied in industries. The method differs from the statistical least squares technique, which bases comparisons relative to an average producer. DEA identifies a "frontier" on which the relative performance of all utilities in the sample can be compared. Bafail et al. [16] comment that DEA identifies efficient and inefficient units where results are considered in contexts unique to the set of DMUs considered. Furthermore, DEA facilitates the comparison of each unit with the most efficient among them. Many researchers such as Thanassoulis [17] consider DEA more appropriate and result-oriented for measuring the efficiency of organizations than multivariate analysis techniques.

When used in a benchmarking environment, the efficient DMUs may not necessarily form a "production frontier", but rather lead to a "best-practice frontier" [18]. According to [19], DEA is a "balanced benchmarking" method. Ranking, scaling, AHP, MDS, TOPSIS, etc. can be used either independently or in conjunction with each other for performance evaluation and benchmarking. The authors of [20] suggested the application of a DEA model for ranking companies at the initial stage, followed by AHP, while Jandaghi and Ramshini [8] applied fuzzy AHP and CCR methods to measure the performance of Iranian petroleum industries. Similarly, Mohaghar et al. [21] used a special version of DEA in comparison to the fuzzy VICAR method for a supplier selection problem, while authors of [22] have used a fuzzy version of DEA in comparison with fuzzy AHP for a facility layout design problem.

2.1. DEA Models

In the DEA model initially developed by Charnes et al. [13], a score of 1 is assigned to a decision-making unit (DMU) only when comparisons with other relevant DMUs do not provide evidence of inefficiency for the same sets of inputs and outputs. DEA assigns an efficiency score less than 1 to (relatively) inefficient units. A score less than 1 means a linear combination of other DMUs. The score reflects the radial distance from the estimated production frontier to the DMU under consideration. There are a number of equivalent formulations for DEA. The most direct formulation of the exposition given above is as follows: Let $x_i \in X$ be the vector of inputs into DMU_i. Let $y_i \in Y$ be the corresponding vector of outputs. Let x_0 be the inputs and y_0 be the outputs. The best combination of $DMUs$ determines DMU_0. This problem can be written as the measure of efficiency for DMU_0 given by the following fractional program:

$$\text{Min} \quad Z = \theta,$$

subject to:

$$\sum \lambda_i x_i \leq \theta x_0,$$
$$\sum \lambda_i y_i \leq y_0,$$

where λ_i is the weight given to DMU_i in its efforts to determine DMU_0, and θ is the efficiency of DMU_0. In general, it should include DMU_0 on the left-hand side of the equations. Then, the optimal θ cannot possibly be more than 1. When someone solves this linear program, the efficiency (θ) of $DMU_0 = 1$, meaning that the unit is efficient.

The efficiency of a DMU is simply the relationship of the outputs to inputs and is constrained to be less than 1. The goal is to find a set of prices and values that puts the target DMU in the best possible light. So the simple LP maximization model is

$$\text{Max} \quad Z = \frac{u^T y_0}{v^T x_0},$$

subject to:

$$\frac{u^T y_j}{v^T x_j} \leq 1 \qquad j = 0, 1, ..., n,$$

$$u^T, v^T \geq 0.$$

Here, u^T and v^T are vectors of prices and values, respectively, which have nothing to do with real prices and values: they are artificial constructs.

2.2. Methodological Extension of DEA

Originally, Charnes et al. [13] developed the DEA technique for evaluating the relative efficiency of production systems of a similar nature. The basic DEA models discuss the way the returns-to-scale, the geometry of the envelopment surface, and the efficient projections are identified. Flexibility is the main advantage of this generic model. If new variables and weights are added, DEA can be refined to reflect managerial or organization factors, sharpen efficiency estimates, and/or overcome inconsistencies. Many researchers, including the authors of [23], through empirical analyses found new applications for DEA in banking, industry, and higher education systems, as well as facility layouts, etc. The main problem was that when the number of inputs and outputs was large, the DEA analysis yielded a large number of most efficient DMUs. Moreover, there were a large number of DMUs attached with 0 weights. In view of the limitations of the original model, investigators attempted to develop more versatile models to suit constant returns and variable returns, namely constant return to scale (CRS) and variable returns to scale (VRS) models. In addition, Talluri [24] discusses some methodological extensions of the original DEA model by [13]. For more literature, see [25,26]. Alternative DEA models such as the multiplicative [27] and the additive [28] models of Charnes et al. [13] were developed later to suit different applications. Banker et al. [29] developed the Banker, Charnes, Cooper (BCC) model to estimate the pure technical efficiency of DMUs with reference to the efficient frontier. It also identifies whether a DMU is operating in increasing, decreasing, or constant returns to scale. Charnes, Cooper, and Rhodes (CCR) models are a specific type of BCC model. Bafail et al. [30] used BCC models to evaluate the relative efficiency of Saudi banks. Their analyses were supported by results obtained from cluster analysis on the data sets.

The choice of a particular DEA model is dependent upon the objective of the study and the expected results. The goal of DEA arises from situations where the productive efficiency of a system or DMU is of importance. DEA measures how well the DMUs convert inputs into outputs, while multiple criteria decision making (MCDM) models have arisen from problems of ranking when there are alternatives that have conflicting criteria. A methodological connection between MCDM is to define maximizing criteria (benefits) as outputs and minimizing criteria (costs) as inputs. A very useful literature survey is provided in [25,26] on the use of selected DEA models for various problem domains. Many other approaches combining DEA with AHP, maximum efficiency ratio models, and a multi-objective linear program are found in literature [8].

2.3. Tie-Breaking Models for Super Efficiency

A major limitation of the basic models is that when the number of inputs and outputs is large, the DEA analysis yields a large number of most efficient or 100% efficient DMUs. The CCR super-efficiency DEA model, the so-called AP model, was first developed under constant returns to scale (CRS) by Andersen and Petersen [31]. The problem with the AP model was unfeasibility and instability when some inputs are close to 0. The authors of [32] discuss super-efficiency models, which were first introduced as a "tie-breaking procedure" for ranking units rated as 100% efficient in conventional DEA models. When a DMU being evaluated is excluded from the reference set of DEA, the resulting DEA model is called a super-efficiency DEA model. It is concluded by [33] that the super-efficiency scores of DMUs enables us to distinguish between the efficient observations. On the other hand, another approach for ranking inefficient DMUs is proposed in [34] to overcome the issues caused by unfeasibility, as a modification of the earlier super-efficiency models. In addition, Ebadi [35] proposes variable returns to scale (VRS) super-efficiency models using input–output orientation, while Salhieh and Al-Harris [36] recommend a super-efficiency model for further discrimination between product concepts. The super-efficiency metric can be calculated for the efficient concepts based on the following model:

$$\text{Min} \quad \delta = \frac{\frac{1}{m}\sum_{i=1}^{m}\frac{\bar{x}_i}{x_{ik}}}{\frac{1}{s}\sum_{r=1}^{s}\frac{\bar{y}_r}{y_{rk}}},$$

subject to

$$\sum_{j=1,j\neq k}^{n}\lambda_j x_j \geq \bar{x}_i \qquad i = 1, 2, .., m,$$

$$\sum_{j=1,j\neq k}^{n}\lambda_j y_j \leq \bar{y}_r \qquad r = 1, 2, .., s,$$

$$\lambda_j \geq 0 \qquad j = 1, 2, .., m \quad j \neq k,$$

$$\bar{x}_i \geq x_{ik} \qquad i = 1, 2, .., m,$$

$$0 \leq \bar{y}_r \leq y_{rk} \qquad r = 1, 2, .., s.$$

A professional software package that makes use of these super-efficiency models to estimate super-efficiencies under various optimization modes is used for support in this work. Technical/standard efficiencies were also computed using the same software.

2.4. Multidimensional Scaling (MDS)

DEA is commonly used to evaluate the performance efficiency of firms. It is worth mentioning here that MDS finds wide applications in marketing research for market positioning. Recent studies have presented hybrid methods involving DEA in combination with multivariate analysis techniques such as principal component analysis (PCA), multidimensional scaling (MDS), discriminant analysis, multiple regression, and cluster analysis to solve such problems. It has been proven by researchers that such an integrated approach for performance evaluation leads to an enhancement of confidence in the DEA.

Furthermore, based on this the DEA results can be used to illustrate the similarities and differences between various models and firms. The use of MDS is reported in [37] to interpret DEA results obtained as efficiency scores. The measure of dissimilarities between two companies is considered as Euclidean distances based on 35 variables. The authors of [38] applied DEA to study healthcare management in 23 Taiwanese cities and counties. They suggested two-dimensional scatter plots from MDS analysis as a reference plane for planning and adjusting improvements. Cho and Park [39] applied a hybrid approach of DEA-PCA to measure the efficiency of mobile content firms to analyze the efficiency and the relative performance of firms. PCA and discriminant analysis (DA) have been used to validate the relationship between the DEA models and mobile content types. The results have been visulized with MDS and analyzed with respect to the data structures. Sagarra et al. [40] have analyzed 55 universities-based data on 33 variables for a seven-year period from 2007 to 2012. Scatter plots for efficiency–teaching–research against subject mix were analyzed. They applied cluster analysis with the aid of a dendrogram to handle the six-dimensional projection on to a two dimensional plane.

In this study, we use the multidimensional scaling approach (MDS) simultaneously with DEA to counter-check the DEA analysis results. From a non-technical point of view, MDS provides a visual representation of proximities in the form of similarities or dissimilarities (distances) among objects of the same type. The method implies working out a matrix of perceived similarities between 10 petrochemical industries, given in the form of financial performance data or KPIs. The variable values represent similarities or distances. Thus, the data to be analyzed is a collection of 10 variables based on which a distance function is defined, $\delta_{i,j}$, as the distance between the ith and j^{th} companies. These distances are the entries of the dissimilarity matrix as given here:

$$\Delta = \begin{pmatrix} \delta_{1,1} & \delta_{1,2} & \cdots & \delta_{1,I} \\ \delta_{2,1} & \delta_{2,2} & \cdots & \delta_{2,I} \\ \vdots & \vdots & \ddots & \vdots \\ \delta_{I,1} & \delta_{I,2} & \cdots & \delta_{I,I} \end{pmatrix},$$

where the main aim is, given Δ, to find I vectors $x_1, .., x_i \in \Re^N$ such that $||x_i - x_j|| \approx \delta_{i,j}$ for $\forall i, j = 1, ..., I$, where $||.||$ is a vector norm or Euclidean distance. In MDS, this norm is normally the Euclidean distance, but in a broader sense, it may also be a metric or arbitrary distance function.

3. Data Analysis

DEA being a data-oriented, non-parametric method, to evaluate relative efficiency based on pre-selected inputs and outputs, it is important that the most appropriate inputs and outputs (I/Os) are chosen for the models. The researchers indicate that the effectiveness of using parametric methods in measuring and comparing the efficiencies of macroeconomic systems, such as stock markets, is debatable because they may not yield comprehensive results. In fact, it is extremely difficult to measure the performance of the petrochemical sector in simple terms such as capital structure, volume of business or investments, total assets or share equities, etc. Furthermore, using multivariate analysis techniques to analyze, classify, or rank different entities based on multiple criteria without consideration for the variability and randomness of influencing factors may result in erratic conclusions. Therefore, DEA is a very useful set of techniques where multiple inputs and outputs are considered to measure the relative performance of organizational units. The efficiency of an organization is computed using DEA by transforming the inputs into the outputs in relation to the several influential I/O factors.

3.1. Data Preparation and Selecting I/O Variables

Data preparation is a crucial stage of experimental and empirical studies, where the data is scanned and pre-assessed for proper use. While collecting data, many times a large number of possible combinations of variables might be accessible. Morita and Avkiran [41] proposed an I/O selection method that uses diagonal layout experiments, which is a statistical approach to find an optimal combination. To assess data from the NIKKEI 500 index, they demonstrate their model using a few discriminating techniques such as ANOVA.

Ten petrochemical companies in the Kingdom of Saudi Arabia were considered for analysis in this proof-of-concept study. The primary reason to select these companies is that they are the only petrochemical companies listed in the Saudi Stock Exchange Market and are found to be the most prominent and leading companies in the sector. Data on the performance measures were collected from Saudi Stock Exchange sources. Obviously, the data needs to be pre-processed in order to suit the DEA input format before performing the DEA. Many generic and sector-specific issues related to data preparation for DEA are summarized by Dyson et al. [42]. The first concern is about the number of DMUs' inputs and outputs. It is believed that the selection of I/O variables for analysis using DEA remains as one of the most challenging issues. The choice of I/O variables depends on how well each can contribute to reveal the discrimination between efficient and inefficient units. There is a clear trade-off at this stage. If a large number of DMUs is chosen, there is a greater probability of capturing high performance units that would improve the discriminatory power. In contrast, a large data set may decrease the homogeneity of the data, where larger data sets demand more computational effort. Many researchers, including the authors of [42–44], have given thumb rules to estimate the number of inputs and outputs and the number of DMUs. According to these recommendations, the number of DMUs to be chosen vary from 10 to 24. In this work, the number of DMUs is decided to be 10 due to technical constraints and difficulties in data access.

The discriminatory power of the analysis can be enhanced by reducing the number of input and output factors. If there are inputs or outputs that correlate highly with one another, the discriminatory power of the model will be lost [31]. Furthermore, many potential errors in data sets prepared after data collection may lead to erratic results. The effects of frequently noticed insufficiencies like imbalances in data sets, presence of negative data values or absence of the positivity requirement of DEA, data scaling or translation errors, missing data, etc. were eliminated or reduced by normalizing the entire data sets. A number of normalization methods in this regard are proposed and discussed by [44–49], which propose a fuzzy mathematical approach.

In this work, we used data on key financial performance measures collected from 10 top Saudi petrochemical companies within the time period of 2010 to 2015, and a snapshot of 2013 data is presented in Table 1. Out of 15 measures collected, 5 input variables and 5 output variables were selected for the purpose of analysis. These were decided upon following the correlation analysis performed to eliminate replication of highly correlated I/O variables.

Table 1. Petrochemicals Input and Output Data for 2013, as on 31/12/2013 [50].

Companies	Inputs					Outputs				
	Avg. Price	General Admin Expenses	Depreciation & Amortization	Total Assets	Owners' Equity	Gross Profit	Book Value	P/B	Gross Profit Margin (%)	Net Cash Flow
SAFCO	151	81140	373936	9459857	8268786	2878615	25	6	68	3272122
YANSAB	58	231851	1080097	4607895	15043331	3224790	27	3	34	4138027
SIPCHEM	23	136535	558071	16688750	5793223	1298580	16	2	32	1739805
SABIC	97	12759672	14283312	339070569	156271417	55344363	52	2	29	56421004
NIC	28	915657	1462237	47270232	12006133	4837264	18	2	27	3962872
SPCO	16	22324	211170	8678361	5794433	432159	13	2	18	426972
PETROCHEM	22	250516	819114	21005725	4120810	726515	9	3	16	788636
SIIG	26	261541	819419	25374069	6331913	726515	14	2	16	783016
KAYAN	12	366843	2291716	46217826	14093625	602333	9	2	6	2562576
PETRORABIGH	17	695240	2196598	45593819	8917457	461093	10	2	1	2593015

SABIC – Saudi Arabian basic Industries;
SAFCO - Saudi Arabian Fertilizer Company;
YANSAB - Yanbu National Petrochemical Company;
SPCO - Sahara Petrochemical Co.;
PETROCHEM – Petrochem Middle East (PME);

SIPCHEM - Saudi International Petrochemical Company;
KAYAN - Saudi Kayan Petrochemical Company;
NIC - National Industrialization Company;
PETRORABHIGH - Rabigh Refining & Petrochemical Co.,
SIIG - Saudi Industrial Investment Group

3.2. Super-Efficiency Model (SEM)

The weakness of the basic DEA model is its inability to discriminate highly efficient DMUs. The efficiency plot displays many 100% DMUs crowded at the DEA frontier, rendering the task of ranking difficult. So it was decided to apply super-efficiency models using the slack-based measures (SBM). While computing super-efficiency, the LP solution is relaxed so that the observed DMU is not used as its own peer, which results in efficiency scores greater than 100%. $x_{i,1}$ and $y_{r,1}$ are dropped from the left-hand side of the input and output inequalities correspondingly, to eliminate the peer effect of DMU1. Hence, the analyst is now able to compare efficient firms with efficiency scores of 100% ($\Theta = 1$) that operate at the frontier. The changes in the low efficiency scores are observed to be less [51].

Data analysis was done with the support of Frontier Analyst (Ver. 4.0, Banxia Software Ltd., Kandel, UK) software. Alternative formulations were also used with the help of the "advanced" tab of the analysis options in the software. Standard formulation, so-called BCC/CCR technical efficiency, and super-efficiency formulations were used in this work, with one optimization mode (minimize input) and two scaling modes (constant returns and variable returns). The software Frontier Analyst also offers additive formulations, allocative (cost-based) formulations, and Malmquist formulations. Each mode allows certain options. With the help of super-efficiency analysis, efficiency improvements for efficient units can be estimated, where interpretation of the results turns out to be more complicated when super-efficiency is enabled.

4. Results and Discussions

A large number of DMUs is effective in improving the discriminatory power of the DEA model. However, we have decided to consider 10 DMUs in this study due to technical constraints and the practical difficulties in data collection and compilation. Similarly, the number of variables used was limited to 10. As argued in [43]), large data sets are likely to reduce the homogeneity of the data. Therefore, following the thumb rules suggested by researchers presented in the previous section, 5 different evaluation models were decided using different input/output combinations out of 5 input and 5 output variables. Since the variables considered/selected are all financial performance measures, the analysis is expected to provide the efficiency scores for the financial performance of the companies. As there were 10 financial measures of petrochemical companies from Saudi Arabia, model variations could be achieved by changing the I/O combinations. Standard efficiency (BCC/CCR technical efficiency) and super-efficiencies for the 10 petrochemical industries were obtained using the DEA approach with the software support, and results are tabulated in Table 2. The technical efficiency has been calculated through the BCC/CCR model introduced above, where the scores fit in the scale of [0...100], while the super-efficiency scores overflow this scale and hence, are normalized as 1/10 of calculated scores.

Model 1 uses all of the 5 input and 5 output variables, where it is observed that 9 out of the 10 petrochemical companies have scored technical efficiency levels of 100% at the DEA frontier, which does not help to discriminate the performances. On the other hand, the super-efficiency computation managed to significantly separate them from one another, noting that the scores in % overflow the scale of 0–100 due to the calculations provided through the super-efficiency models used. However, the results in Table 2 are normalized to fit them in the scale of 0–100%. The Saudi Arabian Fertilizer Company (SAFCO) (Jubail, Saudi Arabia) and Saudi Arabian Basic Industries (SABIC) (Riyadh, Saudi Arabia) achieved 100.0% technical efficiency and super-efficiency with normalized scores. The Yanbu National Petrochemical Company (YANSAB) (Yanbu, Saudi Arabia) and the Sahara Petrochemical Company (SPCO) (Jubail, Saudi Arabia) come next with super-efficiency scores of 73.52% and 50.08%, respectively, followed by PETROCHEM (Jubail, Saudi Arabia) with score of 22.03% and the Saudi International Petrochemical Company (SIPCHEM) (Jubail, Saudi Arabia) with score of 21.82%. The remaining 4 companies performed more or less the same. The Saudi Industrial Investment Group (SIIG) (Riyadh, Saudi Arabia) achieved technical efficiency of 85% and super-efficiencies of 8.5% only. DMUs away from the DEA frontier are expected to show low super-efficiency scores, as clearly indicated in [51]. In **Model 1**, all input variables, including depreciation, were considered to be controlled. When it was considered as an uncontrolled input variable, the performance efficiency scores of YANSAB and SPCO also rose to 100.00%. Thus, it is very clear that under controlled input scenarios, the discrimination of DMUs becomes vague, as a large number of DMUs tends to cluster around the nearest ones.

Models 2, 3, 4, and **5** were constructed by changing the input/output combinations. In all of the model combinations, SAFCO was found to exhibit the highest super-efficiency scores, ranging from 23.74% to 100.00%. The technical and super-efficiency scores of SABIC in **Model 4** were found to be 47.10% and 4.71% only, respectively. In **Models 2** and **5**, the performance of SABIC and YANSAB were found to be very close. In **Model 5**, SABIC achieved an efficiency of 100.00%. and YANSAB performed at 20.89%, compared to 73.52% in **Model 1**. The performance of The Rabigh Refining and Petrochemical Company (PETRORABIGH) (Rabigh, Saudi Arabia) and SIIG remain very low in all the models, and the National Industrialization Company (NIC) (Riyadh, Saudi Arabia) was found to be slightly above them. Meanwhile, the performance of SIPCHEM, PETROCHEM, and the Saudi Kayan Petrochemical Company (KAYAN) (Jubail, Saudi Arabia) were found to be very close together in all models, though SIPCHEM is a little bit ahead of the other two, with super-efficiency scores ranging from 3.56% to 21.82%. It was found that at low technical efficiency levels, when DMUs are far away from the DEA frontier, both the technical and super-efficiency scores tend to be the same or closer. For example, in the case SIIG, technical efficiency is lowest at 1.52% in **Model 4**, and the maximum is 8.5% in **Model 1** and **Model 5**. It may be noted that the super-efficiency scores of SIIG before normalization are the same as those of the technical efficiency scores. The estimated level of improvement that can be achieved for DMUs based on the I/O configuration is tabulated in Table 3; about 35.3% improvement is possible in the gross profit margin %. Average price, an input variable, has a potential of 11.57%. Depreciation records the minimum possibility of improvement, with only 0.83%. Efficiency plots for the 10 companies, with "Gross Profit Margin %", and "Average price" indicated on the x- and y-axes of the plots, respectively, are given in Figure 1a,b. Such efficiency plots from the software demonstrate the relative positioning of the companies with respect to the variables chosen. SAFCO, with a gross profit margin of 64.70%, sits at the top of the list for efficiency and hence is placed at the top right-hand corner of the plot.

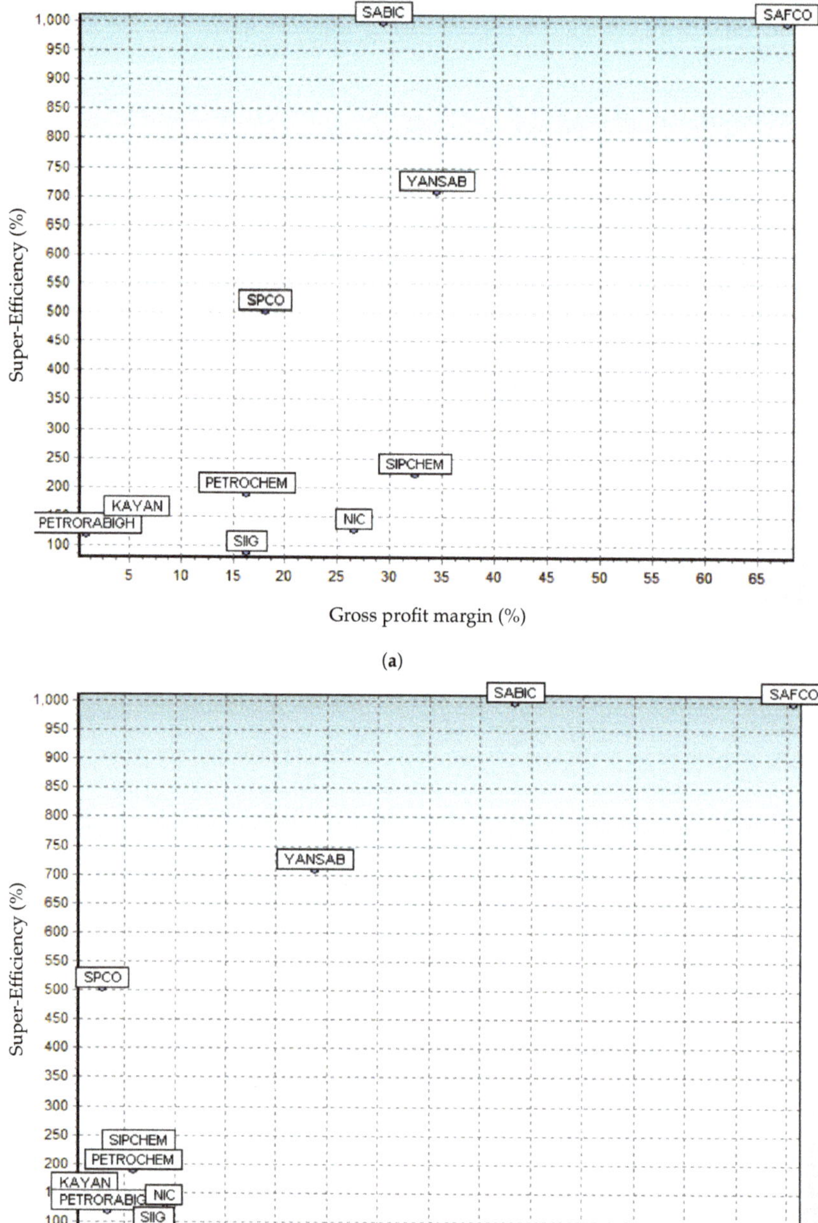

Figure 1. Efficiency plots of the decision-making units (DMUs).

Table 2. Technical and super-efficiencies with 5 different data envelopment analysis (DEA) models.

Models	Model 1		Model 2		Model 3		Model 4		Model 5	
Minimise Inputs	Gen. Admin Exp., Total Assets, Owner's Equity, Depreciation, Avrg. Price		Gen. Admin exp., Total Assets, Depreciation		Gen. Admin exp., Total Assets, Owner's equity		Gen. Admin Exp., Total Assets, Depreciation		Owner's Equity, Depreciation, Avrg. Price	
Outputs	Gross profit, Net Cash Flow, P/B, Book Value, Gross Profit Margin		Net Cash flow, P/B		Gross Profit, Net Cash Flow,		Net Cash Flow, P/B		Book value, Gross Profit Margin	
Company	Tech Eff. (%)	Sup. Eff (%)	Tech. Eff.(%)	Sup. Eff (%)	Tech. Eff (%)	Sup. Eff (%)	Tech. Eff (%)	Sup. Eff (%)	Tech. Eff (%)	Sup. Eff (%)
SAFCO	100.00	100.00	100.00	100.00	100.00	23.74	100.00	28.56	100.00	100.00
SABIC	100.00	100.00	100.00	100.00	96.60	9.66	47.10	4.71	100.00	100.00
YANSAB	100.00	73.52	100.00	32.18	100.00	25.96	100.00	25.96	100.00	20.80
SPCO	100.00	50.00	100.00	36.35	54.60	5.46	100.00	12.12	100.00	25.24
PETROCHEM	100.00	22.03	40.30	4.03	49.50	4.59	22.80	2.28	100.00	14.06
SIPCHEM	100.00	21.82	53.60	5.36	75.90	7.59	35.60	3.56	100.00	19.70
KAYAN	100.00	15.36	19.60	1.96	45.90	4.59	17.30	1.73	100.00	13.33
NIC	100.00	12.92	37.90	3.79	100.00	11.57	31.00	3.10	99.50	9.95
PETRORABIGH	100.00	11.60	19.00	1.90	73.50	7.35	15.40	1.54	88.20	8.82
SIIG	85.00	8.50	33.30	3.33	32.10	3.21	15.20	1.52	85.00	8.50
Optimization Mode	Var. returns		Var. returns		Const. returns		Const. returns		Var. returns	
	Min. inputs		Min. inputs		Min. inputs		Min. inputs		Min. inputs	

Table 3. Total potential improvements.

Input/Output Variables	Improvement %
Average Price	**11.57**
General Admin Expenses	10.38
Total Assets	7.75
Owners' Equity	2.83
Depreciation	0.83
Profit Margin	**35.30**
Gross Profit	14.59
Net Cash Flow	10.61
P/B	3.39
Book value	2.75

Based on the financial performance data, MDS plots the companies on a map such that they are seen as similar organizations to each other and are placed near to each other on the 2-dimensional map in Figure 1, where the companies perceived to be different from each other are placed far away from each other. Referring to Figure 1a,b, the efficiency scores in the DEA efficiency plots reflects the radial distance from the estimated production frontier to the DMU under consideration. Similarly, the MDS plot in Figure 2 displays the distances (or proximities) among companies, which are positioned accordingly. We applied the ALSCAL algorithm within the MDS module of SPSS software to all financial data used in DEA in order to exhibit the positioning of the 10 companies in the scatter plot (Euclidean distance model).

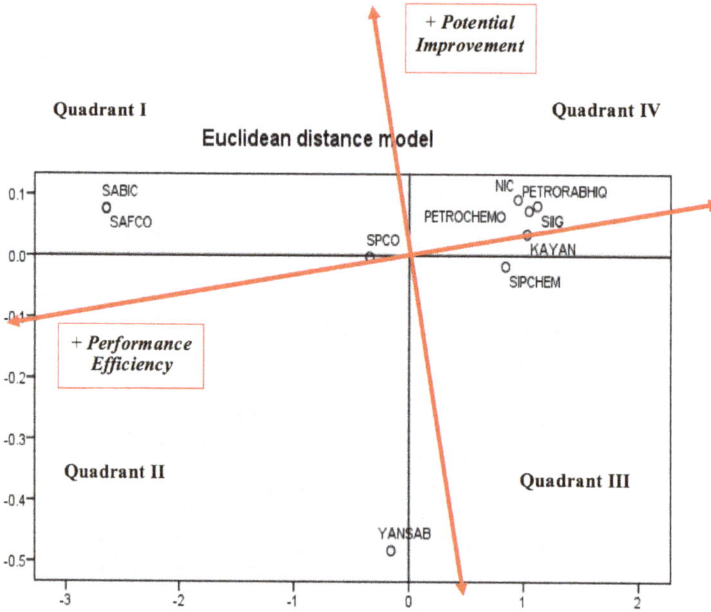

Figure 2. Multidimensional scaling (MDS) scatter plot of petrochemical companies in Saudi Arabia.

The most challenging task in MDS is the interpretation of dimensions of the 2-dimensional scatter plot. The default orientations of the axes in the scatter plots are arbitrary, and they can be rotated in any direction. To facilitate realistic interpretations of the plot, meaningful dimensions, clusters of points, or patterns need to be added. The most reasonable approach in interpreting dimensions is to use multiple regression to regress significant variables on the coordinates for different dimensions.

We used the potential improvement (%) of the I/O variables as obtained from the DEA frontier plot (Table 2) to determine the axes of the MDS plot. Out of the inputs and outputs, gross profit margin (%) and average price have the maximum potential for improvement, 35.3% and 11.57%, respectively. These key variables leading to higher performance efficiency and potential improvement are used as the axes dimensions and for positioning the companies in the DEA efficiency plots. SAFCO, SABIC, YANSAB, and SPCO were found to lead in performance efficiency as well as potential improvement (Figure 1).

In the MDS plot, SAFCO, SABIC, YANSAB, and SPCO lie in the first and second quadrants defined by the new dimensions. The companies KAYAN and SIPCHEM lie in the third quadrant, and PETROCHEM, SIG, NIC, and PETRORABIGH lie in the fourth quadrant. Table 4 displays a comparative analysis of output from both DEA and MDS.

Table 4. Output from DEA and MDS compared.

MDS Plot	Company	Super-Efficiency (%)
Quadrant I and II	SAFCO	100.0
	SABIC	100.0
	YANSAB	73.5
	SPCO	50.1
Quadrant III	SIPCHEM	21.8
	PETROCHEM	22.0
	KAYAN	15.4
Quadrant IV	NIC	12.9
	PETRORABIGH	11.6
	SIIG	8.5

5. Conclusions

The primary contribution of this work is to identify a performance evaluation approach for the unbiased positioning of Saudi Arabian petrochemical companies. Considering the importance of the petrochemical industry in the Saudi economy, it is vital to continuously monitor the efficiency and the performance of the companies in the sector, for the success of the economy. Given the boundaries of data access and the technical constraints of this prototype study, a framework is successfully established based on the DEA–MDS hybrid approach for assessing the efficiency of petrochemical industries. Despite these findings, certain limitations remain. This evaluation is based only on the financial data of the firms considered; non-financial data was not taken into account. In recent years, non-financial data, such as social involvement, environmental interfaces, and employee relations etc., have been increasingly included in the efficiency evaluations of companies.

In view of the intrinsic nature of the energy market and industry, and the vital role of the petrochemical sector in alleviating the related issues, further research needs to be conducted in this direction. The authors feel that this work could be continued with all possible combinatorial approaches. The hybrid approach of DEA–MDS presented here can be extended with more data and in combination with multiple regressions. Other multivariate techniques such as principal component analysis, logistic regression, and discriminant analysis will be extremely useful.

The most challenging task in DEA is discriminating the efficient and inefficient companies in an unambiguous way. The authors expect that econometric model variations and the new method of multi-component DEA can provide better performance measurement models. Since in this study, financial indicators were used for evaluating the performance of companies, the authors suggest that qualitative indicators also be considered in evaluating the performance of companies. The results can be compared. The use of more market indices based on expert opinion to provide performance appraisal models is also recommended.

Author Contributions: All authors have equally contributed to research activities of the project. H.A. was project lead, while the other co-authors were technical contributors. The paper was first drafted by S.A.K., revised by M.E.A., reviewed and confirmed by H.A., A.O.B. and R.A.

Funding: This project was funded by the Deanship of Scientific Research (DSR.) at King Abdulaziz University, Jeddah, Saudi Arabia, under grant no. RG-10-34. The authors therefore acknowledge with thanks the DSR's technical and financial support.

Acknowledgments: The authors would like to acknowledge with thanks the DSR's technical and financial support and the useful feedback/comments of anonymous reviewers.

Conflicts of Interest: The authors declare no conflict of interest.

References

1. SAMA. *Table(9): RESERVE ASSETS*; Technical Report; Saudi Arabian Monetary Agency: Jeddah, Saudi Arabia, 2015.
2. Sg. *Saudi Gazette Report on The story of oil in Saudi Arabia Retrieved on 07-0-2014*; Technical Report; Saudi Gazette: Jeddah, Saudi Arabia, 2014.
3. Emrouznejad, A.; Parker, B.R.; Tavares, G. Evaluation of research in efficiency and productivity: A survey and analysis of the first 30 years of scholarly literature in DEA. *J. Socio-Econ. Plann. Sci.* **2008**, *42*, 151–157. [CrossRef]
4. Oliveira, L.S.M.; Correia, T.C.V.D.; de Mello, J.C.C.B.S. Data Envelopment Analysis Applied to Evaluate the Usage of Oil and Natural Gas: South America Case. In Proceedings of the International Conference on Operational Research on Development ICORD VI, Fortelenza, Brasil, Brazil, 29–31 August 2007; pp. 487–495.
5. Ranjbar, M.H.; Abedini, B.; Afroomand, E. Performance evaluation of petrochemical firms accepted in Tehran stock exchange using DEA (window analysis). *Eur. Online J. Nat. Soc. Sci.* **2013**, *2*, 580–588.
6. Alsahlawi, M.A. Measuring Energy Efficiency in GCC Countries Using Data Envelopment Analysis. *J. Bus. Inq.* **2013**, *12*, 15–30.
7. Shaverdia, M.; Heshmatib, M.R.; Ramezanic, I. Application of Fuzzy AHP Approach for Financial Performance Evaluation of Iranian Petrochemical Sector. *Procedia Comput. Sci.* **2014**, *31*, 995–1004. [CrossRef]
8. Jandaghi, G.; Ramshini, M. A Performance Measurement Model for Automotive and Petrochemical Companies Using FAHP and CCR Method. *Eur. J. Acad. Essays* **2014**, *1*, 80–92.
9. Xian, Y.; Wang, K.; Shi, X.; Zhang, C.; Wei, Y.M.; Huang, Z. Carbon emissions intensity reduction target for China's power industry: An efficiency and productivity perspective. *J. Clean. Prod.* **2018**, *197*, 1022–1034. [CrossRef]
10. Wang, J.; Wang, K.; Shi, X.; Wei, Y.M. Spatial heterogeneity and driving forces of environmental productivity growth in China: Would it help to switch pollutant discharge fees to environmental taxes? *J. Clean. Prod.* **2019**, *223*, 36–44. [CrossRef]
11. Wang, K.; Wu, M.; Sun, Y.; Shi, X.; Sun, A.; Zhang, P. Resource abundance, industrial structure, and regional carbon emissions efficiency in China. *Resour. Policy* **2019**, *60*, 203–214. [CrossRef]
12. Farrell, M.J. The Measurement of Productive Efficiency. *J. R. Stat. Soc.* **1957**, *120*, 253–281. [CrossRef]
13. Charnes, A.; Cooper, W.; Rhodes, E. Measuring efficiency of decision-making units. *Eur. J. Oper. Res.* **1978**, *2*, 428–449. [CrossRef]
14. Toloo, M.; Nalchigar, S. A new integrated DEA model for finding most BCC-efficient DMU. *Appl. Math. Model.* **2009**, *33*, 597–604. [CrossRef]
15. Chen, W.C. Integrating Approaches to Efficiency and Productivity Measurement. Ph.D. Thesis, Georgia Institute of Technology, Atlanta, GA, USA, 2003.
16. Bafail, A.O.; Abdelaal, R.M.; Karuvat, S.A. DEA as an Integrated Approach for measuring Efficiency of a Dynamic Economy—A case study. In Proceedings of the Computational Engineering in Systems Applications (CESA 2003), Lille, France, 9–11 July 2003.
17. Thanassoulis, E. Comparison of regression analysis and data envelopment analysis as alternative methods for performance assessments. *J. Oper. Res. Soc.* **1993**, *4*, 1129–1144. [CrossRef]
18. Cook, W.D.; Tone, K.; Zhu, J. Data envelopment analysis: Prior to choosing a model. *Omega* **2014**, *44*, 1–4. [CrossRef]
19. Sherman, H.D.; Zhu, J. Analyzing performance in service organizations. *Sloan Manag. Rev.* **2013**, *54*, 37–42.

20. Stiakakis, E.; Sifaleras, A. Combining the priority rankings of DEA and AHP methodologies: A case study on an ICT industry. *Int. J. Data Anal. Tech. Strateg.* **2013**, *5*, 101–114. [CrossRef]

21. Mohaghar, A.; Fathi, M.R.; Jafarzadeh, A.H. A supplier selection method using AR-DEA and Fuzzy VIKOR. *Int. J. Ind. Eng.* **2013**, *20*, 387–400.

22. Azadeh, A.; Moradi, B. Simulation and optimization of facility layout design problem with safety and ergonomics factors. *Int. J. Ind. Eng.* **2014**, *21*, 2014.

23. Charnes, A.; Cooper, W.W.; Lewin, A.Y.; Seiford, L.M. *Data Envelopment Analysis: Theory, Methodology and Applications*; Springer: Berlin, Germany, 1997.

24. Talluri, S. Data Envelopment Analysis: Models and Extensions. Production and Operations Management. *Decis. Line* **2000**, *31*, 8–11.

25. Liu, J.S.; Lu, L.Y.Y.; Lu, W.M.; Lin, B.J.Y. Data envelopment analysis 1978–2010: A citation-based literature survey. *Omega* **2013**, *41*, 3–15. [CrossRef]

26. Liu, J.S.; Lu, L.Y.Y.; Lu, W.M.; Lin, B.J.Y. A survey of DEA Applications. *Omega* **2013**, *41*, 893–900. [CrossRef]

27. Charnes, A.; Cooper, W.; Rhodes, E. A multiplicative model for efficiency analysis. *Socio-Econ. Plann. Sci.* **1982**, *16*, 223–224. [CrossRef]

28. Charnes, A.; Cooper, W.W.; Golany, B.L.; Seiford, L.; Stutz, J. Foundations of data envelopment analysis for pareto-koopmans efficient empirical production functions. *J. Econ.* **1985**, *30*, 91–107. [CrossRef]

29. Banker, R.D.; Charnes, A.; Cooper, W.W. Some models for estimating technical and scale inefficiencies in data envelopment analysis. *Manag. Sci.* **1984**, *30*, 1078–1092. [CrossRef]

30. Bafail, A.O.; Abdelaal, R.M.; Karuvat, S.A. A DEA Approach for Measuring Relative Performance of Saudi Banks. In Proceedings of the Efficiency and Productivity Analysis in the 21th Century, Moscow, Russia, 24–26 June 2002.

31. Andersen, P.; Petersen, N.C. A procedure for ranking efficient units in data envelopment analysis. *Manag. Sci.* **1993**, *39*, 1261–1264. [CrossRef]

32. Lovell, C.K.; Rouse, A. Equivalent standard DEA models to provide super-efficiency scores. *J. Oper. Res. Soc.* **2003**, *54*, 101–108. [CrossRef]

33. Yawe, B. Hospital Performance Evaluation in Uganda: A Super-Efficiency Data Envelope Analysis Model. *Zambia Soc. Sci. J.* **2010**, *1*, 6.

34. Minh, N.K.; Khanh, P.V.; Tuan, P.A. A New Approach for Ranking Efficient Units in Data Envelopment Analysis and Application to a Sample of Vietnamese Agricultural Bank Branches. *Am. J. Oper. Res.* **2012**, *2*, 126–136. [CrossRef]

35. Ebadi, S. Using a Super Efficiency Model for Ranking Units in DEA. *Appl. Math. Sci.* **2012**, *6*, 2043–2048.

36. Salhieh, S.M.; Al-Harris, M.Y. New product concept selection: An integrated approach using data envelopment analysis (DEA) and conjoint analysis (CA). *Int. J. Eng. Technol.* **2014**, *3*, 44–55. [CrossRef]

37. Cinca, C.S.; Molinero, C.M.; Callen, Y.F. *The Path of Efficiency in DEA: Multidimensional Scaling as a Tool for Post-Optimality Analysis*; Discussion Papers in Management; University of Southampton: Southampton, UK, 2001; ISSN 1356-3548.

38. Cheng, H.; Shen, S.L.; Lu, Y.C. Resource allocation of city and county under government supervision: An effectiveness analysis of the healthcare system in Taiwan. In Proceedings of the 3rd IEEE International Conference on Innovative Computing Technology (INTECH), London, UK, 29–31 August 2013; pp. 155–161.

39. Cho, E.J.; Park, M.C. Evaluating the Efficiency of Mobile Content Companies Using Data Envelopment Analysis and Principal Component Analysis. *ETRI J.* **2011**, *33*, 443–453. [CrossRef]

40. Sagarra, M.; Agasisti, T.; Molinero, C.M. Exploring the Efficiency of Mexican Universities: Integrating Data Envelopment Analysis and Multidimensional Scaling. *Omega* **2017**, *67*, 123–133. [CrossRef]

41. Morita, H.; Avkiran, N.K. Selecting Inputs And Outputs In Data Envelopment Analysis by Designing Statistical Experiments. *J. Oper. Res. Soc. Jpn.* **2009**, *52*, 163–173. [CrossRef]

42. Dyson, R.G.; Allen, R.; Camanho, A.S.; Podinovski, V.V.; Sarrico, C.S.; Shale, E.A. Pitfalls and Protocols in DEA. *Eur. J. Oper. Res.* **2001**, *132*, 245–259. [CrossRef]

43. Golany, B.; Roll, Y. An Application Procedure for DEA. *Omega* **1989**, *17*, 237–250. [CrossRef]

44. Bowlin, W.F. Measuring Performance: An Introduction to Data Envelopment Analysis (DEA). *J. Cost Anal.* **1998**, *7*, 3–27. [CrossRef]

45. Ali, A.I.; Seiford, L.M. Translation Invariance in Data Envelopment Analysis. *Oper. Res. Lett.* **1990**, *9*, 403–405.

46. Charnes, A.; Cooper, W.W.; Thrall, R.M. A Structure for Characterizing and Classifying Efficiency and Inefficiency in Data Envelopment Analysis. *J. Prod. Anal.* **1991**, *2*, 197–237. [CrossRef]
47. Lewis, H.F.; Sexton, T.R. *Data Envelopment Analysis with Reverse Inputs*; Working Paper; State University of New York at Stony Brook: Stony Brook, NY, USA, 2000.
48. Pastor, J.T. Translation Invariance in DEA: A Generalization. *Ann. Oper. Res.* **1996**, *66*, 93–102. [CrossRef]
49. Kao, C.; Liu, S.T. Data Envelopment Analysis with Missing Data: An Application to University Libraries in Taiwan. *J. Oper. Res. Soc.* **2000**, *51*, 897–905. [CrossRef]
50. GULFBASE. Petrochemical Data for 2013 Retrieved on 19-05-2015. Available online: http://www.gulfbase.com (accessed on 4 June 2019).
51. Erkoc, T.E. Estimation Methodology of Economic Efficiency: Stochastic Frontier Analysis vs Data Envelopment Analysis. *Int. J. Acad. Res. Econ. Manag. Sci.* **2012**, *1*, 1–23.

© 2019 by the authors. Licensee MDPI, Basel, Switzerland. This article is an open access article distributed under the terms and conditions of the Creative Commons Attribution (CC BY) license (http://creativecommons.org/licenses/by/4.0/).

Article

A Meta-Model-Based Multi-Objective Evolutionary Approach to Robust Job Shop Scheduling

Zigao Wu [1,*], Shaohua Yu [2] and Tiancheng Li [3,*]

[1]   Department of Industrial Engineering, Northwestern Polytechnical University, Xi'an 710072, China
[2]   Laboratoire Genie Industriel, CentraleSupélec, Université Paris-Saclay, 91190 Saint-Aubin, France; shaohua.yu@centralesupelec.fr
[3]   Key Laboratory of Information Fusion Technology (Ministry of Education), School of Automation, Northwestern Polytechnical University, Xi'an 710072, China
*   Correspondence: zgwu@mail.nwpu.edu.cn (Z.W.); t.c.li@mail.nwpu.edu.cn or t.c.li@usal.es (T.L.)

Received: 15 May 2019; Accepted: 5 June 2019; Published: 10 June 2019

Abstract: In the real-world manufacturing system, various uncertain events can occur and disrupt the normal production activities. This paper addresses the multi-objective job shop scheduling problem with random machine breakdowns. As the key of our approach, the robustness of a schedule is considered jointly with the makespan and is defined as expected makespan delay, for which a meta-model is designed by using a data-driven response surface method. Correspondingly, a multi-objective evolutionary algorithm (MOEA) is proposed based on the meta-model to solve the multi-objective optimization problem. Extensive experiments based on the job shop benchmark problems are conducted. The results demonstrate that the Pareto solution sets of the MOEA are much better in both convergence and diversity than those of the algorithms based on the existing slack-based surrogate measures. The MOEA is also compared with the algorithm based on Monte Carlo approximation, showing that their Pareto solution sets are close to each other while the MOEA is much more computationally efficient.

Keywords: scheduling; evolutionary algorithm; robustness; multi-objective; machine breakdown

---

1. Introduction

Production scheduling is of great significance in both scientific study and engineering applications [1–5]. Generally, the aim of the job shop scheduling problem (JSS) is to find a schedule that minimizes certain performance objective, given a set of machines and a set of jobs. However, in practice, the execution of a schedule is usually confronted with disruptions and unforeseen events, such as random machine breakdowns (RMDs), which make the actual performance of a schedule hard to predict. Against this background, we will focus on the multi-objective robust JSS under RMDs with the goal of optimizing the makespan and the robustness simultaneously.

In the last few decades, the JSS problems have been extensively studied, most of which have addressed the JSS with makespan as the objective [6–8], such as the hybrid genetic algorithm [9], the genetic algorithm [10] with search area adaptation [11], the global optimization technique which combines tabu search with the ant colony optimization [12], and the memetic algorithm conditioned on a limited set of human operators [13]. However, it is often assumed that the problem parameters about jobs and machines are known and deterministic. This makes it difficult to generate a good schedule for a real-world job shop which is subjected to various uncertainties [14], such as machine breakdowns, variable processing times, and due date changes. Mehta et al. [15] had classified the uncertainties in the practical manufacturing into three main categories: complete unknowns, suspicions about the future, and known uncertainties. Complete unknowns are those unpredictable events, e.g. a sudden strike, about which no a-priori information is available, while suspicions about the future arise from the intuition and experience of the human scheduler. On the other hand, known uncertainties are those events about which some information is available in advance, such as machine breakdowns [16–18] whose frequency and duration may be characterized by probability distributions. Under these uncertainties, a schedule will be difficult to execute as planned, and finally the actual performance of the schedule will deteriorate.

Recently, robust optimization has gained intensive interest [19], where most of existing studies focus on addressing the scheduling problems under known uncertainties. The robustness of a schedule indicates the ability of the schedule to preserve a specified level of solution quality in the presence of uncertainties [20], which is generally measured by the expected deviation of the performance from its initial performance under uncertainties [21]. Liu et al. [22] defined the robust schedule as a schedule that is insensitive to uncertainties, such as that a schedule may degrade its performance to a very small degree under disruptions. Thus, in addition to the makespan, the robustness of a schedule will also be taken as one of the objectives in the robust scheduling. Xiao et al. [23] addressed the stochastic JSS problem with uncertain processing times, and the robustness took the expected relative deviation between the realized makespan and the predictive makespan. Zuo et al. [24] considered both the expectation and standard deviation of the performance of a schedule. Ahmadi et al. [25] defined the robustness of a schedule as the expected deviation of starting and completion time of each job between preschedule and realized schedule under RMD.

With simultaneous consideration of the performance and the robustness of a schedule, the JSS under uncertainties holds a multi-objective nature. Usually, the two objectives are combined by the weight sum, and then the problem with two objectives will be transferred into a single-objective problem, such as that described in [26,27]. However, providing a wide range of solutions to decision-makers might be more useful [20], since decision-makers can make a better trade-off between the performance and the robustness for their schedules. The multi-objective evolutionary algorithm (MOEA), such as NSGA II [28–30], has been successfully solved the classic JSS without considering uncertainties [31]. In addition, Hosseinabadi, et al. [32] proposed a TIME_GELS algorithm that uses the gravitational emulation local search [33] for solving the multiobjective flexible dynamic job-shop scheduling problem. But, when the robustness is considered, a MOEA should further be able to solve the problems in the presence of uncertainties [34].

Because of the intractable complexity of JSS with uncertainties, it is difficult to evaluate the effects of uncertainties on a schedule, and thus the robustness is not available in a closed form. In this case, approximation methods should be applied for fitness evaluation in the MOEA. The simplest way is to employ the Monte Carlo method [35,36] to estimate the robustness, by averaging the objective values over a few randomly sampled uncertainty scenarios. The Monte Carlo method is based on random sampling and has been proven to be a powerful method for estimation [37–39]. However, it may cause potentially expensive fitness evaluations and reduce the computing efficiency. For this reason, the problem approximation that tends to replace the original statement of a problem by one which is simple but easier to solve, has been applied. Mirabi, et al. [40] simplified the machine breakdown by assuming the repair time is constant, while Liu et al. [41] assumed that all possible machine breakdowns during a scheduling horizon are aggregated as one. However, this may lead to a large mismatch between the model and the actual problem. In contrast, the meta-model which is an approximate function of the real fitness, also known as the surrogate measure, may be preferred. However, the design of a meta-model for the robustness is challenging. So far, the available surrogate measures for the robustness measured by the expected makespan delay (EMD) only include the average total slack time of operations in [42] and the sum of free slack times of operations in [26]. Since these surrogate measures ignored uncertainties, their estimation accuracy will be reduced [43].

In comparison with the existing research, the main contributions of our approach are as follows:

- The robustness of a schedule is considered jointly with the makespan and is defined as EMD, for which a meta-model is designed by using a data-driven response surface method.
- A MOEA is proposed based on the meta-model, gaining excellent performance and efficiency.

The remainder of this paper is organized as follow. The multi-objective optimization model for the JSS under RMD is defined in the next section. In Section 3, a meta-model-based MOEA is proposed. The performance of the proposed algorithm is presented in Section 4, with comparison with the Monte Carlo method. In Section 5, we conclude the whole study.

2. Problem Definition

We consider the JSS problem with n jobs ($J = \{J_j | j = 1, 2, \ldots, n\}$) to be processed on m machines ($M = \{M_i | i = 1, 2, \ldots, m\}$). All jobs and machines are available at time zero. The processing of job j on machine i is called operation O_{ij}, $i \in [1, m], j \in [1, n]$, and its processing time p_{ij} is constant and known. Each job includes m operations ($O_j = \{O_{ij} | i = 1, 2, \ldots, m\}$, $j \in [1, n]$) that must be processed in a specified sequence through each machine. A feasible schedule that specifies the starting and completion time of all operations should satisfy the constraints: (1) job splitting is not allowed; (2) an operation is not allowed to be preempted by the others; (3) each operation is performed only once on one machine; and (4) each machine performs only one operation at a time.

In this study, the operation-based machine breakdown model presented in [21] will be applied. More specifically, the machine breakdown is modeled by two parameters: the downtime required to repair the machine after its breakdown, and the breakdown probability during a time interval. The machine breakdown probability $\Pr_{ij}$ when processing operation O_{ij} can be calculated by Equation (1), where λ_0 is the machine failure rate of each machine.

$$\Pr_{ij} = \min\{\lambda_0 p_{ij}, 1\} \tag{1}$$

The downtime D_{ij} of a machine after a breakdown when processing operation O_{ij} is modeled as an exponential distribution, as shown in Equation (2), where β_0 is the expectation of the downtime.

$$f(D_{ij}) = \begin{cases} \frac{1}{\beta_0} e^{-\frac{1}{\beta_0} D_{ij}} & D_{ij} > 0 \\ 0 & D_{ij} \leq 0 \end{cases} \tag{2}$$

In the classic JSS problems without considering uncertainties, the aim of scheduling is generally to find a schedule that minimizes the makespan. The makespan of a schedule is equal to the maximum completion time of all operations in the schedule, which can be determined by Equation (3), where ct_{ij} is the completion time of operation O_{ij}.

$$C_{max}^0 = \max_{i \in M} \{\max_{j \in J} \{ct_{ij}\}\} \tag{3}$$

However, the makespan of a schedule will be affected by RMDs in practice [16–18]. Since RMDs postpone the completion time of operations, the actual makespan C_{max}^r of a schedule will be delayed. As shown in Figure 1a, the makespan C_{max}^0 of the schedule before execution is equal to 10. If a machine breakdown with one unit of downtime occurring on the operation O_{21}, as shown in Figure 1b, its completion time is directly postponed by one unit of time. Then, all the completion times of its subsequent operations $\{O_{22}, O_{11}, O_{23}, O_{12}, O_{31}\}$ are also delayed by one unit of time. Finally, the actual makespan C_{max}^r of the schedule is equal to 11, which has been delayed by one unit of time.

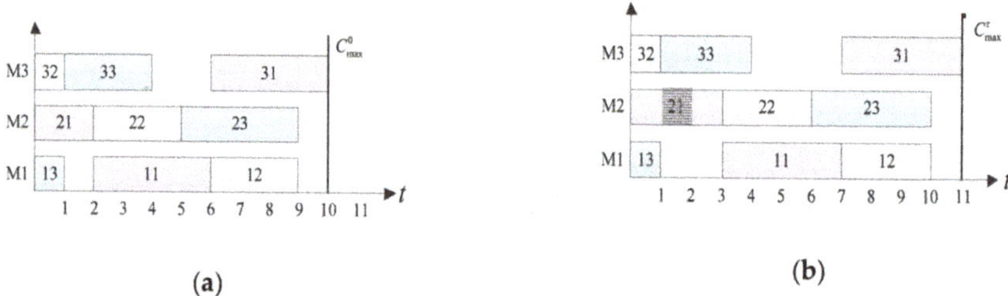

Figure 1. An example for a schedule with 3 machines and 3 jobs under a machine breakdown. (**a**) The schedule before a machine breakdown; (**b**) the schedule after a machine breakdown.

According to the analysis above, the influence of machine breakdowns on the makespan of a schedule can be given by the makespan delay δ_c in Equation (4).

$$\delta_c = \max(C^r_{max} - C^0_{max}, \ 0) \tag{4}$$

Since machine breakdowns take place randomly, the makespan delay of a schedule will vary with the actual scenario of machine breakdowns, which makes the actual makespan very instability. The stochastic change of the actual makespan will reduce the performance of a schedule, such as that it may lead to the products cannot be delivered on time and lose the customer good will. Therefore, it is preferred that the makespan of a schedule is robust under RMDs. To measure the robustness of makespan, the EMD will further be applied, as the Δ_c shown in Equation (5).

$$\Delta_c = E(\delta_c) = \int_0^{+\infty} \delta_c f(\delta_c)d\delta \tag{5}$$

However, a schedule with the minimum makespan is generally very compact, which means that it may be very sensitive to RMD and with a large EMD. Thus, the makespan C^0_{max} of a schedule will conflict with the EMD. Since different decision-makers have various preferences for the makespan and the makespan delay, it is worth providing a wide range of schedules for decision-makers to make the best trade-off. When consider the makespan and the EMD of a schedule at the same time, the JSS under RMD can be modeled as a multi-objective optimization problem. Let A be the set of precedence constraints (O_{ij}, O_{kj}) that require job j to be processed on machine i before it is processed on machine k, the multi-objective optimization model can be provided as follows:

Minimize:

$$F = (C^0_{max}, \ \Delta_c) \tag{6}$$

Subject to:

$$st_{kj} - st_{ij} \geq p_{ij} , \text{ for all } (O_{ij}, O_{kj}) \in A \tag{7}$$

$$st_{ij} - st_{il} \geq p_{il} \text{ or } st_{il} - st_{ij} \geq p_{ij}, \text{ for all } O_{ij} \text{ and } O_{il} \tag{8}$$

$$st_{ij} \geq 0, \text{ for all } O_{ij} \tag{9}$$

$$\lambda_i = \lambda_0, \text{ for all } i \in M \tag{10}$$

$$D_{ij} \sim Exp(1/\beta_0), \text{ for all } O_{ij} \tag{11}$$

3. The Meta-Model Based MOEA

When solving the multi-objective optimization model in Section 2 by a MOEA, the primary task is to evaluate the fitness of each individual in a population. However, the EMD in Equation (5) cannot be analytically calculated for the intractable complexity of JSS under RMD. Although the commonly-used Monte Carlo simulation can be used to approximate the EMD, it will make the MOEA inefficient for it is very time-consuming to evaluate each single individual, especially for the problems with larger scales. In view of this, a meta-model-based MOEA will be proposed to solve the multi-objective optimization model for the robust JSS.

3.1. Framework of The Algorithm

The meta-model-based MOEA is designed according to the basic framework of the classic NSGA-II [28]. As shown in Figure 2, the algorithm begins with an initial population P_0 with N randomly generated individuals. Before executing the following genetic operators, the meta-model Δ_c^a of the Δ_c will be constructed based on the initial population P_0. Then, the selection, crossover, and mutation operators will be applied on the current population P_k to generate new individuals and construct a combined population R_{k+1}. The fitness of individuals in the combined population R_{k+1} will first be evaluated by the makespan $C_{\max}^0$ and the proposed meta-model Δ_c^a, and then the individual-based evolution control will be applied to update the fitness of some individuals. Finally, the next generation population P_{k+1} will be generated according to the ranks of individuals. When the maximum generation number is reached, the algorithm will stop and return the obtained Pareto solution set.

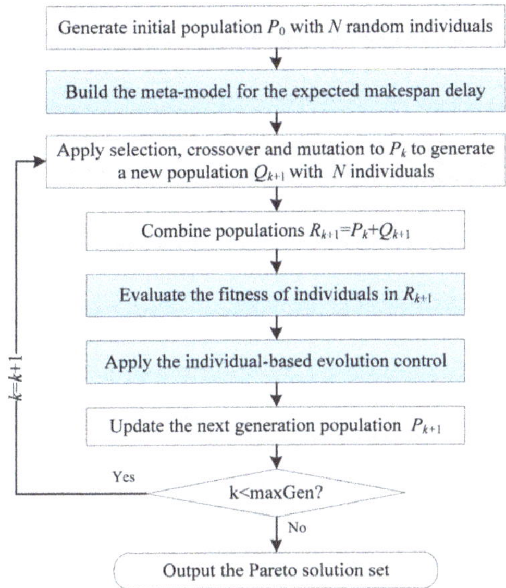

Figure 2. The flow chart of the meta-model-based multi-objective evolutionary algorithm (MOEA).

3.2. Meta-Model of The EMD

The meta-model for the EMD will be constructed based on the response surface methodology. As shown in Equation (12), this method applies a quadratic polynomial $\hat{f}(x)$ to approximate the function relation between the input x and the output y of a system,

$$y \approx \hat{f}(x) = a_0 + Bx + xCx^{\mathrm{T}}, \tag{12}$$

where, x is the input variant vector with v variables as shown in Equation (13), a_0 is the constant term, B is the coefficient vector of the linear term as shown in Equation (14), and C is the coefficient matrix of the quadratic term as shown in Equation (15).

$$x = (x_1, x_2, \ldots, x_v) \tag{13}$$

$$B = (b_1, b_2, \ldots, b_v) \tag{14}$$

$$C = \begin{bmatrix} c_{11} & c_{12} & \cdots & c_{1v} \\ & c_{22} & \cdots & c_{2v} \\ & & \ddots & \vdots \\ & & & c_{vv} \end{bmatrix} \tag{15}$$

However, a schedule cannot be directly taken as an input variant of the EMD, since it cannot be quantified. Therefore, to construct a meta-model by the response surface method for the EMD, the primary task is to extract features related to the EMD from the schedule. To this end, we will further analyze how RMDs affect the makespan of a schedule. As we all know, a feasible job shop schedule is decided by the process constraints and the resource constraints. As a result, machines may have some idle time during a schedule period, as shown in Figure 3a. In the classic JSS problems, we are devoted to reducing the idle time to minimize the makespan for improving the utilities of machines. However, when RMDs are considered, the idle time may be useful for an operation to control the influence of machine breakdowns on the makespan of a schedule and then improve the robustness of the makespan.

The available idle time of operations in a schedule can be classified into two types: the free slack time and the total slack time. The former is the time that an operation can be delayed without delaying the starting of its very next operations, while the latter is the difference between the earliest and latest starting times of an operation without delaying the makespan. Take the schedule in Figure 3a as an example, the free slack time of operations $\{O_{13}, O_{11}, O_{12}, O_{21}, O_{22}, O_{23}, O_{32}, O_{33}, O_{31}\}$ are $\{0, 0, 1, 0, 0, 1, 0, 1, 0\}$, respectively. The earliest and latest starting time of operations $\{O_{13}, O_{11}, O_{12}, O_{21}, O_{22}, O_{23}, O_{32}, O_{33}, O_{31}\}$ without delaying the makespan is $\{0, 2, 6, 0, 2, 5, 0, 1, 6\}$ and $\{1, 2, 7, 0, 3, 6, 2, 3, 6\}$, respectively. Therefore, the total slack time of each operation is $\{1, 0, 1, 0, 1, 1, 2, 2, 0\}$, respectively.

It is clear that not all operations have the free/total slack time. For an operation without slack time, the makespan of a schedule will be directly delayed, when an RMD takes place on it. As shown in Figure 3b, when a RMD with one unit of downtime takes places on the operation O_{11}, the actual makespan $C_{\max}^r$ is changed to 11, which is directly delayed by one unit of time. However, when an RMD takes places on an operation with slack time, the makespan will not be delayed until the slack time of this operation is used up. As shown in Figure 3c, after a RMD with one unit of downtime takes place on the operation O_{33} with two units of free slack time, the makespan is still equal to 10, for the free slack time of this operation is larger than the downtime of the breakdown. On the other hand, although the operation O_{22} has no free slack time, the makespan can also be protected by the total slack time of the operation, as shown in Figure 3d.

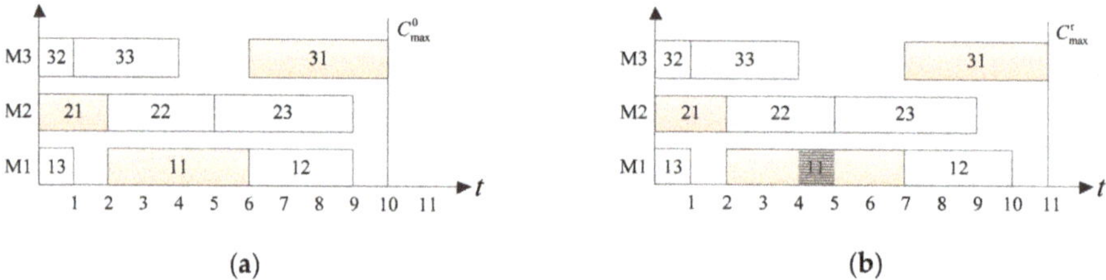

Figure 3. *Cont.*

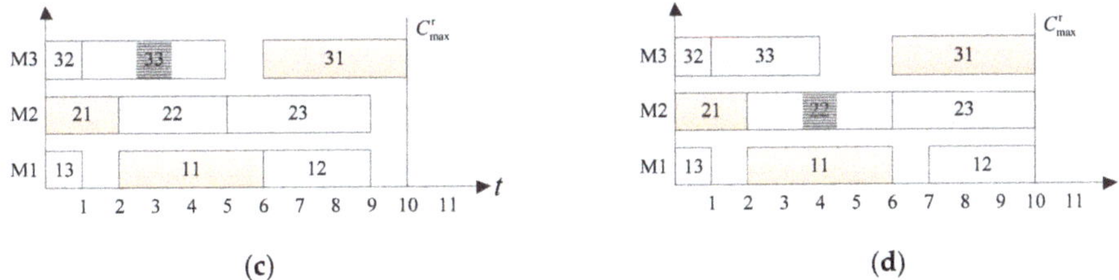

Figure 3. The relationship between the makespan delay and the machine breakdowns on different operations. (**a**) No machine breakdown; (**b**) a machine breakdown on operation O_{11}; (**c**) a machine breakdown on operation O_{33}; (**d**) a machine breakdown on operation O_{22}.

According to the analysis above, it can be found that the makespan delay of a schedule depends on the machine breakdown level, the free slack time and total slack time of each operation. In view of this, we will extract the mathematical features for the schedule under RMDs from these three aspects: the RMDs, the set O_y of operations with slack time and the set O_n of operations without slack time. For the RMDs, the machine failure rate λ_0 and the expected downtime β_0 after a breakdown will be taken. For the operations in the set O_y, all their processing time p_{ij}, free slack time fs_{ij} and total slack time ts_{ij} will be taken. As for the operations in the set O_n, only their processing time p_{ij} will be taken.

However, it is practically impossible to consider all the processing time, free slack time, and total slack time of operations as the input variants of the EMD. To reduce the number of input variants, we will further generalize these basic features into some comprehensive features. Formally, the sum of processing time p_s^y and p_s^n will be used to represent the processing time of operations in the sets O_y and O_n, respectively. For the slack time, the average free slack time fs_a and the average total slack time ts_a of operations in the set O_y will be applied.

Finally, the input variants of the EMD can be listed as follows: the machine failure rate λ_0, the expected downtime β_0, the sum of processing time p_s^n, the sum of processing time p_s^y, the average free slack time fs_a, and the average total slack time ts_a. Then, the input variant vector x of the EMD can be set as $x = (x_1, x_2, x_3, x_4, x_5, x_6) = (\lambda_0, \beta_0, p_s^n, p_s^y, fs_a, ts_a)$, and then the meta-model Δ_c^a of Δ_c can be defined by Equation (16).

$$\Delta_c \approx \Delta_c^a = a_0 + \sum_{i=1}^{6} b_i x_i + \sum_{i=1}^{5} \sum_{j=1}^{6} c_{ij} x_i x_j \tag{16}$$

Then, the coefficients should further be determined to finalize the meta-model Δ_c^a. As shown in Algorithm 1, it takes the initial population P_0 and the input variant vector x as the inputs, and outputs the meta-model Δ_c^a with the determined coefficients. First, a training data set D_c which includes N data instances will first be generated based on the initial population P_0. A data instance $I_i = (x_i, \Delta_c^i)$ is composed of the values of the input variant vector x_i and the corresponding Δ_c^i. The values of the input variant vector x_i can be determined once a schedule s_i is generated based on the ith individual in the initial population P_0. Since the EMD cannot be analytically calculated, it will be evaluated by the Monte Carlo approximation Δ_c^{sim} as shown in Equation (17). After the training data set D_c is constructed, the Multiple Linear Regression will be used to determine the coefficients of the meta-model Δ_c^a for the EMD,

$$\Delta_c^{sim} = \frac{1}{N_s} \sum_{i=1}^{N_s} (C_{max}^i - C_{max}^0), \tag{17}$$

where N_s is the simulation times and C_{max}^i is the makespan of a schedule under the ith simulation.

Algorithm 1 The pseudo-code to finalize the meta-model.

Inputs: the initial population P_0 and the input variant vector x
Outputs: the meta-model Δ_c^a with the determined coefficients

---

1:   Set the training data set $D_c = \varnothing$;
2:   Generate N_s machine breakdown scenarios with the machine failure rate λ_0 and the expected downtime β_0;
3:   for $i = 1$ to N
4:       Generate the schedule s_i based on the ith individual in P_0;
5:       Determine the makespan $C_{\max}^0$ of the schedule s_i by Equation (3);
6:       Calculate the free slack time fs_{ij} and the total slack time ts_{ij} in schedule s_i;
7:       Calculate the sum of processing time p_s^n and p_s^y;
8:       Calculate the average free slack time fs_a and the average total slack time ts_a;
9:       Determine the values x_i of the input variant vector $x = (\lambda_0, \beta_0, p_s^n, p_s^y, fs_a, ts_a)$;
10:      Determined the EMD $\Delta_c^i = \Delta_c^{sim}$ by Equation (17);
11:      Generate the ith data instance $I_i = (x_i, \Delta_c^i)$;
12:      Update the training data set $D_c = D_c \cup I_i$;
13:   end for
14:   Based on the training data set D_c, apply the multiple linear regression to determine the coefficients of the meta-model Δ_c^a in Equation (16);
15:   Return the finalized meta-model Δ_c^a.

3.3. Fitness Evaluation

To compare the fitness of different individuals, the makespan and the EMD of each individual in a population should be evaluated. Once a schedule is generated, the makespan can be directly determined by Equation (3). However, the EMD cannot be analytically calculated for the complexity of the JSS. Although it can be effectively approximated by the time-consuming Monte Carlo approximation in Equation (17), the efficiency of the algorithm will be significantly reduced. In view of this, we will apply the proposed meta-model Δ_c^a in Equation (16) to approximate the EMD. The basic motivation for using the meta-model in the fitness evaluation is to reduce the number of expensive fitness evaluations without degrading the quality of the obtained optimal solution.

Based on the proposed meta-model Δ_c^a, Algorithm 2 can be used to provide the fitness set $F_{k+1} = \left\{ (C_{\max}^0(s_0), \Delta_c^a(s_0)), \ldots, (C_{\max}^0(s_i), \Delta_c^a(s_i)), \ldots, (C_{\max}^0(s_i), \Delta_c^a(s_i)) \right\}$ of the individuals in the combined population P_{k+1}, where $F_{k+1}^i = (C_{\max}^0(s_i), \Delta_c^a(s_i))$ represents the fitness of the ith individual in the combined population P_{k+1} with the makespan $F_{k+1}^i(1) = C_{\max}^0(s_i)$ and the EMD $F_{k+1}^i(2) = \Delta_c^a(s_i)$.

Algorithm 2 Fitness evaluation for the combined population

Inputs: the combined population R_{k+1}
Outputs: the fitness set F_{k+1} of the individuals in R_{k+1}

1:   Set the fitness set $F_{k+1} = \varnothing$;
2:   for $i = 1$ to $2N$
3:       Select the ith individual chm_i from the population R_{k+1};
4:       Generate the schedule s_i based on the individual chm_i;
5:       Evaluate the makespan $C_{\max}^{0,i}$ of the schedule s_i by Equation (3);
6:       Get machine failure rate λ_0 and expected downtime β_0;
7:       Calculate the free slack time fs_{ij} and the total slack time ts_{ij} in the schedule s_i;
8:       Calculate the sum of processing time p_s^n and p_s^y;
9:       Calculate the average free slack time fs_a and the average total slack time ts_a;
10:      Determine the values x_i of the input variant vector $x = (\lambda_0, \beta_0, p_s^n, p_s^y, fs_a, ts_a)$
11:      Evaluate the EMD by $\Delta_c^a(s_i)$ in Equation (16) with the values of x_i;
12:      Update the fitness set $F_{k+1} = F_{k+1} \cup F_{k+1}^i = F_{k+1} \cup (C_{\max}^0(s_i), \Delta_c^a(s_i))$;
13:   end for
14:   Return the fitness set F_{k+1};

3.4. Individual-Based Evolution Control

Generally, the approximate model is assumed to be of high fidelity and, therefore, the real fitness function will be not at all used in the evolution [20]. However, an evolutionary algorithm using meta-models without controlling the evolution using the real fitness function can run the risk of an incorrect convergence [28]. For this reason, the meta-model is combined with the real fitness function in our algorithm, which is often known as evolution control or model management.

As shown in Algorithm 3, the individual-based evolution control framework will be applied, which chooses the best individuals according to the pre-evaluation using the meta-model Δ_c^a for reevaluation using the real fitness function. For this purpose, the fitness set F_{k+1} will first be ranked by the fast non-dominated sorting. And then, the individuals with the rank $rank(F_{k+1}^i) = 1$ will further be reevaluated by the Monte Carlo approximation Δ_c^{sim} in Equation (17). In addition, to avoid the unnecessary simulation computational time, the repeated individuals will only be evaluated once.

Algorithm 3 Individual-based evolution control framework

Inputs: the fitness set F_{k+1} of the combined population R_{k+1}
Outputs: the modified fitness set $\hat{F}_{k+1}$

1:   Generate N_s scenarios with the machine failure rate λ_0 and the expected downtime β_0;
2:   Rank the fitness set F_{k+1} by the fast non-dominated sorting;
3:   Sort the fitness set F_{k+1} in the ascending lexicographic order of the rank, the makespan and the EMD;
4:   Set $\hat{F}_{k+1} = F_{k+1}$;
5:   for $i = 1$ to $2N$
6:      if $rank(F_{k+1}^i) > 1$
7:         Break;
8:      else if $F_{k+1}^i(1) = F_{k+1}^{i-1}(1)$ and $F_{k+1}^i(2) = F_{k+1}^{i-1}(2)$
9:         Update the fitness $\hat{F}_{k+1}^i$ by $\hat{F}_{k+1}^i(2) = \hat{F}_{k+1}^{i-1}(2)$;
10:      else
11:         Generate the schedule s_i of the individual associated by F_{k+1}^i in the R_{k+1};
12:         Evaluate the EMD by the Monte Carlo method Δ_c^{sim} in Equation (17);
13:         Update the fitness $\hat{F}_{k+1}^i$ by $\hat{F}_{k+1}^i(2) = \Delta_c^{sim}$;
14:      end if
15: end for
16: Return the modified fitness set $\hat{F}_{k+1}$.

3.5. Evolutionary Operators

In our algorithm, the preference list representation is applied to code the chromosomes. The chromosome built by this coding method is made up of m substrings corresponding to m machines. Every substring is a preference list of n jobs on the corresponding machine. Supposing the chromosome is [(2 3 1) (1 3 2) (2 1 3)], the substring (2 3 1) is the preference list for machine 1, the substring (1 3 2) for machine 2 and the substring (2 1 3) for machine 3. When decoding, the job the first to appear in every precedence list will be selected firstly. If a selected operation meets the process constraint, the operation will be scheduled, and then it is removed from corresponding preference list. If there are more than one operation can be scheduled, then select one randomly.

The main genetic operators include selection, crossover and mutation. The usual binary tournament selection is used to select parent individuals for generating child solutions. Namely, randomly select two individuals from the population, and choose one of them with better fitness for the subsequent genetic operators. In the crossover operator, the substring crossover which exchanges the substrings of parents between two randomly selected machine numbers is applied. For the mutation operator, the swap-mutation operator to a randomly selected substring is applied.

Before updating the next generation population, the fast non-dominated sorting approach is applied to ranking the solutions in the combined population. Then, the population will be updated by choosing the individuals in the order of their ranks. Since all the previous and current population members are included in the combined population, the elitism can also be ensured.

4. Experimental Analysis

In this section, the performance of the meta-model in evaluating the EMD will first be presented, and then the meta-model-based MOEA will be used to solve the robust JSS problem.

4.1. Experiment Setting

The algorithm is implemented using C++ and run on a 2.8 GHz PC with an Intel Pentium dual-core CPU and 2 GB of RAM. The parameters are listed as follows: the population size is 1024; the generation number is 64; the crossover rate is 0.95; the mutation rate is 0.05; the machine breakdown ratio is 0.005; the expected downtime is 20; and the simulation times are 600.

In the literature, many benchmark problems have been generated by different researchers to test the performance of different algorithms, which are also very useful for this research for they include a wide range of problem instances. In this study, the problem instances La01-La40 with sizes from 10×5 to 30×10 in the benchmark problem set LA (Lawrence in 1984) and the problem instances Ta01-Ta40 with sizes from 15×15 to 30×15 in the benchmark problem set TA (Taillard in 1994) will be applied. In total, there are 80 benchmark problem instances will be used to test the performance of the proposed algorithm.

4.2. Evaluation Performance of The Proposed Mete-Model

To distinguish the robustness of different schedules, an effective meta-model must have high evaluation accuracy and perform a strong linear correlation to the real value of the robustness. To show the accuracy of the proposed meta-model Δ_c^a in evaluating the EMD, the average $\overline{\chi}$ in Equation (18) and standard variance $\sigma(\chi)$ in Equation (19) of the absolute relative deviation $\chi(\Delta_c^a, \Delta_c^{sim})$ from the Monte Carlo approximation Δ_c^{sim} will be applied. In addition, a correlation study will be conducted using IBM SPSS. For each test problem, the R^2 statistic and the significance level $Sig.$ of the linear model ANOVA are recorded. The value of R^2 is used to measure the fitting degree of the meta-model, while the significance level $Sig.$ is used to test whether there is a significant linear correlation between the meta-model and the EMD. Therefore, a good meta-model should be with a large value of R^2 and a small value of $Sig.$ for each test problem.

$$\overline{\chi} = \frac{1}{N} \sum_{i=1}^{N} \chi(\Delta_c^a, \Delta_c^{sim}) = \frac{1}{N} \sum_{i=1}^{N} \left| \frac{\Delta_c^a - \Delta_c^{sim}}{\Delta_c^{sim}} \right| \tag{18}$$

$$\sigma(\chi) = \sqrt{\frac{\sum_{i=1}^{N} \left(\chi(\Delta_c^a, \Delta_c^{sim}) - \overline{\chi} \right)^2}{N-1}} \tag{19}$$

The experimental results have been processed and recorded in Tables 1 and 2 for the problem sets LA and TA, respectively. It can be found that the maximum and minimum values of $\overline{\chi}$ for all problems in LA are equal to 0.041 and 0.020, respectively. And, the maximum and minimum values of $\overline{\chi}$ for all problems in TA are equal to 0.026 and 0.017, respectively. That is, the values of $\overline{\chi}$ for all test problems are less than 0.05 in average. Therefore, the values of the meta-model are very close to that of the Monte Carlo approximation, which indicate that the proposed meta-model have high accuracy in evaluating the EMD for the JSS under RMD. On the other hand, the maximum and minimum values of $\sigma(\chi)$ for all problems in LA are equal to 0.031 and 0.015, respectively. The maximum and minimum values of $\sigma(\chi)$ for all problems in TA are equal to 0.020 and 0.013, respectively. All these results show that the meta-model have a small variance in the absolute relative deviation, which indicate that the performance of the meta-model is also robust in evaluating the EMD.

Table 1. The experimental results of the meta-model in the problem set LA.

Cases	$n \times m$	$\bar{\chi}$	$\sigma(\chi)$	R^2	Sig.	Cases	$n \times m$	$\bar{\chi}$	$\sigma(\chi)$	R^2	Sig.
La01	10×5	0.035	0.027	0.74	<0.01	La21	15×10	0.026	0.019	0.77	<0.01
La02	10×5	0.040	0.029	0.59	<0.01	La22	15×10	0.028	0.021	0.74	<0.01
La03	10×5	0.041	0.031	0.53	<0.01	La23	15×10	0.026	0.020	0.78	<0.01
La04	10×5	0.038	0.028	0.71	<0.01	La24	15×10	0.027	0.021	0.76	<0.01
La05	10×5	0.034	0.026	0.78	<0.01	La25	15×10	0.028	0.021	0.75	<0.01
La06	15×5	0.029	0.022	0.80	<0.01	La26	20×10	0.023	0.018	0.76	<0.01
La07	15×5	0.033	0.024	0.71	<0.01	La27	20×10	0.024	0.018	0.74	<0.01
La08	15×5	0.032	0.024	0.72	<0.01	La28	20×10	0.024	0.019	0.73	<0.01
La09	15×5	0.033	0.026	0.68	<0.01	La29	20×10	0.024	0.018	0.75	<0.01
La10	15×5	0.033	0.024	0.76	<0.01	La30	20×10	0.025	0.019	0.74	<0.01
La11	20×5	0.026	0.020	0.76	<0.01	La31	30×10	0.020	0.016	0.75	<0.01
La12	20×5	0.029	0.023	0.76	<0.01	La32	30×10	0.020	0.015	0.75	<0.01
La13	20×5	0.029	0.023	0.70	<0.01	La33	20×10	0.020	0.015	0.76	<0.01
La14	20×5	0.029	0.023	0.76	<0.01	La34	30×10	0.021	0.016	0.76	<0.01
La15	20×5	0.029	0.023	0.71	<0.01	La35	30×10	0.022	0.017	0.82	<0.01
La16	10×10	0.031	0.024	0.74	<0.01	La36	15×15	0.023	0.018	0.80	<0.01
La17	10×10	0.031	0.024	0.74	<0.01	La37	15×15	0.023	0.018	0.75	<0.01
La18	10×10	0.032	0.025	0.72	<0.01	La38	15×15	0.025	0.019	0.74	<0.01
La19	10×10	0.030	0.023	0.72	<0.01	La39	15×15	0.025	0.018	0.75	<0.01
La20	10×10	0.035	0.026	0.65	<0.01	La40	15×15	0.025	0.019	0.71	<0.01
Aver.	/	**0.032**	**0.025**	**0.71**	**<0.01**	**Aver.**	/	**0.024**	**0.018**	**0.76**	**<0.01**

The results can also be clearly presented by the quartile graphs of the absolute relative deviation $\chi(\Delta_c^a, \Delta_c^{\text{sim}})$, as shown in Figures 4 and 5. In Figure 4, the maximum value of the absolute relative deviations $\chi(\Delta_c^a, \Delta_c^{\text{sim}})$ for all problems in LA is about 0.19, but more than 75% of the values are less than 0.06 for all the problems in the problem set LA. Especially, when the problem scale is larger, such as the problems LA21-LA40, more than 75% of the values of $\chi(\Delta_c^a, \Delta_c^{\text{sim}})$ are less than 0.04. As for the problem set TA in Figure 5, the maximum value of the absolute relative deviation $\chi(\Delta_c^a, \Delta_c^{\text{sim}})$ for all problems is only about 0.12, and more than 75% of the values are less than 0.04 for all the test problems. Therefore, it can be concluded that the proposed meta-model has a very small estimation error for the EMD.

On the other hand, the results of the linear model ANOVA have also been provided in Tables 1 and 2. For the results in Table 1, except for the problems La02, La03, La09, and La20, we can find that all the values of R^2 are larger than 0.70. Especially for the problems La21-La40 with larger problem scales, the average of R^2 even reaches to 0.76. All the values of R^2 are larger than 0.70 for the problems in TA and the average value is about 0.76 as shown in Table 2. In addition, for all the problems in LA and TA, the significance level *Sig.* is less than 0.01. All these results show that the proposed meta-model have a significant linear correlation with the expected makespan.

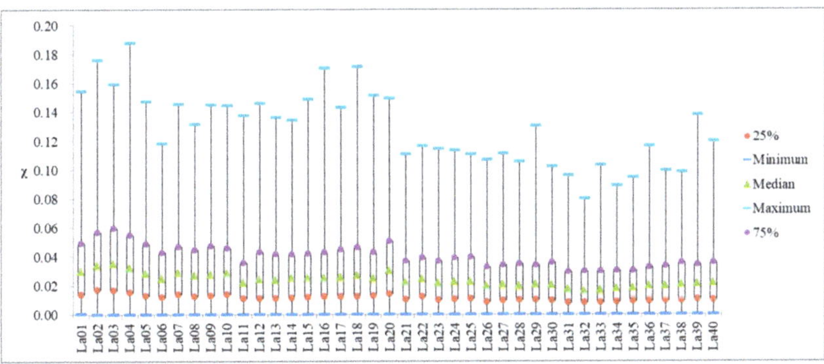

Figure 4. The quartile graph of the absolute relative deviation $\chi(\Delta_c^a, \Delta_c^{\text{sim}})$ in the problem set LA.

Table 2. The experimental results of the meta-model in the problem set TA.

Cases	$n \times m$	$\bar{\chi}$	$\sigma(\chi)$	R^2	Sig.	Cases	$n \times m$	$\bar{\chi}$	$\sigma(\chi)$	R^2	Sig.
Ta01	15×15	0.026	0.019	0.73	<0.01	Ta21	20×20	0.019	0.015	0.78	<0.01
Ta02	15×15	0.023	0.018	0.76	<0.01	Ta22	20×20	0.018	0.014	0.79	<0.01
Ta03	15×15	0.025	0.019	0.74	<0.01	Ta23	20×20	0.018	0.014	0.76	<0.01
Ta04	15×15	0.023	0.018	0.77	<0.01	Ta24	20×20	0.019	0.014	0.77	<0.01
Ta05	15×15	0.023	0.018	0.78	<0.01	Ta25	20×20	0.018	0.014	0.79	<0.01
Ta06	15×15	0.023	0.017	0.76	<0.01	Ta26	20×20	0.018	0.014	0.78	<0.01
Ta07	15×15	0.026	0.020	0.73	<0.01	Ta27	20×20	0.019	0.015	0.75	<0.01
Ta08	15×15	0.024	0.018	0.76	<0.01	Ta28	20×20	0.019	0.014	0.77	<0.01
Ta09	15×15	0.025	0.018	0.74	<0.01	Ta29	20×20	0.019	0.015	0.76	<0.01
Ta10	15×15	0.023	0.018	0.77	<0.01	Ta30	20×20	0.018	0.014	0.79	<0.01
Ta11	20×15	0.022	0.016	0.76	<0.01	Ta31	30×15	0.018	0.014	0.77	<0.01
Ta12	20×15	0.020	0.016	0.78	<0.01	Ta32	30×15	0.019	0.015	0.77	<0.01
Ta13	20×15	0.021	0.016	0.78	<0.01	Ta33	30×15	0.018	0.014	0.74	<0.01
Ta14	20×15	0.020	0.016	0.77	<0.01	Ta34	30×15	0.017	0.014	0.78	<0.01
Ta15	20×15	0.021	0.016	0.73	<0.01	Ta35	30×15	0.018	0.014	0.80	<0.01
Ta16	20×15	0.021	0.015	0.78	<0.01	Ta36	30×15	0.018	0.014	0.76	<0.01
Ta17	20×15	0.023	0.017	0.74	<0.01	Ta37	30×15	0.018	0.014	0.76	<0.01
Ta18	20×15	0.021	0.016	0.75	<0.01	Ta38	30×15	0.018	0.014	0.79	<0.01
Ta19	20×15	0.021	0.017	0.76	<0.01	Ta39	30×15	0.019	0.013	0.81	<0.01
Ta20	20×15	0.021	0.016	0.79	<0.01	Ta40	30×15	0.017	0.013	0.78	<0.01
Aver.	/	**0.023**	**0.017**	**0.76**	**<0.01**	**Aver.**	/	**0.018**	**0.014**	**0.77**	**<0.01**

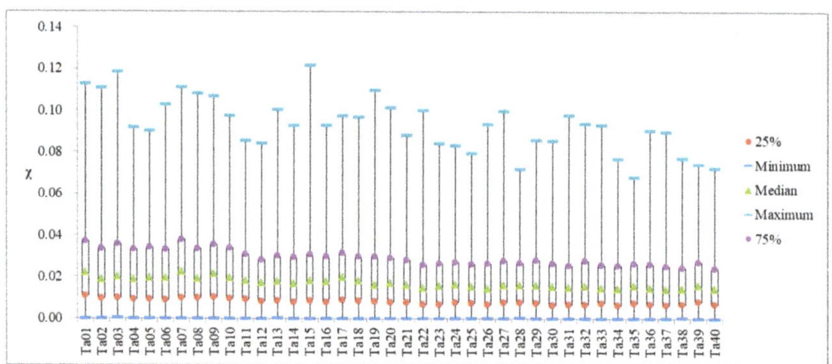

Figure 5. The quartile graph of the absolute relative deviation $\chi(\Delta_c^a, \Delta_c^{sim})$ in the problem set TA.

In summary, the proposed meta-model Δ_c^a have high evaluation accuracy and holds a strong linear correlation to the EMD. This means that the meta-model Δ_c^a is able to effectively distinguish the robustness of different schedules in the evolutionary algorithm.

4.3. Optimization Performance of The Proposed Algorithm

In this section, the performance of the proposed meta-model (MM)-based MOEA in optimizing the makespan and the EMD for the JSS under RMD will be presented. To show the performance of the algorithm, the Monte Carlo approximation in Equation (17)-based MOEA (MC) will be applied. In addition, the proposed meta-meta will also be compared with the existing surrogate measures, and thus the results on the MOEAs with the average total slack time (SM1) in [42] and the sum of free slack time (SM2) in [26] will also be provided. By implementing these algorithms with various approximations of EMD, four Pareto solution sets can be obtained for each problem.

To investigate the performance of MOEAs, many metrics have been developed and applied in the related research [44]. Since a single metric can only provide some specific but incomplete of performance, to comprehensively evaluate the performance of the proposed algorithm, both the average distance and the number of distinct choices will be used in our experiments to measure the convergence and diversity of the algorithm, respectively.

Average distance metric A_d in Equation (20) evaluates the closeness of the obtained Pareto solution set PF_{find} to the true Pareto solution set PF_{true}, where $d(z_i, a_j)$ denotes the Euclidean distance between z_i in PF_{find} and all points a_j in PF_{true}.

$$A_d = \frac{1}{|PF_{find}|} \sum_{i=1}^{|PF_{find}|} \min_{j=1}^{|PF_{true}|} d(z_i, a_j) \tag{20}$$

Number of distinct choices N_μ in Equation (21) focuses on the distribution of solutions, which defines the number of distinct choices for a pre-specified value of μ, $0 < \mu < 1$. In this metric, an m-dimensional objective space will be divided into $1/\mu^m$ number of small grids, where any solutions within the same grid are considered similar to one another. If there are individuals in the obtained Pareto set PF_{find} that fall into the region $T(l_m, \ldots, l_2, l_1)$, $NT(l_m, \ldots, l_2, l_1)$ is equal to one, otherwise zero. In our experiments, the value of μ for the metric N_μ is taken as 0.05.

$$N_\mu(PF_{find}) = \sum_{l_m=0}^{1/\mu-1} \cdots \sum_{l_2=0}^{1/\mu-1} \sum_{l_1=0}^{1/\mu-1} NT(l_m, \ldots, l_2, l_1) \tag{21}$$

Since the NP-hard nature of the JSS under RMD, the true Pareto set PF_{true} is not available for all test problems. In view of this, the approximate Pareto solution set which is provided by all comparison algorithms will be used to represent the true one. That is, under the obtained Pareto fronts $(PF_{find}^1, PF_{find}^2, \ldots, PF_{find}^{n_p})$ by different algorithms for a test problem, the true Pareto solution set can be approximated by Equation (22), where $a_j \prec b_i$ implies that a_j dominates b_i. Besides, in order to reduce the scale difference between different objectives, all Pareto fronts will be normalized by $PF_{find} = (PF_{find} - PF_{true}^{min})/(PF_{true}^{max} - PF_{true}^{min})$ based on the maximum and minimum of objectives of the true Pareto set PF_{true}.

$$PF_{true} \approx \left\{ b_i \middle| \forall b_i, \neg \exists a_j \in (PF_{find}^1 \cup PF_{find}^2 \cup \ldots \cup PF_{find}^{n_p}) \prec b_i \right\} \tag{22}$$

The results on the metric A_d of the algorithms MC, MM, SM1, and SM2 have been provided in Tables 3 and 4 for the problem sets LA and TA, respectively. The results show that the algorithms SM1 and SM2 have the similar performance in the convergence, for the values of A_d of the algorithms SM1 and SM2 are always close to each other. For example, in Table 3, their average values of A_d in the problems La01–La20 are 1.27 and 1.26, and the average values of A_d in the problems La21–La40 are 0.21 and 0.21, respectively. The similar results can also be found in Table 4, the average values of A_d in the problems Ta01–Ta20 are 0.21 and 0.24, while they are 0.22 and 0.23 in the problems Ta21–Ta40.

By comparison, the proposed algorithm MM performs better in the convergence than the algorithms SM1 and SM2. This is because that, except for the problems La39, Ta14, and Ta39, all the values of A_d for the algorithm MM in the problem sets LA and TA are less than that of the algorithms SM1 and SM2. In average, the values of A_d for the algorithm MM in the problems La01–La20 and La21–La40 are 0.16 and 0.08, while the corresponding values of A_d for the algorithms SM1 and SM2 are about 1.26 and 0.21, respectively. And, in the problem set TA, the average values of A_d for the algorithm MM in the problems La01–La20 and La21–La40 are both 0.09, while the average values of A_d for the algorithms SM1 and SM2 are about 0.22.

On the other hand, the results also indicate that the convergence of the algorithm MM can even be very close to that of the algorithm MC. As shown in Table 3, the results show that the proposed algorithm MM can have the best convergence in the problems La04, La05, La07, La10, La12, La13, La14, La22, La24, La26, La27, and La31 from the problem set LA and Ta06, Ta09,Ta19, Ta23, Ta24, Ta37, and Ta37 from the problem set TA. In addition, the average of A_d for the algorithm MM in the problems La01–La21 with smaller problem scales is 0.16, which is 0.12 larger than that of the algorithm MC. But, in the problems with larger problem scales, such as the problems La21–La40 and Ta01–Ta40, the average of A_d for the algorithm MM is only 0.06 larger that of the algorithm MC. Therefore, we conclude that the proposed algorithm MM is better than the algorithms SM1 and SM2 and similar to the algorithm MC in convergence.

Table 3. The values of A_d for the algorithms Monte Carlo approximation-based MOEA (MC), meta-model (MM), total slack time (SM1), and free slack time (SM2) in the problem set LA.

Cases	$n \times m$	MC	MM	SM1	SM2	Cases	$n \times m$	MC	MM	SM1	SM2
La01	10×5	0.01	0.07	0.99	1.57	La21	15×10	0.00	0.10	0.31	0.34
La02	10×5	0.01	0.04	0.26	0.50	La22	15×10	0.03	0.01	0.19	0.13
La03	10×5	0.01	0.05	0.08	0.10	La23	15×10	0.00	0.14	0.19	0.31
La04	10×5	0.07	0.00	0.16	0.36	La24	15×10	0.04	0.03	0.20	0.12
La05	10×5	0.00	0.00	0.00	0.00	La25	15×10	0.00	0.07	0.19	0.30
La06	15×5	0.00	0.92	3.96	7.19	La26	20×10	0.02	0.02	0.23	0.17
La07	15×5	0.52	0.00	1.55	1.22	La27	20×10	0.06	0.02	0.26	0.06
La08	15×5	0.00	0.33	3.70	2.95	La28	20×10	0.00	0.11	0.18	0.15
La09	15×5	0.00	0.35	2.89	1.45	La29	20×10	0.01	0.07	0.17	0.08
La10	15×5	0.00	0.00	0.00	0.00	La30	20×10	0.02	0.04	0.26	0.24
La11	20×5	0.00	0.66	6.06	4.84	La31	30×10	0.15	0.00	0.21	0.25
La12	20×5	0.04	0.03	1.38	0.92	La32	30×10	0.00	0.18	0.24	0.24
La13	20×5	0.06	0.05	1.34	1.22	La33	20×10	0.00	0.17	0.24	0.09
La14	20×5	0.00	0.00	0.00	0.00	La34	30×10	0.00	0.10	0.27	0.08
La15	20×5	0.00	0.30	0.63	1.12	La35	30×10	0.01	0.16	0.18	0.26
La16	10×10	0.09	0.14	1.09	0.36	La36	15×15	0.00	0.12	0.19	0.38
La17	10×10	0.00	0.08	0.20	0.16	La37	15×15	0.00	0.12	0.22	0.25
La18	10×10	0.00	0.15	0.30	0.46	La38	15×15	0.00	0.04	0.18	0.27
La19	10×10	0.01	0.02	0.34	0.13	La39	15×15	0.00	0.13	0.04	0.20
La20	10×10	0.02	0.09	0.40	0.55	La40	15×15	0.00	0.05	0.24	0.31
Aver.	/	**0.04**	**0.16**	**1.27**	**1.26**	Aver.	/	**0.02**	**0.08**	**0.21**	**0.21**

Table 4. The values of A_d for the algorithms MC, MM, SM1 and SM2 in the problem set TA.

Cases	$n \times m$	MC	MM	SM1	SM2	Cases	$n \times m$	MC	MM	SM1	SM2
Ta01	15×15	0.00	0.08	0.19	0.20	Ta21	20×20	0.02	0.05	0.23	0.31
Ta02	15×15	0.00	0.08	0.22	0.33	Ta22	20×20	0.00	0.17	0.28	0.34
Ta03	15×15	0.00	0.04	0.17	0.09	Ta23	20×20	0.09	0.01	0.14	0.19
Ta04	15×15	0.01	0.12	0.21	0.17	Ta24	20×20	0.02	0.01	0.17	0.16
Ta05	15×15	0.00	0.15	0.15	0.24	Ta25	20×20	0.00	0.22	0.29	0.24
Ta06	15×15	0.04	0.01	0.23	0.22	Ta26	20×20	0.00	0.07	0.30	0.22
Ta07	15×15	0.00	0.14	0.24	0.26	Ta27	20×20	0.00	0.09	0.18	0.38
Ta08	15×15	0.01	0.02	0.08	0.11	Ta28	20×20	0.03	0.05	0.30	0.21
Ta09	15×15	0.06	0.03	0.27	0.30	Ta29	20×20	0.02	0.06	0.31	0.34
Ta10	15×15	0.06	0.10	0.29	0.22	Ta30	20×20	0.02	0.05	0.24	0.24
Ta11	20×15	0.00	0.22	0.29	0.30	Ta31	30×15	0.02	0.04	0.16	0.17
Ta12	20×15	0.01	0.04	0.24	0.19	Ta32	30×15	0.05	0.05	0.13	0.14
Ta13	20×15	0.00	0.08	0.14	0.29	Ta33	30×15	0.03	0.05	0.34	0.26
Ta14	20×15	0.01	0.12	0.10	0.17	Ta34	30×15	0.00	0.13	0.33	0.39
Ta15	20×15	0.00	0.21	0.26	0.37	Ta35	30×15	0.00	0.09	0.21	0.14
Ta16	20×15	0.00	0.06	0.23	0.21	Ta36	30×15	0.02	0.03	0.15	0.18
Ta17	20×15	0.03	0.06	0.14	0.33	Ta37	30×15	0.03	0.02	0.17	0.08
Ta18	20×15	0.00	0.07	0.24	0.30	Ta38	30×15	0.00	0.27	0.24	0.35
Ta19	20×15	0.09	0.00	0.30	0.34	Ta39	30×15	0.07	0.16	0.01	0.13
Ta20	20×15	0.00	0.14	0.17	0.19	Ta40	30×15	0.00	0.13	0.19	0.18
Aver.	/	**0.02**	**0.09**	**0.21**	**0.24**	Aver.	/	**0.02**	**0.09**	**0.22**	**0.23**

The results on the metric N_μ of the algorithms MC, MM, SM1 and SM2 are presented in Tables 5 and 6 for the problem sets LA and TA, respectively. The results show that the algorithms SM1 and SM2 also have the similar performance in the diversity. From the results in Table 5, we can found that the average values of N_μ for the algorithms SM1 and SM2 in the problems La01–La20 are 4.7 and 5.2, and the average values of A_d for the problems La21–La40 are 9.3 and 10.5, respectively. That is, the average difference of A_d between the algorithms SM1 and SM2 is only about 1.0 in average. The similar results can also be found in Table 6, the average values of N_μ for the problems Ta01–Ta20 are 10.2 and 9.7, while they are 8.9 and 8.4 in the problems Ta21–Ta40.

Table 5. The values of N_μ for the algorithms MC, MM, SM1, and SM2 in the problem set LA.

Cases	$n \times m$	MC	MM	SM1	SM2	Cases	$n \times m$	MC	MM	SM1	SM2
La01	10×5	11	12	2	4	La21	15×10	15	15	8	13
La02	10×5	12	9	5	4	La22	15×10	16	17	10	9
La03	10×5	11	8	2	6	La23	15×10	13	18	6	12
La04	10×5	9	8	4	4	La24	15×10	12	12	10	15
La05	10×5	1	1	2	3	La25	15×10	14	15	11	9
La06	15×5	3	3	3	6	La26	20×10	15	15	10	10
La07	15×5	11	6	4	8	La27	20×10	12	14	10	13
La08	15×5	5	7	3	9	La28	20×10	14	18	8	7
La09	15×5	7	9	9	7	La29	20×10	13	14	7	11
La10	15×5	1	2	4	2	La30	20×10	13	16	10	9
La11	20×5	4	5	6	5	La31	30×10	15	12	9	8
La12	20×5	11	8	5	7	La32	30×10	13	12	7	12
La13	20×5	12	7	7	6	La33	20×10	11	13	8	9
La14	20×5	1	1	3	1	La34	30×10	13	16	11	10
La15	20×5	12	16	7	6	La35	30×10	11	16	10	7
La16	10×10	9	8	2	2	La36	15×15	12	16	11	11
La17	10×10	10	11	7	7	La37	15×15	14	17	8	9
La18	10×10	12	14	8	4	La38	15×15	15	18	14	13
La19	10×10	12	12	6	6	La39	15×15	18	14	8	12
La20	10×10	12	14	4	7	La40	15×15	14	17	10	11
Aver.	/	8.3	8.1	4.7	5.2	Aver.	/	13.7	15.3	9.3	10.5

Table 6. The values of N_μ for the algorithms MC, MM, SM1, and SM2 in the problem set TA.

Cases	$n \times m$	MC	MM	SM1	SM2	Cases	$n \times m$	MC	MM	SM1	SM2
Ta01	15×15	16	18	9	9	Ta21	20×20	17	14	10	9
Ta02	15×15	14	10	10	12	Ta22	20×20	13	15	6	8
Ta03	15×15	15	19	12	10	Ta23	20×20	15	11	11	11
Ta04	15×15	14	19	11	11	Ta24	20×20	13	13	11	9
Ta05	15×15	17	15	9	11	Ta25	20×20	14	12	11	9
Ta06	15×15	16	14	9	10	Ta26	20×20	16	12	10	10
Ta07	15×15	15	17	12	10	Ta27	20×20	14	8	12	9
Ta08	15×15	17	11	13	9	Ta28	20×20	15	17	7	9
Ta09	15×15	15	14	10	8	Ta29	20×20	17	12	12	10
Ta10	15×15	15	14	11	12	Ta30	20×20	17	10	7	7
Ta11	20×15	13	15	10	10	Ta31	30×15	13	13	9	10
Ta12	20×15	12	14	8	7	Ta32	30×15	13	12	7	8
Ta13	20×15	16	13	8	10	Ta33	30×15	13	11	6	5
Ta14	20×15	13	14	9	10	Ta34	30×15	14	13	10	9
Ta15	20×15	17	18	12	11	Ta35	30×15	10	14	9	7
Ta16	20×15	19	13	8	8	Ta36	30×15	15	10	9	6
Ta17	20×15	15	11	10	9	Ta37	30×15	13	11	6	8
Ta18	20×15	16	15	10	9	Ta38	30×15	9	15	8	10
Ta19	20×15	10	13	12	8	Ta39	30×15	11	14	7	6
Ta20	20×15	16	17	11	10	Ta40	30×15	13	13	9	7
Aver.	/	15.1	14.7	10.2	9.7	Aver.	/	13.8	12.5	8.9	8.4

Compared with the algorithms SM1 and SM2, the proposed algorithm MM performs better in diversity. This is because that the values of A_d for the algorithm MM are larger than that of the algorithms SM1 and SM2 for most of the problems in the problem sets LA and TA as shown in Tables 5 and 6. In addition, the results have also shown the algorithm MM have the similar performance in diversity to that of the algorithm MC. For example, the average values of A_d for the algorithms MM and MC in the problems La01–La20 are about 8.1 and 8.3, while the average values of A_d are 15.3 and 13.7 in the problems La21–La40, respectively. Similar results can be found in Table 6 for the problem set TA. Therefore, we can conclude that the proposed algorithm MM is very close to the algorithm MC in diversity, but it is much better than the algorithms SM1 and SM2.

Taking the problems instances La06, La15, Ta15, and Ta36 as examples, the performance of the algorithms MC, MM, SM1, and SM2 can also be clearly illustrated by the Pareto fronts. As shown in Figure 6, the Pareto front of the algorithm MM is very close to that of the algorithm MC, while the Pareto fronts of the algorithms SM1 and SM2 are very far from that of the algorithm MC. On the other hand, the number of solutions in the Pareto front of the algorithm MM is approximately equal to that of the algorithm MC, while the number of solutions in the Pareto fronts of the algorithms SM1 and SM2 are much less in comparison.

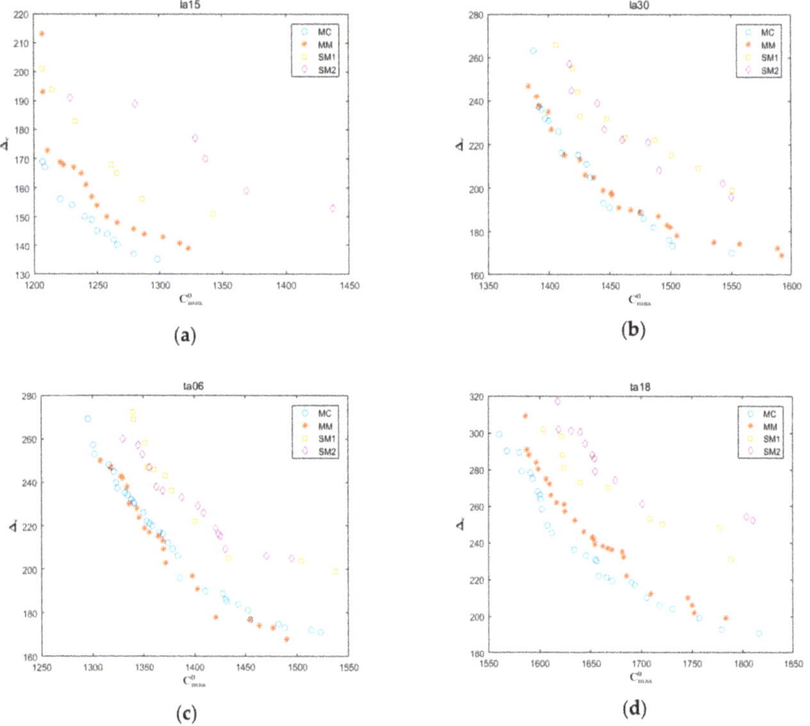

Figure 6. The Pareto fronts of algorithms MC, MM, SM1, and SM2. (**a**) For the problem La15; (**b**) for the problem La30; (**c**) for the problem Ta06; (**d**) for the problem Ta18.

However, as indicated earlier, the algorithm based on the Monte Carlo approximation will be very time-consuming. To investigate computational efforts for each algorithm, the average T_a and the relative value T_r of CPU time have been recorded, where T_r is the ratio of the average T_a of an algorithm (MC, MM, SM1, or SM2) to that of the algorithm MC. In this experiment, the simulation times for the Monte Carlo approximation are set as 30, which is generally the required minimum number of samples in statistical estimation [20]. As the results shown in Table 7, the algorithms SM1 and SM2 always consume the least time, while the algorithm MC consumes the most. By comparison, the time consumption of the proposed algorithm MM is almost equal to that of the algorithms SM1 and SM2, but it is much less than that of the algorithm MC. What is more, the time consumption of the algorithm MC increases significantly with the problem scale. For example, the average T_a of the algorithm MC in the problems La01-La05 is 66, which is close to that of the algorithm MM with the value 53. But, in the problems Ta31-Ta40, the time consumption of the algorithm MC is about 8 times of that of the algorithm MM. In more complex problems or uncertain scenarios, a larger sample size may be needed for the algorithm MC, which will result in a computational time that is unacceptable.

Table 7. Results on the CPU time (in seconds) of algorithms MC, MM, SM1, and SM2.

Cases	$n \times m$	MC		MM		SM1		SM2	
		T_a	T_r	T_a	T_r	T_a	T_r	T_a	T_r
La01–La05	10×5	66	1.00	53	0.80	50	0.76	50	0.76
La06–La10	15×5	83	1.00	54	0.65	52	0.63	50	0.60
La11–La15	20×5	113	1.00	55	0.49	55	0.49	52	0.46
La16–La20	10×10	102	1.00	54	0.53	54	0.53	51	0.50
La21–La25	15×10	153	1.00	56	0.37	55	0.36	54	0.35
La26–La30	20×10	220	1.00	60	0.27	59	0.27	57	0.26
La31–La35	30×10	398	1.00	73	0.18	70	0.18	68	0.17
La36–La40	15×15	238	1.00	65	0.27	62	0.26	60	0.25
Ta01–Ta10	15×15	231	1.00	63	0.27	61	0.26	59	0.26
Ta11–Ta20	20×15	351	1.00	75	0.21	70	0.20	68	0.19
Ta21–Ta30	20×20	543	1.00	85	0.16	82	0.15	83	0.15
Ta31–Ta40	30×15	733	1.00	95	0.13	92	0.13	89	0.12
Aver.	/	**269.3**	**1.00**	**65.7**	**0.24**	**63.5**	**0.24**	**61.8**	**0.23**

5. Conclusions

In this study, we have addressed the robust JSS with RMDs, in which the makespan and the EMD are considered simultaneously. To improve the efficiency of the MOEA, a meta-model has been constructed by using the data-driven response surface method. Then, with the individual-based evolution control, the meta-model-based MOEA has been proposed to solve this problem. The results have shown that regarding the convergence and diversity, the proposed algorithm yields better Pareto solution sets than the algorithms with the existing slack time-based surrogate measures. Moreover, the meta-model has high accuracy in evaluating the EMD similar to the Monte Carlo approximation. Overall, the proposed meta-model-based MOEA can effectively and efficiently solve the robust JSS with RMDs.

Author Contributions: Conceptualization, Z.W.; Formal analysis, S.Y.; Funding acquisition, T.L.; Investigation, S.Y.; Methodology, Z.W.; Supervision, T.L.; Writing—original draft, Z.W.; Writing—review and editing, T.L.

Funding: This research was funded in part by National Natural Science Foundation of China under grant no 51475383 and the Career Start-up Grant of the Northwestern Polytechnical University.

Conflicts of Interest: The authors declare no conflict of interest.

Notations:

JSS	Job shop scheduling problem	RMD	Random machine breakdowns
EMD	Expected makespan delay	MOEA	Multi-objective evolutionary algorithm
n	Number of jobs	m	Number of machines
J_j	Job j, $j = 1, 2, \ldots, n$	M_i	Machine i, $i = 1, 2, \ldots, m$
J	Set of jobs, $\{J_j \mid j = 1, 2, \ldots, n\}$	M	Set of machines $\{M_i \mid i = 1, 2, \ldots, m\}$
O_{ij}	Operation that job j is on machine i	O_j	Set of operations for job j
p_{ij}	Processing time of operation O_{ij}	st_{ij}	Starting time of operation O_{ij}
ct_{ij}	Completion time of operation O_{ij}	fs_{ij}	Free slack time of operation O_{ij}
ts_{ij}	Total slack time of operation O_{ij}	Pr_{ij}	Machine breakdown probability of O_{ij}
D_{ij}	Downtime when processing O_{ij}	C_{max}^0	Makespan of a schedule before execution
C_{max}^r	Actual makespan of a schedule	δ_c	Makespan delay of a schedule
Δ_c	Expression of expected makespan delay	Δ_c^a	Meta-model of Δ_c
Δ_c^{sim}	Monte Carlo approximation of Δ_c	O_n	Set of operations without slack time
O_y	Set of operations with slack time	p_s^n	Sum of processing time in the set O_n
p_s^y	Sum of processing time in the set O_y	fs_a	Average free slack time in the set O_y
ts_a	Average total slack time in the set O_y	λ_0	Machine failure rate of each machine

β_0	Expectation of the downtime		P_0	Initial population
P_k	Current population in generation k		R_{k+1}	Combined population in generation k
x	Input vector $x = (\lambda_0, \beta_0, p_s^n, p_s^y, fs_a, ts_a)$		I_i	A data instance $I_i = (x_i, \Delta_c^i)$
D_c	Training data set		s_i	Schedule of ith individual in R_{k+1}
F_{k+1}^i	Fitness of the ith individual in R_{k+1}		F_{k+1}	Fitness set of the population R_{k+1}

References

1. Sotskov, Y.N.; Egorova, N.G. The optimality region for a single-machine scheduling problem with bounded durations of the jobs and the total completion time objective. *Mathematics* **2019**, *7*, 382. [CrossRef]
2. Gafarov, E.; Werner, F. Two-machine job-shop scheduling with equal processing times on each machine. *Mathematics* **2019**, *7*, 301. [CrossRef]
3. Luan, F.; Cai, Z.; Wu, S.; Jiang, T.; Li, F.; Yang, J. Improved whale algorithm for solving the flexible job shop scheduling problem. *Mathematics* **2019**, *7*, 384. [CrossRef]
4. Turker, A.; Aktepe, A.; Inal, A.; Ersoz, O.; Das, G.; Birgoren, B. A decision support system for dynamic job-shop scheduling using real-time data with simulation. *Mathematics* **2019**, *7*, 278. [CrossRef]
5. Sun, L.; Lin, L.; Li, H.; Gen, M. Cooperative co-evolution algorithm with an MRF-based decomposition strategy for stochastic flexible job shop scheduling. *Mathematics* **2019**, *7*, 318. [CrossRef]
6. Zhang, J.; Ding, G.; Zou, Y.; Qin, S.; Fu, J. Review of job shop scheduling research and its new perspectives under Industry 4.0. *J Intell. Manuf.* **2019**, *30*, 1809–1830. [CrossRef]
7. Potts, C.N.; Strusevich, V.A. Fifty years of scheduling: A survey of milestones. *J. Oper. Res. Soc.* **2009**, *60*, S41–S68. [CrossRef]
8. García-León, A.A.; Dauzère-Pérèsb, S.; Mati, Y. An efficient Pareto approach for solving the multi-objective flexible job-shop scheduling problem with regular criterio. *Comput. Oper. Res.* **2019**, *108*, 187–200. [CrossRef]
9. Zhang, C.; Rao, Y.; Li, P. An effective hybrid genetic algorithm for the job shop scheduling problem. *Int. J. Adv. Manuf. Tech.* **2008**, *39*, 965–974. [CrossRef]
10. Li, L.; Jiao, L.; Stolkin, R.; Liu, F. Mixed second order partial derivatives decomposition method for large scale optimization. *Appl. Soft Comput.* **2017**, *61*, 1013–1021. [CrossRef]
11. Watanabe, M.; Ida, K.; Gen, M. A genetic algorithm with modified crossover operator and search area adaptation for the job-shop scheduling problem. *Comput. Ind. Eng.* **2005**, *48*, 743–752. [CrossRef]
12. Eswaramurthy, V.P.; Tamilarasi, A. Hybridizing tabu search with ant colony optimization for solving job shop scheduling problems. *Int. J. Adv. Manuf. Tech.* **2009**, *40*, 1004–1015. [CrossRef]
13. Mencía, C.; Mencía, R.; Sierra, M.R.; Varela, R. Memetic algorithms for the job shop scheduling problem with operators. *Appl. Soft. Comput.* **2015**, *34*, 94–105. [CrossRef]
14. Buddala, R.; Mahapatra, S.S. Two-stage teaching-learning-based optimization method for flexible job-shop scheduling under machine breakdown. *Int. J. Adv. Manuf. Tech.* **2019**, *100*, 1419–1432. [CrossRef]
15. Mehta, S.V.; Uzsoy, R.M. Predictable scheduling of a job shop subject to breakdowns. *IEEE Trans. Robotic. Autom.* **1998**, *14*, 365–378. [CrossRef]
16. Lei, D. Minimizing makespan for scheduling stochastic job shop with random breakdown. *Appl. Math. Comput.* **2012**, *218*, 11851–11858. [CrossRef]
17. Nouiri, M.; Bekrar, A.; Jemai, A.; Trentesaux, D.; Ammari, A.C.; Niar, S. Two stage particle swarm optimization to solve the flexible job shop predictive scheduling problem considering possible machine breakdowns. *Comput. Ind. Eng.* **2017**, *112*, 595–606. [CrossRef]
18. von Hoyningen-Huene, W.; Kiesmueller, G.P. Evaluation of the expected makespan of a set of non-resumable jobs on parallel machines with stochastic failures. *Eur. J. Oper. Res.* **2015**, *240*, 439–446. [CrossRef]
19. Jamili, A. Robust job shop scheduling problem: Mathematical models, exact and heuristic algorithms. *Expert Syst. Appl.* **2016**, *55*, 341–350. [CrossRef]
20. Xiong, J.; Xing, L.; Chen, Y. Robust scheduling for multi-objective flexible job-shop problems with random machine breakdowns. *Int. J. Prod. Econ.* **2013**, *141*, 112–126. [CrossRef]
21. Wu, Z.; Sun, S.; Xiao, S. Risk measure of job shop scheduling with random machine breakdowns. *Comput. Oper. Res.* **2018**, *99*, 1–12. [CrossRef]
22. Liu, N.; Abdelrahman, M.A.; Ramaswamy, S.R. A Complete Multiagent Framework for Robust and Adaptable Dynamic Job Shop Scheduling. *IEEE Trans. Syst. Man. Cybern. Part C* **2007**, *37*, 904–916. [CrossRef]

23. Xiao, S.; Sun, S.; Jin, J.J. Surrogate Measures for the Robust Scheduling of Stochastic Job Shop Scheduling Problems. *Energies* **2017**, *10*, 543. [CrossRef]

24. Zuo, X.; Mo, H.; Wu, J. A robust scheduling method based on a multi-objective immune algorithm. *Inform Sciences* **2009**, *179*, 3359–3369. [CrossRef]

25. Ahmadi, E.; Zandieh, M.; Farrokh, M.; Emami, S.M. A multi objective optimization approach for flexible job shop scheduling problem under random machine breakdown by evolutionary algorithms. *Comput. Oper. Res.* **2016**, *73*, 56–66. [CrossRef]

26. Al-Fawzan, M.A.; Haouari, M. A bi-objective model for robust resource-constrained project scheduling. *Int. J. Prod. Econ.* **2005**, *96*, 175–187. [CrossRef]

27. Goren, S.; Sabuncuoglu, I. Optimization of schedule robustness and stability under random machine breakdowns and processing time variability. *IIE Trans.* **2010**, *42*, 203–220. [CrossRef]

28. Deb, K.; Pratap, A.; Agarwal, S.; Meyarivan, T. A fast and elitist multiobjective genetic algorithm: NSGA-II. *IEEE Trans. Evol. Comput.* **2002**, *6*, 182–197. [CrossRef]

29. Li, L.; Yao, X.; Stolkin, R.; Gong, M.; He, S. An evolutionary multi-objective approach to sparse reconstruction. *IEEE Trans. Evol. Comput.* **2014**, *18*, 827–845.

30. Zhou, A.; Qu, B.; Li, H.; Zhao, S.; Suganthan, P.N.; Zhang, Q. Multiobjective evolutionary algorithms: A survey of the state of the art. *Swarm Evol. Comput.* **2011**, *1*, 32–49. [CrossRef]

31. Xiong, J.; Tan, X.; Yang, K.; Xing, L.; Chen, Y. A Hybrid Multiobjective Evolutionary Approach for Flexible Job-Shop Scheduling Problems. *Math. Probl. Eng.* **2012**, *2012*, 1–27. [CrossRef]

32. Hosseinabadi, A.A.R.; Siar, H.; Shamshirband, S.; Shojafar, M.; Nasir, M.H.N.M. Using the gravitational emulation local search algorithm to solve the multi-objective flexible dynamic job shop scheduling problem in Small and Medium Enterprises. *Ann. Oper. Res.* **2015**, *229*, 451–474. [CrossRef]

33. Hosseinabadi, A.A.R.; Kardgar, M.; Shojafar, M.; Shamshirband, S.; Abraham, A. GELS-GA: Hybrid metaheuristic algorithm for solving Multiple Travelling Salesman Problem. In Proceedings of the 2014 IEEE 14th International Conference on Intelligent Systems Design and Applications, Okinawa, Japan, 28–30 November 2014.

34. Jin, Y.; Branke, J. Evolutionary Optimization in Uncertain Environments: A Survey. *IEEE T. Evolut. Comput.* **2005**, *9*, 303–317. [CrossRef]

35. Al-Hinai, N.; Elmekkawy, T.Y. Robust and stable flexible job shop scheduling with random machine breakdowns using a hybrid genetic algorithm. *Int. J. Prod. Econ.* **2011**, *132*, 279–291. [CrossRef]

36. Chaari, T.; Chaabane, S.; Loukil, T.; Trentesaux, D. A genetic algorithm for robust hybrid flow shop scheduling. *Int. J. Comput. Integ. M.* **2011**, *24*, 821–833. [CrossRef]

37. Yang, F.; Zheng, L.; Luo, Y. A novel particle filter based on hybrid deterministic and random sampling. *IEEE Access* **2018**, *6*, 67536–67542. [CrossRef]

38. Yang, F.; Luo, Y.; Zheng, L. Double-Layer Cubature Kalman Filter for Nonlinear Estimation. *Sensors* **2019**, *19*, 986. [CrossRef] [PubMed]

39. Wang, X.; Li, T.; Sun, S.; Corchado, J.M. A survey of recent advances in particle filters and remaining challenges for multitarget tracking. *Sensors* **2017**, *17*, 2707. [CrossRef]

40. Mirabi, M.; Ghomi, S.M.T.F.; Jolai, F. A two-stage hybrid flowshop scheduling problem in machine breakdown condition. *J. Intell. Manuf.* **2013**, *24*, 193–199. [CrossRef]

41. Liu, L.; Gu, H.; Xi, Y. Robust and stable scheduling of a single machine with random machine breakdowns. *Int. J. Adv. Manuf. Technol.* **2006**, *31*, 645–654. [CrossRef]

42. Leon, J.; Wu, S.D.; Storer, R.H. Robustness measures and robust scheduling for job shops. *IIE Trans.* **1994**, *26*, 32–43. [CrossRef]

43. Jensen, M.T. Generating robust and flexible job shop schedules using genetic algorithms. *IEEE Trans. Evol. Comput.* **2003**, *7*, 275–288. [CrossRef]

44. Yen, G.G.; He, Z. Performance Metric Ensemble for Multiobjective Evolutionary Algorithms. *IEEE Trans. Evol. Comput.* **2014**, *18*, 131–144. [CrossRef]

 © 2019 by the authors. Licensee MDPI, Basel, Switzerland. This article is an open access article distributed under the terms and conditions of the Creative Commons Attribution (CC BY) license (http://creativecommons.org/licenses/by/4.0/).

Article

Maintenance Optimization for Repairable Deteriorating Systems under Imperfect Preventive Maintenance

Juhyun Lee [1], Byunghoon Kim [2] and Suneung Ahn [2,*]

[1]   Department of Industrial and Management Engineering, Hanyang University, Seoul 04763, Korea
[2]   Department of Industrial and Management Engineering, Hanyang University ERICA, Ansan 15588, Korea
*   Correspondence: sunahn@hanyang.ac.kr; Tel.: +82-031-400-5267

Received: 15 July 2019; Accepted: 5 August 2019; Published: 7 August 2019

Abstract: This study deals with the preventive maintenance optimization problem based on a reliability threshold. The conditional reliability threshold is used instead of the system reliability threshold. Then, the difference between the two thresholds is discussed. The hybrid failure rate model is employed to represent the effect of imperfect preventive maintenance activities. Two maintenance strategies are proposed under two types of reliability constraints. These constraints are set to consider the cost-effective maintenance strategy and to evaluate the balancing point between the expected total maintenance cost rate and the system reliability. The objective of the proposed maintenance strategies is to determine the optimal conditional reliability threshold together with the optimal number of preventive maintenance activities that minimize the expected total maintenance cost per unit time. The optimality conditions of the proposed maintenance strategies are also investigated and shown via four propositions. A numerical example is provided to illustrate the proposed preventive maintenance strategies. Some sensitivity analyses are also conducted to investigate how the parameters of the proposed model affect the optimality of preventive maintenance strategies.

Keywords: preventive maintenance; imperfect repair; optimization; reliability threshold; repairable system

---

1. Introduction

Most engineering systems experience degradation influenced by the usage environment and operation time length such as aircrafts, high-speed railroads, nuclear power plants, and weapons systems, which leads to frequent failures [1]. Once a failure has occurred in the system, it decreases the availability and reliability of the system, which increase unexpected losses. For example, the availability degradation can increase downtime costs and decrease product quality. Additionally, the reliability degradation can affect the safety of systems and reduce the operating life. Therefore, in many industries, maintenance activities are performed to prevent decreasing availability and reliability of the systems due to severe degradation.

Maintenance activities play an important role in reliability theory because they decrease uncertainty of the failure, reduce the operation cost of the system, and increase the functioning life of the system [2]. In practice, due to increasing complexity of the systems, the cost related to the maintenance activities has increased constantly over the past decades, which attributes 15–70% of production costs [3–5]. Therefore, it is important to establish cost-effective maintenance planning for ensuring higher efficiency, quality production, availability of the system, and reliability of the system.

There are two general types of maintenance activities: corrective maintenance (CM) and preventive maintenance (PM). CM is unscheduled and performed when a system fails, and the system is restored from a failed state to a working state [6]. Most CM employs a minimal repair that returns the failed system to its previous working state after repairing the system. Therefore, the minimal repair has been widely used in numerous maintenance models to address maintenance problems such as the failure of a system. There are many studies covering the minimal repair model [7]. On the other hand, PM activities delay the degradation of a system and reduce operational stress. In traditional PM models, a repaired system can end up in the state of either "as good as new" (i.e., perfect PM) or "as bad as old" (i.e., minimal repair) after the PM activities [8]. Thereafter, a more generalized and realistic approach has been proposed, called the imperfect PM model. The imperfect PM model considers that the system state lies somewhere between "as good as new" and "as bad as old" after PM activities.

Many models describing the effect of imperfect PM activities have been discussed in many existing studies. Malik et al. [9] introduced an imperfect PM model based on the age reduction method. Lin et al. [10] proposed a hybrid hazard rate function that combines the age reduction factor and hazard increasing factor. This model can better reflect the general and realistic situations because the model restores the system effective age to a better state and degrades the hazard rate of the system, after PM activities. El-Ferik and Ben-Daya [11] developed an age-based imperfect PM model that uses the hybrid hazard rate model and considers two types of failure modes. Khatab and Abdelhakim [12] developed an imperfect PM model that is subject to random deterioration based on the hybrid hazard rate using a reliability threshold.

In addition, the PM activity is subcategorized into two types of activities: periodic and non-periodic. Periodic PM activities are repeatedly performed over a fixed period of time T. The periodic PM activity is still widely applied to many PM models because it is mathematically convenient [13]. Shue et al. [14] developed an extended periodic imperfect PM model by considering an age-dependent failure type. Some previous studies [15–17] proposed maintenance policies for a leased system using the periodic imperfect PM model. Moreover, other extended studies on periodic PM have been applied to various fields such as warranty policies, production systems, and used systems [18–21]. However, a periodic PM has a shortcoming in that it is hard to prevent failures that frequently occur in the late period of the life of a system based on the periodic PM. Non-periodic PM models have been proposed to handle this shortcoming, which can prevent the inefficient PM activities [22].

In non-periodic PM models, controlling system reliability is an important issue especially when dealing with repairable deteriorating systems. In practice, maintenance based on reliability threshold provides good guidance for decision-makers who want to attain high performance of maintenance for their operating systems [23]. In addition, reliability of an engineering system is a decision variable that can increase investment in production technology [24]. Zhao [25] proposed a PM policy based on the critical reliability level subject to degradation. Zhou et al. [26] proposed a reliability-centered predictive maintenance policy that is composed of scheduled PM and non-scheduled PM using the reliability threshold. They considered the situation in which the system was continuously monitored. Liao et al. [27] developed a sequential imperfect PM model with a reliability threshold for a continuously monitored system that considered both the operational cost and the breakdown cost. Schutz et al. [16] developed a maintenance strategy for leased equipment based on the reliability threshold. Doostparast et al. [28] developed an optimal PM scheduling considering a certain level of reliability for coherent systems. Lin et al. [29] developed an imperfect PM model with a reliability threshold that considered three reliability constraints to help evaluate the maintenance cost. This model used the system reliability threshold that maintains a survival probability until the system replacement. Khatab et al. [30] developed an imperfect PM model for failure-prone second-hand systems considering the conditional reliability threshold and upgrade action. Khatab [31] improved Liao's model [27] by adding mathematical properties and remedying inconsistencies in the maintenance optimization model.

Traditional PM models using the reliability threshold assume that PM activities are performed when the reliability reaches a threshold value, and that the system is constantly monitored and maintained [29]. However, the system reliability may not be maintained at a constant level when following the foregoing assumption. This is because there are two types of reliability threshold in the

PM model: the conditional reliability threshold and the system reliability threshold. The conditional reliability threshold refers to the survival probability of each PM cycle, whereas the system reliability threshold refers to the survival probability until the replacement. The performance of the PM model (e.g., the expected maintenance cost rate and the system operating time until replacement) is not robust to the selection of reliability threshold in constructing the PM model. Therefore, it is important to clearly differentiate between the conditional reliability threshold and the system reliability threshold. This is the motivation of this study.

In addition, it is important to consider the trade-off between reliability and cost when establishing PM models. Some existing models employ reliability thresholds to address the trade-off. However, the existing models assume that the threshold value is same for every PM cycle. This assumption is biased and unrealistic toward reliability. Therefore, the existing models find it difficult to reflect the cost-effectiveness accurately. This is another motivation of this study.

The main contributions of this study are (i) to identify and define the differences between the system reliability threshold and the conditional reliability threshold and (ii) to develop a novel PM model that minimizes the expected maintenance cost rate with consideration of two reliability constraints. Table 1 highlights the contributions of different authors in the field of PM model focusing on the reliability threshold. It can be observed in Table 1 that the literature uses the system reliability threshold more, rather than the conditional reliability threshold. In addition, the literature assumes a fixed level of the reliability threshold.

Table 1. Author contribution in the reliability-based PM model based on reliability constraints, maintenance costs, and imperfect maintenance model.

Research	Types of Reliability Threshold	Reliability Constraints		Maintenance Cost		Other Cost	Imperfect Maintenance Model		
		Fixed	Unfixed	CM	PM	Breakdown	Age Reduction Factor	Hazard Increasing Factor	Other
Zhao [25]	Conditional	√		√	√				√
Zhou et al. [26]	System	√		√	√		√	√	
Liao et al. [27]	Conditional	√		√	√	√	√	√	
Schutz et al. [16]	System	√		√	√		√		
Doostparast et al. [28]	System	√		√	√	√			√
Lin et al. [29]	System	√	√	√	√		√		
Khatab et al. [30]	Conditional	√		√	√	√	√	√	
Khatab [31]	Conditional	√		√	√	√	√	√	
Proposed model	Conditional	√	√	√	√	√	√	√	

In this study, we use the conditional reliability as a criterion of PM activities instead of using system reliability because the conditional reliability threshold can consider the situation in which the system reliability decreases as the frequency of PM activities increase. The hybrid failure rate model is used to represent the effect of imperfect PM activities. In addition, we analyzed the proposed model through two strategies that can help evaluate the trade-off between maintenance cost and system reliability. We discussed the optimal conditions for each strategy via four propositions and one lemma that show the existence and uniqueness of the optimal solution.

The remainder of this paper is organized as follows. Section 2 explains the assumptions to establish the proposed PM model, and derives the function of the expected maintenance cost rate. Section 3 defines two strategies and provides an algorithm to find the optimal solution. Section 4 provides a numerical example to illustrate the proposed PM model and conducts sensitivity analyses to investigate critical elements. Finally, Section 5 discusses the conclusions of this research.

2. Imperfect Preventive Maintenance Model

We consider the situation in which a new system is subject to stochastic deterioration over time and will be replaced at the end of the Nth PM cycle by a new one. The system experiences PM whenever the conditional reliability threshold reaches a certain level R_i. PM is imperfect and modeled via the hybrid hazard rate model of Lin et al. [10]. Minimal repair is taken for any failures prior to the PM actions. The goal of this study is to determine the optimal number of PM activities and the optimal conditional reliability thresholds for each PM cycle that minimizes the expected maintenance cost rate. To consider the above-mentioned goal and assumptions, we compose the long-term expected maintenance cost rate in this study, which is given as:

$$ECR(\underline{R} = (R_1, R_2, \ldots, R_N), N) = \frac{E[MC] + E[BC]}{E[T]}, \tag{1}$$

where $E[MC]$ denotes the expected total maintenance cost, $E[BC]$ denotes the expected total breakdown cost, and $E[T]$ denotes the expected system replacement cycle. It should be noted that $E[MC]$ is composed of the cost for minimal repair, PM, and replacement. These costs are assumed to be constant and can be obtained from a field investigation. As they may also be considered the expected values for the random variables, it is reasonable to assume $max\ (C_{MR}, C_{PM}, C_{BD}) < C_{RP}$.

2.1. Notations and Assumptions

The notations and detailed assumptions in this study are:

Notations

α	Scale parameter of hazard intensity functions where $\alpha > 0$
β	Deterioration parameter of hazard intensity functions where $\beta > 2$
a_i	Age reduction factor where $0 = a_0 < a_1 < \cdots < a_i < \cdots < 1$
b_i	Hazard increasing factor where $1 = b_0 \leq b_1 \leq \cdots \leq b_i \leq \cdots$
x_i	The ith PM interval, where $x_0 = 0$ and $x_1 = Y_1$.
Y_i	Effective age of the system before the ith PM activity
T_O	System lifetime until replacement
R_i	Conditional reliability threshold where $I = 1, 2, \ldots, N$
C_{MR}	Cost of minimal repair
C_{PM}	Cost of PM activity
C_{RP}	Cost of system replacement
C_{BD}	Cost of system breakdown
$R_{SYS,i}$	System reliability up to the ith PM cycle
$R_{CON,i}$	Conditional reliability at the ith PM cycle
$R_{RES,i}$	Restored reliability after the ith PM activity
$h_i(t)$	Hazard intensity functions during the ith PM cycle

Assumptions

1. The planning horizon is infinite.
2. The system is replaced at the Nth PM cycle, and $(N - 1)$ imperfect PMs are performed.
3. The durations of minimal repair, PM, and replacement are negligible.
4. The deterioration process of the system follows the non-homogeneous Poisson process (NHPP), and their hazard intensity function is given as

$$h(t) = \alpha\beta t^{\beta-1}, \text{ where } t \geq 0. \tag{2}$$

5. The hazard rate of the system varies with the hybrid model after imperfect PM activities that are performed whenever the conditional reliability threshold reaches a certain value.
6. Costs for minimal repair, PM, replacement, and breakdown are assumed constant.

2.2. Hybrid Hazard Rate Model under Imperfect PM

Now, we describe the effect of imperfect PM using the hybrid hazard rate model. This model adjusts the hazard rate of the system using the age reduction factor and the hazard increasing factor after PM activity. Let us consider the situation in which a system's age is reduced by the age reduction factor after PM. If the first PM activity is performed at the effective age Y_1, the age of the system is updated from Y_1 to $a_1 Y_1$. The effective age Y_2 is obtained as the sum of $a_1 Y_1$ and x_2, where x_2 denotes the length of the second PM cycle, and the effective age Y_i can then be generalized as

$$Y_i = x_i + a_{i-1} Y_{i-1}, \tag{3}$$

where $Y_0 = 0$, $x_1 = Y_1$, and a_i denotes the age reduction factor for $0 = a_0 < a_1 < \cdots < a_i < \cdots < 1$.

The hybrid hazard rate model adjusts both the effective age of the system and the hazard rate of the system after PM activity. Now, we describe the hazard rate of the system in each PM cycle. The hazard rate of the system in the first PM cycle is the same as Equation (2) within $[0, Y_1)$. After the first PM, the hazard rate of the system is adjusted from $h(Y_1)$ to $b_1 h(a_1 Y_1)$, and thus, the hazard rate in the second PM cycle becomes $b_1 h(a_1 Y_1 + t)$ and $t \in [0, x_2)$. After the second PM, the hazard rate of the system is adjusted from $b_1 h(Y_2)$ to $b_2 b_1 h(a_2 Y_2)$. The hazard rate in the third PM cycle, then becomes $b_2 b_1 h(a_2 Y_2 + t)$ and $t \in [0, x_3)$. According to these procedures, the hazard rate in the ith PM cycle can be generalized as

$$h_i(t) = B_{i-1} h(a_{i-1} Y_{i-1} + t), \tag{4}$$

where $i \geq 1$, $t \in [0, x_i)$, and $B_{i-1} = \Pi_{j=1}^{i-1} b_j$ with $1 = b_0 \leq b_1 \leq \cdots \leq b_i \leq \cdots$.

2.3. Conditional Reliability Threshold

Here, we try to derive the system reliability under imperfect PM. According to the theory of reliability engineering, the reliability is defined as the probability that a system will survive over time period t [32]. The system reliability without PM activity can be defined as:

$$
\begin{aligned}
R_{SYS}(t) &= \Pr\{T \geq t\} \\
&= \exp\left(-\int_0^t h(u)\,du\right).
\end{aligned}
\tag{5}
$$

Let $R_{SYS,i}(t)$ be the system reliability at the ith PM cycle, $R_{RES,i-1}(t)$ be the restored reliability after the $(i-1)$th PM, and $R_{CON,i}(t|a_{i-1} Y_{i-1})$ be the conditional reliability at the ith PM cycle. According to the hybrid hazard rate model, after the first PM activity, the effective age is adjusted from Y_1 to $a_1 Y_1$ and the hazard rate of the system is adjusted from $h(Y_1)$ to $B_1 h(a_1 Y_1)$. Based on these procedures, the restored reliability after the first PM can be inferred as

$$R_{RES,1}(a_1 Y_1) = \exp(-B_1 H(a_1 Y_1)). \tag{6}$$

The conditional reliability at the second PM cycle, $R_{CON,2}(t|a_1 Y_1)$, is the survival probability of the system at time t given that the system undergoes the first PM at Y_1; it follows that

$$R_{CON,2}(t|a_1 Y_1) = \exp\left(-B_1 \int_{a_1 Y_1}^t h(u)\,du\right), \tag{7}$$

where $t \in [a_1 Y_1, Y_2)$. The system reliability after the first PM at time period t is the survival probability over the period $Y_1 + (t - a_1 Y_1)$; it can then be computed as the product of the restored reliability and the conditional reliability, which is given as

$$R_{SYS,2}(t) = R_{RES,1}(a_1 Y_1) R_{CON,2}(t|a_1 Y_1). \tag{8}$$

Based on these procedures, the system reliability after the $(i-1)$th PM cycle can obviously be determined as:

$$R_{SYS,i}(t) = R_{RES,i-1}(a_{i-1}Y_{i-1})R_{CON,i}(t|a_{i-1}Y_{i-1}).\tag{9}$$

According to the aforementioned discussion, we can describe the difference between the conditional reliability threshold and the system reliability threshold through two definitions.

Definition 1. *If imperfect PM activities are performed under the following constraints*

$$R_i = R_{CON,i}(Y_i|a_{i-1}Y_{i-1}),\tag{10}$$

then R_i is called the conditional reliability threshold of the ith PM cycle.

Definition 2. *If imperfect PM activities are performed under the following constraints*

$$\begin{aligned}R_S &= R_{SYS,j}(Y_j)\\&= R_{RES,j-1}(a_{j-1}Y_{j-1})R_{CON,j}(Y_j|a_{j-1}Y_{j-1}),\end{aligned}\tag{11}$$

then R_S is called a system reliability threshold.

Details of the PM model using the system reliability threshold have been reported in the literature [29]. As the conditional reliability threshold can reflect the situation in which the system reliability decreases as the frequency of PM activities increases while maintaining the reliability above a certain level, it is used to model the present work. Moreover, as the system undergoes PM whenever the conditional reliability threshold reaches the certain level, the corresponding reliability constraints can be given as:

$$R_i = \exp\left(-B_{i-1}\int_{a_{i-1}Y_{i-1}}^{Y_i} h(t)dt\right),\tag{12}$$

for $i = 1, 2, \ldots, N$. Equation (12) is inferred as:

$$\ln R_i = -B_{i-1}\int_{a_{i-1}Y_{i-1}}^{Y_i} h(t)dt.\tag{13}$$

Solving Equation (13) with respect to Y_i, the effective age at the ith PM activity can be determined as:

$$Y_i = \left(\frac{-B_{i-1}^{-1}\ln R_i - \sum_{j=1}^{i-1}\left(\prod_{k=j}^{i-1}\rho_k^\beta\right)B_{j-1}^{-1}\ln R_j}{\alpha}\right)^{1/\beta},\tag{14}$$

where $i = 2, \ldots, N$ and $Y_1 = (-\ln R_1/\alpha)^{1/\beta}$.

2.4. Long-Term Expected Maintenance Cost Rate

The expected total maintenance cost until the Nth PM cycle is composed of the replacement cost, PM cost, and minimal repair cost, which is given as:

$$\begin{aligned}E[MC] &= C_{RP} + (N-1)C_{PM} + C_{MR}\sum_{i=1}^{N}\left(\int_0^{Y_i}h_i(t)dt\right)\\&= C_{RP} + (N-1)C_{PM} + C_{MR}\sum_{i=1}^{N}\left(B_{i-1}\int_{a_{i-1}Y_{i-1}}^{Y_i}h(t)dt\right).\end{aligned}\tag{15}$$

Through Equation (13), Equation (15) becomes:

$$E[MC] = C_{RP} + (N-1)C_{PM} - C_{MR}\sum_{i=1}^{N} \ln R_i. \tag{16}$$

Real engineering systems suffer from frequent breakdowns after long-term use due to the severe degradation of the system. At this point, it may be more economical to replace the system rather than performing PM. Liao et al. [27] pointed out the importance of breakdown cost when constructing the PM models and used the breakdown cost in their PM model. They considered that the breakdown costs are charged once per PM cycle. Unlike the computation approach used in Liao et al. [27], Khatab [31] considered the breakdown cost that occurs whenever the system undergoes maintenance activities, such as a minimal repair, PM, and replacement. He discussed that the breakdown costs are computed as the product of the sum of all maintenance activities until the Nth PM cycle and the value of the breakdown cost. This study employs the approach proposed by Khatab [31] in calculating the breakdown cost. The total breakdown cost in this study is given as

$$\begin{aligned} E[BC] &= C_{BD}\left(N + \sum_{i=1}^{N}\int_{0}^{Y_i} h_i(t)dt\right) \\ &= C_{BD}\left(N - \sum_{i=1}^{N}\ln R_i\right). \end{aligned} \tag{17}$$

The replacement cycle of the system can be calculated by the sum of the durations up to the Nth PM cycle, which is given as:

$$T_O = Y_N + \sum_{j=1}^{N-1}(1-a_j)Y_j. \tag{18}$$

Through Equations (15)–(18), the long-term expected maintenance cost rate becomes:

$$ECR(\underline{R}=(R_1,R_2,\ldots,R_N),N) = \frac{C_{RP} + (N-1)C_{PM} - C_{MR}\sum_{i=1}^{N}\ln R_i + C_{BD}\left(N - \sum_{i=1}^{N}\ln R_i\right)}{Y_N + \sum_{j=1}^{N-1}(1-a_j)Y_j}. \tag{19}$$

3. Maintenance Optimization

In this section, we introduce the proposed PM model by discussing two strategies with different reliability constraints. The optimization problem of the proposed PM model is to find the decision variables that minimize the expected maintenance cost rate. In this optimization problem, the decision variables are defined as the optimal number of PM activities and the optimal conditional reliability thresholds.

Strategy 1: The PM activities are performed with the same level of conditional reliability threshold, which is given as:

$$R_1 = \exp\left(-B_{i-1}\int_{a_{i-1}Y_{i-1}}^{Y_i} h(t)dt\right), \text{ for } i = 1,2,\ldots,N. \tag{20}$$

Strategy 2: The PM activities are performed at the different values of conditional reliability thresholds on all PM cycles. This PM strategy relaxes the reliability constraints of strategy 1 and is assumed to follow Equation (12). The system reliability of strategy 2 is lower than that of strategy 1, but it can lower the expected total maintenance cost rate.

3.1. Strategy 1: Same Level of Conditional Reliability Threshold

This strategy assumes that the conditional reliability thresholds on all PM cycles have the same value. The PM activity of this strategy is performed whenever the conditional reliability threshold reaches a certain level. The decision variables in this strategy are the optimal number of PM activities and the optimal conditional reliability threshold. These values minimize the expected total maintenance cost rate. In this strategy, the reliability constraints for performing PM activities follow Equation (20). Solving Equation (20) with respect to Y_i, the effective age at the ith PM activity can be determined as:

$$Y_i = \left(\frac{-B_{i-1}^{-1} \ln R_1 - \sum_{j=1}^{i-1} \left(\prod_{k=j}^{i-1} \rho_k^\beta \right) B_{j-1}^{-1} \ln R_1}{\alpha} \right)^{1/\beta}, \tag{21}$$

where $i = 2, \ldots, N$ and $Y_1 = (-\ln R_1 / \alpha)^{1/\beta}$.

The expected total maintenance cost rate of strategy 1 becomes:

$$ECR_1(R_1, N) = \frac{C_{RP} + (N-1)C_{PM} - NC_{MR} \ln R_1 + C_{BD}(N - N \ln R_1)}{Y_N + \sum_{j=1}^{N-1} (1 - a_j)Y_j}. \tag{22}$$

Now, we try to find the optimal number of PM activities that minimizes Equation (22). The inequalities are:

$$ECR_1(R_1, N-1) > ECR_1(R_1, N) \text{ and } ECR_1(R_1, N) \le ECR_1(R_1, N+1), \tag{23}$$

if and only if

$$K_1(N-1) < C_{RP} - C_{PM} \text{ and } K_1(N) \ge C_{RP} - C_{PM}, \tag{24}$$

where

$$K_1(N) = \left(\frac{\sum_{i=1}^N x_i}{x_{N+1}} - N \right) (C_{PM} + C_{BD} - (C_{MR} + C_{BD}) \ln R_1). \tag{25}$$

Inequalities (23) and (24) are a necessary condition to find where Equation (22) is a convex function with respect to N when R_1 is fixed. Through the following Proposition 1, we can see that there exists an optimal finite and unique N^*.

Proposition 1. *When R_1 is fixed, if Inequalities (22) and (23) are satisfied, Equation (22) is a convex function and there exists an optimal finite and unique N^* that minimizes Equation (22).*

Proof. See Appendix A. □

Through following Proposition 2, we discuss the optimal value of the conditional reliability threshold when N is fixed.

Proposition 2. *When N is fixed, the optimal value for the conditional reliability threshold is given as:*

$$-\ln R_1^* = \frac{N(C_{PM} + C_{BD}) - C_{PM} + C_{RP}}{(\beta - 1)N(C_{MR} + C_{BD})}. \tag{26}$$

Proof. See Appendix A. □

The optimal solutions for strategy 1 can be obtained via the following algorithm that is given as the pseudo-code and is followed as Table 2.

Table 2. Algorithm to obtain the optimal solution for strategy 1.

Algorithm Pseudo-Code for the Minimization of Equation (22)
1: Input data: $C_M, C_R, C_P, C_{BD}, \alpha, \beta, a_i, b_i$;
2: Set $N = 1, M = 2$;
3: Compute $R_1(N), R_1(M)$ given in Equation (26);
4: Compute $ECR_1(N|R_1(N)), ECR_1(M|R_1(M))$ given in Equation (22)
5: **while** $(ECR_1(R_1(N), N) > ECR_1(R_1(M), M))$
6:      $N = N + 1, M = M + 1$;
7:      Compute $R_1(N), R_1(M)$ given in Equation (26);
8:      Compute $ECR_1(R_1(N), N), ECR_1(R_1(M), M)$ given in Equation (22);
9: **end while**
10: Set $N^* = N, R^*_1 = R_1(N^*)$;
11: Using N^* and R^*_1, compute Y_i given in Equation (21);

3.2. Strategy 2: Different Level of Conditional Reliability Threshold

Unlike strategy 1, this strategy assumes that the conditional reliability thresholds on all PM cycles have difference values. The structural properties of this strategy follow Section 2. The aim of this strategy is to find the optimal number of PM activities N^* and the optimal conditional reliability thresholds $\underline{R} = (R_1, R_2, \ldots, R_N)$. The expected total maintenance cost rate is the same as Equation (19). Solving Equation (14) with respect to Y_i, the effective age at the ith PM activity can be determined as:

$$Y_i = \left(\frac{-B_{i-1}^{-1} \ln R_i - \sum_{j=1}^{i-1} \left(\prod_{k=j}^{i-1} \rho_k^\beta \right) B_{j-1}^{-1} \ln R_j}{\alpha} \right)^{1/\beta}, \tag{27}$$

where $i = 2, \ldots, N$ and $Y_1 = (-\ln R_1/\alpha)^{1/\beta}$.

The result of following Lemma 1 will be used in Propositions 3 and 4. The lemma derives the N conditional reliability thresholds to a single conditional reliability threshold.

Lemma 1. *When N is fixed, the conditional reliability thresholds that minimize Equation (17) have the following relationship:*

$$-\ln R_i = \begin{cases} B_{i-1}\left(\frac{A_{i-1}}{A_i} - a_{i-1}^\beta\right)\left(\frac{A_1}{A_{i-1}}\right)(-\ln R_1) & , \text{if } 2 \leq i \leq N-1, \\ B_{N-1}\left(A_{N-1} - a_{N-1}^\beta\right)\left(\frac{A_1}{A_{N-1}}\right)(-\ln R_1) & , \text{if } i = N, \end{cases} \tag{28}$$

where

$$A_i = \left[\frac{1}{B_{N-1}} \left(\frac{B_{i-1} - a_i^\beta B_i}{1 - a_i} \right) \right]^{\beta/(\beta-1)} \quad \text{for all } i. \tag{29}$$

Proof. See Appendix A. $\square$

Through the result of the following Lemma, Equation (19) becomes:

$$ECR_2(R_1, N) = \frac{(C_{RP} + C_{BD}) + (N-1)(C_{PM} + C_{BD}) - (C_{MR} + C_{BD})Mr(N)\ln R_1}{Y_N + \sum_{j=1}^{N-1} (1 - a_j)Y_j}, \tag{30}$$

where

$$Mr(N) = 1 + \sum_{i=2}^{N-1} B_{i-1}\left(\frac{A_{i-1}}{A_i} - a_{i-1}^\beta\right)\left(\frac{A_1}{A_{i-1}}\right) + B_{N-1}\left(A_{N-1} - a_{N-1}^\beta\right)\left(\frac{A_1}{A_{N-1}}\right), \tag{31}$$

where $Mr(1) = 1$ and $Mr(2) = 1 + B_1(A_1 - a_1^\beta)$.

Through following Proposition 3, we can see that there exists an optimal finite and unique N^*.

Proposition 3. *When R_1 is fixed, N^* that satisfies Inequality (32) is the unique and finite optimal value that minimizes Equation (30).*

$$K_2(N-1) < C_{RP} - C_{PM} \text{ and } K_2(N) \geq C_{RP} - C_{PM}, \tag{32}$$

where

$$K_2(N) = \left(\frac{\sum_{i=1}^{N} x_i}{x_{N+1}} - N\right)(C_P + C_{BD}) - (C_M + C_{BD})\frac{\ln R_1}{x_{N+1}}\left(Mr(N+1)\sum_{i=1}^{N} x_i - Mr(N)\sum_{i=1}^{N+1} x_i\right). \tag{33}$$

Proof. See Appendix A. □

The following Proposition provides the optimal conditional reliability threshold of strategy 2.

Proposition 4. *When N is fixed, the solution to the optimal conditional reliability threshold that minimizes Equation (30) is given as:*

$$- \ln R_1^* = \frac{N(C_{PM} + C_{BD}) - C_{PM} + C_{RP}}{(\beta - 1)Mr(N)(C_{MR} + C_{BD})}. \tag{34}$$

Proof. See Appendix A. □

The optimal solutions for strategy 2 can be obtained via the following algorithm that is given as pseudo-code and is followed as Table 3.

Table 3. Algorithm to obtain the optimal solution for strategy 2.

Algorithm Pseudo-Code for the Minimization of Equation (30)
1: Input data: C_M, C_R, C_P, C_{BD}, α, β, a_i, b_i;
2: Set $N = 1$, $M = 2$;
3:       Compute $R_1(N)$, $R_1(M)$ given in Equation (34);
4:       Compute $Mr(N)$, $Mr(M)$ given in Equation (31);
5:       Compute $ECR_2(R_1(N), N)$, $ECR_2(R_1(M), M)$ given in Equation (30);
6: **while** $(ECR_2(R_1(N), N) > ECR_2(R_1(M), M))$
7:       $N = N + 1$, $M = M + 1$;
8:       Compute $R_1(N)$, $R_1(M)$ given in Equation (34);
9:       Compute $Mr(N)$, $Mr(M)$ given in Equation (31);
10:      Compute $ECR_2(R_1(N), N)$, $ECR_2(R_1(M), M)$ given in Equation (30);
11: **end while**
12: Set $N^* = N$, $R^*_1 = R_1(N^*)$;
13: Using N^* and R^*_1, compute $R_N{}^* = (R_1, \ldots, R_N{}^*)$ given in Equation (28);
14: Using N^* and R^*_1, compute Y_i given in Equation (27);

4. Numerical Illustration

In this section, we illustrate the proposed PM strategy based on a numerical example. To conduct the numerical illustration, we used Python 3.7. The procedures of the numerical example are followed as:

Step 1. Input data related to maintenance costs and parameters of hybrid hazard rate function.

Step 2. Compute the conditional reliability threshold and the expected maintenance cost rate.

Step 3. Repeat the step 2 until satisfying $ECR_k(R_1(N), N) > ECR_k(R_1(M), M)$ for $k = 1,2$.

Step 4. Compute PM schedule, the optimal number of PM activities, and the optimal conditional reliability threshold.

In the numerical illustration, two additional PM strategies are considered that can help evaluate the effect of the trade-off between reliability and cost. In the additional PM strategies, the conditional reliability threshold is provided to the decision maker, in which this is given as $R_1 = 0.9$ (PM strategy 3) and $R_1 = 0.3$ (PM strategy 4). Sensitivity analyses are also conducted to find how the parameters affect the proposed PM strategy. Table 4 sets out the parameters for conducting the numerical illustration. The functions of the age reduction factor and the hazard increasing factor are set as follows:

$$a_i = \frac{i}{2i+2}, \ b_i = \frac{13i+4}{12i+4}. \tag{35}$$

Table 4. Setting of parameters.

C_{MR}/C_{PM}	C_{RP}/C_{PM}	C_{BD}/C_{PM}	α	β
3.0	5.0	0.3	2.6	3.2

4.1. Results of the Numerical Illustration

Figure 1 shows the curves for the expected total maintenance cost rate under the increasing PM activities. Table 5 summarizes the results of this study.

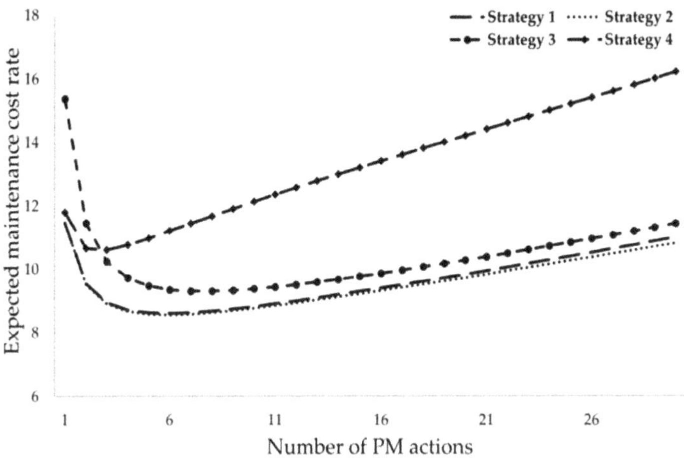

Figure 1. Cost curve under increasing the number of preventive maintenance (PM) activities.

Table 5. The optimal solutions and their system performances.

No. PM	PM Strategy 1		PM Strategy 2		PM Strategy 3		PM Strategy 4	
	x_i	R_S (R^*_1)	x_i	R_S (R^*_1)	x_i	R_S (R^*_1)	x_i	R_S (R^*_1)
1	0.4933	0.7627 (0.7627)	0.5194	0.7265 (0.7265)	0.3672	0.9000 (0.9000)	0.7862	0.3000 (0.3000)
2	0.3626	0.7601 (0.7627)	0.3534	0.7637 (0.7668)	0.2699	0.8988 (0.9000)	0.5779	0.2955 (0.3000)
3	0.3164	0.7561 (0.7627)	0.2968	0.7841 (0.7901)	0.2356	0.8969 (0.9000)	0.5043	0.2886 (0.3000)
4	0.2905	0.7528 (0.7627)	0.2655	0.7981 (0.8072)	0.2163	0.8955 (0.9000)		
5	0.2728	0.7504 (0.7627)	0.2441	0.8091 (0.8196)	0.2031	0.8943 (0.9000)		
6	0.2593	0.7485 (0.7627)	0.3271	0.6660 (0.6753)	0.1931	0.8934 (0.9000)		
7					0.1848	0.8928 (0.9000)		
8					0.1777	0.8922 (0.9000)		
N^*	6		6		8		3	
T_O	1.9950		2.0065		1.8477		1.8684	
R^*_1	0.7627		0.7265					
$EMC\,(N^*, R^*_1)$	8.6034		8.5542		9.2988		10.6077	

From the results of this study, we can see that the larger the conditional reliability threshold that we set, the shorter PM cycle and the more PM activities the optimal solution needs. As shown in Table 5, the optimal value N of PM strategy 3 is higher than the optimal ones obtained from strategies 1, 2, or 4. In addition, as R_1 for the PM strategy 3 is set to a high value, the reliability of the system remains high. However, this leads to high PM cost and breakdown cost, since there are frequent PM activities. As shown in Figure 1, the expected total maintenance cost rate of PM strategy 3 is higher than that of PM strategies 1 and 2.

In the cases where the conditional reliability threshold is set to a low value, the length of the PM cycle increases. The result of this is many system breakdowns and only minimal repairs are allowed, and thus, the expected total maintenance cost rate dramatically increases over time. Moreover, we can observe that a trade-off between the reliability threshold and the expected total maintenance cost rate. Therefore, appropriate conditional reliability thresholds are required to establish a cost-effective maintenance strategy.

Unlike the results of PM strategies 3 and 4, PM strategies 1 and 2 focus on finding a balance between the conditional reliability threshold and the expected total maintenance cost rate. Hence, we can see that the results of PM strategies 1 and 2 are more cost-effective than the results of PM strategies 3 and 4. Furthermore, we can see that PM strategy 2 is more cost-effective than PM strategy 1 because PM strategy 2 relaxed the reliability constraints of PM strategy 1. From Table 5, we can see that the optimal conditional reliability thresholds of PM strategy 2 tend to increase slightly except for the value that corresponds to N^*. This means that it is more reasonable to perform PM activities frequently than to replace the system up to $(N^* - 1)$ PM cycles. Since the system is then replaced at the N^*th PM activity, the conditional reliability threshold of the last cycle decreases. This implies that operating the system for as long as possible in the last PM cycle is more efficient than performing imperfect PM activity.

4.2. Sensitivity Analyses

Several sensitivity analyses were conducted to investigate how the parameters affected the proposed models. First, to determine how the parameters of the hazard intensity function influence the proposed models, sensitivity analysis was conducted for the situation in which α was adjusted from 1.92 to 2.88 and β was adjusted from 2.88 to 4.32. The results are summarized in Table 6. It is shown that changing of α had no effect on the optimal solutions because the change in α could not reflect the trend in the deterioration process. However, since α was used to calculate the PM cycle, the system lifetime and the expected total maintenance cost rate were affected by the change in α.

The change in β affected the optimal solutions unlike the results of the sensitivity analyses for α because β reflects the shape of the hazard intensity. For example, when β received a low value, the hazard intensity increased from the initial period, resulting in shorter PM intervals. As shown in

Table 6, N^* decreased as the value of β was low. Moreover, R_1^* increases as β increases, which implies that the system degradation accelerates when β is high.

Table 6. Results of sensitivity analysis related to the failure rate's parameters.

α	β	Strategy 1					Strategy 2				
		R_1^*	N^*	R_S	T_O	$EMC_1 (N^*, R_1^*)$	R_1^*	N^*	R_S	T_O	$EMC_2 (N^*, R_1^*)$
2.08	3.20	0.7627	6	0.7485	2.1391	8.0239	0.7265	6	0.6660	2.1514	7.7980
2.34	3.20	0.7627	6	0.7485	2.0618	8.3248	0.7265	6	0.6660	2.0736	8.2771
2.60	3.20	0.7627	6	0.7485	1.9550	8.6034	0.7265	6	0.6660	2.0065	8.5542
2.86	3.20	0.7627	6	0.7485	1.9364	8.8635	0.7265	6	0.6660	1.9476	8.8128
3.12	3.20	0.7627	6	0.7485	1.8845	9.1079	0.7265	6	0.6660	1.8953	9.0557
2.60	2.56	0.6650	5	0.6351	1.7113	10.0687	0.6251	5	0.5486	1.7213	10.0101
2.60	2.88	0.7128	5	0.6933	1.7342	9.2751	0.6831	5	0.6090	1.7441	9.2224
2.60	3.20	0.7627	6	0.7485	1.9550	8.6034	0.7265	6	0.6660	2.0065	8.5542
2.60	3.52	0.7894	6	0.7798	2.0492	8.0434	0.7595	6	0.7021	2.0604	7.9998
2.60	3.84	0.8190	7	0.8119	2.3400	7.5694	0.7857	7	0.7379	2.3523	7.5298

In addition, we conducted sensitivity analyses to investigate how the change in related costs for maintenance activity affected the proposed model. The results are summarized in Table 7. As the replacement cost increases or the PM cost decreases, the optimal number of PM activities increases. This implies that it is more economical to perform PM activities than to replace the system. As the minimal repair cost increases, the optimal conditional reliability threshold increases. However, the optimal number of PM activities remains unchanged. As the conditional reliability threshold increases, the length of the PM cycle shortens, thereby increasing the expected total maintenance cost rate. Indeed, as shown in Table 7, the optimal conditional reliability threshold and the expected total maintenance cost rate increases as the minimal repair cost increases. Moreover, for both PM strategies, the value of the optimal conditional reliability threshold is significantly sensitive to changes in the minimal repair cost. Therefore, decision-makers should pay close attention to the estimation of minimal repair cost.

Table 7. Results of sensitivity analysis related to maintenance costs.

C_{MR}	C_{PM}	C_{RP}	Strategy 1					Strategy 2				
			R_1^*	N^*	R_S	T_O	$EMC_1 (N^*, R_1^*)$	R_1^*	N^*	R_S	T_O	$EMC_2 (N^*, R_1^*)$
2.40	1.00	5.00	0.7181	6	0.7018	2.1241	8.0805	0.6766	6	0.6085	2.1363	8.0342
2.70	1.00	5.00	0.7423	6	0.7271	2.0553	8.3510	0.7036	6	0.6395	2.0671	8.3032
3.00	1.00	5.00	0.7627	6	0.7485	1.9550	8.6034	0.7265	6	0.6660	2.0065	8.5542
3.30	1.00	5.00	0.7801	6	0.7668	1.9415	8.8406	0.7461	6	0.6890	1.9526	8.7900
3.60	1.00	5.00	0.7952	6	0.7826	1.8935	9.0645	0.7631	6	0.7090	1.9044	9.0126
3.00	0.80	5.00	0.7912	7	0.7772	2.1434	8.0756	0.7472	7	0.6991	2.1562	8.0276
3.00	0.90	5.00	0.7715	6	0.7577	1.9682	8.3511	0.7364	6	0.6776	1.9795	8.3033
3.00	1.00	5.00	0.7627	6	0.7485	1.9550	8.6034	0.7265	6	0.6660	2.0065	8.5542
3.00	1.10	5.00	0.7406	5	0.7274	1.7924	8.8454	0.7169	5	0.6453	1.8023	8.7970
3.00	1.20	5.00	0.7325	5	0.7189	1.8127	9.0673	0.7082	5	0.6350	1.8227	9.0177
3.00	1.00	4.00	0.7697	5	0.7577	1.7170	8.0477	0.7482	5	0.6827	1.7265	8.0037
3.00	1.00	4.50	0.7592	5	0.7467	1.7448	8.3365	0.7368	5	0.6691	1.7544	8.2910
3.00	1.00	5.00	0.7627	6	0.7485	1.9550	8.6034	0.7265	6	0.6660	2.0065	8.5542
3.00	1.00	5.50	0.7652	7	0.7497	2.2347	8.8521	0.7168	7	0.6643	2.2481	8.7994
3.00	1.00	6.00	0.7577	7	0.7418	2.2601	9.0746	0.7081	7	0.6543	2.2736	9.0206

5. Conclusions

In this study, we established a cost-effective PM strategy for a repairable system subject to stochastic deteriorations. Two PM strategies were proposed and modeled by two reliability constraints that help evaluate the trade-off between the expected total maintenance cost rate and the system reliability. Strategy 1 constricted the conditional reliability threshold with the same level, whereas

Strategy 2 relaxed the reliability constraints of Strategy 1 by using different level of the conditional reliability threshold. Moreover, this study discussed the difference between the conditional reliability threshold and the system reliability threshold via two definitions. The conditional reliability threshold was used as a condition variable that represents the decreasing system reliability as the number of PM activities increases.

This study employed a hybrid hazard rate model to represent imperfect PM activities. The function of the expected total maintenance cost rate was determined, including the costs of PM activity, replacement, minimal repair, and breakdown. We provided the structural properties of the proposed PM strategies via four propositions and proved their existence and uniqueness of the propositions. Two algorithms were proposed to find optimal solutions. A numerical example was conducted to illustrate the proposed PM strategy. Sensitivity analyses were also conducted to investigate how the parameters of the proposed PM strategy affect the optimal solutions.

5.1. Contributions to Theory

The contributions of this study are as follows:

(1) This study identifies and defines the differences between the system reliability threshold and the conditional reliability threshold. This contributes to the reliability theory.
(2) This study develops a cost-effective PM model using the conditional reliability threshold that considers the decreasing system reliability as the number of PM activities increases, which can relax the trade-off between the reliability and the cost.
(3) This study provides the optimal conditions for each strategy through four propositions and one lemma that show the uniqueness and existence of the optimal solution.

5.2. Implications for the Decision-Makers

Decision makers need to decide whether they undertake all the maintenance activities to keep the system in good condition or reduce maintenance activities in planning PM policy to cut down the costs. For example, if they reduce the costs related to maintenance activities in a PM strategy, it is difficult to obtain high quality of the system due to the insufficient PM activities. On the other hand, frequent PM activities result in higher maintenance costs and poorer system availability. The proposed model can be a good alternative in solving these problems. Based on the proposed model, they can avoid both the extremely frequent maintenance activities and the insufficient maintenance ones effectively.

5.3. Limitations and Further Research

This study has some limitations. First, the parameters of hazards rate function are not estimated based on data but predetermined manually. Second, we need more constraints to apply the model in real word applications because the proposed model is applied to complex and expensive systems. For example, the optimal PM strategies in this work may be expanded to the system under lease, considering the various constraints for the reliability such as the maximum reliability, the minimum reliability, and the reliability at the end of lease. To tackle these limitations, we can estimate the parameters based on the data and add more constraints to handle the complex and difficult situations in real world applications.

For the next step of the research, we can also extend the proposed model to apply to many different areas that cover multi component systems. The expected result is a novel algorithm that not only solves the integrated optimization problem of the proposed PM strategies but also maintains production costs and requirements. In addition, another next step of this study includes solving the two dimensional warranty problem of the proposed model while considering the warranty cost and the numerous uncertainties.

Author Contributions: Conceptualization, writing—original draft preparation, formal analysis, visualization, J.L.; methodology, J.L and S.A.; supervision, writing—review and editing, B.K. and S.A.; data curation, B.K. and J.L.; funding acquisition, project administration, S.A.

Funding: This work was supported by the National Research Foundation of Korea (NRF) grant funded by the Korea government (MSIP) (No. NRF-2018R1A2B6003232).

Conflicts of Interest: The authors declare no conflict of interest.

Appendix A

Appendix A.1 Proof of Proposition 1

When R_1 is fixed, the expected total maintenance cost rate becomes a univariate function. Inequalities (23)–(24) are a necessary condition to obtain the optimal number of PM activities. Inequality (24) is obtained from substituting Equation (22) into Inequality (23). If N^* satisfies Inequality (24), it becomes a finite and unique optimal solution such that $Min(ECR_1(N), N = 1,2,\ldots) = ECR_1(N^*)$. Now, to prove Proposition 1, we recall $K_1(N)$, which is as follows:

$$K_1(N) = \left(\frac{\sum_{i=1}^{N} x_i}{x_{N+1}} - N \right)(C_{PM} + C_{BD} - (C_{MR} + C_{BD}) \ln R_1). \tag{A1}$$

The left hand side of $K_1(N)$ increases as N increases because x_{N+1} decrease to zero. Moreover, the right hand side of $K_1(N)$ is a positive because $-\ln R_1$ is a positive value. Hence, $K_1(N)$ in Inequality (24) is the increasing function with respect to N, and N^* that satisfies Inequality (24) is then a finite and unique optimal solution that minimizes Equation (22).

Appendix A.2 Proof of Proposition 2

When the number of PM activities is fixed, the condition to obtain the optimal conditional reliability threshold that minimizes Equation (22) is as follows:

$$\frac{dECR_1(R_1|N)}{dR_1} = 0. \tag{A2}$$

Solving Equation (A2) with R_1, we obtained Equation (26). Moreover, Equation (26) is a finite and unique solution minimizes Equation (22) because $d^2ECR_1(R_1|N)/d^2R_1 > 0$.

Appendix A.3 Proof of Lemma 1

When the number of PM activities is fixed, the conditions to obtain the optimal conditional reliability threshold that minimizes Equation (17) are as follows:

$$\begin{cases} \dfrac{dECR_2(\underline{R}=(R_1,R_2,\cdots,R_N)|N)}{dR_1} = 0 \\ \qquad\vdots \\ \dfrac{dECR_2(\underline{R}=(R_1,R_2,\cdots,R_N)|N)}{dR_N} = 0 \end{cases}. \tag{A3}$$

Equation (A3) yields the following relationships:

$$\alpha\beta(C_m + C_{bd})B_{i-1}\left(\frac{dT_o}{dR_i}\left(Y_N + \sum_{i=1}^{N-1}(1 - a_i Y_i) \right) \right) = ECR_2(\underline{R} = (R_1, R_2, \cdots, R_N)|N), \tag{A4}$$

for $i = 1, 2, \ldots, N$. Solving the simultaneous equation in Equation (A4), we have the following relationships:

$$
\begin{aligned}
S(N) \quad &= \left[\frac{1}{B_{N-1}}\frac{B_{N-2}-a_{N-1}^{\beta}B_{N-1}}{1-a_{N-1}}\right]^{\frac{\beta}{\beta-1}} S(N-1) \\[4pt]
&\quad\vdots \\[4pt]
&= \left[\frac{1}{B_{N-1}}\frac{B_{1}-a_{2}^{\beta}B_{2}}{1-a_{2}}\right]^{\frac{\beta}{\beta-1}} S(2) \\[4pt]
&= \left[\frac{1}{B_{N-1}}\frac{1-a_{1}^{\beta}B_{1}}{1-a_{1}}\right]^{\frac{\beta}{\beta-1}} S(1)
\end{aligned}
\tag{A5}
$$

where $S(1) = -\ln R_1$ and

$$
S(i) = -B_{i-1}^{-1}\ln R_i - \sum_{j=1}^{i-1}\left(\prod_{k=j}^{i-1}\rho_k^{\beta}\right)B_{j-1}^{-1}\ln R_j, \text{ for } i = 2,3,\ldots,N.
\tag{A6}
$$

Using the relationships in Equation (A5), we obtained Equation (28).

Appendix A.4 Proof of Proposition 3

Using the results of Lemma 1, Equation (17) becomes Equation (30). To prove Proposition 3, we redefine $K_2(N)$ in Inequality (33) as follows:

$$
K_2(N) = A(N) + B(N)C(N),
\tag{A7}
$$

where

$$
\begin{aligned}
A(N) &= \left(\frac{\sum_{i=1}^{N}x_i}{x_{N+1}} - N\right)(C_{PM} + C_{BD}), \\
B(N) &= -(C_{MR} + C_{BD})\frac{\ln R_1}{x_{N+1}}, \\
C(N) &= \sum_{i=1}^{N}x_i(Mr(N+1) - Mr(N)) - Mr(N)x_{N+1}.
\end{aligned}
\tag{A8}
$$

$A(N)$ and $B(N)$ increase as N increase, and $C(N)$ increases as N increases because x_{N+1} decreases to zero. Hence, there exists a finite and unique N^* that minimizes Equation (30).

Appendix A.5 Proof of Proposition 4

Using the results of Lemma 1, Equation (17) becomes Equation (30), and then, Equation (34) is a finite and unique solution that minimizes Equation (30) because $d^2 ECR_2(R_1|N)/d^2 R_1 > 0$.

References

1. Lee, J.; Park, J.; Ahn, S. On determining a non-periodic preventive maintenance schedule using the failure rate threshold for a repairable system. *Smart Struct. Syst.* **2018**, *22*, 151–159.
2. Kilic, E.; Ali, S.S.; Weber, G.W.; Dubey, R. A value-adding approach to reliability under preventive maintenance costs and its applications. *Optimization* **2014**, *63*, 1805–1816. [CrossRef]
3. Velmurugan, R.; Dhingra, T. Maintenance strategy selection and its impact in maintenance function: A conceptual framework. *Int. J. Oper. Prod. Manag.* **2015**, *35*, 1622–1661. [CrossRef]
4. Ilangkumaran, M.; Kumanan, S. Application of hybrid VIKOR model in selection of maintenance strategy. *Int. J. Inf. Syst. Supply Chain Manag.* **2012**, *5*, 59–81. [CrossRef]
5. Lee, J.; Ni, J.; Djurdjanovic, D.; Qiu, H.; Liao, H. Intelligent prognostics tools and e-maintenance. *Comput. Ind.* **2006**, *57*, 476–489. [CrossRef]
6. Ahn, S.; Kim, W. On determination of the preventive maintenance interval guaranteeing system availability under a periodic maintenance policy. *Struct. Infrastruct. Eng.* **2011**, *7*, 307–314. [CrossRef]
7. Murthy, D. A note on minimal repair. *IEEE Trans. Reliab.* **1991**, *40*, 245–246. [CrossRef]
8. Yanez, M.; Joglar, F.; Modarres, M. Generalized renewal process for analysis of repairable systems with limited failure experience. *Reliab. Eng. Syst. Saf.* **2002**, *77*, 167–180. [CrossRef]

9. Malik, M.A.K. Reliable preventive maintenance scheduling. *AIIE Trans.* **1979**, *11*, 221–228. [CrossRef]
10. Lin, D.; Zuo, M.J.; Yam, R.C. Sequential imperfect preventive maintenance models with two categories of failure modes. *Nav. Res. Logist.* **2001**, *48*, 172–183. [CrossRef]
11. El-Ferik, S.; Ben-Daya, M. Age-based hybrid model for imperfect preventive maintenance. *IIE Trans.* **2006**, *38*, 365–375. [CrossRef]
12. Khatab, A. Hybrid hazard rate model for imperfect preventive maintenance of systems subject to random deterioration. *J. Intell. Manuf.* **2015**, *26*, 601–608. [CrossRef]
13. Huang, Y.-S.; Chen, E.; Ho, J.-W. Two-dimensional warranty with reliability-based preventive maintenance. *IEEE Trans. Reliab.* **2013**, *62*, 898–907. [CrossRef]
14. Sheu, S.-H.; Chang, C.-C. An extended periodic imperfect preventive maintenance model with age-dependent failure type. *IEEE Trans. Reliab.* **2009**, *58*, 397–405. [CrossRef]
15. Yeh, R.H.; Kao, K.-C.; Chang, W.L. Preventive-maintenance policy for leased products under various maintenance costs. *Expert Syst. Appl.* **2011**, *38*, 3558–3562. [CrossRef]
16. Schutz, J.; Rezg, N. Maintenance strategy for leased equipment. *Comput. Ind. Eng.* **2013**, *66*, 593–600. [CrossRef]
17. Mabrouk, A.B.; Chelbi, A.; Radhoui, M. Optimal imperfect maintenance strategy for leased equipment. *Int. J. Prod. Econ.* **2016**, *178*, 57–64. [CrossRef]
18. Jung, K.M.; Park, M.; Park, D.H. Cost optimization model following extended renewing two-phase warranty. *Comput. Ind. Eng.* **2015**, *79*, 188–194. [CrossRef]
19. Liao, G.-L. Optimal economic production quantity policy for a parallel system with repair, rework, free-repair warranty and maintenance. *Int. J. Prod. Econ.* **2016**, *54*, 6265–6280. [CrossRef]
20. Park, M.; Jung, K.M.; Park, D.H. Optimization of periodic preventive maintenance policy following the expiration of two-dimensional warranty. *Reliab. Eng. Syst. Saf.* **2018**, *170*, 1–9. [CrossRef]
21. Sett, B.K.; Sarkar, S.; Sarkar, B. Optimal buffer inventory and inspection errors in an imperfect production system with preventive maintenance. *Int. J. Adv. Manuf. Technol.* **2017**, *90*, 545–560. [CrossRef]
22. Nakagawa, T. Sequential imperfect preventive maintenance policies. *IEEE Trans. Reliab.* **1988**, *37*, 295–298. [CrossRef]
23. Sarkar, B.; Sana, S.S.; Chaudhuri, K. Optimal reliability, production lotsize and safety stock: An economic manufacturing quantity model. *Int. J. Ind. Eng. Manag. Sci.* **2010**, *5*, 192–202. [CrossRef]
24. Sarkar, B.; Sana, S.S.; Chaudhuri, K. Optimal reliability, production lot size and safety stock in an imperfect production system. *Int. J. Math. Oper. Res.* **2010**, *2*, 467–490. [CrossRef]
25. Zhao, Y. On preventive maintenance policy of a critical reliability level for system subject to degradation. *Reliab. Eng. Syst. Saf.* **2003**, *79*, 301–308. [CrossRef]
26. Zhou, X.; Xi, L.; Lee, J. Reliability-centered predictive maintenance scheduling for a continuously monitored system subject to degradation. *Reliab. Eng. Syst. Saf.* **2007**, *92*, 530–534. [CrossRef]
27. Liao, W.; Pan, E.; Xi, L. Preventive maintenance scheduling for repairable system with deterioration. *J. Intell. Manuf.* **2010**, *21*, 875–884. [CrossRef]
28. Doostparast, M.; Kolahan, F.; Doostparast, M. A reliability-based approach to optimize preventive maintenance scheduling for coherent systems. *Reliab. Eng. Syst. Saf.* **2014**, *126*, 98–106. [CrossRef]
29. Lin, Z.-L.; Huang, Y.-S.; Fang, C.-C. Non-periodic preventive maintenance with reliability thresholds for complex repairable systems. *Reliab. Eng. Syst. Saf.* **2015**, *136*, 145–156. [CrossRef]
30. Khatab, A.; Diallo, C.; Sidibe, I. Optimizing upgrade and imperfect preventive maintenance in failure-prone second-hand systems. *J. Manuf. Syst.* **2017**, *43*, 58–78. [CrossRef]
31. Khatab, A. Maintenance optimization in failure-prone systems under imperfect preventive maintenance. *J. Intell. Manuf.* **2018**, *29*, 707–717. [CrossRef]
32. Ebeling, C.E. *An Introduction to Reliability and Maintainability Engineering*; Tata McGraw-Hill Education: New York, NY, USA, 2004.

 © 2019 by the authors. Licensee MDPI, Basel, Switzerland. This article is an open access article distributed under the terms and conditions of the Creative Commons Attribution (CC BY) license (http://creativecommons.org/licenses/by/4.0/).

Article

Dynamic Agile Distributed Development Method

Muhammad Asgher Nadeem and Scott Uk-Jin Lee *

Department of Computer Science and Engineering, Hanyang University, Ansan, Gyeonggi-do 15588, Korea;
engr.nadeem18@gmail.com
* Correspondence: scottlee@hanyang.ac.kr

Received: 23 July 2019; Accepted: 24 September 2019; Published: 13 October 2019

Abstract: "Agile" is an effective software engineering model with a high trust and acceptance rate among its users. The term agility comes from the concept of rapid development and working in a team for better results and a faster competition rate when compared with any other software engineering model. In this study, an assessment of the different patterns, frameworks, and application program interfaces available for distributed development in an agile model is given. After analyzing the state-of-the-art distributed models, a novel framework of a dynamic agile distributed development method (DADDM) is introduced in this paper. Many researchers have worked on global software development using the agile approach; however, we are presenting the idea of incorporating the agile benefits with dynamic distributed software development. The applicability of the proposed model is checked via two selected parameters: a feasibility study and a business study. The complete DADDM development life cycle is presented in the methodology section. The techniques used in DADDM and team members' roles and responsibilities in DADDM are defined in this study. This study reflects all pillars of planning, controlling, organizing, and management of leadership. The use of DADDM in distributed agile development encourages future researchers to use this proposed framework for comparison and testing of their models and to check the effectiveness through a comparison with DADDM.

Keywords: distributed development; agile development; application program interfaces in agile distributed projects; framework for agile distributed development

---

1. Introduction

In distributed agile development, the agility rate is much higher in a team working with 40 people. The main focus of distributed agile development is the ability to respond to changing requirements and environments [1]. The ability to meet changing project deadlines with a satisfied customer base is the core concept of agile development.

There are different kinds of agile methods that exist. Agile-distributed development procedures consist of various frameworks that focus on a wide range of project management methods. Major types of agile development are dynamic systems development, crystal methods, agile, and extreme programming [2]. The highly recommended and most effective methods are extreme programming and agile. Agile works on the principle of an empirical process control model. This difference makes agile more effective and a better choice for system development projects. The agile method for distributed development involves agile teams consisting of the product owner and agile master, and the definitions of events, rules, and artifacts. Agile events are used to complete the processes on time, create regularity in tasks, and reduce the time required for unnecessary meetings [3].

Currently, the value of agile development is increasing because of its intuitive functionality. Projects are being completed on time and according to the specific needs of customers. The salient features of the dynamic agile-distributed method are innovation, flexibility, and effective management.

Project development is central to the responsive and disciplinary approach, as it has the program's baseline constraints as control objectives.

Isolation exists in both customary prerequisite project building and dynamic agile distributed development method (DADDM) procedures. It works through absolute requirements by appropriately assembling, arranging, and archiving all prerequisites while avoiding any conflicts. In DADDM, reliable creation is due to the correspondence issues among clients. However, it gives the developer a chance to create prerequisites, for example, the use of cases. These utilization cases are written in common language and are given useful prerequisites. Numerous distinctions exist and are of prime significance, particularly for programming improvement ventures.

1.1. DADDM Methodologies to Client

In the past, prerequisite issues were less significant. This was because there were not many prerequisites or changes required by the customer. Many organizations are trying to improve their products and are working towards complex requirements to improve customer satisfaction. These include the prerequisites of over-burden, no client commitment, the absence of compelling correspondences, the fulfilment of time constraints, spending issues, and substantially more inconveniences. To conquer these issues, computer science specialists presented the agile manifesto in distributed development in agile modeling. The DADDM methodology essentially focuses on programming advancement rather than documentation; it effectively works when clients' requirements change to ensure clients' satisfaction, provide fast advancement, and much more.

Many techniques exist, including:

1. XP, generally known as Extreme Programming;
2. Approach of Agile Modeling;
3. Nimble AGILE Methodology;
4. Nimble Feature Driven Development;
5. Nimble Automated Software Engineering;
6. Deft Lean Systems Engineering, Lean Software Development, Adaptive Software Development of Agile.

There is typically an organized structure for the improvement of enormous and complex frameworks. The three useful systems being embraced are theory, cooperation, and learning. Theory relates to arranging the board requirements, which allows experts to prepare for future prospects. Cooperation is similar to client association, as clients and colleagues work together among themselves to discover fundamental clashes and issues with important understandings. Learning deals with preparing or change of prerequisites during the improvement process, which implies that designers or clients work together and collaborate in a team for better results. The main objective of this is to create fundamental models applicable to programming improvements that utilize tests and conceptualization.

1.2. Agile Artifact

Agile operates on a pattern that is designed with a single purpose that maximizes transparency and provides adaptation and opportunity for interoperability. A product background is updated regularly as the product gets larger. The development phase requires more improvements and involves irregularities in preferences. That is, generating a breakthrough requires certain changes be made in the product backup.

1.2.1. Role of the Agile Team

Agile works on pre-determined characteristics for effective participation and the fast delivery of the product [4]. These are as follows:

Owner of product

It is the owner's job to accurately understand the metrics and needs of the market and business values.

Product Owner Responsibilities

- The development team should work in a specific order to ensure rapid work completion;
- The purpose of the product must be retained within the team by defining and updating product requirements;
- Feedback must be collected from customers, clients, and partners, and this information must be transferred to the team;
- Optimized work must be carried out in every stage;
- The quality of the product and delivery method must be considered;
- A product release plan must be constructed [5].

An overview of the team members is shown in Figure 1.

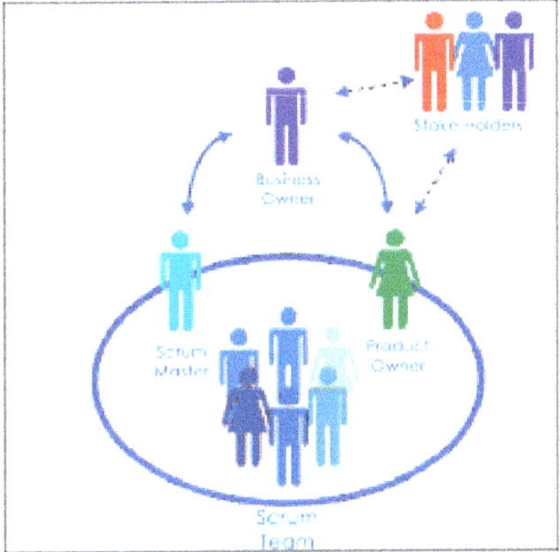

Figure 1. Overview of agile team members.

1.2.2. Agile Master

To help the team and owner, the agile master works as a trainer for understanding the approaches used in the management process of an agile project. This role helps streamline the workflow to adjust the agile project management protocols. In the project, the agile master acts like a backbone. The agile master fully understands the jobs of the team and is also aware of the requirements needed for getting a job done. In addition, the agile master knows all the details of the project and wears multiple hats in the project.

Role of the Agile Master

- Helps the team understand the values and approaches of the agile process;
- Helps the product owner update, check, and document product requirements;
- Makes sure the opportunity and goals of the product are well understood by every team member;
- Is responsible for properly conducting agile events;
- Minimizes team clashes and other weaknesses that are barriers to the on-time delivery of product;
- Manages the team in the right way to attain maximum production;
- Keeps a record of implementation of every successful sprint.

1.2.3. Development Team

The team consists of different divisions, including developers, testers, and engineers. The agile team is well organized and works in collaboration by helping each other to achieve targets. Every member of the team updates and shares their work with other members to keep everyone informed about the progress of the project [3].

Responsibilities of Agile Development Team

- Teams are independent, and the product owner or agile master do not put pressure to give any output at the end but check their work consistently with light touch;
- Groups do not separate, and each team member works according to their experience;
- Groups are cross-functional and cooperate to check the progress after each iteration.

1.3. Agile Framework

Agile frameworks work on a group of approved and synchronized procedures. Every event is time-boxed and very similar to sprints in which the entire development of the product takes place. The life cycle of every agile event is dependent on the objective success status for the stage.

1.4. Types of Agile Meetings

a. Sprint Planning Meeting
The first step is to arrange a planning meeting before the project starts. In this phase, the product owner plans the product requirements and identifies the most important things that are mandatory and must be executed quickly. In turn, the team strategizes to implement the checklist provided by the product owner. In this meeting, the team plans the deadlines and decides how work will be done efficiently. All aims and goals are planned during the sprint meeting.

b. Daily Agile Meeting
Daily meetings of teams begin once the sprint planning conference has started. These are very short meetings, not more than fifteen minutes, in which challenges, issues, progress, and feedback of the project are discussed among team members. New solutions and different strategies are also discussed on a daily basis.

c. Sprint Review Meeting
This review meeting is held at the end of the sprint to discuss what the team has achieved at the end of the iteration. Stakeholders and clients participate in this meeting. The team discusses the next plan that will be executed in the next phase. This review meeting emphases cost, timeline, and available resources.

1.5. Extreme Programming

This procedure of agile project management emphases changes in response to client and customer preferences. It was created by Kent Back in 1999 as a new perspective for software development with a basic emphasis on how the code is written effectively, rapidly, and error free. The main idea is that a unified method of XP involves the development of a system based on teamwork and elements being developed including coding, testing, and design screening [6].

1.5.1. Extreme Programming Value

Extreme programming is not just a multi-functioning method; there are many positive things to help and guide the team's behavior and to promote extremely high programming for better project management.

There are five key values for high-level project management:

- Simplicity—Workload is divided into small components to eliminate the complexity. The team only focuses on part of the work at a specific time and meets it.
- Effective communication—The team combines the collaborative work, including every outcome and error; daily meetings are known as the interaction.
- Continuous Impact—The team's focus and their workflow center around the opinions and changes from customers.
- Respect—Each member of the team is recognized for his feedback and takes part in the effort. Despite the organizational status, the administration admires every team member and gives respect, and the developing team does the same with the customer's demands.
- Courage—The groups are stimulated instead of accepting failure during these projects, and the focus is not on their propaganda.

1.5.2. Extreme Programming Rules

Dawn Wells [6] published the first XP rules in 1999 to clarify the key components of software development rules and list the activities that gradually became the working part of the XP model.

1.6. Management

- The Project Manager establishes a workshop to start work for their team;
- The Project Manager determines the project's speed;
- Stand-up Meetings are held on a daily basis;
- The performance of every team member is checked and reinstated when any procedural or dispute problems arise;
- The development approach changes if the XP method does not properly incorporate the team member [7].

1.7. Designing

- Start with a simple design to avoid delay in completion;
- Functional increases in design should be saved later;
- A class responsibility collaboration (CRC) card is used.

1.8. Coding

- Collective code ownership is applied such that any developer can change the code, detect or repair the errors.
- Using system residences (each team member including clients can explain how the system works).
- Programming is paired to write code and send it to production.
- Integration of code is done every few hours and is stored [8].

1.9. Testing

- Acceptance test requirements are repeated.
- Unit testing is performed regularly before issuing the code.

1.10. Kanban

Kanban is best recognized as a scheduling system, which has a great advantage over production inventory. Kanban limits the amount of extraordinary work. This prevents businesses and teams from being neutral and chaotic, which damages inventory track records. Kanban only uses the time-consuming approach in which the project managers enable them to use their resources [9].

1.10.1. Kanban Project Management Principles

The Kanban technique works on four elementary principles [10] that describe Kanban's process. These principles are as follows.

1.10.2. Start Now

Kanban does not work in a phase and does not work in the cross-ending method. It can be applied to any workflow in the project life cycle. Kanban is already under development and can be applied to a project.

1.10.3. Save Current Processes, Role, and Responsibilities

Contrary to other methods, there are certain rules and frameworks that must work in the project. Kanban teams can continue to work with the same responsibilities in their current roles before they accept and synchronize them. This involves old business practices as well as deliberately fraudulent frameworks.

1.10.4. Leadership Actions

Kanban's flexible model allows transparency in the working frame. Since there is no such classification for the Kanban perspective, every member of the team knows how to access the host for the top and bottom of the project. Kanban promotes leadership power over all stages with its power mode, which is available for every team member. This requires the team to retain a mentality of continuous upgrading on a discrete level.

1.10.5. Kanban Practice

The team follows actions that must be completed for the successful Kanban process.

1.10.6. Workflow

Kanban encourages teams to work in projects and tasks within the work of quick and efficient workflow.

1.10.7. Adjustable Boards

The Kanban Board is a pictorial demonstration of the workflow at each phase of the project life cycle. The board has three parts: development, completion, and application. It serves as a real-time inventory that is constantly updated to include any obstacles, problems, and constraints that are potential threats.

1.10.8. Kanban Cards

A section of the Kanban board is made of Kanban cards. Cards have information such as job details and their development assigned to it as well as its history. These cards serve as a center of information that is accessible to the entire team. One of the most impressive and interesting features of the Kanban cards is the ability to move cards from one side to another, with smooth drag and drop functionality.

1.10.9. Limit Work in Progress

The best use of Kanban is to bind WIP (work in progress). Tasks are assigned extreme limits (which are located in the column of the counting board in this case). Number of tasks in progress cannot exceed the number of jobs requested. In a second step through a Kanban system, teams must work together to narrow down the limitations.

1.10.10. Manage Flow

Workflow through the Kanban system should be free from any roadblocks. Efficient workflow flows from one phase to another phase in a continuous mode, providing evidence that team efficiency and work metrics are fruitful and have business value.

1.10.11. Continuous Improvement

Kanban does not promise successful implementation. The teams need to work on their weaknesses and improve their strengths to maintain stability and expedite quick delivery.

1.11. Application Software Development

Appropriate software development is a framework consisting of adaptive modes, which adopt an adaptive process while building large and complex products via mutual cooperation. The best practices of application software development are emphasized to accommodate adaptive, growing patterns, so that we can dynamically cope with unusual conditions when they occur.

1.11.1. What Is Distributed Agile?

Project development with distributed agility is an approach to software development that connects the team members from different countries, time zones, and language. Good communication and creating a highly functional environment are the fundamental elements for better operation and good results from the team. Some principles of distributed agile need to be integrated with the agile software development to make it distributed agile development:

- Interactions among team members and a better understanding of the processes and tools;
- Working software with comprehensive documentation;
- Contract negotiation and customer collaboration;
- How to respond to changes in plans.

1.11.2. Distributed Agile Software Development

Distributed agile software development combines the features of an agile model with distributed computing technology and upgrading the agile model to the next level. In distributed agile software development, we make use of external resources and high-profile competent in-house teams. The highlighted outcome for the approach is the minimum time required for the project completion. We can meet deadlines easily and never need to work in high-risk situations. In this development model, we integrate team members across the world and provide the best utilization of this extended team. One of the most appropriate ways of uniting them and working efficiently is to develop an effective communication environment.

1.11.3. Significance of DADDM

- Improves the productivity of dynamic agile programming by presenting intuitive and semiautomatic strategies in various periods for improving requirement engineering;
- Decreases the intellectual multifaceted nature of light-footed programming advances and determines complex choice circumstances effectively by defining scientific models;
- Improves the correspondence and coordination of dispersed agile programming advancement groups by presenting semi-mechanized strategies to the appropriate groups;
- It improves the nature of agile programming resulting in lower levels of risk by considering all significant factors (for example conditions, limits) in numerical streamlining models;
- Agile programming improvement focuses on the best arrangement for a particular setting, and clients' criticisms are addressed by modifying requirements, limits, and needs.

2. Literature Review

Extensive remote software development programs are complex and have a low tolerance rate that creates many challenges. Despite the challenges of handling multiple development teams, the methods of agility are being adopted based on the quick process to support the team's agreement. Artifacts are solid products of the software development process aimed at the teams' perspective on the same development program. The purpose of the study of Bass [11] was to focus on the development process to meet the requirements of remote software development programs concentrating on the infrastructure used in the development process. Using practitioner interviews, 46 software development teams were adopted in nine international companies combined with a documentary approach, documentary sources, and observations. The data analysis used was an open coding memo consistent comparison. This study identified 25 architectures of five types including features, sprints, releases, products, and corporate governance. It was discovered that with the plans of conventional angels to encourage the rules, strategies associated with planning-based methods were encouraged. Study-related experimental evidence has been used to identify the main owner of every artifact and generate a graphic map of specific activities. Finally, the programs developed in the study created a dynamic and plan-oriented architecture to improve the performance of the enterprise's quality and technology strategy, while the risk of failure was also reduced in the agile development environment [11]. It worked well to satisfy the requirements of remote software development programs. However, its applicability is smaller in agile development, which highlights the need for our work on designing and developing dynamic agile distributed development methods.

A combined animated approach and distribution software development (DSD) were used to promote better quality software solutions in less time and at lower cost. The method helps in both the agility and distribution and reduces major challenges and risks. This work was intended to create a broad set of risk factors affecting the output of distributed freeware projects and indicate risk management practices [12]. Their work was unique because it integrated the simulation method with DSD, and it will be more useful if applied in dynamic project development.

The study provided continuous comparisons to analyze deeper interviews of twenty projects from three practitioners and thirteen different information technology (IT) organizations. Field experience was supported by extensive research literature on the management of risk in agile and distributed development. Interviews and project work documents showed the five risk factors of the group under five main risk categories. The risk category was mapped for organizational changes to facilitate results of the real world. Risk factors can be attributed to the DSD, and the development-related issues in agile are presented. Besides, some new risk factors have been tested using this process, and further investigation is needed. Distributed agile development (DAD) is being adopted in organizations to meet the changing business needs caused by global transformation. This study reduced the risk impact and enhanced the possible options to mitigate the influence to a maximum extent [12].

Software development has become mainstream. The authors [13] worked on the Scaled Agile Framework (SAFE) and showed the best practices for studying two industry case studies [13].

Knowledge management (KM) is mandatory for any software project, but especially in global software development, where members of the team are separated from time to time. The knowledge management software is organizing knowledge in various ways to enhance transparency and improve the performance. One way to evaluate these strategies is presented in Reference [14], as they defined seven Knowledge Management areas. The purpose of their research was to study the methods of knowledge and sharing in many distributed ages. This is done by managing a series of interviews during a specified time period. If we apply their findings for the agile development platform, then we can get a high customer satisfaction level. Their results showed that sharing knowledge in remote areas and trusted fiscal projects created a sound base of successful project completion [14].

Adapting mass agility often requires the integration of agile methods and non-dynamic growth elements to build hybrid adaptive procedures. The challenge is to determine what elements or components (freelance or unauthorized) of hybrid evolution procedures are relevant to develop a reference architecture. The authors [15] addressed this important challenge and used a hybrid creative experiment to produce a hybrid adaptive model. They proved that there are a lot of chances for improving adaptive hybrid and agile-based development approaches using controversial engineering approaches [15].

The options available for project management methods have increased significantly; there are many questions for project managers considering alternatives. Project rating systems and standards are not met with logical project goals, features or environments. The main reason for this is that many software projects do not receive time, budget or client's attention. The purpose of the paper [16] was to identify and categorize the important success factors (CSFs) and to develop a controversial approach based on traditional project-oriented and freelance models. After reviewing the previous work, 37 CSFs were identified from 148 articles for software development projects. These were then ranked in three main CSFs categories: organizational, team, and customer factors. An unsatisfactory model highlighted the need to meet project properties and project management procedures on CSFs. This study suggested a model and helped to develop rules to assess the role of CSF for the success of the project. Although future experimental testing of this conceptual model is necessary, it provided an initial step in collecting the number of databases, provided a detailed experimental analysis for comparative studies, and improved the description of CSF [16]. The importance of critical success factors will become more impactful when it comes with dynamic project development, where the customer is allowed to make changes at any level of the development phased. So, if we take the work on this research as the initial point for our research and integrate it with the dynamic agile development method, we should achieve a high rate of customer satisfaction.

In the past, it was assumed that the global software development and the agile methods are incompatible with each other. The author of Reference [17] explained some of the core difficulties and benefits of using the agile method for global software development. They suggested applying this on an industrial level. This point is the base line of our work and we adopted their approach for the compatibility design of the software development in the distributed agile environment.

Agile distributed development models are currently widely used by most of the software organizations due to the fact of limited time and cost, but this method also creates many additional problems for software development teams. The result was an unstable software architecture. To address this problem, a situational ADSD (agile distributed software development) framework [18] was proposed, which can handle different requirement challenges for different situations. The gaps in situational ADSD also motivated us to propose a dynamic and generic DADDM, which applies to all kinds of distributed software development processes to complete on time and budget.

3. Methodology

3.1. Assessment of the Patterns for Agile-Based Distributed Development

3.1.1. Agile Integration

Organizations need to put things like business segments or administrations together rapidly. At its core, agile development is an approach to complete programming tasks more rapidly. Presently, agile development can be used in conjunction with small groups and a gradual methodology to determine what is required and how to achieve it more successfully instead of making it into a multi-year plan. Thus, agile is not just about software development or external software acquisition. The agile philosophy can also apply to broader concepts of such as integration and infrastructure. However, it brought change to the overall tech industry, driving the fast-moving needs of the customers.

3.1.2. Distributed Environment

Integration has been around for decades, but this approach is closely related to technical concepts and organizational patterns. Both are required in the modern context. The integrated technology pattern around the ESB (enterprise service bus), where a centralized software component performs integrations to backend systems. The goal of integration was to make the logic easy and accept the changes. In the past, integration was performed by the most prominent people, who were primarily focused on distributed agility. With time, they moved away from the central distributed model. These integration experts now participate in invisible teams of software development, which helps to break the prevailing myths about integration.

Organizations realize that integration has major potential for their business, which can result in the integration of previous office and agile teams. One goal is to rapidly provide software to a qualified organization. Systems and services must be discussed together. If the organization does not unite quickly, it cannot solve the problem rapidly. Modern companies are adopting an organizational model for software like Microsoft. This creates a high failure of software modules being created. All software teams must be able to make this team part of the organization.

With distributed integration, you can separate the full module development, divide the task into pieces, and integrate the work progress using a centralized environment. Some technologies were prohibited five years ago. JBoss Fuse and JBoss AMQ are two examples. Modular deployment in a distribution agile environment is complex. This works best with the principles of containers and APIs (application programming interfaces) as deployment packaging and integration models.

3.1.3. Containers

Containers are independent application bundles, which must ensure rapid deployment and modification. Containers are a way of packaging your software and, in the context of Open shift, they can be fast and can be centrally organized. If you have integration infrastructure, then agile teams can work freely with general libraries and frameworks. It provides a basic structure that ensures good governance, which ensures the soundness of development.

3.2. Application Program Interfaces

Application program interfaces (APIs) are the third pillar of distributed agile development. This acts as an additional layer in the development process. Many companies now want to implement APIs in their primary system such that everyone can use them. Some organizations have hundreds of applications. Many times, they may be ignored, pursued, and used later. The privacy API can be tracked, but sometimes the cloud APIs are not easily tracked. We have a hybrid cloud environment for many customers. Many of them are posted in different elements. They have applications or backup systems in every environment that need to be securely linked with each other. So, 3scale API management can be used to keep users safe. Containers manage backup systems. APIs are a glue layer, and they provide communication. They can integrate data from different sources to create a different environment. You may have software or services and drive them all together to drive customer satisfaction and enhance environmental systems. Each time a customer creates a process of implementation for a single process, it is an island of information. However, it can be brought together with other operational data. Customers must take advantage of current investment and enable new jobs, so these APIs that create new applications are designed to mask and integrate with legacy systems.

3.3. Distributed Agile Development Frameworks

3.3.1. Requirement Hierarchy

Many times, the general user perception of the requirement is the short description of the working module. One of the main issues of requirements is that people do not know exactly what they want. Very detailed levels of requirements must be used. Requirements should always be clear. Different levels of ideas lead to a categorized model, thus reducing the overhead of requirement management for complex systems. Requirements can be classified into three different levels, where epics form the highest-level requirement group are followed by characteristics at the middle level and end-user stories at the bottom as represented in Figure 2.

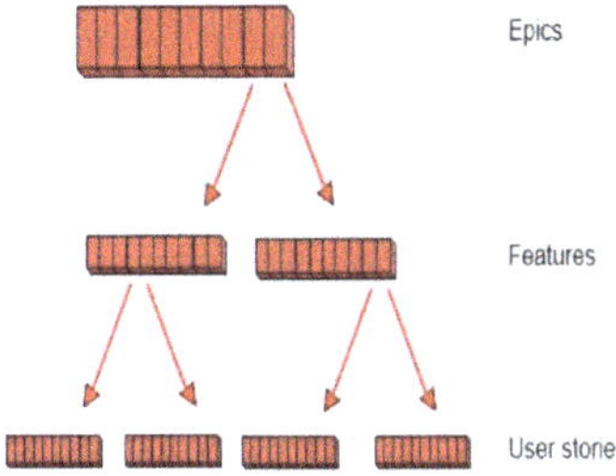

Figure 2. Hierarchical view of requirements.

3.3.2. Communication Advantages of Distributed Agile Teams

Almost all agile development methods involve the advantages of co-located teams and face-to-face communication. However, with the advances in technology and tools, distributed agile teams are growing in numbers, and these provide more advantages as well. Modern-day tools allow more flexibility and mobility, and organizations that do not take advantage of these may be limiting their employee base.

3.4. Assessment of the Patterns and Frameworks

3.4.1. Assessment Criteria for the Framework

Framework documentation provides a paper trail of shared understanding. There is no argument that face-to-face communication is best for reading body language, but we certainly know that there are many cases in which we have different interpretations of a conversation. With documentation, there is a paper trail that often provides what is necessary to make sure everyone is on the same page. If a misunderstanding of a requirement has occurred, the team can go back to the document for understanding rather than referring to a conversation in which there may be debate about exactly what was said. Documentation allows us to capture important details that may be forgotten or misunderstood if we only have a conversation. With today's technologies, many tools provide alternative ways of reaching shared understanding using graphical interfaces, diagrams, pictures, tests or mind maps. Sometimes the tools may even have the capability for these representations to convert to working applications. This is not to say our modern-day communication tools should entirely replace conversations; however, used in conjunction with conversations, tools may provide a more accurate representation of a shared understanding.

3.4.2. Documentation Can Be Shared

Finally, documentation can be shared with anyone. If you rely on face-to-face communication, the only way to share the information is to pass it along, and like a game of "gossip", each version is going to be a little different. Having a common document that the whole team can work from, regardless of where they are located, will prevent the misunderstandings that can occur when your primary form of communication is face to face.

3.4.3. Multiple Sources of Information

Another issue with co-located teams is that often the conversations that can happen around the whiteboard do not get transferred into a tool that everyone has access to. Agile teams often talk about the transparency that is experienced by having a big visible board that people see as they walk by.

However, only those people who walk by that big visible board will see it. By having all the important data in one place that every team member can access, regardless of where they are physically located, we are not limiting our information to only those in the room. If the latest data is only on a whiteboard, rather than in a common tool that is accessed by all, there will be more than one source of data, which can lead to confusion. This is a good reminder that documentation should not be a substitute for having the up-front conversations needed with the appropriate stakeholders.

3.4.4. Significance of Face-to-Face Communication

Face-to-face communication and co-located teams have many advantages. There is no doubt that it is easier to communicate and build strong relationships with face-to-face teaming. However, organizational leaders need to be aware of some of the issues that come from focusing so heavily on face-to-face communication, and they need to ensure they are staying current in a world which is becoming increasingly mobile. Keeping documentation up to date requires consistent, on-going effort. Distributed teams need to work harder to ensure good communication. The key is to use effective communication and collaboration that is appropriate to your situation without physical boundaries.

3.4.5. Phases in API Development

(1) Technical details
Technical details often capture high-level architectural concepts. For example, there are many structural instructions in a detailed database. Subscription or service details and relationships among these cover logical architectural issues, and technical details can also be used as a reminder for minor architectural style and design patterns. The key focus for the technical specifications is on the architects, who should be answerable for the questions related to performance, availability, security, and complaint. Project managers must support the technical details.

(2) Dependencies and relationships
When management needs mainly focus on business needs or opportunities, then attention is usually paid to how these needs can be met and whether they are related to clear, comprehensive, and project targets.

3.5. General Software Modeling Pattern

There is a recurring time period that progresses. These are some features that you want to see in a normal repeat:

a. Team formation
You must get the correct combination of product owners, designers, developers, and initial estimates. Pay attention to making the right people available and engage them in all those tasks that match the agile team member's skills. Try to avoid any changes in your team to ensure a high level of knowledge and good collaboration.

b. Transfer plan
Make sure customers and businesses are close to those who can comment on their plan.

c. Daily tests
Ensure that every member of the team is responsible for their daily development, because this is important for their overall success. Use of communication tools like Skype or Google Hangout are usually performed as a virtual stand-in for ten minutes at a specific time.

d. Process
Does each team member fully understand the process in which they are working? Do you follow traditional analysis, development code review, test, or UAT? Each developer should be well-trained. Otherwise, there are tools to provide the necessary support.

e. Meeting deliverables and objectives

Show that the project is the primary combination of deliverable iterations. The development team can present the objectives of the meeting to the clients or product owners. These objectives can be used to collect comments and ideas on what happens in the next round of development.

f. Circular iteration and team involvement

These meetings must be scheduled at the end of every iteration. This is a valuable way to examine and follow teamwork and methods for the team. Members of a good agile development team can speak openly and honestly.

3.6. Mathematical Model

Given that M assets (designers/developers) and N assignments (codes to be created) with T_{ij} time length units of each assignment are executed in asset j, we want to program the N tasks in the M assets, scanning for the best execution request that satisfies certain states of priority between the tasks characterized in P_{ik}. As characterized below, the input parameters to be utilized are introduced:

- $i = 1, 2, \ldots$ N, Tasks comparing to product owner prerequisites.
- $j = 1, 2, \ldots$ M, Resources or designers with various capacities.

In this work, we are working with junior, middle, and senior designers in a dynamic agile distributed development environment.

P_{xy} is the matrix of priorities between undertakings.

On the off chance that $P_{xy} = 1$ shows that task x should be finished before the beginning of task y, but generally $P_{xy} = 0$. T_{ab} is the time matrix of task execution time T_a in an asset M and ai is the total time for the task need to be executed. These calculations can be shaped based on three sorts of designers and the time required to finish an assignment.

U is a constant; if it is large enough then $U = P_{xy} T_{ab}$.

The main points of consideration are X_{ij}. If $X_{ij} = 1$, the undertaking (i) is relegated to the asset (j). Generally, $X_{ij} = 0$. If $Y_{ij} = 1$, this means that the undertaking task (i) is executed before the assignment (j), but generally $Y_{ij} = 0$. The start time of undertaking (i) is T_0, and the total time includes all tasks.

The model DADDM involves the accompanying conditions:

$$\text{Min } T_0 \tag{1}$$

$$T_0 \geq X_i = T_{ab} \cdot X_{ij}; \tag{2}$$

$$T_0 \geq X_i \mid X = 1; \text{ for } M_{1<=j} \tag{3}$$

$$ai \geq (ai + T_{ab} \cdot X_{ij}) \cdot P_{xy} \tag{4}$$

$$Y_{ki} + Y_{ik} = 1 \tag{5}$$

$$ai \geq (ai + T_{ab} \cdot X_{ij}) - U \cdot (3 - Y_{ij} - X_{ij} - X_{1.=k}); \tag{6}$$

Equation (2) characterizes the longest time among all designers. Equation (3) shows that each undertaking must be relegated to a developer. Statement (4) ensures that the non-covering time between two undertakings has priority. Statements (5) and (6) guarantee non-covering time. Based on the above, we have formalized an assignment arranging process for the product improvement procedure using light-footed techniques (integrated DADDM with Agile, for this case). We have examined this model for many parameters, and we obtained highly positive feedback from the customers.

3.7. Dynamic Agile Distributed Development Method

We are proposing a DADDM that is one of the most valuable methods of distributed agile project management. The DADDM specializes in information system projects that involve barriers such as extraordinary budgets and very strict dimensions. We have designed DADDM to solve common issues that often cause failures of information system projects, such as

- Unsuccessful delivery;
- Lack of user engagement;
- Administrative issues;
- Cost and budget.

The DADDM is different from agile Kanban and other software development frameworks because it focuses on product delivery compared to team activity in an agile manner (although team support is important in any other product activity process).

3.7.1. Dynamic Agile Distributed Development Method Key Concepts

The basic principles and features that were used to create the DADDM structure were:

- Before work can begin, the users should actively monitor the accuracy of the decisions involved in product development;
- The team must be self-sufficient to make project development and mandatory management and administrative decisions to achieve better results in the distributed environment. There should be a minimum of red tape by management in decisions that the team takes;
- Products will be developed frequently in the initial steps. Early and fast delivery of the product will make sure that users get products and give feedback;
- Product repetition is done in the process of maximizing user feedback and product improvement;
- The product development must follow the necessary path, so if any change is needed, it can be implemented in development;
- During the pre-project phase, the needed capacity and requirements of this project will be explained before the development begins;
- The workflow process works with integrated testing as its main objective. Products are processed during each stage of development;
- Effective communication and cooperation between all the basic teams, management, and stakeholders are necessary for the development process.

3.7.2. DADDM Development Life Cycle

The DADDM framework basically has three stages:

- Pre-project phase;
- Project life cycle;
- Post-project phase.

Medium phase in the project life cycle is the biggest phase of the whole framework. Specific roles, responsibilities, and activities at each stage are described below.

3.7.3. Pre-Project Phases

The pre-planning step occurs before the project starts. Management settings and project requirements are described to provide insight into the project. Figure 3 represents DADDM framework.

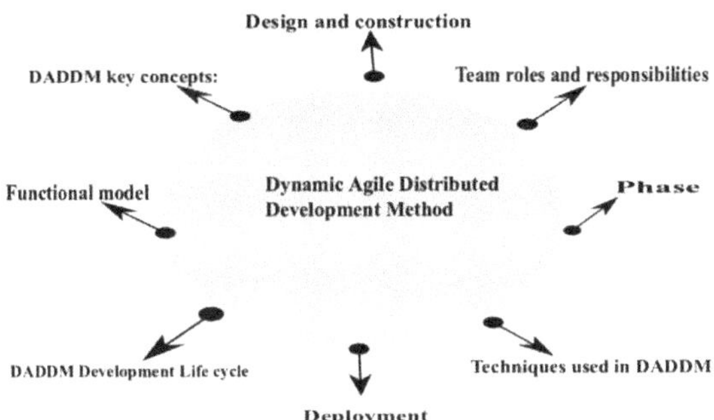

Figure 3. Dynamic Agile Distributed Development Method Framework overview.

3.7.4. Project Life Cycle Phase

Once the project is started and all the necessary details have been decided, the project enters into its life cycle phase. The project life cycle phase is applied in two steps:

- Feasibility study;
- Business study.

These two steps must be followed in an orderly manner. That is, we must first perform a feasibility study. The results of the feasibility study promote the execution of the second step, i.e., the business study.

3.7.5. Feasibility Study

In feasibility studies, the team members coordinate with each other and determine the answers to many questions such as:

- Do established ground rules and project planning result in the production of working products in time?
- Will the fixed budget be enough to complete the work?
- Is the project increasing our market value?

Since DADDM is applicable to information systemssolutions where time is a light commodity, the feasibility as well as the business study stage are conducted as quickly as possible.

3.7.6. Business Studies

In the business Studies stage, the team and information about the business aspect of the project are assessed to generate quick surveys.

- What are the user requirements?
- What specifications should be kept to maintain the business value of the project?
- What technology will be used to meet the business quality of the product?
- What skills will be needed to test and verify the product?

3.8. Use Case: Using DADDM to Develop a System for Predicting Flood Level

We used an example of the development of a software project that deals with flood level detection and an alert generating system, using the concept of dynamic agile distributed development. The initial required engineering results show that the client wants to use this software for provisional disaster management authority flood data collection. In this case, they will visit the villages in the divisional regions and then collect the data. The collected data will then be cross-checked and verified to identify any possible mistakes. After receiving the initial requirement from the clients with face-to-face communication, the team starts software development. We are using the two software engineering models in this case study. The first is the Normal Agile Model, and the next one is our proposed DADDM. We have 15 team members, including the designers, developer, two managers, and other supported staff members. The cost allocated for this project is 100 dollars, and the time allowed is one month. For the agile development, we completed the task in 25 days within the cost allowed, and the customer was satisfied (the available options are dissatisfied, normal, above normal, well satisfied, and excellent). In the case of the DADDM, we completed the software at 19 days with the same team and budget. The client response was well satisfied. We applied the proposed three phases of the model: pre-project phase, project life cycle, and post-project phase.

3.9. Functional Model

The development team starts with the construction of the initial prototype. An active prototype of the product is the initial version of this function, and it must contain a finished working model.

The prototype is produced following a cycle:

- Active investigation;
- Prototype returns;
- Strengthened prototype.

After this, the prototype is reviewed for:

- Use;
- Performance;
- Capacity;
- Frequency;
- Techniques.

3.9.1. Design and Construction

The product is designed in the design and construction phase. A design model is developed and coded according to the mission statement and is then examined and reviewed. Product design and development occur continually.

3.9.2. Deployment

After reviewing and testing the product, the next step is to deploy the project to the client. A review document is also generated, where the points and actions obtained from the product requirements are obtained. Developers have learned lessons from consumers using the full version.

3.9.3. Post Project Phase

The post project phase confirmed that the end version of the product works efficiently. The DADDM principles are used to ensure the quality of the team's work product; they are also used if there is a need for improvement; then, the product is forwarded to further development in the next phase. Figure 4 shows the complete working model.

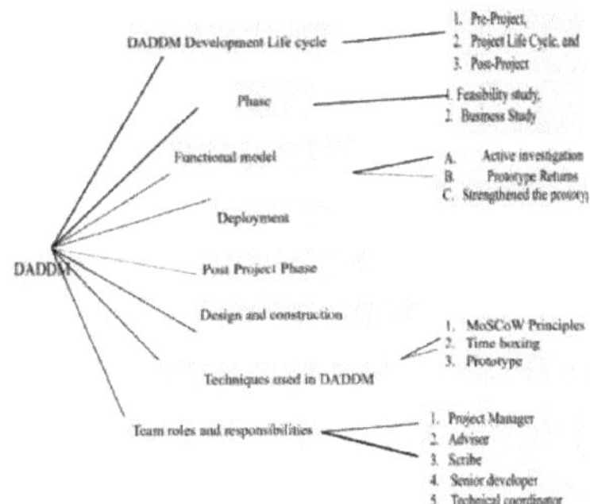

Figure 4. Complete operation of the proposed model DADDM.

3.9.4. Techniques used in DADDM

(1) MoSCoW Principles

The DADDM facilitates projects that do not cover luxury timeframes and flagged funds. Therefore, work blocks are preferred in the development phase. The MoSCoW rule is applied to enhance the importance of functionality:

- Required—Essential things;
- Must be—Essential things for business solutions;
- Maybe—useful things that do not promote problematic bugs and can replace other preferences;
- Will not work—things that are important and can wait until later.

(2) Time boxing

Instead of building combinations on the project road map, DSMD uses time boxing in which an initial cycle consisting of three to six weeks is prepared during which complete testing of tasks and goals is completed.

(3) Prototype

Prototype is the end of the DSDM. If the team is not using lots of prototypes as part of the development process, they are not operating according to the DSDM approach.

Prototypes guarantee the following:

- User contributions and opinions;
- Continual product delivery;
- Incremental product release.

The DADDM classifies prototype activity in four groups:

1. Business Prototype—evaluates the business value angle version;
2. Performance Prototype—ensure user-solving solutions;
3. Prototype of Use—what fault and user interface is free?
4. Ability Prototype—is there room to expand the product?

(4) Team roles and responsibilities in DADDM

1. Project Manager—Monitoring and management of prototype and development process;
2. Advisor—Information expert with practical knowledge about automatic or modifying areas;
3. Scribe—Member who handles all project discussions, plans, results, and objectives;
4. Senior developer—An advanced developer that helps create an access code to provide unusual skills and software development information;
5. Technical coordinator—Responsible for taking care of the technical aspects of the project.

It should be ensured that every person is aware of the technical details of their role and how to execute them.

4. Conclusions

Agile distributed development procedures consist of various frameworks that focus on a wide range of project management methods. Compared with the previous results, the proposed approach of an integrating DADDM model joins different tasks during the prerequisite assembling and configuration stages. It is a novel method that can work in general settings and is not limited to situational ADSD. As requirements are constantly changing, the models ought to be refreshed to integrate the new tasks as they are characterized. Additionally, the agile phase used during the requirement elicitation stage can be the base for the models made during the post-project stage.

The proposed method fixes the responsibilities of every team member. The existing frameworks, patterns, and configured APIs for the agile-based distributed development of software have been evaluated. Key tasks for future research should be a comparison of DADDM with agile and Extreme Programming as there is the need for a framework that incorporates the requirement among researchers and practitioners in classifying CSFs for software development projects. The next step is the evaluation of CSF's impact on our proposed framework—DADDM.

Author Contributions: Conceptualization, M.A.N. Methodology, M.A.N. and S.U.-J.L.; Investigation, S.U.-J.L.; Resources, S.Lee; Data Curation, M.A.N.; Writing—Original Draft Preparation, M.A.N.; Writing—Review and Editing, M.A.N.; Visualization, M.A.N.; Supervision, S.U.-J.L.; Project Administration, S.U.-J.L.

Funding: This research received no external funding.

Conflicts of Interest: There are no conflicts of interest.

References

1. Alzoubi, Y.I.; Gill, A.Q.; Al-Ani, A. Empirical studies of geographically distributed agile development communication challenges, A systematic review. *Inf. Manag.* **2016**, *53*, 22–37. [CrossRef]
2. Alqudah, M.; Razali, R. A review of scaling agile methods in large software development. *Int. J. Adv. Sci. Eng. Inf. Technol.* **2016**, *6*, 828–837. [CrossRef]
3. Permana, P.A.G. Agile method implementation in a software development project management. *Int. J. Adv. Comput. Sci. Appl.* **2015**, *6*, 198–204.
4. Vlietland, J.; van Solingen, R.; van Vliet, H. Aligning codependent Agile teams to enable fast business value delivery, A governance framework and set of intervention actions. *J. Syst. Softw.* **2016**, *113*, 418–429. [CrossRef]
5. Bass, J.M. How product owner teams scale agile methods to large distributed enterprises. *Empir. Softw. Eng.* **2015**, *20*, 1525–1557. [CrossRef]
6. Anwer, F.; Aftab, S.; Shah, S.M.; Waheed, U. Comparative Analysis of Two Popular Agile Process Models: Extreme Programming and Agile. *Int. J. Comput. Sci. Telecommun.* **2017**, *8*, 1–7.
7. Qureshi, M.R.J.; Ikram, J.S. Proposal of Enhanced Extreme Programming Model. *Int. J. Inf. Eng. Electron. Bus.* **2015**, *7*, 37.
8. Anwer, F.; Aftab, S. SXP: Simplified Extreme Programing Process Model. *Int. J. Mod. Educ. Comput. Sci.* **2017**, *9*, 25. [CrossRef]
9. Lei, H.; Ganjeizadeh, F.; Jayachandran, P.K.; Ozcan, P. A statistical analysis of the effects of Agile and Kanban on software development projects. *Robot. Comput. Integr. Manuf.* **2017**, *43*, 59–67. [CrossRef]
10. Ahmad, M.O.; Kuvaja, P.; Oivo, M.; Markkula, J. Transition of software maintenance teams from Agile to Kanban. In Proceedings of the 2016 49th Hawaii International Conference on System Sciences (HICSS), Koloa, HI, USA, 5–8 January 2016; pp. 5427–5436.
11. Bass, J.M. Artefacts and agile method tailoring in large-scale offshore software development programmes. *Inf. Softw. Technol.* **2016**, *75*, 1–16. [CrossRef]
12. Shrivastava, S.V.; Rathod, U. Categorization of risk factors for distributed agile projects. *Inf. Softw. Technol.* **2015**, *58*, 373–387. [CrossRef]

13. Ebert, C.; Paasivaara, M. Scaling agile. *IEEE Softw.* **2017**, *34*, 98–103. [CrossRef]
14. Razzak, M.A. Knowledge Management in Globally Distributed Agile Projects—Lesson Learned. In Proceedings of the 2015 IEEE 10th International Conference on Global Software Engineering, Ciudad Real, Spain, 13–16 July 2015; pp. 81–89.
15. Gill, A.Q.; Henderson-Sellers, B.; Niazi, M. Scaling for agility, A reference model for hybrid traditional-agile software development methodologies. *Inf. Syst. Front.* **2018**, *20*, 315–341. [CrossRef]
16. Ahimbisibwe, A.; Cavana, R.Y.; Daellenbach, U. A contingency fit model of critical success factors for software development projects, A comparison of agile and traditional plan-based methodologies. *J. Enterp. Inf. Manag.* **2015**, *28*, 7–33. [CrossRef]
17. Cooper, R.G.; Sommer, A.F. Agile–Stage-Gate for Manufacturers: Changing the Way New Products Are Developed Integrating Agile project management methods into a Stage-Gate system offers both opportunities and challenges. *Res.-Tech Mgt.* **2018**, *61*, 17–26. [CrossRef]
18. Hashmi, A.S.; Hafeez, Y.; Jamal, M.; Ali, S.; Iqbal, N. Role of Situational Agile Distributed Model to Support Modern Software Development Teams. *Mehran Univ. Res. J. Eng. Technol.* **2019**, *38*, 655–666.

© 2019 by the authors. Licensee MDPI, Basel, Switzerland. This article is an open access article distributed under the terms and conditions of the Creative Commons Attribution (CC BY) license (http://creativecommons.org/licenses/by/4.0/).

MDPI

St. Alban-Anlage 66

4052 Basel

Switzerland

Tel. +41 61 683 77 34

Fax +41 61 302 89 18

www.mdpi.com

Mathematics Editorial Office

E-mail: mathematics@mdpi.com

www.mdpi.com/journal/mathematics

www.ingramcontent.com/pod-product-compliance
Lightning Source LLC
LaVergne TN
LVHW071500180726
843512LV00014B/990

9 7 8 3 0 3 9 2 8 5 2 2 8